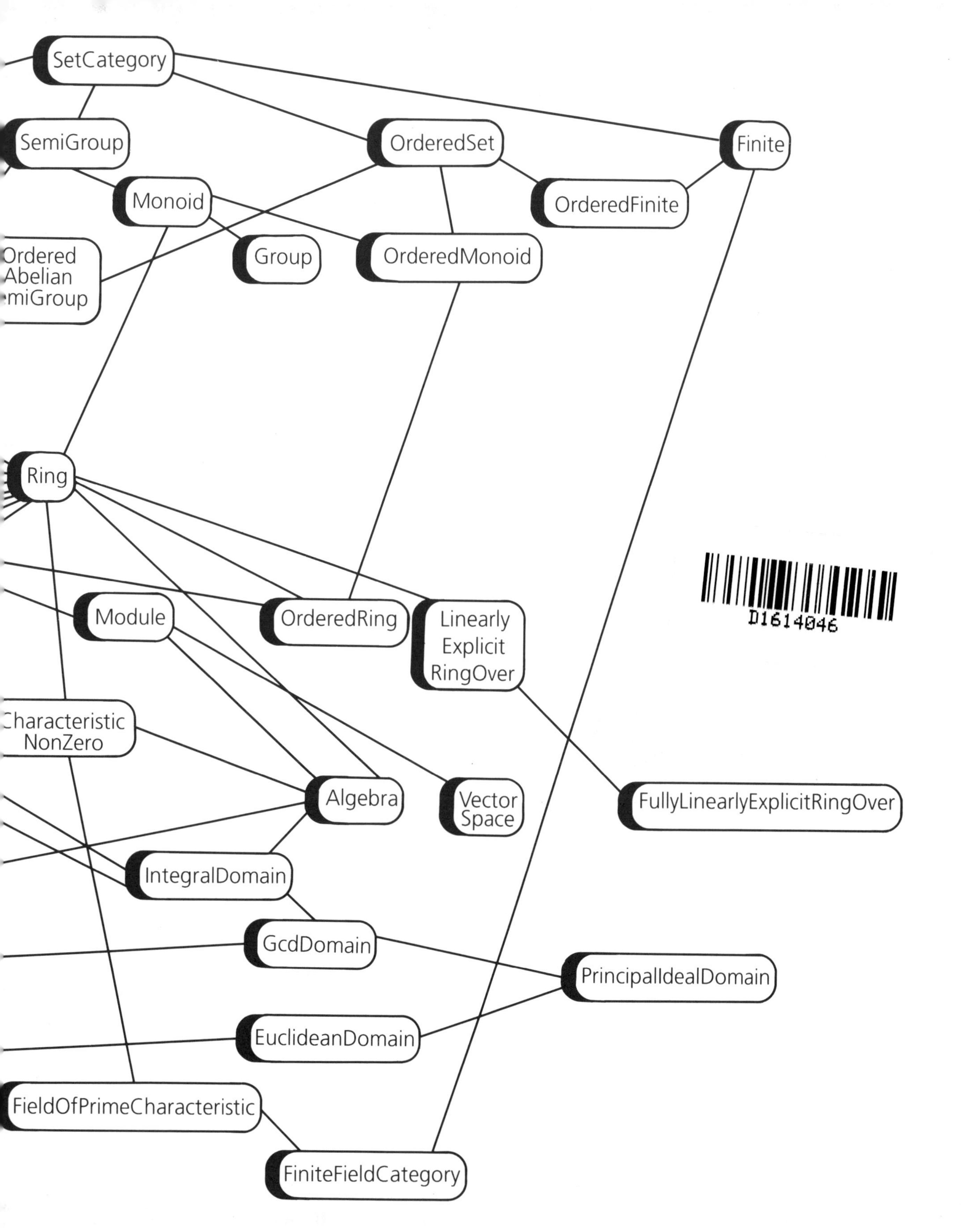

SetCategory
SemiGroup
OrderedSet
Finite
Monoid
OrderedFinite
Ordered
Abelian
miGroup
Group
OrderedMonoid
Ring
Module
OrderedRing
Linearly
Explicit
RingOver
haracteristic
NonZero
Algebra
Vector
Space
FullyLinearlyExplicitRingOver
IntegralDomain
GcdDomain
PrincipalIdealDomain
EuclideanDomain
FieldOfPrimeCharacteristic
FiniteFieldCategory
D1614046

The Scientific Computation System

Richard D. Jenks Robert S. Sutor

The Scientific Computation System

With Contributions From

Scott C. Morrison
Jonathan M. Steinbach
Barry M. Trager
Stephen M. Watt
Manuel Bronstein
William H. Burge
Timothy P. Daly
Michael Dewar
Patrizia Gianni
Johannes Grabmeier
William Sit
Clifton J. Williamson

NAG®

The Numerical Algorithms Group Limited

Springer-Verlag
New York Berlin Heidelberg London Paris
Tokyo Hong Kong Barcelona Budapest

Richard D. Jenks
Robert S. Sutor
IBM Thomas J. Watson Research Center
Route 134/Taconic Parkway
P.O. Box 218
Yorktown Heights, NY 10598
U.S.A.

The AXIOM system is marketed and distributed by The Numerical Algorithms Group, Ltd., Oxford, UK, and Numerical Algorithms Group, Inc., Downer's Grove, Illinois, USA.

Library of Congress Cataloging-in-Publication Data
Jenks, Richard D.
AXIOM : the scientific computation system / Richard D. Jenks
Robert S. Sutor.
p. cm.
Includes bibliographical references and index.
ISBN 0-387-97855-0
1. AXIOM (Computer file) 2. Mathematics—Data processing.
I. Sutor, Robert S. II. Title.
QA76.95.J46 1992
510'.285'5133—dc20 92-14454

Printed on acid-free paper.

Production supervised by Kenneth Dreyhaupt; manufacturing coordinated by Rhea Talbert.
Photocomposed copy prepared from the authors' LaTeX file.
Printed and bound by R.R. Donnelley & Sons, Harrisonburg, Virginia.
Printed in the United States of America.

9 8 7 6 5 4 3 2 1

ISBN 0-387-97855-0 Springer-Verlag New York Berlin Heidelberg
ISBN 3-540-97855-0 Springer-Verlag Berlin Heidelberg New York

To my children, Douglas, Daniel, and Susan,
for their love, support, and understanding over the years.
R.D.J.

To Judith and Kate,
to whom my debt is beyond computation.
R.S.S.

Foreword

You are holding in your hands an unusual book. Winston Churchill once said that the empires of the future will be empires of the mind. This book might hold an electronic key to such an empire.

When computers were young and slow, the emerging computer science developed dreams of Artificial Intelligence and Automatic Theorem Proving in which theorems can be proved by machines instead of mathematicians. Now, when computer hardware has matured and become cheaper and faster, there is not too much talk of putting the burden of formulating and proving theorems on the computer's shoulders. Moreover, even in those cases when computer programs do prove theorems, or establish counter-examples (for example, the solution of the four color problem, the non-existence of projective planes of order 10, the disproof of the Mertens conjecture), humans carry most of the burden in the form of programming and verification.

It is the language of computer programming that has turned out to be the crucial instrument of productivity in the evolution of scientific computing. The original Artificial Intelligence efforts gave birth to the first symbolic manipulation systems based on LISP. The first complete symbolic manipulation or, as they are called now, computer algebra packages tried to imbed the development programming and execution of mathematical problems into a framework of familiar symbolic notations, operations and conventions. In the third decade of symbolic computations, a couple of these early systems—REDUCE and MACSYMA—still hold their own among faithful users.

AXIOM was born in the mid-70's as a system called Scratchpad developed by IBM researchers. Scratchpad/AXIOM was born big—its original platform was an IBM mainframe 3081, and later a 3090. The system was growing and learning during the decade of the 80's, and its development and progress influenced the field of computer algebra. During this period, the first commercially available computer algebra packages for mini and and microcomputers made their debut. By now, our readers are aware of Mathematica, Maple, Derive, and Macsyma. These systems (as well as a few special purpose computer algebra packages in academia) emphasize ease of operation and standard scientific conventions, and come with a prepared set of mathematical solutions for typical tasks confronting an applied scientist or an engineer. These features brought a recognition of the enormous benefits of computer algebra to the widest circles of scientists and engineers.

The Scratchpad system took its time to blossom into the beautiful AXIOM product. There is no rival to this powerful environment in its scope and, most importantly, in its structure and organization. AXIOM contains the basis for any comprehensive and elaborate mathematical development. It gives the user all Foundation and Algebra instruments necessary to develop a computer realization of sophisticated mathematical objects in exactly the way a mathematician would do it. AXIOM is also the basis of a complete scientific cyberspace—it provides an environment for mathematical objects used in scientific computation, and the means of controlling and communicating between these objects. Knowledge of only a few AXIOM language features and operating principles is all that is required to make impressive progress in a given domain of interest. The system is powerful. It is not an interactive interpretive environment operating only in response to one line commands—it is a complete language with rich syntax and a full compiler. Mathematics can be developed and explored with ease by the user of AXIOM. In fact, during AXIOM's growth cycle, many detailed mathematical domains were constructed. Some of them are a part of AXIOM's core and are described in this book. For a bird's eye view of the algebra hierarchy of AXIOM, glance inside the book cover.

The crucial strength of AXIOM lies in its excellent structural features and unlimited expandability—it is open, modular system designed to support an ever growing number of facilities with minimal increase in structural complexity. Its design also supports the integration of other computation tools such as numerical software libraries written in Fortran and C. While AXIOM is already a very powerful system, the prospect of scientists using the system to develop their own fields of Science is truly exciting—the day is still young for AXIOM.

Over the last several years Scratchpad/AXIOM has scored many successes in theoretical mathematics, mathematical physics, combinatorics, digital signal processing, cryptography and parallel processing. We have to confess that we enjoyed using Scratchpad/AXIOM. It provided us with an excellent environment for our research, and allowed us to solve problems intractable on other systems. We were able to prove new diophantine results for π; establish the Grothendieck conjecture for certain classes of linear differential equations; study the arithmetic properties of the uniformization of hyperelliptic and other algebraic curves; construct new factorization algorithms based on formal groups; within Scratchpad/AXIOM we were able to obtain new identities needed for quantum field theory (elliptic genus formula and double scaling limit for quantum gravity), and classify period relations for CM varieties in terms of hypergeometric series.

The AXIOM system is now supported and distributed by NAG, the group that is well known for its high quality software products for numerical and statistical computations. The development of AXIOM in IBM was conducted at IBM T.J. Watson Research Center at Yorktown, New York by a symbolic computation group headed by Richard D. Jenks. Shmuel Winograd of IBM was instrumental in the progress of symbolic research at IBM.

This book opens the wonderful world of AXIOM, guiding the reader and user through AXIOM's definitions, rules, applications and interfaces. A variety of fully developed areas of mathematics are presented as packages, and the user is well advised to take advantage of the sophisticated realization of familiar mathematics. The AXIOM book is easy to read and the AXIOM system is easy to use. It possesses all the features required of a modern computer environment (for example, windowing, integration of operating system features, and interactive graphics). AXIOM comes with a detailed hypertext interface (HyperDoc), an elaborate browser, and complete on-line documentation. The HyperDoc allows novices to solve their problems in a straightforward way, by providing menus for step-by-step interactive entry.

The appearance of AXIOM in the scientific market moves symbolic computing into a higher plane, where scientists can formulate their statements in their own language and receive computer assistance in their proofs. AXIOM's performance on workstations is truly impressive, and users of AXIOM will get more from them than we, the early users, got from mainframes. AXIOM provides a powerful scientific environment for easy construction of mathematical tools and algorithms; it is a symbolic manipulation system, and a high performance numerical system, with full graphics capabilities. We expect every (computer) power hungry scientist will want to take full advantage of AXIOM.

David V. Chudnovsky Gregory V. Chudnovsky

Contents

Contributors

The design and development of AXIOM was led by the Symbolic Computation Group of the Mathematical Sciences Department, IBM Thomas J. Watson Research Center, Yorktown Heights, New York. The current implementation of AXIOM is the product of many people. The primary contributors are:

Richard D. Jenks (IBM, Yorktown) received a Ph.D. from the University of Illinois and was a principal architect of the **Scratchpad** computer algebra system (1971). In 1977, Jenks initiated the AXIOM effort with the design of MODLISP, inspired by earlier work with Rüdiger Loos (Tübingen), James Griesmer (IBM, Yorktown), and David Y. Y. Yun (Hawaii). Joint work with David R. Barton (Berkeley, California) and James Davenport led to the design and implementation of prototypes and the concept of categories (1980). More recently, Jenks led the effort on user interface software for AXIOM.

Barry M. Trager (IBM, Yorktown) received a Ph.D. from MIT while working in the **MACSYMA** computer algebra group. Trager's thesis laid the groundwork for a complete theory for closed-form integration of elementary functions and its implementation in AXIOM. Trager and Richard Jenks are responsible for the original abstract datatype design and implementation of the programming language with its current MODLISP-based compiler and run-time system. Trager is also responsible for the overall design of the current AXIOM library and for the implementation of many of its components.

Stephen M. Watt (IBM, Yorktown) received a Ph.D. from the University of Waterloo and is one of the original authors of the **Maple** computer algebra system. Since joining IBM in 1984, he has made central contributions to the AXIOM language and system design, as well as numerous contributions to the library. He is the principal architect of the new AXIOM compiler, planned for Release 2.

Robert S. Sutor (IBM, Yorktown) received a Ph.D. in mathematics from Princeton University and has been involved with the design and implementation of the system interpreter, system commands, and documentation since 1984. Sutor's contributions to the AXIOM library include factored objects, partial fractions, and the original implementation of finite field extensions. Recently, he has devised technology for producing automatic hard-copy and on-line documentation from single source files.

Scott C. Morrison (IBM, Yorktown) received an M.S. from the University of California, Berkeley, and is a principal person responsible for the design and implementation of the AXIOM interface, including the interpreter, HyperDoc, and applications of the computer graphics system.

Manuel Bronstein (ETH, Zürich) received a Ph.D. in mathematics from the University of California, Berkeley, completing the theoretical work on closed-form integration by Barry Trager. Bronstein designed and implemented the algebraic structures and algorithms in the AXIOM library for integration, closed form solution of differential equations, operator algebras, and manipulation of top-level mathematical expressions. He also designed (with Richard Jenks) and implemented the current pattern match facility for AXIOM.

William H. Burge (IBM, Yorktown) received a Ph.D. from Cambridge University, implemented the AXIOM parser, designed (with Stephen Watt) and implemented the stream and power series structures, and numerous algebraic facilities including those for data structures, power series, and combinatorics.

Timothy P. Daly (IBM, Yorktown) is pursuing a Ph.D. in computer science at Brooklyn Polytechnic Institute and is responsible for porting, testing, performance, and system support work for AXIOM.

James Davenport (Bath) received a Ph.D. from Cambridge University, is the author of several computer algebra textbooks, and has long recognized the need for AXIOM's generality for computer algebra. He was involved with the early prototype design of system internals and the original category hierarchy for AXIOM (with David R. Barton). More recently, Davenport and Barry Trager designed the algebraic category hierarchy currently used in AXIOM. Davenport is Hebron and Medlock Professor of Information Technology at Bath University.

Michael Dewar (Bath) received a Ph.D. from the University of Bath for his work on the IRENA system (an interface between the **REDUCE** computer algebra system and the NAG Library of numerical subprograms), and work on interfacing algebraic and numerical systems in general. He has contributed code to produce FORTRAN output from AXIOM, and is currently developing a comprehensive foreign language interface and a link to the NAG Library for release 2 of AXIOM.

Albrecht Fortenbacher (IBM Scientific Center, Heidelberg) received a doctorate from the University of Karlsruhe and is a designer and implementer of the type-inferencing code in the AXIOM interpreter. The result of research by Fortenbacher on type coercion by rewrite rules will soon be incorporated into AXIOM.

Patrizia Gianni (Pisa) received a Laurea in mathematics from the University of Pisa and is the prime author of the polynomial and rational function component of the AXIOM library. Her contributions include algorithms for greatest common divisors, factorization, ideals, Gröbner bases, solutions of polynomial systems, and linear algebra. She is currently Associate Professor of Mathematics at the University of Pisa.

Johannes Grabmeier (IBM Scientific Center, Heidelberg) received a Ph.D. from University Bayreuth (Bavaria) and is responsible for many AXIOM packages, including those for representation theory (with Holger Gollan (Essen)), permutation groups (with Gerhard Schneider (Essen)), finite fields (with Alfred Scheerhorn), and non-associative algebra (with Robert Wisbauer (Düsseldorf)).

Larry Lambe received a Ph.D. from the University of Illinois (Chicago) and has been using AXIOM for research in homological algebra. Lambe contributed facilities for Lie ring and exterior algebra calculations and has worked with Scott Morrison on various graphics applications.

Michael Monagan (ETH, Zürich) received a Ph.D. from the University of Waterloo and is a principal contributor to the **Maple** computer algebra system. He designed and implemented the category hierarchy and domains for data structures (with Stephen Watt), multi-precision floating point arithmetic, code for polynomials modulo a prime, and also worked on the new compiler.

William Sit (CCNY) received a Ph.D. from Columbia University. He has been using AXIOM for research in differential algebra, and contributed operations for differential polynomials (with Manuel Bronstein).

Jonathan M. Steinbach (IBM, Yorktown) received a B.A. degree from Ohio State University and has responsibility for the AXIOM computer graphics facility. He has modified and extended this facility from the original design by Jim Wen. Steinbach is currently involved in the new compiler effort.

Jim Wen, a graduate student in computer graphics at Brown University, designed and implemented the original computer graphics system for AXIOM with pop-up control panels for interactive manipulation of graphic objects.

Clifton J. Williamson (Cal Poly) received a Ph.D. in Mathematics from the University of California, Berkeley. He implemented the power series (with William Burge and Stephen Watt), matrix, and limit facilities in the library and made numerous contributions to the HyperDoc documentation and algebraic side of the computer graphics facility. Williamson is currently an Assistant Professor of Mathematics at California Polytechnic State University, San Luis Obispo.

Contributions to the current AXIOM system were also made by: Yurij Baransky (IBM Research, Yorktown), David R. Barton, Bruce Char (Drexel), Korrinn Fu, Rüdiger Gebauer, Holger Gollan (Essen), Steven J. Gortler, Michael Lucks, Victor Miller (IBM Research, Yorktown), C. Andrew Neff (IBM Research, Yorktown), H. Michael Möller (Hagen), Simon Robinson, Gerhard Schneider (Essen), Thorsten Werther (Bonn), John M. Wiley, Waldemar Wiwianka (Paderborn), David Y. Y. Yun (Hawaii).

Other group members, visitors and contributors to AXIOM include Richard Anderson, George Andrews, David R. Barton, Alexandre Bouyer, Martin Brock, Florian Bundschuh, Cheekai Chin, David V. Chudnovsky, Gregory V. Chudnovsky, Josh Cohen, Gary Cornell, Jean Della Dora, Claire DiCrescendo, Dominique Duval, Lars Erickson, Timothy Freeman, Marc Gaetano, Vladimir A. Grinberg, Florian Bundschuh, Oswald Gschnitzer, Klaus Kusche, Bernhard Kutzler, Mohammed Mobarak, Julian A. Padget, Michael Rothstein, Alfred Scheerhorn, William F. Schelter, Martin Schönert, Fritz Schwarz, Christine J. Sundaresan, Moss E. Sweedler, Themos T. Tsikas, Berhard Wall, Robert Wisbauer, and Knut Wolf.

This book has contributions from several people in addition to its principal authors. Scott Morrison is responsible for the computer graphics gallery and the programs in Appendix F. Jonathan Steinbach wrote the original version of Chapter 7. Michael Dewar contributed material on the FORTRAN interface in Chapter 4. Manuel Bronstein, Clifton Williamson, Patricia Gianni, Johannes Grabmeier, and Barry Trager, and Stephen Watt contributed to Chapters 8 and 9 and Appendix E. William Burge, Timothy Daly, Larry Lambe, and William Sit contributed material to Chapter 9.

The authors would like to thank the production staff at Springer-Verlag for their guidance in the preparation of this book, and Jean K. Rivlin of IBM Yorktown Heights for her assistance in producing the camera-ready copy. Also, thanks to Robert F. Caviness, James H. Davenport, Sam Dooley, Richard J. Fateman, Stuart I. Feldman, Stephen J. Hague, John A. Nelder, Eugene J. Surowitz, Themos T. Tsikas, James W. Thatcher, and Richard E. Zippel for their constructive suggestions on drafts of the book.

Introduction to AXIOM

Welcome to the world of AXIOM. We call AXIOM a scientific computation system: a self-contained toolbox designed to meet your scientific programming needs, from symbolics, to numerics, to graphics.

This introduction is a quick overview of what AXIOM offers.

Symbolic computation

AXIOM provides a wide range of simple commands for symbolic mathematical problem solving. Do you need to solve an equation, to expand a series, or to obtain an integral? If so, just ask AXIOM to do it.

Integrate $\frac{1}{(x^3\ (a+bx)^{1/3})}$ with respect to x.

```
integrate(1/(x**3 * (a+b*x)**(1/3)),x)
```

$$\frac{\left(\begin{array}{l} -2\ b^2\ x^2\ \sqrt{3}\ \log\left(\sqrt[3]{a}\ \sqrt[3]{b\ x+a}^{\,2}+\sqrt[3]{a}^{\,2}\ \sqrt[3]{b\ x+a}+a\right)+ \\ 4\ b^2\ x^2\ \sqrt{3}\ \log\left(\sqrt[3]{a}^{\,2}\ \sqrt[3]{b\ x+a}-a\right)+ \\ 12\ b^2\ x^2\ \arctan\left(\frac{2\ \sqrt{3}\ \sqrt[3]{a}^{\,2}\ \sqrt[3]{b\ x+a}+a\ \sqrt{3}}{3\ a}\right)+ \\ (12\ b\ x-9\ a)\ \sqrt{3}\ \sqrt[3]{a}\ \sqrt[3]{b\ x+a}^{\,2} \end{array}\right)}{18\ a^2\ x^2\ \sqrt{3}\ \sqrt[3]{a}} \qquad (1)$$

Type: Union(Expression Integer, ...)

AXIOM provides state-of-the-art algebraic machinery to handle your most advanced symbolic problems. For example, AXIOM's integrator gives you the answer when an answer exists. If one does not, it provides a proof that there is no answer. Integration is just one of a multitude of symbolic operations that AXIOM provides.

Numeric computation

AXIOM has a numerical library that includes operations for linear algebra, solution of equations, and special functions. For many of these operations, you can select any number of floating point digits to be carried out in the computation.

Solve $x^{49} - 49x^4 + 9$ to 49 digits of accuracy.

```
solve(x**49-49*x**4+9 = 0,1.e-49)
```

$$\begin{aligned}[x &= -0.6546536706904271136718122105095984761851224331556, \\ x &= 1.086921395653859508493939035954893289009213388763, \\ x &= 0.6546536707255271739694686066136764835361487607661]\end{aligned} \tag{2}$$

Type: List Equation Polynomial Float

The output of a computation can be converted to FORTRAN to be used in a later numerical computation. Besides floating point numbers, AXIOM provides literally dozens of kinds of numbers to compute with. These range from various kinds of integers, to fractions, complex numbers, quaternions, continued fractions, and to numbers represented with an arbitrary base.

What is `10` to the `100` th power in base `32`?

```
radix(10**100,32)
```

```
4I9LKIP9GRSTC5IF164PO5V72ME827226JSLAP462585Q7H
00000000000000000000                                  (3)
```

Type: RadixExpansion 32

Graphics

You may often want to visualize a symbolic formula or draw a graph from a set of numerical values. To do this, you can call upon the AXIOM graphics capability.

Draw $J_0(\sqrt{x^2+y^2})$ for $-20 \le x, y \le 20$.

```
draw(5*besselJ(0,sqrt(x**2+y**2)), x=-20..20, y=-20..20)
```

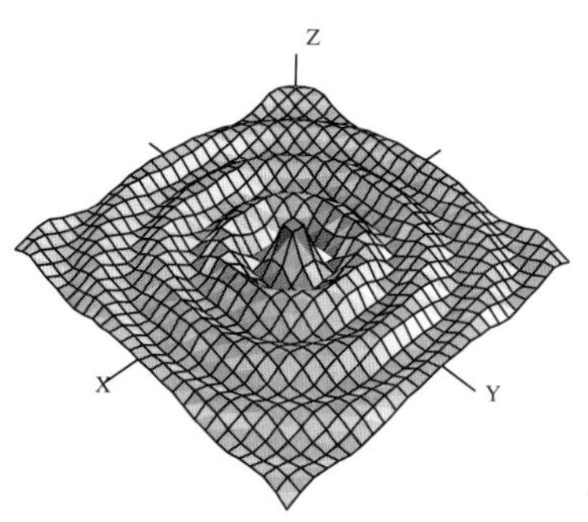

Graphs in AXIOM are interactive objects you can manipulate with your mouse. Just click on the graph, and a control panel pops up. Using this mouse and the control panel, you can translate, rotate, zoom, change the coloring, lighting, shading, and perspective on the picture. You can also generate a PostScript copy of your graph to produce hard-copy output.

HyperDoc

HyperDoc presents you windows on the world of AXIOM, offering on-line help, examples, tutorials, a browser, and reference material. HyperDoc gives you on-line access to this book in a "hypertext" format. Words that appear in a different font (for example, Matrix, **factor**, and *category*) are generally mouse-active; if you click on one with your mouse, HyperDoc shows you a new window for that word.

As another example of a HyperDoc facility, suppose that you want to compute the roots of $x^{49} - 49x^4 + 9$ to 49 digits (as in our previous example) and you don't know how to tell AXIOM to do this. The "basic command" facility of HyperDoc leads the way. Through the series of HyperDoc windows shown in Figure 1 and the specified mouse clicks, you and HyperDoc generate the correct command to issue to compute the answer.

Interactive Programming

AXIOM's interactive programming language lets you define your own functions. A simple example of a user-defined function is one that computes the successive Legendre polynomials. AXIOM lets you define these polynomials in a piece-wise way.

The first Legendre polynomial.

```
p(0) == 1
```

Type: Void

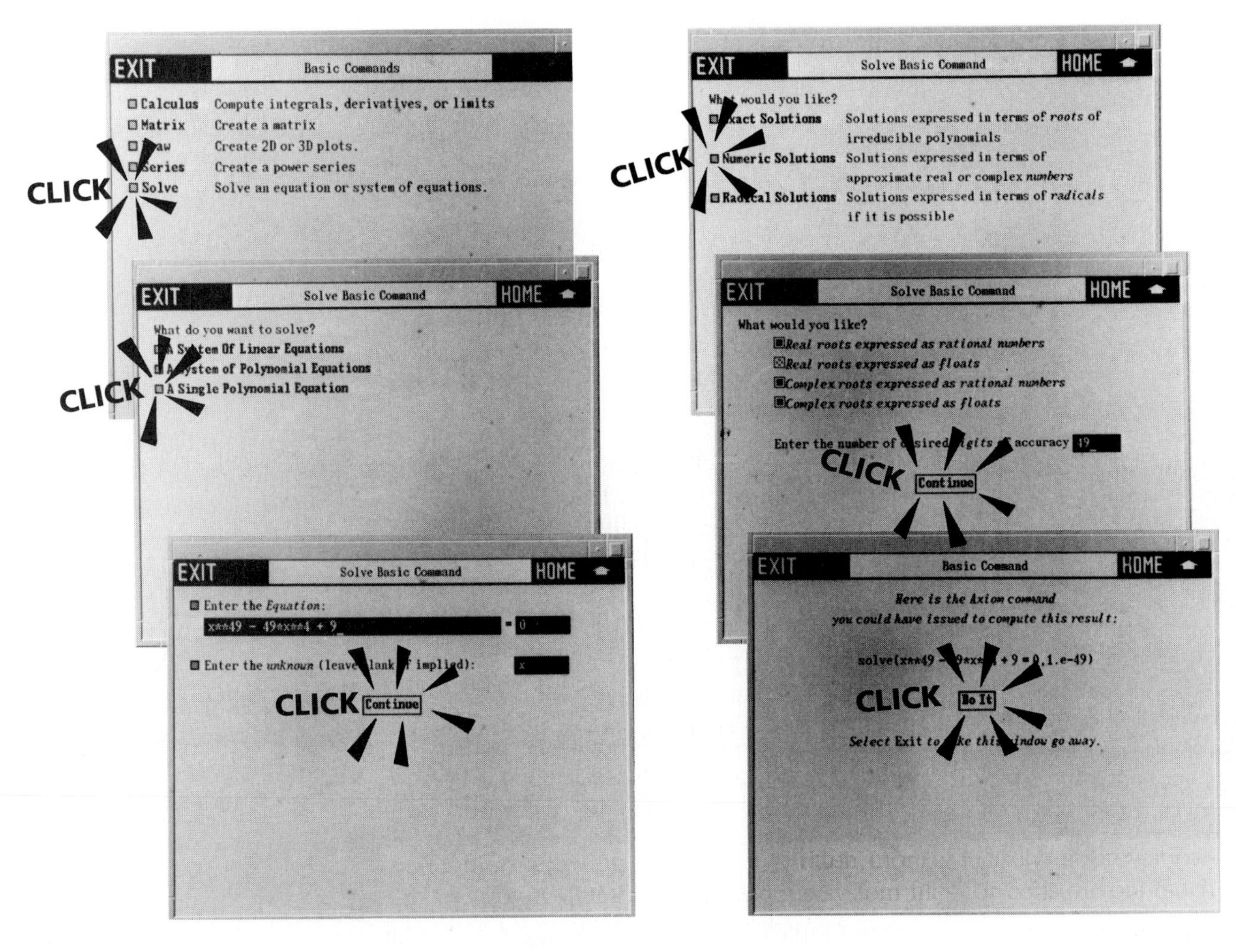

Figure 1: Computing the roots of $x^{49} - 49x^4 + 9$.

The second Legendre polynomial.

```
p(1) == x
```

Type: Void

The nth Legendre polynomial for $(n > 1)$.

```
p(n) == ((2*n-1)*x*p(n-1) - (n-1) * p(n-2))/n
```

Type: Void

In addition to letting you define simple functions like this, the interactive language can be used to create entire application packages. All the graphs in the AXIOM Images section in the center of the book, for example, were created by programs written in the interactive language.

The above definitions for p do no computation—they simply tell AXIOM how to compute p(k) for some positive integer k. To actually get a value of a Legendre polynomial, you ask for it.

What is the tenth Legendre polynomial?

```
p(10)
```

```
Compiling function p with type Integer -> Polynomial
Fraction Integer
Compiling function p as a recurrence relation.
```

$$\frac{46189}{256}\,x^{10} - \frac{109395}{256}\,x^{8} + \frac{45045}{128}\,x^{6} - \frac{15015}{128}\,x^{4} + \frac{3465}{256}\,x^{2} - \frac{63}{256} \tag{7}$$

Type: Polynomial Fraction Integer

AXIOM applies the above pieces for p to obtain the value of p(10). But it does more: it creates an optimized, compiled function for p. The function is formed by putting the pieces together into a single piece of code. By *compiled*, we mean that the function is translated into basic machine-code. By *optimized*, we mean that certain transformations are performed on that code to make it run faster. For p, AXIOM actually translates the original definition that is recursive (one that calls itself) to one that is iterative (one that consists of a simple loop).

What is the coefficient of x^{90} in p(90)?

```
coefficient(p(90),x,90)
```

$$\frac{5688265542052017822223458237426581853561497449095175}{77371252455336267181195264} \tag{8}$$

Type: Polynomial Fraction Integer

In general, a user function is type-analyzed and compiled on first use. Later, if you use it with a different kind of object, the function is recompiled if necessary.

Data Structures

A variety of data structures are available for interactive use. These include strings, lists, vectors, sets, multisets, and hash tables. A particularly useful structure for interactive use is the infinite stream:

Create the infinite stream of derivatives of Legendre polynomials

```
[D(p(i),x) for i in 1..]
```

$$\left[1,\ 3\,x,\ \frac{15}{2}\,x^{2} - \frac{3}{2},\ \frac{35}{2}\,x^{3} - \frac{15}{2}\,x,\ \frac{315}{8}\,x^{4} - \frac{105}{4}\,x^{2} + \frac{15}{8},\right.$$
$$\left.\frac{693}{8}\,x^{5} - \frac{315}{4}\,x^{3} + \frac{105}{8}\,x,\ \frac{3003}{16}\,x^{6} - \frac{3465}{16}\,x^{4} + \frac{945}{16}\,x^{2} - \frac{35}{16},\ \ldots\right] \tag{9}$$

Type: Stream Polynomial Fraction Integer

Streams display only a few of their initial elements. Otherwise, they are "lazy": they only compute elements when you ask for them.

Data structures are an important component for building application software. Advanced users can represent data for applications in optimal fashion. In all, AXIOM offers over forty kinds of aggregate data structures, ranging from mutable structures (such as cyclic lists and flexible arrays) to storage efficient structures (such as bit vectors). As an example, streams are used as the internal data structure for power series.

What is the series expansion of $\log(\cot(x))$ about $x = \pi/2$?

```
series(log(cot(x)),x = %pi/2)
```

$$\log\left(\frac{2\ x - \%pi}{2}\right) + \log(-1) + \frac{1}{3}\left(x - \frac{\%pi}{2}\right)^2 + \frac{7}{90}\left(x - \frac{\%pi}{2}\right)^4 + \frac{62}{2835}\left(x - \frac{\%pi}{2}\right)^6 + O\left(\left(x - \frac{\%pi}{2}\right)^8\right) \tag{10}$$

Type: UnivariatePuiseuxSeries(Expression Integer, x, pi/2)

Series and streams make no attempt to compute *all* their elements! Rather, they stand ready to deliver elements on demand.

What is the coefficient of the 50 th term of this series?

```
coefficient(%,50)
```

$$\frac{44590788901016030052447242300856550965644}{7131469286438669111584090881309360354581359130859375} \tag{11}$$

Type: Expression Integer

Mathematical Structures

AXIOM also has many kinds of mathematical structures. These range from simple ones (like polynomials and matrices) to more esoteric ones (like ideals and Clifford algebras). Most structures allow the construction of arbitrarily complicated "types."

Even a simple input expression can result in a type with several levels.

```
matrix [[x + %i,0], [1,-2]]
```

$$\begin{bmatrix} x + \%i & 0 \\ 1 & -2 \end{bmatrix} \tag{12}$$

Type: Matrix Polynomial Complex Integer

The AXIOM interpreter builds types in response to user input. Often, the type of the result is changed in order to be applicable to an operation.

The inverse operation requires that elements of the above matrices are fractions.

```
inverse(%)
```

$$\begin{bmatrix} \frac{1}{x+\%i} & 0 \\ \frac{1}{2\ x+2\ \%i} & -\frac{1}{2} \end{bmatrix} \tag{13}$$

Type: Union(Matrix Fraction Polynomial Complex Integer, ...)

Pattern Matching

A convenient facility for symbolic computation is "pattern matching." Suppose you have a trigonometric expression and you want to transform it to some equivalent form. Use a `rule` command to describe the transformation rules you need. Then give the rules a name and apply that name as a function to your trigonometric expression.

Introduce two rewrite rules.

```
sinCosExpandRules := rule
  sin(x+y)  ==  sin(x)*cos(y) + sin(y)*cos(x)
  cos(x+y)  ==  cos(x)*cos(y) - sin(x)*sin(y)
  sin(2*x)  ==  2*sin(x)*cos(x)
  cos(2*x)  ==  cos(x)**2 - sin(x)**2
```

$$\{\sin(y+x) == \cos(x)\,\sin(y) + \cos(y)\,\sin(x),$$
$$\cos(y+x) == -\sin(x)\,\sin(y) + \cos(x)\,\cos(y), \tag{14}$$
$$\sin(2\ x) == 2\ \cos(x)\,\sin(x),\ \cos(2\ x) == -\sin(x)^2 + \cos(x)^2\}$$

Type: Ruleset(Integer, Integer, Expression Integer)

Apply the rules to a simple trigonometric expression.

```
sinCosExpandRules(sin(a+2*b+c))
```

$$\left(-\cos(a)\sin(b)^2 - 2\cos(b)\sin(a)\sin(b) + \cos(a)\cos(b)^2\right)\sin(c)-$$
$$\cos(c)\sin(a)\sin(b)^2 + 2\cos(a)\cos(b)\cos(c)\sin(b)+ \tag{15}$$
$$\cos(b)^2\cos(c)\sin(a)$$

Type: Expression Integer

Using input files, you can create your own library of transformation rules relevant to your applications, then selectively apply the rules you need.

Polymorphic Algorithms

All components of the AXIOM algebra library are written in the AXIOM library language. This language is similar to the interactive language except for protocols that authors are obliged to follow. The library language permits you to write "polymorphic algorithms," algorithms defined to work in their most natural settings and over a variety of types.

Define a system of polynomial equations `S`.

```
S := [3*x**3 + y + 1 = 0,y**2 = 4]
```

$$\left[y + 3\ x^3 + 1 = 0,\ y^2 = 4\right] \tag{16}$$

Type: List Equation Polynomial Integer

Solve the system S using rational number arithmetic and 30 digits of accuracy.

```
solve(S,1/10**30)
```

$$\left[\left[y = -2,\ x = \frac{1757879671211184245283070414507}{2535301200456458802993406410752}\right],\right. \tag{17}$$

$$\left.[y = 2,\ x = -1]\right]$$

Type: List List Equation Polynomial Fraction Integer

Solve S with the solutions expressed in radicals.

```
radicalSolve(S)
```

$$\left[[y = 2,\ x = -1],\ \left[y = 2,\ x = \frac{-\sqrt{-3}+1}{2}\right],\right.$$

$$\left[y = 2,\ x = \frac{\sqrt{-3}+1}{2}\right],\ \left[y = -2,\ x = \frac{1}{\sqrt[3]{3}}\right], \tag{18}$$

$$\left.\left[y = -2,\ x = \frac{\sqrt{-1}\,\sqrt{3}-1}{2\,\sqrt[3]{3}}\right],\ \left[y = -2,\ x = \frac{-\sqrt{-1}\,\sqrt{3}-1}{2\,\sqrt[3]{3}}\right]\right]$$

Type: List List Equation Expression Integer

While these solutions look very different, the results were produced by the same internal algorithm! The internal algorithm actually works with equations over any "field." Examples of fields are the rational numbers, floating point numbers, rational functions, power series, and general expressions involving radicals.

Extensibility

Users and system developers alike can augment the AXIOM library, all using one common language. Library code, like interpreter code, is compiled into machine binary code for run-time efficiency.

Using this language, you can create new computational types and new algorithmic packages. All library code is polymorphic, described in terms of a database of algebraic properties. By following the language protocols, there is an automatic, guaranteed interaction between your code and that of colleagues and system implementers.

A Technical Introduction to AXIOM

AXIOM has both an *interactive language* for user interactions and a *programming language* for building library modules. Like Modula 2, PASCAL, FORTRAN, and Ada, the programming language emphasizes strict type-checking. Unlike these languages, types in AXIOM are dynamic objects: they are created at run-time in response to user commands.

Here is the idea of the AXIOM programming language in a nutshell. AXIOM types range from algebraic ones (like polynomials, matrices, and power series) to data structures (like lists, dictionaries, and input files). Types combine in any meaningful way. You can build polynomials of matrices, matrices of polynomials of power series, hash tables with symbolic keys and rational function entries, and so on.

Categories define algebraic properties to ensure mathematical correctness. They ensure, for example, that matrices of polynomials are OK, but matrices of input files are not. Through categories, programs can discover that polynomials of continued fractions have a commutative multiplication whereas polynomials of matrices do not.

Categories allow algorithms to be defined in their most natural setting. For example, an algorithm can be defined to solve polynomial equations over *any* field. Likewise a greatest common divisor can compute the "gcd" of two elements from *any* Euclidean domain. Categories foil attempts to compute meaningless "gcds", for example, of two hashtables. Categories also enable algorithms to be compiled into machine code that can be run with arbitrary types.

The AXIOM interactive language is oriented towards ease-of-use. The AXIOM interpreter uses type-inferencing to deduce the type of an object from user input. Type declarations can generally be omitted for common types in the interactive language.

So much for the nutshell. Here are these basic ideas described by ten design principles:

Types are Defined by Abstract Datatype Programs

Basic types are called *domains of computation*, or, simply, *domains*. Domains are defined by AXIOM programs of the form:

```
Name(...): Exports == Implementation
```

Each domain has a capitalized `Name` that is used to refer to the class of its members. For example, Integer denotes "the class of integers," Float, "the class of floating point numbers," and String, "the class of strings."

The "`...`" part following `Name` lists zero or more parameters to the constructor. Some basic ones like Integer take no parameters. Others, like Matrix, Polynomial and List, take a single parameter that again must be a domain. For example, Matrix(Integer) denotes "matrices over the integers," Polynomial (Float) denotes "polynomial with floating point coefficients," and List (Matrix (Polynomial (Integer))) denotes "lists of matrices of polynomials over the integers." There is no restriction on the number or type of parameters of a domain constructor.

The `Exports` part specifies operations for creating and manipulating objects of the domain. For example, type Integer exports constants `0` and `1`, and operations "+", "`-`", and "`*`". While these operations are common, others such as **odd?** and **bit?** are not.

The `Implementation` part defines functions that implement the exported operations of the domain. These functions are frequently described in terms of another lower-level domain used to represent the objects of the domain.

The Type of Basic Objects is a Domain or Subdomain

Every AXIOM object belongs to a *unique* domain. The domain of an object is also called its *type*. Thus the integer 7 has type `Integer` and the string `"daniel"` has type `String`.

The type of an object, however, is not unique. The type of integer 7 is not only Integer but `NonNegativeInteger`, `PositiveInteger`, and possibly, in general, any other "subdomain" of the domain `Integer`. A *subdomain* is a domain with a "membership predicate". `PositiveInteger` is a subdomain of `Integer` with the predicate "is the integer > 0?".

Subdomains with names are defined by abstract datatype programs similar to those for domains. The *Export* part of a subdomain, however, must list a subset of the exports of the domain. The `Implementation` part optionally gives special definitions for subdomain objects.

Domains Have Types Called Categories

Domain and subdomains in AXIOM are themselves objects that have types. The type of a domain or subdomain is called a *category*. Categories are described by programs of the form:

```
Name(...): Category == Exports
```

The type of every category is the distinguished symbol `Category`. The category `Name` is used to designate the class of domains of that type. For example, category Ring designates the class of all rings. Like domains, categories can take zero or more parameters as indicated by the "`...`" part following `Name`. Two examples are Module(R) and MatrixCategory(R,Row,Col).

The `Exports` part defines a set of operations. For example, Ring exports the operations "`0`", "`1`", "`+`", "`-`", and "`*`". Many algebraic domains such as Integer and Polynomial (Float) are rings. String and List (R) (for any domain R) are not.

Categories serve to ensure the type-correctness. The definition of matrices states `Matrix(R: Ring)` requiring its single parameter `R` to be a ring. Thus a "matrix of polynomials" is allowed, but "matrix of lists" is not.

Operations Can Refer To Abstract Types

All operations have prescribed source and target types. Types can be denoted by symbols that stand for domains, called "symbolic domains." The following lines of AXIOM code use a symbolic domain `R`:

```
R: Ring
power: (R, NonNegativeInteger): R -> R
power(x, n) == x ** n
```

Line 1 declares the symbol `R` to be a ring. Line 2 declares the type of `power` in terms of `R`. From the definition on line 3, `power(3,2)` produces 9 for `x = 3` and `R =` Integer. Also, `power(3.0,2)` produces `9.0` for `x = 3.0` and `R =` Float. `power("oxford",2)` however fails since `"oxford"` has type String which is not a ring.

Using symbolic domains, algorithms can be defined in their most natural or general setting.

Categories Form Hierarchies

Categories form hierarchies (technically, directed-acyclic graphs). A simplified hierarchical world of algebraic categories is shown below in Figure 2. At the top of this world is SetCategory, the class of algebraic sets. The notions of parents, ancestors, and descendants is clear. Thus ordered sets (domains of category OrderedSet) and rings are also algebraic sets. Likewise, fields and integral domains are rings and algebraic sets. However fields and integral domains are not ordered sets.

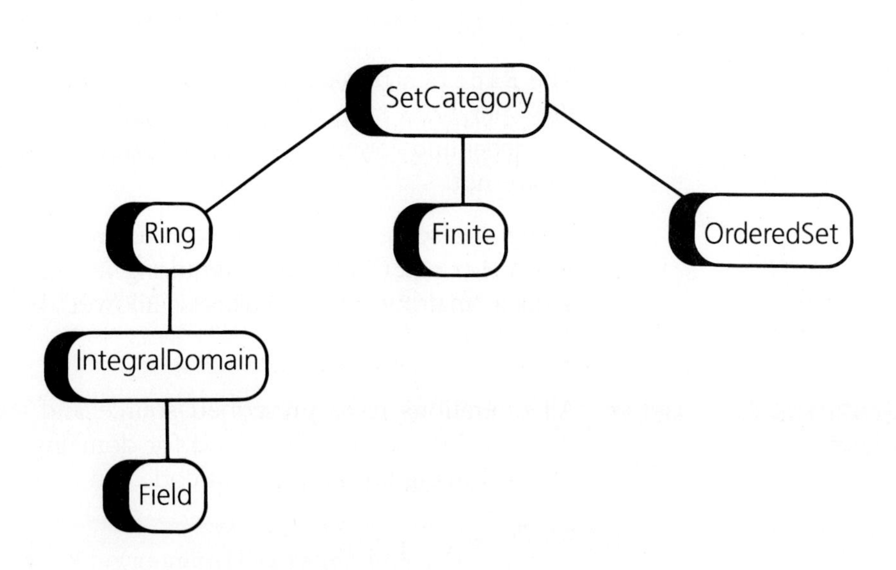

Figure 2: A simplified category hierarchy.

Domains Belong to Categories by Assertion

A category designates a class of domains. Which domains? You might think that Ring designates the class of all domains that export `0`, `1`, "+", "`-`", and "`*`". But this is not so. Each domain must *assert* which categories it belongs to.

The `Export` part of the definition for Integer reads, for example:

```
Join(OrderedSet, IntegralDomain,  ...) with ...
```

This definition asserts that Integer is both an ordered set and an integral domain. In fact, Integer does not explicitly export constants `0` and `1` and operations "+", "`-`" and "`*`" at all: it inherits them all from `Ring`! Since IntegralDomain is a descendant of `Ring`, Integer is therefore also a ring.

Assertions can be conditional. For example, Complex(R) defines its exports by:

```
Ring with ... if R has Field then Field ...
```

Thus Complex(Float) is a field but Complex(Integer) is not since Integer is not a field.

You may wonder: "Why not simply let the set of operations determine whether a domain belongs to a given category?". AXIOM allows operation names (for example, **norm**) to have very different meanings in different contexts. The meaning of an operation in AXIOM is determined by context. By associating operations with categories, operation names can be reused whenever appropriate or convenient to do so. As a simple example, the operation "<" might be used to denote lexicographic-comparison in an algorithm. However, it is wrong to use the same "<" with this definition of absolute-value: `abs(x) == if x < 0 then -x else x`. Such a definition for `abs` in AXIOM is protected by context: argument `x` is required to be a member of a domain of category OrderedSet.

Packages Are Clusters of Polymorphic Operations

In AXIOM, facilities for symbolic integration, solution of equations, and the like are placed in "packages". A *package* is a special kind of domain: one whose exported operations depend solely on the parameters of the constructor and/or explicit domains.

If you want to use AXIOM, for example, to define some algorithms for solving equations of polynomials over an arbitrary field `F`, you can do so with a package of the form:

```
MySolve(F: Field): Exports == Implementation
```

where `Exports` specifies the **solve** operations you wish to export and `Implementation` defines functions for implementing your algorithms.

Once AXIOM has compiled your package, your algorithms can then be used for any `F`: floating-point numbers, rational numbers, complex rational functions, and power series, to name a few.

The Interpreter Builds Domains Dynamically

The AXIOM interpreter reads user input then builds whatever types it needs to perform the indicated computations. For example, to create the matrix

$$M = \begin{pmatrix} x^2+1 & 0 \\ 0 & x/2 \end{pmatrix}$$

the interpreter first loads the modules Matrix, Polynomial, Fraction, and Integer from the library, then builds the *domain tower* "matrices of polynomials of rational numbers (fractions of integers)".

Once a domain tower is built, computation proceeds by calling operations down the tower. For example, suppose that the user asks to square the above matrix. To do this, the function "`*`" from Matrix is passed `M` to compute `M * M`. The function is also passed an environment containing `R` that, in this case, is Polynomial (Fraction (Integer)). This results in the successive calling of the "`*`" operations from Polynomial, then from Fraction, and then finally from Integer before a result is passed back up the tower.

Categories play a policing role in the building of domains. Because the argument of Matrix is required to be a ring, AXIOM will not build nonsensical types such as "matrices of input files".

AXIOM Code is Compiled

AXIOM programs are statically compiled to machine code, then placed into library modules. Categories provide an important role in obtaining efficient object code by enabling:

- static type-checking at compile time;
- fast linkage to operations in domain-valued parameters;
- optimization techniques to be used for partially specified types (operations for "vectors of `R`", for instance, can be open-coded even though `R` is unknown).

AXIOM is Extensible

Users and system implementers alike use the AXIOM language to add facilities to the AXIOM library. The entire AXIOM library is in fact written in the AXIOM source code and available for user modification and/or extension.

AXIOM's use of abstract datatypes clearly separates the exports of a domain (what operations are defined) from its implementation (how the objects are represented and operations are defined). Users of a domain can thus only create and manipulate objects through these exported operations. This allows implementers to "remove and replace" parts of the library safely by newly upgraded (and, we hope, correct) implementations without consequence to its users.

Categories protect names by context, making the same names available for use in other contexts. Categories also provide for code-economy. Algorithms can be parameterized categorically to characterize their correct and most general context. Once compiled, the same machine code is applicable in all such contexts.

Finally, AXIOM provides an automatic, guaranteed interaction between new and old code. For example:

- if you write a new algorithm that requires a parameter to be a field, then your algorithm will work automatically with every field defined in the system; past, present, or future.
- if you introduce a new domain constructor that produces a field, then the objects of that domain can be used as parameters to any algorithm using field objects defined in the system; past, present, or future.

These are the key ideas. For further information, we particularly recommend your reading chapters 11, 12, and 13, where these ideas are explained in greater detail.

PART I

Basic Features of AXIOM

CHAPTER 1

An Overview of AXIOM

Welcome to the AXIOM environment for interactive computation and problem solving. Consider this chapter a brief, whirlwind tour of the AXIOM world. We introduce you to AXIOM's graphics and the AXIOM language. Then we give a sampling of the large variety of facilities in the AXIOM system, ranging from the various kinds of numbers, to data types (like lists, arrays, and sets) and mathematical objects (like matrices, integrals, and differential equations). We conclude with the discussion of system commands and an interactive "undo."

Before embarking on the tour, we need to brief those readers working interactively with AXIOM on some details. Others can skip right immediately to Section 1.2 on page 21.

1.1 Starting Up and Winding Down

You need to know how to start the AXIOM system and how to stop it. We assume that AXIOM has been correctly installed on your machine (as described in another AXIOM document).

To begin using AXIOM, issue the command **axiom** to the operating system shell. There is a brief pause, some start-up messages, and then one or more windows appear.

If you are not running AXIOM under the X Window System, there is only one window (the console). At the lower left of the screen there is a prompt that looks like

```
(1) ->
```

When you want to enter input to AXIOM, you do so on the same line after the prompt. The "1" in "(1)" is the computation step number and is incremented after you enter AXIOM statements. Note, however, that a system command such as `)clear all` may change the step number in other ways. We talk about step numbers more when we discuss system commands and the workspace history facility.

If you are running AXIOM under the X Window System, there may be two windows: the console window (as just described) and the HyperDoc main menu. HyperDoc is a multiple-window hypertext system that lets you view AXIOM documentation and examples on-line, execute AXIOM expressions, and generate graphics. If you are in a graphical windowing environment, it is usually started automatically when AXIOM begins. If it is not running, issue `)hd` to start it. We discuss the basics of HyperDoc in Chapter 3.

To interrupt an AXIOM computation, hold down the **Ctrl** (control) key and press **c**. This brings you back to the AXIOM prompt.

To exit from AXIOM, move to the console window, type `)quit` at the input prompt and press the **Enter** key. You will probably be prompted with the following message:

Please enter **y** or **yes** if you really want to leave the interactive environment and return to the operating system

You should respond **yes**, for example, to exit AXIOM.

We are purposely vague in describing exactly what your screen looks like or what messages AXIOM displays. AXIOM runs on a number of different machines, operating systems and window environments, and these differences all affect the physical look of the system. You can also change the way that AXIOM behaves via *system commands* described later in this chapter and in Appendix A. System commands are special commands, like `)set`, that begin with a closing parenthesis and are used to change your environment. For example, you can set a system variable so that you are not prompted for confirmation when you want to leave AXIOM.

1.1.1 Clef

If you are using AXIOM under the X Window System, the Clef command line editor is probably available and installed. With this editor you can recall previous lines with the up and down arrow keys (↑ and ↓). To move forward and backward on a line, use the right and left arrows (→ and ←). You can use the **Insert** key to toggle insert mode on or off. When you are in insert mode, the cursor appears as a large block and if you type anything, the characters are inserted into the line without deleting the previous ones.

If you press the **Home** key, the cursor moves to the beginning of the line and if you press the **End** key, the cursor moves to the end of the line. Pressing **Ctrl**-**End** deletes all the text from the cursor to the end of the line.

Clef also provides AXIOM operation name completion for a limited set of operations. If you enter a few letters and then press the **Tab** key, Clef tries to use those letters as the prefix of an AXIOM operation name. If a name appears and it is not what you want, press **Tab** again to see another name.

You are ready to begin your journey into the world of AXIOM. Proceed to the first stop.

1.2 Typographic Conventions

In this book we have followed these typographical conventions:

- Categories, domains and packages are displayed in a sans-serif typeface: Ring, Integer, DiophantineSolutionPackage.
- Prefix operators, infix operators, and punctuation symbols in the AXIOM language are displayed in the text like this: "`+`", "`$`", "`+->`".
- AXIOM expressions or expression fragments are displayed in a monospace typeface: `inc(x) == x + 1`.
- For clarity of presentation, TeX is often used to format expressions: $g(x) = x^2 + 1$.
- Function names and HyperDoc button names are displayed in the text in a bold typeface: **factor**, **integrate**, **Lighting**.
- Italics are used for emphasis and for words defined in the glossary: *category*.

This book contains over 2500 examples of AXIOM input and output. All examples were run though AXIOM and their output was created in TeX form for this book by the AXIOM TexFormat package. We have deleted system messages from the example output if those messages are not important for the discussions in which the examples appear.

1.3 The AXIOM Language

The AXIOM language is a rich language for performing interactive computations and for building components of the AXIOM library. Here we present only some basic aspects of the language that you need to know for the rest of this chapter. Our discussion here is intentionally informal, with details unveiled on an "as needed" basis. For more information on a particular construct, we suggest you consult the index at the back of the book.

1.3.1 Arithmetic Expressions

For arithmetic expressions, use the "+" and "-" *operators* as in mathematics. Use "*" for multiplication, and "**" for exponentiation. To create a fraction, use "/". When an expression contains several operators, those of highest *precedence* are evaluated first. For arithmetic operators, "**" has highest precedence, "*" and "/" have the next highest precedence, and "+" and "-" have the lowest precedence.

AXIOM puts implicit parentheses around operations of higher precedence, and groups those of equal precedence from left to right.

```
1 + 2 - 3 / 4 * 3 ** 2 - 1
```

$$-\frac{19}{4} \tag{1}$$

Type: Fraction Integer

The above expression is equivalent to this.

```
((1 + 2) - ((3 / 4) * (3 ** 2))) - 1
```

$$-\frac{19}{4} \tag{2}$$

Type: Fraction Integer

If an expression contains subexpressions enclosed in parentheses, the parenthesized subexpressions are evaluated first (from left to right, from inside out).

```
1 + 2 - 3/ (4 * 3 ** (2 - 1))
```

$$\frac{11}{4} \tag{3}$$

Type: Fraction Integer

1.3.2 Previous Results

Use the percent sign ("%") to refer to the last result. Also, use "%%" to refer to previous results. `%%(-1)` is equivalent to "%", `%%(-2)` returns the next to the last result, and so on. `%%(1)` returns the result from step number 1, `%%(2)` returns the result from step number 2, and so on. `%%(0)` is not defined.

This is ten to the tenth power.

```
10 ** 10
```

$$10000000000 \tag{1}$$

Type: PositiveInteger

This is the last result minus one.

```
% - 1
```

$$9999999999 \tag{2}$$

Type: PositiveInteger

This is the last result.

```
%%(-1)
```

$$9999999999 \tag{3}$$

Type: PositiveInteger

This is the result from step number 1.

```
%%(1)
```

$$10000000000 \tag{4}$$

Type: PositiveInteger

1.3.3 Some Types

Everything in AXIOM has a type. The type determines what operations you can perform on an object and how the object can be used. An entire chapter of this book (Chapter 2) is dedicated to the interactive use of types. Several of the final chapters discuss how types are built and how they are organized in the AXIOM library.

Positive integers are given type PositiveInteger.

```
8
```

$$8 \tag{1}$$

Type: PositiveInteger

Negative ones are given type Integer. This fine distinction is helpful to the AXIOM interpreter.

```
-8
```

$$-8 \tag{2}$$

Type: Integer

Here a positive integer exponent gives a polynomial result.

```
x**8
```

$$x^8 \tag{3}$$

Type: Polynomial Integer

Here a negative integer exponent produces a fraction.

```
x**(-8)
```

$$\frac{1}{x^8} \tag{4}$$

Type: Fraction Polynomial Integer

1.3.4 Symbols, Variables, Assignments, and Declarations

A *symbol* is a literal used for the input of things like the "variables" in polynomials and power series.

We use the three symbols `x`, `y`, and `z` in entering this polynomial.

```
(x - y*z)**2
```

$$y^2\ z^2 - 2\ x\ y\ z + x^2 \tag{1}$$

Type: Polynomial Integer

A symbol has a name beginning with an uppercase or lowercase alphabetic character, "%", or "!". Successive characters (if any) can be any of the above, digits, or "?". Case is distinguished: the symbol `points` is different from the symbol `Points`.

A symbol can also be used in AXIOM as a *variable*. A variable refers to a value. To *assign* a value to a variable, the operator ":=" is used.[1] A variable initially has no restrictions on the kinds of values to which it can refer.

This assignment gives the value 4 (an integer) to a variable named `x`.

```
x := 4
```

$$4 \tag{2}$$

Type: PositiveInteger

This gives the value `z + 3/5` (a polynomial) to `x`.

```
x := z + 3/5
```

$$z + \frac{3}{5} \tag{3}$$

Type: Polynomial Fraction Integer

To restrict the types of objects that can be assigned to a variable, use a *declaration*

```
y : Integer
```

Type: Void

After a variable is declared to be of some type, only values of that type can be assigned to that variable.

```
y := 89
```

$$89 \tag{5}$$

Type: Integer

The declaration for `y` forces values assigned to `y` to be converted to integer values.

```
y := sin %pi
```

$$0 \tag{6}$$

Type: Integer

If no such conversion is possible, AXIOM refuses to assign a value to `y`.

```
y := 2/3
```

```
Cannot convert right-hand side of assignment
2
-
3

to an object of the type Integer of the left-hand
side.
```

[1]AXIOM actually has two forms of assignment: *immediate* assignment, as discussed here, and *delayed assignment*. See Section 5.1 on page 109 for details.

A type declaration can also be given together with an assignment. The declaration can assist AXIOM in choosing the correct operations to apply.

```
f : Float := 2/3
```

$$0.6666666666666666666666666666666666666667 \qquad (7)$$

Type: Float

Any number of expressions can be given on input line. Just separate them by semicolons. Only the result of evaluating the last expression is displayed.

These two expressions have the same effect as the previous single expression.

```
f : Float; f := 2/3
```

$$0.6666666666666666666666666666666666666667 \qquad (8)$$

Type: Float

The type of a symbol is either Symbol or Variable(*name*) where *name* is the name of the symbol.

By default, the interpreter gives this symbol the type Variable(q).

```
q
```

$$q \qquad (9)$$

Type: Variable q

When multiple symbols are involved, Symbol is used.

```
[q, r]
```

$$[q, r] \qquad (10)$$

Type: List Symbol

What happens when you try to use a symbol that is the name of a variable?

```
f
```

$$0.6666666666666666666666666666666666666667 \qquad (11)$$

Type: Float

Use a single quote ("'") before the name to get the symbol.

```
'f
```

$$f \qquad (12)$$

Type: Variable f

Quoting a name creates a symbol by preventing evaluation of the name as a variable. Experience will teach you when you are most likely going to need to use a quote. We try to point out the location of such trouble spots.

1.3.5 Conversion

Objects of one type can usually be "converted" to objects of several other types. To *convert* an object to a new type, use the "::" infix operator.[2] For example, to display an object, it is necessary to convert the object to type OutputForm.

[2]Conversion is discussed in detail in Section 2.7 on page 78.

This produces a polynomial with rational number coefficients.

```
p := r**2 + 2/3
```

$$r^2 + \frac{2}{3} \tag{1}$$

Type: Polynomial Fraction Integer

Create a quotient of polynomials with integer coefficients by using "`::`".

```
p :: Fraction Polynomial Integer
```

$$\frac{3\ r^2 + 2}{3} \tag{2}$$

Type: Fraction Polynomial Integer

Some conversions can be performed automatically when AXIOM tries to evaluate your input. Others conversions must be explicitly requested.

1.3.6 Calling Functions

As we saw earlier, when you want to add or subtract two values, you place the arithmetic operator "+" or "`-`" between the two *arguments* denoting the values. To use most other AXIOM operations, however, you use another syntax: write the name of the operation first, then an open parenthesis, then each of the arguments separated by commas, and, finally, a closing parenthesis. If the operation takes only one argument and the argument is a number or a symbol, you can omit the parentheses.

This calls the operation **factor** with the single integer argument `120`.

```
factor(120)
```

$$2^3\ 3\ 5 \tag{1}$$

Type: Factored Integer

This is a call to **divide** with the two integer arguments `125` and `7`.

```
divide(125,7)
```

$$[quotient = 17,\ remainder = 6] \tag{2}$$

Type: Record(quotient: Integer, remainder: Integer)

This calls **quatern** with four floating-point arguments.

```
quatern(3.4,5.6,2.9,0.1)
```

$$3.4 + 5.6\ i + 2.9\ j + 0.1\ k \tag{3}$$

Type: Quaternion Float

This is the same as `factorial(10)`.

```
factorial 10
```

$$3628800 \tag{4}$$

Type: PositiveInteger

1.3.7 Some Predefined Macros

AXIOM provides several *macros* for your convenience.[3] Macros are names (or forms) that expand to larger expressions for commonly used values.

%i	The square root of -1.
%e	The base of the natural logarithm.
%pi	π.
%infinity	∞.
%plusInfinity	$+\infty$.
%minusInfinity	$-\infty$.

1.3.8 Long Lines

When you enter AXIOM expressions from your keyboard, there will be times when they are too long to fit on one line. AXIOM does not care how long your lines are, so you can let them continue from the right margin to the left side of the next line.

Alternatively, you may want to enter several shorter lines and have AXIOM glue them together. To get this glue, put an underscore (_) at the end of each line you wish to continue.

```
2_
+_
3
```

is the same as if you had entered

```
2+3
```

If you are putting your AXIOM statements in an input file (see Section 4.1 on page 99), you can use indentation to indicate the structure of your program. (see Section 5.2 on page 112).

1.3.9 Comments

The comment beginning with -- is ignored by AXIOM.

Comment statements begin with two consecutive hyphens or two consecutive plus signs and continue until the end of the line.

```
2 + 3 -- this is rather simple, no?
```

$$5 \tag{1}$$

Type: PositiveInteger

There is no way to write long multi-line comments other than starting each line with "--" or "++".

[3]See Section 6.2 on page 136 for a discussion on how to write your own macros.

1.4 Graphics

AXIOM has a two- and three-dimensional drawing and rendering package that allows you to draw, shade, color, rotate, translate, map, clip, scale and combine graphic output of AXIOM computations. The graphics interface is capable of plotting functions of one or more variables and plotting parametric surfaces. Once the graphics figure appears in a window, move your mouse to the window and click. A control panel appears immediately and allows you to interactively transform the object.

This is an example of AXIOM's two-dimensional plotting. From the 2D Control Panel you can rescale the plot, turn axes and units on and off and save the image, among other things. This PostScript image was produced by clicking on the **PS** 2D Control Panel button.

```
draw(cos(5*t/8), t=0..16*%pi, coordinates==polar)
```

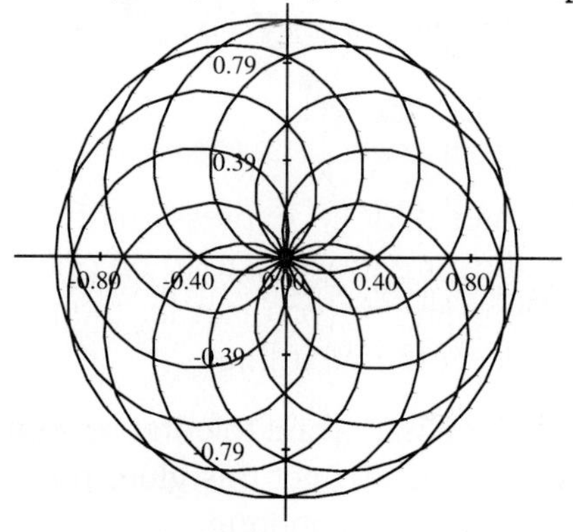

This is an example of AXIOM's three-dimensional plotting. It is a monochrome graph of the complex arctangent function. The image displayed was rotated and had the "shade" and "outline" display options set from the 3D Control Panel. The PostScript output was produced by clicking on the **save** 3D Control Panel button and then clicking on the **PS** button. See Section 8.1 on page 227 for more details and examples of AXIOM's numeric and graphics capabilities.

```
draw((x,y) +-> real atan complex(x,y), -%pi..%pi, -
  %pi..%pi, colorFunction == (x,y) +-> argument atan
  complex(x,y))
```

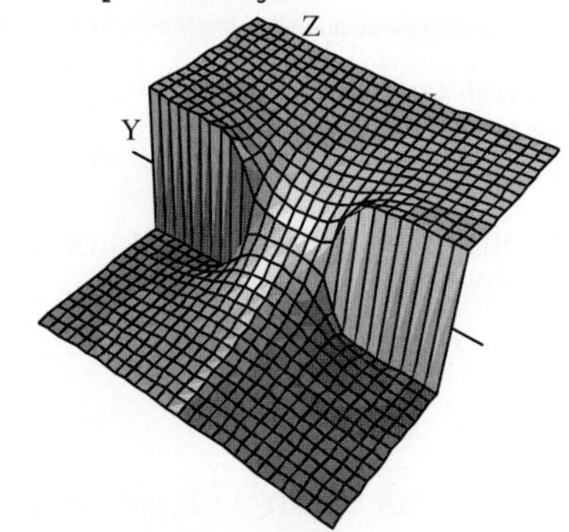

An exhibit of AXIOM Images is given in the center section of this book. For a description of the commands and programs that produced these figures, see Appendix F. PostScript output is available so that AXIOM images can be printed.[4] See Chapter 7 for more examples and details about using AXIOM's graphics facilities.

[4]PostScript is a trademark of Adobe Systems Incorporated, registered in the United States.

1.5 Numbers

AXIOM distinguishes very carefully between different kinds of numbers, how they are represented and what their properties are. Here are a sampling of some of these kinds of numbers and some things you can do with them.

Integer arithmetic is always exact.

```
11**13 * 13**11 * 17**7 - 19**5 * 23**3
```

25387751112538918594666224484237298 (1)

Type: PositiveInteger

Integers can be represented in factored form.

```
factor 643238070748569023720594412551704344145570763243
```

$11^{13}\ 13^{11}\ 17^{7}\ 19^{5}\ 23^{3}\ 29^{2}$ (2)

Type: Factored Integer

Results stay factored when you do arithmetic. Note that the 12 is automatically factored for you.

```
% * 12
```

$2^{2}\ 3\ 11^{13}\ 13^{11}\ 17^{7}\ 19^{5}\ 23^{3}\ 29^{2}$ (3)

Type: Factored Integer

Integers can also be displayed to bases other than 10. This is an integer in base 11.

```
radix(25937424601,11)
```

10000000000 (4)

Type: RadixExpansion 11

Roman numerals are also available for those special occasions.

```
roman(1992)
```

MCMXCII (5)

Type: RomanNumeral

Rational number arithmetic is also exact.

```
r := 10 + 9/2 + 8/3 + 7/4 + 6/5 + 5/6 + 4/7 + 3/8 + 2/9
```

$$\frac{55739}{2520} \quad (6)$$

Type: Fraction Integer

To factor fractions, you have to map **factor** onto the numerator and denominator.

```
map(factor,r)
```

$$\frac{139\ 401}{2^{3}\ 3^{2}\ 5\ 7} \quad (7)$$

Type: Fraction Factored Integer

Type SmallInteger refers to machine word-length integers. In English, this expression means "11 as a small integer".

```
11@SmallInteger
```

11 (8)

Type: SmallInteger

Machine double-precision floating-point numbers are also available for numeric and graphical applications.

```
123.21@SmallFloat
```

123.21000000000001 (9)

Type: SmallFloat

The normal floating-point type in AXIOM, Float, is a software implementation of floating-point numbers in which the exponent and the mantissa may have any number of digits.[5] The types Complex(Float) and Complex(SmallFloat) are the corresponding software implementations of complex floating-point numbers.

This is a floating-point approximation to about twenty digits. The "::" is used here to change from one kind of object (here, a rational number) to another (a floating-point number).

```
r :: Float
```

$$22.118650793650793650793650793650793650793650793651079\ldots$$

22.11865079365079365079365079365079365079 (10)

Type: Float

Use **digits** to change the number of digits in the representation. This operation returns the previous value so you can reset it later.

```
digits(22)
```

40 (11)

Type: PositiveInteger

To 22 digits of precision, the number $e^{\pi\sqrt{163.0}}$ appears to be an integer.

```
exp(%pi * sqrt 163.0)
```

262537412640768744.0 (12)

Type: Float

Increase the precision to forty digits and try again.

```
digits(40); exp(%pi * sqrt 163.0)
```

262537412640768743.9999999999992500725976 (13)

Type: Float

Here are complex numbers with rational numbers as real and imaginary parts.

```
(2/3 + %i)**3
```

$$-\frac{46}{27}+\frac{1}{3}\ \%i \qquad (14)$$

Type: Complex Fraction Integer

The standard operations on complex numbers are available.

```
conjugate %
```

$$-\frac{46}{27}-\frac{1}{3}\ \%i \qquad (15)$$

Type: Complex Fraction Integer

You can factor complex integers.

```
factor(89 - 23 * %i)
```

$$-(1+\%i)\ (2+\%i)^2\ (3+2\ \%i)^2 \qquad (16)$$

Type: Factored Complex Integer

Complex numbers with floating point parts are also available.

```
exp(%pi/4.0 * %i)
```

0.7071067811865475244008443621048490392849+

0.7071067811865475244008443621048490392848 %i (17)

Type: Complex Float

[5]See 'Float' on page 368 and 'SmallFloat' on page 452 for additional information on floating-point types.

Every rational number has an exact representation as a repeating decimal expansion (see 'DecimalExpansion' on page 350).

```
decimal(1/352)
```

$$0.0028\overline{409} \tag{18}$$

Type: DecimalExpansion

A rational number can also be expressed as a continued fraction (see 'ContinuedFraction' on page 335).

```
continuedFraction(6543/210)
```

$$31 + \frac{1|}{|6} + \frac{1|}{|2} + \frac{1|}{|1} + \frac{1|}{|3} \tag{19}$$

Type: ContinuedFraction Integer

Also, partial fractions can be used and can be displayed in a compact ...

```
partialFraction(1,factorial(10))
```

$$\frac{159}{2^8} - \frac{23}{3^4} - \frac{12}{5^2} + \frac{1}{7} \tag{20}$$

Type: PartialFraction Integer

or expanded format (see 'PartialFraction' on page 433).

```
padicFraction(%)
```

$$\frac{1}{2} + \frac{1}{2^4} + \frac{1}{2^5} + \frac{1}{2^6} + \frac{1}{2^7} + \frac{1}{2^8} - \frac{2}{3^2} - \frac{1}{3^3} - \frac{2}{3^4} - \frac{2}{5} - \frac{2}{5^2} + \frac{1}{7} \tag{21}$$

Type: PartialFraction Integer

Like integers, bases (radices) other than ten can be used for rational numbers (see 'RadixExpansion' on page 444). Here we use base eight.

```
radix(4/7, 8)
```

$$0.\overline{4} \tag{22}$$

Type: RadixExpansion 8

Of course, there are complex versions of these as well. AXIOM decides to make the result a complex rational number.

```
% + 2/3*%i
```

$$\frac{4}{7} + \frac{2}{3}\ \%i \tag{23}$$

Type: Complex Fraction Integer

You can also use AXIOM to manipulate fractional powers.

```
(5 + sqrt 63 + sqrt 847)**(1/3)
```

$$\sqrt[3]{\sqrt{847} + 3\ \sqrt{7} + 5} \tag{24}$$

Type: AlgebraicNumber

You can also compute with integers modulo a prime.

```
x : PrimeField 7 := 5
```

$$5 \tag{25}$$

Type: PrimeField 7

Arithmetic is then done modulo 7.

```
x**3
```

$$6 \tag{26}$$

Type: PrimeField 7

Since 7 is prime, you can invert nonzero values.

```
1/x
```

$$3 \tag{27}$$

Type: PrimeField 7

You can also compute modulo an integer that is not a prime.

```
y : IntegerMod 6 := 5
```

$$5 \tag{28}$$

Type: IntegerMod 6

All of the usual arithmetic operations are available.

```
y**3
```

$$5 \tag{29}$$

Type: IntegerMod 6

Inversion is not available if the modulus is not a prime number. Modular arithmetic and prime fields are discussed in Section 8.11.1 on page 276.

```
1/y
```

```
Cannot find a definition or library operation named /
with argument types
                        PositiveInteger
                          IntegerMod 6
```

This defines a to be an algebraic number, that is, a root of a polynomial equation.

```
a := rootOf(a**5 + a**3 + a**2 + 3,a)
```

$$a \tag{30}$$

Type: Expression Integer

Computations with a are reduced according to the polynomial equation.

```
(a + 1)**10
```

$$-85\ a^4 - 264\ a^3 - 378\ a^2 - 458\ a - 287 \tag{31}$$

Type: Expression Integer

Define b to be an algebraic number involving a.

```
b := rootOf(b**4 + a,b)
```

$$b \tag{32}$$

Type: Expression Integer

Do some arithmetic.

```
2/(b - 1)
```

$$\frac{2}{b-1} \tag{33}$$

Type: Expression Integer

To expand and simplify this, call **ratDenom** to rationalize the denominator.

```
ratDenom(%)
```

$$\left(a^4 - a^3 + 2\ a^2 - a + 1\right) b^3 + \left(a^4 - a^3 + 2\ a^2 - a + 1\right) b^2 + \left(a^4 - a^3 + 2\ a^2 - a + 1\right) b + a^4 - a^3 + 2\ a^2 - a + 1 \tag{34}$$

Type: Expression Integer

If we do this, we should get b.

```
2/%+1
```

$$\frac{\left(\left(a^4-a^3+2\,a^2-a+1\right)b^3+\left(a^4-a^3+2\,a^2-a+1\right)b^2+\left(a^4-a^3+2\,a^2-a+1\right)b+a^4-a^3+2\,a^2-a+3\right)}{\left(\left(a^4-a^3+2\,a^2-a+1\right)b^3+\left(a^4-a^3+2\,a^2-a+1\right)b^2+\left(a^4-a^3+2\,a^2-a+1\right)b+a^4-a^3+2\,a^2-a+1\right)} \quad (35)$$

Type: Expression Integer

But we need to rationalize the denominator again.

```
ratForm(%)
```

$$b \quad (36)$$

Type: Expression Integer

Types Quaternion and Octonion are also available. Multiplication of quaternions is non-commutative, as expected.

```
q:=quatern(1,2,3,4)*quatern(5,6,7,8) -
  quatern(5,6,7,8)*quatern(1,2,3,4)
```

$$-8\,i+16\,j-8\,k \quad (37)$$

Type: Quaternion Integer

1.6 Data Structures

AXIOM has a large variety of data structures available. Many data structures are particularly useful for interactive computation and others are useful for building applications. The data structures of AXIOM are organized into *category hierarchies* as shown on the inside back cover.

A *list* is the most commonly used data structure in AXIOM for holding objects all of the same type.[6] The name *list* is short for "linked-list of nodes." Each node consists of a value (**first**) and a link (**rest**) that *points* to the next node, or to a distinguished value denoting the empty list. To get to, say, the third element, AXIOM starts at the front of the list, then traverses across two links to the third node.

Write a list of elements using square brackets with commas separating the elements.

```
u := [1,-7,11]
```

$$[1,\ -7,\ 11] \quad (1)$$

Type: List Integer

This is the value at the third node. Alternatively, you can say `u.3`.

```
first rest rest u
```

$$11 \quad (2)$$

Type: PositiveInteger

Many operations are defined on lists, such as: **empty?**, to test that a list has no elements; **cons**`(x,l)`, to create a new list with **first** element `x`

[6]Lists are discussed in 'List' on page 404 and in Section 5.5 on page 130.

and **rest** 1; **reverse**, to create a new list with elements in reverse order; and **sort**, to arrange elements in order.

An important point about lists is that they are "mutable": their constituent elements and links can be changed "in place." To do this, use any of the operations whose names end with the character "!".

The operation **concat!**(u,v) replaces the last link of the list u to point to some other list v. Since u refers to the original list, this change is seen by u.

```
concat!(u,[9,1,3,-4]); u
```

$$[1, -7, 11, 9, 1, 3, -4] \quad (3)$$

Type: List Integer

A *cyclic list* is a list with a "cycle": a link pointing back to an earlier node of the list. To create a cycle, first get a node somewhere down the list.

```
lastnode := rest(u,3)
```

$$[9, 1, 3, -4] \quad (4)$$

Type: List Integer

Use **setrest!** to change the link emanating from that node to point back to an earlier part of the list.

```
setrest!(lastnode,rest(u,2)); u
```

$$[1, -7, \overline{11, 9}] \quad (5)$$

Type: List Integer

A *stream* is a structure that (potentially) has an infinite number of distinct elements.[7] Think of a stream as an "infinite list" where elements are computed successively.

Create an infinite stream of factored integers. Only a certain number of initial elements are computed and displayed.

```
[factor(i) for i in 2.. by 2]
```

$$\left[2,\ 2^2,\ 2\ 3,\ 2^3,\ 2\ 5,\ 2^2\ 3,\ 2\ 7,\ \ldots\right] \quad (6)$$

Type: Stream Factored Integer

AXIOM represents streams by a collection of already-computed elements together with a function to compute the next element "on demand." Asking for the n th element causes elements 1 through n to be evaluated.

```
%.36
```

$$2^3\ 3^2 \quad (7)$$

Type: Factored Integer

Streams can also be finite or cyclic. They are implemented by a linked list structure similar to lists and have many of the same operations. For example, **first** and **rest** are used to access elements and successive nodes of a stream.

A *one-dimensional array* is another data structure used to hold objects of the same type.[8] Unlike lists, one-dimensional arrays are inflexible—they are implemented using a fixed block of storage. Their advantage is that they give quick and equal access time to any element.

[7] Streams are discussed in 'Stream' on page 457 and in Section 5.5 on page 130.
[8] See 'OneDimensionalArray' on page 425 for details.

A simple way to create a one-dimensional array is to apply the operation **oneDimensionalArray** to a list of elements.

```
a := oneDimensionalArray [1, -7, 3, 3/2]
```

$$\left[1,\ -7,\ 3,\ \frac{3}{2}\right] \tag{8}$$

Type: OneDimensionalArray Fraction Integer

One-dimensional arrays are also mutable: you can change their constituent elements "in place."

```
a.3 := 11; a
```

$$\left[1,\ -7,\ 11,\ \frac{3}{2}\right] \tag{9}$$

Type: OneDimensionalArray Fraction Integer

However, one-dimensional arrays are not flexible structures. You cannot destructively **concat!** them together.

```
concat!(a,oneDimensionalArray [1,-2])
```

```
Cannot find a definition or library operation named
concat! with argument types
          OneDimensionalArray Fraction Integer
              OneDimensionalArray Integer
```

Examples of datatypes similar to OneDimensionalArray are: Vector (vectors are mathematical structures implemented by one-dimensional arrays), String (arrays of "characters," represented by byte vectors), and Bits (represented by "bit vectors").

A vector of 32 bits, each representing the Boolean value `true`.

```
bits(32,true)
```

"11111111111111111111111111111111" (10)

Type: Bits

A *flexible array* is a cross between a list and a one-dimensional array.[9] Like a one-dimensional array, a flexible array occupies a fixed block of storage. Its block of storage, however, has room to expand! When it gets full, it grows (a new, larger block of storage is allocated); when it has too much room, it contracts.

Create a flexible array of three elements.

```
f := flexibleArray [2, 7, -5]
```

$$[2,\ 7,\ -5] \tag{11}$$

Type: FlexibleArray Integer

Insert some elements between the second and third elements.

```
insert!(flexibleArray [11, -3],f,2)
```

$$[2,\ 11,\ -3,\ 7,\ -5] \tag{12}$$

Type: FlexibleArray Integer

Flexible arrays are used to implement "heaps." A *heap* is an example of a data structure called a *priority queue*, where elements are ordered with respect to one another.[10] A heap is organized so as to optimize insertion

[9]See 'FlexibleArray' on page 366 for details.

[10]See 'Heap' on page 378 for more details. Heaps are also examples of data structures

and extraction of maximum elements. The **extract!** operation returns the maximum element of the heap, after destructively removing that element and reorganizing the heap so that the next maximum element is ready to be delivered.

An easy way to create a heap is to apply the operation **heap** to a list of values.

```
h := heap [-4,7,11,3,4,-7]
```

$$[11, 4, 7, -4, 3, -7] \tag{13}$$

Type: Heap Integer

This loop extracts elements one-at-a-time from h until the heap is exhausted, returning the elements as a list in the order they were extracted.

```
[extract!(h) while not empty?(h)]
```

$$[11, 7, 4, 3, -4, -7] \tag{14}$$

Type: List Integer

A *binary tree* is a "tree" with at most two branches per node: it is either empty, or else is a node consisting of a value, and a left and right subtree (again, binary trees).[11]

A *binary search tree* is a binary tree such that, for each node, the value of the node is greater than to all values (if any) in the left subtree, and less than or equal all values (if any) in the right subtree.

```
binarySearchTree [5,3,2,9,4,7,11]
```

$$[[2, 3, 4], 5, [7, 9, 11]] \tag{15}$$

Type: BinarySearchTree PositiveInteger

A *balanced binary tree* is useful for doing modular computations. Given a list lm of moduli, **modTree**(a,lm) produces a balanced binary tree with the values $a \bmod m$ at its leaves.

```
modTree(8,[2,3,5,7])
```

$$[0, 2, 3, 1] \tag{16}$$

Type: List Integer

A *set* is a collection of elements where duplication and order is irrelevant.[12] Sets are always finite and have no corresponding structure like streams for infinite collections.

Create sets using braces ("{" and "}") rather than brackets.

```
fs := {1/3,4/5,-1/3,4/5}
```

$$\left\{-\frac{1}{3}, \frac{1}{3}, \frac{4}{5}\right\} \tag{17}$$

Type: Set Fraction Integer

A *multiset* is a set that keeps track of the number of duplicate values.[13]

---

called *bags*. Other bag data structures are Stack, Queue, and Dequeue.

[11]Example of binary tree types are BinarySearchTree (see 'BinarySearchTree' on page 313, PendantTree, TournamentTree, and BalancedBinaryTree (see 'BalancedBinaryTree' on page 311).

[12]See 'Set' on page 449 for more details.

[13]See 'MultiSet' on page 420 for details.

For all the primes p between 2 and 1000, find the distribution of $p \bmod 5$.

```
multiset [x rem 5 for x in primes(2,1000)]
```

$$\{0,\ 40 : 1,\ 47 : 2,\ 42 : 3,\ 38 : 4\} \tag{18}$$

Type: Multiset Integer

A *table* is conceptually a set of "key–value" pairs and is a generalization of a multiset.[14] The domain Table(Key, Entry) provides a general-purpose type for tables with *values* of type `Entry` indexed by *keys* of type `Key`.

Compute the above distribution of primes using tables. First, let t denote an empty table of keys and values, each of type Integer.

```
t : Table(Integer,Integer) := empty()
```

$$\mathrm{table}() \tag{19}$$

Type: Table(Integer, Integer)

We define a function **howMany** to return the number of values of a given modulus `k` seen so far. It calls **search**`(k,t)` which returns the number of values stored under the key `k` in table `t`, or `"failed"` if no such value is yet stored in `t` under `k`.

In English, this says "Define `howMany(k)` as follows. First, let n be the value of **search**(k, t). Then, if n has the value "$failed$", return the value 1; otherwise return $n + 1$."

```
howMany(k) == (n:=search(k,t); n case "failed" => 1; n+1)
```

Type: Void

Run through the primes to create the table, then print the table. The expression `t.m := howMany(m)` updates the value in table t stored under key m.

```
for p in primes(2,1000) repeat (m:= p rem 5; t.m:=
  howMany(m)); t
```

$$\mathrm{table}\left(4 = 38,\ 3 = 42,\ 2 = 47,\ 1 = 40,\ 0 = 1\right) \tag{21}$$

Type: Table(Integer, Integer)

A *record* is an example of an inhomogeneous collection of objects.[15] A record consists of a set of named *selectors* that can be used to access its components.

Declare that `daniel` can only be assigned a record with two prescribed fields.

```
daniel : Record(age : Integer, salary : Float)
```

Type: Void

Give `daniel` a value, using square brackets to enclose the values of the fields.

```
daniel := [28, 32005.12]
```

$$\left[age = 28,\ salary = 32005.12\right] \tag{23}$$

Type: Record(age: Integer, salary: Float)

[14]For examples of tables, see AssociationList ('AssociationList' on page 309), HashTable, KeyedAccessFile ('KeyedAccessFile' on page 390), Library ('Library' on page 393), SparseTable ('SparseTable' on page 455), StringTable ('StringTable' on page 462), and Table ('Table' on page 465).

[15]See Section 2.4 on page 71 for details.

Give daniel a raise.

```
daniel.salary := 35000; daniel
```

$[age = 28,\ salary = 35000]\ 35000.0$ (24)

Type: Record(age: Integer, salary: Float)

A *union* is a data structure used when objects have multiple types.[16]

Let dog be either an integer or a string value.

```
dog: Union(licenseNumber: Integer, name: String)
```

Type: Void

Give dog a name.

```
dog := "Whisper"
```

"Whisper" (26)

Type: Union(name: String, ...)

All told, there are over forty different data structures in AXIOM. Using the domain constructors described in Chapter 13, you can add your own data structure or extend an existing one. Choosing the right data structure for your application may be the key to obtaining good performance.

1.7 Expanding to Higher Dimensions

To get higher dimensional aggregates, you can create one-dimensional aggregates with elements that are themselves aggregates, for example, lists of lists, one-dimensional arrays of lists of multisets, and so on. For applications requiring two-dimensional homogeneous aggregates, you will likely find *two-dimensional arrays* and *matrices* most useful.

The entries in TwoDimensionalArray and Matrix objects are all the same type, except that those for Matrix must belong to a Ring. You create and access elements in roughly the same way. Since matrices have an understood algebraic structure, certain algebraic operations are available for matrices but not for arrays. Because of this, we limit our discussion here to Matrix, that can be regarded as an extension of TwoDimensionalArray.[17]

You can create a matrix from a list of lists, where each of the inner lists represents a row of the matrix.

```
m := matrix([[1,2], [3,4]])
```

$$\begin{bmatrix} 1 & 2 \\ 3 & 4 \end{bmatrix} \quad (1)$$

Type: Matrix Integer

[16] See Section 2.5 on page 73 for details.

[17] See 'TwoDimensionalArray' on page 469 for more information about arrays. For more information about AXIOM's linear algebra facilities, see 'Matrix' on page 414, 'Permanent' on page 436, 'SquareMatrix' on page 456, 'Vector' on page 478, Section 8.4 on page 241(computation of eigenvalues and eigenvectors), and Section 8.5 on page 244(solution of linear and polynomial equations).

The "collections" construct (see Section 5.5 on page 130) is useful for creating matrices whose entries are given by formulas.

```
matrix([[1/(i + j - x) for i in 1..4] for j in 1..4])
```

$$\begin{bmatrix} -\frac{1}{x-2} & -\frac{1}{x-3} & -\frac{1}{x-4} & -\frac{1}{x-5} \\ -\frac{1}{x-3} & -\frac{1}{x-4} & -\frac{1}{x-5} & -\frac{1}{x-6} \\ -\frac{1}{x-4} & -\frac{1}{x-5} & -\frac{1}{x-6} & -\frac{1}{x-7} \\ -\frac{1}{x-5} & -\frac{1}{x-6} & -\frac{1}{x-7} & -\frac{1}{x-8} \end{bmatrix} \quad (2)$$

Type: Matrix Fraction Polynomial Integer

Let vm denote the three by three Vandermonde matrix.

```
vm := matrix [[1,1,1], [x,y,z], [x*x,y*y,z*z]]
```

$$\begin{bmatrix} 1 & 1 & 1 \\ x & y & z \\ x^2 & y^2 & z^2 \end{bmatrix} \quad (3)$$

Type: Matrix Polynomial Integer

Use this syntax to extract an entry in the matrix.

```
vm(3,3)
```

$$z^2 \quad (4)$$

Type: Polynomial Integer

You can also pull out a **row** or a column.

```
column(vm,2)
```

$$\left[1,\ y,\ y^2\right] \quad (5)$$

Type: Vector Polynomial Integer

You can do arithmetic.

```
vm * vm
```

$$\begin{bmatrix} x^2+x+1 & y^2+y+1 & z^2+z+1 \\ x^2\ z+x\ y+x & y^2\ z+y^2+x & z^3+y\ z+x \\ x^2\ z^2+x\ y^2+x^2 & y^2\ z^2+y^3+x^2 & z^4+y^2\ z+x^2 \end{bmatrix} \quad (6)$$

Type: Matrix Polynomial Integer

You can perform operations such as **transpose**, **trace**, and **determinant**.

```
factor determinant vm
```

$$(y-x)\ (z-y)\ (z-x) \quad (7)$$

Type: Factored Polynomial Integer

1.8 Writing Your Own Functions

AXIOM provides you with a very large library of predefined operations and objects to compute with. You can use the AXIOM library of constructors to create new objects dynamically of quite arbitrary complexity. For example, you can make lists of matrices of fractions of polynomials with complex floating point numbers as coefficients. Moreover, the library provides a wealth of operations that allow you to create and manipulate these objects.

For many applications, you need to interact with the interpreter and write some AXIOM programs to tackle your application. AXIOM allows you to write functions interactively, thereby effectively extending the system library. Here we give a few simple examples, leaving the details to Chapter 6.

We begin by looking at several ways that you can define the "factorial" function in AXIOM. The first way is to give a piece-wise definition of the function. This method is best for a general recurrence relation since the pieces are gathered together and compiled into an efficient iterative function. Furthermore, enough previously computed values are automatically saved so that a subsequent call to the function can pick up from where it left off.

Define the value of **fact** at 0.

```
fact(0) == 1
```

Type: Void

Define the value of `fact(n)` for general `n`.

```
fact(n) == n*fact(n-1)
```

Type: Void

Ask for the value at `50`. The resulting function created by AXIOM computes the value by iteration.

```
fact(50)
```

```
Compiling function fact with type Integer -> Integer
Compiling function fact as a recurrence relation.
```

$$304140932017133780436126081660647688443776415689605120000000000000 \tag{3}$$

Type: PositiveInteger

A second definition uses an `if-then-else` and recursion.

```
fac(n) == if n < 3 then n else n * fac(n - 1)
```

Type: Void

This function is less efficient than the previous version since each iteration involves a recursive function call.

```
fac(50)
```

$$304140932017133780436126081660647688443776415689605120000000000000 \tag{5}$$

Type: PositiveInteger

A third version directly uses iteration.

```
fa(n) == (a := 1; for i in 2..n repeat a := a*i; a)
```

Type: Void

This is the least space-consumptive version.

```
fa(50)
```

$$304140932017133780436126081660647688443776415689605120000000000000 \tag{7}$$

Type: PositiveInteger

A final version appears to construct a large list and then reduces over it with multiplication.

```
f(n) == reduce(*,[i for i in 2..n])
```

Type: Void

In fact, the resulting computation is optimized into an efficient iteration loop equivalent to that of the third version.

```
f(50)
```

$$30414093201713378043612608166064768844377641568960512000000000000 \quad (9)$$

Type: PositiveInteger

The library version uses an algorithm that is different from the four above because it highly optimizes the recurrence relation definition of **factorial**.

```
factorial(50)
```

$$30414093201713378043612608166064768844377641568960512000000000000 \quad (10)$$

Type: PositiveInteger

You are not limited to one-line functions in AXIOM. If you place your function definitions in **.input** files (see Section 4.1 on page 99), you can have multi-line functions that use indentation for grouping.

Given `n` elements, **diagonalMatrix** creates an `n` by `n` matrix with those elements down the diagonal. This function uses a permutation matrix that interchanges the `i`th and `j`th rows of a matrix by which it is right-multiplied.

This function definition shows a style of definition that can be used in **.input** files. Indentation is used to create *blocks*: sequences of expressions that are evaluated in sequence except as modified by control statements such as `if-then-else` and `return`.

```
permMat(n, i, j) ==
  m := diagonalMatrix
    [(if i = k or j = k then 0 else 1)
      for k in 1..n]
  m(i,j) := 1
  m(j,i) := 1
  m
```

Type: Void

This creates a four by four matrix that interchanges the second and third rows.

```
p := permMat(4,2,3)
```

$$\begin{bmatrix} 1 & 0 & 0 & 0 \\ 0 & 0 & 1 & 0 \\ 0 & 1 & 0 & 0 \\ 0 & 0 & 0 & 1 \end{bmatrix} \quad (12)$$

Type: Matrix Integer

Create an example matrix to permute.

```
m := matrix [[4*i + j for j in 1..4] for i in 0..3]
```

$$\begin{bmatrix} 1 & 2 & 3 & 4 \\ 5 & 6 & 7 & 8 \\ 9 & 10 & 11 & 12 \\ 13 & 14 & 15 & 16 \end{bmatrix} \quad (13)$$

Type: Matrix Integer

Interchange the second and third rows of m.

```
permMat(4,2,3) * m
```

$$\begin{bmatrix} 1 & 2 & 3 & 4 \\ 9 & 10 & 11 & 12 \\ 5 & 6 & 7 & 8 \\ 13 & 14 & 15 & 16 \end{bmatrix} \qquad (14)$$

Type: Matrix Integer

A function can also be passed as an argument to another function, which then applies the function or passes it off to some other function that does. You often have to declare the type of a function that has functional arguments.

This declares **t** to be a two-argument function that returns a Float. The first argument is a function that takes one Float argument and returns a Float.

```
t : (Float -> Float, Float) -> Float
```

Type: Void

This is the definition of **t**.

```
t(fun, x) == fun(x)**2 + sin(x)**2
```

Type: Void

We have not defined a **cos** in the workspace. The one from the AXIOM library will do.

```
t(cos, 5.2058)
```

1.0 (17)

Type: Float

Here we define our own (user-defined) function.

```
cosinv(y) == cos(1/y)
```

Type: Void

Pass this function as an argument to **t**.

```
t(cosinv, 5.2058)
```

1.739223724180051649254147684772932520785 (19)

Type: Float

AXIOM also has pattern matching capabilities for simplification of expressions and for defining new functions by rules. For example, suppose that you want to apply regularly a transformation that groups together products of radicals:

$$\sqrt{a}\,\sqrt{b} \mapsto \sqrt{ab}, \quad (\forall a)(\forall b)$$

Note that such a transformation is not generally correct. AXIOM never uses it automatically.

Give this rule the name **groupSqrt**.

```
groupSqrt := rule(sqrt(a) * sqrt(b) == sqrt(a*b))
```

$$\%D\ \sqrt{a}\ \sqrt{b} == \%D\ \sqrt{a\ b} \tag{20}$$

Type: RewriteRule(Integer, Integer, Expression Integer)

Here is a test expression.

```
a := (sqrt(x) + sqrt(y) + sqrt(z))**4
```

$$\left(\left(4\ z+4\ y+12\ x\right)\ \sqrt{y}+\left(4\ z+12\ y+4\ x\right)\ \sqrt{x}\right)\ \sqrt{z}+ \\ \left(12\ z+4\ y+4\ x\right)\ \sqrt{x}\ \sqrt{y}+z^2+\left(6\ y+6\ x\right)\ z+y^2+6\ x\ y+x^2 \tag{21}$$

Type: Expression Integer

The rule **groupSqrt** successfully simplifies the expression.

```
groupSqrt a
```

$$\left(4\ z+4\ y+12\ x\right)\ \sqrt{y\ z}+\left(4\ z+12\ y+4\ x\right)\ \sqrt{x\ z}+ \\ \left(12\ z+4\ y+4\ x\right)\ \sqrt{x\ y}+z^2+\left(6\ y+6\ x\right)\ z+y^2+6\ x\ y+x^2 \tag{22}$$

Type: Expression Integer

1.9 Polynomials

Polynomials are the commonly used algebraic types in symbolic computation. Interactive users of AXIOM generally only see one type of polynomial: Polynomial(R). This type represents polynomials in any number of unspecified variables over a particular coefficient domain R. This type represents its coefficients *sparsely*: only terms with non-zero coefficients are represented.

In building applications, many other kinds of polynomial representations are useful. Polynomials may have one variable or multiple variables, the variables can be named or unnamed, the coefficients can be stored sparsely or densely. So-called "distributed multivariate polynomials" store polynomials as coefficients paired with vectors of exponents. This type is particularly efficient for use in algorithms for solving systems of non-linear polynomial equations.

The polynomial constructor most familiar to the interactive user is Polynomial.

```
(x**2 - x*y**3 +3*y)**2
```

$$x^2\ y^6-6\ x\ y^4-2\ x^3\ y^3+9\ y^2+6\ x^2\ y+x^4 \tag{1}$$

Type: Polynomial Integer

If you wish to restrict the variables used, UnivariatePolynomial provides polynomials in one variable.

```
p: UP(x,INT) := (3*x-1)**2 * (2*x + 8)
```

$$18\ x^3+60\ x^2-46\ x+8 \tag{2}$$

Type: UnivariatePolynomial(x, Integer)

The constructor MultivariatePolynomial provides polynomials in one or more specified variables.

```
m: MPOLY([x,y],INT) := (x**2-x*y**3+3*y)**2
```

$$x^4-2\ y^3\ x^3+\left(y^6+6\ y\right)\ x^2-6\ y^4\ x+9\ y^2 \tag{3}$$

Type: MultivariatePolynomial([x, y], Integer)

You can change the way the polynomial appears by modifying the variable ordering in the explicit list.

```
m :: MPOLY([y,x],INT)
```

$$x^2\ y^6 - 6\ x\ y^4 - 2\ x^3\ y^3 + 9\ y^2 + 6\ x^2\ y + x^4 \qquad (4)$$

Type: MultivariatePolynomial([y, x], Integer)

The constructor DistributedMultivariatePolynomial provides polynomials in one or more specified variables with the monomials ordered lexicographically.

```
m :: DMP([y,x],INT)
```

$$y^6\ x^2 - 6\ y^4\ x - 2\ y^3\ x^3 + 9\ y^2 + 6\ y\ x^2 + x^4 \qquad (5)$$

Type: DistributedMultivariatePolynomial([y, x], Integer)

The constructor Homogeneous-DistributedMultivariatePolynomial is similar except that the monomials are ordered by total order refined by reverse lexicographic order.

```
m :: HDMP([y,x],INT)
```

$$y^6\ x^2 - 2\ y^3\ x^3 - 6\ y^4\ x + x^4 + 6\ y\ x^2 + 9\ y^2 \qquad (6)$$

Type: HomogeneousDistributedMultivariatePolynomial([y, x], Integer)

More generally, the domain constructor GeneralDistributedMultivariatePolynomial allows the user to provide an arbitrary predicate to define his own term ordering. These last three constructors are typically used in Gröbner basis applications and when a flat (that is, non-recursive) display is wanted and the term ordering is critical for controlling the computation.

1.10 Limits

AXIOM's **limit** function is usually used to evaluate limits of quotients where the numerator and denominator both tend to zero or both tend to infinity. To find the limit of an expression `f` as a real variable `x` tends to a limit value `a`, enter `limit(f, x=a)`. Use **complexLimit** if the variable is complex. Additional information and examples of limits are in Section 8.6 on page 249.

You can take limits of functions with parameters.

```
g := csc(a*x) / csch(b*x)
```

$$\frac{\csc\left(a\ x\right)}{\operatorname{csch}\left(b\ x\right)} \qquad (1)$$

Type: Expression Integer

As you can see, the limit is expressed in terms of the parameters.

```
limit(g,x=0)
```

$$\frac{b}{a} \qquad (2)$$

Type: Union(OrderedCompletion Expression Integer, ...)

A variable may also approach plus or minus infinity:

```
h := (1 + k/x)**x
```

$$\frac{x + k}{x}^{x} \qquad (3)$$

Type: Expression Integer

Use %plusInfinity and %minusInfinity to denote ∞ and $-\infty$.

```
limit(h,x=%plusInfinity)
```

$$\%e^k \qquad (4)$$

Type: Union(OrderedCompletion Expression Integer, ...)

A function can be defined on both sides of a particular value, but may tend to different limits as its variable approaches that value from the left and from the right.

```
limit(sqrt(y**2)/y,y = 0)
```

$$[leftHandLimit = -1, \; rightHandLimit = 1] \qquad (5)$$

Type: Union(Record(leftHandLimit: Union(OrderedCompletion Expression Integer, "failed"), rightHandLimit: Union(OrderedCompletion Expression Integer, "failed")), ...)

As x approaches 0 along the real axis, exp(-1/x**2) tends to 0.

```
limit(exp(-1/x**2),x = 0)
```

$$0 \qquad (6)$$

Type: Union(OrderedCompletion Expression Integer, ...)

However, if x is allowed to approach 0 along any path in the complex plane, the limiting value of exp(-1/x**2) depends on the path taken because the function has an essential singularity at x=0. This is reflected in the error message returned by the function.

```
complexLimit(exp(-1/x**2),x = 0)
```

"failed" (7)

Type: Union("failed", ...)

1.11 Series

AXIOM also provides power series. By default, AXIOM tries to compute and display the first ten elements of a series. Use)set streams calculate to change the default value to something else. For the purposes of this book, we have used this system command to display fewer than ten terms. For more information about working with series, see Section 8.9 on page 255.

You can convert a functional expression to a power series by using the operation **series**. In this example, sin(a*x) is expanded in powers of (x - 0), that is, in powers of x.

```
series(sin(a*x),x = 0)
```

$$a\,x - \frac{a^3}{6}\,x^3 + \frac{a^5}{120}\,x^5 - \frac{a^7}{5040}\,x^7 + O\left(x^9\right) \qquad (1)$$

Type: UnivariatePuiseuxSeries(Expression Integer, x, 0)

This expression expands sin(a*x) in powers of (x - %pi/4).

```
series(sin(a*x),x = %pi/4)
```

$$\sin\left(\frac{a\ \%pi}{4}\right)+a\ \cos\left(\frac{a\ \%pi}{4}\right)\ \left(x-\frac{\%pi}{4}\right)-$$
$$\frac{a^2\ \sin\left(\frac{a\ \%pi}{4}\right)}{2}\ \left(x-\frac{\%pi}{4}\right)^2-\frac{a^3\ \cos\left(\frac{a\ \%pi}{4}\right)}{6}\ \left(x-\frac{\%pi}{4}\right)^3+$$
$$\frac{a^4\ \sin\left(\frac{a\ \%pi}{4}\right)}{24}\ \left(x-\frac{\%pi}{4}\right)^4+\frac{a^5\ \cos\left(\frac{a\ \%pi}{4}\right)}{120}\ \left(x-\frac{\%pi}{4}\right)^5- \tag{2}$$
$$\frac{a^6\ \sin\left(\frac{a\ \%pi}{4}\right)}{720}\ \left(x-\frac{\%pi}{4}\right)^6-\frac{a^7\ \cos\left(\frac{a\ \%pi}{4}\right)}{5040}\ \left(x-\frac{\%pi}{4}\right)^7+$$
$$O\left(\left(x-\frac{\%pi}{4}\right)^8\right)$$

Type: UnivariatePuiseuxSeries(Expression Integer, x, pi/4)

AXIOM provides *Puiseux series:* series with rational number exponents. The first argument to **series** is an in-place function that computes the n th coefficient. (Recall that the "+->" is an infix operator meaning "maps to.")

```
series(n +-> (-1)**((3*n - 4)/6)/factorial(n - 1/3),x =
  0,4/3..,2)
```

$$x^{\frac{4}{3}}-\frac{1}{6}\ x^{\frac{10}{3}}+O\left(x^4\right) \tag{3}$$

Type: UnivariatePuiseuxSeries(Expression Integer, x, 0)

Once you have created a power series, you can perform arithmetic operations on that series. We compute the Taylor expansion of 1/(1-x).

```
f := series(1/(1-x),x = 0)
```

$$1+x+x^2+x^3+x^4+x^5+x^6+x^7+O\left(x^8\right) \tag{4}$$

Type: UnivariatePuiseuxSeries(Expression Integer, x, 0)

Compute the square of the series.

```
f ** 2
```

$$1+2\ x+3\ x^2+4\ x^3+5\ x^4+6\ x^5+7\ x^6+8\ x^7+O\left(x^8\right) \tag{5}$$

Type: UnivariatePuiseuxSeries(Expression Integer, x, 0)

The usual elementary functions (**log**, **exp**, trigonometric functions, and so on) are defined for power series.

```
f := series(1/(1-x),x = 0)
```

$$1+x+x^2+x^3+x^4+x^5+x^6+x^7+O\left(x^8\right) \tag{6}$$

Type: UnivariatePuiseuxSeries(Expression Integer, x, 0)

```
g := log(f)
```

$$x+\frac{1}{2}\ x^2+\frac{1}{3}\ x^3+\frac{1}{4}\ x^4+\frac{1}{5}\ x^5+\frac{1}{6}\ x^6+\frac{1}{7}\ x^7+\frac{1}{8}\ x^8+O\left(x^9\right) \tag{7}$$

Type: UnivariatePuiseuxSeries(Expression Integer, x, 0)

```
exp(g)
```

$$1+x+x^2+x^3+x^4+x^5+x^6+x^7+O\left(x^8\right) \quad (8)$$

Type: UnivariatePuiseuxSeries(Expression Integer, x, 0)

Here is a way to obtain numerical approximations of `e` from the Taylor series expansion of `exp(x)`. First create the desired Taylor expansion.

```
f := taylor(exp(x))
```

$$1+x+\frac{1}{2}\,x^2+\frac{1}{6}\,x^3+\frac{1}{24}\,x^4+\frac{1}{120}\,x^5+\frac{1}{720}\,x^6+\frac{1}{5040}\,x^7+ O\left(x^8\right) \quad (9)$$

Type: UnivariateTaylorSeries(Expression Integer, x, 0)

Evaluate the series at the value `1.0`. As you see, you get a sequence of partial sums.

```
eval(f,1.0)
```

$$[1.0,\ 2.0,\ 2.5,\ 2.666666666666666666666666666666666666667,\ 2.708333333333333333333333333333333333333,\ 2.716666666666666666666666666666666666667,\ 2.718055555555555555555555555555555555556,\ \ldots] \quad (10)$$

Type: Stream Expression Float

1.12 Derivatives

Use the AXIOM function **D** to differentiate an expression.

To find the derivative of an expression `f` with respect to a variable `x`, enter `D(f, x)`.

```
f := exp exp x
```

$$\%e^{\%e^{x}} \quad (1)$$

Type: Expression Integer

```
D(f, x)
```

$$\%e^{x}\ \%e^{\%e^{x}} \quad (2)$$

Type: Expression Integer

An optional third argument `n` in **D** asks AXIOM for the `n` th derivative of `f`. This finds the fourth derivative of `f` with respect to `x`.

```
D(f, x, 4)
```

$$\left(\%e^{x\,4}+6\ \%e^{x\,3}+7\ \%e^{x\,2}+\%e^{x}\right)\ \%e^{\%e^{x}} \quad (3)$$

Type: Expression Integer

You can also compute partial derivatives by specifying the order of differentiation.

```
g := sin(x**2 + y)
```

$$\sin\left(y+x^2\right) \quad (4)$$

Type: Expression Integer

```
D(g, y)
```

$$\cos\left(y + x^2\right) \tag{5}$$

Type: Expression Integer

```
D(g, [y, y, x, x])
```

$$4\ x^2\ \sin\left(y + x^2\right) - 2\ \cos\left(y + x^2\right) \tag{6}$$

Type: Expression Integer

AXIOM can manipulate the derivatives (partial and iterated) of expressions involving formal operators. All the dependencies must be explicit.

This returns 0 since F (so far) does not explicitly depend on x.

```
D(F,x)
```

$$0 \tag{7}$$

Type: Polynomial Integer

Suppose that we have F a function of x, y, and z, where x and y are themselves functions of z.

Start by declaring that F, x, and y are operators.

```
F := operator 'F; x := operator 'x; y := operator 'y
```

$$y \tag{8}$$

Type: BasicOperator

You can use F, x, and y in expressions.

```
a := F(x z, y z, z**2) + x y(z+1)
```

$$x\,(y\,(z+1)) + F\left(x\,(z),\ y\,(z),\ z^2\right) \tag{9}$$

Type: Expression Integer

Differentiate formally with respect to z. The formal derivatives appearing in dadz are not just formal symbols, but do represent the derivatives of x, y, and F.

```
dadz := D(a, z)
```

$$\begin{array}{l} 2\ z\ F_{,3}\left(x\,(z),\ y\,(z),\ z^2\right) + y'\,(z)\ F_{,2}\left(x\,(z),\ y\,(z),\ z^2\right) + \\ x'\,(z)\ F_{,1}\left(x\,(z),\ y\,(z),\ z^2\right) + x'\,(y\,(z+1))\ y'\,(z+1) \end{array} \tag{10}$$

Type: Expression Integer

You can evaluate the above for particular functional values of F, x, and y. If x(z) is exp(z) and y(z) is log(z+1), then this evaluates dadz.

```
eval(eval(dadz, 'x, z +-> exp z), 'y, z +-> log(z+1))
```

$$\frac{\left(\begin{array}{l} \left(2\ z^2 + 2\ z\right)\ F_{,3}\left(\%e^z,\ \log(z+1),\ z^2\right) + \\ F_{,2}\left(\%e^z,\ \log(z+1),\ z^2\right) + \\ (z+1)\ \%e^z\ F_{,1}\left(\%e^z,\ \log(z+1),\ z^2\right) + z + 1 \end{array}\right)}{z+1} \tag{11}$$

Type: Expression Integer

You obtain the same result by first evaluating a and then differentiating.

```
eval(eval(a, 'x, z +-> exp z), 'y, z +-> log(z+1))
```

$$F\left(\%e^{z},\ \log(z+1),\ z^{2}\right)+z+2 \qquad (12)$$

Type: Expression Integer

```
D(%, z)
```

$$\frac{\left(\begin{array}{l}\left(2\ z^{2}+2\ z\right)\ F_{,3}\left(\%e^{z},\ \log(z+1),\ z^{2}\right)+\\ F_{,2}\left(\%e^{z},\ \log(z+1),\ z^{2}\right)+\\ (z+1)\ \%e^{z}\ F_{,1}\left(\%e^{z},\ \log(z+1),\ z^{2}\right)+z+1\end{array}\right)}{z+1} \qquad (13)$$

Type: Expression Integer

1.13 Integration

AXIOM has extensive library facilities for integration.

The first example is the integration of a fraction with denominator that factors into a quadratic and a quartic irreducible polynomial. The usual partial fraction approach used by most other computer algebra systems either fails or introduces expensive unneeded algebraic numbers.

We use a factorization-free algorithm.

```
integrate((x**2+2*x+1)/((x+1)**6+1),x)
```

$$\frac{\arctan\left(x^{3}+3\ x^{2}+3\ x+1\right)}{3} \qquad (1)$$

Type: Union(Expression Integer, ...)

When real parameters are present, the form of the integral can depend on the signs of some expressions.

Rather than query the user or make sign assumptions, AXIOM returns all possible answers.

```
integrate(1/(x**2 + a),x)
```

$$\left[\frac{\log\left(\left(x^{2}-a\right)\ \sqrt{-a}+2\ a\ x\right)-\log\left(x^{2}+a\right)}{2\ \sqrt{-a}},\ \frac{\arctan\left(\frac{x}{\sqrt{a}}\right)}{\sqrt{a}}\right] \qquad (2)$$

Type: Union(List Expression Integer, ...)

The **integrate** operation generally assumes that all parameters are real. The only exception is when the integrand has complex valued quantities.

If the parameter is complex instead of real, then the notion of sign is undefined and there is a unique answer. You can request this answer by "prepending" the word "complex" to the command name:

```
complexIntegrate(1/(x**2 + a),x)
```

$$\frac{\log\left(\frac{-\sqrt{-a}+x}{2\ \sqrt{-a}}\right)-\log\left(\frac{-\sqrt{-a}-x}{2\ \sqrt{-a}}\right)}{2\ \sqrt{-a}} \qquad (3)$$

Type: Expression Integer

The following two examples illustrate the limitations of table-based approaches. The two integrands are very similar, but the answer to one of them requires the addition of two new algebraic numbers.

This one is the easy one. The next one looks very similar but the answer is much more complicated.

```
integrate(x**3 / (a+b*x)**(1/3),x)
```

$$\frac{\left(120\ b^3\ x^3-135\ a\ b^2\ x^2+162\ a^2\ b\ x-243\ a^3\right)\ \sqrt[3]{b\ x+a}^{\,2}}{440\ b^4} \tag{4}$$

Type: Union(Expression Integer, ...)

Only an algorithmic approach is guaranteed to find what new constants must be added in order to find a solution.

```
integrate(1 / (x**3 * (a+b*x)**(1/3)),x)
```

$$\frac{\left(\begin{array}{l}-2\ b^2\ x^2\ \sqrt{3}\ \log\left(\sqrt[3]{a}\ \sqrt[3]{b\ x+a}^{\,2}+\sqrt[3]{a}^{\,2}\ \sqrt[3]{b\ x+a}+a\right)+\\ 4\ b^2\ x^2\ \sqrt{3}\ \log\left(\sqrt[3]{a}^{\,2}\ \sqrt[3]{b\ x+a}-a\right)+\\ 12\ b^2\ x^2\ \arctan\left(\frac{2\ \sqrt{3}\ \sqrt[3]{a}^{\,2}\ \sqrt[3]{b\ x+a}+a\ \sqrt{3}}{3\ a}\right)+\\ (12\ b\ x-9\ a)\ \sqrt{3}\ \sqrt[3]{a}\ \sqrt[3]{b\ x+a}^{\,2}\end{array}\right)}{18\ a^2\ x^2\ \sqrt{3}\ \sqrt[3]{a}} \tag{5}$$

Type: Union(Expression Integer, ...)

Some computer algebra systems use heuristics or table-driven approaches to integration. When these systems cannot determine the answer to an integration problem, they reply "I don't know." AXIOM uses a algorithm for integration. that conclusively proves that an integral cannot be expressed in terms of elementary functions.

When AXIOM returns an integral sign, it has proved that no answer exists as an elementary function.

```
integrate(log(1 + sqrt(a*x + b)) / x,x)
```

$$\int^{x} \frac{\log\left(\sqrt{b+\%Q\ a}+1\right)}{\%Q}\ d\%Q \tag{6}$$

Type: Union(Expression Integer, ...)

AXIOM can handle complicated mixed functions much beyond what you can find in tables.

Whenever possible, AXIOM tries to express the answer using the functions present in the integrand.

```
integrate((sinh(1+sqrt(x+b))+2*sqrt(x+b)) / (sqrt(x+b) *
  (x + cosh(1+sqrt(x + b)))), x)
```

$$2\ \log\left(\frac{-2\ \cosh\left(\sqrt{x+b}+1\right)-2\ x}{\sinh\left(\sqrt{x+b}+1\right)-\cosh\left(\sqrt{x+b}+1\right)}\right)-2\ \sqrt{x+b}-4 \tag{7}$$

Type: Union(Expression Integer, ...)

A strong structure-checking algorithm in AXIOM finds hidden algebraic relationships between functions.

```
integrate(tan(atan(x)/3),x)
```

$$\frac{\left(\begin{array}{l} 8\log\left(3\tan\left(\frac{\arctan(x)}{3}\right)^2-1\right)-3\tan\left(\frac{\arctan(x)}{3}\right)^2+ \\ 18\,x\tan\left(\frac{\arctan(x)}{3}\right)+16 \end{array}\right)}{18} \tag{8}$$

Type: Union(Expression Integer, ...)

The discovery of this algebraic relationship is necessary for correct integration of this function. Here are the details:

1. If $x = \tan t$ and $g = \tan(t/3)$ then the following algebraic relation is true:
$$g^3 - 3xg^2 - 3g + x = 0$$
2. Integrate g using this algebraic relation; this produces:
$$\frac{(24g^2-8)\log(3g^2-1)+(81x^2+24)g^2+72xg-27x^2-16}{54g^2-18}$$
3. Rationalize the denominator, producing:
$$\frac{8\log(3g^2-1)-3g^2+18xg+16}{18}$$
Replace g by the initial definition $g = \tan(\arctan(x)/3)$ to produce the final result.

This is an example of a mixed function where the algebraic layer is over the transcendental one.

```
integrate((x + 1) / (x*(x + log x) ** (3/2)), x)
```

$$-\frac{2\sqrt{\log(x)+x}}{\log(x)+x} \tag{9}$$

Type: Union(Expression Integer, ...)

While incomplete for non-elementary functions, AXIOM can handle some of them.

```
integrate(exp(-x**2) * erf(x) / (erf(x)**3 - erf(x)**2 -
  erf(x) + 1),x)
```

$$\frac{\left(\begin{array}{l} (-\operatorname{erf}(x)+1)\sqrt{\%pi}\log(\operatorname{erf}(x)+1)+ \\ (\operatorname{erf}(x)-1)\sqrt{\%pi}\log(\operatorname{erf}(x)-1)-2\sqrt{\%pi} \end{array}\right)}{8\operatorname{erf}(x)-8} \tag{10}$$

Type: Union(Expression Integer, ...)

More examples of AXIOM's integration capabilities are discussed in Section 8.8 on page 252.

1.14 Differential Equations

The general approach used in integration also carries over to the solution of linear differential equations.

Let's solve some differential equations. Let y be the unknown function in terms of x.

```
y := operator 'y
```

$$y \tag{1}$$

Type: BasicOperator

Here we solve a third order equation with polynomial coefficients.

```
deq := x**3 * D(y x, x, 3) + x**2 * D(y x, x, 2) - 2 * x *
  D(y x, x) + 2 * y x = 2 * x**4
```

$$x^3\ y'''(x) + x^2\ y''(x) - 2\ x\ y'(x) + 2\ y(x) = 2\ x^4 \tag{2}$$

Type: Equation Expression Integer

```
solve(deq, y, x)
```

$$\left[particular = \frac{x^4}{15},\ basis = \left[\frac{1}{x},\ x,\ x^2\right]\right] \tag{3}$$

Type: Union(Record(particular: Expression Integer, basis: List Expression Integer), ...)

Here we find all the algebraic function solutions of the equation.

```
deq := (x**2 + 1) * D(y x, x, 2) + 3 * x * D(y x, x) + y x
  = 0
```

$$\left(x^2+1\right)\ y''(x) + 3\ x\ y'(x) + y(x) = 0 \tag{4}$$

Type: Equation Expression Integer

```
solve(deq, y, x)
```

$$\left[particular = 0,\ basis = \left[\frac{1}{\sqrt{x^2+1}},\ \frac{\log\left(\sqrt{x^2+1}-x\right)}{\sqrt{x^2+1}}\right]\right] \tag{5}$$

Type: Union(Record(particular: Expression Integer, basis: List Expression Integer), ...)

Coefficients of differential equations can come from arbitrary constant fields. For example, coefficients can contain algebraic numbers.

This example has solutions whose logarithmic derivative is an algebraic function of degree two.

```
eq := 2*x**3 * D(y x,x,2) + 3*x**2 * D(y x,x) - 2 * y x
```

$$2\ x^3\ y''(x) + 3\ x^2\ y'(x) - 2\ y(x) \tag{6}$$

Type: Expression Integer

```
solve(eq,y,x).basis
```

$$\left[\%e^{\left(-\frac{2}{\sqrt{x}}\right)},\ \%e^{\frac{2}{\sqrt{x}}}\right] \tag{7}$$

Type: List Expression Integer

Here's another differential equation to solve.

```
deq := D(y x, x) = y(x) / (x + y(x) * log y x)
```

$$y'(x) = \frac{y(x)}{y(x)\log(y(x)) + x} \tag{8}$$

Type: Equation Expression Integer

```
solve(deq, y, x)
```

$$\frac{y(x)\log(y(x))^2 - 2\,x}{2\,y(x)} \tag{9}$$

Type: Union(Expression Integer, ...)

Rather than attempting to get a closed form solution of a differential equation, you instead might want to find an approximate solution in the form of a series.

Let's solve a system of nonlinear first order equations and get a solution in power series. Tell AXIOM that `x` is also an operator.

```
x := operator 'x
```

$$x \tag{10}$$

Type: BasicOperator

Here are the two equations forming our system.

```
eq1 := D(x(t), t) = 1 + x(t)**2
```

$$x'(t) = x(t)^2 + 1 \tag{11}$$

Type: Equation Expression Integer

```
eq2 := D(y(t), t) = x(t) * y(t)
```

$$y'(t) = x(t)\,y(t) \tag{12}$$

Type: Equation Expression Integer

We can solve the system around `t = 0` with the initial conditions `x(0) = 0` and `y(0) = 1`. Notice that since we give the unknowns in the order `[x, y]`, the answer is a list of two series in the order `[series for x(t), series for y(t)]`.

```
seriesSolve([eq2, eq1], [x, y], t = 0, [y(0) = 1, x(0) =
  0])
```

$$\left[t + \frac{1}{3}\,t^3 + \frac{2}{15}\,t^5 + \frac{17}{315}\,t^7 + O\left(t^8\right),\right.$$
$$\left. 1 + \frac{1}{2}\,t^2 + \frac{5}{24}\,t^4 + \frac{61}{720}\,t^6 + O\left(t^8\right)\right] \tag{13}$$

Type: List UnivariateTaylorSeries(Expression Integer, t, 0)

1.15 Solution of Equations

AXIOM also has state-of-the-art algorithms for the solution of systems of polynomial equations. When the number of equations and unknowns is the same, and you have no symbolic coefficients, you can use **solve** for real roots and **complexSolve** for complex roots. In each case, you tell AXIOM how accurate you want your result to be. All operations in the **solve** family return answers in the form of a list of solution sets, where each solution set is a list of equations.

A system of two equations involving a symbolic parameter t.

```
S(t) == [x**2-2*y**2 - t,x*y-y-5*x + 5]
```

Type: Void

Find the real roots of S(19) with rational arithmetic, correct to within $1/10^{20}$.

```
solve(S(19),1/10**20)
```

$$\left[\left[y = 5,\ x = -\frac{245168263225309344251 1}{295147905179352825856}\right],\ \left[y = 5,\ x = \frac{2451682632253093442511}{295147905179352825856}\right]\right] \tag{2}$$

Type: List List Equation Polynomial Fraction Integer

Find the complex roots of S(19) with floating point coefficients to 20 digits accuracy in the mantissa.

```
complexSolve(S(19),10.e-20)
```

$$[[y = -3.0\ \%i,\ x = 1.0],\ [y = 3.0\ \%i,\ x = 1.0],\ [y = 5.0,\ x = 8.306623862918074852575221582451359125798],\ [y = 5.0,\ x = -8.306623862918074852575221582451359125798]] \tag{3}$$

Type: List List Equation Polynomial Complex Float

If a system of equations has symbolic coefficients and you want a solution in radicals, try **radicalSolve**.

```
radicalSolve(S(a),[x,y])
```

$$\left[\left[x = -\sqrt{a+50},\ y = 5\right],\ \left[x = \sqrt{a+50},\ y = 5\right],\ \left[x = 1,\ y = \sqrt{\frac{-a+1}{2}}\right],\ \left[x = 1,\ y = -\sqrt{\frac{-a+1}{2}}\right]\right] \tag{4}$$

Type: List List Equation Expression Integer

For systems of equations with symbolic coefficients, you can apply **solve**, listing the variables that you want AXIOM to solve for. For polynomial equations, a solution cannot usually be expressed solely in terms of the other variables. Instead, the solution is presented as a "triangular" system of equations, where each polynomial has coefficients involving only the succeeding variables. This is analogous to converting a linear system of equations to "triangular form".

A system of three equations in five variables.

```
eqns := [x**2 - y + z,x**2*z + x**4 - b*y, y**2 *z - a -
  b*x]
```

$$\left[z - y + x^2,\ x^2\ z - b\ y + x^4,\ y^2\ z - b\ x - a\right] \tag{5}$$

Type: List Polynomial Integer

Solve the system for unknowns $[x, y, z]$, reducing the solution to triangular form.

```
solve(eqns,[x,y,z])
```

$$\left[\left[x=-\frac{a}{b},\ y=0,\ z=-\frac{a^2}{b^2}\right],\right.$$

$$\left[x=\frac{z^3+2\ b\ z^2+b^2\ z-a}{b},\ y=z+b,\right.$$

$$\left.\left.z^6+4\ b\ z^5+6\ b^2\ z^4+\left(4\ b^3-2\ a\right)\ z^3+\left(b^4-4\ a\ b\right)\ z^2-2\ a\ b^2\ z-b^3+a^2=0\right]\right] \quad (6)$$

Type: List List Equation Fraction Polynomial Integer

1.16 System Commands

We conclude our tour of AXIOM with a brief discussion of *system commands*. System commands are special statements that start with a closing parenthesis (“`)`”). They are used to control or display your AXIOM environment, start the HyperDoc system, issue operating system commands and leave AXIOM. For example, `)system` is used to issue commands to the operating system from AXIOM. Here is a brief description of some of these commands. For more information on specific commands, see Appendix A.

Perhaps the most important user command is the `)clear all` command that initializes your environment. Every section and subsection in this book has an invisible `)clear all` that is read prior to the examples given in the section. `)clear all` gives you a fresh, empty environment with no user variables defined and the step number reset to `1`. The `)clear` command can also be used to selectively clear values and properties of system variables.

Another useful system command is `)read`. A preferred way to develop an application in AXIOM is to put your interactive commands into a file, say **my.input** file. To get AXIOM to read this file, you use the system command `)read my.input`. If you need to make changes to your approach or definitions, go into your favorite editor, change **my.input**, then `)read my.input` again.

Other system commands include: `)history`, to display previous input and/or output lines; `)display`, to display properties and values of workspace variables; and `)what`.

Issue)what to get a list of AXIOM objects that contain a given substring in their name.

```
)what operations integrate

Operations whose names satisfy the above pattern(s):

HermiteIntegrate       algintegrate
complexIntegrate       expintegrate
extendedIntegrate      fintegrate
infieldIntegrate       integrate
internalIntegrate      lazyGintegrate
lazyIntegrate          lfintegrate
limitedIntegrate       palgintegrate
primintegrate          primintegratefrac

To get more information about an operation such as
HermiteIntegrate , issue the command )display op
HermiteIntegrate
```

A useful system command is)undo. Sometimes while computing interactively with AXIOM, you make a mistake and enter an incorrect definition or assignment. Or perhaps you need to try one of several alternative approaches, one after another, to find the best way to approach an application. For this, you will find the *undo* facility of AXIOM helpful.

System command)undo n means "undo back to step n"; it restores the values of user variables to those that existed immediately after input expression n was evaluated. Similarly,)undo -n undoes changes caused by the last n input expressions. Once you have done an)undo, you can continue on from there, or make a change and **redo** all your input expressions from the point of the)undo forward. The)undo is completely general: it changes the environment like any user expression. Thus you can)undo any previous undo.

Here is a sample dialogue between user and AXIOM.

"Let me define two mutually dependent functions f and g piece-wise."

```
f(0) == 1; g(0) == 1
```

(1)

Type: Void

"Here is the general term for f."

```
f(n) == e/2*f(n-1) - x*g(n-1)
```

(2)

Type: Void

"And here is the general term for g."

```
g(n) == -x*f(n-1) + d/3*g(n-1)
```

(3)

Type: Void

"What is value of f(3)?"

```
f(3)
```

$$-x^3 + \left(e + \frac{1}{3}\, d\right) x^2 + \left(-\frac{1}{4}\, e^2 - \frac{1}{6}\, d\, e - \frac{1}{9}\, d^2\right) x + \frac{1}{8}\, e^3 \qquad (4)$$

Type: Polynomial Fraction Integer

"Hmm, I think I want to define f differently. Undo to the environment right after I defined f."

```
)undo 2
```

"Here is how I think I want f to be defined instead."

```
f(n) == d/3*f(n-1) - x*g(n-1)
```

(5)

Type: Void

Redo the computation from expression 3 forward.

```
)undo )redo
```

"And here is the general term for g."

```
g(n) == -x*f(n-1) + d/3*g(n-1)
```

(6)

Type: Void

"What is value of f(3)?"

```
f(3)
```

$$-x^3 + d\, x^2 - \frac{1}{3}\, d^2\, x + \frac{1}{27}\, d^3 \qquad (7)$$

Type: Polynomial Fraction Integer

"I want my old definition of f after all. Undo the undo and restore the environment to that immediately after (4)."

```
)undo 4
```

"Check that the value of f(3) is restored."

```
f(3)
```

$$-x^3 + \left(e + \frac{1}{3}\, d\right) x^2 + \left(-\frac{1}{4}\, e^2 - \frac{1}{6}\, d\, e - \frac{1}{9}\, d^2\right) x + \frac{1}{8}\, e^3 \qquad (8)$$

Type: Polynomial Fraction Integer

After you have gone off on several tangents, then backtracked to previous points in your conversation using `)undo`, you might want to save all the "correct" input commands you issued, disregarding those undone. The system command `)history )write mynew.input` writes a clean straight-line program onto the file **mynew.input** on your disk.

This concludes your tour of AXIOM. To disembark, issue the system command `)quit` to leave AXIOM and return to the operating system.

CHAPTER 2

Using Types and Modes

In this chapter we look at the key notion of *type* and its generalization *mode*. We show that every AXIOM object has a type that determines what you can do with the object. In particular, we explain how to use types to call specific functions from particular parts of the library and how types and modes can be used to create new objects from old. We also look at Record and Union types and the special type Any. Finally, we give you an idea of how AXIOM manipulates types and modes internally to resolve ambiguities.

2.1 The Basic Idea

The AXIOM world deals with many kinds of objects. There are mathematical objects such as numbers and polynomials, data structure objects such as lists and arrays, and graphics objects such as points and graphic images. Functions are objects too.

AXIOM organizes objects using the notion of *domain of computation*, or simply *domain*. Each domain denotes a class of objects. The class of objects it denotes is usually given by the name of the domain: Integer for the integers, Float for floating-point numbers, and so on. The convention is that the first letter of a domain name is capitalized. Similarly, the domain Polynomial(Integer) denotes "polynomials with integer coefficients." Also, Matrix(Float) denotes "matrices with floating-point entries."

Every basic AXIOM object belongs to a unique domain. The integer 3 belongs to the domain Integer and the polynomial `x + 3` belongs to the domain Polynomial(Integer). The domain of an object is also called its *type*.

Thus we speak of "the type Integer" and "the type Polynomial(Integer)."

After an AXIOM computation, the type is displayed toward the right-hand side of the page (or screen).

```
-3
```

$$-3 \tag{1}$$

Type: Integer

Here we create a rational number but it looks like the last result. The type however tells you it is different. You cannot identify the type of an object by how AXIOM displays the object.

```
-3/1
```

$$-3 \tag{2}$$

Type: Fraction Integer

When a computation produces a result of a simpler type, AXIOM leaves the type unsimplified. Thus no information is lost.

```
x + 3 - x
```

$$3 \tag{3}$$

Type: Polynomial Integer

This seldom matters since AXIOM retracts the answer to the simpler type if it is necessary.

```
factorial(%)
```

$$6 \tag{4}$$

Type: Expression Integer

When you issue a positive number, the type PositiveInteger is printed. Surely, 3 also has type Integer! The curious reader may now have two questions. First, is the type of an object not unique? Second, how is PositiveInteger related to Integer? Read on!

```
3
```

$$3 \tag{5}$$

Type: PositiveInteger

Any domain can be refined to a *subdomain* by a membership *predicate*.[1] For example, the domain Integer can be refined to the subdomain PositiveInteger, the set of integers `x` such that `x > 0`, by giving the AXIOM predicate `x +-> x > 0`. Similarly, AXIOM can define subdomains such as "the subdomain of diagonal matrices," "the subdomain of lists of length two," "the subdomain of monic irreducible polynomials in `x`," and so on. Trivially, any domain is a subdomain of itself.

While an object belongs to a unique domain, it can belong to any number of subdomains. Any subdomain of the domain of an object can be used as the *type* of that object. The type of 3 is indeed both Integer and PositiveInteger as well as any other subdomain of integer whose predicate is satisfied, such as "the prime integers," "the odd positive integers between 3 and 17," and so on.

[1]A predicate is a function that, when applied to an object of the domain, returns either `true` or `false`.

2.1.1 Domain Constructors

In AXIOM, domains are objects. You can create them, pass them to functions, and, as we'll see later, test them for certain properties.

In AXIOM, you ask for a value of a function by applying its name to a set of arguments.

To ask for "the factorial of 7" you enter this expression to AXIOM. This applies the function `factorial` to the value 7 to compute the result.

```
factorial(7)
```

5040 (1)

Type: PositiveInteger

Enter the type Polynomial (Integer) as an expression to AXIOM. This looks much like a function call as well. It is! The result is appropriately stated to be of type Domain, which according to our usual convention, denotes the class of all domains.

```
Polynomial(Integer)
```

Polynomial Integer (2)

Type: Domain

The most basic operation involving domains is that of building a new domain from a given one. To create the domain of "polynomials over the integers," AXIOM applies the function Polynomial to the domain Integer. A function like Polynomial is called a *domain constructor* or, more simply, a *constructor*. A domain constructor is a function that creates a domain. An argument to a domain constructor can be another domain or, in general, an arbitrary kind of object. Polynomial takes a single domain argument while SquareMatrix takes a positive integer as an argument to give its dimension and a domain argument to give the type of its components.

What kinds of domains can you use as the argument to Polynomial or SquareMatrix or List? Well, the first two are mathematical in nature. You want to be able to perform algebraic operations like "+" and "*" on polynomials and square matrices, and operations such as **determinant** on square matrices. So you want to allow polynomials of integers *and* polynomials of square matrices with complex number coefficients and, in general, anything that "makes sense." At the same time, you don't want AXIOM to be able to build nonsense domains such as "polynomials of strings!"

In contrast to algebraic structures, data structures can hold any kind of object. Operations on lists such as **insert**, **delete**, and **concat** just manipulate the list itself without changing or operating on its elements. Thus you can build List over almost any datatype, including itself.

Create a complicated algebraic domain.

```
List (List (Matrix (Polynomial (Complex (Fraction
  (Integer))))))
```

List List Matrix Polynomial Complex Fraction Integer (3)

Type: Domain

Try to create a meaningless domain.

```
Polynomial(String)
```

```
Polynomial String is not a valid type.
```

Evidently from our last example, AXIOM has some mechanism that tells what a constructor can use as an argument. This brings us to the notion of *category*. As domains are objects, they too have a domain. The domain of a domain is a category. A category is simply a type whose members are domains.

A common algebraic category is Ring, the class of all domains that are "rings." A ring is an algebraic structure with constants `0` and `1` and operations "`+`", "`-`", and "`*`". These operations are assumed "closed" with respect to the domain, meaning that they take two objects of the domain and produce a result object also in the domain. The operations are understood to satisfy certain "axioms," certain mathematical principles providing the algebraic foundation for rings. For example, the *additive inverse axiom* for rings states:

> Every element `x` has an additive inverse `y` such that `x + y = 0`.

The prototypical example of a domain that is a ring is the integers. Keep them in mind whenever we mention Ring.

Many algebraic domain constructors such as Complex, Polynomial, Fraction, take rings as arguments and return rings as values. You can use the infix operator "`has`" to ask a domain if it belongs to a particular category.

All numerical types are rings. Domain constructor Polynomial builds "the ring of polynomials over any other ring."

```
Polynomial(Integer) has Ring
```

true (4)

Type: Boolean

Constructor List never produces a ring.

```
List(Integer) has Ring
```

false (5)

Type: Boolean

The constructor Matrix(R) builds "the domain of all matrices over the ring R." This domain is never a ring since the operations "`+`", "`-`", and "`*`" on matrices of arbitrary shapes are undefined.

```
Matrix(Integer) has Ring
```

false (6)

Type: Boolean

Thus you can never build polynomials over matrices.

```
Polynomial(Matrix(Integer))
```

```
Polynomial Matrix Integer is not a valid type.
```

Use SquareMatrix(n,R) instead. For any positive integer n, it builds "the ring of n by n matrices over R."

```
Polynomial(SquareMatrix(7,Complex(Integer)))
```

Polynomial SquareMatrix (7, Complex Integer) (7)

Type: Domain

Another common category is Field, the class of all fields. A field is a ring with additional operations. For example, a field has commutative multiplication and a closed operation "/" for the division of two elements. Integer is not a field since, for example, 3/2 does not have an integer result. The prototypical example of a field is the rational numbers, that is, the domain Fraction(Integer). In general, the constructor Fraction takes a ring as an argument and returns a field.[2] Other domain constructors, such as Complex, build fields only if their argument domain is a field.

The complex integers (often called the "Gaussian integers") do not form a field.

```
Complex(Integer) has Field
```

false (8)

Type: Boolean

But fractions of complex integers do.

```
Fraction(Complex(Integer)) has Field
```

true (9)

Type: Boolean

The algebraically equivalent domain of complex rational numbers is a field since domain constructor Complex produces a field whenever its argument is a field.

```
Complex(Fraction(Integer)) has Field
```

true (10)

Type: Boolean

The most basic category is Type. It denotes the class of all domains and subdomains.[3] Domain constructor List is able to build "lists of elements from domain D" for arbitrary D simply by requiring that D belong to category Type.

Now, you may ask, what exactly is a category? Like domains, categories can be defined in the AXIOM language. A category is defined by three components:

1. a name (for example, Ring), used to refer to the class of domains that the category represents;
2. a set of operations, used to refer to the operations that the domains of this class support (for example, "+", "-", and "*" for rings); and
3. an optional list of other categories that this category extends.

---

[2]Actually, the argument domain must have some additional properties so as to belong to category IntegralDomain.

[3]Type does not denote the class of all types. The type of all categories is Category. The type of Type itself is undefined.

This last component is a new idea. And it is key to the design of AXIOM! Because categories can extend one another, they form hierarchies. Detailed charts showing the category hierarchies in AXIOM are displayed in the endpages of this book. There you see that all categories are extensions of Type and that Field is an extension of Ring.

The operations supported by the domains of a category are called the *exports* of that category because these are the operations made available for system-wide use. The exports of a domain of a given category are not only the ones explicitly mentioned by the category. Since a category extends other categories, the operations of these other categories—and all categories these other categories extend—are also exported by the domains.

For example, polynomial domains belong to PolynomialCategory. This category explicitly mentions some twenty-nine operations on polynomials, but it extends eleven other categories (including Ring). As a result, the current system has over one hundred operations on polynomials.

If a domain belongs to a category that extends, say, Ring, it is convenient to say that the domain exports Ring. The name of the category thus provides a convenient shorthand for the list of operations exported by the category. Rather than listing operations such as "+" and "*" of Ring each time they are needed, the definition of a type simply asserts that it exports category Ring.

The category name, however, is more than a shorthand. The name Ring, in fact, implies that the operations exported by rings are required to satisfy a set of "axioms" associated with the name Ring.[4]

Why is it not correct to assume that some type is a ring if it exports all of the operations of Ring? Here is why. Some languages such as **APL** denote the Boolean constants `true` and `false` by the integers `1` and `0` respectively, then use "+" and "*" to denote the logical operators **or** and **and**. But with these definitions Boolean is not a ring since the additive inverse axiom is violated.[5] This alternative definition of Boolean can be easily and correctly implemented in AXIOM, since Boolean simply does not assert that it is of category Ring. This prevents the system from building meaningless domains such as Polynomial(Boolean) and then wrongfully applying algorithms that presume that the ring axioms hold.

Enough on categories. To learn more about them, see Chapter 12. We

---

[4]This subtle but important feature distinguishes AXIOM from other abstract datatype designs.

[5]There is no inverse element `a` such that `1 + a = 0`, or, in the usual terms: `true or a = false`.

now return to our the discussion of domains.

Domains *export* a set of operations to make them available for system-wide use. Integer, for example, exports the operations "+" and "=" given by the *signatures* "+": (Integer,Integer) → Integer and "=": (Integer,Integer) → Boolean, respectively. Each of these operations takes two Integer arguments. The "+" operation also returns an Integer but "=" returns a Boolean: `true` or `false`. The operations exported by a domain usually manipulate objects of the domain—but not always.

The operations of a domain may actually take as arguments, and return as values, objects from any domain. For example, Fraction (Integer) exports the operations "/": (Integer,Integer) → Fraction(Integer) and **characteristic**: → NonNegativeInteger.

Suppose all operations of a domain take as arguments and return as values, only objects from *other* domains. This kind of domain is what AXIOM calls a *package*.

A package does not designate a class of objects at all. Rather, a package is just a collection of operations. Actually the bulk of the AXIOM library of algorithms consists of packages. The facilities for factorization; integration; solution of linear, polynomial, and differential equations; computation of limits; and so on, are all defined in packages. Domains needed by algorithms can be passed to a package as arguments or used by name if they are not "variable." Packages are useful for defining operations that convert objects of one type to another, particularly when these types have different parameterizations. As an example, the package PolynomialFunction2(R,S) defines operations that convert polynomials over a domain `R` to polynomials over `S`. To convert an object from Polynomial(Integer) to Polynomial(Float), AXIOM builds the package PolynomialFunctions2(Integer,Float) in order to create the required conversion function. (This happens "behind the scenes" for you: see Section 2.7 on page 78 for details on how to convert objects.)

AXIOM categories, domains and packages and all their contained functions are written in the AXIOM programming language and have been compiled into machine code. This is what comprises the AXIOM *library*. In the rest of this book we show you how to use these domains and their functions and how to write your own functions.

2.2 Writing Types and Modes

We have already seen in the last section several examples of types. Most of these examples had either no arguments (for example, Integer) or one argument (for example, Polynomial (Integer)). In this section we give details about writing arbitrary types. We then define *modes* and discuss how to write them. We conclude the section with a discussion on constructor abbreviations.

When might you need to write a type or mode? You need to do so when you declare variables.

```
a : PositiveInteger
```

Type: Void

You need to do so when you declare functions (Section 2.3 on page 69),

```
f : Integer -> String
```

Type: Void

You need to do so when you convert an object from one type to another (Section 2.7 on page 78).

```
factor(2 :: Complex(Integer))
```

$$-\%i\ (1+\%i)^2 \tag{3}$$

Type: Factored Complex Integer

```
(2 = 3)$Integer
```

false (4)

Type: Boolean

You need to do so when you give computation target type information (Section 2.9 on page 83).

```
(2 = 3)@Boolean
```

false (5)

Type: Boolean

2.2.1 Types with No Arguments

A constructor with no arguments can be written either with or without trailing opening and closing parentheses ("()").

Boolean() is the same as Boolean	Integer() is the same as Integer
String() is the same as String	Void() is the same as Void

It is customary to omit the parentheses.

2.2.2 Types with One Argument

A constructor with one argument can frequently be written with no parentheses. Types nest from right to left so that Complex Fraction Polynomial Integer is the same as Complex (Fraction (Polynomial (Integer))). You need to use parentheses to force the application of a constructor to the correct argument, but you need not use any more than is necessary to remove ambiguities.

Here are some guidelines for using parentheses (they are possibly slightly more restrictive than they need to be).

If the argument is an expression like 2 + 3 then you must enclose the argument in parentheses.

```
e : PrimeField(2 + 3)
```

Type: Void

If the type is to be used with package calling then you must enclose the argument in parentheses.

```
content(2)$Polynomial(Integer)
```

$$2 \tag{2}$$

Type: Integer

Alternatively, you can write the type without parentheses then enclose the whole type expression with parentheses.

```
content(2)$(Polynomial Complex Fraction Integer)
```

$$2 \tag{3}$$

Type: Complex Fraction Integer

If you supply computation target type information (Section 2.9 on page 83) then you should enclose the argument in parentheses.

```
(2/3)@Fraction(Polynomial(Integer))
```

$$\frac{2}{3} \tag{4}$$

Type: Fraction Polynomial Integer

If the type itself has parentheses around it and we are not in the case of the first example above, then the parentheses can usually be omitted.

```
(2/3)@Fraction(Polynomial Integer)
```

$$\frac{2}{3} \tag{5}$$

Type: Fraction Polynomial Integer

If the type is used in a declaration and the argument is a single-word type, integer or symbol, then the parentheses can usually be omitted.

```
(d,f,g) : Complex Polynomial Integer
```

Type: Void

2.2.3 Types with More Than One Argument

If a constructor has more than one argument, you must use parentheses. Some examples are

UnivariatePolynomial(x, Float)
MultivariatePolynomial([z,w,r], Complex Float)
SquareMatrix(3, Integer)
FactoredFunctions2(Integer,Fraction Integer)

2.2.4 Modes

A *mode* is a type that possibly is a question mark ("?") or contains one in an argument position. For example, the following are all modes.

?	Polynomial ?
Matrix Polynomial ?	SquareMatrix(3,?)
Integer	OneDimensionalArray(Float)

As is evident from these examples, a mode is a type with a part that is not specified (indicated by a question mark). Only one "?" is allowed per mode and it must appear in the most deeply nested argument that is a type. Thus ?(Integer), Matrix(? (Polynomial)), SquareMatrix(?, Integer) and SquareMatrix(?, ?) are all invalid. The question mark must take the place of a domain, not data (for example, the integer that is the dimension of a square matrix). This rules out, for example, the two SquareMatrix expressions.

Modes can be used for declarations (Section 2.3 on page 69) and conversions (Section 2.7 on page 78). However, you cannot use a mode for package calling or giving target type information.

2.2.5 Abbreviations

Every constructor has an abbreviation that you can freely substitute for the constructor name. In some cases, the abbreviation is nothing more than the capitalized version of the constructor name.

Aside from allowing types to be written more concisely, abbreviations are used by AXIOM to name various system files for constructors (such as library filenames, test input files and example files). Here are some common abbreviations.

COMPLEX abbreviates Complex	EXPR abbreviates Expression
FLOAT abbreviates Float	FRAC abbreviates Fraction
INT abbreviates Integer	MATRIX abbreviates Matrix
NNI abbreviates NonNegativeInteger	PI abbreviates PositiveInteger
POLY abbreviates Polynomial	SF abbreviates SmallFloat
STRING abbreviates String	UP abbreviates UnivariatePolynomial

You can combine both full constructor names and abbreviations in a type expression. Here are some types using abbreviations.

POLY INT is the same as Polynomial(INT)
POLY(Integer) is the same as Polynomial(Integer)
POLY(Integer) is the same as Polynomial(INT)

FRAC(COMPLEX(INT)) is the same as Fraction Complex Integer
FRAC(COMPLEX(INT)) is the same as FRAC(Complex Integer)

There are several ways of finding the names of constructors and their abbreviations. For a specific constructor, use `)abbreviation query`. You can also use the `)what` system command to see the names and abbreviations of constructors. For more information about `)what`, see Section A.26 on page 594.

`)abbreviation query` can be abbreviated (no pun intended) to `)abb q`.

```
)abb q Integer
```

```
INT abbreviates domain Integer
```

The `)abbreviation query` command lists the constructor name if you give the abbreviation. Issue `)abb q` if you want to see the names and abbreviations of all AXIOM constructors.

```
)abb q DMP
```

```
DMP abbreviates domain
DistributedMultivariatePolynomial
```

Issue this to see all packages whose names contain the string "ode".

```
)what packages ode
```

```
-------------------- Packages --------------------

Packages with names matching patterns:
  ode

 EXPRODE   ExpressionSpaceODESolver
 GRAY      GrayCode
 NODE1     NonLinearFirstOrderODESolver
 ODECONST  ConstantLODE
 ODEEF     ElementaryFunctionODESolver
 ODEINT    ODEIntegration
 ODEPAL    PureAlgebraicLODE
 ODERAT    RationalLODE              ODERED    ReduceLODE
 ODESYS    SystemODESolver           ODETOOLS  ODETools
 RTODETLS  RatODETools
 UTSODE    UnivariateTaylorSeriesODESolver
```

2.3 Declarations

A *declaration* is an expression used to restrict the type of values that can be assigned to variables. A colon ("`:`") is always used after a variable or list of variables to be declared.

For a single variable, the syntax for declaration is

variableName : *typeOrMode*

For multiple variables, the syntax is

(*variableName*$_1$, *variableName*$_2$, ... *variableName*$_N$): *typeOrMode*

You can always combine a declaration with an assignment. When you do, it is equivalent to first giving a declaration statement, then giving an assignment. For more information on assignment, see Section 1.3.4 on page 23 and Section 5.1 on page 109. To see how to declare your own functions, see Section 6.4 on page 140.

This declares one variable to have a type.

```
a : Integer
```

Type: Void

This declares several variables to have a type.

```
(b,c) : Integer
```

Type: Void

a, b and c can only hold integer values.

```
a := 45
```

45 (3)

Type: Integer

If a value cannot be converted to a declared type, an error message is displayed.

```
b := 4/5
Cannot convert right-hand side of assignment
4
-
5

to an object of the type Integer of the left-hand
side.
```

This declares a variable with a mode.

```
n : Complex ?
```

Type: Void

This declares several variables with a mode.

```
(p,q,r) : Matrix Polynomial ?
```

Type: Void

This complex object has integer real and imaginary parts.

```
n := -36 + 9 * %i
```

$$-36+9\ \%i \tag{6}$$

Type: Complex Integer

This complex object has fractional symbolic real and imaginary parts.

```
n := complex(4/(x + y),y/x)
```

$$\frac{4}{y+x}+\frac{y}{x}\ \%i \tag{7}$$

Type: Complex Fraction Polynomial Integer

This matrix has entries that are polynomials with integer coefficients.

```
p := [[1,2],[3,4],[5,6]]
```

$$\begin{bmatrix} 1 & 2 \\ 3 & 4 \\ 5 & 6 \end{bmatrix} \tag{8}$$

Type: Matrix Polynomial Integer

This matrix has a single entry that is a polynomial with rational number coefficients.

```
q := [[x - 2/3]]
```

$$\begin{bmatrix} x-\frac{2}{3} \end{bmatrix} \tag{9}$$

Type: Matrix Polynomial Fraction Integer

This matrix has entries that are polynomials with complex integer coefficients.

```
r := [[1-%i*x,7*y+4*%i]]
```

$$\begin{bmatrix} -\%i\ x+1 & 7\ y+4\ \%i \end{bmatrix} \tag{10}$$

Type: Matrix Polynomial Complex Integer

Note the difference between this and the next example. This is a complex object with polynomial real and imaginary parts.

```
f : COMPLEX POLY ? := (x + y*%i)**2
```

$$-y^2+x^2+2\ x\ y\ \%i \tag{11}$$

Type: Complex Polynomial Integer

This is a polynomial with complex integer coefficients. The objects are convertible from one to the other. See Section 2.7 on page 78 for more information.

```
g : POLY COMPLEX ? := (x + y*%i)**2
```

$$-y^2+2\ \%i\ x\ y+x^2 \tag{12}$$

Type: Polynomial Complex Integer

2.4 Records

A Record is an object composed of one or more other objects, each of which is referenced with a *selector*. Components can all belong to the same type or each can have a different type.

> The syntax for writing a Record type is
>
> Record($selector_1$: $type_1$, $selector_2$: $type_2$, ..., $selector_N$: $type_N$)
>
> You must be careful if a selector has the same name as a variable in the workspace. If this occurs, precede the selector name by a single quote.

Record components are implicitly ordered. All the components of a record can be set at once by assigning the record a bracketed *tuple* of values of the proper length (for example, `r : Record(a: Integer, b: String) := [1, "two"]`). To access a component of a record `r`, write the name `r`, followed by a period, followed by a selector.

The object returned by this computation is a record with two components: a `quotient` part and a `remainder` part.

```
u := divide(5,2)
```

$$[quotient = 2,\ remainder = 1] \quad (1)$$

Type: Record(quotient: Integer, remainder: Integer)

This is the quotient part.

```
u.quotient
```

$$2 \quad (2)$$

Type: PositiveInteger

This is the remainder part.

```
u.remainder
```

$$1 \quad (3)$$

Type: PositiveInteger

You can use selector expressions on the left-hand side of an assignment to change destructively the components of a record.

```
u.quotient := 8978
```

$$8978 \quad (4)$$

Type: PositiveInteger

The selected component `quotient` has the value `8978`, which is what is returned by the assignment. Check that the value of `u` was modified.

```
u
```

$$[quotient = 8978,\ remainder = 1] \quad (5)$$

Type: Record(quotient: Integer, remainder: Integer)

Selectors are evaluated. Thus you can use variables that evaluate to selectors instead of the selectors themselves.

```
s := 'quotient
```

$$quotient \quad (6)$$

Type: Variable quotient

Be careful! A selector could have the same name as a variable in the workspace. If this occurs, precede the selector name by a single quote, as in `u.'quotient`.

```
divide(5,2).s
```

$$2 \quad (7)$$

Type: PositiveInteger

Here we declare that the value of `bd` has two components: a string, to be accessed via `name`, and an integer, to be accessed via `birthdayMonth`.

```
bd : Record(name : String, birthdayMonth : Integer)
```

Type: Void

You must initially set the value of the entire Record at once.

```
bd := ["Judith", 3]
```

$$[name = \texttt{"Judith"},\ birthdayMonth = 3] \quad (9)$$

Type: Record(name: String, birthdayMonth: Integer)

Once set, you can change any of the individual components.

```
bd.name := "Katie"
```

"Katie" (10)

Type: String

Records may be nested and the selector names can be shared at different levels.

```
r : Record(a : Record(b: Integer, c: Integer), b: Integer)
```

Type: Void

The record `r` has a `b` selector at two different levels. Here is an initial value for `r`.

```
r := [[1,2],3]
```

$$[a = [b = 1,\ c = 2],\ b = 3] \tag{12}$$

Type: Record(a: Record(b: Integer, c: Integer), b: Integer)

This extracts the `b` component from the `a` component of `r`.

```
r.a.b
```

$$1 \tag{13}$$

Type: PositiveInteger

This extracts the `b` component from `r`.

```
r.b
```

$$3 \tag{14}$$

Type: PositiveInteger

You can also use spaces or parentheses to refer to Record components. This is the same as `r.a`.

```
r(a)
```

$$[b = 1,\ c = 2] \tag{15}$$

Type: Record(b: Integer, c: Integer)

This is the same as `r.b`.

```
r b
```

$$3 \tag{16}$$

Type: PositiveInteger

This is the same as `r.b := 10`.

```
r(b) := 10
```

$$10 \tag{17}$$

Type: PositiveInteger

Look at `r` to make sure it was modified.

```
r
```

$$[a = [b = 1,\ c = 2],\ b = 10] \tag{18}$$

Type: Record(a: Record(b: Integer, c: Integer), b: Integer)

2.5 Unions

Type Union is used for objects that can be of any of a specific finite set of types. Two versions of unions are available, one with selectors (like records) and one without.

2.5.1 Unions Without Selectors

The declaration `x : Union(Integer, String, Float)` states that `x` can have values that are integers, strings or "big" floats. If, for example, the Union object is an integer, the object is said to belong to the Integer *branch* of the Union.[6]

> The syntax for writing a Union type without selectors is
>
> Union($type_1$, $type_2$, ..., $type_N$)
>
> The types in a union without selectors must be distinct.

It is possible to create unions like Union(Integer, PositiveInteger) but they are difficult to work with because of the overlap in the branch types. See below for the rules AXIOM uses for converting something into a union object.

The `case` infix operator returns a Boolean and can be used to determine the branch in which an object lies.

This function displays a message stating in which branch of the Union the object (defined as `x` above) lies.

```
sayBranch(x : Union(Integer,String,Float)) : Void  ==
  output
    x case Integer => "Integer branch"
    x case String  => "String branch"
    "Float branch"
```

Type: Void

This tries **sayBranch** with an integer.

```
sayBranch 1

Integer branch
```

Type: Void

This tries **sayBranch** with a string.

```
sayBranch "hello"

String branch
```

Type: Void

This tries **sayBranch** with a floating-point number.

```
sayBranch 2.718281828

Float branch
```

Type: Void

There are two things of interest about this particular example to which we would like to draw your attention.

1. AXIOM normally converts a result to the target value before passing

[6]Note that we are being a bit careless with the language here. Technically, the type of `x` is always Union(Integer, String, Float). If it belongs to the Integer branch, `x` may be converted to an object of type Integer.

it to the function. If we left the declaration information out of this function definition then the `sayBranch` call would have been attempted with an Integer rather than a Union, and an error would have resulted.

2. The types in a Union are searched in the order given. So if the type were given as
`sayBranch(x: Union(String,Integer,Float,Any)): Void`
then the result would have been "String branch" because there is a conversion from Integer to String.

Sometimes Union types can sometimes have extremely long names. AXIOM therefore abbreviates the names of unions by printing the type of the branch first within the Union and then eliding the remaining types with an ellipsis ("...").

Here the Integer branch is displayed first. Use "::" to create a Union object from an object.

```
78 :: Union(Integer,String)
```

78 (5)

Type: Union(Integer, ...)

Here the String branch is displayed first.

```
s := "string" :: Union(Integer,String)
```

"string" (6)

Type: Union(String, ...)

Use `typeOf` to see the full and actual Union type.

```
typeOf s
```

Union (Integer , String) (7)

Type: Domain

A common operation that returns a union is **exquo** which returns the "exact quotient" if the quotient is exact,...

```
three := exquo(6,2)
```

3 (8)

Type: Union(Integer, ...)

and `"failed"` if the quotient is not exact.

```
exquo(5,2)
```

"failed" (9)

Type: Union("failed", ...)

A union with a `"failed"` is frequently used to indicate the failure or lack of applicability of an object. As another example, assign an integer a variable `r` declared to be a rational number.

```
r: FRAC INT := 3
```

3 (10)

Type: Fraction Integer

The operation **retractIfCan** tries to retract the fraction to the underlying domain Integer. It produces a union object. Here it succeeds.

```
retractIfCan(r)
```

3 (11)

Type: Union(Integer, ...)

Assign it a rational number.

```
r := 3/2
```

$$\frac{3}{2} \tag{12}$$

Type: Fraction Integer

Here the retraction fails.

```
retractIfCan(r)
```

"failed" (13)

Type: Union("failed", ...)

2.5.2 Unions With Selectors

Like records (Section 2.4 on page 71), you can write Union types with selectors.

The syntax for writing a Union type with selectors is

Union($selector_1$: $type_1$, $selector_2$: $type_2$, ..., $selector_N$: $type_N$)

You must be careful if a selector has the same name as a variable in the workspace. If this occurs, precede the selector name by a single quote. It is an error to use a selector that does not correspond to the branch of the Union in which the element actually lies.

Be sure to understand the difference between records and unions with selectors. Records can have more than one component and the selectors are used to refer to the components. Unions always have one component but the type of that one component can vary. An object of type Record(a: Integer, b: Float, c: String) contains an integer *and* a float *and* a string. An object of type Union(a: Integer, b: Float, c: String) contains an integer *or* a float *or* a string.

Here is a version of the **sayBranch** function (cf. Section 2.5.1 on page 74) that works with a union with selectors. It displays a message stating in which branch of the Union the object lies.

```
sayBranch(x:Union(i:Integer,s:String,f:Float)):Void==
  output
    x case i => "Integer branch"
    x case s  => "String branch"
    "Float branch"
```

Note that `case` uses the selector name as its right-hand argument. If you accidentally use the branch type on the right-hand side of `case`, `false` will be returned.

Declare variable u to have a union type with selectors.

```
u : Union(i : Integer, s : String)
```

Type: Void

Give an initial value to u.

```
u := "good morning"
```

"good morning" (2)

Type: Union(s: String, ...)

Use case to determine in which branch of a Union an object lies.

```
u case i
```

false (3)

Type: Boolean

```
u case s
```

true (4)

Type: Boolean

To access the element in a particular branch, use the selector.

```
u.s
```

"good morning" (5)

Type: String

2.6 The "Any" Domain

With the exception of objects of type Record, all AXIOM data structures are homogenous, that is, they hold objects all of the same type. If you need to get around this, you can use type Any. Using Any, for example, you can create lists whose elements are integers, rational numbers, strings, and even other lists.

Declare u to have type Any.

```
u: Any
```

Type: Void

Assign a list of mixed type values to u

```
u := [1, 7.2, 3/2, x**2, "wally"]
```

$$\left[1,\ 7.2,\ \frac{3}{2},\ x^2,\ \texttt{"wally"}\right] \quad (2)$$

Type: List Any

When we ask for the elements, AXIOM displays these types.

```
u.1
```

1 (3)

Type: PositiveInteger

Actually, these objects belong to Any but AXIOM automatically converts them to their natural types for you.

```
u.3
```

$$\frac{3}{2} \quad (4)$$

Type: Fraction Integer

Since type Any can be anything, it can only belong to type Type. Therefore it cannot be used in algebraic domains.

```
v : Matrix(Any)
```

```
Matrix Any is not a valid type.
```

Perhaps you are wondering how AXIOM internally represents objects of type Any. An object of type Any consists not only a data part representing its normal value, but also a type part (a *badge*) giving its type. For example, the value 1 of type PositiveInteger as an object of type Any internally looks like [1,PositiveInteger()].

2.7 Conversion

Conversion is the process of changing an object of one type into an object of another type. The syntax for conversion is:

object :: *newType*

By default, 3 has the type PositiveInteger.

```
3
```

$$3 \tag{1}$$

Type: PositiveInteger

We can change this into an object of type Fraction Integer by using "::".

```
3 :: Fraction Integer
```

$$3 \tag{2}$$

Type: Fraction Integer

A *coercion* is a special kind of conversion that AXIOM is allowed to do automatically when you enter an expression. Coercions are usually somewhat safer than more general conversions. The AXIOM library contains operations called **coerce** and **convert**. Only the **coerce** operations can be used by the interpreter to change an object into an object of another type unless you explicitly use a "::".

By now you will be quite familiar with what types and modes look like. It is useful to think of a type or mode as a pattern for what you want the result to be.

Let's start with a square matrix of polynomials with complex rational number coefficients.

```
m : SquareMatrix(2,POLY COMPLEX FRAC INT)
```

Type: Void

```
m := matrix [[x-3/4*%i,z*y**2+1/2],[3/7*%i*y**4 - x,12-
  %i*9/5]]
```

$$\begin{bmatrix} x - \frac{3}{4}\,\%i & y^2\,z + \frac{1}{2} \\ \frac{3}{7}\,\%i\,y^4 - x & 12 - \frac{9}{5}\,\%i \end{bmatrix} \quad (4)$$

Type: SquareMatrix(2, Polynomial Complex Fraction Integer)

We first want to interchange the Complex and Fraction layers. We do the conversion by doing the interchange in the type expression.

```
m1 := m :: SquareMatrix(2,POLY FRAC COMPLEX INT)
```

$$\begin{bmatrix} x - \frac{3\,\%i}{4} & y^2\,z + \frac{1}{2} \\ \frac{3\,\%i}{7}\,y^4 - x & \frac{60-9\,\%i}{5} \end{bmatrix} \quad (5)$$

Type: SquareMatrix(2, Polynomial Fraction Complex Integer)

Interchange the Polynomial and the Fraction levels.

```
m2 := m1 :: SquareMatrix(2,FRAC POLY COMPLEX INT)
```

$$\begin{bmatrix} \frac{4\,x-3\,\%i}{4} & \frac{2\,y^2\,z+1}{2} \\ \frac{3\,\%i\,y^4-7\,x}{7} & \frac{60-9\,\%i}{5} \end{bmatrix} \quad (6)$$

Type: SquareMatrix(2, Fraction Polynomial Complex Integer)

Interchange the Polynomial and the Complex levels.

```
m3 := m2 :: SquareMatrix(2,FRAC COMPLEX POLY INT)
```

$$\begin{bmatrix} \frac{4\,x-3\,\%i}{4} & \frac{2\,y^2\,z+1}{2} \\ \frac{-7\,x+3\,y^4\,\%i}{7} & \frac{60-9\,\%i}{5} \end{bmatrix} \quad (7)$$

Type: SquareMatrix(2, Fraction Complex Polynomial Integer)

All the entries have changed types, although in comparing the last two results only the entry in the lower left corner looks different. We did all the intermediate steps to show you what AXIOM can do.

In fact, we could have combined all these into one conversion.

```
m :: SquareMatrix(2,FRAC COMPLEX POLY INT)
```

$$\begin{bmatrix} \frac{4\,x-3\,\%i}{4} & \frac{2\,y^2\,z+1}{2} \\ \frac{-7\,x+3\,y^4\,\%i}{7} & \frac{60-9\,\%i}{5} \end{bmatrix} \quad (8)$$

Type: SquareMatrix(2, Fraction Complex Polynomial Integer)

There are times when AXIOM is not be able to do the conversion in one step. You may need to break up the transformation into several conversions in order to get an object of the desired type.

We cannot move either Fraction or Complex above (or to the left of, depending on how you look at it) SquareMatrix because each of these levels requires that its argument type have commutative multiplication, whereas SquareMatrix does not.[7] The Integer level did not move anywhere because it does not allow any arguments. We also did not move the SquareMatrix part anywhere, but we could have.

[7]Fraction requires that its argument belong to the category IntegralDomain and Complex requires that its argument belong to CommutativeRing. See Section 2.1 on page 59 for a brief discussion of categories.

Recall that m looks like this.

```
m
```

$$\begin{bmatrix} x - \frac{3}{4}\,\%i & y^2\,z + \frac{1}{2} \\ \frac{3}{7}\,\%i\,y^4 - x & 12 - \frac{9}{5}\,\%i \end{bmatrix} \quad (9)$$

Type: SquareMatrix(2, Polynomial Complex Fraction Integer)

If we want a polynomial with matrix coefficients rather than a matrix with polynomial entries, we can just do the conversion.

```
m :: POLY SquareMatrix(2,COMPLEX FRAC INT)
```

$$\begin{bmatrix} 0 & 1 \\ 0 & 0 \end{bmatrix} y^2\,z + \begin{bmatrix} 0 & 0 \\ \frac{3}{7}\,\%i & 0 \end{bmatrix} y^4 + \begin{bmatrix} 1 & 0 \\ -1 & 0 \end{bmatrix} x + \begin{bmatrix} -\frac{3}{4}\,\%i & \frac{1}{2} \\ 0 & 12 - \frac{9}{5}\,\%i \end{bmatrix} \quad (10)$$

Type: Polynomial SquareMatrix(2, Complex Fraction Integer)

We have not yet used modes for any conversions. Modes are a great shorthand for indicating the type of the object you want. Instead of using the long type expression in the last example, we could have simply said this.

```
m :: POLY ?
```

$$\begin{bmatrix} 0 & 1 \\ 0 & 0 \end{bmatrix} y^2\,z + \begin{bmatrix} 0 & 0 \\ \frac{3}{7}\,\%i & 0 \end{bmatrix} y^4 + \begin{bmatrix} 1 & 0 \\ -1 & 0 \end{bmatrix} x + \begin{bmatrix} -\frac{3}{4}\,\%i & \frac{1}{2} \\ 0 & 12 - \frac{9}{5}\,\%i \end{bmatrix} \quad (11)$$

Type: Polynomial SquareMatrix(2, Complex Fraction Integer)

We can also indicate more structure if we want the entries of the matrices to be fractions.

```
m :: POLY SquareMatrix(2,FRAC ?)
```

$$\begin{bmatrix} 0 & 1 \\ 0 & 0 \end{bmatrix} y^2\,z + \begin{bmatrix} 0 & 0 \\ \frac{3\,\%i}{7} & 0 \end{bmatrix} y^4 + \begin{bmatrix} 1 & 0 \\ -1 & 0 \end{bmatrix} x + \begin{bmatrix} -\frac{3\,\%i}{4} & \frac{1}{2} \\ 0 & \frac{60-9\,\%i}{5} \end{bmatrix} \quad (12)$$

Type: Polynomial SquareMatrix(2, Fraction Complex Integer)

2.8 Subdomains Again

A *subdomain* S of a domain D is a domain consisting of

1. those elements of D that satisfy some *predicate* (that is, a test that returns `true` or `false`) and
2. a subset of the operations of D.

Every domain is a subdomain of itself, trivially satisfying the membership test: `true`.

Currently, there are only two system-defined subdomains in AXIOM that receive substantial use. PositiveInteger and NonNegativeInteger are subdomains of Integer. An element x of NonNegativeInteger is an integer that is greater than or equal to zero, that is, satisfies `x >= 0`. An element `x` of PositiveInteger is a nonnegative integer that is, in fact, greater than zero, that is, satisfies `x > 0`. Not all operations from Integer are available for these subdomains. For example, negation and subtraction are not provided since the subdomains are not closed under those operations. When you use an integer in an expression, AXIOM assigns to it the type that is the most specific subdomain whose predicate is satisfied.

This is a positive integer.

```
5
```

5 (1)

Type: PositiveInteger

This is a nonnegative integer.

```
0
```

0 (2)

Type: NonNegativeInteger

This is neither of the above.

```
-5
```

-5 (3)

Type: Integer

Furthermore, unless you are assigning an integer to a declared variable or using a conversion, any integer result has as type the most specific subdomain.

```
(-2) - (-3)
```

1 (4)

Type: PositiveInteger

```
0 :: Integer
```

0 (5)

Type: Integer

```
x : NonNegativeInteger := 5
```

5 (6)

Type: NonNegativeInteger

When necessary, AXIOM converts an integer object into one belonging to a less specific subdomain. For example, in `3-2`, the arguments to "`-`" are both elements of PositiveInteger, but this type does not provide a subtraction operation. Neither does NonNegativeInteger, so 3 and 2 are viewed as elements of Integer, where their difference can be calculated. The result is `1`, which AXIOM then automatically assigns the type PositiveInteger.

Certain operations are very sensitive to the subdomains to which their arguments belong. This is an element of PositiveInteger.

```
2 ** 2
```

4 (7)

Type: PositiveInteger

This is an element of Fraction Integer.

```
2 ** (-2)
```

$$\frac{1}{4} \tag{8}$$

Type: Fraction Integer

It makes sense then that this is a list of elements of PositiveInteger.

```
[10**i for i in 2..5]
```

$$[100, 1000, 10000, 100000] \tag{9}$$

Type: List PositiveInteger

What should the type of `[10**(i-1) for i in 2..5]` be? On one hand, `i-1` is always an integer greater than zero as `i` ranges from 2 to 5 and so `10**i` is also always a positive integer. On the other, `i-1` is a very simple function of `i`. AXIOM does not try to analyze every such function over the index's range of values to determine whether it is always positive or nowhere negative. For an arbitrary AXIOM function, this analysis is not possible.

So, to be consistent no such analysis is done and we get this.

```
[10**(i-1) for i in 2..5]
```

$$[10, 100, 1000, 10000] \tag{10}$$

Type: List Fraction Integer

To get a list of elements of PositiveInteger instead, you have two choices. You can use a conversion.

```
[10**((i-1) :: PI) for i in 2..5]
```

$$[10, 100, 1000, 10000] \tag{11}$$

Type: List PositiveInteger

Or you can use `pretend`.

```
[10**((i-1) pretend PI) for i in 2..5]
```

$$[10, 100, 1000, 10000] \tag{12}$$

Type: List PositiveInteger

The operation `pretend` is used to defeat the AXIOM type system. The expression `object pretend D` means "make a new object (without copying) of type `D` from `object`." If `object` were an integer and you told AXIOM to pretend it was a list, you would probably see a message about a fatal error being caught and memory possibly being damaged. Lists do not have the same internal representation as integers!

You use `pretend` at your peril.

Use `pretend` with great care! AXIOM trusts you that the value is of the specified type.

```
(2/3) pretend Complex Integer
```

$$2 + 3\,\%i \tag{13}$$

Type: Complex Integer

2.9 Package Calling and Target Types

AXIOM works hard to figure out what you mean by an expression without your having to qualify it with type information. Nevertheless, there are times when you need to help it along by providing hints (or even orders!) to get AXIOM to do what you want.

We saw in Section 2.3 on page 69 that declarations using types and modes control the type of the results produced. For example, we can either produce a complex object with polynomial real and imaginary parts or a polynomial with complex integer coefficients, depending on the declaration).

Package calling is how you tell AXIOM to use a particular function from a particular part of the library.

Use the "/" from Fraction Integer to create a fraction of two integers.

```
2/3
```

$$\frac{2}{3} \tag{1}$$

Type: Fraction Integer

If we wanted a floating point number, we can say "use the "/" in Float."

```
(2/3)$Float
```

$$0.66666666666666666667 \tag{2}$$

Type: Float

Perhaps we actually wanted a fraction of complex integers.

```
(2/3)$Fraction(Complex Integer)
```

$$\frac{2}{3} \tag{3}$$

Type: Fraction Complex Integer

In each case, AXIOM used the indicated operations, sometimes first needing to convert the two integers into objects of an appropriate type. In these examples, "/" is written as an infix operator.

> To use package calling with an infix operator, use the following syntax:
>
> (arg_1 *op* arg_1)$*type*

We used, for example, `(2/3)$Float`. The expression `2 + 3 + 4` is equivalent to `(2+3) + 4`. Therefore in the expression `(2 + 3 + 4)$Float` the second "+" comes from the Float domain. Can you guess whether the first "+" comes from Integer or Float?[8]

[8]Float, because the package call causes AXIOM to convert `(2 + 3)` and `4` to type Float. Before the sum is converted, it is given a target type (see below) of Float by AXIOM and then evaluated. The target type causes the "+" from Float to be used.

For an operator written before its arguments, you must use parentheses around the arguments (even if there is only one), and follow the closing parenthesis by a "$" and then the type.

fun (arg_1 , arg_1 , ... , arg_N)$*type*

For example, to call the "minimum" function from SmallFloat on two integers, you could write `min(4,89)$SmallFloat`. Another use of package calling is to tell AXIOM to use a library function rather than a function you defined. We discuss this in Section 6.9 on page 145.

Sometimes rather than specifying where an operation comes from, you just want to say what type the result should be. We say that you provide a *target type* for the expression. Instead of using a "$", use a "@" to specify the requested target type. Otherwise, the syntax is the same. Note that giving a target type is not the same as explicitly doing a conversion. The first says "try to pick operations so that the result has such-and-such a type." The second says "compute the result and then convert to an object of such-and-such a type."

Sometimes it makes sense, as in this expression, to say "choose the operations in this expression so that the final result is a Float."

```
(2/3)@Float
```

0.66666666666666666667 (4)

Type: Float

Here we used "@" to say that the target type of the left-hand side was Float. In this simple case, there was no real difference between using "$" and "@". You can see the difference if you try the following.

This says to try to choose "+" so that the result is a string. AXIOM cannot do this.

```
(2 + 3)@String
```

```
An expression involving @ String actually evaluated
to one of type PositiveInteger. Perhaps you should
use :: String.
```

This says to get the "+" from String and apply it to the two integers. AXIOM also cannot do this because there is no "+" exported by String.

```
(2 + 3)$String
```

```
The function + is not implemented in String.
```

(By the way, the operation **concat** or juxtaposition is used to concatenate two strings.)

When we have more than one operation in an expression, the difference is even more evident. The following two expressions show that AXIOM uses the target type to create different objects. The "+", "*" and "**" operations are all chosen so that an object of the correct final type is created.

This says that the operations should be chosen so that the result is a Complex object.

```
((x + y * %i)**2)@(Complex Polynomial Integer)
```

$$-y^2 + x^2 + 2\ x\ y\ \%i \tag{5}$$

Type: Complex Polynomial Integer

This says that the operations should be chosen so that the result is a Polynomial object.

```
((x + y * %i)**2)@(Polynomial Complex Integer)
```

$$-y^2 + 2\ \%i\ x\ y + x^2 \tag{6}$$

Type: Polynomial Complex Integer

What do you think might happen if we left off all target type and package call information in this last example?

```
(x + y * %i)**2
```

$$-y^2 + 2\ \%i\ x\ y + x^2 \tag{7}$$

Type: Polynomial Complex Integer

We can convert it to Complex as an afterthought. But this is more work than just saying making what we want in the first place.

```
% :: Complex ?
```

$$-y^2 + x^2 + 2\ x\ y\ \%i \tag{8}$$

Type: Complex Polynomial Integer

Finally, another use of package calling is to qualify fully an operation that is passed as an argument to a function.

Start with a small matrix of integers.

```
h := matrix [[8,6],[-4,9]]
```

$$\begin{bmatrix} 8 & 6 \\ -4 & 9 \end{bmatrix} \tag{9}$$

Type: Matrix Integer

We want to produce a new matrix that has for entries the multiplicative inverses of the entries of h. One way to do this is by calling **map** with the **inv** function from Fraction (Integer).

```
map(inv$Fraction(Integer),h)
```

$$\begin{bmatrix} \frac{1}{8} & \frac{1}{6} \\ -\frac{1}{4} & \frac{1}{9} \end{bmatrix} \tag{10}$$

Type: Matrix Fraction Integer

We could have been a bit less verbose and used abbreviations.

```
map(inv$FRAC(INT),h)
```

$$\begin{bmatrix} \frac{1}{8} & \frac{1}{6} \\ -\frac{1}{4} & \frac{1}{9} \end{bmatrix} \tag{11}$$

Type: Matrix Fraction Integer

As it turns out, AXIOM is smart enough to know what we mean anyway. We can just say this.

```
map(inv,h)
```

$$\begin{bmatrix} \frac{1}{8} & \frac{1}{6} \\ -\frac{1}{4} & \frac{1}{9} \end{bmatrix} \tag{12}$$

Type: Matrix Fraction Integer

2.10 Resolving Types

In this section we briefly describe an internal process by which AXIOM determines a type to which two objects of possibly different types can be converted. We do this to give you further insight into how AXIOM takes your input, analyzes it, and produces a result.

What happens when you enter `x + 1` to AXIOM? Let's look at what you get from the two terms of this expression.

This is a symbolic object whose type indicates the name.

```
x
```

$$x \tag{1}$$

Type: Variable x

This is a positive integer.

```
1
```

$$1 \tag{2}$$

Type: PositiveInteger

There are no operations in PositiveInteger that add positive integers to objects of type Variable(x) nor are there any in Variable(x). Before it can add the two parts, AXIOM must come up with a common type to which both **x** and 1 can be converted. We say that AXIOM must *resolve* the two types into a common type. In this example, the common type is Polynomial(Integer).

Once this is determined, both parts are converted into polynomials, and the addition operation from Polynomial(Integer) is used to get the answer.

```
x + 1
```

$$x + 1 \tag{3}$$

Type: Polynomial Integer

AXIOM can always resolve two types: if nothing resembling the original types can be found, then Any is be used. This is fine and useful in some cases.

```
["string",3.14159]
```

$$["string", 3.14159] \tag{4}$$

Type: List Any

In other cases objects of type Any can't be used by the operations you specified.

```
"string" + 3.14159
```

```
Cannot find a definition or library operation named +
with argument types
                              String
                              Float
```

Although this example was contrived, your expressions may need to be qualified slightly to help AXIOM resolve the types involved. You may need to declare a few variables, do some package calling, provide some target type information or do some explicit conversions.

We suggest that you just enter the expression you want evaluated and see what AXIOM does. We think you will be impressed with its ability to "do what I mean." If AXIOM is still being obtuse, give it some hints. As you work with AXIOM, you will learn where it needs a little help to analyze quickly and perform your computations.

2.11 Exposing Domains and Packages

In this section we discuss how AXIOM makes some operations available to you while hiding others that are meant to be used by developers or only in rare cases. If you are a new user of AXIOM, it is likely that everything you need is available by default and you may want to skip over this section on first reading.

Every domain and package in the AXIOM library is either *exposed* (meaning that you can use its operations without doing anything special) or it is *hidden* (meaning you have to either package call (see Section 2.9 on page 83) the operations it contains or explicitly expose it to use the operations). The initial exposure status for a constructor is set in the file **INTERP.EXPOSED** (see the *Installer's Note* for AXIOM if you need to know the location of this file). Constructors are collected together in *exposure groups*. Categories are all in the exposure group "categories" and the bulk of the basic set of packages and domains that are exposed are in the exposure group "basic." Here is an abbreviated sample of the file:

```
basic
        AlgebraicNumber                              AN
        AlgebraGivenByStructuralConstants            ALGSC
        Any                                          ANY
        AnyFunctions1                                ANY1
        BinaryExpansion                              BINARY
        Boolean                                      BOOLEAN
        CardinalNumber                               CARD
        CartesianTensor                              CARTEN
        Character                                    CHAR
        CharacterClass                               CCLASS
        CliffordAlgebra                              CLIF
        Color                                        COLOR
        Complex                                      COMPLEX
        ContinuedFraction                            CONTFRAC
        DecimalExpansion                             DECIMAL
        ...

categories
        AbelianGroup                                 ABELGRP
        AbelianMonoid                                ABELMON
        AbelianMonoidRing                            AMR
        AbelianSemiGroup                             ABELSG
        Aggregate                                    AGG
        Algebra                                      ALGEBRA
        AlgebraicallyClosedField                     ACF
        AlgebraicallyClosedFunctionSpace             ACFS
```

```
ArcHyperbolicFunctionCategory            AHYP
...
```

For each constructor in a group, the full name and the abbreviation is given. There are other groups in **INTERP.EXPOSED** but initially only the constructors in exposure group "basic" are exposed.

As an interactive user of AXIOM, you do not need to modify this file. Instead, use `)set expose` to expose, hide or query the exposure status of an individual constructor or exposure group. The reason for having exposure groups is to be able to expose or hide multiple constructors with a single command. For example, you might group together into exposure group "quantum" a number of domains and packages useful for quantum mechanical computations. These probably should not be available to every user, but you want an easy way to make the whole collection visible to AXIOM when it is looking for operations to apply.

If you wanted to hide all the basic constructors available by default, you would issue `)set expose drop group basic`. We do not recommend that you do this. If, however, you discover that you have hidden all the basic constructors, you should issue `)set expose add group basic` to restore your default environment.

It is more likely that you would want to expose or hide individual constructors. In Section 6.19 on page 170 we use several operations from OutputForm, a domain usually hidden. To avoid package calling every operation from OutputForm, we expose the domain and let AXIOM conclude that those operations should be used. Use `)set expose add constructor` and `)set expose drop constructor` to expose and hide a constructor, respectively. You should use the constructor name, not the abbreviation. The `)set expose` command guides you through these options.

If you expose a previously hidden constructor, AXIOM exhibits new behavior (that was your intention) though you might not expect the results that you get. OutputForm is, in fact, one of the worst offenders in this regard. This domain is meant to be used by other domains for creating a structure that AXIOM knows how to display. It has functions like "+" that form output representations rather than do mathematical calculations. Because of the order in which AXIOM looks at constructors when it is deciding what operation to apply, OutputForm might be used instead of what you expect.

This is a polynomial.

```
x + x
```

$$2\,x \qquad (1)$$

Type: Polynomial Integer

Expose OutputForm.

```
)set expose add constructor OutputForm

OutputForm is now explicitly exposed in frame G1077
```

This is what we get when OutputForm is automatically available.

```
x + x
```

$$x + x \tag{2}$$

Type: OutputForm

Hide OutputForm so we don't run into problems with any later examples!

```
)set expose drop constructor OutputForm

OutputForm is now explicitly hidden in frame G1077
```

Finally, exposure is done on a frame-by-frame basis. A *frame* (see Section A.10 on page 581) is one of possibly several logical AXIOM workspaces within a physical one, each having its own environment (for example, variables and function definitions). If you have several AXIOM workspace windows on your screen, they are all different frames, automatically created for you by HyperDoc. Frames can be manually created, made active and destroyed by the `)frame` system command. They do not share exposure information, so you need to use `)set expose` in each one to add or drop constructors from view.

2.12 Commands for Snooping

To conclude this chapter, we introduce you to some system commands that you can use for getting more information about domains, packages, categories, and operations. The most powerful AXIOM facility for getting information about constructors and operations is the Browse component of HyperDoc. This is discussed in Chapter 14.

Use the `)what` system command to see lists of system objects whose name contain a particular substring (uppercase or lowercase is not significant).

Issue this to see a list of all operations with "`complex`" in their names.

```
)what operation complex

Operations whose names satisfy the above pattern(s):

complex                   complexEigenvalues
complexEigenvectors       complexElementary
complexExpand             complexIntegrate
complexLimit              complexNormalize
complexNumeric            complexRoots
complexSolve              complexZeros

To get more information about an operation such as
complex, issue the command )display op complex
```

If you want to see all domains with "`matrix`" in their names, issue this.

```
)what domain matrix
-------------------- Domains ---------------------

Domains with names matching patterns:
  matrix

 DHMATRIX DenavitHartenbergMatrix
 DPMM     DirectProductMatrixModule
 IMATRIX  IndexedMatrix
 LSQM     LieSquareMatrix
 MATCAT-  MatrixCategory&        MATRIX   Matrix
 RMATCAT- RectangularMatrixCategory&
 RMATRIX  RectangularMatrix
 SMATCAT- SquareMatrixCategory&
 SQMATRIX SquareMatrix
```

Similarly, if you wish to see all packages whose names contain "`gauss`", enter this.

```
)what package gauss
-------------------- Packages --------------------

Packages with names matching patterns:
  gauss

 GAUSSFAC GaussianFactorizationPackage
```

This command shows all the operations that Any provides. Wherever "`$`" appears, it means "Any".

```
)show Any
Any is a domain constructor.
Abbreviation for Any is ANY
This constructor is exposed in this frame.
Issue )edit any.spad to see algebra source code for
ANY

------------------- Operations -------------------

?=? : ($,$) -> Boolean
coerce : $ -> OutputForm
dom : $ -> SExpression
domainOf : $ -> OutputForm
obj : $ -> None
objectOf : $ -> OutputForm
any : (SExpression,None) -> $
showTypeInOutput : Boolean -> String
```

This displays all operations with the name **complex**.

```
)display operation complex
There is one exposed function called complex :
[1] (D1,D1) -> D from D if D1 has COMRING and D has
      COMPCAT D1
```

Let's analyze this output.

First we find out what some of the abbreviations mean.

```
)abbreviation query COMPCAT

COMPCAT abbreviates category ComplexCategory


)abbreviation query COMRING

COMRING abbreviates category CommutativeRing
```

So if D1 is a commutative ring (such as the integers or floats) and D belongs to ComplexCategory D1, then there is an operation called **complex** that takes two elements of D1 and creates an element of D. The primary example of a constructor implementing domains belonging to ComplexCategory is Complex. See 'Complex' on page 333 for more information on that and see Section 6.4 on page 140 for more information on function types.

CHAPTER 3

Using HyperDoc

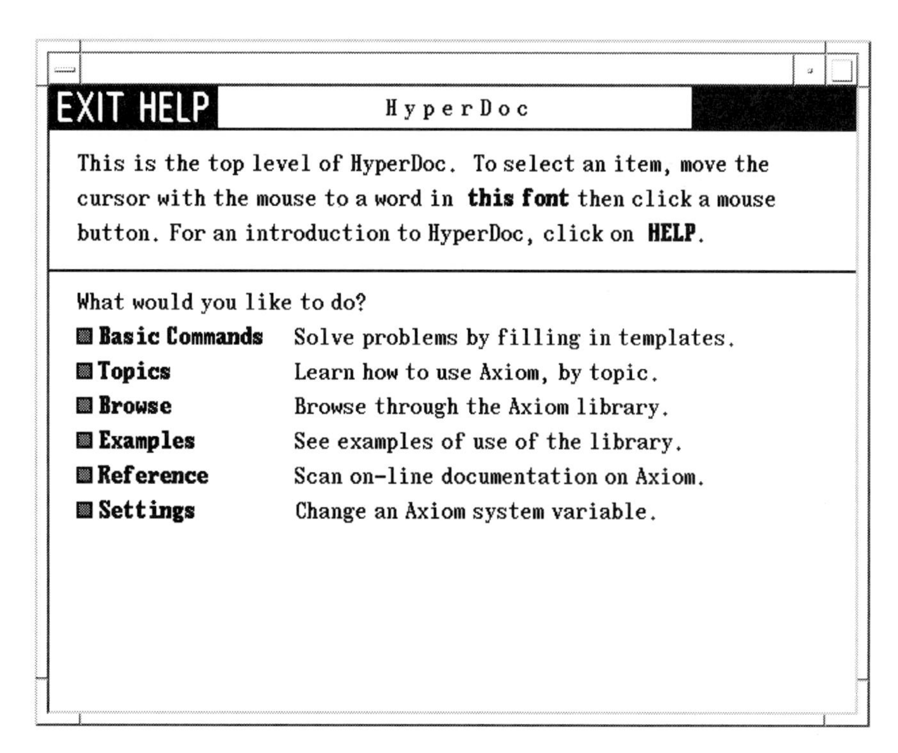

Figure 3.1: The HyperDoc root window page.

HyperDoc is the gateway to AXIOM. It's both an on-line tutorial and an on-line reference manual. It also enables you to use AXIOM simply by using the mouse and filling in templates. HyperDoc is available to you if you are running AXIOM under the X Window System.

Pages usually have active areas, marked in **this font** (bold face). As you move the mouse pointer to an active area, the pointer changes from a filled dot to an open circle. The active areas are usually linked to other pages. When you click on an active area, you move to the linked page.

3.1 Headings

Most pages have a standard set of buttons at the top of the page. This is what they mean:

HELP Click on this to get help. The button only appears if there is specific help for the page you are viewing. You can get *general* help for HyperDoc by clicking the help button on the home page.

[up-arrow button] Click here to go back one page. By clicking on this button repeatedly, you can go back several pages and then take off in a new direction.

HOME Go back to the home page, that is, the page on which you started. Use HyperDoc to explore, to make forays into new topics. Don't worry about how to get back. HyperDoc remembers where you came from. Just click on this button to return.

EXIT From the root window (the one that is displayed when you start the system) this button leaves the HyperDoc program, and it must be restarted if you want to use it again. From any other HyperDoc window, it just makes that one window go away. You *must* use this button to get rid of a window. If you use the window manager "Close" button, then all of HyperDoc goes away.

The buttons are not be displayed if they are not applicable to the page you are viewing. For example, there is no **HOME** button on the top-level menu.

3.2 Scroll Bars

Whenever there is too much text to fit on a page, a *scroll bar* automatically appears along the right side.

With a scroll bar, your page becomes an aperture, that is, a window into a larger amount of text than can be displayed at one time. The scroll bar lets you move up and down in the text to see different parts. It also shows where the aperture is relative to the whole text. The aperture is indicated by a strip on the scroll bar.

Move the cursor with the mouse to the "down-arrow" at the bottom of the scroll bar and click. See that the aperture moves down one line. Do it several times. Each time you click, the aperture moves down one line. Move the mouse to the "up-arrow" at the top of the scroll bar and click. The aperture moves up one line each time you click.

Next move the mouse to any position along the middle of the scroll bar and click. HyperDoc attempts to move the top of the aperture to this point in the text.

You cannot make the aperture go off the bottom edge. When the aperture is about half the size of text, the lowest you can move the aperture is halfway down.

To move up or down one screen at a time, use the **PageUp** and **PageDown** keys on your keyboard. They move the visible part of the region up and down one page each time you press them.

3.3 Input Areas

Input areas are boxes where you can put data.

Suppose you have more than one input area on a page. First, move your mouse cursor to somewhere within the page. Note that only the first input area has an underscore cursor. Characters that you type are inserted in front of the underscore. This means that when you type characters at your keyboard, they go into this first input area.

The input area grows to accommodate as many characters as you type. Use the **Backspace** key to erase characters to the left. To modify what you type, use the right-arrow → and left-arrow keys ← and the keys **Insert**, **Delete**, **Home** and **End**. These keys are found immediately on the right of the standard IBM keyboard.

If you press the **Home** key, the cursor moves to the beginning of the line and if you press the **End** key, the cursor moves to the end of the line. Pressing **Ctrl**-**End** deletes all the text from the cursor to the end of the line.

To type information into another input area, use the **Enter** or **Tab** key to move from one input area to another. To move in the reverse order, use **Shift**-**Tab**.

You can also move from one input area to another using your mouse. Notice that each input area is active. Click on one of the areas. As you can see, the underscore cursor moves to that window.

3.4 Buttons

Some pages have *radio buttons* and *toggles*. Radio buttons are a group of buttons like those on car radios: you can select only one at a time. Once you have selected a button, an "×" appears in the button. To change the selection, move the cursor with the mouse to a different radio button and click.

A toggle is a square button you can either select (it has an "×") or not (it has no "×"). Unlike radio buttons, you can set a group of them any way you like. To change the selections, move the cursor with the mouse to a toggle and click.

3.5 Search Strings

A *search string* is used for searching some database. To learn about search strings, we suggest that you bring up the HyperDoc glossary. To do this from the top-level page of Browse:

1. Click on **Reference**, bringing up the AXIOM Reference page.
2. Click on **Glossary**, bringing up the glossary.

The glossary has an input area at its bottom. We review the various kinds of search strings you can enter to search the glossary.

The simplest search string is a word, for example, `operation`. A word only matches an entry having exactly that spelling. Enter the word `operation` into the input area above then click on **Search**. As you can see, `operation` matches only one entry, namely with `operation` itself.

Normally matching is insensitive to whether the alphabetic characters of your search string are in uppercase or lowercase. Thus `operation` and `OperAtion` both have the same effect.

You will very often want to use the wildcard "`*`" in your search string so as to match multiple entries in the list. The search key "`*`" matches every entry in the list. You can also use "`*`" anywhere within a search string to match an arbitrary substring. Try `cat*` for example: enter `cat*` into the input area and click on **Search**. This matches several entries.

You use any number of wildcards in a search string as long as they are not adjacent. Try search strings such as `*dom*`. As you see, this search string matches `domain`, `domain constructor`, `subdomain`, and so on.

3.5.1 Logical Searches

For more complicated searches, you can use "`and`", "`or`", and "`not`" with basic search strings; write logical expressions using these three operators just as in the AXIOM language. For example, `domain or package` matches the two entries `domain` and `package`. Similarly, `dom* and` `*con*` matches `domain constructor` and others. Also `not *a*` matches every entry that does not contain the letter a somewhere.

Use parentheses for grouping. For example, `dom* and (not *con*)` matches `domain` but not `domain constructor`.

There is no limit to how complex your logical expression can be. For example,

```
a* or b* or c* or d* or e* and (not *a*)
```

is a valid expression.

3.6 Example Pages

Many pages have AXIOM example commands. Each command has an active "button" along the left margin. When you click on this button, the output for the command is "pasted-in." Click again on the button and you see that the pasted-in output disappears.

Maybe you would like to run an example? To do so, just click on any part of its text! When you do, the example line is copied into a new interactive AXIOM buffer for this HyperDoc page.

Sometimes one example line cannot be run before you run an earlier one. Don't worry—HyperDoc automatically runs all the necessary lines in the right order!

The new interactive AXIOM buffer disappears when you leave HyperDoc. If you want to get rid of it beforehand, use the **Cancel** button of the X Window manager or issue the AXIOM system command `)close`.

3.7 X Window Resources for HyperDoc

You can control the appearance of HyperDoc while running under Version 11 of the X Window System by placing the following resources in the file **.Xdefaults** in your home directory. In what follows, *font* is any valid X11 font name (for example, `Rom14`) and *color* is any valid X11 color specification (for example, `NavyBlue`). For more information about fonts and colors, refer to the X Window documentation for your system.

`Axiom.hyperdoc.RmFont:` *font*
: This is the standard text font. The default value is `"Rom14"`.

`Axiom.hyperdoc.RmColor:` *color*
: This is the standard text color. The default value is `"black"`.

`Axiom.hyperdoc.ActiveFont:` *font*
: This is the font used for HyperDoc link buttons. The default value is `"Bld14"`.

`Axiom.hyperdoc.ActiveColor:` *color*
: This is the color used for HyperDoc link buttons. The default value is `"black"`.

`Axiom.hyperdoc.AxiomFont:` *font*
: This is the font used for active AXIOM commands.[1] The default value is `"Bld14"`.

`Axiom.hyperdoc.AxiomColor:` *color*
: This is the color used for active AXIOM commands.[2] The default value is `"black"`.

`Axiom.hyperdoc.BoldFont:` *font*
: This is the font used for bold face. The default value is `"Bld14"`.

`Axiom.hyperdoc.BoldColor:` *color*
: This is the color used for bold face. The default value is `"black"`.

`Axiom.hyperdoc.TtFont:` *font*
: This is the font used for AXIOM output in HyperDoc. This font must be fixed-width. The default value is `"Rom14"`.

`Axiom.hyperdoc.TtColor:` *color*
: This is the color used for AXIOM output in HyperDoc. The default value is `"black"`.

`Axiom.hyperdoc.EmphasizeFont:` *font*
: This is the font used for italics. The default value is `"Itl14"`.

`Axiom.hyperdoc.EmphasizeColor:` *color*
: This is the color used for italics. The default value is `"black"`.

`Axiom.hyperdoc.InputBackground:` *color*
: This is the color used as the background for input areas. The default value is `"black"`.

`Axiom.hyperdoc.InputForeground:` *color*
: This is the color used as the foreground for input areas. The default value is `"white"`.

`Axiom.hyperdoc.BorderColor:` *color*
: This is the color used for drawing border lines. The default value is `"black"`.

`Axiom.hyperdoc.Background:` *color*
: This is the color used for the background of all windows. The default value is `"white"`.

---

[1]This was called `Axiom.hyperdoc.SpadFont` in early versions of AXIOM.

[2]This was called `Axiom.hyperdoc.SpadColor` in early versions of AXIOM.

CHAPTER 4

Input Files and Output Styles

In this chapter we discuss how to collect AXIOM statements and commands into files and then read the contents into the workspace. We also show how to display the results of your computations in several different styles including TeX, FORTRAN and monospace two-dimensional format.[1]

The printed version of this book uses the AXIOM TeX output formatter. When we demonstrate a particular output style, we will need to turn TeX formatting off and the output style on so that the correct output is shown in the text.

4.1 Input Files

In this section we explain what an *input file* is and why you would want to know about it. We discuss where AXIOM looks for input files and how you can direct it to look elsewhere. We also show how to read the contents of an input file into the *workspace* and how to use the *history* facility to generate an input file from the statements you have entered directly into the workspace.

An *input* file contains AXIOM expressions and system commands. Anything that you can enter directly to AXIOM can be put into an input file. This is how you save input functions and expressions that you wish to

[1] TeX is a trademark of the American Mathematical Society.

read into AXIOM more than one time.

To read an input file into AXIOM, use the `)read` system command. For example, you can read a file in a particular directory by issuing

```
)read /spad/src/input/matrix.input
```

The "**.input**" is optional; this also works:

```
)read /spad/src/input/matrix
```

What happens if you just enter `)read matrix.input` or even `)read matrix`? AXIOM looks in your current working directory for input files that are not qualified by a directory name. Typically, this directory is the directory from which you invoked AXIOM. To change the current working directory, use the `)cd` system command. The command `)cd` by itself shows the current working directory. To change it to the `src/input` subdirectory for user "babar", issue

```
)cd /u/babar/src/input
```

AXIOM looks first in this directory for an input file. If it is not found, it looks in the system's directories, assuming you meant some input file that was provided with AXIOM.

> If you have the AXIOM history facility turned on (which it is by default), you can save all the lines you have entered into the workspace by entering
>
> `)history )write`
> AXIOM tells you what input file to edit to see your statements. The file is in your home directory or in the directory you specified with `)cd`.

In Section 5.2 on page 112 we discuss using indentation in input files to group statements into *blocks*.

4.2 The axiom.input File

When AXIOM starts up, it tries to read the input file **axiom.input** from your home directory. It there is no **axiom.input** in your home directory, it reads the copy located in its own **src/input** directory. The file usually contains system commands to personalize your AXIOM environment. In the remainder of this section we mention a few things that users frequently place in their **axiom.input** files.

In order to have FORTRAN output always produced from your computations, place the system command `)set output fortran on` in **axiom.input**. If you do not want to be prompted for confirmation when you issue the `)quit` system command, place `)set quit unprotected` in **axiom.input**. If you then decide that you do want to be prompted, issue `)set quit protected`. This is the default setting so that new users do not leave AXIOM inadvertently.[2]

To see the other system variables you can set, issue `)set` or use the HyperDoc **Settings** facility to view and change AXIOM system variables.

4.3 Common Features of Using Output Formats

In this section we discuss how to start and stop the display of the different output formats and how to send the output to the screen or to a file. To fix ideas, we use FORTRAN output format for most of the examples.

You can use the `)set output` system command to toggle or redirect the different kinds of output. The name of the kind of output follows "output" in the command. The names are

fortran for FORTRAN output.
algebra for monospace two-dimensional mathematical output.
tex for TeX output.
script for IBM Script Formula Format output.

For example, issue `)set output fortran on` to turn on FORTRAN format and issue `)set output fortran off` to turn it off. By default, `algebra` is `on` and all others are `off`. When output is started, it is sent to the screen. To send the output to a file, give the file name without directory or extension. AXIOM appends a file extension depending on the kind of output being produced.

Issue this to redirect FORTRAN output to, for example, the file **linalg.sfort**.

```
)set output fortran linalg
```

```
FORTRAN output will be written to file
```

You must *also* turn on the creation of FORTRAN output. The above just says where it goes if it is created.

```
)set output fortran on
```

In what directory is this output placed? It goes into the directory from which you started AXIOM, or if you have used the `)cd` system command, the one that you specified with `)cd`. You should use `)cd` before you send the output to the file.

You can always direct output back to the screen by issuing this.

```
)set output fortran console
```

[2]The system command `)pquit` always prompts you for confirmation.

Let's make sure FORTRAN formatting is off so that nothing we do from now on produces FORTRAN output.

```
)set output fortran off
```

We also delete the demonstrated output file we created.

```
)system rm linalg.sfort
```

You can abbreviate the words "on," "off" and "console" to the minimal number of characters needed to distinguish them. Because of this, you cannot send output to files called **on.sfort, off.sfort, of.sfort, console.sfort, consol.sfort** and so on.

The width of the output on the page is set by `)set output length` for all formats except FORTRAN. Use `)set fortran fortlength` to change the FORTRAN line length from its default value of `72`.

4.4 Monospace Two-Dimensional Mathematical Format

This is the default output format for AXIOM. It is usually on when you start the system.

If it is not, issue this.

```
)set output algebra on
```

Since the printed version of this book (as opposed to the HyperDoc version) shows output produced by the TeX output formatter, let us temporarily turn off TeX output.

```
)set output tex off
```

Here is an example of what it looks like.

```
matrix [[i*x**i + j*%i*y**j for i in 1..2] for j in 3..4]
```

```
        +      3            3      2+
        |3%i y  + x   3%i y  + 2x  |
   (1)  |                          |
        |      4            4      2|
        +4%i y  + x   4%i y  + 2x  +
```

(0)

Type: Matrix Polynomial Complex Integer

Issue this to turn off this kind of formatting.

```
)set output algebra off
```

Turn TeX output on again.

```
)set output tex on
```

The characters used for the matrix brackets above are rather ugly. You get this character set when you issue `)set output characters plain`. This character set should be used when you are running on a machine that does not support the IBM extended ASCII character set. If you are running on an IBM workstation, for example, issue `)set output characters default` to get better looking output.

4.5 TEX Format

AXIOM can produce TEX output for your expressions. The output is produced using macros from the LATEX document preparation system by Leslie Lamport.[3] The printed version of this book was produced using this formatter.

To turn on TEX output formatting, issue this.

```
)set output tex on
```

Here is an example of its output.

```
matrix [[i*x**i + j*\%i*y**j for i in 1..2] for j in 3..4]
```

```
\[
\left[
\begin{array}{cc}
\displaystyle
{{3 \  \%i \ {y^3}}+ x}&
\displaystyle
{{3 \  \%i \ {y^3}}+{2 \ {x^2}}}
\\
\displaystyle
{{4 \  \%i \ {y^4}}+ x}&
\displaystyle
{{4 \  \%i \ {y^4}}+{2 \ {x^2}}}
\end{array}
\right] \leqno (3)
\]
```

This formats as

$$\left[\begin{array}{cc} 3\ \%i\ y^3 + x & 3\ \%i\ y^3 + 2\ x^2 \\ 4\ \%i\ y^4 + x & 4\ \%i\ y^4 + 2\ x^2 \end{array}\right]$$

To turn TEX output formatting off, issue `)set output tex off`. The LATEX macros in the output generated by AXIOM are all standard except for the following definitions:

```
\def\csch{\mathop{\rm csch}\nolimits}

\def\erf{\mathop{\rm erf}\nolimits}

\def\zag#1#2{
  {{\hfill \left. {#1} \right|}
   \over
   {\left| {#2} \right. \hfill}
  }
}
```

[3]See Leslie Lamport, *LATEX: A Document Preparation System,* Reading, Massachusetts: Addison-Wesley Publishing Company, Inc., 1986.

4.6 IBM Script Formula Format

AXIOM can produce IBM Script Formula Format output for your expressions.

To turn IBM Script Formula Format on, issue this.

```
)set output script on
```

Here is an example of its output.

```
matrix [[i*x**i + j*%i*y**j for i in 1..2] for j in 3..4]
```

```
.eq set blank @
:df.
<left lb <<<<3 @@ %i @@ <y sup 3>>+x> here <<3 @@ %i @@
<y sup 3>>+<2 @@ <x sup 2>>>> habove <<<4 @@ %i @@
<y sup 4>>+x> here <<4 @@ %i @@ <y sup 4>>+<2 @@
<x up 2>>>>> right rb>
:edf.
```

To turn IBM Script Formula Format output formatting off, issue this.

```
)set output script off
```

4.7 FORTRAN Format

In addition to turning FORTRAN output on and off and stating where the output should be placed, there are many options that control the appearance of the generated code. In this section we describe some of the basic options. Issue `)set fortran` to see a full list with their current settings.

The output FORTRAN expression usually begins in column 7. If the expression needs more than one line, the ampersand character "&" is used in column 6. Since some versions of FORTRAN have restrictions on the number of lines per statement, AXIOM breaks long expressions into segments with a maximum of 1320 characters (20 lines of 66 characters) per segment. If you want to change this, say, to 660 characters, issue the system command `)set fortran explength 660`. You can turn off the line breaking by issuing `)set fortran segment off`. Various code optimization levels are available.

FORTRAN output is produced after you issue this.

```
)set output fortran on
```

For the initial examples, we set the optimization level to 0, which is the lowest level.

```
)set fortran optlevel 0
```

The output is usually in columns 7 through 72, although fewer columns are used in the following examples so that the output fits nicely on the page.

```
)set fortran fortlength 60
```

By default, the output goes to the screen and is displayed before the standard AXIOM two-dimensional output. In this example, an assignment to the variable R1 was generated because this is the result of step 1.

```
(x+y)**3
```

```
   REAL T2,T1
   R1=y**3+3.*x*y*y+3.*x*x*y+x**3
```

$$y^3 + 3\ x\ y^2 + 3\ x^2\ y + x^3 \tag{1}$$

Type: Polynomial Integer

Here is an example that illustrates the line breaking.

```
(x+y+z)**3
```

```
R2=z**3+(3.*y+3.*x)*z*z+(3.*y*y+6.*x*y+3.*x*x)*z+y**3+
  &3.*x*y*y+3.*x*x*y+x**3
```

$$z^3 + (3\ y + 3\ x)\ z^2 + \left(3\ y^2 + 6\ x\ y + 3\ x^2\right)\ z + y^3 + 3\ x\ y^2 + 3\ x^2\ y + x^3 \tag{2}$$

Type: Polynomial Integer

Note in the above examples that integers are generally converted to floating point numbers, except in exponents. This is the default behavior but can be turned off by issuing `)set fortran ints2floats off`. The rules governing when the conversion is done are:

1. If an integer is an exponent, convert it to a floating point number if it is greater than 32767 in absolute value, otherwise leave it as an integer.
2. Convert all other integers in an expression to floating point numbers.

These rules only govern integers in expressions. Numbers generated by AXIOM for `DIMENSION` statements are also integers.

To set the type of generated FORTRAN data, use one of the following:

```
)set fortran defaulttype REAL
)set fortran defaulttype INTEGER
)set fortran defaulttype COMPLEX
)set fortran defaulttype LOGICAL
)set fortran defaulttype CHARACTER
```

When temporaries are created, they are given a default type of REAL. Also, the REAL versions of functions are used by default.

```
sin(x)
```

```
   R3=SIN(x)
```

$$\sin(x) \tag{3}$$

Type: Expression Integer

At optimization level 1, AXIOM removes common subexpressions.

```
)set fortran optlevel 1
```

```
(x+y+z)**3
```

```
   REAL T2,T1
   T1=y*y
   T2=x*x
R4=z**3+(3.*y+3.*x)*z*z+(3.*T1+6.*x*y+3.*T2)*z+y**3+3.
  &*x*T1+3.*T2*y+x**3
```

$$z^3+(3\ y+3\ x)\ z^2+\left(3\ y^2+6\ x\ y+3\ x^2\right)\ z+y^3+3\ x\ y^2+ \\ 3\ x^2\ y+x^3 \tag{4}$$

Type: Polynomial Integer

This changes the precision to DOUBLE. Substitute `single` for `double` to return to single precision.

```
)set fortran precision double
```

Complex constants display the precision.

```
2.3 + 5.6*%i
```

```
   R5=(2.3D0,5.6D0)
```

$$2.3+5.6\ \%i \tag{5}$$

Type: Complex Float

The function names that AXIOM generates depend on the chosen precision.

```
sin %e
```

```
   R6=DSIN(DEXP(1.0D0))
```

$$\sin(\%e) \tag{6}$$

Type: Expression Integer

Reset the precision to `single` and look at these two examples again.

```
)set fortran precision single
```

```
2.3 + 5.6*%i
```

```
   R7=(2.3,5.6)
```

$$2.3+5.6\ \%i \tag{7}$$

Type: Complex Float

```
sin %e
```

```
   R8=SIN(EXP(1.))
```

$$\sin(\%e) \tag{8}$$

Type: Expression Integer

Expressions that look like lists, streams, sets or matrices cause array code to be generated.

```
[x+1,y+1,z+1]

   REAL T1
   DIMENSION T1(0:2)
   T1(0)=x+1.
   T1(1)=y+1.
   T1(2)=z+1.
   R9=T1
```

$$[x+1,\ y+1,\ z+1] \tag{9}$$

Type: List Polynomial Integer

A temporary variable is generated to be the name of the array. This may have to be changed in your particular application.

```
{2,3,4,3,5}

   REAL T1
   DIMENSION T1(0:3)
   T1(0)=2.
   T1(1)=3.
   T1(2)=4.
   T1(3)=5.
   R10=T1
```

$$\{2,\ 3,\ 4,\ 5\} \tag{10}$$

Type: Set PositiveInteger

By default, the starting index for generated FORTRAN arrays is 0.

```
matrix [[2.3,9.7],[0.0,18.778]]

   REAL T1
   DIMENSION T1(0:1,0:1)
   T1(0,0)=2.3
   T1(0,1)=9.7
   T1(1,0)=0.0
   T1(1,1)=18.778
   T1
```

$$\begin{bmatrix} 2.3 & 9.7 \\ 0.0 & 18.778 \end{bmatrix} \tag{11}$$

Type: Matrix Float

To change the starting index for generated FORTRAN arrays to be 1, issue this. This value can only be 0 or 1.

```
)set fortran startindex 1
```

Look at the code generated for the matrix again.

```
matrix [[2.3,9.7],[0.0,18.778]]
```

```
   REAL T1
   DIMENSION T1(1:2,1:2)
   T1(1,1)=2.3
   T1(1,2)=9.7
   T1(2,1)=0.0
   T1(2,2)=18.778
   T1
```

$$\begin{bmatrix} 2.3 & 9.7 \\ 0.0 & 18.778 \end{bmatrix} \qquad (12)$$

Type: Matrix Float

CHAPTER 5

Introduction to the AXIOM Interactive Language

In this chapter we look at some of the basic components of the AXIOM language that you can use interactively. We show how to create a *block* of expressions, how to form loops and list iterations, how to modify the sequential evaluation of a block and how to use `if-then-else` to evaluate parts of your program conditionally. We suggest you first read the boxed material in each section and then proceed to a more thorough reading of the chapter.

5.1 Immediate and Delayed Assignments

A *variable* in AXIOM refers to a value. A variable has a name beginning with an uppercase or lowercase alphabetic character, "%", or "!". Successive characters (if any) can be any of the above, digits, or "?". Case is distinguished. The following are all examples of valid, distinct variable names:

```
a           tooBig?     a1B2c3%!?
A           %j          numberOfPoints
beta6       %J          numberofpoints
```

The "`:=`" operator is the immediate *assignment* operator. Use it to associate a value with a variable.

> The syntax for immediate assignment for a single variable is
>
> *variable* `:=` *expression*
>
> The value returned by an immediate assignment is the value of *expression.*

The right-hand side of the expression is evaluated, yielding `1`. This value is then assigned to `a`.

```
a := 1
```

1 (1)

Type: PositiveInteger

The right-hand side of the expression is evaluated, yielding `1`. This value is then assigned to `b`. Thus `a` and `b` both have the value `1` after the sequence of assignments.

```
b := a
```

1 (2)

Type: PositiveInteger

What is the value of `b` if `a` is assigned the value `2`?

```
a := 2
```

2 (3)

Type: PositiveInteger

As you see, the value of `b` is left unchanged.

```
b
```

1 (4)

Type: PositiveInteger

This is what we mean when we say this kind of assignment is *immediate*; `b` has no dependency on `a` after the initial assignment. This is the usual notion of assignment found in programming languages such as C, PASCAL and FORTRAN.

AXIOM provides delayed assignment with "`==`". This implements a delayed evaluation of the right-hand side and dependency checking.

> The syntax for delayed assignment is
>
> *variable* `==` *expression*
>
> The value returned by a delayed assignment is the unique value of Void.

Using a and b as above, these are the corresponding delayed assignments.

```
a == 1
```

Type: Void

```
b == a
```

Type: Void

The right-hand side of each delayed assignment is left unevaluated until the variables on the left-hand sides are evaluated. Therefore this evaluation and ...

```
a
Compiling body of rule a to compute value of type
PositiveInteger
```

1 (7)

Type: PositiveInteger

this evaluation seem the same as before.

```
b
Compiling body of rule b to compute value of type
PositiveInteger
```

1 (8)

Type: PositiveInteger

If we change a to 2

```
a == 2
Compiled code for a has been cleared.
Compiled code for b has been cleared.
1 old definition(s) deleted for function or rule a
```

Type: Void

then a evaluates to 2, as expected, but

```
a
Compiling body of rule a to compute value of type
PositiveInteger
```

2 (10)

Type: PositiveInteger

the value of b reflects the change to a.

```
b
Compiling body of rule b to compute value of type
PositiveInteger
```

2 (11)

Type: PositiveInteger

It is possible to set several variables at the same time by using a *tuple* of variables and a tuple of expressions.[1]

[1]A *tuple* is a collection of things separated by commas, often surrounded by parentheses.

> The syntax for multiple immediate assignments is
>
> (var_1, var_2, ..., var_N) := ($expr_1$, $expr_2$, ..., $expr_N$)
>
> The value returned by an immediate assignment is the value of $expr_N$.

This sets x to 1 and y to 2.

```
(x,y) := (1,2)
```

2 (12)

Type: PositiveInteger

Multiple immediate assigments are parallel in the sense that the expressions on the right are all evaluated before any assignments on the left are made. However, the order of evaluation of these expressions is undefined.

You can use multiple immediate assignment to swap the values held by variables.

```
(x,y) := (y,x)
```

1 (13)

Type: PositiveInteger

x has the previous value of y.

```
x
```

2 (14)

Type: PositiveInteger

y has the previous value of x.

```
y
```

1 (15)

Type: PositiveInteger

There is no syntactic form for multiple delayed assignments. See the discussion in Section 6.8 on page 144 about how AXIOM differentiates between delayed assignments and user functions of no arguments.

5.2 Blocks

A *block* is a sequence of expressions evaluated in the order that they appear, except as modified by control expressions such as `leave`, `return`, `iterate` and `if-then-else` constructions. The value of a block is the value of the expression last evaluated in the block.

To leave a block early, use "=>". For example, `i < 0 => x`. The expression before the "=>" must evaluate to `true` or `false`. The expression following the "=>" is the return value for the block.

A block can be constructed in two ways:

1. the expressions can be separated by semicolons and the resulting expression surrounded by parentheses, and
2. the expressions can be written on succeeding lines with each line

indented the same number of spaces (which must be greater than zero). A block entered in this form is called a *pile*.

Only the first form is available if you are entering expressions directly to AXIOM. Both forms are available in **.input** files.

> The syntax for a simple block of expressions entered interactively is
>
> $$(\; expression_1 \; ; \; expression_2 \; ; \; \ldots \; ; \; expression_N \;)$$
>
> The value returned by a block is the value of an "=>" expression, or $expression_N$ if no "=>" is encountered.

In **.input** files, blocks can also be written using *piles*. The examples throughout this book are assumed to come from **.input** files.

In this example, we assign a rational number to a using a block consisting of three expressions. This block is written as a pile. Each expression in the pile has the same indentation, in this case two spaces to the right of the first line.

```
a :=
  i := gcd(234,672)
  i := 3*i**5 - i + 1
  1 / i
```

$$\frac{1}{23323} \tag{1}$$

Type: Fraction Integer

Here is the same block written on one line. This is how you are required to enter it at the input prompt.

```
a := (i := gcd(234,672); i := 3*i**5 - i + 1; 1 / i)
```

$$\frac{1}{23323} \tag{2}$$

Type: Fraction Integer

Blocks can be used to put several expressions on one line. The value returned is that of the last expression.

```
(a := 1; b := 2; c := 3; [a,b,c])
```

$$[1, 2, 3] \tag{3}$$

Type: List PositiveInteger

AXIOM gives you two ways of writing a block and the preferred way in an **.input** file is to use a pile. Roughly speaking, a pile is a block whose constituent expressions are indented the same amount. You begin a pile by starting a new line for the first expression, indenting it to the right of the previous line. You then enter the second expression on a new line, vertically aligning it with the first line. And so on. If you need to enter an inner pile, further indent its lines to the right of the outer pile. AXIOM knows where a pile ends. It ends when a subsequent line is indented to the left of the pile or the end of the file.

Blocks can be used to perform several steps before an assignment (immediate or delayed) is made.

```
d :=
   c := a**2 + b**2
   sqrt(c * 1.3)
```

2.5495097567963924 (4)

Type: Float

Blocks can be used in the arguments to functions. (Here `h` is assigned `2.1 + 3.5`.)

```
h := 2.1 +
   1.0
   3.5
```

5.6 (5)

Type: Float

Here the second argument to **eval** is `x = z`, where the value of `z` is computed in the first line of the block starting on the second line.

```
eval(x**2 - x*y**2,
     z := %pi/2.0 - exp(4.1)
     x = z
   )
```

$58.769491270567072878\ y^2 + 3453.853104201259382$ (6)

Type: Polynomial Float

Blocks can be used in the clauses of `if-then-else` expressions (see Section 5.3 on page 115).

```
if h > 3.1 then 1.0 else (z := cos(h); max(z,0.5))
```

1.0 (7)

Type: Float

This is the pile version of the last block.

```
if h > 3.1 then
    1.0
  else
    z := cos(h)
    max(z,0.5)
```

1.0 (8)

Type: Float

Blocks can be nested.

```
a := (b := factorial(12); c := (d := eulerPhi(22);
  factorial(d));b+c)
```

482630400 (9)

Type: PositiveInteger

This is the pile version of the last block.

```
a :=
  b := factorial(12)
  c :=
    d := eulerPhi(22)
    factorial(d)
  b+c
```

482630400 (10)

Type: PositiveInteger

Since c + d does equal 3628855, a has the value of c and the last line is never evaluated.

```
a :=
  c := factorial 10
  d := fibonacci 10
  c + d = 3628855 => c
  d
```

$$3628800 \qquad (11)$$

Type: PositiveInteger

5.3 if-then-else

Like many other programming languages, AXIOM uses the three keywords `if`, `then` and `else` to form conditional expressions. The `else` part of the conditional is optional. The expression between the `if` and `then` keywords is a *predicate*: an expression that evaluates to or is convertible to either `true` or `false`, that is, a Boolean.

> The syntax for conditional expressions is
>
> `if` *predicate* `then` $expression_1$ `else` $expression_2$
>
> where the `else` $expression_2$ part is optional. The value returned from a conditional expression is $expression_1$ if the predicate evaluates to `true` and $expression_2$ otherwise. If no `else` clause is given, the value is always the unique value of Void.

An `if-then-else` expression always returns a value. If the `else` clause is missing then the entire expression returns the unique value of Void. If both clauses are present, the type of the value returned by `if` is obtained by resolving the types of the values of the two clauses. See Section 2.10 on page 86 for more information.

The predicate must evaluate to, or be convertible to, an object of type Boolean: `true` or `false`. By default, the equal sign "=" creates an equation.

This is an equation. In particular, it is an object of type Equation Polynomial Integer.

```
x + 1 = y
```

$$x + 1 = y \qquad (1)$$

Type: Equation Polynomial Integer

However, for predicates in `if` expressions, AXIOM places a default target type of Boolean on the predicate and equality testing is performed. Thus you need not qualify the "=" in any way. In other contexts you may need to tell AXIOM that you want to test for equality rather than create an equation. In those cases, use "@" and a target type of Boolean. See Section 2.9 on page 83 for more information.

The compound symbol meaning "not equal" in AXIOM is "~=". This can be used directly without a package call or a target specification. The expression `a ~= b` is directly translated into `not (a = b)`.

Many other functions have return values of type Boolean. These include <, <=, >, >=, ~= and `member?`. By convention, operations with names ending in "?" return Boolean values.

The usual rules for piles are suspended for conditional expressions. In **.input** files, the `then` and `else` keywords can begin in the same column as the corresponding `if` but may also appear to the right. Each of the following styles of writing `if-then-else` expressions is acceptable:

```
if i>0 then output("positive") else output("nonpositive")

if i > 0 then output("positive")
  else output("nonpositive")

if i > 0 then output("positive")
else output("nonpositive")

if i > 0
then output("positive")
else output("nonpositive")

if i > 0
  then output("positive")
  else output("nonpositive")
```

A block can follow the `then` or `else` keywords. In the following two assignments to `a`, the `then` and `else` clauses each are followed by two-line piles. The value returned in each is the value of the second line.

```
a :=
  if i > 0 then
    j := sin(i * pi())
    exp(j + 1/j)
  else
    j := cos(i * 0.5 * pi())
    log(abs(j)**5 + 1)

a :=
  if i > 0
    then
      j := sin(i * pi())
      exp(j + 1/j)
    else
      j := cos(i * 0.5 * pi())
      log(abs(j)**5 + 1)
```

These are both equivalent to the following:

```
a :=
  if i > 0 then (j := sin(i * pi()); exp(j + 1/j))
  else (j := cos(i * 0.5 * pi()); log(abs(j)**5 + 1))
```

5.4 Loops

A *loop* is an expression that contains another expression, called the *loop body*, which is to be evaluated zero or more times. All loops contain the `repeat` keyword and return the unique value of Void. Loops can contain inner loops to any depth.

The most basic loop is of the form

`repeat` *loopBody*

Unless *loopBody* contains a `leave` or `return` expression, the loop repeats forever. The value returned by the loop is the unique value of Void.

5.4.1 Compiling vs. Interpreting Loops

AXIOM tries to determine completely the type of every object in a loop and then to translate the loop body to LISP or even to machine code. This translation is called *compilation.*

If AXIOM decides that it cannot compile the loop, it issues a message stating the problem and then the following message:

We will attempt to step through and interpret the code.

It is still possible that AXIOM can evaluate the loop but in *interpret-code mode.* See Section 6.10 on page 146 where this is discussed in terms of compiling versus interpreting functions.

5.4.2 return in Loops

A `return` expression is used to exit a function with a particular value. In particular, if a `return` is in a loop within the function, the loop is terminated whenever the `return` is evaluated.

Suppose we start with this.

```
f() ==
  i := 1
  repeat
    if factorial(i) > 1000 then return i
    i := i + 1
```

Type: Void

When `factorial(i)` is big enough, control passes from inside the loop all the way outside the function, returning the value of `i` (or so we think).

```
f()
```

Type: Void

What went wrong? Isn't it obvious that this function should return an integer? Well, AXIOM makes no attempt to analyze the structure of a loop to determine if it always returns a value because, in general, this is impossible. So AXIOM has this simple rule: the type of the function is determined by the type of its body, in this case a block. The normal value of a block is the value of its last expression, in this case, a loop. And the value of every loop is the unique value of Void! So the return type of **f** is Void.

There are two ways to fix this. The best way is for you to tell AXIOM what the return type of `f` is. You do this using giving `f` a declaration `f: () -> Integer` prior to calling for its value. This tells AXIOM: "trust me—an integer is returned." We'll explain more about this in the next chapter. Another clumsy way is to add a dummy expression as follows.

Since we want an integer, let's stick in a dummy final expression that is an integer and will never be evaluated.

```
f() ==
  i := 1
  repeat
    if factorial(i) > 1000 then return i
    i := i + 1
  0
```

Type: Void

When we try **f** again we get what we wanted. See Section 6.15 on page 159 for more information.

```
f()
```

7 (4)

Type: PositiveInteger

5.4.3 leave in Loops

The `leave` keyword is often more useful in terminating a loop. A `leave` causes control to transfer to the expression immediately following the loop. As loops always return the unique value of Void, you cannot return a value with `leave`. That is, `leave` takes no argument.

This example is a modification of the last example in the previous section. Instead of using `return`, we'll use `leave`.

```
f() ==
  i := 1
  repeat
    if factorial(i) > 1000 then leave
    i := i + 1
  i
```

Type: Void

The loop terminates when `factorial(i)` gets big enough, the last line of the function evaluates to the corresponding "good" value of `i`, and the function terminates, returning that value.

```
f()
```

7 (2)

Type: PositiveInteger

You can only use `leave` to terminate the evaluation of one loop. Let's consider a loop within a loop, that is, a loop with a nested loop. First, we initialize two counter variables.

```
(i,j) := (1, 1)
```

1 (3)

Type: PositiveInteger

Nested loops must have multiple `leave` expressions at the appropriate nesting level. How would you rewrite this so `(i + j) > 10` is only evaluated once?

```
repeat
  repeat
    if (i + j) > 10 then leave
    j := j + 1
  if (i + j) > 10 then leave
  i := i + 1
```

Type: Void

5.4.4 leave vs. => in Loop Bodies

Compare the following two loops:

```
i := 1
repeat
  i := i + 1
  i > 3 => i
  output(i)
```

```
i := 1
repeat
  i := i + 1
  if i > 3 then leave
  output(i)
```

In the example on the left, the values 2 and 3 for `i` are displayed but then the "=>" does not allow control to reach the call to **output** again. The loop will not terminate until you run out of space or interrupt the execution. The variable `i` will continue to be incremented because the "=>" only means to leave the *block,* not the loop.

In the example on the right, upon reaching 4, the `leave` will be executed, and both the block and the loop will terminate. This is one of the reasons why both "=>" and `leave` are provided. Using a `while` clause (see below) with the "=>" lets you simulate the action of `leave`.

5.4.5 More Examples of leave

Here we give four examples of `repeat` loops that terminate when a value exceeds a given bound.

First, initialize `i` as the loop counter.

```
i := 0
```

0 (1)

Type: NonNegativeInteger

Here is the first loop. When the square of `i` exceeds `100`, the loop terminates.

```
repeat
  i := i + 1
  if i**2 > 100 then leave
```

Type: Void

Upon completion, i should have the value 11.

```
i
```

$$11 \tag{3}$$

Type: NonNegativeInteger

Do the same thing except use "=>" instead an if-then expression.

```
i := 0
```

$$0 \tag{4}$$

Type: NonNegativeInteger

```
repeat
  i := i + 1
  i**2 > 100 => leave
```

Type: Void

```
i
```

$$11 \tag{6}$$

Type: NonNegativeInteger

As a third example, we use a simple loop to compute n!.

```
(n, i, f) := (100, 1, 1)
```

$$1 \tag{7}$$

Type: PositiveInteger

Use i as the iteration variable and f to compute the factorial.

```
repeat
  if i > n then leave
  f := f * i
  i := i + 1
```

Type: Void

Look at the value of f.

```
f
```

$$\begin{array}{l}93326215443944152681699238856266700490715968264381621468\\59296389521759999322991560894146397615651828625369792082\\7223758251185210916864000000000000000000000000\end{array} \tag{9}$$

Type: PositiveInteger

Finally, we show an example of nested loops. First define a four by four matrix.

```
m := matrix [[21,37,53,14], [8,-24,22,-16], [2,10,15,14],
  [26,33,55,-13]]
```

$$\begin{bmatrix} 21 & 37 & 53 & 14 \\ 8 & -24 & 22 & -16 \\ 2 & 10 & 15 & 14 \\ 26 & 33 & 55 & -13 \end{bmatrix} \tag{10}$$

Type: Matrix Integer

Next, set row counter r and column counter c to 1. Note: if we were writing a function, these would all be local variables rather than global workspace variables.

```
(r, c) := (1, 1)
```

$$1 \tag{11}$$

Type: PositiveInteger

Also, let lastrow and lastcol be the final row and column index.

```
(lastrow, lastcol) := (nrows(m), ncols(m))
```

4 (12)

Type: PositiveInteger

Scan the rows looking for the first negative element. We remark that you can reformulate this example in a better, more concise form by using a for clause with repeat. See Section 5.4.8 on page 123 for more information.

```
repeat
  if r > lastrow then leave
  c := 1
  repeat
    if c > lastcol then leave
    if elt(m,r,c) < 0 then
      output [r, c, elt(m,r,c)]
      r := lastrow
      leave      -- don't look any further
    c := c + 1
  r := r + 1
```

```
[2,2,- 24]
```

Type: Void

5.4.6 iterate in Loops

AXIOM provides an iterate expression that skips over the remainder of a loop body and starts the next loop iteration.

We first initialize a counter.

```
i := 0
```

0 (1)

Type: NonNegativeInteger

Display the even integers from 2 to 5.

```
repeat
  i := i + 1
  if i > 5 then leave
  if odd?(i) then iterate
  output(i)
```

```
2
4
```

Type: Void

5.4.7 while Loops

The repeat in a loop can be modified by adding one or more while clauses. Each clause contains a *predicate* immediately following the while keyword. The predicate is tested *before* the evaluation of the body of the loop. The loop body is evaluated whenever the predicate in a while clause is true.

The syntax for a simple loop using `while` is

`while` *predicate* `repeat` *loopBody*

The *predicate* is evaluated before *loopBody* is evaluated. A `while` loop terminates immediately when *predicate* evaluates to `false` or when a `leave` or `return` expression is evaluated in *loopBody*. The value returned by the loop is the unique value of Void.

Here is a simple example of using `while` in a loop. We first initialize the counter.

```
i := 1
```

1 (1)

Type: PositiveInteger

The steps involved in computing this example are (1) set `i` to `1`, (2) test the condition `i < 1` and determine that it is not true, and (3) do not evaluate the loop body and therefore do not display `"hello"`.

```
while i < 1 repeat
  output "hello"
  i := i + 1
```

Type: Void

If you have multiple predicates to be tested use the logical `and` operation to separate them. AXIOM evaluates these predicates from left to right.

```
(x, y) := (1, 1)
```

1 (3)

Type: PositiveInteger

```
while x < 4 and y < 10 repeat
  output [x,y]
  x := x + 1
  y := y + 2
```

```
[1,1]
[2,3]
[3,5]
```

Type: Void

A `leave` expression can be included in a loop body to terminate a loop even if the predicate in any `while` clauses are not `false`.

```
(x, y) := (1, 1)
```

1 (5)

Type: PositiveInteger

This loop has multiple `while` clauses and the loop terminates before any one of their conditions evaluates to `false`.

```
while x < 4 while y < 10 repeat
  if x + y > 7 then leave
  output [x,y]
  x := x + 1
  y := y + 2
```

```
[1,1]
[2,3]
```

Type: Void

Here's a different version of the nested loops that looked for the first negative element in a matrix.

```
m := matrix [[21,37,53,14], [8,-24,22,-16], [2,10,15,14],
  [26,33,55,-13]]
```

$$\begin{bmatrix} 21 & 37 & 53 & 14 \\ 8 & -24 & 22 & -16 \\ 2 & 10 & 15 & 14 \\ 26 & 33 & 55 & -13 \end{bmatrix} \qquad (7)$$

Type: Matrix Integer

Initialized the row index to 1 and get the number of rows and columns. If we were writing a function, these would all be local variables.

```
r := 1
```

$$1 \qquad (8)$$

Type: PositiveInteger

```
(lastrow, lastcol) := (nrows(m), ncols(m))
```

$$4 \qquad (9)$$

Type: PositiveInteger

Scan the rows looking for the first negative element.

```
while r <= lastrow repeat
  c := 1  -- index of first column
  while c <= lastcol repeat
    if elt(m,r,c) < 0 then
      output [r, c, elt(m,r,c)]
      r := lastrow
      leave     -- don't look any further
    c := c + 1
  r := r + 1
```

```
[2,2,- 24]
```

Type: Void

5.4.8 for Loops

AXIOM provides the `for` and `in` keywords in `repeat` loops, allowing you to iterate across all elements of a list, or to have a variable take on integral values from a lower bound to an upper bound. We shall refer to these modifying clauses of `repeat` loops as `for` clauses. These clauses can be present in addition to `while` clauses. As with all other types of `repeat` loops, `leave` can be used to prematurely terminate the evaluation of the loop.

The syntax for a simple loop using `for` is

`for` *iterator* `repeat` *loopBody*

The *iterator* has several forms. Each form has an end test which is evaluated before *loopBody* is evaluated. A `for` loop terminates immediately when the end test succeeds (evaluates to `true`) or when a `leave` or `return` expression is evaluated in *loopBody*. The value returned by the loop is the unique value of Void.

5.4.9 for i in n..m repeat

If `for` is followed by a variable name, the `in` keyword and then an integer segment of the form `n..m`, the end test for this loop is the predicate `i > m`. The body of the loop is evaluated `m-n+1` times if this number is greater than 0. If this number is less than or equal to 0, the loop body is not evaluated at all.

The variable `i` has the value `n, n+1, ..., m` for successive iterations of the loop body. The loop variable is a *local variable* within the loop body: its value is not available outside the loop body and its value and type within the loop body completely mask any outer definition of a variable with the same name.

This loop prints the values of 10^3, 11^3, and 12^3:

```
for i in 10..12 repeat output(i**3)
1000
1331
1728
```

Type: Void

Here is a sample list.

```
a := [1,2,3]
```

$[1, 2, 3]$ (2)

Type: List PositiveInteger

Iterate across this list, using "`.`" to access the elements of a list and the # operation to count its elements.

```
for i in 1..#a repeat output(a.i)
1
2
3
```

Type: Void

This type of iteration is applicable to anything that uses "`.`". You can also use it with functions that use indices to extract elements.

Define m to be a matrix.

```
m := matrix [[1,2],[4,3],[9,0]]
```

$$\begin{bmatrix} 1 & 2 \\ 4 & 3 \\ 9 & 0 \end{bmatrix} \qquad (4)$$

Type: Matrix Integer

Display the rows of m.

```
for i in 1..nrows(m) repeat output row(m,i)
```

```
[1,2]
[4,3]
[9,0]
```

Type: Void

You can use `iterate` with `for`-loops.

Display the even integers in a segment.

```
for i in 1..5 repeat
  if odd?(i) then iterate
  output(i)
```

```
2
4
```

Type: Void

See 'Segment' on page 447 for more information about segments.

5.4.10 for i in n..m by s repeat

By default, the difference between values taken on by a variable in loops such as `for i in n..m repeat ...` is 1. It is possible to supply another, possibly negative, step value by using the `by` keyword along with `for` and `in`. Like the upper and lower bounds, the step value following the `by` keyword must be an integer. Note that the loop `for i in 1..2 by 0 repeat output(i)` will not terminate by itself, as the step value does not change the index from its initial value of 1.

This expression displays the odd integers between two bounds.

```
for i in 1..5 by 2 repeat output(i)
```

```
1
3
5
```

Type: Void

Use this to display the numbers in reverse order.

```
for i in 5..1 by -2 repeat output(i)
```

```
5
3
1
```

Type: Void

5.4.11 for i in n.. repeat

If the value after the ".." is omitted, the loop has no end test. A potentially infinite loop is thus created. The variable is given the successive values `n, n+1, n+2, ...` and the loop is terminated only if a `leave` or `return` expression is evaluated in the loop body. However you may also add some other modifying clause on the `repeat` (for example, a `while` clause) to stop the loop.

This loop displays the integers greater than or equal to 15 and less than the first prime greater than 15.

```
for i in 15.. while not prime?(i) repeat output(i)
```

```
15
16
```

Type: Void

5.4.12 for x in l repeat

Another variant of the `for` loop has the form:

`for` *x* `in` *list* `repeat` *loopBody*

This form is used when you want to iterate directly over the elements of a list. In this form of the `for` loop, the variable `x` takes on the value of each successive element in `l`. The end test is most simply stated in English: "are there no more `x` in `l`?"

If `l` is this list,

```
l := [0,-5,3]
```

$$[0, -5, 3] \tag{1}$$

Type: List Integer

display all elements of `l`, one per line.

```
for x in l repeat output(x)
```

```
0
- 5
3
```

Type: Void

Since the list constructing expression `expand [n..m]` creates the list `[n, n+1, ..., m]`[2], you might be tempted to think that the loops

```
for i in n..m repeat output(i)
```

and

```
for x in expand [n..m] repeat output(x)
```

are equivalent. The second form first creates the list `expand [n..m]` (no matter how large it might be) and then does the iteration. The first form potentially runs in much less space, as the index variable `i` is simply incremented once per loop and the list is not actually created. Using the first form is much more efficient.

[2]This list is empty if `n > m`.

Of course, sometimes you really want to iterate across a specific list. This displays each of the factors of 2400000.

```
for f in factors(factor(2400000)) repeat output(f)
```

```
[factor= 2,exponent= 8]
[factor= 3,exponent= 1]
[factor= 5,exponent= 5]
```

Type: Void

5.4.13 "Such that" Predicates

A `for` loop can be followed by a "|" and then a predicate. The predicate qualifies the use of the values from the iterator following the `for`. Think of the vertical bar "|" as the phrase "such that."

This loop expression prints out the integers n in the given segment such that n is odd.

```
for n in 0..4 | odd? n repeat output n
```

```
1
3
```

Type: Void

A `for` loop can also be written

`for` *iterator* | *predicate* `repeat` *loopBody*

which is equivalent to:

`for` *iterator* `repeat if` *predicate* `then` *loopBody* `else iterate`

The predicate need not refer only to the variable in the `for` clause: any variable in an outer scope can be part of the predicate.

In this example, the predicate on the inner `for` loop uses i from the outer loop and the j from the `for` clause that it directly modifies.

```
for i in 1..50 repeat
  for j in 1..50 | factorial(i+j) < 25 repeat
    output [i,j]
```

```
[1,1]
[1,2]
[1,3]
[2,1]
[2,2]
[3,1]
```

Type: Void

5.4.14 Parallel Iteration

The last example of the previous section gives an example of *nested iteration*: a loop is contained in another loop. Sometimes you want to iterate across two lists in parallel, or perhaps you want to traverse a list while incrementing a variable.

The general syntax of a repeat loop is

$$\mathit{iterator}_1 \;\; \mathit{iterator}_2 \; \ldots \mathit{iterator}_N \;\; \texttt{repeat} \;\; \mathit{loopBody}$$

where each *iterator* is either a `for` or a `while` clause. The loop terminates immediately when the end test of any *iterator* succeeds or when a `leave` or `return` expression is evaluated in *loopBody*. The value returned by the loop is the unique value of Void.

Here we write a loop to iterate across two lists, computing the sum of the pairwise product of elements. Here is the first list.

```
l := [1,3,5,7]
```

$[1, 3, 5, 7]$ (1)

Type: List PositiveInteger

And the second.

```
m := [100,200]
```

$[100, 200]$ (2)

Type: List PositiveInteger

The initial value of the sum counter.

```
sum := 0
```

0 (3)

Type: NonNegativeInteger

The last two elements of `l` are not used in the calculation because `m` has two fewer elements than `l`.

```
for x in l for y in m repeat
    sum := sum + x*y
```

Type: Void

Display the "dot product."

```
sum
```

700 (5)

Type: NonNegativeInteger

Next, we write a loop to compute the sum of the products of the loop elements with their positions in the loop.

```
l := [2,3,5,7,11,13,17,19,23,29,31,37]
```

$[2, 3, 5, 7, 11, 13, 17, 19, 23, 29, 31, 37]$ (6)

Type: List PositiveInteger

The initial sum.

```
sum := 0
```

0 (7)

Type: NonNegativeInteger

Here looping stops when the list `l` is exhausted, even though the `for i in 0..` specifies no terminating condition.

```
for i in 0.. for x in l repeat sum := i * x
```

Type: Void

Display this weighted sum.

```
sum
```

407 (9)

Type: NonNegativeInteger

When "|" is used to qualify any of the for clauses in a parallel iteration, the variables in the predicates can be from an outer scope or from a for clause in or to the left of a modified clause.

This is correct:

```
for i in 1..10 repeat
  for j in 200..300 | odd? (i+j) repeat
    output [i,j]
```

This is not correct since the variable j has not been defined outside the inner loop.

```
for i in 1..10 | odd? (i+j) repeat  -- wrong, j not defined
  for j in 200..300 repeat
    output [i,j]
```

This example shows that it is possible to mix several of the forms of repeat modifying clauses on a loop.

```
for i in 1..10
    for j in 151..160 | odd? j
      while i + j < 160 repeat
        output [i,j]
```

```
[1,151]
[3,153]
```

Type: Void

Here are useful rules for composing loop expressions:

1. while predicates can only refer to variables that are global (or in an outer scope) or that are defined in for clauses to the left of the predicate.
2. A "such that" predicate (something following "|") must directly follow a for clause and can only refer to variables that are global (or in an outer scope) or defined in the modified for clause or any for clause to the left.

5.5 Creating Lists and Streams with Iterators

All of what we did for loops in Section 5.4 on page 117 can be transformed into expressions that create lists and streams. The `repeat`, `leave` or `iterate` words are not used but all the other ideas carry over. Before we give you the general rule, here are some examples which give you the idea.

This creates a simple list of the integers from `1` to `10`.

```
list := [i for i in 1..10]
```

$$[1, 2, 3, 4, 5, 6, 7, 8, 9, 10] \quad (1)$$

Type: List PositiveInteger

Create a stream of the integers greater than or equal to `1`.

```
stream := [i for i in 1..]
```

$$[1, 2, 3, 4, 5, 6, 7, \ldots] \quad (2)$$

Type: Stream PositiveInteger

This is a list of the prime integers between `1` and `10`, inclusive.

```
[i for i in 1..10 | prime? i]
```

$$[2, 3, 5, 7] \quad (3)$$

Type: List PositiveInteger

This is a stream of the prime integers greater than or equal to `1`.

```
[i for i in 1.. | prime? i]
```

$$[2, 3, 5, 7, 11, 13, 17, \ldots] \quad (4)$$

Type: Stream PositiveInteger

This is a list of the integers between `1` and `10`, inclusive, whose squares are less than `700`.

```
[i for i in 1..10 while i*i < 700]
```

$$[1, 2, 3, 4, 5, 6, 7, 8, 9, 10] \quad (5)$$

Type: List PositiveInteger

This is a stream of the integers greater than or equal to `1` whose squares are less than `700`.

```
[i for i in 1.. while i*i < 700]
```

$$[1, 2, 3, 4, 5, 6, 7, \ldots] \quad (6)$$

Type: Stream PositiveInteger

Got the idea? Here is the general rule.

> The general syntax of a collection is
>
> [*collectExpression* $iterator_1$ $iterator_2$ ... $iterator_N$]
>
> where each $iterator_i$ is either a `for` or a `while` clause. The loop terminates immediately when the end test of any $iterator_i$ succeeds or when a `return` expression is evaluated in *collectExpression*. The value returned by the collection is either a list or a stream of elements, one for each iteration of the *collectExpression*.

Be careful when you use `while` to create a stream. By default, AXIOM tries to compute and display the first ten elements of a stream. If the `while` condition is not satisfied quickly, AXIOM can spend a long (possibly infinite) time trying to compute the elements. Use `)set streams calculate` to change the default to something else. This also affects the number of terms computed and displayed for power series. For the purposes of this book, we have used this system command to display fewer than ten terms.

Use nested iterators to create lists of lists which can then be given as an argument to **matrix**.

```
matrix [[x**i+j for i in 1..3] for j in 10..12]
```

$$\begin{bmatrix} x+10 & x^2+10 & x^3+10 \\ x+11 & x^2+11 & x^3+11 \\ x+12 & x^2+12 & x^3+12 \end{bmatrix} \tag{7}$$

Type: Matrix Polynomial Integer

You can also create lists of streams, streams of lists and streams of streams. Here is a stream of streams.

```
[[i/j for i in j+1..] for j in 1..]
```

$$\left[[2, 3, 4, 5, 6, 7, 8, \ldots], \left[\frac{3}{2}, 2, \frac{5}{2}, 3, \frac{7}{2}, 4, \frac{9}{2}, \ldots\right], \left[\frac{4}{3}, \frac{5}{3}, 2, \frac{7}{3}, \frac{8}{3}, 3, \frac{10}{3}, \ldots\right], \left[\frac{5}{4}, \frac{3}{2}, \frac{7}{4}, 2, \frac{9}{4}, \frac{5}{2}, \frac{11}{4}, \ldots\right], \left[\frac{6}{5}, \frac{7}{5}, \frac{8}{5}, \frac{9}{5}, 2, \frac{11}{5}, \frac{12}{5}, \ldots\right], \left[\frac{7}{6}, \frac{4}{3}, \frac{3}{2}, \frac{5}{3}, \frac{11}{6}, 2, \frac{13}{6}, \ldots\right], \left[\frac{8}{7}, \frac{9}{7}, \frac{10}{7}, \frac{11}{7}, \frac{12}{7}, \frac{13}{7}, 2, \ldots\right], \ldots\right] \tag{8}$$

Type: Stream Stream Fraction Integer

You can use parallel iteration across lists and streams to create new lists.

```
[i/j for i in 3.. by 10 for j in 2..]
```

$$\left[\frac{3}{2}, \frac{13}{3}, \frac{23}{4}, \frac{33}{5}, \frac{43}{6}, \frac{53}{7}, \frac{63}{8}, \ldots\right] \tag{9}$$

Type: Stream Fraction Integer

Iteration stops if the end of a list or stream is reached.

```
[i**j for i in 1..7 for j in 2.. ]
```

$$[1, 8, 81, 1024, 15625, 279936, 5764801] \tag{10}$$

Type: Stream Integer

As with loops, you can combine these modifiers to make very complicated conditions.

```
[[[i,j] for i in 10..15 | prime? i] for j in 17..22 | j =
  squareFreePart j]
```

$$[[[11, 17], [13, 17]], [[11, 19], [13, 19]], [[11, 21], [13, 21]], [[11, 22], [13, 22]]] \tag{11}$$

Type: List List List PositiveInteger

See 'List' on page 404 and 'Stream' on page 457 for more information on creating and manipulating lists and streams, respectively.

5.6 An Example: Streams of Primes

We conclude this chapter with an example of the creation and manipulation of infinite streams of prime integers. This might be useful for experiments with numbers or other applications where you are using sequences of primes over and over again. As for all streams, the stream of primes is only computed as far out as you need. Once computed, however, all the primes up to that point are saved for future reference.

Two useful operations provided by the AXIOM library are **prime?** and **nextPrime**. A straight-forward way to create a stream of prime numbers is to start with the stream of positive integers `[2,..]` and filter out those that are prime.

Create a stream of primes.

```
primes : Stream Integer := [i for i in 2.. | prime? i]
```

$$[2, 3, 5, 7, 11, 13, 17, \ldots] \tag{1}$$

Type: Stream Integer

A more elegant way, however, is to use the **generate** operation from Stream. Given an initial value `a` and a function `f`, **generate** constructs the stream `[a, f(a), f(f(a)), ...]`. This function gives you the quickest method of getting the stream of primes.

This is how you use **generate** to generate an infinite stream of primes.

```
primes := generate(nextPrime,2)
```

$$[2, 3, 5, 7, 11, 13, 17, \ldots] \tag{2}$$

Type: Stream Integer

Once the stream is generated, you might only be interested in primes starting at a particular value.

```
smallPrimes := [p for p in primes | p > 1000]
```

$$[1009, 1013, 1019, 1021, 1031, 1033, 1039, \ldots] \tag{3}$$

Type: Stream Integer

Here are the first 11 primes greater than 1000.

```
[p for p in smallPrimes for i in 1..11]
```

$$[1009, 1013, 1019, 1021, 1031, 1033, 1039, \ldots] \tag{4}$$

Type: Stream Integer

Here is a stream of primes between 1000 and 1200.

```
[p for p in smallPrimes while p < 1200]
```

$$[1009, 1013, 1019, 1021, 1031, 1033, 1039, \ldots] \tag{5}$$

Type: Stream Integer

To get these expanded into a finite stream, you call **complete** on the stream.

```
complete %
```

$$[1009, 1013, 1019, 1021, 1031, 1033, 1039, \ldots] \tag{6}$$

Type: Stream Integer

Twin primes are consecutive odd number pairs which are prime. Here is the stream of twin primes.

```
twinPrimes := [[p,p+2] for p in primes | prime?(p + 2)]
```

$$[[3, 5], [5, 7], [11, 13], [17, 19], [29, 31], [41, 43], [59, 61], \ldots] \quad (7)$$

Type: Stream List Integer

Since we already have the primes computed we can avoid the call to **prime?** by using a double iteration. This time we'll just generate a stream of the first of the twin primes.

```
firstOfTwins:= [p for p in primes for q in rest primes |
  q=p+2]
```

$$[3, 5, 11, 17, 29, 41, 59, \ldots] \quad (8)$$

Type: Stream Integer

Let's try to compute the infinite stream of triplet primes, the set of primes `p` such that `[p,p+2,p+4]` are primes. For example, `[3,5,7]` is a triple prime. We could do this by a triple `for` iteration. A more economical way is to use **firstOfTwins**. This time however, put a semicolon at the end of the line.

Put a semicolon at the end so that no elements are computed.

```
firstTriplets := [p for p in firstOfTwins for q in rest
  firstOfTwins | q = p+2];
```

(9)

Type: Stream Integer

What happened? As you know, by default AXIOM displays the first ten elements of a stream when you first display it. And, therefore, it needs to compute them! If you want *no* elements computed, just terminate the expression by a semicolon ("`;`").[3]

Compute the first triplet prime.

```
firstTriplets.1
```

$$3 \quad (10)$$

Type: PositiveInteger

If you want to compute another, just ask for it. But wait a second! Given three consecutive odd integers, one of them must be divisible by 3. Thus there is only one triplet prime. But suppose that you did not know this and wanted to know what was the tenth triplet prime.

```
firstTriples.10
```

To compute the tenth triplet prime, AXIOM first must compute the second, the third, and so on. But since there isn't even a second triplet prime, AXIOM will compute forever. Nonetheless, this effort can produce a useful result. After waiting a bit, hit **Ctrl**-**c**. The system responds as follows.

```
>> System error:
Console interrupt.
```

[3]Why does this happen? The semi-colon prevents the display of the result of evaluating the expression. Since no stream elements are needed for display (or anything else, so far), none are computed.

```
You are being returned to the top level of
the interpreter.
```

Let's say that you want to know how many primes have been computed. Issue

```
numberOfComputedEntries primes
```

and, for this discussion, let's say that the result is `2045`.

How big is the `2045` th prime?

```
primes.2045
```

17837 (11)

Type: PositiveInteger

What you have learned is that there are no triplet primes between 5 and 17837. Although this result is well known (some might even say trivial), there are many experiments you could make where the result is not known. What you see here is a paradigm for testing of hypotheses. Here our hypothesis could have been: "there is more than one triplet prime." We have tested this hypothesis for 17837 cases. With streams, you can let your machine run, interrupt it to see how far you've gotten, then start it up and let it continue from where it left off.

CHAPTER 6

User-Defined Functions, Macros and Rules

In this chapter we show you how to write functions and macros, and we explain how AXIOM looks for and applies them. We show some simple one-line examples of functions, together with larger ones that are defined piece-by-piece or through the use of piles.

6.1 Functions vs. Macros

A function is a program to perform some computation. Most functions have names so that it is easy to refer to them. A simple example of a function is one named **abs** which computes the absolute value of an integer.

This is a use of the "absolute value" library function for integers.

```
abs(-8)
```

8 (1)

Type: PositiveInteger

This is an unnamed function that does the same thing, using the "maps-to" syntax "+->" that we discuss in Section 6.17 on page 165.

```
(x +-> if x < 0 then -x else x)(-8)
```

8 (2)

Type: PositiveInteger

Functions can be used alone or serve as the building blocks for larger programs. Usually they return a value that you might want to use in the next stage of a computation, but not always (for example, see 'Exit' on page 355 and 'Void' on page 480). They may also read data from your keyboard, move information from one place to another, or format and display results on your screen.

In AXIOM, as in mathematics, functions are usually *parameterized*. Each time you *call* (some people say *apply* or *invoke*) a function, you give values to the parameters (variables). Such a value is called an *argument* of the function. AXIOM uses the arguments for the computation. In this way you get different results depending on what you "feed" the function.

Functions can have local variables or refer to global variables in the workspace. AXIOM can often *compile* functions so that they execute very efficiently. Functions can be passed as arguments to other functions.

Macros are textual substitutions. They are used to clarify the meaning of constants or expressions and to be templates for frequently used expressions. Macros can be parameterized but they are not objects that can be passed as arguments to functions. In effect, macros are extensions to the AXIOM expression parser.

6.2 Macros

A *macro* provides general textual substitution of an AXIOM expression for a name. You can think of a macro as being a generalized abbreviation. You can only have one macro in your workspace with a given name, no matter how many arguments it has.

> The two general forms for macros are
>
> `macro` *name* == *body*
> `macro` *name(arg1,...)* == *body*
>
> where the body of the macro can be any AXIOM expression.

For example, suppose you decided that you like to use `df` for **D**. You define the macro `df` like this.

```
macro df == D
```

Type: Void

Whenever you type `df`, the system expands it to **D**.

```
df(x**2 + x + 1,x)
```

$$2\,x+1 \qquad (2)$$

Type: Polynomial Integer

Macros can be parameterized and so can be used for many different kinds of objects.

```
macro ff(x) == x**2 + 1
```

Type: Void

Apply it to a number, a symbol, or an expression.

```
ff z
```

$$z^2 + 1 \tag{4}$$

Type: Polynomial Integer

Macros can also be nested, but you get an error message if you run out of space because of an infinite nesting loop.

```
macro gg(x) == ff(2*x - 2/3)
```

Type: Void

This new macro is fine as it does not produce a loop.

```
gg(1/w)
```

$$\frac{13\ w^2 - 24\ w + 36}{9\ w^2} \tag{6}$$

Type: Fraction Polynomial Integer

This, however, loops since `gg` is defined in terms of `ff`.

```
macro ff(x) == gg(-x)
```

Type: Void

The body of a macro can be a block.

```
macro next == (past := present; present := future; future
  := past + present)
```

Type: Void

Before entering `next`, we need values for `present` and `future`.

```
present : Integer := 0
```

$$0 \tag{9}$$

Type: Integer

```
future : Integer := 1
```

$$1 \tag{10}$$

Type: Integer

Repeatedly evaluating `next` produces the next Fibonacci number.

```
next
```

$$1 \tag{11}$$

Type: Integer

And the next one.

```
next
```

$$2 \tag{12}$$

Type: Integer

Here is the infinite stream of the rest of the Fibonacci numbers.

```
[next for i in 1..]
```

$$[3, 5, 8, 13, 21, 34, 55, \ldots] \tag{13}$$

Type: Stream Integer

Bundle all the above lines into a single macro.

```
macro fibStream ==
  present : Integer := 1
  future : Integer := 1
  [next for i in 1..] where
    macro next ==
      past := present
      present := future
      future := past + present
```

Type: Void

Use **concat** to start with the first two Fibonacci numbers.

```
concat([0,1],fibStream)
```

$$[0, 1, 2, 3, 5, 8, 13, \ldots] \tag{15}$$

Type: Stream Integer

An easier way to compute these numbers is to use the library operation **fibonacci**.

```
[fibonacci i for i in 1..]
```

$$[1, 1, 2, 3, 5, 8, 13, \ldots] \tag{16}$$

Type: Stream Integer

6.3 Introduction to Functions

Each name in your workspace can refer to a single object. This may be any kind of object including a function. You can use interactively any function from the library or any that you define in the workspace. In the library the same name can have very many functions, but you can have only one function with a given name, although it can have any number of arguments that you choose.

If you define a function in the workspace that has the same name and number of arguments as one in the library, then your definition takes precedence. In fact, to get the library function you must *package-call* it (see Section 2.9 on page 83).

To use a function in AXIOM, you apply it to its arguments. Most functions are applied by entering the name of the function followed by its argument or arguments.

```
factor(12)
```

$$2^2\ 3 \tag{1}$$

Type: Factored Integer

Some functions like "+" have *infix operators* as names.

```
3 + 4
```

7 (2)

Type: PositiveInteger

The function "+" has two arguments. When you give it more than two arguments, AXIOM groups the arguments to the left. This expression is equivalent to `(1 + 2) + 7`.

```
1 + 2 + 7
```

10 (3)

Type: PositiveInteger

All operations, including infix operators, can be written in prefix form, that is, with the operation name followed by the arguments in parentheses. For example, `2 + 3` can alternatively be written as `+(2,3)`. But `+(2,3,4)` is an error since "+" takes only two arguments.

Prefix operations are generally applied before the infix operation. Thus `factorial 3 + 1` means `factorial(3) + 1` producing `7`, and `- 2 + 5` means `(-2) + 5` producing `3`. An example of a prefix operator is prefix "`-`". For example, `- 2 + 5` converts to `(- 2) + 5` producing the value `3`. Any prefix function taking two arguments can be written in an infix manner by putting an ampersand ("&") before the name. Thus `D(2*x,x)` can be written as `2*x &D x` returning `2`.

Every function in AXIOM is identified by a *name* and *type*.[1] The type of a function is always a mapping of the form Source → Target where `Source` and `Target` are types. To enter a type from the keyboard, enter the arrow by using a hyphen "`-`" followed by a greater-than sign "`>`", e.g. `Integer -> Integer`.

Let's go back to "+". There are many "+" functions in the AXIOM library: one for integers, one for floats, another for rational numbers, and so on. These "+" functions have different types and thus are different functions. You've seen examples of this *overloading* before—using the same name for different functions. Overloading is the rule rather than the exception. You can add two integers, two polynomials, two matrices or two power series. These are all done with the same function name but with different functions.

[1]An exception is an "anonymous function" discussed in Section 6.17 on page 165.

6.4 Declaring the Type of Functions

In Section 2.3 on page 69 we discussed how to declare a variable to restrict the kind of values that can be assigned to it. In this section we show how to declare a variable that refers to function objects.

> A function is an object of type
>
> Source → Type
>
> where `Source` and `Target` can be any type. A common type for `Source` is Tuple(T_1, ..., T_n), usually written (T_1, ..., T_n), to indicate a function of n arguments.

If g takes an Integer, a Float and another Integer, and returns a String, the declaration is written this way.

```
g: (Integer,Float,Integer) -> String
```

Type: Void

The types need not be written fully; using abbreviations, the above declaration is:

```
g: (INT,FLOAT,INT) -> STRING
```

Type: Void

It is possible for a function to take no arguments. If h takes no arguments but returns a Polynomial Integer, any of the following declarations is acceptable.

```
h: () -> POLY INT
```

Type: Void

```
h: () -> Polynomial INT
```

Type: Void

```
h: () -> POLY Integer
```

Type: Void

> Functions can also be declared when they are being defined. The syntax for combined declaration/definition is:
>
> *functionName*(*parm*$_1$: *parmType*$_1$, ..., *parm*$_N$: *parmType*$_N$): *functionReturnType*

The following definition fragments show how this can be done for the functions g and h above.

```
g(arg1: INT, arg2: FLOAT, arg3: INT): STRING == ...
```

```
h(): POLY INT == ...
```

A current restriction on function declarations is that they must involve fully specified types (that is, cannot include modes involving explicit or implicit "?"). For more information on declaring things in general, see Section 2.3 on page 69.

6.5 One-Line Functions

As you use AXIOM, you will find that you will write many short functions to codify sequences of operations that you often perform. In this section we write some simple one-line functions.

This is a simple recursive factorial function for positive integers.

```
fac n == if n < 3 then n else n * fac(n-1)
```

Type: Void

```
fac 10
```

$$3628800 \tag{2}$$

Type: PositiveInteger

This function computes `1 + 1/2 + 1/3 + ... + 1/n`.

```
s n == reduce(+,[1/i for i in 1..n])
```

Type: Void

```
s 50
```

$$\frac{13943237577224054960759}{3099044504245996706400} \tag{4}$$

Type: Fraction Integer

This function computes a Mersenne number, several of which are prime.

```
mersenne i == 2**i - 1
```

Type: Void

If you type `mersenne`, AXIOM shows you the function definition.

```
mersenne
```

$$\text{mersenne } i \; == \; 2^i - 1 \tag{6}$$

Type: FunctionCalled mersenne

Generate a stream of Mersenne numbers.

```
[mersenne i for i in 1..]
```

$$[1, 3, 7, 15, 31, 63, 127, \ldots] \tag{7}$$

Type: Stream Integer

Create a stream of those values of i such that mersenne(i) is prime.

```
mersenneIndex := [n for n in 1.. | prime?(mersenne(n))]
```

$$[2, 3, 5, 7, 13, 17, 19, \ldots] \tag{8}$$

Type: Stream PositiveInteger

Finally, write a function that returns the n th Mersenne prime.

```
mersennePrime n == mersenne mersenneIndex(n)
```

Type: Void

```
mersennePrime 5
```

$$8191 \tag{10}$$

Type: PositiveInteger

6.6 Declared vs. Undeclared Functions

If you declare the type of a function, you can apply it to any data that can be converted to the source type of the function.

Define **f** with type Integer → Integer.

```
f(x: Integer): Integer == x + 1
```

```
Function declaration f : Integer -> Integer has been
added to workspace.
```

Type: Void

The function **f** can be applied to integers, ...

```
f 9
```

```
Compiling function f with type Integer -> Integer
```

$$10 \tag{2}$$

Type: PositiveInteger

and to values that convert to integers, ...

```
f(-2.0)
```

$$-1 \tag{3}$$

Type: Integer

but not to values that cannot be converted to integers.

```
f(2/3)
```

```
Conversion failed in the compiled user function f.

Cannot convert from type Fraction Integer to Integer
for value
2
-
3
```

To make the function over a wide range of types, do not declare its type.

Give the same definition with no declaration.

```
g x == x + 1
```

Type: Void

If x + 1 makes sense, you can apply **g** to x.

```
g 9
```

```
Compiling function g with type PositiveInteger ->
PositiveInteger
```

10 (5)

Type: PositiveInteger

A version of **g** with different argument types get compiled for each new kind of argument used.

```
g(2/3)
```

```
Compiling function g with type Fraction Integer ->
Fraction Integer
```

$$\frac{5}{3} \qquad (6)$$

Type: Fraction Integer

Here x+1 for x = "axiom" makes no sense.

```
g("axiom")
```

```
Cannot find a definition or library operation named +
with argument types
                         String
                     PositiveInteger

We will attempt to step through and interpret the
code.

Cannot find a definition or library operation named +
with argument types
                         String
                     PositiveInteger
```

As you will see in Chapter 12, AXIOM has a formal idea of categories for what "makes sense."

6.7 Functions vs. Operations

A function is an object that you can create, manipulate, pass to, and return from functions (for some interesting examples of library functions that manipulate functions, see 'MappingPackage1' on page 411). Yet, we often seem to use the term *operation* and function interchangeably in AXIOM. What is the distinction?

First consider values and types associated with some variable n in your workspace. You can make the declaration n : Integer, then assign n an integer value. You then speak of the integer n. However, note that the integer is not the name n itself, but the value that you assign to n.

Similarly, you can declare a variable `f` in your workspace to have type Integer → Integer, then assign `f`, through a definition or an assignment of an anonymous function. You then speak of the function `f`. However, the function is not `f`, but the value that you assign to `f`.

A function is a value, in fact, some machine code for doing something. Doing what? Well, performing some *operation*. Formally, an operation consists of the constituent parts of `f` in your workspace, excluding the value; thus an operation has a name and a type. An operation is what domains and packages export. Thus Ring exports one operation "+". Every ring also exports this operation. Also, the author of every ring in the system is obliged under contract (see Section 11.3 on page 504) to provide an implementation for this operation.

This chapter is all about functions—how you create them interactively and how you apply them to meet your needs. In Chapter 11 you will learn how to create them for the AXIOM library. Then in Chapter 12, you will learn about categories and exported operations.

6.8 Delayed Assignments vs. Functions with No Arguments

In Section 5.1 on page 109 we discussed the difference between immediate and delayed assignments. In this section we show the difference between delayed assignments and functions of no arguments.

A function of no arguments is sometimes called a *nullary function*.

```
sin24() == sin(24.0)
```

Type: Void

You must use the parentheses ("()") to evaluate it. Like a delayed assignment, the right-hand-side of a function evaluation is not evaluated until the left-hand-side is used.

```
sin24()
Compiling function sin24 with type () -> Float
```

$$-0.90557836200662384514 \qquad (2)$$

Type: Float

If you omit the parentheses, you just get the function definition.

```
sin24
```

$$\mathrm{sin}\,24\;() \;==\; \sin\left(float\,(240,\,-1,\,10)\$\mathrm{Float}\,\right) \qquad (3)$$

Type: FunctionCalled sin24

You do not use the parentheses "()" in a delayed assignment...

```
cos24 == cos(24.0)
```

Type: Void

nor in the evaluation.

```
cos24
Compiling body of rule cos24 to compute value of type
Float
```

0.42417900733699697594 (5)

Type: Float

The only syntactic difference between delayed assignments and nullary functions is that you use "`()`" in the latter case.

6.9 How AXIOM Determines What Function to Use

What happens if you define a function that has the same name as a library function? Well, if your function has the same name and number of arguments (we sometimes say *arity*) as another function in the library, then your function covers up the library function. If you want then to call the library function, you will have to package-call it. AXIOM can use both the functions you write and those that come from the library. Let's do a simple example to illustrate this.

Suppose you (wrongly!) define **sin** in this way.

```
sin x == 1.0
```

Type: Void

The value `1.0` is returned for any argument.

```
sin 4.3
Compiling function sin with type Float -> Float
```

1.0 (2)

Type: Float

If you want the library operation, we have to package-call it (see Section 2.9 on page 83 for more information).

```
sin(4.3)$Float
```

−0.91616593674945498404 (3)

Type: Float

```
sin(34.6)$Float
```

−0.042468034716950101543 (4)

Type: Float

Even worse, say we accidentally used the same name as a library function in the function.

```
sin x == sin x
```

Type: Void

Then AXIOM definitely does not understand us.

```
sin 4.3
We cannot determine the type of sin because cannot
analyze the non-recursive part, if it exists. We
suggest that you declare the function.
```

Again, we could package-call the inside function.

```
sin x == sin(x)$Float
```

Type: Void

```
sin 4.3
```

$$-0.9161659367494549840 4 \qquad (7)$$

Type: Float

Of course, you are unlikely to make such obvious errors. It is more probable that you would write a function and in the body use a function that you think is a library function. If you had also written a function by that same name, the library function would be invisible.

How does AXIOM determine what library function to call? It very much depends on the particular example, but the simple case of creating the polynomial `x + 2/3` will give you an idea.

1. The **x** is analyzed and its default type is Variable(x).
2. The 2 is analyzed and its default type is PositiveInteger.
3. The 3 is analyzed and its default type is PositiveInteger.
4. Because the arguments to "/" are integers, AXIOM gives the expression 2/3 a default target type of Fraction(Integer).
5. AXIOM looks in PositiveInteger for "/". It is not found.
6. AXIOM looks in Fraction(Integer) for "/". It is found for arguments of type Integer.
7. The 2 and 3 are converted to objects of type Integer (this is trivial) and "/" is applied, creating an object of type Fraction(Integer).
8. No "+" for arguments of types Variable(x) and Fraction(Integer) are found in either domain.
9. AXIOM resolves (see Section 2.10 on page 86) the types and gets Polynomial (Fraction (Integer)).
10. The **x** and the 2/3 are converted to objects of this type and "+" is applied, yielding the answer, an object of type Polynomial (Fraction (Integer)).

6.10 Compiling vs. Interpreting

When possible, AXIOM completely determines the type of every object in a function, then translates the function definition to Common LISP or to machine code (see next section). This translation, called *compilation*, happens the first time you call the function and results in a computational delay. Subsequent function calls with the same argument types use the compiled version of the code without delay.

If AXIOM cannot determine the type of everything, the function may still be executed but in *interpret-code mode*: each statement in the function is analyzed and executed as the control flow indicates. This process is slower than executing a compiled function, but it allows the execution of code that may involve objects whose types change.

If AXIOM decides that it cannot compile the code, it issue a message stating the problem and then the following message:

We will attempt to step through and interpret the code.

This is not a time to panic. Rather, it just means that what you gave to AXIOM is somehow ambiguous: either it is not specific enough to be analyzed completely, or it is beyond AXIOM's present interactive compilation abilities.

This function runs in interpret-code mode, but it does not compile.

```
varPolys(vars) ==
  for var in vars repeat
    output(1 :: UnivariatePolynomial(var,Integer))
```

Type: Void

For `vars` equal to `['x, 'y, 'z]`, this function displays 1 three times.

```
varPolys ['x,'y,'z]

Cannot compile conversion for types involving local
variables. In particular, could not compile the
expression involving ::
UnivariatePolynomial(var,Integer)
We will attempt to step through and interpret the
code.

1
1
1
```

Type: Void

The type of the argument to **output** changes in each iteration, so AXIOM cannot compile the function. In this case, even the inner loop by itself would have a problem:

```
for var in ['x,'y,'z] repeat
  output(1 :: UnivariatePolynomial(var,Integer))

Cannot compile conversion for types involving local
variables. In particular, could not compile the
expression involving ::
UnivariatePolynomial(var,Integer)
We will attempt to step through and interpret the
code.

1
1
1
```

Type: Void

Sometimes you can help a function to compile by using an extra conversion or by using `pretend`. See Section 2.8 on page 80 for details.

When a function is compilable, you have the choice of whether it is compiled to Common LISP and then interpreted by the Common LISP interpreter or then further compiled from Common LISP to machine code. The option is controlled via `)set functions compile`. Issue `)set functions compile on` to compile all the way to machine code. With the default setting `)set functions compile off`, AXIOM has its Common LISP code interpreted because the overhead of further compilation is larger than the run-time of most of the functions our users have defined. You may find that selectively turning this option on and off will give you the best performance in your particular application. For example, if you are writing functions for graphics applications where hundreds of points are being computed, it is almost certainly true that you will get the best performance by issuing `)set functions compile on`.

6.11 Piece-Wise Function Definitions

To move beyond functions defined in one line, we introduce in this section functions that are defined piece-by-piece. That is, we say "use this definition when the argument is such-and-such and use this other definition when the argument is that-and-that."

6.11.1 A Basic Example

There are many other ways to define a factorial function for nonnegative integers. You might say factorial of `0` is `1`, otherwise factorial of `n` is `n` times factorial of `n-1`. Here is one way to do this in AXIOM.

Here is the value for `n = 0`.

```
fact(0) == 1
```

Type: Void

Here is the value for `n > 0`. The vertical bar "`|`" means "such that".

```
fact(n | n > 0) == n * fact(n - 1)
```

Type: Void

What is the value for `n = 3`?

```
fact(3)
```

6 (3)

Type: PositiveInteger

What is the value for `n = -3`?

```
fact(-3)
```

```
You did not define fact for argument -3 .
```

Now for a second definition. Here is the value for n = 0.

```
facto(0) == 1
```

Type: Void

Give an error message if n < 0.

```
facto(n | n < 0) == error "arguments to fact must be non-
  negative"
```

Type: Void

Here is the value otherwise.

```
facto(n) == n * facto(n - 1)
```

Type: Void

What is the value for n = 7?

```
facto(3)
```

$$6 \tag{7}$$

Type: PositiveInteger

What is the value for n = -7?

```
facto(-3)
```

```
Error signalled from user code in function facto:
arguments to fact must be non-negative
```

To see the current piece-wise definition of a function, use `)display value`.

```
)display value facto
```

```
Definition:
  facto 0 == 1
  facto (n | n < 0) ==
    error(arguments to fact must be non-negative)
  facto n == n facto(n - 1)
```

In general a *piece-wise definition* of a function consists of two or more parts. Each part gives a "piece" of the entire definition. AXIOM collects the pieces of a function as you enter them. When you ask for a value of the function, it then "glues" the pieces together to form a function.

The two piece-wise definitions for the factorial function are examples of recursive functions, that is, functions that are defined in terms of themselves. Here is an interesting doubly-recursive function. This function returns the value 11 for all positive integer arguments.

Here is the first of two pieces.

```
eleven(n | n < 1) == n + 11
```

Type: Void

And the general case.

```
eleven(m) == eleven(eleven(m - 12))
```

Type: Void

Compute elevens, the infinite stream of values of eleven.

```
elevens := [eleven(i) for i in 0..]
```

[11, 11, 11, 11, 11, 11, 11, ...] (10)

Type: Stream Integer

What is the value at n = 200?

```
elevens 200
```

11 (11)

Type: PositiveInteger

What is the AXIOM's definition of eleven?

```
)display value eleven
```

```
Definition:
  eleven (m | m < 1) == m + 11
  eleven m == eleven(eleven(m - 12))
```

6.11.2 Picking Up the Pieces

Here are the details about how AXIOM creates a function from its pieces. AXIOM converts the ith piece of a function definition into a conditional expression of the form: `if` $pred_i$ `then` $expression_i$. If any new piece has a $pred_i$ that is identical[2] to an earlier $pred_j$, the earlier piece is removed. Otherwise, the new piece is always added at the end.

> If there are n pieces to a function definition for f, the function defined f is:
>
> `if` $pred_1$ `then` $expression_1$ `else`
>
> ...
>
> `if` $pred_n$ `then` $expression_n$ `else`
>
> `error "You did not define f for argument <arg>."`

You can give definitions of any number of mutually recursive function definitions, piece-wise or otherwise. No computation is done until you ask for a value. When you do ask for a value, all the relevant definitions are gathered, analyzed, and translated into separate functions and compiled.

Let's recall the definition of **eleven** from the previous section.

```
eleven(n | n < 1) == n + 11
```

Type: Void

[2]after all variables are uniformly named

```
eleven(m) == eleven(eleven(m - 12))
```

Type: Void

A similar doubly-recursive function below produces -11 for all negative positive integers. If you haven't worked out why or how **eleven** works, the structure of this definition gives a clue.

This definition we write as a block.

```
minusEleven(n) ==
  n >= 0 => n - 11
  minusEleven (5 + minusEleven(n + 7))
```

Type: Void

Define s(n) to be the sum of plus and minus "eleven" functions divided by n. Since 11 - 11 = 0, we define s(0) to be 1.

```
s(0) == 1
```

Type: Void

And the general term.

```
s(n) == (eleven(n) + minusEleven(n))/n
```

Type: Void

What are the first ten values of s?

```
[s(n) for n in 0..]
```

$$[1,\ 1,\ 1,\ 1,\ 1,\ 1,\ 1,\ \ldots] \tag{6}$$

Type: Stream Fraction Integer

AXIOM can create infinite streams in the positive direction (for example, for index values 0,1, ...) or negative direction (for example, for index values 0,-1,-2, ...). Here we would like a stream of values of s(n) that is infinite in both directions. The function t(n) below returns the nth term of the infinite stream [s(0), s(1), s(-1), s(2), s(-2), ...]. Its definition has three pieces.

Define the initial term.

```
t(1) == s(0)
```

Type: Void

The even numbered terms are the s(i) for positive i. We use "quo" rather than "/" since we want the result to be an integer.

```
t(n | even?(n)) == s(n quo 2)
```

Type: Void

Finally, the odd numbered terms are the s(i) for negative i. In piece-wise definitions, you can use different variables to define different pieces. AXIOM will not get confused.

```
t(p) == s(- p quo 2)
```

Type: Void

Look at the definition of t. In the first piece, the variable n was used; in the second piece, p. AXIOM always uses your last variable to display your definitions back to you.

```
)display value t
Definition:
  t 1 == s(0)
  t (p | even?(p)) == s(p quo 2)
  t p == s(- p quo 2)
```

Create a series of values of s applied to alternating positive and negative arguments.

```
[t(i) for i in 1..]
```

$$[1, 1, 1, 1, 1, 1, 1, \ldots] \tag{10}$$

Type: Stream Fraction Integer

Evidently t(n) = 1 for all i. Check it at n= 100.

```
t(100)
```

$$1 \tag{11}$$

Type: Fraction Integer

6.11.3 Predicates

We have already seen some examples of predicates (Section 6.11.1 on page 148). Predicates are Boolean-valued expressions and AXIOM uses them for filtering collections (see Section 5.5 on page 130) and for placing constraints on function arguments. In this section we discuss their latter usage.

The simplest use of a predicate is one you don't see at all.

```
opposite 'right == 'left
```

Type: Void

Here is a longer way to give the "opposite definition."

```
opposite (x | x = 'left) == 'right
```

Type: Void

Try it out.

```
for x in ['right,'left,'inbetween] repeat output opposite
  x
left
right

The function opposite is not defined for the given
argument(s).
```

Explicit predicates tell AXIOM that the given function definition piece is to be applied if the predicate evaluates to true for the arguments to the function. You can use such "constant" arguments only for integers, strings, quoted symbols, and the Boolean values `true` and `false`. The following are all valid function definition fragments using constant arguments.

```
a(1) == ...
b("unramified") == ...
c('untested) == ...
d(true) == ...
```

If a function has more than one argument, each argument can have its own predicate. However, if a predicate involves two or more arguments, it must be given *after* all the arguments mentioned in the predicate have been given. You are always safe to give a single predicate at the end of the argument list.

A function involving predicates on two arguments.

```
inFirstHalfQuadrant(x | x > 0,y | y < x) == true
```

Type: Void

This is incorrect as it gives a predicate on y before the argument y is given.

```
inFirstHalfQuadrant(x | x > 0 and y < x,y) == true
```

Type: Void

It is always correct to write the predicate at the end.

```
inFirstHalfQuadrant(x,y | x > 0 and y < x) == true
```

Type: Void

Here is the rest of the definition.

```
inFirstHalfQuadrant(x,y) == false
```

Type: Void

Try it out.

```
[inFirstHalfQuadrant(i,3) for i in 1..5]
```

$$[\text{false}, \text{false}, \text{false}, \text{true}, \text{true}] \qquad (7)$$

Type: List Boolean

6.12 Caching Previously Computed Results

By default, AXIOM does not save the values of any function. You can cause it to save values and not to recompute unnecessarily by using `)set functions cache`. This should be used before the functions are defined or, at least, before they are executed. The word following "cache" should be `0` to turn off caching, a positive integer `n` to save the last `n` computed values or "all" to save all computed values. If you then give a list of names of functions, the caching only affects those functions. Use no list of names or "all" when you want to define the default behavior for functions not specifically mentioned in other `)set functions cache` statements. If you give no list of names, all functions will have the caching behavior. If you explicitly turn on caching for one or more names, you must explicitly turn off caching for those names when you want to stop saving their values.

This causes the functions **f** and **g** to have the last three computed values saved.

```
)set functions cache 3 f g

function f will cache the last 3 values.
function g will cache the last 3 values.
```

This is a sample definition for **f**.

```
f x == factorial(2**x)
```

Type: Void

A message is displayed stating what **f** will cache.

```
f(4)

Compiling function f with type PositiveInteger ->
Integer
f will cache 3 most recently computed value(s).
```

20922789888000 (2)

Type: PositiveInteger

This causes all other functions to have all computed values saved by default.

```
)set functions cache all

In general, interpreter functions will cache all
values.
```

This causes all functions that have not been specifically cached in some way to have no computed values saved.

```
)set functions cache 0

 In general, functions will cache no returned values.
```

We also make **f** and **g** uncached.

```
)set functions cache 0 f g

Caching for function f is turned off
Caching for function g is turned off
```

Be careful about caching functions that have *side effects*. Such a function might destructively modify the elements of an array or issue a **draw** command, for example. A function that you expect to execute every time it is called should not be cached. Also, it is highly unlikely that a function with no arguments should be cached.

6.13 Recurrence Relations

One of the most useful classes of function are those defined via a "recurrence relation." A *recurrence relation* makes each successive value depend on some or all of the previous values. A simple example is the ordinary "factorial" function:

```
fact(0) == 1
fact(n | n > 0) == n * fact(n-1)
```

The value of `fact(10)` depends on the value of `fact(9)`, `fact(9)` on `fact(8)`, and so on. Because it depends on only one previous value, it is usually called a *first order recurrence relation*. You can easily imagine a function based on two, three or more previous values. The Fibonacci numbers are probably the most famous function defined by a second order recurrence relation.

The library function **fibonacci** computes Fibonacci numbers. It is obviously optimized for speed.

```
[fibonacci(i) for i in 0..]
```

$$[0,\ 1,\ 1,\ 2,\ 3,\ 5,\ 8,\ \ldots] \tag{1}$$

Type: Stream Integer

Define the Fibonacci numbers ourselves using a piece-wise definition.

```
fib(1) == 1
```

Type: Void

```
fib(2) == 1
```

Type: Void

```
fib(n) == fib(n-1) + fib(n-2)
```

Type: Void

As defined, this recurrence relation is obviously doubly-recursive. To compute `fib(10)`, we need to compute `fib(9)` and `fib(8)`. And to `fib(9)`, we need to compute `fib(8)` and `fib(7)`. And so on. It seems that to compute `fib(10)` we need to compute `fib(9)` once, `fib(8)` twice, `fib(7)` three times. Look familiar? The number of function calls needed to compute *any* second order recurrence relation in the obvious way is exactly `fib(n)`. These numbers grow! For example, if AXIOM actually did this, then `fib(500)` requires more than 10^{104} function calls. And, given all this, our definition of **fib** obviously could not be used to calculate the five-hundredth Fibonacci number.

Let's try it anyway.

```
fib(500)
Compiling function fib as a recurrence relation.
```

$$13942322456169788013972438287040728395007025658769730726410\\8962948325571622863290691557658876222521294125 \quad (5)$$

Type: PositiveInteger

Since this takes a short time to compute, it obviously didn't do as many as 10^{104} operations! By default, AXIOM transforms any recurrence relation it recognizes into an iteration. Iterations are efficient. To compute the value of the n th term of a recurrence relation using an iteration requires only n function calls.[3]

To turn off this special recurrence relation compilation, issue

```
)set functions recurrence off
```

To turn it back on, substitute "on" for "off".

The transformations that AXIOM uses for **fib** caches the last two values.[4] If, after computing a value for **fib**, you ask for some larger value, AXIOM picks up the cached values and continues computing from there. See Section 6.16 on page 162 for an example of a function definition that has this same behavior. Also see Section 6.12 on page 153 for a more general discussion of how you can cache function values.

Recurrence relations can be used for defining recurrence relations involving polynomials, rational functions, or anything you like. Here we compute the infinite stream of Legendre polynomials.

The Legendre polynomial of degree 0.

```
p(0) == 1
```

Type: Void

The Legendre polynomial of degree 1.

```
p(1) == x
```

Type: Void

The Legendre polynomial of degree n.

```
p(n) == ((2*n-1)*x*p(n-1) - (n-1)*p(n-2))/n
```

Type: Void

---

[3]If you compare the speed of our **fib** function to the library function, our version is still slower. This is because the library **fibonacci** uses a "powering algorithm" with a computing time proportional to $\log^3(n)$ to compute fibonacci(n).

[4]For a more general k th order recurrence relation, AXIOM caches the last k values.

Compute the Legendre polynomial of degree 6.

```
p(6)
Compiling function p as a recurrence relation.
```

$$\frac{231}{16} x^6 - \frac{315}{16} x^4 + \frac{105}{16} x^2 - \frac{5}{16} \quad (9)$$

Type: Polynomial Fraction Integer

6.14 Making Functions from Objects

There are many times when you compute a complicated expression and then wish to use that expression as the body of a function. AXIOM provides an operation called **function** to do this. It creates a function object and places it into the workspace. There are several versions, depending on how many arguments the function has. The first argument to **function** is always the expression to be converted into the function body, and the second is always the name to be used for the function. For more information, see 'MakeFunction' on page 409.

Start with a simple example of a polynomial in three variables.

```
p := -x + y**2 - z**3
```

$$-z^3 + y^2 - x \quad (1)$$

Type: Polynomial Integer

To make this into a function of no arguments that simply returns the polynomial, use the two argument form of **function**.

```
function(p,'f0)
```

$$f0 \quad (2)$$

Type: Symbol

To avoid possible conflicts (see below), it is a good idea to quote always this second argument.

```
f0
```

$$f0\ () \ == \ -z^3 + y^2 - x \quad (3)$$

Type: FunctionCalled f0

This is what you get when you evaluate the function.

```
f0()
```

$$-z^3 + y^2 - x \quad (4)$$

Type: Polynomial Integer

To make a function in **x**, use a version of **function** that takes three arguments. The last argument is the name of the variable to use as the parameter. Typically, this variable occurs in the expression and, like the function name, you should quote it to avoid possible confusion.

```
function(p,'f1,'x)
```

$$f1 \quad (5)$$

Type: Symbol

This is what the new function looks like.

```
f1
```

$$f1\ x \ == \ -z^3 + y^2 - x \quad (6)$$

Type: FunctionCalled f1

This is the value of **f1** at `x = 3`. Notice that the return type of the function is Polynomial (Integer), the same as `p`.

```
f1(3)
```

$$-z^3 + y^2 - 3 \tag{7}$$

Type: Polynomial Integer

To use `x` and `y` as parameters, use the four argument form of **function**.

```
function(p,'f2,'x,'y)
```

$$f2 \tag{8}$$

Type: Symbol

```
f2
```

$$f2\,(x,\,y)\ ==\ -z^3 + y^2 - x \tag{9}$$

Type: FunctionCalled f2

Evaluate `f2` at `x = 3` and `y = 0`. The return type of **f2** is still Polynomial(Integer) because the variable `z` is still present and not one of the parameters.

```
f2(3,0)
```

$$-z^3 - 3 \tag{10}$$

Type: Polynomial Integer

Finally, use all three variables as parameters. There is no five argument form of **function**, so use the one with three arguments, the third argument being a list of the parameters.

```
function(p,'f3,['x,'y,'z])
```

$$f3 \tag{11}$$

Type: Symbol

Evaluate this using the same values for `x` and `y` as above, but let `z` be `-6`. The result type of **f3** is Integer.

```
f3
```

$$f3\,(x,\,y,\,z)\ ==\ -z^3 + y^2 - x \tag{12}$$

Type: FunctionCalled f3

```
f3(3,0,-6)
```

$$213 \tag{13}$$

Type: PositiveInteger

The four functions we have defined via `p` have been undeclared. To declare a function whose body is to be generated by **function**, issue the declaration *before* the function is created.

```
g: (Integer, Integer) -> Float
```

Type: Void

```
D(sin(x-y)/cos(x+y),x)
```

$$\frac{\sin(-y+x)\,\sin(y+x) + \cos(-y+x)\,\cos(y+x)}{\cos(y+x)^2} \tag{15}$$

Type: Expression Integer

```
function(%,'g,'x,'y)
```

$$g \tag{16}$$

Type: Symbol

```
g
```

$$g\ (x,\ y)\ ==\ \frac{\sin(-y+x)\ \sin(y+x)+\cos(-y+x)\ \cos(y+x)}{\cos(y+x)^2} \tag{17}$$

Type: FunctionCalled g

It is an error to use `g` without the quote in the penultimate expression since `g` had been declared but did not have a value. Similarly, since it is common to overuse variable names like `x`, `y`, and so on, you avoid problems if you always quote the variable names for **function**. In general, if `x` has a value and you use `x` without a quote in a call to **function**, then AXIOM does not know what you are trying to do.

What kind of object is allowable as the first argument to **function**? Let's use the Browse facility of HyperDoc to find out. At the main Browse menu, enter the string `function` and then click on **Operations.** The exposed operations called **function** all take an object whose type belongs to category ConvertibleTo InputForm. What domains are those? Go back to the main Browse menu, erase `function`, enter `ConvertibleTo` in the input area, and click on **categories** on the **Constructors** line. At the bottom of the page, enter `InputForm` in the input area following **S =**. Click on **Cross Reference** and then on **Domains**. The list you see contains over forty domains that belong to the category ConvertibleTo InputForm. Thus you can use **function** for Integer, Float, String, Complex, Expression, and so on.

6.15 Functions Defined with Blocks

You need not restrict yourself to functions that only fit on one line or are written in a piece-wise manner. The body of the function can be a block, as discussed in Section 5.2 on page 112.

Here is a short function that swaps two elements of a list, array or vector.

```
swap(m,i,j) ==
  temp := m.i
  m.i := m.j
  m.j := temp
```

Type: Void

The significance of **swap** is that it has a destructive effect on its first argument.

```
k := [1,2,3,4,5]
```

$$[1,\ 2,\ 3,\ 4,\ 5] \tag{2}$$

Type: List PositiveInteger

```
swap(k,2,4)
```

2 (3)

Type: PositiveInteger

You see that the second and fourth elements are interchanged.

```
k
```

[1, 4, 3, 2, 5] (4)

Type: List PositiveInteger

Using this, we write a couple of different sort functions. First, a simple bubble sort. The operation "#" returns the number of elements in an aggregate.

```
bubbleSort(m) ==
  n := #m
  for i in 1..(n-1) repeat
    for j in n..(i+1) by -1 repeat
      if m.j < m.(j-1) then swap(m,j,j-1)
  m
```

Type: Void

Let this be the list we want to sort.

```
m := [8,4,-3,9]
```

[8, 4, −3, 9] (6)

Type: List Integer

This is the result of sorting.

```
bubbleSort(m)
```

[−3, 4, 8, 9] (7)

Type: List Integer

Moreover, `m` is destructively changed to be the sorted version.

```
m
```

[−3, 4, 8, 9] (8)

Type: List Integer

This function implements an insertion sort. The basic idea is to traverse the list and insert the `i` th element in its correct position among the `i-1` previous elements. Since we start at the beginning of the list, the list elements before the `i` th element have already been placed in ascending order.

```
insertionSort(m) ==
  for i in 2..#m repeat
    j := i
    while j > 1 and m.j < m.(j-1) repeat
      swap(m,j,j-1)
      j := j - 1
  m
```

Type: Void

As with our bubble sort, this is a destructive function.

```
m := [8,4,-3,9]
```

[8, 4, −3, 9] (10)

Type: List Integer

```
insertionSort(m)
```

[−3, 4, 8, 9] (11)

Type: List Integer

m

$$[-3, 4, 8, 9] \tag{12}$$

Type: List Integer

Neither of the above functions is efficient for sorting large lists since they reference elements by asking for the j th element of the structure m.

Here is a more efficient bubble sort for lists.

```
bubbleSort2(m: List Integer): List Integer ==
  null m => m
  l := m
  while not null (r := l.rest) repeat
     r := bubbleSort2 r
     x := l.first
     if x < r.first then
       l.first := r.first
       r.first := x
     l.rest := r
     l := l.rest
  m
```

Type: Void

Try it out.

```
bubbleSort2 [3,7,2]
```

$$[7, 3, 2] \tag{14}$$

Type: List Integer

This definition is both recursive and iterative, and is tricky! Unless you are *really* curious about this definition, we suggest you skip immediately to the next section.

Here are the key points in the definition. First notice that if you are sorting a list with less than two elements, there is nothing to do: just return the list. This definition returns immediately if there are zero elements, and skips the entire `while` loop if there is just one element.

The second point to realize is that on each outer iteration, the bubble sort ensures that the minimum element is propagated leftmost. Each iteration of the `while` loop calls **bubbleSort2** recursively to sort all but the first element. When finished, the minimum element is either in the first or second position. The conditional expression ensures that it comes first. If it is in the second, then a swap occurs. In any case, the **rest** of the original list must be updated to hold the result of the recursive call.

6.16 Free and Local Variables

When you want to refer to a variable that is not local to your function, use a "free" declaration. Variables declared to be free are assumed to be defined globally in the workspace.

This is a global workspace variable.

```
counter := 0
```

0 (1)

Type: NonNegativeInteger

This function refers to the global counter.

```
f() ==
  free counter
  counter := counter + 1
```

Type: Void

The global counter is incremented by 1.

```
f()
```

1 (3)

Type: PositiveInteger

```
counter
```

1 (4)

Type: NonNegativeInteger

Usually AXIOM can tell that you mean to refer to a global variable and so free isn't always necessary. However, for clarity and the sake of self-documentation, we encourage you to use it.

Declare a variable to be "local" when you do not want to refer to a global variable by the same name.

This function uses counter as a local variable.

```
g() ==
  local counter
  counter := 7
```

Type: Void

Apply the function.

```
g()
```

7 (6)

Type: PositiveInteger

Check that the global value of counter is unchanged.

```
counter
```

1 (7)

Type: NonNegativeInteger

Parameters to a function are local variables in the function. Even if you issue a free declaration for a parameter, it is still be local.

What happens if you do not declare that a variable x in the body of your function is `local` or `free`? Well, AXIOM decides on this basis:

1. AXIOM scans your function line-by-line, from top-to-bottom. The right-hand side of an assignment is looked at before the left-hand side.
2. If **x** is referenced before it is assigned a value, it is a `free` (global) variable.
3. If **x** is assigned a value before it is referenced, it is a `local` variable.

Set two global variables to 1.

```
a := b := 1
```

1 (8)

Type: PositiveInteger

Refer to a before it is assigned a value, but assign a value to b before it is referenced.

```
h() ==
  b := a + 1
  a := b + a
```

Type: Void

Can you predict this result?

```
h()
```

3 (10)

Type: PositiveInteger

How about this one?

```
[a, b]
```

[3, 1] (11)

Type: List PositiveInteger

What happened? In the first line of the function body for `h`, `a` is referenced on the right-hand side of the assignment. Thus `a` is a free variable. The variable `b` is not referenced in that line, but it is assigned a value. Thus `b` is a local variable and is given the value `a + 1 = 2`. In the second line, the free variable `a` is assigned the value `b + a = 2 + 1 = 3.` This is the value returned by the function. Since `a` was free in **h**, the global variable `a` has value `3.` Since `b` was local in **h**, the global variable `b` is unchanged—it still has the value `1.`

It is good programming practice always to declare global variables. However, by far the most common situation is to have local variables in your functions. No declaration is needed for this situation, but be sure to initialize their values.

AXIOM does not allow *fluid variables*, that is, variables *bound* by a function `f` that can be referenced by functions called by `f`.

Values are passed to functions by *reference*: a pointer to the value is passed rather than a copy of the value or a pointer to a copy.

This is a global variable that is bound to a record object.

```
r : Record(i : Integer) := [1]
```

$[i = 1]$ (12)

Type: Record(i: Integer)

This function first modifies the one component of its record argument and then rebinds the parameter to another record.

```
resetRecord rr ==
  rr.i := 2
  rr := [10]
```

Type: Void

Pass `r` as an argument to **resetRecord**.

```
resetRecord r
```

$[i = 10]$ (14)

Type: Record(i: Integer)

The value of `r` was changed by the expression `rr.i := 2` but not by `rr := [10]`.

```
r
```

$[i = 2]$ (15)

Type: Record(i: Integer)

To conclude this section, we give an iterative definition of a function that computes Fibonacci numbers. This definition approximates the definition into which AXIOM transforms the recurrence relation definition of **fib** in Section 6.13 on page 155.

Global variables `past` and `present` are used to hold the last computed Fibonacci numbers.

```
past := present := 1
```

1 (16)

Type: PositiveInteger

Global variable `index` gives the current index of `present`.

```
index := 2
```

2 (17)

Type: PositiveInteger

Here is a recurrence relation defined in terms of these three global variables.

```
fib(n) ==
  free past, present, index
  n < 3 => 1
  n = index - 1 => past
  if n < index-1 then
    (past,present) := (1,1)
    index := 2
  while (index < n) repeat
    (past,present) := (present, past+present)
    index := index + 1
  present
```

Type: Void

Compute the infinite stream of Fibonacci numbers.

```
fibs := [fib(n) for n in 1..]
```

$$[1, 1, 2, 3, 5, 8, 13, \ldots] \tag{19}$$

Type: Stream PositiveInteger

What is the 1000th Fibonacci number?

```
fibs 1000
```

$$\begin{array}{l}43466557686937456435688527675040625802564660517371780402481\\ 72908953655541794905189040387984007925516929592259308032263\\ 47752096896232398733224711616429964409065331879382989696499\\ 28516003704476137795166849228875\end{array} \tag{20}$$

Type: PositiveInteger

As an exercise, we suggest you write a function in an iterative style that computes the value of the recurrence relation $p(n) = p(n-1) - 2\,p(n-2) + 4\,p(n-3)$ having the initial values $p(1) = 1$, $p(2) = 3$ and $p(3) = 9$. How would you write the function using an element OneDimensionalArray or Vector to hold the previously computed values?

6.17 Anonymous Functions

An *anonymous function* is a function that is defined by giving a list of parameters, the "maps-to" compound symbol "+->" (from the mathematical symbol $\mapsto$), and by an expression involving the parameters, the evaluation of which determines the return value of the function.

(*parm*$_1$, *parm*$_2$, ..., *parm*$_N$) +-> *expression*

You can apply an anonymous function in several ways.

1. Place the anonymous function definition in parentheses directly followed by a list of arguments.
2. Assign the anonymous function to a variable and then use the variable name when you would normally use a function name.
3. Use "==" to use the anonymous function definition as the arguments and body of a regular function definition.
4. Have a named function contain a declared anonymous function and use the result returned by the named function.

6.17.1 Some Examples

Anonymous functions are particularly useful for defining functions "on the fly." That is, they are handy for simple functions that are used only in one place. In the following examples, we show how to write some simple anonymous functions.

This is a simple absolute value function.

```
x +-> if x < 0 then -x else x
```

$$x \mapsto \text{if } x < 0 \begin{array}{l} \text{then} \quad -x \\ \text{else} \quad x \end{array} \tag{1}$$

Type: AnonymousFunction

```
abs1 := %
```

$$x \mapsto \text{if } x < 0 \begin{array}{l} \text{then} \quad -x \\ \text{else} \quad x \end{array} \tag{2}$$

Type: AnonymousFunction

This function returns `true` if the absolute value of the first argument is greater than the absolute value of the second, `false` otherwise.

```
(x,y) +-> abs1(x) > abs1(y)
```

$$(x,\ y) \mapsto \text{abs1}\,(y) < \text{abs1}\,(x) \tag{3}$$

Type: AnonymousFunction

We use the above function to "sort" a list of integers.

```
sort(%,[3,9,-4,10,-3,-1,-9,5])
```

$$[10,\ -9,\ 9,\ 5,\ -4,\ -3,\ 3,\ -1] \tag{4}$$

Type: List Integer

This function returns `1` if `i + j` is even, `-1` otherwise.

```
ev := ( (i,j) +-> if even?(i+j) then 1 else -1)
```

$$(i,\ j) \mapsto \text{if } \text{even?}\,(i+j) \begin{array}{l} \text{then} \quad 1 \\ \text{else} \quad -1 \end{array} \tag{5}$$

Type: AnonymousFunction

We create a four-by-four matrix containing `1` or `-1` depending on whether the row plus the column index is even or not.

```
matrix([[ev(row,col) for row in 1..4] for col in 1..4])
```

$$\begin{bmatrix} 1 & -1 & 1 & -1 \\ -1 & 1 & -1 & 1 \\ 1 & -1 & 1 & -1 \\ -1 & 1 & -1 & 1 \end{bmatrix} \tag{6}$$

Type: Matrix Integer

This function returns `true` if a polynomial in `x` has multiple roots, `false` otherwise. It is defined and applied in the same expression.

```
( p +-> not one?(gcd(p,D(p,x))) )(x**2+4*x+4)
```

$$\text{true} \tag{7}$$

Type: Boolean

This and the next expression are equivalent.

```
g(x,y,z) == cos(x + sin(y + tan(z)))
```

Type: Void

The one you use is a matter of taste.

```
g == (x,y,z) +-> cos(x + sin(y + tan(z)))
```

Type: Void

6.17.2 Declaring Anonymous Functions

If you declare any of the arguments you must declare all of them. Thus,

```
(x: INT,y): FRAC INT +-> (x + 2*y)/(y - 1)
```

is not legal.

This is an example of a fully declared anonymous function. The output shown just indicates that the object you created is a particular kind of map, that is, function.

```
(x: INT,y: INT): FRAC INT +-> (x + 2*y)/(y - 1)
```

theMap (...) (1)

Type: ((Integer, Integer) → Fraction Integer)

AXIOM allows you to declare the arguments and not declare the return type.

```
(x: INT,y: INT) +-> (x + 2*y)/(y - 1)
```

theMap (...) (2)

Type: ((Integer, Integer) → Fraction Integer)

The return type is computed from the types of the arguments and the body of the function. You cannot declare the return type if you do not declare the arguments. Therefore,

```
(x,y): FRAC INT +-> (x + 2*y)/(y - 1)
```

is not legal.

This and the next expression are equivalent.

```
h(x: INT,y: INT): FRAC INT == (x + 2*y)/(y - 1)
```

```
Function declaration h : (Integer,Integer) -> Fraction

Integer has been added to workspace.
```

Type: Void

The one you use is a matter of taste.

```
h == (x: INT,y: INT): FRAC INT +-> (x + 2*y)/(y - 1)
```

```
Function declaration h : (Integer,Integer) -> Fraction

Integer has been added to workspace.
```

Type: Void

When should you declare an anonymous function?

1. If you use an anonymous function and AXIOM can't figure out what you are trying to do, declare the function.
2. If the function has nontrivial argument types or a nontrivial return type that AXIOM may be able to determine eventually, but you are not willing to wait that long, declare the function.
3. If the function will only be used for arguments of specific types and it is not too much trouble to declare the function, do so.
4. If you are using the anonymous function as an argument to another function (such as **map** or **sort**), consider declaring the function.
5. If you define an anonymous function inside a named function, you *must* declare the anonymous function.

This is an example of a named function for integers that returns a function.

```
addx x == ((y: Integer): Integer +-> x + y)
```

Type: Void

We define **g** to be a function that adds 10 to its argument.

```
g := addx 10
```

```
Compiling function addx with type PositiveInteger ->
(Integer -> Integer)
```

theMap (...) (6)

Type: (Integer → Integer)

Try it out.

```
g 3
```

13 (7)

Type: PositiveInteger

```
g(-4)
```

6 (8)

Type: PositiveInteger

An anonymous function cannot be recursive: since it does not have a name, you cannot even call it within itself! If you place an anonymous function inside a named function, the anonymous function must be declared.

6.18 Example: A Database

This example shows how you can use AXIOM to organize a database of lineage data and then query the database for relationships.

The database is entered as "assertions" that are really pieces of a function definition.

```
children("albert") == ["albertJr","richard","diane"]
```

Type: Void

Each piece `children(x) == y` means "the children of **x** are **y**".

```
children("richard") == ["douglas","daniel","susan"]
```

Type: Void

This family tree thus spans four generations.

```
children("douglas") == ["dougie","valerie"]
```

Type: Void

Say "no one else has children."

```
children(x) == []
```

Type: Void

We need some functions for computing lineage. Start with `childOf`.

```
childOf(x,y) == member?(x,children(y))
```

Type: Void

To find the `parentOf` someone, you have to scan the database of people applying `children`.

```
parentOf(x) ==
  for y in people repeat
    (if childOf(x,y) then return y)
  "unknown"
```

Type: Void

And a grandparent of **x** is just a parent of a parent of **x**.

```
grandParentOf(x) == parentOf parentOf x
```

Type: Void

The grandchildren of **x** are the people **y** such that **x** is a grandparent of **y**.

```
grandchildren(x) == [y for y in people | grandParentOf(y)
  = x]
```

Type: Void

Suppose you want to make a list of all great-grandparents. Well, a great-grandparent is a grandparent of a person who has children.

```
greatGrandParents == [x for x in people |
  reduce(_or,[not empty? children(y) for y in
grandchildren(x)],false)]
```

Type: Void

Define `descendants` to include the parent as well.

```
descendants(x) ==
  kids := children(x)
  null kids => [x]
  concat(x,reduce(concat,[descendants(y)
    for y in kids],[]))
```

Type: Void

Finally, we need a list of people. Since all people are descendants of "albert", let's say so.

```
people == descendants "albert"
```

Type: Void

We have used "==" to define the database and some functions to query the database. But no computation is done until we ask for some information. Then, once and for all, the functions are analyzed and compiled to machine code for run-time efficiency. Notice that no types are given anywhere in this example. They are not needed.

Who are the grandchildren of "richard"?

```
grandchildren "richard"
```

["dougie", "valerie"] (12)

Type: List String

Who are the great-grandparents?

```
greatGrandParents
```

["albert"] (13)

Type: List String

6.19 Example: A Famous Triangle

In this example we write some functions that display Pascal's triangle. It demonstrates the use of piece-wise definitions and some output operations you probably haven't seen before.

To make these output operations available, we have to *expose* the domain OutputForm. See Section 2.11 on page 87 for more information about exposing domains and packages.

```
)set expose add constructor OutputForm
```

```
OutputForm is now explicitly exposed in frame G1077
```

Define the values along the first row and any column `i`.

```
pascal(1,i) == 1
```

Type: Void

Define the values for when the row and column index `i` are equal. Repeating the argument name indicates that the two index values are equal.

```
pascal(n,n) == 1
```

Type: Void

```
pascal(i,j | 1 < i and i < j) ==
   pascal(i-1,j-1)+pascal(i,j-1)
```

Type: Void

Now that we have defined the coefficients in Pascal's triangle, let's write a couple of one-liners to display it.

First, define a function that gives the nth row.

```
pascalRow(n) == [pascal(i,n) for i in 1..n]
```

Type: Void

Next, we write the function **displayRow** to display the row, separating entries by blanks and centering.

```
displayRow(n) == output center blankSeparate pascalRow(n)
```

Type: Void

Here we have used three output operations. Operation **output** displays the printable form of objects on the screen, **center** centers a printable form in the width of the screen, and **blankSeparate** takes a list of printable forms and inserts a blank between successive elements.

Look at the result.

```
for i in 1..7 repeat displayRow i
```

```
                 1
               1  1
             1  2  1
           1  3  3  1
         1  4  6  4  1
       1  5  10  10  5  1
     1  6  15  20  15  6  1
```

Type: Void

Being purists, we find this less than satisfactory. Traditionally, elements of Pascal's triangle are centered between the left and right elements on the line above.

To fix this misalignment, we go back and redefine **pascalRow** to right adjust the entries within the triangle within a width of four characters.

```
pascalRow(n) == [right(pascal(i,n),4) for i in 1..n]
```

Type: Void

Finally let's look at our purely reformatted triangle.

```
for i in 1..7 repeat displayRow i
```

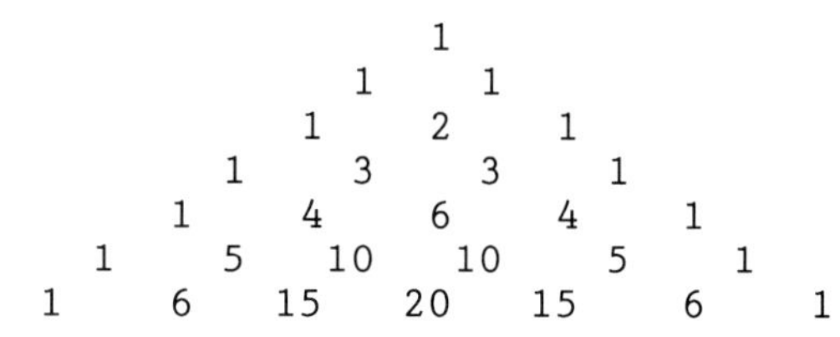

Type: Void

Unexpose OutputForm so we don't get unexpected results later.

```
)set expose drop constructor OutputForm
```

```
OutputForm is now explicitly hidden in frame G1077
```

6.20 Example: Testing for Palindromes

In this section we define a function **pal?** that tests whether its argument is a *palindrome*, that is, something that reads the same backwards and forwards. For example, the string "Madam I'm Adam" is a palindrome (excluding blanks and punctuation) and so is the number `123454321`. The definition works for any datatype that has `n` components that are accessed by the indices `1...n`.

Here is the definition for **pal?**. It is simply a call to an auxiliary function called **palAux?**. We are following the convention of ending a function's name with "?" if the function returns a Boolean value.

```
pal? s == palAux?(s,1,#s)
```

Type: Void

Here is **palAux?**. It works by comparing elements that are equidistant from the start and end of the object.

```
palAux?(s,i,j) ==
  j > i =>
    (s.i = s.j) and palAux?(s,i+1,i-1)
  true
```

Type: Void

Try **pal?** on some examples. First, a string.

```
pal? "Oxford"
```

false (3)

Type: Boolean

A list of polynomials.

```
pal? [4,a,x-1,0,x-1,a,4]
```

true (4)

Type: Boolean

A list of integers from the example in the last section.

```
pal? [1,6,15,20,15,6,1]
```

true (5)

Type: Boolean

To use **pal?** on an integer, first convert it to a string.

```
pal?(1441::String)
```

true (6)

Type: Boolean

Compute an infinite stream of decimal numbers, each of which is an obvious palindrome.

```
ones := [reduce(+,[10**j for j in 0..i]) for i in 1..]
```

[11, 111, 1111, 11111, 111111, 1111111, 11111111, ...] (7)

Type: Stream PositiveInteger

How about their squares?

```
squares := [x**2 for x in ones]
```

[121, 12321, 1234321, 123454321, 12345654321, 1234567654321,
123456787654321, 12345678987654321, 1234567900987654321, ...] (8)

Type: Stream PositiveInteger

Well, let's test them all!

```
[pal?(x::String) for x in squares]
```

[true, true, true, true, true, true, true, true, true, ...] (9)

Type: Stream Boolean

6.21 Rules and Pattern Matching

A common mathematical formula is

$$\log(x) + \log(y) = \log(xy) \quad \forall\, x \text{ and } y.$$

The presence of "∀" indicates that **x** and **y** can stand for arbitrary mathematical expressions in the above formula. You can use such mathematical formulas in AXIOM to specify "rewrite rules". Rewrite rules are objects in AXIOM that can be assigned to variables for later use, often for the purpose of simplification. Rewrite rules look like ordinary function definitions except that they are preceded by the reserved word `rule`. For example, a rewrite rule for the above formula is:

```
rule log(x) + log(y) == log(x * y)
```

Like function definitions, no action is taken when a rewrite rule is issued. Think of rewrite rules as functions that take one argument. When a rewrite rule `A = B` is applied to an argument `f`, its meaning is: "rewrite every subexpression of `f` that *matches* `A` by `B`." The left-hand side of a rewrite rule is called a *pattern*; its right-side side is called its *substitution*.

Create a rewrite rule named **logrule**. The generated symbol beginning with a "%" is a place-holder for any other terms that might occur in the sum.

```
logrule := rule log(x) + log(y) == log(x * y)
```

$$\log(y) + \log(x) + \%B == \log(x\ y) + \%B \tag{1}$$

Type: RewriteRule(Integer, Integer, Expression Integer)

Create an expression with logarithms.

```
f := log sin x + log x
```

$$\log(\sin(x)) + \log(x) \tag{2}$$

Type: Expression Integer

Apply **logrule** to `f`.

```
logrule f
```

$$\log(x \sin(x)) \tag{3}$$

Type: Expression Integer

The meaning of our example rewrite rule is: "for all expressions **x** and `y`, rewrite `log(x) + log(y)` by `log(x * y)`." Patterns generally have both operation names (here, **log** and "+") and variables (here, **x** and `y`). By default, every operation name stands for itself. Thus **log** matches only "`log`" and not any other operation such as **sin**. On the other hand, variables do not stand for themselves. Rather, a variable denotes a *pattern variable* that is free to match any expression whatsoever.

When a rewrite rule is applied, a process called *pattern matching* goes to work by systematically scanning the subexpressions of the argument. When a subexpression is found that "matches" the pattern, the subexpression is replaced by the right-hand side of the rule. The details of what happens will be covered later.

The customary AXIOM notation for patterns is actually a shorthand for a longer, more general notation. Pattern variables can be made explicit by using a percent ("%") as the first character of the variable name. To say that a name stands for itself, you can prefix that name with a quote operator ("'"). Although the current AXIOM parser does not let you quote an operation name, this more general notation gives you an alternate way of giving the same rewrite rule:

```
rule log(%x) + log(%y) == log(x * y)
```

This longer notation gives you patterns that the standard notation won't handle. For example, the rule

```
rule %f(c * 'x) ==  c*%f(x)
```

means "for all `f` and `c`, replace `f(y)` by `c * f(x)` when `y` is the product of `c` and the explicit variable `x`."

Thus the pattern can have several adornments on the names that appear there. Normally, all these adornments are dropped in the substitution on the right-hand side.

To summarize:

> To enter a single rule in AXIOM, use the following syntax:
>
> `rule` *leftHandSide* `==` *rightHandSide*
>
> The *leftHandSide* is a pattern to be matched and the *rightHandSide* is its substitution. The rule is an object of type RewriteRule that can be assigned to a variable and applied to expressions to transform them.

Rewrite rules can be collected into rulesets so that a set of rules can be applied at once. Here is another simplification rule for logarithms.

$$y \log(x) = \log(x^y) \quad \forall\, x \text{ and } y.$$

If instead of giving a single rule following the reserved word `rule` you give a "pile" of rules, you create what is called a *ruleset*. Like rules, rulesets are objects in AXIOM and can be assigned to variables. You will find it useful to group commonly used rules into input files, and read them in as needed.

Create a ruleset named `logrules`.

```
logrules := rule
  log(x) + log(y)  == log(x * y)
  y * log x        == log(x ** y)
```

$$\{\log(y) + \log(x) + \%C == \log(x\ y) + \%C,\ y\ \log(x) == \log\left(x^y\right)\} \qquad (4)$$

Type: Ruleset(Integer, Integer, Expression Integer)

Again, create an expression f containing logarithms.

```
f := a * log(sin x) - 2 * log x
```

$$a \log(\sin(x)) - 2 \log(x) \tag{5}$$

Type: Expression Integer

Apply the ruleset **logrules** to f.

```
logrules f
```

$$\log\left(\frac{\sin(x)^a}{x^2}\right) \tag{6}$$

Type: Expression Integer

We have allowed pattern variables to match arbitrary expressions in the above examples. Often you want a variable only to match expressions satisfying some predicate. For example, we may want to apply the transformation

$$y \log(x) = \log(x^y)$$

only when y is an integer. The way to restrict a pattern variable y by a predicate f(y) is by using a vertical bar "|", which means "such that," in much the same way it is used in function definitions. You do this only once, but at the earliest (meaning deepest and leftmost) part of the pattern.

This restricts the logarithmic rule to create integer exponents only.

```
logrules2 := rule
  log(x) + log(y)              == log(x * y)
  (y | integer? y) * log x == log(x ** y)
```

$$\{\log(y) + \log(x) + \%E == \log(x\ y) + \%E,\ y \log(x) == \log\left(x^y\right)\} \tag{7}$$

Type: Ruleset(Integer, Integer, Expression Integer)

Compare this with the result of applying the previous set of rules.

```
f
```

$$a \log(\sin(x)) - 2 \log(x) \tag{8}$$

Type: Expression Integer

```
logrules2 f
```

$$a \log(\sin(x)) + \log\left(\frac{1}{x^2}\right) \tag{9}$$

Type: Expression Integer

This is an example of some of the usual identities involving products of sines and cosines.

```
sinCosProducts == rule
  sin(x) * sin(y) == (cos(x-y) - cos(x + y))/2
  cos(x) * cos(y) == (cos(x-y) + cos(x+y))/2
  sin(x) * cos(y) == (sin(x-y) + sin(x + y))/2
```

Type: Void

```
g := sin(a)*sin(b) + cos(b)*cos(a) + sin(2*a)*cos(2*a)
```

$$\sin(a) \sin(b) + \cos(2\ a) \sin(2\ a) + \cos(a) \cos(b) \tag{11}$$

Type: Expression Integer

```
sinCosProducts g
```

```
Compiling body of rule sinCosProducts to compute value
of type Ruleset(Integer,Integer,Expression Integer)
```

$$\frac{\sin(4\ a) + 2\ \cos(-b + a)}{2} \tag{12}$$

Type: Expression Integer

Another qualification you will often want to use is to allow a pattern to match an identity element. Using the pattern `x + y`, for example, neither `x` nor `y` matches the expression `0`. Similarly, if a pattern contains a product `x*y` or an exponentiation `x**y`, then neither `x` or `y` matches `1`.

If identical elements were matched, pattern matching would generally loop. Here is an expansion rule for exponentials.

```
exprule := rule exp(a + b) == exp(a) * exp(b)
```

$$\%e^{(b+a)} == \%e^{a}\ \%e^{b} \tag{13}$$

Type: RewriteRule(Integer, Integer, Expression Integer)

This rule would cause infinite rewriting on this if either `a` or `b` were allowed to match `0`.

```
exprule exp x
```

$$\%e^{x} \tag{14}$$

Type: Expression Integer

There are occasions when you do want a pattern variable in a sum or product to match `0` or `1`. If so, prefix its name with a "`?`" whenever it appears in a left-hand side of a rule. For example, consider the following rule for the exponential integral:

$$\int \left(\frac{y + e^{x}}{x}\right)\ dx = \int \frac{y}{x}\ dx + \mathrm{Ei}(x) \quad \forall\, x \text{ and } y.$$

This rule is valid for `y = 0`. One solution is to create a Ruleset with two rules, one with and one without `y`. A better solution is to use an "optional" pattern variable.

Define rule `eirule` with a pattern variable `?y` to indicate that an expression may or may not occur.

```
eirule := rule integral((?y + exp x)/x,x) ==
  integral(y/x,x) + Ei x
```

$$\int^{x} \frac{\%e^{\%J} + y}{\%J}\ d\%J == {'integral}\left(\frac{y}{x},\ x\right) + {'Ei}(x) \tag{15}$$

Type: RewriteRule(Integer, Integer, Expression Integer)

Apply rule `eirule` to an integral without this term.

```
eirule integral(exp u/u, u)
```

$$Ei(u) \tag{16}$$

Type: Expression Integer

Apply rule `eirule` to an integral with this term.

```
eirule integral(sin u + exp u/u, u)
```

$$\int^{u} \sin(\%J)\ d\%J + Ei(u) \tag{17}$$

Type: Expression Integer

Here is one final adornment you will find useful. When matching a pattern of the form `x + y` to an expression containing a long sum of the form `a +...+ b`, there is no way to predict in advance which subset of the sum matches `x` and which matches `y`. Aside from efficiency, this is generally unimportant since the rule holds for any possible combination of matches for `x` and `y`. In some situations, however, you may want to say which pattern variable is a sum (or product) of several terms, and which should match only a single term. To do this, put a prefix colon "`:`" before the pattern variable that you want to match multiple terms.

The remaining rules involve operators `u` and `v`.

```
u := operator 'u
```

$$u \tag{18}$$

Type: BasicOperator

These definitions tell AXIOM that `u` and `v` are formal operators to be used in expressions.

```
v := operator 'v
```

$$v \tag{19}$$

Type: BasicOperator

First define `myRule` with no restrictions on the pattern variables `x` and `y`.

```
myRule := rule u(x + y) == u x + v y
```

$$u(y+x) == {'}v(y) + {'}u(x) \tag{20}$$

Type: RewriteRule(Integer, Integer, Expression Integer)

Apply `myRule` to an expression.

```
myRule u(a + b + c + d)
```

$$v(d+c+b)+u(a) \tag{21}$$

Type: Expression Integer

Define `myOtherRule` to match several terms so that the rule gets applied recursively.

```
myOtherRule := rule u(:x + y) == u x + v y
```

$$u(y+x) == {'}v(y) + {'}u(x) \tag{22}$$

Type: RewriteRule(Integer, Integer, Expression Integer)

Apply `myOtherRule` to the same expression.

```
myOtherRule u(a + b + c + d)
```

$$v(c)+v(b)+v(a)+u(d) \tag{23}$$

Type: Expression Integer

Summary of pattern variable adornments:

`(x | predicate?(x))`	means that the substutution `s` for `x` must satisfy `predicate?(s) = true.`
`?x`	means that `x` can match an identity element (0 or 1).
`:x`	means that `x` can match several terms in a sum.

Here are some final remarks on pattern matching. Pattern matching provides a very useful paradigm for solving certain classes of problems, namely, those that involve transformations of one form to another and back. However, it is important to recognize its limitations.

First, pattern matching slows down as the number of rules you have to apply increases. Thus it is good practice to organize the sets of rules you use optimally so that irrelevant rules are never included.

Second, careless use of pattern matching can lead to wrong answers. You should avoid using pattern matching to handle hidden algebraic relationships that can go undetected by other programs. As a simple example, a symbol such as "J" can easily be used to represent the square root of `-1` or some other important algebraic quantity. Many algorithms branch on whether an expression is zero or not, then divide by that expression if it is not. If you fail to simplify an expression involving powers of `J` to `-1`, algorithms may incorrectly assume an expression is non-zero, take a wrong branch, and produce a meaningless result.

Pattern matching should also not be used as a substitute for a domain. In AXIOM, objects of one domain are transformed to objects of other domains using well-defined **coerce** operations. Pattern matching should be used on objects that are all the same type. Thus if your application can be handled by type Expression in AXIOM and you think you need pattern matching, consider this choice carefully. You may well be better served by extending an existing domain or by building a new domain of objects for your application.

CHAPTER 7

Graphics

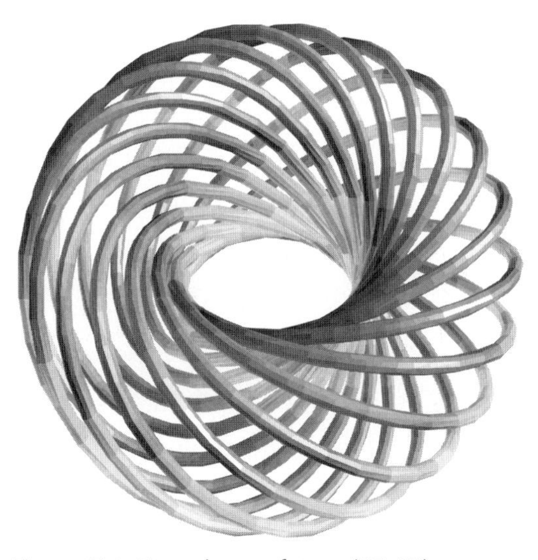

Figure 7.1: Torus knot of type (15,17).

This chapter shows how to use the AXIOM graphics facilities under the X Window System. AXIOM has two-dimensional and three-dimensional drawing and rendering packages that allow the drawing, coloring, transforming, mapping, clipping, and combining of graphic output from AXIOM computations. This facility is particularly useful for investigating problems in areas such as topology. The graphics package is capable of plotting functions of one or more variables or plotting parametric surfaces and curves. Various coordinate systems are also available, such as polar and spherical.

A graph is displayed in a viewport window and it has a control-panel that uses interactive mouse commands. PostScript and other output forms are available so that AXIOM images can be printed or used by other programs.[1]

7.1 Two-Dimensional Graphics

The AXIOM two-dimensional graphics package provides the ability to display

- curves defined by functions of a single real variable
- curves defined by parametric equations
- implicit non-singular curves defined by polynomial equations
- planar graphs generated from lists of point components.

These graphs can be modified by specifying various options, such as calculating points in the polar coordinate system or changing the size of the graph viewport window.

7.1.1 Plotting Two-Dimensional Functions of One Variable

The first kind of two-dimensional graph is that of a curve defined by a function `y = f(x)` over a finite interval of the `x` axis.

The general format for drawing a function defined by a formula `f(x)` is:

`draw(f(x), x = a..b,` *options*`)`

where `a..b` defines the range of `x`, and where *options* prescribes zero or more options as described in Section 7.1.4 on page 185. An example of an option is `curveColor == bright red()`. An alternative format involving functions `f` and `g` is also available.

A simple way to plot a function is to use a formula. The first argument is the formula. For the second argument, write the name of the independent variable (here, `x`), followed by an "=", and the range of values.

[1]PostScript is a trademark of Adobe Systems Incorporated, registered in the United States.

Display this formula over the range $0 \leq x \leq 6$. AXIOM converts your formula to a compiled function so that the results can be computed quickly and efficiently.

```
draw(sin(tan(x)) - tan(sin(x)),x = 0..6)
```

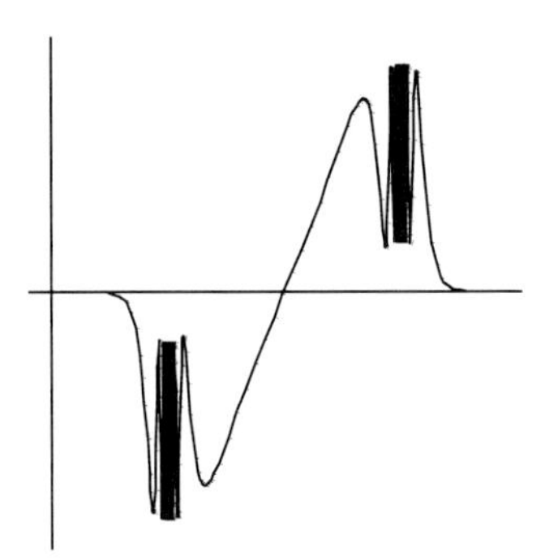

Notice that AXIOM compiled the function before the graph was put on the screen.

Here is the same graph on a different interval. This time we give the graph a title.

```
draw(sin(tan(x)) - tan(sin(x)),x = 10..16)
```

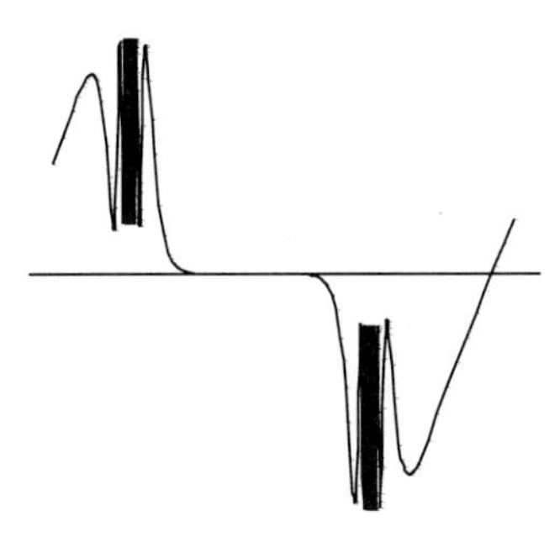

Once again the formula is converted to a compiled function before any points were computed. If you want to graph the same function on several intervals, it is a good idea to define the function first so that the function has to be compiled only once.

This time we first define the function.

```
f(x) == (x-1)*(x-2)*(x-3)
```

Type: Void

To draw the function, the first argument is its name and the second is just the range with no independent variable.

```
draw(f, 0..4)
```

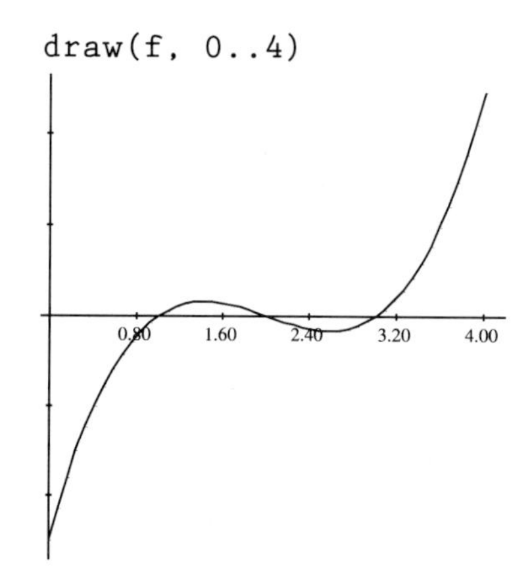

7.1.2 Plotting Two-Dimensional Parametric Plane Curves

The second kind of two-dimensional graph is that of curves produced by parametric equations. Let `x = f(t)` and `y = g(t)` be formulas or two functions `f` and `g` as the parameter `t` ranges over an interval `[a,b]`. The function **curve** takes the two functions `f` and `g` as its parameters.

> The general format for drawing a two-dimensional plane curve defined by parametric formulas `x = f(t)` and `y = g(t)` is:
>
> `draw(curve(f(t), g(t)), t = a..b,` *options*`)`
>
> where `a..b` defines the range of the independent variable `t`, and where *options* prescribes zero or more options as described in Section 7.2.4 on page 200. An example of an option is `curveColor == bright red()`.

Here's an example:

Define a parametric curve using a range involving %pi, AXIOM's way of saying π. For parametric curves, AXIOM compiles two functions, one for each of the functions f and g.

```
draw(curve(sin(t)*sin(2*t)*sin(3*t),
  sin(4*t)*sin(5*t)*sin(6*t)), t = 0..2*%pi)
```

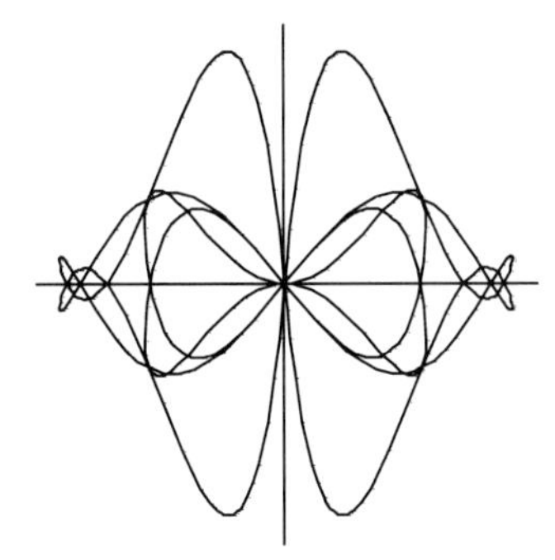

The title may be an arbitrary string and is an optional argument to the **draw** command.

```
draw(curve(cos(t), sin(t)), t = 0..2*%pi)
```

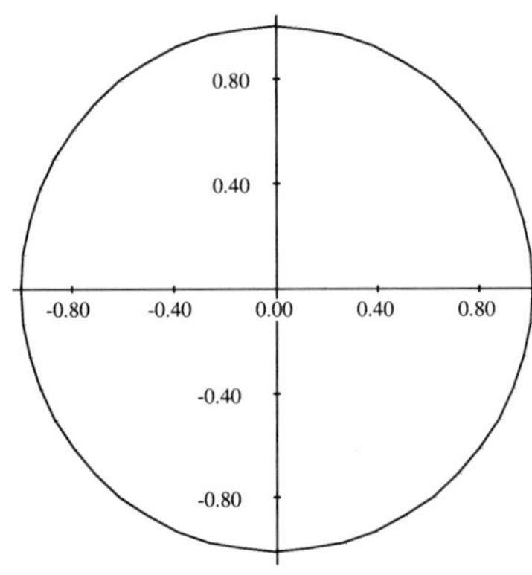

If you plan on plotting x = f(t), y = g(t) as t ranges over several intervals, you may want to define functions f and g first, so that they need not be recompiled every time you create a new graph. Here's an example:

As before, you can first define the functions you wish to draw.

```
f(t:SF):SF == sin(3*t/4)
```

Type: Void

AXIOM compiles them to map SmallFloat values to SmallFloat values.

```
g(t:SF):SF == sin(t)
```

Type: Void

Give to `curve` the names of the functions, then write the range without the name of the independent variable.

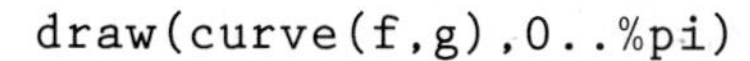

```
draw(curve(f,g),0..%pi)
```

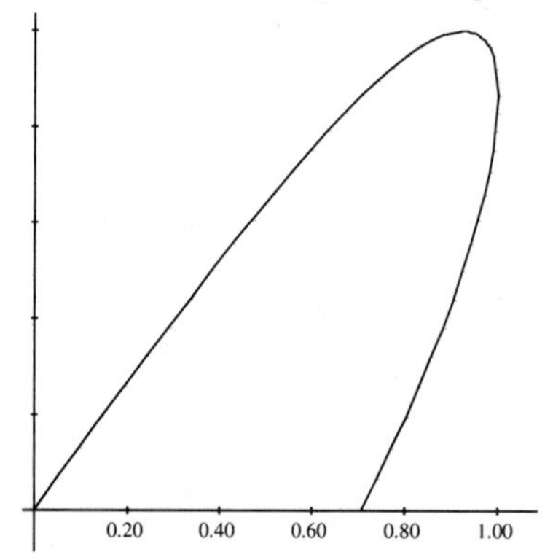

Here is another look at the same curve but over a different range. Notice that `f` and `g` are not recompiled. Also note that AXIOM provides a default title based on the first function specified in **curve**.

```
draw(curve(f,g),-4*%pi..4*%pi)
```

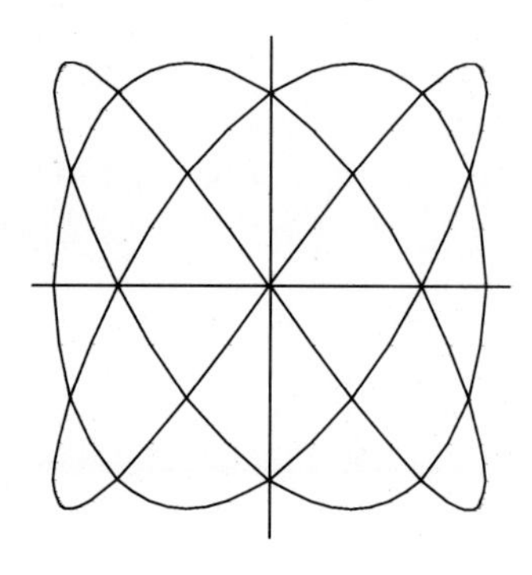

7.1.3 Plotting Plane Algebraic Curves

A third kind of two-dimensional graph is a non-singular "solution curve" in a rectangular region of the plane. A solution curve is a curve defined by a polynomial equation `p(x,y) = 0`. Non-singular means that the curve is "smooth" in that it does not cross itself or come to a point (cusp). Algebraically, this means that for any point `(x,y)` on the curve, that is, a point such that `p(x,y) = 0`, the partial derivatives $\frac{\partial p}{\partial x}(x, y)$ and $\frac{\partial p}{\partial y}(x, y)$ are not both zero.

The general format for drawing a non-singular solution curve given by a polynomial of the form `p(x,y) = 0` is:

```
draw(p(x,y) = 0, x, y, range == [a..b, c..d],
                options)
```

where the second and third arguments name the first and second independent variables of `p`. A `range` option is always given to designate a bounding rectangular region of the plane $a \le x \le b, c \le y \le d$. Zero or more additional options as described in Section 7.1.4 on page 185 may be given.

We require that the polynomial has rational or integral coefficients. Here is an algebraic curve example ("Cartesian ovals"):

```
p := ((x**2 + y**2 + 1) - 8*x)**2 - (8*(x**2 + y**2 + 1)-
  4*x-1)
```

$$y^4 + \left(2\,x^2 - 16\,x - 6\right)\,y^2 + x^4 - 16\,x^3 + 58\,x^2 - 12\,x - 6 \qquad (1)$$

Type: Polynomial Integer

The first argument is always expressed as an equation of the form `p = 0` where `p` is a polynomial.

```
draw(p = 0, x, y, range == [-1..11, -7..7])
```

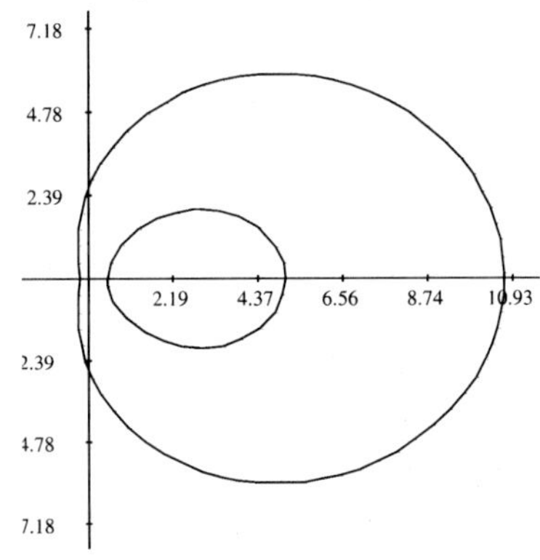

7.1.4 Two-Dimensional Options

The **draw** commands take an optional list of options, such as `title` shown above. Each option is given by the syntax: *name* == *value*. Here is a list of the available options in the order that they are described below.

adaptive	clip	unit
clip	curveColor	range
toScale	pointColor	coordinates

The `adaptive` option turns adaptive plotting on or off. Adaptive plotting uses an algorithm that traverses a graph and computes more points for those parts of the graph with high curvature. The higher the curvature of a region is, the more points the algorithm computes.

The adaptive option is normally on. Here we turn it off.

```
draw(sin(1/x),x=-2*%pi..2*%pi, adaptive == false)
```

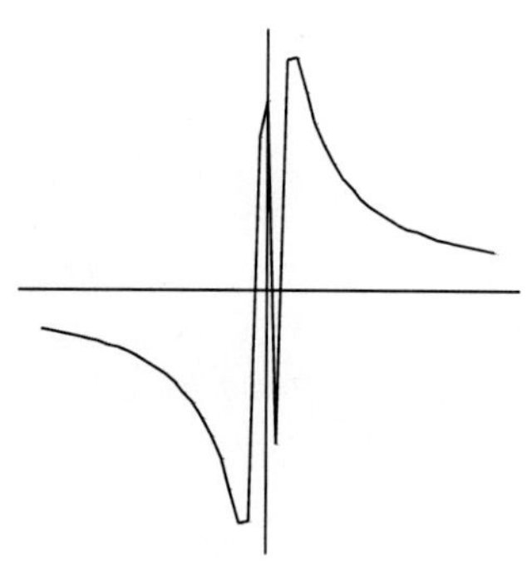

The clip option turns clipping on or off. If on, large values are cut off according to **clipPointsDefault**.

```
draw(tan(x),x=-2*%pi..2*%pi, clip == true)
```

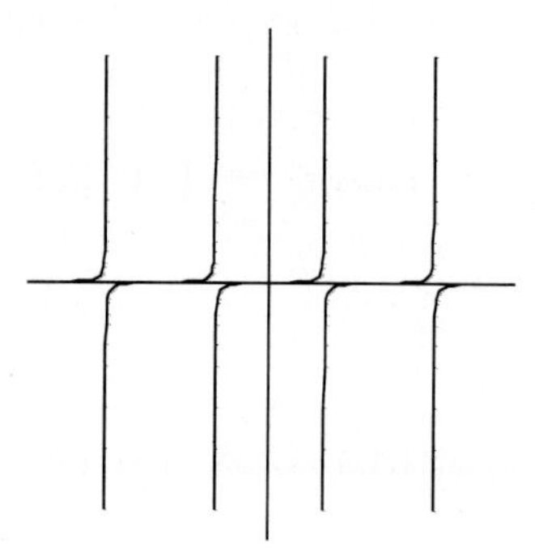

Option toScale does plotting to scale if true or uses the entire viewport if false. The default can be determined using **drawToScale**.

```
draw(sin(x),x=-%pi..%pi, toScale == true, unit ==
  [1.0,1.0])
```

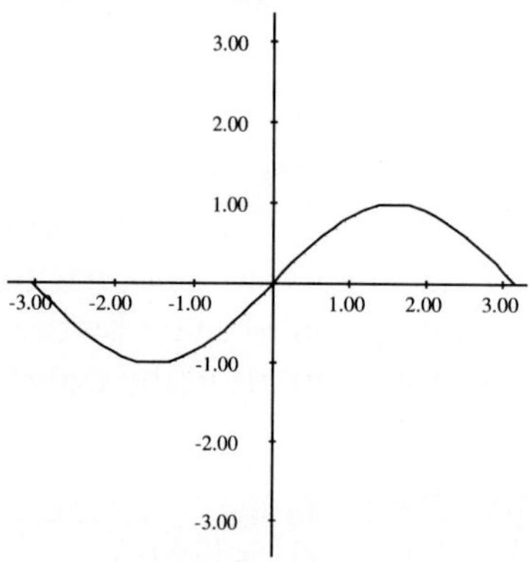

Option `clip` with a range sets point clipping of a graph within the ranges specified in the list `[x range,y range]`. If only one range is specified, clipping applies to the y-axis.

```
draw(sec(x),x=-2*%pi..2*%pi, clip == [-2*%pi..2*%pi,-
  %pi..%pi], unit == [1.0,1.0])
```

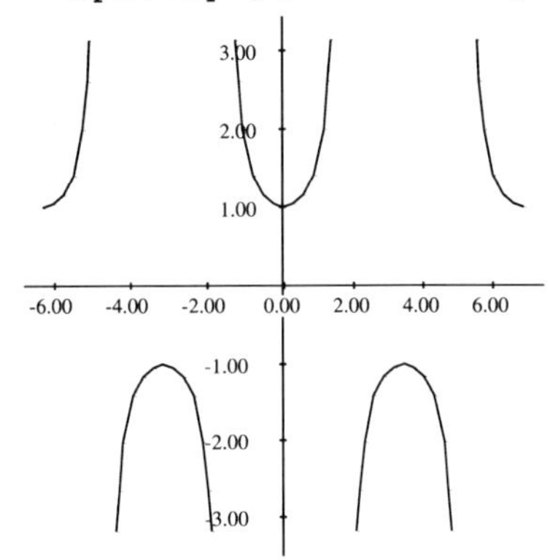

Option `curveColor` sets the color of the graph curves or lines to be the indicated palette color (see Section 7.1.5 on page 189 and Section 7.1.6 on page 190).

```
draw(sin(x),x=-%pi..%pi, curveColor == bright red())
```

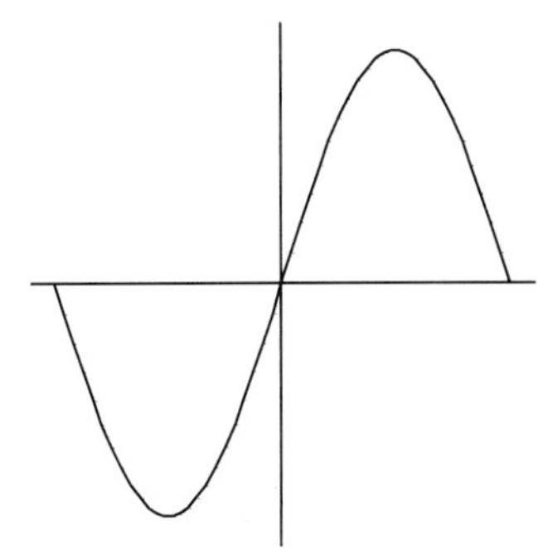

Option `pointColor` sets the color of the graph points to the indicated palette color (see Section 7.1.5 on page 189 and Section 7.1.6 on page 190).

```
draw(sin(x),x=-%pi..%pi, pointColor == pastel yellow())
```

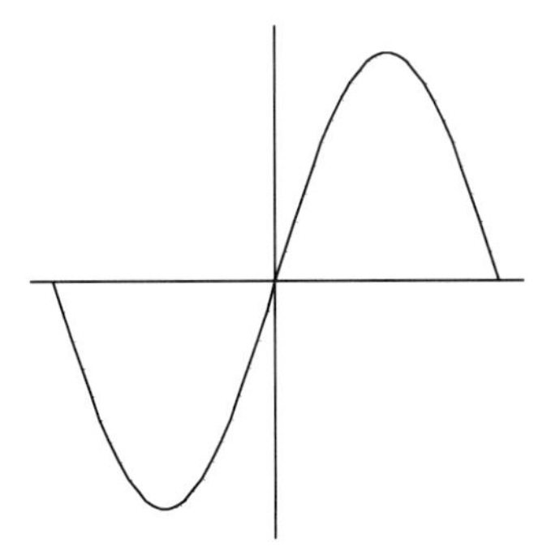

Option `unit` sets the intervals at which the axis units are plotted according to the indicated steps [x interval, y interval].

```
draw(curve(9*sin(3*t/4),8*sin(t)), t = -4*%pi..4*%pi, unit
  == [2.0,1.0])
```

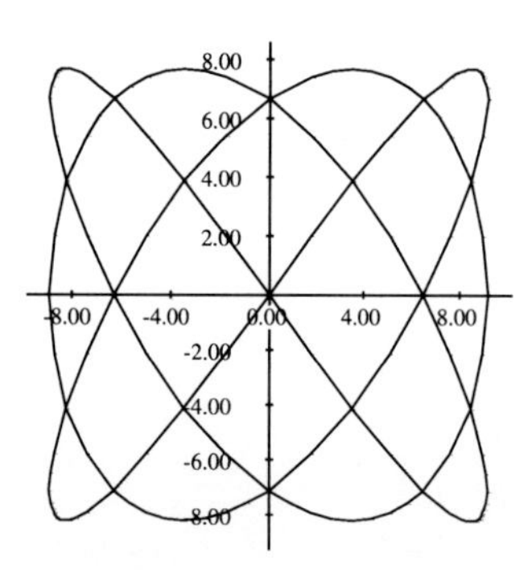

Option `range` sets the range of variables in a graph to be within the ranges for solving plane algebraic curve plots.

```
draw(y**2 + y - (x**3 - x) = 0, x, y, range == [-2..2,-
  2..1], unit==[1.0,1.0])
```

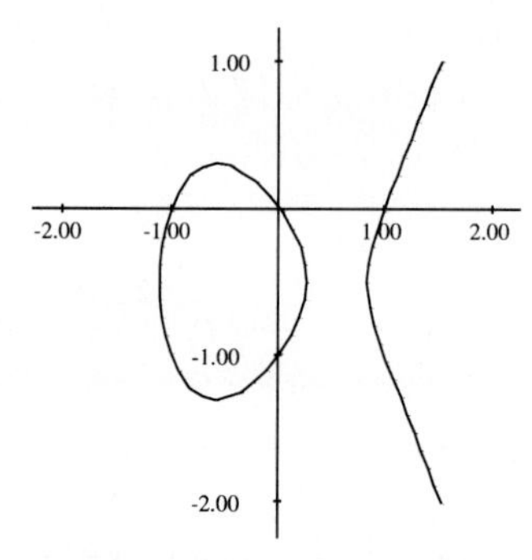

A second example of a solution plot.

```
draw(x**2 + y**2 = 1, x, y, range == [-3/2..3/2,-
  3/2..3/2], unit==[0.5,0.5])
```

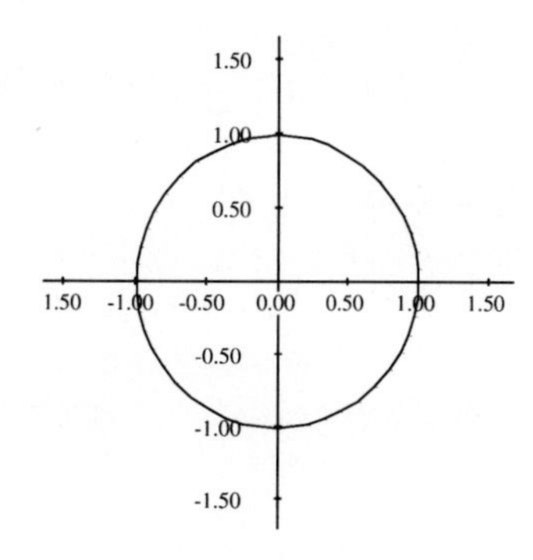

Option `coordinates` indicates the coordinate system in which the graph is plotted. The default is to use the Cartesian coordinate system. For more details, see Section 7.2.7 on page 210 .

```
draw(curve(sin(5*t),t),t=0..2*%pi, coordinates == polar)
```

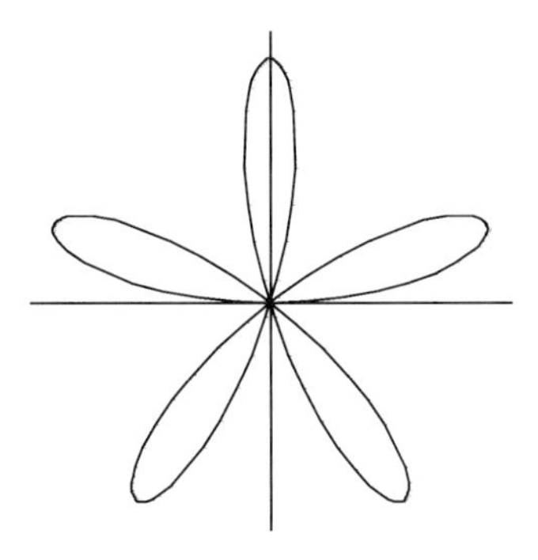

7.1.5 Color

The domain Color provides operations for manipulating colors in two-dimensional graphs. Colors are objects of Color. Each color has a *hue* and a *weight*. Hues are represented by integers that range from `1` to the **numberOfHues()**, normally `27`. Weights are floats and have the value `1.0` by default.

color (*integer*)
  creates a color of hue *integer* and weight `1.0`.

hue (*color*)
  returns the hue of *color* as an integer.

red ()
  , **blue**(), **green**(), and **yellow**() create colors of that hue with weight `1.0`.

color$_1$ + *color*$_2$ returns the color that results from additively combining the indicated *color*$_1$ and *color*$_2$. Color addition is not commutative: changing the order of the arguments produces different results.

integer * *color* changes the weight of *color* by *integer* without affecting its hue. For example, `red() + 3*yellow()` produces a color closer to yellow than to red. Color multiplication is not associative: changing the order of grouping produces different results.

These functions can be used to change the point and curve colors for two- and three-dimensional graphs. Use the `pointColor` option for points.

```
draw(x**2,x=-1..1,pointColor == green())
```

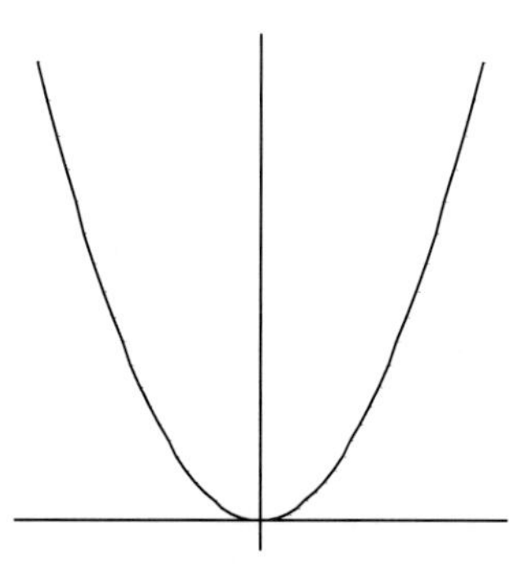

Use the `curveColor` option for curves.

```
draw(x**2,x=-1..1,curveColor == color(13) + 2*blue())
```

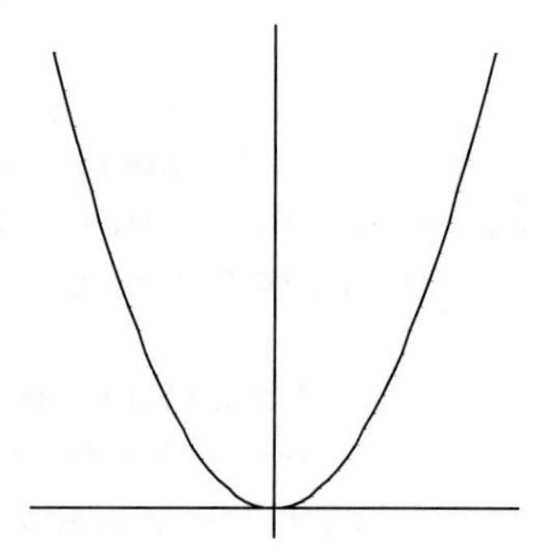

7.1.6 Palette

Domain Palette is the domain of shades of colors: **dark**, **dim**, **bright**, **pastel**, and **light**, designated by the integers 1 through 5, respectively.

Colors are normally "bright."

```
shade red()
```

3 (1)

Type: PositiveInteger

To change the shade of a color, apply the name of a shade to it.

```
myFavoriteColor := dark blue()
```

[Hue: 22, Weight: 1.0] from the Dark palette (2)

Type: Palette

The expression `shade(color)` returns the value of a shade of `color`.

```
shade myFavoriteColor
```

1 (3)

Type: PositiveInteger

The expression hue(color) returns its hue.

```
hue myFavoriteColor
```

Hue: 22, Weight: 1.0 (4)

Type: Color

Palettes can be used in specifying colors in two-dimensional graphs.

```
draw(x**2,x=-1..1,curveColor == dark blue())
```

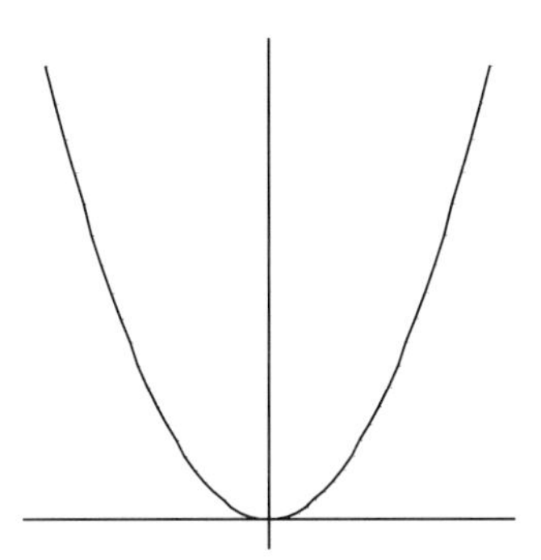

7.1.7 Two-Dimensional Control-Panel

Once you have created a viewport, move your mouse to the viewport and click with your left mouse button to display a control-panel. The panel is displayed on the side of the viewport closest to where you clicked. Each of the buttons which toggle on and off show the current state of the graph.

Transformations

Object transformations are executed from the control-panel by mouse-activated potentiometer windows.

Scale: To scale a graph, click on a mouse button within the **Scale** window in the upper left corner of the control-panel. The axes along which the scaling is to occur are indicated by setting the toggles above the arrow. With `X On` and `Y On` appearing, both axes are selected and scaling is uniform. If either is not selected, for example, if `X Off` appears, scaling is non-uniform.

Translate: To translate a graph, click the mouse in the **Translate** window in the direction you wish the graph to move. This window is located in the upper right corner of the control-panel. Along the top of the **Translate** window are two buttons for selecting the direction of translation. Translation along both coordinate axes results when `X On` and `Y On` appear or along one axis when one is on, for example, `X On` and `Y Off` appear.

Messages

The window directly below the transformation potentiometer windows is used to display system messages relating to the viewport and the control-panel. The following format is displayed:

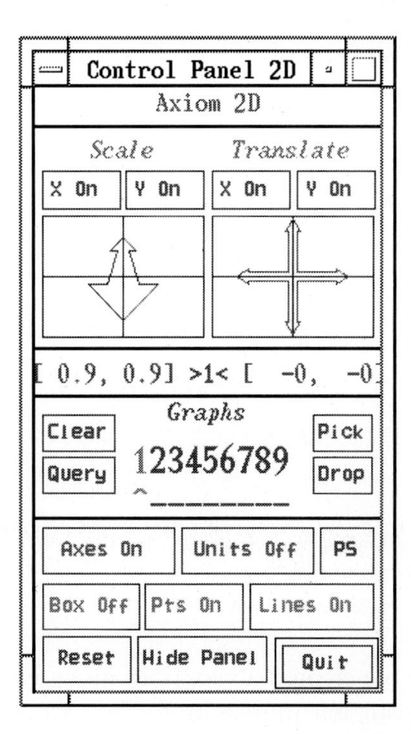

Figure 7.2: Two-dimensional control-panel.

[scaleX, scaleY] >graph< [translateX, translateY]

The two values to the left show the scale factor along the X and Y coordinate axes. The two values to the right show the distance of translation from the center in the X and Y directions. The number in the center shows which graph in the viewport this data pertains to. When multiple graphs exist in the same viewport, the graph must be selected (see "Multiple Graphs," below) in order for its transformation data to be shown, otherwise the number is 1.

Multiple Graphs

The **Graphs** window contains buttons that allow the placement of two-dimensional graphs into one of nine available slots in any other two-dimensional viewport. In the center of the window are numeral buttons from one to nine that show whether a graph is displayed in the viewport. Below each number button is a button showing whether a graph that is present is selected for application of some transformation. When the caret symbol is displayed, then the graph in that slot will be manipulated. Initially, the graph for which the viewport is created occupies the first slot, is displayed, and is selected.

Clear: The **Clear** button deselects every viewport graph slot. A graph slot is reselected by selecting the button below its number.

Query: The **Query** button is used to display the scale and translate data for the indicated graph. When this button is selected the message "Click on the graph to query" appears. Select a slot number button from the **Graphs** window. The scaling factor and translation offset of the graph are then displayed in the message window.

Pick: The **Pick** button is used to select a graph to be placed or dropped into the indicated viewport. When this button is selected, the message "Click on the graph to pick" appears. Click on the slot with the graph number of the desired graph. The graph information is held waiting for you to execute a **Drop** in some other graph.

Drop: Once a graph has been picked up using the **Pick** button, the **Drop** button places it into a new viewport slot. The message "Click on the graph to drop" appears in the message window when the **Drop** button is selected. By selecting one of the slot number buttons in the **Graphs** window, the graph currently being held is dropped into this slot and displayed.

Buttons

Axes turns the coordinate axes on or off.

Units turns the units along the **x** and **y** axis on or off.

Box encloses the area of the viewport graph in a bounding box, or removes the box if already enclosed.

Pts turns on or off the display of points.

Lines turns on or off the display of lines connecting points.

PS writes the current viewport contents to a file **axiom2D.ps** or to a name specified in the user's **.Xdefaults** file. The file is placed in the directory from which AXIOM or the **viewAlone** program was invoked.

Reset resets the object transformation characteristics and attributes back to their initial states.

Hide makes the control-panel disappear.

Quit queries whether the current viewport session should be terminated.

7.1.8 Operations for Two-Dimensional Graphics

Here is a summary of useful AXIOM operations for two-dimensional graphics. Each operation name is followed by a list of arguments. Each argument is written as a variable informally named according to the type of the argument (for example, *integer*). If appropriate, a default value for an argument is given in parentheses immediately following the name.

adaptive ([*boolean*(true)])
sets or indicates whether graphs are plotted according to the adaptive refinement algorithm.

axesColorDefault ([*color*(dark blue())])
sets or indicates the default color of the axes in a two-dimensional graph viewport.

clipPointsDefault ([*boolean*(false)])
sets or indicates whether point clipping is to be applied as the default for graph plots.

drawToScale ([*boolean*(false)])
sets or indicates whether the plot of a graph is "to scale" or uses the entire viewport space as the default.

lineColorDefault ([*color*(pastel yellow())])
sets or indicates the default color of the lines or curves in a two-dimensional graph viewport.

maxPoints ([*integer*(500)])
sets or indicates the default maximum number of possible points to be used when constructing a two-dimensional graph.

minPoints ([*integer*(21)])
sets or indicates the default minimum number of possible points to be used when constructing a two-dimensional graph.

pointColorDefault ([*color*(bright red())])
sets or indicates the default color of the points in a two-dimensional graph viewport.

pointSizeDefault ([*integer*(5)])
sets or indicates the default size of the dot used to plot points in a two-dimensional graph.

screenResolution ([*integer*(600)])
sets or indicates the default screen resolution constant used in setting the computation limit of adaptively generated curve plots.

unitsColorDefault ([*color*(dim green())])
sets or indicates the default color of the unit labels in a two-dimensional graph viewport.

viewDefaults ()
resets the default settings for the following attributes: point color, line color, axes color, units color, point size, viewport upper left-hand corner position, and the viewport size.

viewPosDefault ([*list*([100,100])])
sets or indicates the default position of the upper left-hand corner of a two-dimensional viewport, relative to the display root window. The upper left-hand corner of the display is considered to be at the

(0, 0) position.

viewSizeDefault ([*list*([200,200])])
sets or indicates the default size in which two dimensional viewport windows are shown. It is defined by a width and then a height.

viewWriteAvailable ([*list*(["pixmap", "bitmap", "postscript", "image"])])
indicates the possible file types that can be created with the **write** function.

viewWriteDefault ([*list*([])])
sets or indicates the default types of files, in addition to the **data** file, that are created when a **write** function is executed on a viewport.

units (*viewport, integer*(1), *string*("off"))
turns the units on or off for the graph with index *integer*.

axes (*viewport, integer*(1), *string*("on"))
turns the axes on or off for the graph with index *integer*.

close (*viewport*)
closes *viewport*.

connect (*viewport, integer*(1), *string*("on"))
declares whether lines connecting the points are displayed or not.

controlPanel (*viewport, string*("off"))
declares whether the two-dimensional control-panel is automatically displayed or not.

graphs (*viewport*)
returns a list describing the state of each graph. If the graph state is not being used this is shown by "undefined", otherwise a description of the graph's contents is shown.

graphStates (*viewport*)
displays a list of all the graph states available for *viewport*, giving the values for every property.

key (*viewport*)
returns the process ID number for *viewport*.

move (*viewport, $integer_x$(viewPosDefault), $integer_y$(viewPosDefault)*)
moves *viewport* on the screen so that the upper left-hand corner of *viewport* is at the position *(x,y)*.

options (*viewport*)
returns a list of all the DrawOptions used by *viewport*.

points (*viewport, integer*(1), *string*("on"))
specifies whether the graph points for graph *integer* are to be displayed or not.

region (*viewport, integer*(1), *string*("off"))
declares whether graph *integer* is or is not to be displayed with a

bounding rectangle.

reset (*viewport*)
: resets all the properties of *viewport*.

resize (*viewport*, $integer_{width}$, $integer_{height}$)
: resizes *viewport* with a new *width* and *height*.

scale (*viewport*, $integer_n$`(1)`, $integer_x$`(0.9)`, $integer_y$`(0.9)`)
: scales values for the x and y coordinates of graph n.

show (*viewport*, $integer_n$`(1)`, *string*`("on")`)
: indicates if graph n is shown or not.

title (*viewport*, *string*`("Axiom 2D")`)
: designates the title for *viewport*.

translate (*viewport*, $integer_n$`(1)`, $float_x$`(0.0)`, $float_y$`(0.0)`)
: causes graph n to be moved x and y units in the respective directions.

write (*viewport*, $string_{directory}$, [*strings*])
: if no third argument is given, writes the **data** file onto the directory with extension **data**. The third argument can be a single string or a list of strings with some or all the entries `"pixmap"`, `"bitmap"`, `"postscript"`, and `"image"`.

7.2 Three-Dimensional Graphics

The AXIOM three-dimensional graphics package provides the ability to

- generate surfaces defined by a function of two real variables
- generate space curves and tubes defined by parametric equations
- generate surfaces defined by parametric equations

These graphs can be modified by using various options, such as calculating points in the spherical coordinate system or changing the polygon grid size of a surface.

7.2.1 Plotting Three-Dimensional Functions of Two Variables

The simplest three-dimensional graph is that of a surface defined by a function of two variables, `z = f(x,y)`.

> The general format for drawing a surface defined by a formula `f(x,y)` of two variables `x` and `y` is:
>
> `draw(f(x,y), x = a..b, y = c..d,` *options*)
>
> where `a..b` and `c..d` define the range of `x` and `y`, and where *options* prescribes zero or more options as described in Section 7.2.4 on page 200. An example of an option is `title == "Title of Graph"`. An alternative format involving a function `f` is also available.

The simplest way to plot a function of two variables is to use a formula. With formulas you always precede the range specifications with the variable name and an "=" sign.

```
draw(cos(x*y),x=-3..3,y=-3..3)
```

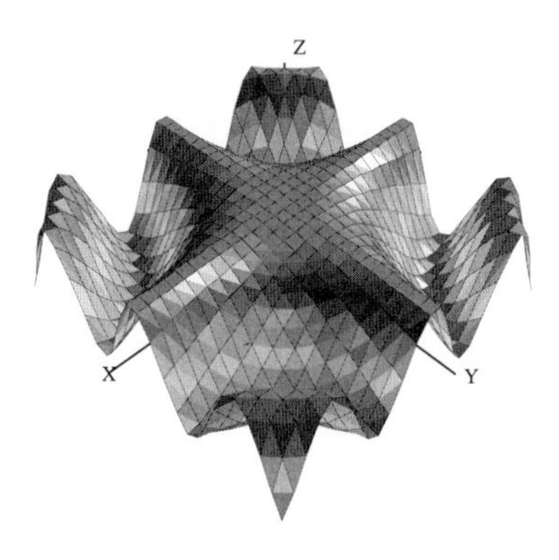

If you intend to use a function more than once, or it is long and complex, then first give its definition to AXIOM.

```
f(x,y) == sin(x)*cos(y)
```

Type: Void

To draw the function, just give its name and drop the variables from the range specifications. AXIOM compiles your function for efficient computation of data for the graph. Notice that AXIOM uses the text of your function as a default title.

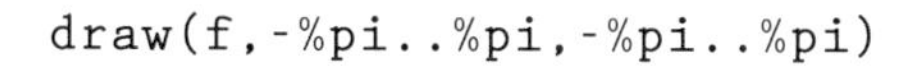

```
draw(f,-%pi..%pi,-%pi..%pi)
```

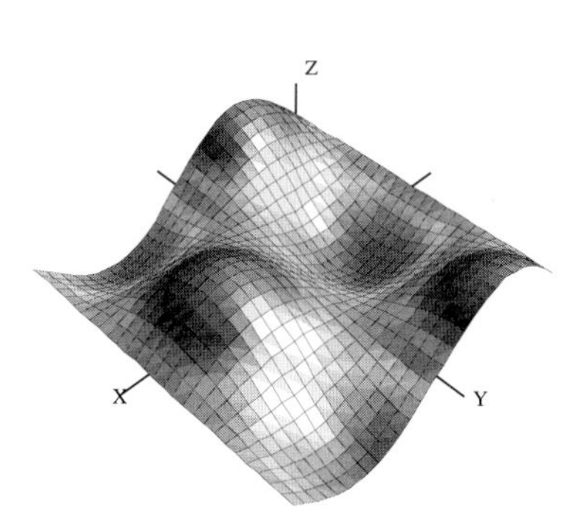

7.2.2 Plotting Three-Dimensional Parametric Space Curves

A second kind of three-dimensional graph is a three-dimensional space curve defined by the parametric equations for `x(t)`, `y(t)`, and `z(t)` as a function of an independent variable `t`.

The general format for drawing a three-dimensional space curve defined by parametric formulas `x = f(t)`, `y = g(t)`, and `z = h(t)` is:

```
draw(curve(f(t),g(t),h(t)), t = a..b, options)
```

where `a..b` defines the range of the independent variable `t`, and where *options* prescribes zero or more options as described in Section 7.2.4 on page 200. An example of an option is `title == "Title of Graph"`. An alternative format involving functions `f`, `g` and `h` is also available.

If you use explicit formulas to draw a space curve, always precede the range specification with the variable name and an "=" sign.

```
draw(curve(5*cos(t), 5*sin(t),t), t=-12..12)
```

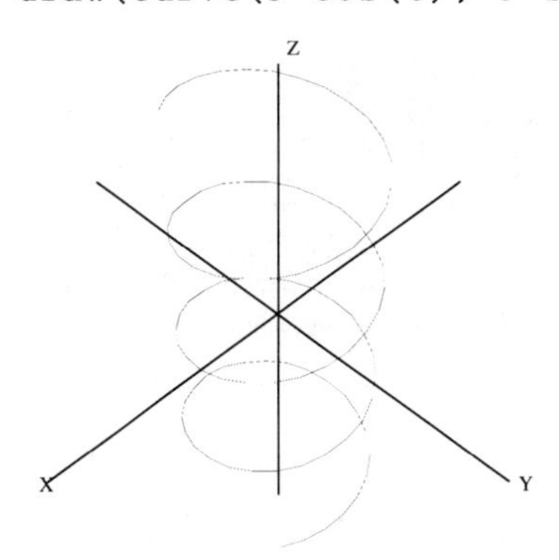

Alternatively, you can draw space curves by referring to functions.

```
i1(t:SF):SF == sin(t)*cos(3*t/5)
```

Type: Void

This is useful if the functions are to be used more than once ...

```
i2(t:SF):SF == cos(t)*cos(3*t/5)
```

Type: Void

or if the functions are long and complex.

```
i3(t:SF):SF == cos(t)*sin(3*t/5)
```

Type: Void

Give the names of the functions and drop the variable name specification in the second argument. Again, AXIOM supplies a default title.

```
draw(curve(i1,i2,i3),0..15*%pi)
```

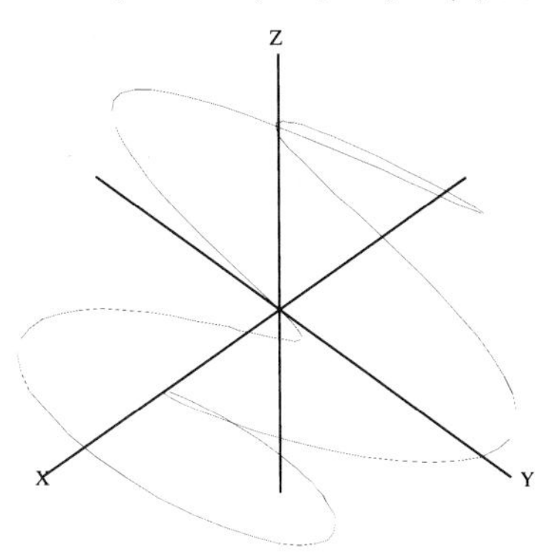

7.2.3 Plotting Three-Dimensional Parametric Surfaces

A third kind of three-dimensional graph is a surface defined by parametric equations for x(u,v), y(u,v), and z(u,v) of two independent variables u and v.

> The general format for drawing a three-dimensional graph defined by parametric formulas x = f(u,v), y = g(u,v), and z = h(u,v) is:
>
> draw(surface(f(u,v),g(u,v),h(u,v)), u = a..b, v = c..d, *options*)
>
> where a..b and c..d define the range of the independent variables u and v, and where *options* prescribes zero or more options as described in Section 7.2.4 on page 200. An example of an option is title == "Title of Graph". An alternative format involving functions f, g and h is also available.

This example draws a graph of a surface plotted using the parabolic cylindrical coordinate system option. The values of the functions supplied to **surface** are interpreted in coordinates as given by a coordinates option, here as parabolic cylindrical coordinates (see Section 7.2.7 on page 210).

```
draw(surface(u*cos(v), u*sin(v), v*cos(u)), u=-4..4,
  v=0..%pi, coordinates== parabolicCylindrical)
```

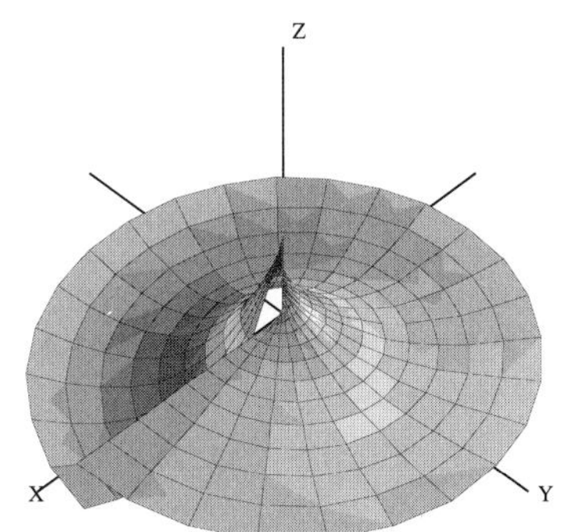

Again, you can graph these parametric surfaces using functions, if the functions are long and complex.

Here we declare the types of arguments and values to be of type SmallFloat.

```
n1(u:SF,v:SF):SF == u*cos(v)
```

Type: Void

As shown by previous examples, these declarations are necessary.

```
n2(u:SF,v:SF):SF == u*sin(v)
```

Type: Void

In either case, AXIOM compiles the functions when needed to graph a result.

```
n3(u:SF,v:SF):SF == u
```

Type: Void

Without these declarations, you have to suffix floats with `@SF` to get a SmallFloat result. However, a call here with an unadorned float produces a SmallFloat.

```
n3(0.5,1.0)
```

0.5 (4)

Type: SmallFloat

Draw the surface by referencing the function names, this time choosing the toroidal coordinate system.

```
draw(surface(n1,n2,n3), 1..4, 1..2*%pi, coordinates ==
  toroidal(1$SF))
```

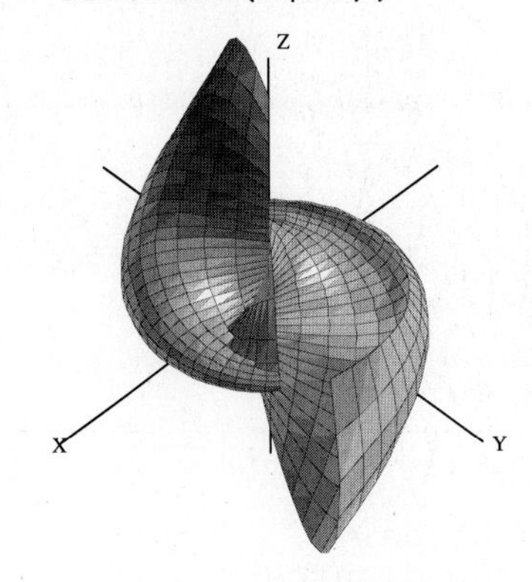

7.2.4 Three-Dimensional Options

The **draw** commands optionally take an optional list of options such as `coordinates` as shown in the last example. Each option is given by the syntax: `name == value`. Here is a list of the available options in the order that they are described below:

title	coordinates	var1Steps
style	tubeRadius	var2Steps
colorFunction	tubePoints	space

Torus knot of type (15,17).

AXIOM Images

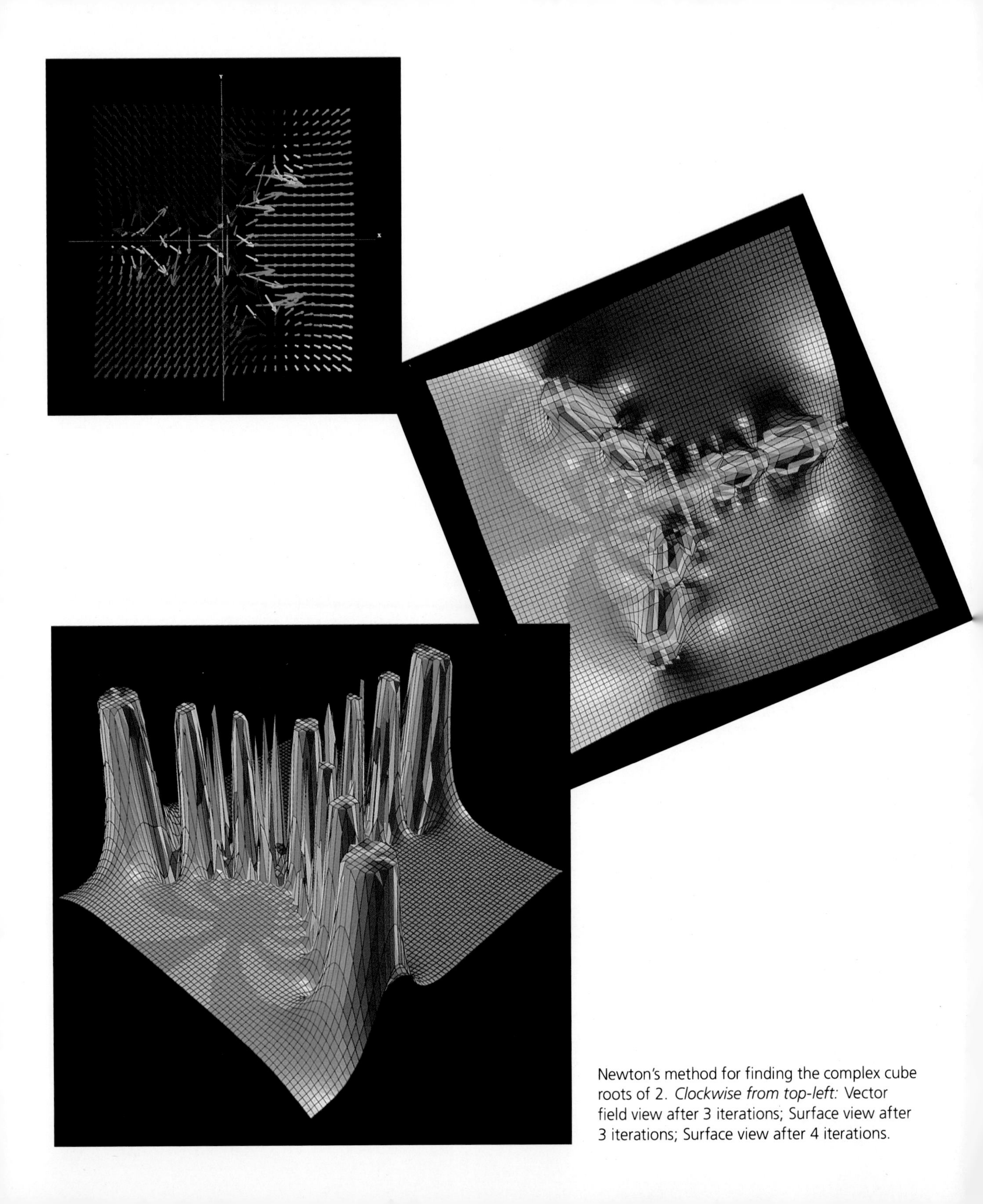

Newton's method for finding the complex cube roots of 2. *Clockwise from top-left:* Vector field view after 3 iterations; Surface view after 3 iterations; Surface view after 4 iterations.

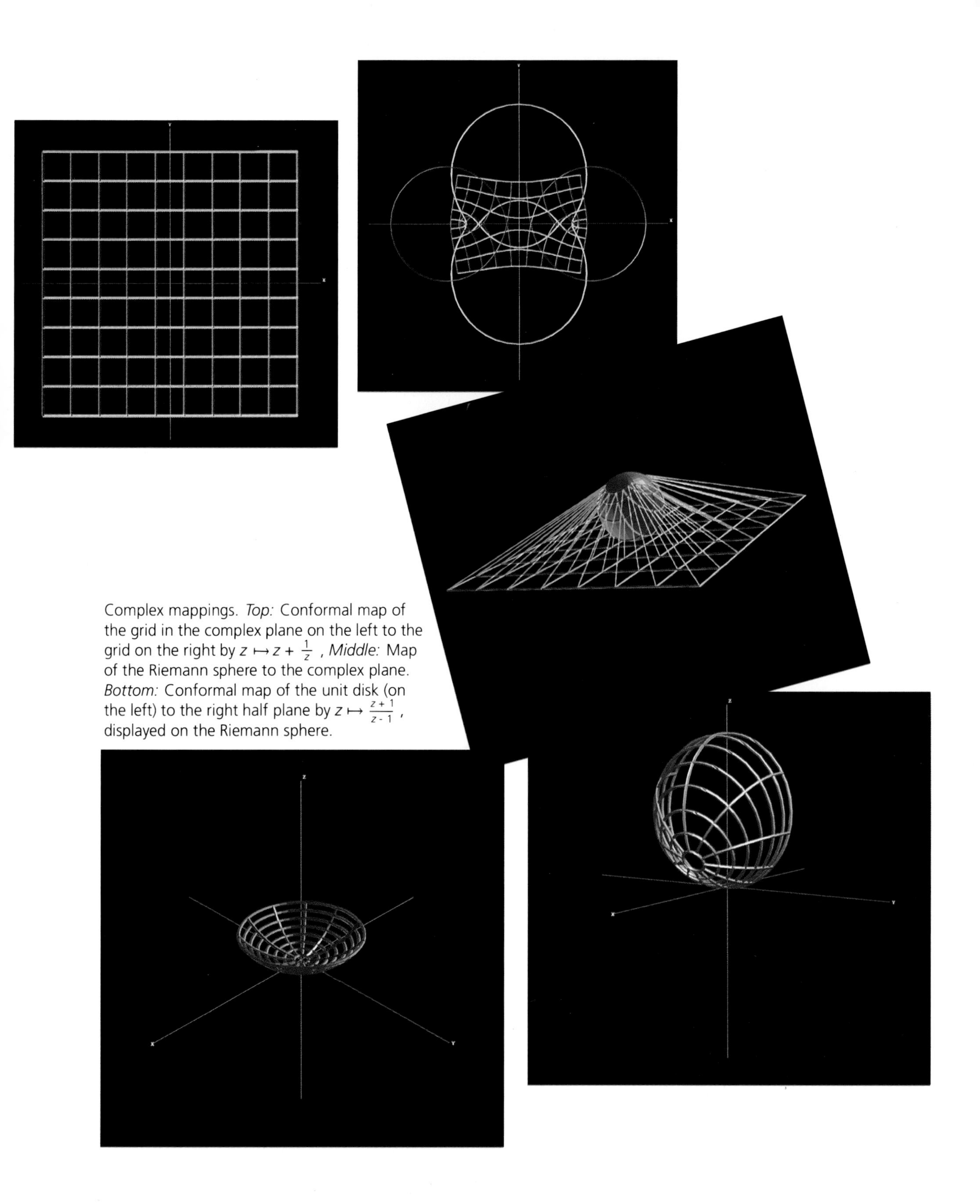

Complex mappings. *Top:* Conformal map of the grid in the complex plane on the left to the grid on the right by $z \mapsto z + \frac{1}{z}$, *Middle:* Map of the Riemann sphere to the complex plane. *Bottom:* Conformal map of the unit disk (on the left) to the right half plane by $z \mapsto \frac{z+1}{z-1}$, displayed on the Riemann sphere.

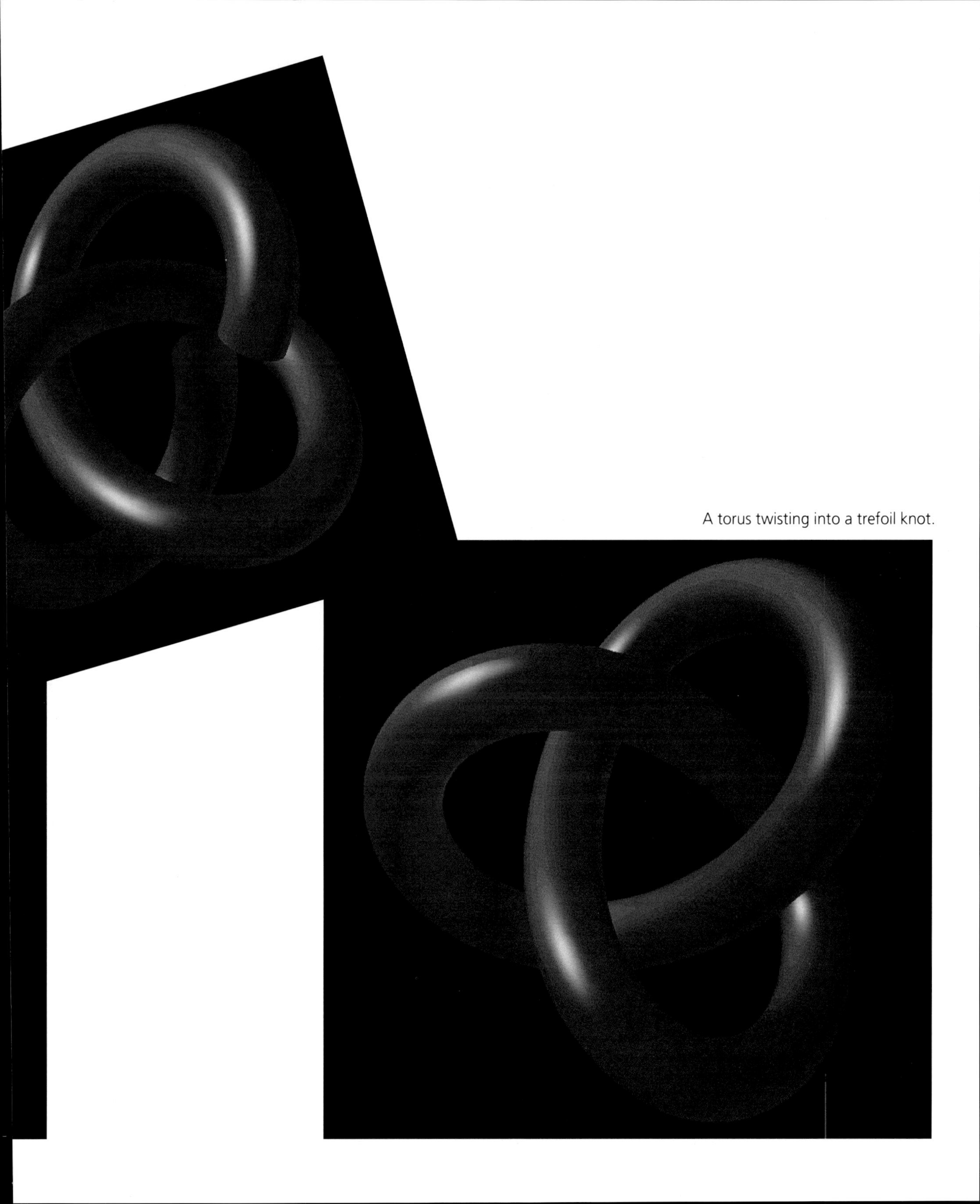

A torus twisting into a trefoil knot.

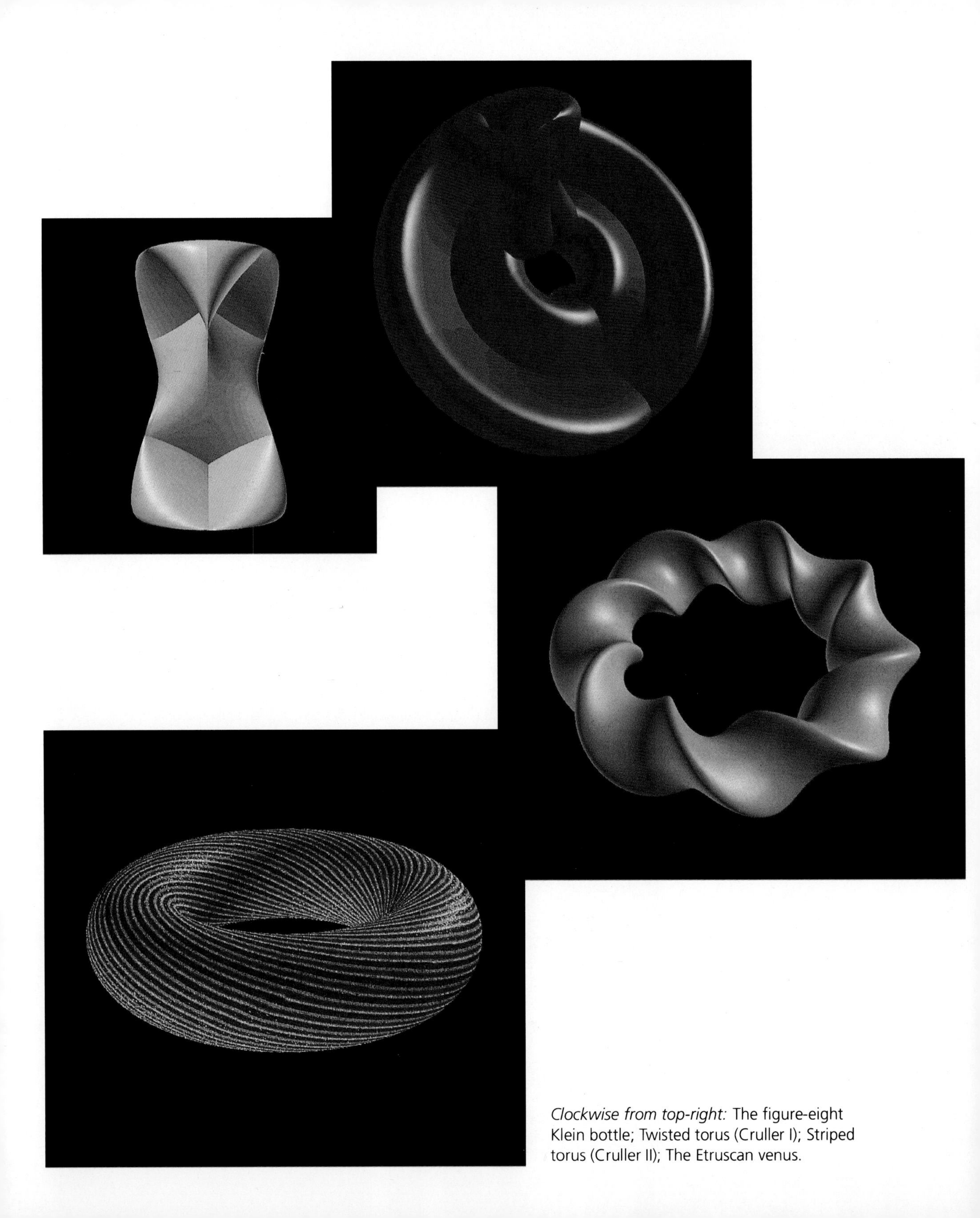

Clockwise from top-right: The figure-eight Klein bottle; Twisted torus (Cruller I); Striped torus (Cruller II); The Etruscan venus.

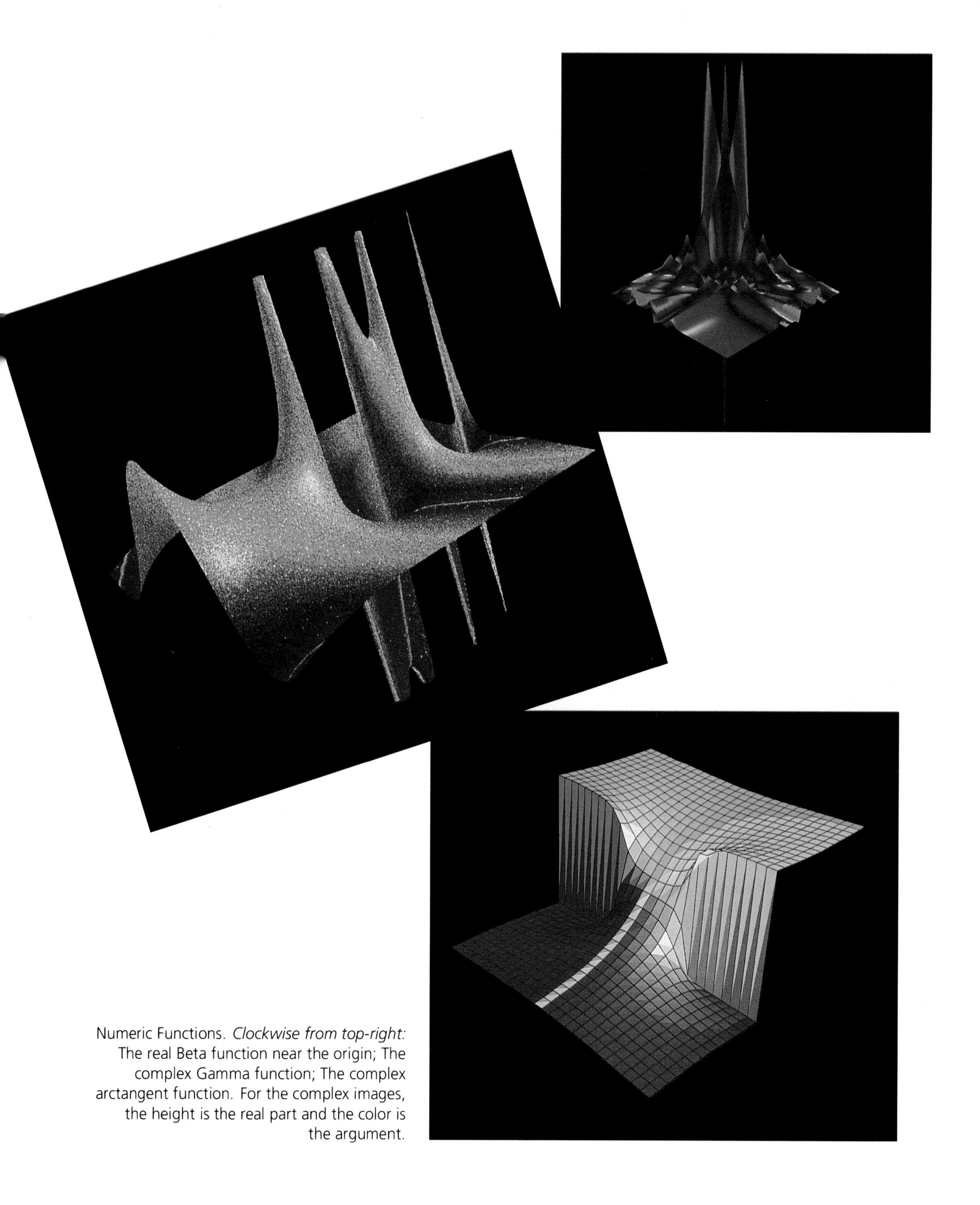

Numeric Functions. *Clockwise from top-right:* The real Beta function near the origin; The complex Gamma function; The complex arctangent function. For the complex images, the height is the real part and the color is the argument.

Clockwise from top-left: Sierpinsky's tetrahedron; Antoine's necklace; Scherk's minimal surface, defined implicitly by $e^z \cos(x) = \cos(y)$; Ribbon plot of x, x^2, x^3, x^4, and x^5.

See Appendix F
for Axiom programs to produce these images.

The option `title` gives your graph a title.

```
draw(cos(x*y),x=0..2*%pi,y=0..%pi,title == "Title of
  Graph")
```

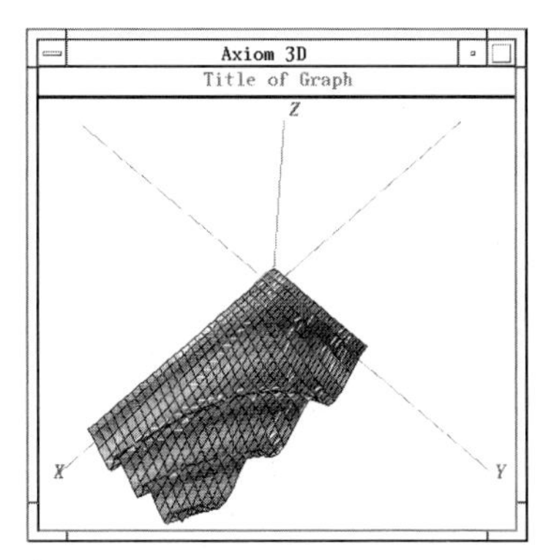

The `style` determines which of four rendering algorithms is used for the graph. The choices are `"wireMesh"`, `"solid"`, `"shade"`, and `"smooth"`.

```
draw(cos(x*y),x=-3..3,y=-3..3, style=="smooth",
  title=="Smooth Option")
```

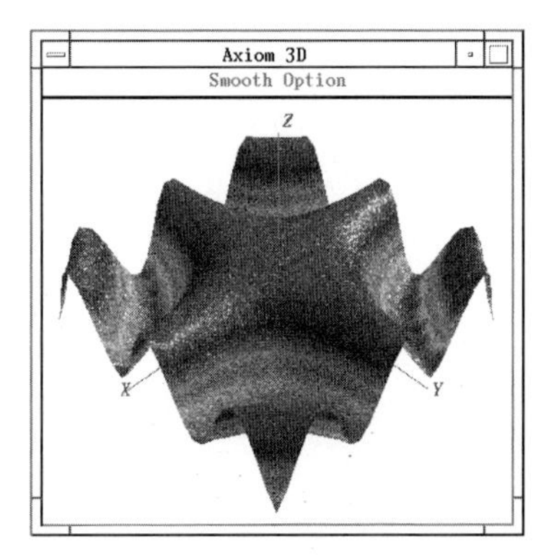

In all but the wire-mesh style, polygons in a surface or tube plot are normally colored in a graph according to their z-coordinate value. Space curves are colored according to their parametric variable value. To change this, you can give a coloring function. The coloring function is sampled across the range of its arguments, then normalized onto the standard AXIOM colormap.

A function of one variable makes the color depend on the value of the parametric variable specified for a tube plot.

```
color1(t) == t
```

Type: Void

```
draw(curve(sin(t), cos(t),0), t=0..2*%pi, tubeRadius ==
  .3, colorFunction == color1)
```

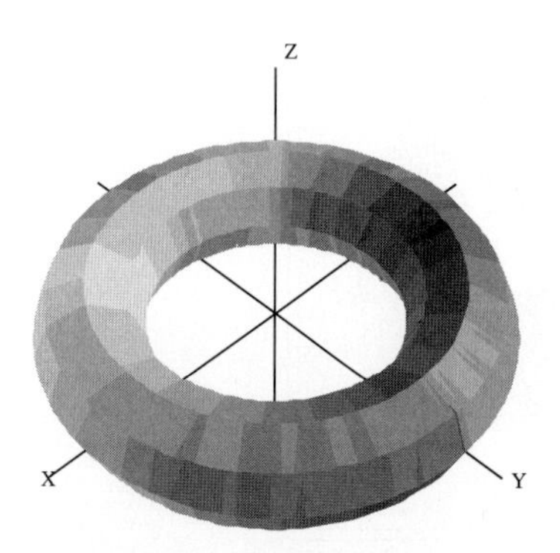

A function of two variables makes the color depend on the values of the independent variables.

```
color2(u,v) == u**2 - v**2
```

Type: Void

Use the option `colorFunction` for special coloring.

```
draw(cos(u*v), u=-3..3, v=-3..3, colorFunction == color2)
```

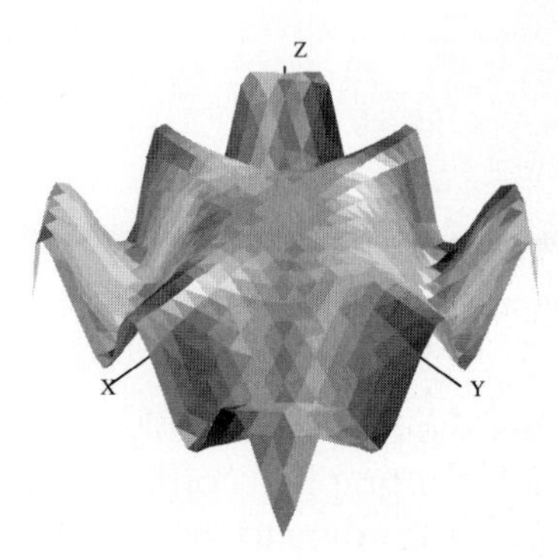

With a three variable function, the color also depends on the value of the function.

```
color3(x,y,fxy) == sin(x*fxy) + cos(y*fxy)
```

Type: Void

```
draw(cos(x*y), x=-3..3, y=-3..3, colorFunction == color3)
```

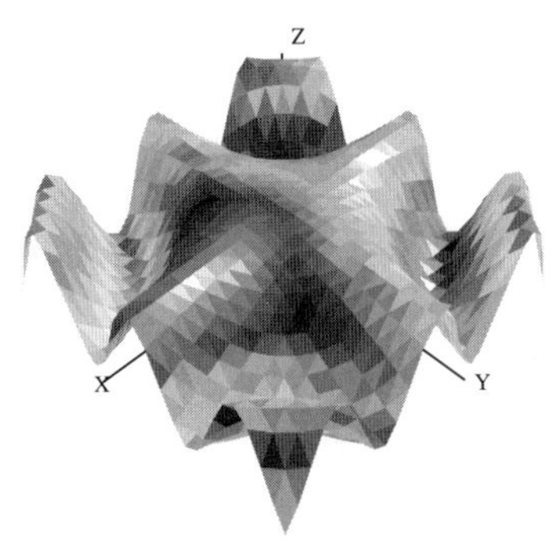

Normally the Cartesian coordinate system is used. To change this, use the `coordinates` option. For details, see Section 7.2.7 on page 210.

```
m(u:SF,v:SF):SF == 1
```

Type: Void

Use the spherical coordinate system.

```
draw(m, 0..2*%pi,0..%pi, coordinates == spherical,
  style=="shade")
```

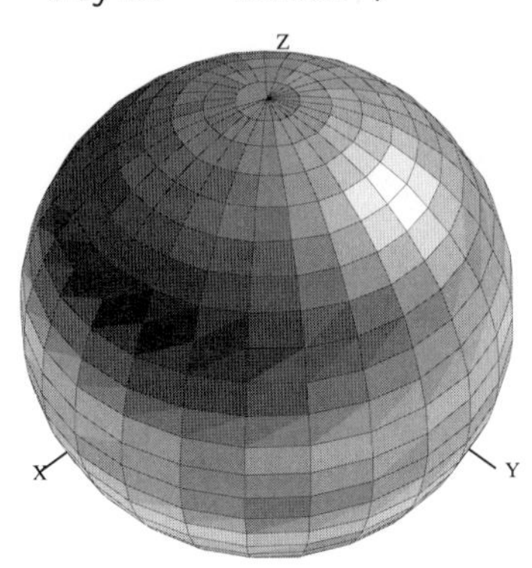

Space curves may be displayed as tubes with polygonal cross sections. Two options, `tubeRadius` and `tubePoints`, control the size and shape of this cross section.

The tubeRadius option specifies the radius of the tube that encircles the specified space curve.

```
draw(curve(sin(t),cos(t),0),t=0..2*%pi, style=="shade",
  tubeRadius == .3)
```

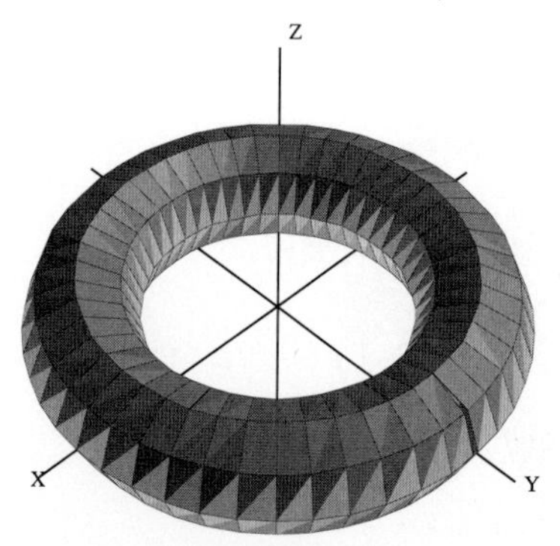

The tubePoints option specifies the number of vertices defining the polygon that is used to create a tube around the specified space curve. The larger this number is, the more cylindrical the tube becomes.

```
draw(curve(sin(t), cos(t), 0), t=0..2*%pi, style=="shade",
  tubeRadius == .25, tubePoints == 3)
```

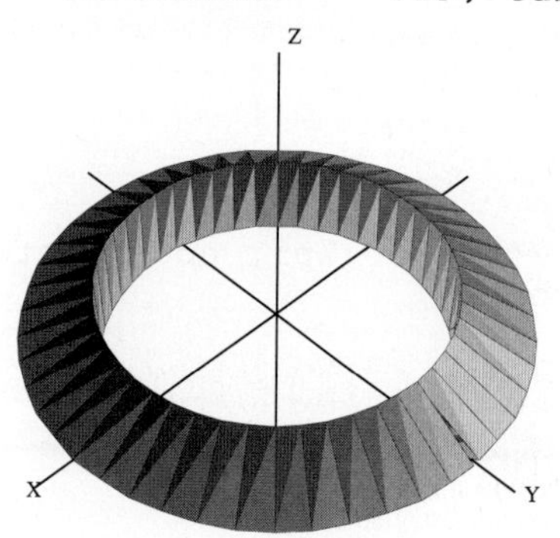

Options **var1Steps** and **var2Steps** specify the number of intervals into which the grid defining a surface plot is subdivided with respect to the first and second parameters of the surface function(s).

```
draw(cos(x*y),x=-3..3,y=-3..3, style=="shade", var1Steps
  == 30, var2Steps == 30)
```

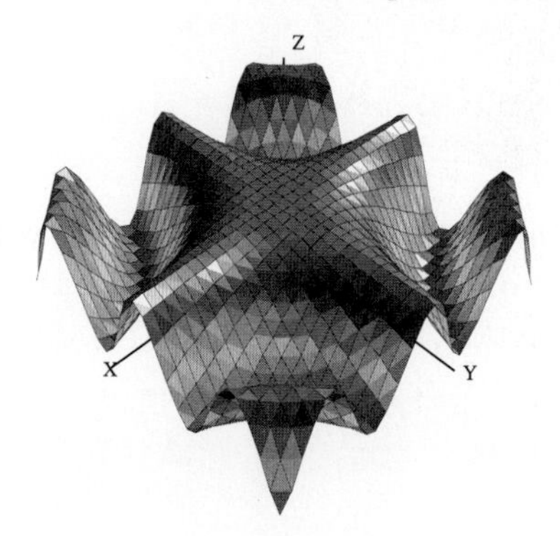

The space option of a **draw** command lets you build multiple graphs in three space. To use this option, first create an empty three-space object, then use the space option thereafter. There is no restriction as to the number or kinds of graphs that can be combined this way.

Create an empty three-space object.

```
s := create3Space()$(ThreeSpace SF)
```

3-Space with 0 components (5)

Type: ThreeSpace SmallFloat

```
m(u:SF,v:SF):SF == 1
```

Type: Void

Add a graph to this three-space object. The new graph destructively inserts the graph into s.

```
draw(m,0..%pi,0..2*%pi, coordinates == spherical, space ==
  s)
```

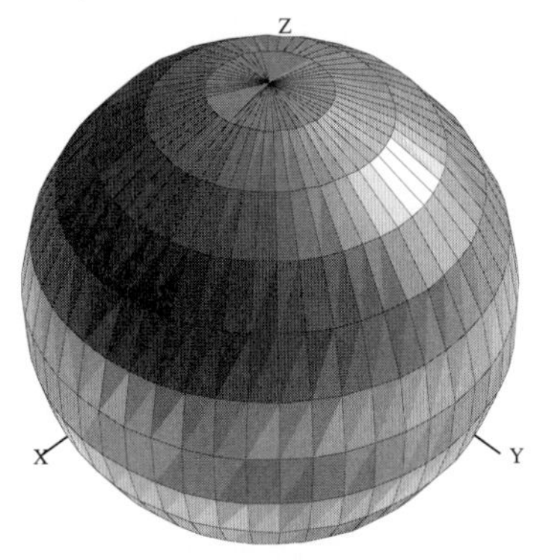

Add a second graph to s.

```
v := draw(curve(1.5*sin(t), 1.5*cos(t),0), t=0..2*%pi,
  tubeRadius == .25, space == s)
```

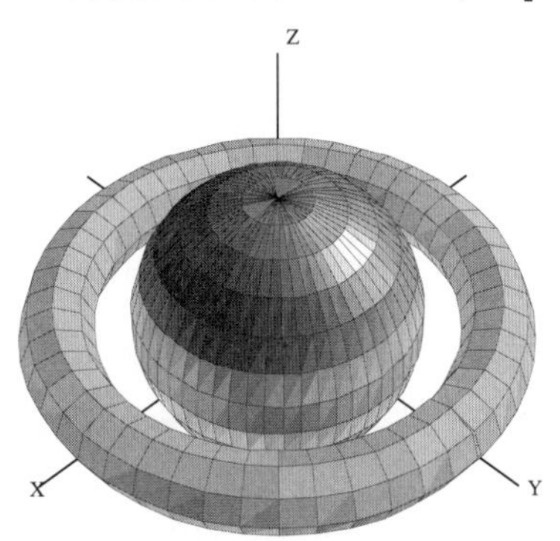

A three-space object can also be obtained from an existing three-dimensional viewport using the **subspace** command. You can then use **makeViewport3D** to create a viewport window.

Assign to subsp the three-space object in viewport v.

```
subsp := subspace v
```

3-Space with 1 components (8)

Type: ThreeSpace SmallFloat

Reset the space component of v to the value of subsp.

```
subspace(v, subsp)
```

ThreeDimensionalViewport : "..." (9)

Type: ThreeDimensionalViewport

Create a viewport window from a three-space object.

```
makeViewport3D(subsp,"Graphs")
```

7.2.5 The makeObject Command

An alternate way to create multiple graphs is to use **makeObject**. The **makeObject** command is similar to the **draw** command, except that it returns a three-space object rather than a ThreeDimensionalViewport. In fact, **makeObject** is called by the **draw** command to create the ThreeSpace then **makeViewport3D** to create a viewport window.

```
m(u:SF,v:SF):SF == 1
```

Type: Void

Do the last example a new way. First use **makeObject** to create a three-space object sph.

```
sph := makeObject(m, 0..%pi, 0..2*%pi,
  coordinates==spherical)
```

Add a second object to sph.

```
makeObject(curve(1.5*sin(t), 1.5*cos(t), 0), t=0..2*%pi,
  space == sph, tubeRadius == .25)
```

Create and display a viewport containing sph.

```
makeViewport3D(sph,"Multiple Objects")
```

Note that an undefined ThreeSpace parameter declared in a **makeObject** or **draw** command results in an error. Use the **create3Space** function to define a ThreeSpace, or obtain a ThreeSpace that has been previously generated before including it in a command line.

7.2.6 Building Three-Dimensional Objects From Primitives

Rather than using the **draw** and **makeObject** commands, you can create three-dimensional graphs from primitives. Operation **create3Space** creates a three-space object to which points, curves and polygons can be added using the operations from the ThreeSpace domain. The resulting object can then be displayed in a viewport using **makeViewport3D**.

Create the empty three-space object space.

```
space := create3Space()$(ThreeSpace SF)
```

3-Space with 0 components (1)

Type: ThreeSpace SmallFloat

Objects can be sent to this space using the operations exported by the ThreeSpace domain. The following examples place curves into space.

Add these eight curves to the space.

```
closedCurve(space,[[0,30,20], [0,30,30], [0,40,30],
  [0,40,100], [0,30,100],[0,30,110], [0,60,110],
  [0,60,100], [0,50,100], [0,50,30], [0,60,30],
  [0,60,20]])
```

3-Space with 1 components (2)

Type: ThreeSpace SmallFloat

```
closedCurve(space,[[80,0,30], [80,0,100], [70,0,110],
  [40,0,110], [30,0,100], [30,0,90], [40,0,90], [40,0,95],
  [45,0,100], [65,0,100], [70,0,95], [70,0,35]])
```

3-Space with 2 components (3)

Type: ThreeSpace SmallFloat

```
closedCurve(space,[[70,0,35], [65,0,30], [45,0,30],
  [40,0,35], [40,0,60], [50,0,60], [50,0,70], [30,0,70],
  [30,0,30], [40,0,20], [70,0,20], [80,0,30]])
```

3-Space with 3 components (4)

Type: ThreeSpace SmallFloat

```
closedCurve(space,[[0,70,20], [0,70,110], [0,110,110],
  [0,120,100], [0,120,70], [0,115,65], [0,120,60],
  [0,120,30], [0,110,20], [0,80,20], [0,80,30],
  [0,80,20]])
```

3-Space with 4 components (5)

Type: ThreeSpace SmallFloat

```
closedCurve(space,[[0,105,30], [0,110,35], [0,110,55],
  [0,105,60], [0,80,60], [0,80,70], [0,105,70],
  [0,110,75], [0,110,95], [0,105,100], [0,80,100],
  [0,80,20], [0,80,30]])
```

3-Space with 5 components (6)

Type: ThreeSpace SmallFloat

```
closedCurve(space,[[140,0,20], [140,0,110], [130,0,110],
  [90,0,20], [101,0,20],[114,0,50], [130,0,50],
  [130,0,60], [119,0,60], [130,0,85], [130,0,20]])
```

3-Space with 6 components (7)

Type: ThreeSpace SmallFloat

```
closedCurve(space,[[0,140,20], [0,140,110],
  [0,150,110], [0,170,50], [0,190,110], [0,200,110],
  [0,200,20], [0,190,20], [0,190,75], [0,175,35],
  [0,165,35],[0,150,75], [0,150,20]])
```

3-Space with 7 components (8)

Type: ThreeSpace SmallFloat

```
closedCurve(space,[[200,0,20], [200,0,110], [189,0,110],
  [160,0,45], [160,0,110], [150,0,110], [150,0,20],
  [161,0,20], [190,0,85], [190,0,20]])
```

3-Space with 8 components (9)

Type: ThreeSpace SmallFloat

Create and display the viewport using **makeViewport3D**. Options may also be given but here are displayed as a list with values enclosed in parentheses.

```
makeViewport3D(space, title == "Letters")
```

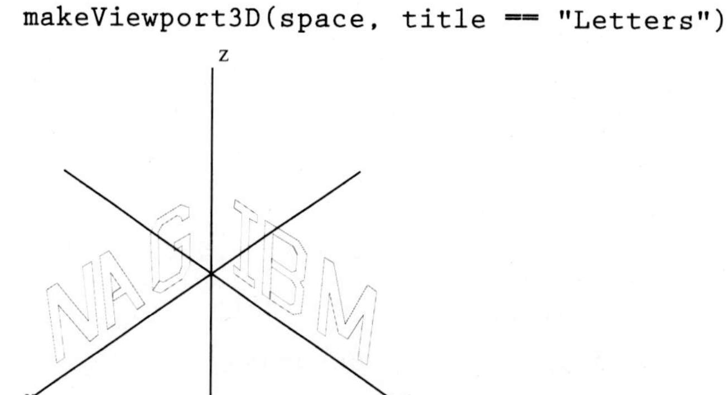

Cube Example

As a second example of the use of primitives, we generate a cube using a polygon mesh. It is important to use a consistent orientation of the polygons for correct generation of three-dimensional objects.

Again start with an empty three-space object.

```
spaceC := create3Space()$(ThreeSpace SF)
```

3-Space with 0 components (10)

Type: ThreeSpace SmallFloat

For convenience, give SmallFloat values +1 and -1 names.

```
x: SF := 1
```

1.0 (11)

Type: SmallFloat

```
y: SF := -1
```

-1.0 (12)

Type: SmallFloat

Define the vertices of the cube.

```
a := point [x,x,y,1::SF]$(Point SF)
```

$[1.0,\ 1.0,\ -1.0,\ 1.0]$ (13)

Type: Point SmallFloat

```
b := point [y,x,y,4::SF]$(Point SF)
```

$[-1.0,\ 1.0,\ -1.0,\ 4.0]$ (14)

Type: Point SmallFloat

```
c := point [y,x,x,8::SF]$(Point SF)
```

$$[-1.0,\ 1.0,\ 1.0,\ 8.0] \tag{15}$$

Type: Point SmallFloat

```
d := point [x,x,x,12::SF]$(Point SF)
```

$$[1.0,\ 1.0,\ 1.0,\ 12.0] \tag{16}$$

Type: Point SmallFloat

```
e := point [x,y,y,16::SF]$(Point SF)
```

$$[1.0,\ -1.0,\ -1.0,\ 16.0] \tag{17}$$

Type: Point SmallFloat

```
f := point [y,y,y,20::SF]$(Point SF)
```

$$[-1.0,\ -1.0,\ -1.0,\ 20.0] \tag{18}$$

Type: Point SmallFloat

```
g := point [y,y,x,24::SF]$(Point SF)
```

$$[-1.0,\ -1.0,\ 1.0,\ 24.0] \tag{19}$$

Type: Point SmallFloat

```
h := point [x,y,x,27::SF]$(Point SF)
```

$$[1.0,\ -1.0,\ 1.0,\ 27.0] \tag{20}$$

Type: Point SmallFloat

Add the faces of the cube as polygons to the space using a consistent orientation.

```
polygon(spaceC,[d,c,g,h])
```

3-Space with 1 components (21)

Type: ThreeSpace SmallFloat

```
polygon(spaceC,[d,h,e,a])
```

3-Space with 2 components (22)

Type: ThreeSpace SmallFloat

```
polygon(spaceC,[c,d,a,b])
```

3-Space with 3 components (23)

Type: ThreeSpace SmallFloat

```
polygon(spaceC,[g,c,b,f])
```

3-Space with 4 components (24)

Type: ThreeSpace SmallFloat

```
polygon(spaceC,[h,g,f,e])
```

3-Space with 5 components (25)

Type: ThreeSpace SmallFloat

```
polygon(spaceC,[e,f,b,a])
```

3-Space with 6 components (26)

Type: ThreeSpace SmallFloat

Create and display the viewport.

```
makeViewport3D(spaceC, title == "Cube")
```

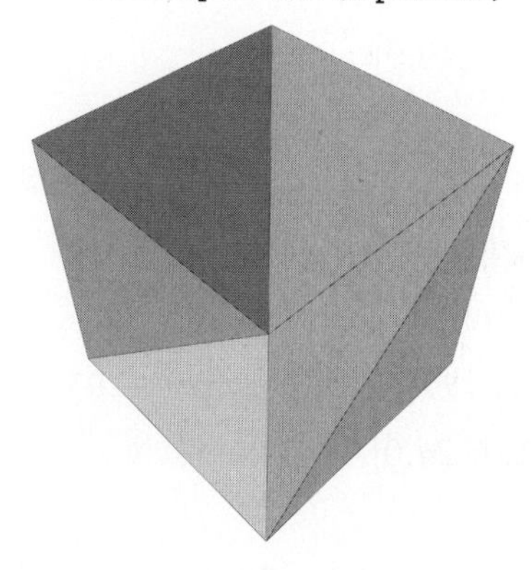

7.2.7 Coordinate System Transformations

The CoordinateSystems package provides coordinate transformation functions that map a given data point from the coordinate system specified into the Cartesian coordinate system. The default coordinate system, given a triplet `(f(u,v), u, v)`, assumes that `z = f(u, v)`, `x = u` and `y = v`, that is, reads the coordinates in `(z, x, y)` order.

```
m(u:SF,v:SF):SF == u**2
```

Type: Void

Graph plotted in default coordinate system.

```
draw(m,0..3,0..5)
```

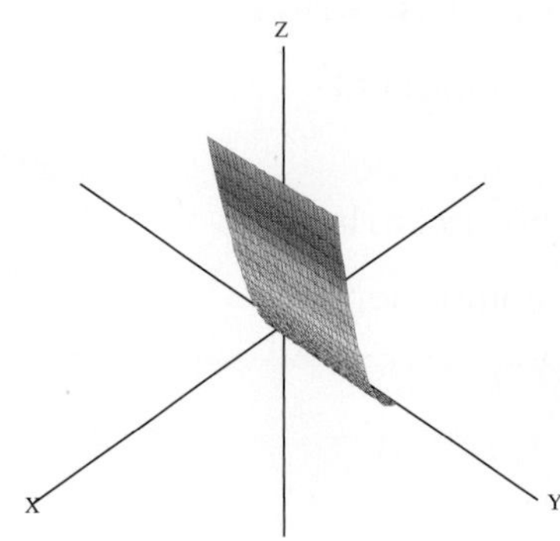

The z coordinate comes first since the first argument of the **draw** command gives its values. In general, the coordinate systems AXIOM provides, or any that you make up, must provide a map to an (x, y, z) triplet in order to be compatible with the **coordinates** DrawOption. Here is an example.

Define the identity function.

```
cartesian(point:Point SF):Point SF == point
```

Type: Void

Pass `cartesian` as the **coordinates** parameter to the **draw** command.

```
draw(m,0..3,0..5,coordinates==cartesian)
```

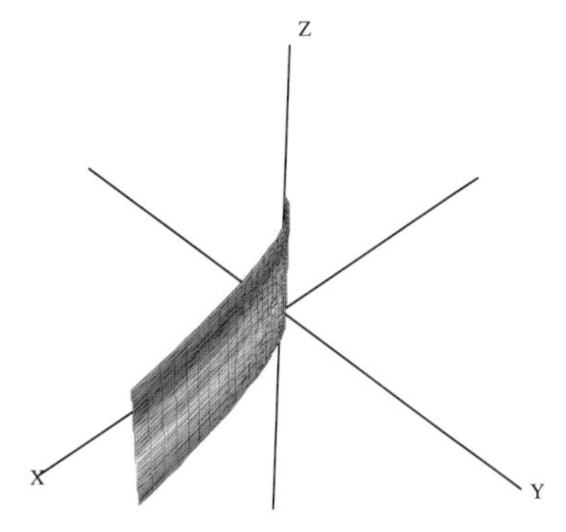

What happened? The option `coordinates == cartesian` directs AXIOM to treat the dependent variable `m` defined by $m = u^2$ as the **x** coordinate. Thus the triplet of values (m, u, v) is transformed to coordinates (x, y, z) and so we get the graph of $x = y^2$.

Here is another example. The **cylindrical** transform takes input of the form (w,u,v), interprets it in the order (r,θ,z) and maps it to the Cartesian coordinates $x = r\cos(\theta)$, $y = r\sin(\theta)$, $z = z$ in which r is the radius, θ is the angle and z is the z-coordinate.

An example using the **cylindrical** coordinates for the constant `r = 3`.

```
f(u:SF,v:SF):SF == 3
```

Type: Void

Graph plotted in cylindrical coordinates.

```
draw(f,0..%pi,0..6,coordinates==cylindrical)
```

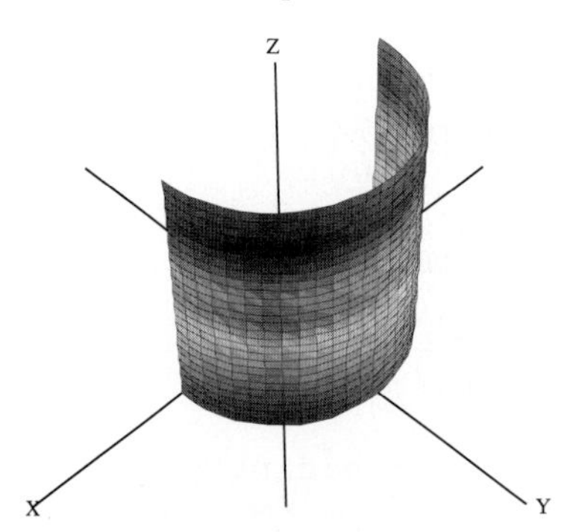

Suppose you would like to specify z as a function of r and θ instead of just r? Well, you still can use the **cylindrical** AXIOM transformation but we have to reorder the triplet before passing it to the transformation.

First, let's create a point to work with and call it `pt` with some color `col`.

```
col := 5
```

5 (4)

Type: PositiveInteger

```
pt := point[1,2,3,col]$(Point SF)
```

[1.0, 2.0, 3.0, 5.0] (5)

Type: Point SmallFloat

The reordering you want is (z, r, θ) to (r, θ, z) so that the first element is moved to the third element, while the second and third elements move forward and the color element does not change.

Define a function **reorder** to reorder the point elements.

```
reorder(p:Point SF):Point SF == point[p.2, p.3, p.1, p.4]
```

Type: Void

The function moves the second and third elements forward but the color does not change.

```
reorder pt
```

[2.0, 3.0, 1.0, 5.0] (7)

Type: Point SmallFloat

The function **newmap** converts our reordered version of the cylindrical coordinate system to the standard (x, y, z) Cartesian system.

```
newmap(pt:Point SF):Point SF == cylindrical(reorder pt)
```

Type: Void

```
newmap pt
```

[−1.9799849932008908, 0.28224001611973443, 1.0, 5.0] (9)

Type: Point SmallFloat

Graph the same function f using the coordinate mapping of the function newmap, so it is now interpreted as $z = 3$:

```
draw(f,0..3,0..2*%pi,coordinates==newmap)
```

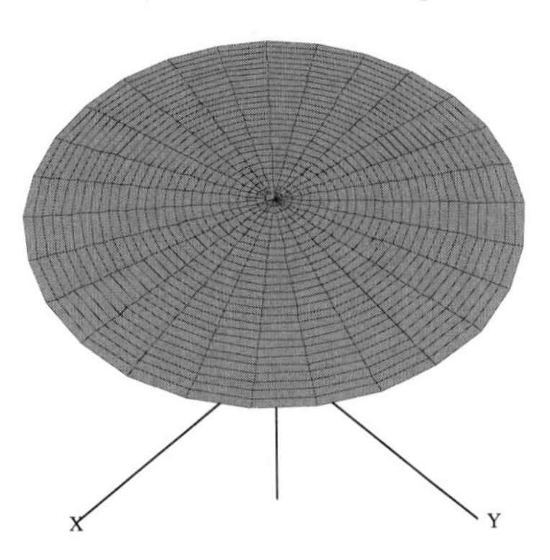

The CoordinateSystems package exports the following operations: **bipolar**, **bipolarCylindrical**, **cartesian**, **conical**, **cylindrical**, **elliptic**, **ellipticCylindrical**, **oblateSpheroidal**, **parabolic**, **parabolicCylindrical**, **paraboloidal**, **polar**, **prolateSpheroidal**, **spherical**, and **toroidal**. Use Browse or the `)show` system command to get more information.

7.2.8 Three-Dimensional Clipping

A three-dimensional graph can be explicitly clipped within the **draw** command by indicating a minimum and maximum threshold for the given function definition. These thresholds can be defined using the AXIOM **min** and **max** functions.

```
gamma(x,y) ==
  g := Gamma complex(x,y)
  point [x, y, max( min(real g, 4), -4), argument g]
```

Type: Void

Here is an example that clips the gamma function in order to eliminate the extreme divergence it creates.

```
draw(gamma, -%pi..%pi, -%pi..%pi, var1Steps == 50,
  var2Steps == 50)
```

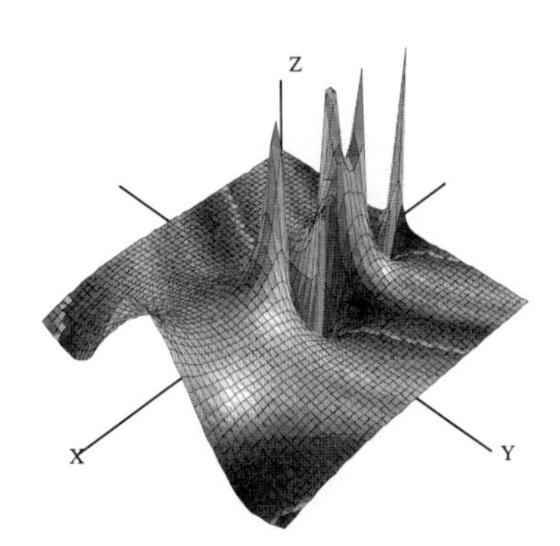

7.2.9 Three-Dimensional Control-Panel

Once you have created a viewport, move your mouse to the viewport and click with your left mouse button. This displays a control-panel on the side of the viewport that is closest to where you clicked.

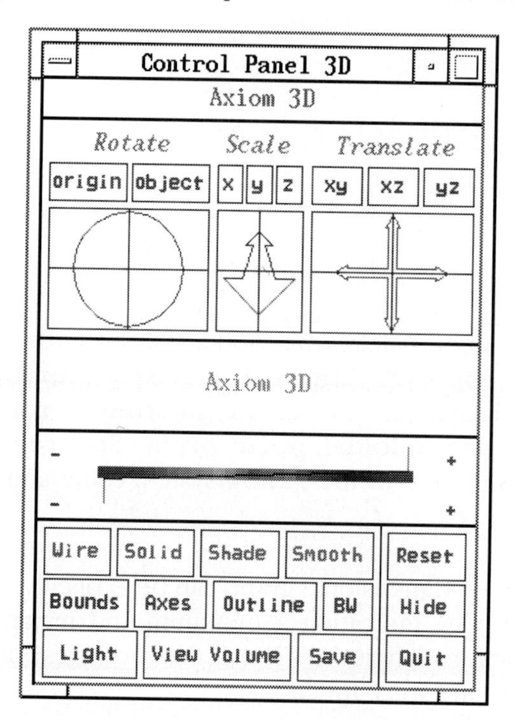

Figure 7.3: Three-dimensional control-panel.

Transformations

We recommend you first select the **Bounds** button while executing transformations since the bounding box displayed indicates the object's position as it changes.

Rotate: A rotation transformation occurs by clicking the mouse within the **Rotate** window in the upper left corner of the control-panel. The rotation is computed in spherical coordinates, using the horizontal mouse position to increment or decrement the value of the longitudinal angle θ within the range of 0 to 2π and the vertical mouse position to increment or decrement the value of the latitudinal angle ϕ within the range of $-\pi$ to π. The active mode of rotation is displayed in green on a color monitor or in clear text on a black and white monitor, while the inactive mode is displayed in red for color display or a mottled pattern for black and white.

origin: The **origin** button indicates that the rotation is to occur with respect to the origin of the viewing space, that is indicated by the axes.

object: The **object** button indicates that the rotation is to occur with respect to the center of volume of the object, independent of the axes' origin position.

Scale: A scaling transformation occurs by clicking the mouse within the **Scale** window in the upper center of the control-panel, containing a zoom arrow. The axes along which the scaling is to occur are indicated by selecting the appropriate button above the zoom arrow window. The selected axes are displayed in green on a color monitor or in clear text on a black and white monitor, while the unselected axes are displayed in red for a color display or a mottled pattern for black and white.

uniform: Uniform scaling along the `x`, `y` and `z` axes occurs when all the axes buttons are selected.

non-uniform: If any of the axes buttons are not selected, non-uniform scaling occurs, that is, scaling occurs only in the direction of the axes that are selected.

Translate: Translation occurs by indicating with the mouse in the **Translate** window the direction you want the graph to move. This window is located in the upper right corner of the control-panel and contains a potentiometer with crossed arrows pointing up, down, left and right. Along the top of the **Translate** window are three buttons (**XY**, **XZ**, and **YZ**) indicating the three orthographic projection planes. Each orientates the group as a view into that plane. Any translation of the graph occurs only along this plane.

Messages

The window directly below the potentiometer windows for transformations is used to display system messages relating to the viewport, the control-panel and the current graph displaying status.

Colormap

Directly below the message window is the colormap range indicator window. The AXIOM Colormap shows a sampling of the spectrum from which hues can be drawn to represent the colors of a surface. The Colormap is composed of five shades for each of the hues along this spectrum. By moving the markers above and below the Colormap, the range of hues that are used to color the existing surface are set. The bottom marker shows the hue for the low end of the color range and the top marker shows the hue for the upper end of the range. Setting the bottom and top markers at the same hue results in monochromatic smooth shading of the graph when **Smooth** mode is selected. At each end of the Colormap are **+** and **-** buttons. When clicked on, these increment or decrement the top or bottom marker.

Buttons

Below the Colormap window and to the left are located various buttons that determine the characteristics of a graph. The buttons along the bottom and right hand side all have special meanings; the remaining buttons in the first row indicate the mode or style used to display the graph. The second row are toggles that turn on or off a property of the graph. On a color monitor, the property is on if green (clear text, on a monochrome monitor) and off if red (mottled pattern, on a monochrome monitor). Here is a list of their functions.

Wire displays surface and tube plots as a wireframe image in a single color (blue) with no hidden surfaces removed, or displays space curve plots in colors based upon their parametric variables. This is the fastest mode for displaying a graph. This is very useful when you want to find a good orientation of your graph.

Solid displays the graph with hidden surfaces removed, drawing each polygon beginning with the furthest from the viewer. The edges of the polygons are displayed in the hues specified by the range in the Colormap window.

Shade displays the graph with hidden surfaces removed and with the polygons shaded, drawing each polygon beginning with the furthest from the viewer. Polygons are shaded in the hues specified by the range in the Colormap window using the Phong illumination model.

Smooth displays the graph using a renderer that computes the graph one line at a time. The location and color of the graph at each visible point on the screen are determined and displayed using the Phong illumination model. Smooth shading is done in one of two ways, depending on the range selected in the colormap window and the number of colors available from the hardware and/or window manager. When the top and bottom markers of the colormap range are set to different hues, the graph is rendered by dithering between the transitions in color hue. When the top and bottom markers of the colormap range are set to the same hue, the graph is rendered using the Phong smooth shading model. However, if enough colors cannot be allocated for this purpose, the renderer reverts to the color dithering method until a sufficient color supply is available. For this reason, it may not be possible to render multiple Phong smooth shaded graphs at the same time on some systems.

Bounds encloses the entire volume of the viewgraph within a bounding box, or removes the box if previously selected. The region that encloses the entire volume of the viewport graph is displayed.

Axes displays Cartesian coordinate axes of the space, or turns them off if previously selected.

Outline causes quadrilateral polygons forming the graph surface to be outlined in black when the graph is displayed in **Shade** mode.

BW converts a color viewport to black and white, or vice-versa. When this button is selected the control-panel and viewport switch to an immutable colormap composed of a range of grey scale patterns or tiles that are used wherever shading is necessary.

Light takes you to a control-panel described below.

ViewVolume takes you to another control-panel as described below.

Save creates a menu of the possible file types that can be written using the control-panel. The **Exit** button leaves the save menu. The **Pixmap** button writes an AXIOM pixmap of the current viewport contents. The file is called **axiom3D.pixmap** and is located in the directory from which AXIOM or **viewAlone** was started. The **PS** button writes the current viewport contents to PostScript output rather than to the viewport window. By default the file is called **axiom3D.ps**; however, if a file name is specified in the user's **.Xdefaults** file it is used. The file is placed in the directory from which the AXIOM or **viewAlone** session was begun. See also the **write** function.

Reset returns the object transformation characteristics back to their initial states.

Hide causes the control-panel for the corresponding viewport to disappear from the screen.

Quit queries whether the current viewport session should be terminated.

Light

The **Light** button changes the control-panel into the **Lighting Control-Panel**. At the top of this panel, the three axes are shown with the same orientation as the object. A light vector from the origin of the axes shows the current position of the light source relative to the object. At the bottom of the panel is an **Abort** button that cancels any changes to the lighting that were made, and a **Return** button that carries out the current set of lighting changes on the graph.

XY: The **XY** lighting axes window is below the **Lighting Control-Panel** title and to the left. This changes the light vector within the **XY** view plane.

Z: The **Z** lighting axis window is below the **Lighting Control-Panel** title and in the center. This changes the **Z** location of the light vector.

Intensity: Below the **Lighting Control-Panel** title and to the right is the light intensity meter. Moving the intensity indicator down decreases the amount of light emitted from the light source. When the indicator is at the top of the meter the light source is emitting at 100% intensity. At the bottom of the meter the light source is emitting at a level slightly above ambient lighting.

View Volume

The **View Volume** button changes the control-panel into the **Viewing Volume Panel**. At the bottom of the viewing panel is an **Abort** button that cancels any changes to the viewing volume that were made and a *Return* button that carries out the current set of viewing changes to the graph.

Eye Reference: At the top of this panel is the **Eye Reference** window. It shows a planar projection of the viewing pyramid from the eye of the viewer relative to the location of the object. This has a bounding region represented by the rectangle on the left. Below the object rectangle is the **Hither** window. By moving the slider in this window the hither clipping plane sets the front of the view volume. As a result of this depth clipping all points of the object closer to the eye than this hither plane are not shown. The **Eye Distance** slider to the right of the **Hither** slider is used to change the degree of perspective in the image.

Clip Volume: The **Clip Volume** window is at the bottom of the **Viewing Volume Panel**. On the right is a **Settings** menu. In this menu are buttons to select viewing attributes. Selecting the **Perspective** button computes the image using perspective projection. The **Show Region** button indicates whether the clipping region of the volume is to be drawn in the viewport and the **Clipping On** button shows whether the view volume clipping is to be in effect when the image is drawn. The left side of the **Clip Volume** window shows the clipping boundary of the graph. Moving the knobs along the **X**, **Y**, and **Z** sliders adjusts the volume of the clipping region accordingly.

7.2.10 Operations for Three-Dimensional Graphics

Here is a summary of useful AXIOM operations for three-dimensional graphics. Each operation name is followed by a list of arguments. Each argument is written as a variable informally named according to the type of the argument (for example, *integer*). If appropriate, a default value for an argument is given in parentheses immediately following the name.

adaptive3D? `()`
tests whether space curves are to be plotted according to the adaptive refinement algorithm.

axes `(`*viewport*, *string*`("on"))`
turns the axes on and off.

close (*viewport*)
: closes the viewport.

colorDef (*viewport, $color_1$* (1), *$color_2$* (27))
: sets the colormap range to be from $color_1$ to $color_2$.

controlPanel (*viewport, string*("off"))
: declares whether the control-panel for the viewport is to be displayed or not.

diagonals (*viewport, string*("off"))
: declares whether the polygon outline includes the diagonals or not.

drawStyle (*viewport, style*)
: selects which of four drawing styles are used: `"wireMesh"`, `"solid"`, `"shade"`, or `"smooth"`.

eyeDistance (*viewport,float*(500))
: sets the distance of the eye from the origin of the object for use in the **perspective**.

key (*viewport*)
: returns the operating system process ID number for the viewport.

lighting (*viewport, $float_x$*(-0.5), *$float_y$*(0.5), *$float_z$*(0.5))
: sets the Cartesian coordinates of the light source.

modifyPointData (*viewport,integer,point*)
: replaces the coordinates of the point with the index *integer* with *point*.

move (*viewport, $integer_x$*(viewPosDefault), *$integer_y$*(viewPosDefault))
: moves the upper left-hand corner of the viewport to screen position ($integer_x$, $integer_y$).

options (*viewport*)
: returns a list of all current draw options.

outlineRender (*viewport, string*("off"))
: turns polygon outlining off or on when drawing in `"shade"` mode.

perspective (*viewport, string*("on"))
: turns perspective viewing on and off.

reset (*viewport*)
: resets the attributes of a viewport to their initial settings.

resize (*viewport, $integer_{width}$* (viewSizeDefault), *$integer_{height}$* (viewSizeDefault))
: resets the width and height values for a viewport.

rotate (*viewport, $number_\theta$*(viewThetaDefault), *$number_\phi$*(viewPhiDefault))
: rotates the viewport by rotation angles for longitude (θ) and latitude (ϕ). Angles designate radians if given as floats, or degrees if given as integers.

setAdaptive3D (*boolean*(true))

sets whether space curves are to be plotted according to the adaptive refinement algorithm.

setMaxPoints3D (*integer*(1000))
: sets the default maximum number of possible points to be used when constructing a three-dimensional space curve.

setMinPoints3D (*integer*(49))
: sets the default minimum number of possible points to be used when constructing a three-dimensional space curve.

setScreenResolution3D (*integer*(500))
: sets the default screen resolution constant used in setting the computation limit of adaptively generated three-dimensional space curve plots.

showRegion (*viewport, string*("off"))
: declares whether the bounding box of a graph is shown or not.

subspace (*viewport*)
: returns the space component.

subspace (*viewport, subspace*)
: resets the space component to *subspace*.

title (*viewport, string*)
: gives the viewport the title *string*.

translate (*viewport*, $float_x$(viewDeltaXDefault), $float_y$(viewDeltaYDefault))
: translates the object horizontally and vertically relative to the center of the viewport.

intensity (*viewport,float*(1.0))
: resets the intensity I of the light source, $0 \leq I \leq 1$.

tubePointsDefault ([*integer*(6)])
: sets or indicates the default number of vertices defining the polygon that is used to create a tube around a space curve.

tubeRadiusDefault ([*float*(0.5)])
: sets or indicates the default radius of the tube that encircles a space curve.

var1StepsDefault ([*integer*(27)])
: sets or indicates the default number of increments into which the grid defining a surface plot is subdivided with respect to the first parameter declared in the surface function.

var2StepsDefault ([*integer*(27)])
: sets or indicates the default number of increments into which the grid defining a surface plot is subdivided with respect to the second parameter declared in the surface function.

viewDefaults ([$integer_{point}$, $integer_{line}$, $integer_{axes}$, $integer_{units}$, $float_{point}$, *list*$_{\text{position}}$, *list*$_{\text{size}}$])

resets the default settings for the point color, line color, axes color, units color, point size, viewport upper left-hand corner position, and the viewport size.

viewDeltaXDefault ([*float*(0)])

resets the default horizontal offset from the center of the viewport, or returns the current default offset if no argument is given.

viewDeltaYDefault ([*float*(0)])

resets the default vertical offset from the center of the viewport, or returns the current default offset if no argument is given.

viewPhiDefault ([*float*(-π/4)])

resets the default latitudinal view angle, or returns the current default angle if no argument is given. ϕ is set to this value.

viewpoint (*viewport*, *float*$_X$, *float*$_Y$, *float*$_Z$)

sets the viewing position in Cartesian coordinates.

viewpoint (*viewport*, *float*$_\theta$, *Float*$_\phi$)

sets the viewing position in spherical coordinates.

viewpoint (*viewport*, *Float*$_\theta$, *Float*$_\phi$, *Float*$_{\text{scaleFactor}}$, *Float*$_{\text{xOffset}}$, *Float*$_{\text{yOffset}}$)

sets the viewing position in spherical coordinates, the scale factor, and offsets. θ (longitude) and ϕ (latitude) are in radians.

viewPosDefault ([*list*([0,0])])

sets or indicates the position of the upper left-hand corner of a two-dimensional viewport, relative to the display root window (the upper left-hand corner of the display is [0, 0]).

viewSizeDefault ([*list*([400,400])])

sets or indicates the width and height dimensions of a viewport.

viewThetaDefault ([*float*(π/4)])

resets the default longitudinal view angle, or returns the current default angle if no argument is given. When a parameter is specified, the default longitudinal view angle θ is set to this value.

viewWriteAvailable ([*list*(["pixmap", "bitmap", "postscript", "image"])])

indicates the possible file types that can be created with the **write** function.

viewWriteDefault ([*list*([])])

sets or indicates the default types of files that are created in addition to the **data** file when a **write** command is executed on a viewport.

viewScaleDefault ([*float*])

sets the default scaling factor, or returns the current factor if no argument is given.

write (*viewport*, *directory*, [*option*])

writes the file **data** for *viewport* in the directory *directory*. An optional third argument specifies a file type (one of `pixmap`, `bitmap`, `postscript`, or `image`), or a list of file types. An additional file is written for each file type listed.

scale (*viewport*, *float*(2.5))
specifies the scaling factor.

7.2.11 Customization using .Xdefaults

Both the two-dimensional and three-dimensional drawing facilities consult the **.Xdefaults** file for various defaults. The list of defaults that are recognized by the graphing routines is discussed in this section. These defaults are preceded by `Axiom.3D.` for three-dimensional viewport defaults, `Axiom.2D.` for two-dimensional viewport defaults, or `Axiom*` (no dot) for those defaults that are acceptable to either viewport type.

`Axiom*buttonFont:` *font*
This indicates which font type is used for the button text on the control-panel. The default value is `"Rom11"`.

`Axiom.2D.graphFont:` *font* (2D only)
This indicates which font type is used for displaying the graph numbers and slots in the **Graphs** section of the two-dimensional control-panel. The default value is `"Rom22"`.

`Axiom.3D.headerFont:` *font*
This indicates which font type is used for the axes labels and potentiometer header names on three-dimensional viewport windows. This is also used for two-dimensional control-panels for indicating which font type is used for potentionmeter header names and multiple graph title headers. The default value is `"Itl14"`.

`Axiom*inverse:` *switch*
This indicates whether the background color is to be inverted from white to black. If `on`, the graph viewports use black as the background color. If `off` or no declaration is made, the graph viewports use a white background. The default value is `"off"`.

`Axiom.3D.lightingFont:` *font* (3D only)
This indicates which font type is used for the **x**, **y**, and **z** labels of the two lighting axes potentiometers, and for the **Intensity** title on the lighting control-panel. The default value is `"Rom10"`.

`Axiom.2D.messageFont, Axiom.3D.messageFont:` *font*
These indicate the font type to be used for the text in the control-panel message window. The default value is `"Rom14"`.

`Axiom*monochrome:` *switch*
This indicates whether the graph viewports are to be displayed as

if the monitor is black and white, that is, a 1 bit plane. If `on` is specified, the viewport display is black and white. If `off` is specified, or no declaration for this default is given, the viewports are displayed in the normal fashion for the monitor in use. The default value is `"off"`.

`Axiom.2D.postScript:` *filename*
: This specifies the name of the file that is generated when a 2D PostScript graph is saved. The default value is `"axiom2D.ps"`.

`Axiom.3D.postScript:` *filename*
: This specifies the name of the file that is generated when a 3D PostScript graph is saved. The default value is `"axiom3D.ps"`.

`Axiom*titleFont` *font*
: This indicates which font type is used for the title text and, for three-dimensional graphs, in the lighting and viewing-volume control-panel windows. The default value is `"Rom14"`.

`Axiom.2D.unitFont:` *font* (2D only)
: This indicates which font type is used for displaying the unit labels on two-dimensional viewport graphs. The default value is `"6x10"`.

`Axiom.3D.volumeFont:` *font* (3D only)
: This indicates which font type is used for the **x**, **y**, and **z** labels of the clipping region sliders; for the **Perspective**, **Show Region**, and **Clipping On** buttons under **Settings**, and above the windows for the **Hither** and **Eye Distance** sliders in the **Viewing Volume Panel** of the three-dimensional control-panel. The default value is `"Rom8"`.

PART II

Advanced Problem Solving and Examples

CHAPTER 8

Advanced Problem Solving

In this chapter we describe techniques useful in solving advanced problems with AXIOM.

8.1 Numeric Functions

AXIOM provides two basic floating-point types: Float and SmallFloat. This section describes how to use numerical operations defined on these types and the related complex types. As we mentioned in Chapter 1, the Float type is a software implementation of floating-point numbers in which the exponent and the significand may have any number of digits. See 'Float' on page 368 for detailed information about this domain. The SmallFloat (see 'SmallFloat' on page 452) is usually a hardware implementation of floating point numbers, corresponding to machine double precision. The types Complex Float and Complex SmallFloat are the corresponding software implementations of complex floating-point numbers. In this section the term *floating-point type* means any of these four types. The floating-point types implement the basic elementary functions. These include (where "$" means SmallFloat, Float, Complex SmallFloat, or Complex Float):

exp, **log**: `$ -> $`
sin, **cos**, **tan**, **cot**, **sec**, **csc**: `$ -> $`
sin, **cos**, **tan**, **cot**, **sec**, **csc**: `$ -> $`
asin, **acos**, **atan**, **acot**, **asec**, **acsc**: `$ -> $`

sinh, **cosh**, **tanh**, **coth**, **sech**, **csch**: `$ -> $`
asinh, **acosh**, **atanh**, **acoth**, **asech**, **acsch**: `$ -> $`
pi: `() -> $`
sqrt: `$ -> $`
nthRoot: `($, Integer) -> $`
******: `($, Fraction Integer) -> $`
******: `($,$) -> $`

The handling of roots depends on whether the floating-point type is real or complex: for the real floating-point types, SmallFloat and Float, if a real root exists the one with the same sign as the radicand is returned; for the complex floating-point types, the principal value is returned. Also, for real floating-point types the inverse functions produce errors if the results are not real. This includes cases such as `asin(1.2)`, `log(-3.2)`, `sqrt(-1.1)`.

The default floating-point type is Float so to evaluate functions using Float or Complex Float, just use normal decimal notation.

```
exp(3.1)
```

$$22.197951281441633405 \qquad (1)$$

Type: Float

```
exp(3.1 + 4.5 * %i)
```

$$-4.6792348860969899118 - 21.699165928071731864\ \%i \qquad (2)$$

Type: Complex Float

To evaluate functions using SmallFloat or Complex SmallFloat, a declaration or conversion is required.

```
r: SF := 3.1; t: SF := 4.5; exp(r + t*%i)
```

$$-4.6792348860969906 - 21.699165928071732\ \%i \qquad (3)$$

Type: Complex SmallFloat

```
exp(3.1::SF + 4.5::SF * %i)
```

$$-4.6792348860969906 - 21.699165928071732\ \%i \qquad (4)$$

Type: Complex SmallFloat

A number of special functions are provided by the package SmallFloatSpecialFunctions for the machine-precision floating-point types. The special functions provided are listed below, where `F` stands for the types SmallFloat and Complex SmallFloat. The real versions of the functions yield an error if the result is not real.

Gamma: `F -> F`
`Gamma(z)` is the Euler gamma function, (z), defined by

$$(z) = \int_0^\infty t^{z-1} e^{-t} dt.$$

Beta: `F -> F`
`Beta(u, v)` is the Euler Beta function, $B(u, v)$, defined by

$$B(u, v) = \int_0^1 t^{u-1}(1-t)^{v-1} dt.$$

This is related to (z) by

$$B(u, v) = \frac{(u)(v)}{(u+v)}.$$

logGamma: `F -> F`
`logGamma(z)` is the natural logarithm of (z). This can often be computed even if (z) cannot.

digamma: `F -> F`
`digamma(z)`, also called `psi(z)`, is the function $\psi(z)$, defined by

$$\psi(z) = '(z)/(z).$$

polygamma: `(NonNegativeInteger, F) -> F`
`polygamma(n, z)` is the `n`th derivative of $\psi(z)$, written $\psi^{(n)}(z)$.

besselJ: `(F,F) -> F`
`besselJ(v,z)` is the Bessel function of the first kind, $J_\nu(z)$. This function satisfies the differential equation

$$z^2 w''(z) + z w'(z) + (z^2 - \nu^2) w(z) = 0.$$

besselY: `(F,F) -> F`
`besselY(v,z)` is the Bessel function of the second kind, $Y_\nu(z)$. This function satisfies the same differential equation as **besselJ**. The implementation simply uses the relation

$$Y_\nu(z) = \frac{J_\nu(z)\cos(\nu\pi) - J_{-\nu}(z)}{\sin(\nu\pi)}.$$

besselI: `(F,F) -> F`
`besselI(v,z)` is the modified Bessel function of the first kind, $I_\nu(z)$. This function satisfies the differential equation

$$z^2 w''(z) + z w'(z) - (z^2 + \nu^2) w(z) = 0.$$

besselK: `(F,F) -> F`
`besselK(v,z)` is the modified Bessel function of the second kind, $K_\nu(z)$. This function satisfies the same differential equation as **besselI**. The implementation simply uses the relation

$$K_\nu(z) = \pi \frac{I_{-\nu}(z) - I_\nu(z)}{2\sin(\nu\pi)}.$$

airyAi: `F -> F`
`airyAi(z)` is the Airy function $Ai(z)$. This function satisfies the differential equation $w''(z) - zw(z) = 0$. The implementation simply uses the relation

$$Ai(-z) = \frac{1}{3}\sqrt{z}(J_{-1/3}(\frac{2}{3}z^{3/2}) + J_{1/3}(\frac{2}{3}z^{3/2})).$$

airyBi: `F -> F`
`airyBi(z)` is the Airy function $Bi(z)$. This function satisfies the same differential equation as **airyAi**. The implementation simply uses the relation

$$Bi(-z) = \frac{1}{3}\sqrt{3z}(J_{-1/3}(\frac{2}{3}z^{3/2}) - J_{1/3}(\frac{2}{3}z^{3/2})).$$

hypergeometric0F1: `(F,F) -> F`
`hypergeometric0F1(c,z)` is the hypergeometric function ${}_0F_1(;c;z)$.

The above special functions are defined only for small floating-point types. If you give Float arguments, they are converted to SmallFloat by AXIOM.

```
Gamma(0.5)**2
```

3.1415926535897944 (5)

Type: SmallFloat

```
a := 2.1; b := 1.1; besselI(a + %i*b, b*a + 1)
```

2.4894824175473493 − 2.3658460381469562 %i (6)

Type: Complex SmallFloat

A number of additional operations may be used to compute numerical values. These are special polynomial functions that can be evaluated for values in any commutative ring R, and in particular for values in any floating-point type. The following operations are provided by the package OrthogonalPolynomialFunctions:

chebyshevT: `(NonNegativeInteger, R) -> R`
`chebyshevT(n,z)` is the `n` th Chebyshev polynomial of the first kind, $T_n(z)$. These are defined by

$$\frac{1 - tz}{1 - 2tz + t^2} = \sum_{n=0}^{\infty} T_n(z)t^n.$$

chebyshevU: `(NonNegativeInteger, R) -> R`
`chebyshevU(n,z)` is the `n` th Chebyshev polynomial of the second kind, $U_n(z)$. These are defined by

$$\frac{1}{1-2tz+t^2} = \sum_{n=0}^{\infty} U_n(z)t^n.$$

hermiteH: `(NonNegativeInteger, R) -> R`
`hermiteH(n,z)` is the `n` th Hermite polynomial, $H_n(z)$. These are defined by

$$e^{2tz-t^2} = \sum_{n=0}^{\infty} H_n(z)\frac{t^n}{n!}.$$

laguerreL: `(NonNegativeInteger, R) -> R`
`laguerreL(n,z)` is the `n` th Laguerre polynomial, $L_n(z)$. These are defined by

$$\frac{e^{-\frac{tz}{1-t}}}{1-t} = \sum_{n=0}^{\infty} L_n(z)\frac{t^n}{n!}.$$

laguerreL: `(NonNegativeInteger,NonNegativeInteger,R)->R`
`laguerreL(m,n,z)` is the associated Laguerre polynomial, $L_n^m(z)$. This is the `m` th derivative of $L_n(z)$.

legendreP: `(NonNegativeInteger, R) -> R`
`legendreP(n,z)` is the `n` th Legendre polynomial, $P_n(z)$. These are defined by

$$\frac{1}{\sqrt{1-2tz+t^2}} = \sum_{n=0}^{\infty} P_n(z)t^n.$$

These operations require non-negative integers for the indices, but otherwise the argument can be given as desired.

```
[chebyshevT(i, z) for i in 0..5]
```

$$\left[1,\ z,\ 2\ z^2-1,\ 4\ z^3-3\ z,\ 8\ z^4-8\ z^2+1,\ 16\ z^5-20\ z^3+5\ z\right] \qquad (7)$$

Type: List Polynomial Integer

The expression `chebyshevT(n,z)` evaluates to the `n` th Chebyshev polynomial of the first kind.

```
chebyshevT(3, 5.0 + 6.0*%i)
```

$$-1675.0+918.0\ \%i \qquad (8)$$

Type: Complex Float

```
chebyshevT(3, 5.0::SmallFloat)
```

$$485.0 \qquad (9)$$

Type: SmallFloat

The expression `chebyshevU(n,z)` evaluates to the `n` th Chebyshev polynomial of the second kind.

```
[chebyshevU(i, z) for i in 0..5]
```

$$\left[1,\ 2\ z,\ 4\ z^2-1,\ 8\ z^3-4\ z,\ 16\ z^4-12\ z^2+1,\ 32\ z^5-32\ z^3+6\ z\right] \qquad (10)$$

Type: List Polynomial Integer

```
chebyshevU(3, 0.2)
```

$$-0.736 \tag{11}$$

Type: Float

The expression `hermiteH(n,z)` evaluates to the n th Hermite polynomial.

```
[hermiteH(i, z) for i in 0..5]
```

$$\left[1,\ 2\ z,\ 4\ z^2-2,\ 8\ z^3-12\ z,\ 16\ z^4-48\ z^2+12,\ 32\ z^5-160\ z^3+120\ z\right] \tag{12}$$

Type: List Polynomial Integer

```
hermiteH(100, 1.0)
```

$$-0.1448706729337934088E93 \tag{13}$$

Type: Float

The expression `laguerreL(n,z)` evaluates to the n th Laguerre polynomial.

```
[laguerreL(i, z) for i in 0..4]
```

$$\left[1,\ -z+1,\ z^2-4\ z+2,\ -z^3+9\ z^2-18\ z+6,\ z^4-16\ z^3+72\ z^2-96\ z+24\right] \tag{14}$$

Type: List Polynomial Integer

```
laguerreL(4, 1.2)
```

$$-13.0944 \tag{15}$$

Type: Float

```
[laguerreL(j, 3, z) for j in 0..4]
```

$$\left[-z^3+9\ z^2-18\ z+6,\ -3\ z^2+18\ z-18,\ -6\ z+18,\ -6,\ 0\right] \tag{16}$$

Type: List Polynomial Integer

```
laguerreL(1, 3, 2.1)
```

$$6.57 \tag{17}$$

Type: Float

The expression `legendreP(n,z)` evaluates to the n th Legendre polynomial,

```
[legendreP(i,z) for i in 0..5]
```

$$\left[1,\ z,\ \frac{3}{2}\ z^2-\frac{1}{2},\ \frac{5}{2}\ z^3-\frac{3}{2}\ z,\ \frac{35}{8}\ z^4-\frac{15}{4}\ z^2+\frac{3}{8},\ \frac{63}{8}\ z^5-\frac{35}{4}\ z^3+\frac{15}{8}\ z\right] \tag{18}$$

Type: List Polynomial Fraction Integer

```
legendreP(3, 3.0*%i)
```

$-72.0\ \%i$ (19)

Type: Complex Float

Finally, three number-theoretic polynomial operations may be evaluated. The following operations are provided by the package NumberTheoreticPolynomialFunctions. .

bernoulliB: `(NonNegativeInteger, R) -> R`
`bernoulliB(n,z)` is the `n` th Bernoulli polynomial, $B_n(z)$. These are defined by

$$\frac{te^{zt}}{e^t - 1} = \sum_{n=0}^{\infty} B_n(z)\frac{t^n}{n!}.$$

eulerE: `(NonNegativeInteger, R) -> R`
`eulerE(n,z)` is the `n` th Euler polynomial, $E_n(z)$. These are defined by

$$\frac{2e^{zt}}{e^t + 1} = \sum_{n=0}^{\infty} E_n(z)\frac{t^n}{n!}.$$

cyclotomic: `(NonNegativeInteger, R) -> R`
`cyclotomic(n,z)` is the `n` th cyclotomic polynomial $_n(z)$. This is the polynomial whose roots are precisely the primitive `n` th roots of unity. This polynomial has degree given by the Euler totient function $\phi(n)$.

The expression `bernoulliB(n,z)` evaluates to the `n` th Bernoulli polynomial.

```
bernoulliB(3, z)
```

$$z^3 - \frac{3}{2}\ z^2 + \frac{1}{2}\ z \qquad (20)$$

Type: Polynomial Fraction Integer

```
bernoulliB(3, 0.7 + 0.4 * %i)
```

$-0.138 - 0.116\ \%i$ (21)

Type: Complex Float

The expression `eulerE(n,z)` evaluates to the `n` th Euler polynomial.

```
eulerE(3, z)
```

$$z^3 - \frac{3}{2}\ z^2 + \frac{1}{4} \qquad (22)$$

Type: Polynomial Fraction Integer

```
eulerE(3, 0.7 + 0.4 * %i)
```

$-0.238 - 0.316\ \%i$ (23)

Type: Complex Float

The expression cyclotomic(n,z) evaluates to the n th cyclotomic polynomial.

```
cyclotomic(3, z)
```

$$z^2 + z + 1 \qquad (24)$$

Type: Polynomial Integer

```
cyclotomic(3, (-1.0 + 0.0 * %i)**(2/3))
```

$$0.0 \qquad (25)$$

Type: Complex Float

Drawing complex functions in AXIOM is presently somewhat awkward compared to drawing real functions. It is necessary to use the **draw** operations that operate on functions rather than expressions.

This is the complex exponential function (rotated interactively). When this is displayed in color, the height is the value of the real part of the function and the color is the imaginary part. Red indicates large negative imaginary values, green indicates imaginary values near zero and blue/violet indicates large positive imaginary values.

```
draw((x,y)+-> real exp complex(x,y), -2..2, -2*%pi..2*%pi,
  colorFunction == (x, y) +-> imag exp complex(x,y),
  title=="exp(x+%i*y)")
```

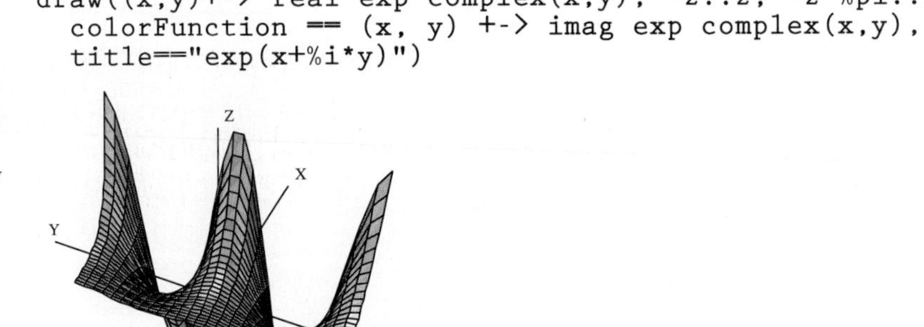

This is the complex arctangent function. Again, the height is the real part of the function value but here the color indicates the function value's phase. The position of the branch cuts are clearly visible and one can see that the function is real only for a real argument.

```
vp := draw((x,y) +-> real atan complex(x,y), -%pi..%pi,
  -%pi..%pi, colorFunction==(x,y) +->argument atan
  complex(x,y), title=="atan(x+%i*y)", style=="shade");
  rotate(vp,-160,-45); vp
```

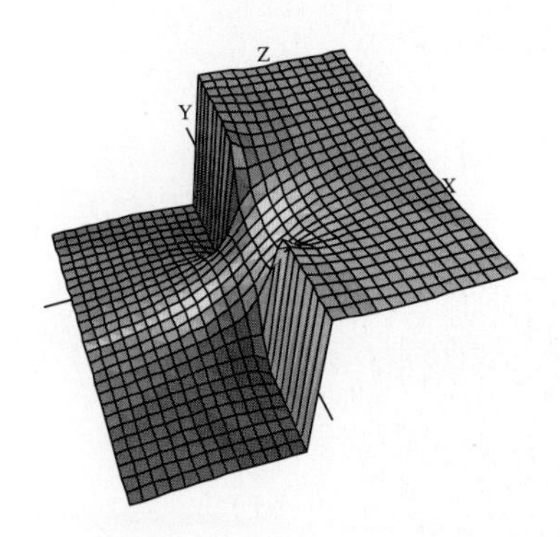

This is the complex Gamma function.

```
draw((x,y) +-> max(min(real Gamma complex(x,y),4),-4),
  -%pi..%pi, -%pi..%pi, style=="shade", colorFunction
  == (x,y) +-> argument Gamma complex(x,y), title ==
  "Gamma(x+%i*y)", var1Steps == 50, var2Steps== 50)
```

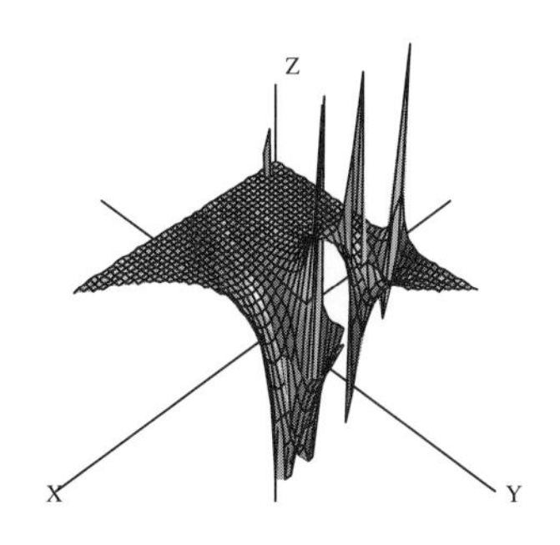

This shows the real Beta function near the origin.

```
draw(Beta(x,y)/100, x=-1.6..1.7, y = -1.6..1.7,
  style=="shade", title=="Beta(x,y)", var1Steps==40,
  var2Steps==40)
```

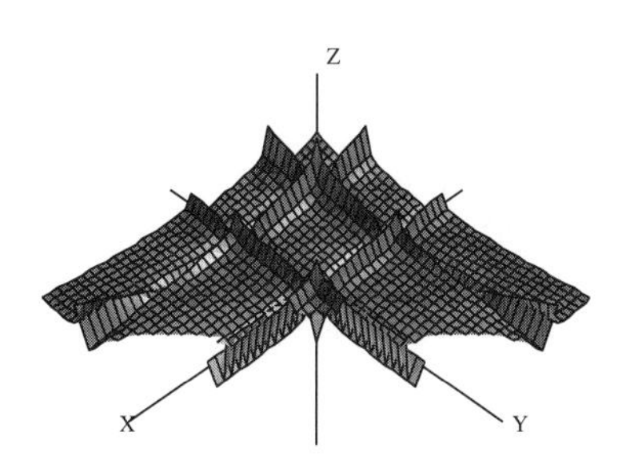

This is the Bessel function $J_\alpha(x)$ for index α in the range -6..4 and argument x in the range 2..14.

```
draw((alpha,x) +-> min(max(besselJ(alpha, x+8), -6), 6),
  -6..4, -6..6, title=="besselJ(alpha,x)", style=="shade",
  var1Steps==40, var2Steps==40)
```

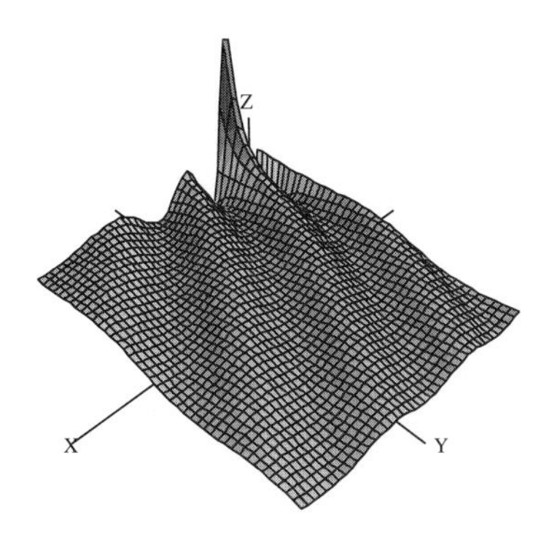

This is the modified Bessel function $I_\alpha(x)$ evaluated for various real values of the index α and fixed argument $x = 5$.

```
draw(besselI(alpha, 5), alpha = -12..12, unit==[5,20])
```

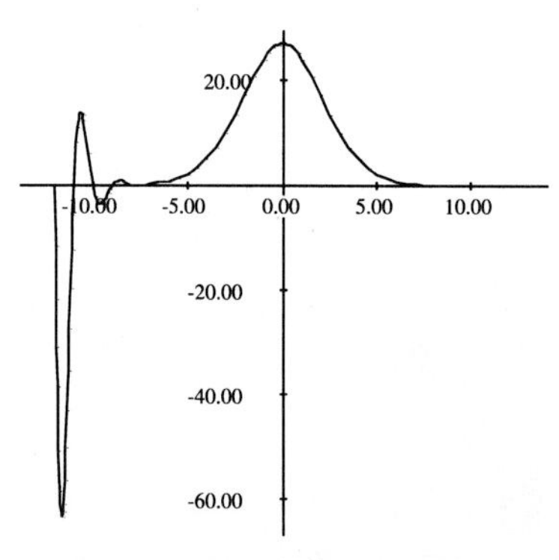

This is similar to the last example except the index α takes on complex values in a `6 x 6` rectangle centered on the origin.

```
draw((x,y) +-> real besselI(complex(x/20, y/20),5), -
  60..60, -60..60, colorFunction == (x,y)+-> argument
  besselI(complex(x/20,y/20),5),
  title=="besselI(x+i*y,5)", style=="shade")
```

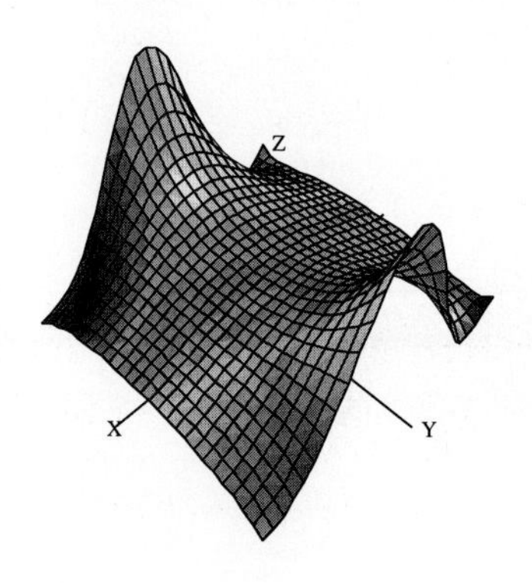

8.2 Polynomial Factorization

The AXIOM polynomial factorization facilities are available for all polynomial types and a wide variety of coefficient domains. Here are some examples.

8.2.1 Integer and Rational Number Coefficients

Polynomials with integer coefficients can be be factored.

```
v := (4*x**3+2*y**2+1)*(12*x**5-x**3*y+12)
```

$$-2\ x^3\ y^3 + \left(24\ x^5 + 24\right)\ y^2 + \left(-4\ x^6 - x^3\right)\ y + 48\ x^8 + 12\ x^5 + 48\ x^3 + 12 \quad (1)$$

Type: Polynomial Integer

```
factor v
```

$$-\left(x^3\ y - 12\ x^5 - 12\right)\ \left(2\ y^2 + 4\ x^3 + 1\right) \qquad (2)$$

Type: Factored Polynomial Integer

Also, AXIOM can factor polynomials with rational number coefficients.

```
w := (4*x**3+(2/3)*x**2+1)*(12*x**5-(1/2)*x**3+12)
```

$$48\ x^8 + 8\ x^7 - 2\ x^6 + \frac{35}{3}\ x^5 + \frac{95}{2}\ x^3 + 8\ x^2 + 12 \qquad (3)$$

Type: Polynomial Fraction Integer

```
factor w
```

$$48\ \left(x^3 + \frac{1}{6}\ x^2 + \frac{1}{4}\right)\ \left(x^5 - \frac{1}{24}\ x^3 + 1\right) \qquad (4)$$

Type: Factored Polynomial Fraction Integer

8.2.2 Simple Algebraic Extension Field Coefficients

Polynomials with coefficients in simple algebraic extensions of the rational numbers can be factored.

Here, aa and bb are symbolic roots of polynomials.

```
aa := rootOf(aa**2+aa+1)
```

$$aa \qquad (1)$$

Type: AlgebraicNumber

```
p:=(x**3+aa**2*x+y)*(aa*x**2+aa*x+aa*y**2)**2
```

$$(-aa - 1)\ y^5 + \left((-aa - 1)\ x^3 + aa\ x\right)\ y^4 +$$

$$\left((-2\ aa - 2)\ x^2 + (-2\ aa - 2)\ x\right)\ y^3 +$$

$$\left((-2\ aa - 2)\ x^5 + (-2\ aa - 2)\ x^4 + 2\ aa\ x^3 + 2\ aa\ x^2\right)\ y^2 + \qquad (2)$$

$$\left((-aa - 1)\ x^4 + (-2\ aa - 2)\ x^3 + (-aa - 1)\ x^2\right)\ y +$$

$$(-aa - 1)\ x^7 + (-2\ aa - 2)\ x^6 - x^5 + 2\ aa\ x^4 + aa\ x^3$$

Type: Polynomial AlgebraicNumber

Note that the second argument to factor can be a list of algebraic extensions to factor over.

```
factor(p,[aa])
```

$$(-aa - 1)\ \left(y + x^3 - \frac{aa}{aa + 1}\ x\right)\ \left(y^2 + x^2 + x\right)^2 \qquad (3)$$

Type: Factored Polynomial AlgebraicNumber

This factors `x**2+3` over the integers.

```
factor(x**2+3)
```

$$x^2 + 3 \tag{4}$$

Type: Factored Polynomial Integer

Factor the same polynomial over the field obtained by adjoining `aa` to the rational numbers.

```
factor(x**2+3, [aa])
```

$$(x - 2\ aa - 1)\ (x + 2\ aa + 1) \tag{5}$$

Type: Factored Polynomial AlgebraicNumber

Factor `x**6+108` over the same field.

```
factor(x**6+108, [aa])
```

$$\left(x^3 - 12\ aa - 6\right)\ \left(x^3 + 12\ aa + 6\right) \tag{6}$$

Type: Factored Polynomial AlgebraicNumber

```
bb:=rootOf(bb**3-2)
```

$$bb \tag{7}$$

Type: AlgebraicNumber

```
factor(x**6+108, [bb])
```

$$\left(x^2 - 3\ bb\ x + 3\ bb^2\right)\ \left(x^2 + 3\ bb^2\right)\ \left(x^2 + 3\ bb\ x + 3\ bb^2\right) \tag{8}$$

Type: Factored Polynomial AlgebraicNumber

Factor again over the field obtained by adjoining both `aa` and `bb` to the rational numbers.

```
factor(x**6+108, [aa,bb])
```

$$\left(x + \frac{(4\ aa + 5)\ bb}{2\ aa - 1}\right)\ \left(x + \frac{(-5\ aa - 4)\ bb}{2\ aa + 3}\right)\ \left(x + \frac{(aa - 4)\ bb}{aa + 3}\right) \cdot \left(x + \frac{(37\ aa + 17)\ bb}{aa + 19}\right)\ \left(x + \frac{(13\ aa + 2)\ bb}{5\ aa - 3}\right)\ \left(x + \frac{(2\ aa + 1)\ bb}{aa + 1}\right) \tag{9}$$

Type: Factored Polynomial AlgebraicNumber

8.2.3 Factoring Rational Functions

Since fractions of polynomials form a field, every element (other than zero) divides any other, so there is no useful notion of irreducible factors. Thus the **factor** operation is not very useful for fractions of polynomials.

There is, instead, a specific operation **factorFraction** that separately factors the numerator and denominator and returns a fraction of the factored results.

```
factorFraction((x**2-4)/(y**2-4))
```

$$\frac{(x - 2)\ (x + 2)}{(y - 2)\ (y + 2)} \tag{1}$$

Type: Fraction Factored Polynomial Integer

You can also use **map**. This expression applies the **factor** operation to the numerator and denominator.

```
map(factor, (x**2-4)/(y**2-4))
```

$$\frac{(x - 2)\ (x + 2)}{(y - 2)\ (y + 2)} \tag{2}$$

Type: Fraction Factored Polynomial Integer

8.3 Manipulating Symbolic Roots of a Polynomial

In this section we show you how to work with one root or all roots of a polynomial. These roots are represented symbolically (as opposed to being numeric approximations). See Section 8.5.2 on page 245 and Section 8.5.3 on page 247 for information about solving for the roots of one or more polynomials.

8.3.1 Using a Single Root of a Polynomial

Use **rootOf** to get a symbolic root of a polynomial: `rootOf(p, x)` returns a root of `p(x)`.

This creates an algebraic number `a`.

```
a := rootOf(a**4+1,a)
```

$$a \tag{1}$$

Type: Expression Integer

To find the algebraic relation that defines `a`, use **definingPolynomial**.

```
definingPolynomial a
```

$$a^4 + 1 \tag{2}$$

Type: Expression Integer

You can use `a` in any further expression, including a nested **rootOf**.

```
b := rootOf(b**2-a-1,b)
```

$$b \tag{3}$$

Type: Expression Integer

Higher powers of the roots are automatically reduced during calculations.

```
a + b
```

$$b + a \tag{4}$$

Type: Expression Integer

```
% ** 5
```

$$\left(10\ a^3 + 11\ a^2 + 2\ a - 4\right)\ b + 15\ a^3 + 10\ a^2 + 4\ a - 10 \tag{5}$$

Type: Expression Integer

The operation **zeroOf** is similar to **rootOf**, except that it may express the root using radicals in some cases.

```
rootOf(c**2+c+1,c)
```

$$c \tag{6}$$

Type: Expression Integer

```
zeroOf(d**2+d+1,d)
```

$$\frac{\sqrt{-3} - 1}{2} \tag{7}$$

Type: Expression Integer

```
rootOf(e**5-2,e)
```

$$e \tag{8}$$

Type: Expression Integer

```
zeroOf(f**5-2,f)
```

$$\sqrt[5]{2} \tag{9}$$

Type: Expression Integer

8.3.2 Using All Roots of a Polynomial

Use **rootsOf** to get all symbolic root of a polynomial: `rootsOf(p, x)` returns a list of all the roots of `p(x)`. If `p(x)` has a multiple root of order `n`, then that root appears `n` times in the list.

Compute all the roots of `x**4 + 1`.

```
l := rootsOf(x**4+1,x)
```

$$[\%x0,\ \%x1,\ \%x2,\ -\%x2 - \%x1 - \%x0] \tag{1}$$

Type: List Expression Integer

As a side effect, the variables `%x0, %x1` and `%x2` are bound to the first three roots of `x**4+1`.

```
%x0**5
```

$$-\%x0 \tag{2}$$

Type: Expression Integer

Although they all satisfy `x**4 + 1 = 0, %x0, %x1,` and `%x2` are different algebraic numbers. To find the algebraic relation that defines each of them, use **definingPolynomial**.

```
definingPolynomial %x0
```

$$\%x0^4 + 1 \tag{3}$$

Type: Expression Integer

```
definingPolynomial %x1
```

$$\%x0^3 + \%x1\ \%x0^2 + \%x1^2\ \%x0 + \%x1^3 \tag{4}$$

Type: Expression Integer

```
definingPolynomial %x2
```

$$\%x1^2 + (\%x0 + \%x2)\ \%x1 + \%x0^2 + \%x2\ \%x0 + \%x2^2 \tag{5}$$

Type: Expression Integer

We can check that the sum and product of the roots of `x**4+1` are its trace and norm.

```
x3 := last l
```

$$-\%x2 - \%x1 - \%x0 \tag{6}$$

Type: Expression Integer

```
%x0 + %x1 + %x2 + x3
```

$$0 \tag{7}$$

Type: Expression Integer

```
%x0 * %x1 * %x2 * x3
```

$$1 \tag{8}$$

Type: Expression Integer

Corresponding to the pair of operations **rootOf/zeroOf** in Section 8.5.2 on page 245, there is an operation **zerosOf** that, like **rootsOf**, computes all the roots of a given polynomial, but which expresses some of them in terms of radicals.

```
zerosOf(y**4+1,y)
```

$$\left[\sqrt[4]{-1},\ \%y1,\ \frac{\sqrt{-3\ \%y1^2 - 2\ \sqrt[4]{-1}\ \%y1 - 3\ \sqrt[4]{-1}^2} - \%y1 - \sqrt[4]{-1}}{2},\right.$$
$$\left.\frac{-\sqrt{-3\ \%y1^2 - 2\ \sqrt[4]{-1}\ \%y1 - 3\ \sqrt[4]{-1}^2} - \%y1 - \sqrt[4]{-1}}{2}\right] \qquad (9)$$

Type: List Expression Integer

As you see, only one implicit algebraic number was created (%y1), and its defining equation is this. The other three roots are expressed in radicals.

```
definingPolynomial %y1
```

$$\sqrt[4]{-1}^3 + \%y1\ \sqrt[4]{-1}^2 + \%y1^2\ \sqrt[4]{-1} + \%y1^3 \qquad (10)$$

Type: Expression Integer

8.4 Computation of Eigenvalues and Eigenvectors

In this section we show you some of AXIOM's facilities for computing and manipulating eigenvalues and eigenvectors, also called characteristic values and characteristic vectors, respectively.

Let's first create a matrix with integer entries.

```
m1 := matrix [[1,2,1],[2,1,-2],[1,-2,4]]
```

$$\begin{bmatrix} 1 & 2 & 1 \\ 2 & 1 & -2 \\ 1 & -2 & 4 \end{bmatrix} \qquad (1)$$

Type: Matrix Integer

To get a list of the *rational* eigenvalues, use the operation **eigenvalues**.

```
leig := eigenvalues(m1)
```

$$[5] \qquad (2)$$

Type: List Fraction Polynomial Fraction Integer

Given an explicit eigenvalue, **eigenvector** computes the eigenvectors corresponding to it.

```
eigenvector(first(leig),m1)
```

$$\left[\begin{bmatrix} 0 \\ -\frac{1}{2} \\ 1 \end{bmatrix}\right] \qquad (3)$$

Type: List Matrix Fraction Polynomial Fraction Integer

The operation **eigenvectors** returns a list of pairs of values and vectors. When an eigenvalue is rational, AXIOM gives you the value explicitly; otherwise, its minimal polynomial is given, (the polynomial of lowest degree with the eigenvalues as roots), together with a parametric representation of the eigenvector using the eigenvalue. This means that if you ask AXIOM to **solve** the minimal polynomial, then you can substitute these roots into the parametric form of the corresponding eigenvectors.

You must be aware that unless an exact eigenvalue has been computed, the eigenvector may be badly in error.

```
eigenvectors(m1)
```

$$\left[\left[eigval = 5, \ eigmult = 1, \ \ eigvec = \left[\begin{bmatrix} 0 \\ -\frac{1}{2} \\ 1 \end{bmatrix} \right] \right], \right.$$

$$\left. \left[algrel = \%C^2 - \%C - 5, \ algmult = 1, \ \ algvec = \left[\begin{bmatrix} \%C \\ 2 \\ 1 \end{bmatrix} \right] \right] \right] \tag{4}$$

Type: List Union(Record(algrel: Fraction Polynomial Fraction Integer, algmult: Integer, algvec: List Matrix Fraction Polynomial Fraction Integer), Record(eigval: Fraction Polynomial Fraction Integer, eigmult: Integer, eigvec: List Matrix Fraction Polynomial Fraction Integer))

Another possibility is to use the operation **radicalEigenvectors** tries to compute explicitly the eigenvectors in terms of radicals.

```
radicalEigenvectors(m1)
```

$$\left[\left[radval = 5, \ radmult = 1, \ \ radvect = \left[\begin{bmatrix} 0 \\ -\frac{1}{2} \\ 1 \end{bmatrix} \right] \right], \right.$$

$$\left[radval = \frac{1}{2}\sqrt{21} + \frac{1}{2}, \ radmult = 1, \right.$$

$$\left. radvect = \left[\begin{bmatrix} \frac{1}{2}\sqrt{21} + \frac{1}{2} \\ 2 \\ 1 \end{bmatrix} \right] \right], \tag{5}$$

$$\left[radval = -\frac{1}{2}\sqrt{21} + \frac{1}{2}, \ radmult = 1, \right.$$

$$\left. \left. radvect = \left[\begin{bmatrix} -\frac{1}{2}\sqrt{21} + \frac{1}{2} \\ 2 \\ 1 \end{bmatrix} \right] \right] \right]$$

Type: List Record(radval: Expression Fraction Integer, radmult: Integer, radvect: List Matrix Expression Fraction Integer)

Alternatively, AXIOM can compute real or complex approximations to the eigenvectors and eigenvalues using the operations **realEigenvectors** or **complexEigenvectors**. They each take an additional argument ϵ to specify the "precision" required. In the real case, this means that each approximation will be within $\pm\epsilon$ of the actual result. In the complex case, this means that each approximation will be within $\pm\epsilon$ of the actual result in each of the real and imaginary parts.

The precision can be specified as a Float if the results are desired in floating-point notation, or as Fraction Integer if the results are to be expressed using rational (or complex rational) numbers.

```
realEigenvectors(m1,1/1000)
```

$$\left[\left[outval = 5,\ outmult = 1,\ \ outvect = \left[\left[\begin{array}{c} 0 \\ -\frac{1}{2} \\ 1 \end{array}\right]\right]\right],\right.$$

$$\left[outval = \frac{5717}{2048},\ outmult = 1,\ \ outvect = \left[\left[\begin{array}{c} \frac{5717}{2048} \\ 2 \\ 1 \end{array}\right]\right]\right], \tag{6}$$

$$\left.\left[outval = -\frac{3669}{2048},\ outmult = 1,\ \ outvect = \left[\left[\begin{array}{c} -\frac{3669}{2048} \\ 2 \\ 1 \end{array}\right]\right]\right]\right]$$

Type: List Record(outval: Fraction Integer, outmult: Integer, outvect: List Matrix Fraction Integer)

If an n by n matrix has n distinct eigenvalues (and therefore n eigenvectors) the operation **eigenMatrix** gives you a matrix of the eigenvectors.

```
eigenMatrix(m1)
```

$$\left[\begin{array}{ccc} 0 & \frac{1}{2}\sqrt{21}+\frac{1}{2} & -\frac{1}{2}\sqrt{21}+\frac{1}{2} \\ -\frac{1}{2} & 2 & 2 \\ 1 & 1 & 1 \end{array}\right] \tag{7}$$

Type: Union(Matrix Expression Fraction Integer, ...)

```
m2 := matrix [[-5,-2],[18,7]]
```

$$\left[\begin{array}{cc} -5 & -2 \\ 18 & 7 \end{array}\right] \tag{8}$$

Type: Matrix Integer

```
eigenMatrix(m2)
```

"failed" (9)

Type: Union("failed", ...)

If a symmetric matrix has a basis of orthonormal eigenvectors, then **orthonormalBasis** computes a list of these vectors.

```
m3 := matrix [[1,2],[2,1]]
```

$$\left[\begin{array}{cc} 1 & 2 \\ 2 & 1 \end{array}\right] \tag{10}$$

Type: Matrix Integer

```
orthonormalBasis(m3)
```

$$\left[\left[\begin{array}{c} -\frac{1}{\sqrt{2}} \\ \frac{1}{\sqrt{2}} \end{array}\right], \left[\begin{array}{c} \frac{1}{\sqrt{2}} \\ \frac{1}{\sqrt{2}} \end{array}\right]\right] \tag{11}$$

Type: List Matrix Expression Fraction Integer

8.5 Solution of Linear and Polynomial Equations

In this section we discuss the AXIOM facilities for solving systems of linear equations, finding the roots of polynomials and solving systems of polynomial equations. For a discussion of the solution of differential equations, see Section 8.10 on page 269.

8.5.1 Solution of Systems of Linear Equations

You can use the operation **solve** to solve systems of linear equations.

The operation **solve** takes two arguments, the list of equations and the list of the unknowns to be solved for. A system of linear equations need not have a unique solution.

To solve the linear system:

$$\begin{array}{rcrcrcr} x & + & y & + & z & = & 8 \\ 3x & - & 2y & + & z & = & 0 \\ x & + & 2y & + & 2z & = & 17 \end{array}$$

evaluate this expression.

```
solve([x+y+z=8,3*x-2*y+z=0,x+2*y+2*z=17],[x,y,z])
```

$$[[x = -1,\ y = 2,\ z = 7]] \tag{1}$$

Type: List List Equation Fraction Polynomial Integer

Parameters are given as new variables starting with a percent sign and "%" and the variables are expressed in terms of the parameters. If the system has no solutions then the empty list is returned.

When you solve the linear system

$$\begin{array}{rcrcrcr} x & + & 2y & + & 3z & = & 2 \\ 2x & + & 3y & + & 4z & = & 2 \\ 3x & + & 4y & + & 5z & = & 2 \end{array}$$

with this expression you get a solution involving a parameter.

```
solve([x+2*y+3*z=2,2*x+3*y+4*z=2,3*x+4*y+5*z=2],[x,y,z])
```

$$[[x = \%I - 2,\ y = -2\ \%I + 2,\ z = \%I]] \tag{2}$$

Type: List List Equation Fraction Polynomial Integer

The system can also be presented as a matrix and a vector. The matrix contains the coefficients of the linear equations and the vector contains the numbers appearing on the right-hand sides of the equations. You may input the matrix as a list of rows and the vector as a list of its elements.

To solve the system:

$$\begin{array}{rcrcrcr} x & + & y & + & z & = & 8 \\ 3x & - & 2y & + & z & = & 0 \\ x & + & 2y & + & 2z & = & 17 \end{array}$$

in matrix form you would evaluate this expression.

```
solve([[1,1,1],[3,-2,1],[1,2,2]],[8,0,17])
```

$$[particular = [-1,\ 2,\ 7],\ basis = [[0,\ 0,\ 0]]] \tag{3}$$

Type: Record(particular: Union(Vector Fraction Integer, "failed"), basis: List Vector Fraction Integer)

The solutions are presented as a Record with two components: the component *particular* contains a particular solution of the given system or the item `"failed"` if there are no solutions, the component *basis* contains a list of vectors that are a basis for the space of solutions of the corresponding homogeneous system. If the system of linear equations does not have a unique solution, then the *basis* component contains non-trivial vectors.

This happens when you solve the linear system

$$\begin{array}{ccccccc} x & + & 2y & + & 3z & = & 2 \\ 2x & + & 3y & + & 4z & = & 2 \\ 3x & + & 4y & + & 5z & = & 2 \end{array}$$

with this command.

```
solve([[1,2,3],[2,3,4],[3,4,5]],[2,2,2])
```

$$\left[particular = [-2,\ 2,\ 0],\ basis = [[1,\ -2,\ 1]]\right] \quad (4)$$

Type: Record(particular: Union(Vector Fraction Integer, "failed"), basis: List Vector Fraction Integer)

All solutions of this system are obtained by adding the particular solution with a linear combination of the *basis* vectors.

When no solution exists then "failed" is returned as the *particular* component, as follows:

```
solve([[1,2,3],[2,3,4],[3,4,5]],[2,3,2])
```

$$\left[particular = \texttt{"failed"},\ basis = [[1,\ -2,\ 1]]\right] \quad (5)$$

Type: Record(particular: Union(Vector Fraction Integer, "failed"), basis: List Vector Fraction Integer)

When you want to solve a system of homogeneous equations (that is, a system where the numbers on the right-hand sides of the equations are all zero) in the matrix form you can omit the second argument and use the **nullSpace** operation.

This computes the solutions of the following system of equations:

$$\begin{array}{ccccccc} x & + & 2y & + & 3z & = & 0 \\ 2x & + & 3y & + & 4z & = & 0 \\ 3x & + & 4y & + & 5z & = & 0 \end{array}$$

The result is given as a list of vectors and these vectors form a basis for the solution space.

```
nullSpace([[1,2,3],[2,3,4],[3,4,5]])
```

$$[[1,\ -2,\ 1]] \quad (6)$$

Type: List Vector Integer

8.5.2 Solution of a Single Polynomial Equation

AXIOM can solve polynomial equations producing either approximate or exact solutions. Exact solutions are either members of the ground field or can be presented symbolically as roots of irreducible polynomials.

This returns the one rational root along with an irreducible polynomial describing the other solutions.

```
solve(x**3 = 8,x)
```

$$\left[x = 2,\ x^2 + 2\ x + 4 = 0\right] \quad (1)$$

Type: List Equation Fraction Polynomial Integer

If you want solutions expressed in terms of radicals you would use this instead.

```
radicalSolve(x**3 = 8,x)
```

$$\left[x = -\sqrt{-3} - 1,\ x = \sqrt{-3} - 1,\ x = 2\right] \quad (2)$$

Type: List Equation Expression Integer

The **solve** command always returns a value but **radicalSolve** returns only the solutions that it is able to express in terms of radicals.

If the polynomial equation has rational coefficients you can ask for approximations to its real roots by calling solve with a second argument that specifies the "precision" ϵ. This means that each approximation will be within $\pm\epsilon$ of the actual result.

Notice that the type of second argument controls the type of the result.

```
solve(x**4 - 10*x**3 + 35*x**2 - 50*x + 25,.0001)
```

$$[x = 3.618011474609375,\ x = 1.381988525390625] \quad (3)$$

Type: List Equation Polynomial Float

If you give a floating-point precision you get a floating-point result; if you give the precision as a rational number you get a rational result.

```
solve(x**3-2,1/1000)
```

$$\left[x = \frac{2581}{2048}\right] \quad (4)$$

Type: List Equation Polynomial Fraction Integer

If you want approximate complex results you should use the command **complexSolve** that takes the same precision argument ϵ.

```
complexSolve(x**3-2,.0001)
```

$$[x = 1.259918212890625,$$

$$x = -0.62989432795395613131 - 1.091094970703125\ \%i, \quad (5)$$

$$x = -0.62989432795395613131 + 1.091094970703125\ \%i]$$

Type: List Equation Polynomial Complex Float

Each approximation will be within $\pm\epsilon$ of the actual result in each of the real and imaginary parts.

```
complexSolve(x**2-2*%i+1,1/100)
```

$$\left[x = -\frac{13028925}{16777216} - \frac{325}{256}\ \%i,\ x = \frac{13028925}{16777216} + \frac{325}{256}\ \%i\right] \quad (6)$$

Type: List Equation Polynomial Complex Fraction Integer

Note that if you omit the "=" from the first argument AXIOM generates an equation by equating the first argument to zero. Also, when only one variable is present in the equation, you do not need to specify the variable to be solved for, that is, you can omit the second argument.

AXIOM can also solve equations involving rational functions. Solutions where the denominator vanishes are discarded.

```
radicalSolve(1/x**3 + 1/x**2 + 1/x = 0,x)
```

$$\left[x = \frac{-\sqrt{-3} - 1}{2},\ x = \frac{\sqrt{-3} - 1}{2}\right] \quad (7)$$

Type: List Equation Expression Integer

8.5.3 Solution of Systems of Polynomial Equations

Given a system of equations of rational functions with exact coefficients:

$$p_1(x_1, \ldots, x_n)$$
$$\vdots$$
$$p_m(x_1, \ldots, x_n)$$

AXIOM can find numeric or symbolic solutions. The system is first split into irreducible components, then for each component, a triangular system of equations is found that reduces the problem to sequential solution of univariate polynomials resulting from substitution of partial solutions from the previous stage.

$$q_1(x_1, \ldots, x_n)$$
$$\vdots$$
$$q_m(x_n)$$

Symbolic solutions can be presented using "implicit" algebraic numbers defined as roots of irreducible polynomials or in terms of radicals. AXIOM can also find approximations to the real or complex roots of a system of polynomial equations to any user-specified accuracy.

The operation **solve** for systems is used in a way similar to **solve** for single equations. Instead of a polynomial equation, one has to give a list of equations and instead of a single variable to solve for, a list of variables. For solutions of single equations see Section 8.5.2 on page 245.

Use the operation **solve** if you want implicitly presented solutions.

```
solve([3*x**3 + y + 1,y**2 -4],[x,y])
```

$$\left[\left[x=-1,\ y=2\right],\ \left[x^2-x+1=0,\ y=2\right],\ \left[3\ x^3-1=0,\ y=-2\right]\right] \quad (1)$$

Type: List List Equation Fraction Polynomial Integer

```
solve([x = y**2-19,y = z**2+x+3,z = 3*x],[x,y,z])
```

$$\left[\left[x=\frac{z}{3},\ y=\frac{3\ z^2+z+9}{3},\ 9\ z^4+6\ z^3+55\ z^2+15\ z-90=0\right]\right] \quad (2)$$

Type: List List Equation Fraction Polynomial Integer

Use **radicalSolve** if you want your solutions expressed in terms of radicals.

```
radicalSolve([3*x**3 + y + 1,y**2 -4],[x,y])
```

$$\left[\left[x = \frac{\sqrt{-3}+1}{2}, \ y = 2\right], \ \left[x = \frac{-\sqrt{-3}+1}{2}, \ y = 2\right], \right.$$
$$\left[x = \frac{-\sqrt{-1}\ \sqrt{3}-1}{2\ \sqrt[3]{3}}, \ y = -2\right], \ \left[x = \frac{\sqrt{-1}\ \sqrt{3}-1}{2\ \sqrt[3]{3}}, \ y = -2\right], \tag{3}$$
$$\left.\left[x = \frac{1}{\sqrt[3]{3}}, \ y = -2\right], \ [x = -1, \ y = 2]\right]$$

Type: List List Equation Expression Integer

To get numeric solutions you only need to give the list of equations and the precision desired. The list of variables would be redundant information since there can be no parameters for the numerical solver.

If the precision is expressed as a floating-point number you get results expressed as floats.

```
solve([x**2*y - 1,x*y**2 - 2],.01)
```

$$[[y = 1.5859375, \ x = 0.79296875]] \tag{4}$$

Type: List List Equation Polynomial Float

To get complex numeric solutions, use the operation **complexSolve**, which takes the same arguments as in the real case.

```
complexSolve([x**2*y - 1,x*y**2 - 2],1/1000)
```

$$\left[\left[y = \frac{1625}{1024}, \ x = \frac{1625}{2048}\right], \right.$$
$$\left[y = -\frac{435445573689}{549755813888} - \frac{1407}{1024}\ \%i, \right.$$
$$\left. x = -\frac{435445573689}{1099511627776} - \frac{1407}{2048}\ \%i\right], \tag{5}$$
$$\left[y = -\frac{435445573689}{549755813888} + \frac{1407}{1024}\ \%i, \right.$$
$$\left.\left. x = -\frac{435445573689}{1099511627776} + \frac{1407}{2048}\ \%i\right]\right]$$

Type: List List Equation Polynomial Complex Fraction Integer

It is also possible to solve systems of equations in rational functions over the rational numbers. Note that `[x = 0.0, a = 0.0]` is not returned as a solution since the denominator vanishes there.

```
solve([x**2/a = a,a = a*x],.001)
```

$$[[x = 1.0, \ a = -1.0], \ [x = 1.0, \ a = 1.0]] \tag{6}$$

Type: List List Equation Polynomial Float

When solving equations with denominators, all solutions where the denominator vanishes are discarded.

```
radicalSolve([x**2/a + a + y**3 - 1,a*y + a + 1],[x,y])
```

$$\left[\left[x=-\sqrt{\frac{-a^4+2\,a^3+3\,a^2+3\,a+1}{a^2}},\ y=\frac{-a-1}{a}\right],\right.$$
$$\left.\left[x=\sqrt{\frac{-a^4+2\,a^3+3\,a^2+3\,a+1}{a^2}},\ y=\frac{-a-1}{a}\right]\right] \quad (7)$$

Type: List List Equation Expression Integer

8.6 Limits

To compute a limit, you must specify a functional expression, a variable, and a limiting value for that variable. If you do not specify a direction, AXIOM attempts to compute a two-sided limit.

Issue this to compute the limit

$$\lim_{x\to 1} \frac{x^2-3x+2}{x^2-1}.$$

```
limit((x**2 - 3*x + 2)/(x**2 - 1),x = 1)
```

$$-\frac{1}{2} \quad (1)$$

Type: Union(OrderedCompletion Fraction Polynomial Integer, ...)

Sometimes the limit when approached from the left is different from the limit from the right and, in this case, you may wish to ask for a one-sided limit. Also, if you have a function that is only defined on one side of a particular value, you can compute a one-sided limit.

The function `log(x)` is only defined to the right of zero, that is, for `x > 0`. Thus, when computing limits of functions involving `log(x)`, you probably want a "right-hand" limit.

```
limit(x * log(x),x = 0,"right")
```

$$0 \quad (2)$$

Type: Union(OrderedCompletion Expression Integer, ...)

When you do not specify `"right"` or `"left"` as the optional fourth argument, **limit** tries to compute a two-sided limit. Here the limit from the left does not exist, as AXIOM indicates when you try to take a two-sided limit.

```
limit(x * log(x),x = 0)
```

$$[leftHandLimit = \texttt{"failed"},\ rightHandLimit = 0] \quad (3)$$

Type: Union(Record(leftHandLimit: Union(OrderedCompletion Expression Integer, "failed"), rightHandLimit: Union(OrderedCompletion Expression Integer, "failed")), ...)

A function can be defined on both sides of a particular value, but tend to different limits as its variable approaches that value from the left and from the right. We can construct an example of this as follows: Since $\sqrt{y^2}$ is simply the absolute value of y, the function $\sqrt{y^2}/y$ is simply the sign (+1 or -1) of the nonzero real number y. Therefore, $\sqrt{y^2}/y = -1$ for `y < 0` and $\sqrt{y^2}/y = +1$ for `y > 0`.

This is what happens when we take the limit at `y = 0`. The answer returned by AXIOM gives both a "left-hand" and a "right-hand" limit.

```
limit(sqrt(y**2)/y,y = 0)
```

$$[leftHandLimit = -1,\ rightHandLimit = 1] \qquad (4)$$

Type: Union(Record(leftHandLimit: Union(OrderedCompletion Expression Integer, "failed"), rightHandLimit: Union(OrderedCompletion Expression Integer, "failed")), ...)

Here is another example, this time using a more complicated function.

```
limit(sqrt(1 - cos(t))/t,t = 0)
```

$$\left[leftHandLimit = -\frac{1}{\sqrt{2}},\ rightHandLimit = \frac{1}{\sqrt{2}}\right] \qquad (5)$$

Type: Union(Record(leftHandLimit: Union(OrderedCompletion Expression Integer, "failed"), rightHandLimit: Union(OrderedCompletion Expression Integer, "failed")), ...)

You can compute limits at infinity by passing either $+\infty$ or $-\infty$ as the third argument of **limit**.

To do this, use the constants `%plusInfinity` and `%minusInfinity`.

```
limit(sqrt(3*x**2 + 1)/(5*x),x = %plusInfinity)
```

$$\frac{\sqrt{3}}{5} \qquad (6)$$

Type: Union(OrderedCompletion Expression Integer, ...)

```
limit(sqrt(3*x**2 + 1)/(5*x),x = %minusInfinity)
```

$$-\frac{\sqrt{3}}{5} \qquad (7)$$

Type: Union(OrderedCompletion Expression Integer, ...)

You can take limits of functions with parameters. As you can see, the limit is expressed in terms of the parameters.

```
limit(sinh(a*x)/tan(b*x),x = 0)
```

$$\frac{a}{b} \qquad (8)$$

Type: Union(OrderedCompletion Expression Integer, ...)

When you use **limit**, you are taking the limit of a real function of a real variable.

When you compute this, AXIOM returns `0` because, as a function of a real variable, `sin(1/z)` is always between `-1` and `1`, so `z * sin(1/z)` tends to `0` as `z` tends to `0`.

```
limit(z * sin(1/z),z = 0)
```

$$0 \qquad (9)$$

Type: Union(OrderedCompletion Expression Integer, ...)

However, as a function of a *complex* variable, `sin(1/z)` is badly behaved near `0` (one says that `sin(1/z)` has an *essential singularity* at `z = 0`).

When viewed as a function of a complex variable, `z * sin(1/z)` does not approach any limit as `z` tends to `0` in the complex plane. AXIOM indicates this when we call **complexLimit**.

```
complexLimit(z * sin(1/z),z = 0)
```

"failed" (10)

Type: Union("failed", ...)

You can also take complex limits at infinity, that is, limits of a function of `z` as `z` approaches infinity on the Riemann sphere. Use the symbol `%infinity` to denote "complex infinity."

As above, to compute complex limits rather than real limits, use **complexLimit**.

```
complexLimit((2 + z)/(1 - z),z = %infinity)
```

$$-1 \quad (11)$$

Type: OnePointCompletion Fraction Polynomial Integer

In many cases, a limit of a real function of a real variable exists when the corresponding complex limit does not. This limit exists.

```
limit(sin(x)/x,x = %plusInfinity)
```

$$0 \quad (12)$$

Type: Union(OrderedCompletion Expression Integer, ...)

But this limit does not.

```
complexLimit(sin(x)/x,x = %infinity)
```

"failed" (13)

Type: Union("failed", ...)

8.7 Laplace Transforms

AXIOM can compute some forward Laplace transforms, mostly of elementary functions not involving logarithms, although some cases of special functions are handled.

To compute the forward Laplace transform of `F(t)` with respect to `t` and express the result as `f(s)`, issue the command `laplace(F(t), t, s)`.

```
laplace(sin(a*t)*cosh(a*t)-cos(a*t)*sinh(a*t), t, s)
```

$$\frac{4\ a^3}{s^4 + 4\ a^4} \quad (1)$$

Type: Expression Integer

Here are some other non-trivial examples.

```
laplace((exp(a*t) - exp(b*t))/t, t, s)
```

$$-\log(s - a) + \log(s - b) \quad (2)$$

Type: Expression Integer

```
laplace(2/t * (1 - cos(a*t)), t, s)
```

$$\log\left(s^2 + a^2\right) - 2\ \log(s) \quad (3)$$

Type: Expression Integer

```
laplace(exp(-a*t) * sin(b*t) / b**2, t, s)
```

$$\frac{1}{b\ s^2+2\ a\ b\ s+b^3+a^2\ b} \tag{4}$$

Type: Expression Integer

```
laplace((cos(a*t) - cos(b*t))/t, t, s)
```

$$\frac{\log\left(s^2+b^2\right)-\log\left(s^2+a^2\right)}{2} \tag{5}$$

Type: Expression Integer

AXIOM also knows about a few special functions.

```
laplace(exp(a*t+b)*Ei(c*t), t, s)
```

$$\frac{\%e^b\ \log\left(\frac{s+c-a}{c}\right)}{s-a} \tag{6}$$

Type: Expression Integer

```
laplace(a*Ci(b*t) + c*Si(d*t), t, s)
```

$$\frac{a\ \log\left(\frac{s^2+b^2}{b^2}\right)+2\ c\ \arctan\left(\frac{d}{s}\right)}{2\ s} \tag{7}$$

Type: Expression Integer

When AXIOM does not know about a particular transform, it keeps it as a formal transform in the answer.

```
laplace(sin(a*t) - a*t*cos(a*t) + exp(t**2), t, s)
```

$$\frac{\left(s^4+2\ a^2\ s^2+a^4\right)\ \mathrm{laplace}\left(\%e^{t^2},\ t,\ s\right)+2\ a^3}{s^4+2\ a^2\ s^2+a^4} \tag{8}$$

Type: Expression Integer

8.8 Integration

Integration is the reverse process of differentiation, that is, an *integral* of a function `f` with respect to a variable `x` is any function `g` such that `D(g,x)` is equal to `f`.

The package FunctionSpaceIntegration provides the top-level integration operation, **integrate**, for integrating real-valued elementary functions.

```
integrate(cosh(a*x)*sinh(a*x), x)
```

$$\frac{\sinh(a\ x)^2+\cosh(a\ x)^2}{4\ a} \tag{1}$$

Type: Union(Expression Integer, ...)

Unfortunately, antiderivatives of most functions cannot be expressed in terms of elementary functions.

```
integrate(log(1 + sqrt(a * x + b)) / x, x)
```

$$\int^x \frac{\log\left(\sqrt{b+\%W\ a}+1\right)}{\%W}\ d\%W \tag{2}$$

Type: Union(Expression Integer, ...)

Given an elementary function to integrate, AXIOM returns a formal integral as above only when it can prove that the integral is not elementary and not when it cannot determine the integral. In this rare case it prints a message that it cannot determine if an elementary integral exists.

Similar functions may have antiderivatives that look quite different because the form of the antiderivative depends on the sign of a constant that appears in the function.

```
integrate(1/(x**2 - 2),x)
```

$$\frac{\log\left(\left(x^2+2\right)\sqrt{2}-4\ x\right)-\log\left(x^2-2\right)}{2\ \sqrt{2}} \tag{3}$$

Type: Union(Expression Integer, ...)

```
integrate(1/(x**2 + 2),x)
```

$$\frac{\arctan\left(\frac{x}{\sqrt{2}}\right)}{\sqrt{2}} \tag{4}$$

Type: Union(Expression Integer, ...)

If the integrand contains parameters, then there may be several possible antiderivatives, depending on the signs of expressions of the parameters.

In this case AXIOM returns a list of answers that cover all the possible cases. Here you use the answer involving the square root of a when a > 0 and the answer involving the square root of -a when a < 0.

```
integrate(x**2 / (x**4 - a**2), x)
```

$$\left[\frac{\log\left(\left(x^2+a\right)\sqrt{a}-2\ a\ x\right)+2\ \arctan\left(\frac{x}{\sqrt{a}}\right)-\log\left(x^2-a\right)}{4\ \sqrt{a}},\right.$$
$$\left.\frac{\log\left(\left(x^2-a\right)\sqrt{-a}+2\ a\ x\right)+2\ \arctan\left(\frac{x}{\sqrt{-a}}\right)-\log\left(x^2+a\right)}{4\ \sqrt{-a}}\right] \tag{5}$$

Type: Union(List Expression Integer, ...)

If the parameters and the variables of integration can be complex numbers rather than real, then the notion of sign is not defined. In this case all the possible answers can be expressed as one complex function. To get that function, rather than a list of real functions, use **complexIntegrate**, which is provided by the package FunctionSpaceComplexIntegration.

This operation is used for integrating complex-valued elementary functions.

```
complexIntegrate(x**2 / (x**4 - a**2), x)
```

$$\frac{\left(\begin{array}{l}\sqrt{4\ a}\log\left(\frac{-\sqrt{-4\ a}+2\ x}{4\ \sqrt{-4\ a}}\right)+\sqrt{-4\ a}\log\left(\frac{-\sqrt{4\ a}+2\ x}{4\ \sqrt{4\ a}}\right)- \\ \sqrt{-4\ a}\log\left(\frac{-\sqrt{4\ a}-2\ x}{4\ \sqrt{4\ a}}\right)-\sqrt{4\ a}\log\left(\frac{-\sqrt{-4\ a}-2\ x}{4\ \sqrt{-4\ a}}\right)\end{array}\right)}{2\ \sqrt{-4\ a}\ \sqrt{4\ a}} \tag{6}$$

Type: Expression Integer

As with the real case, antiderivatives for most complex-valued functions cannot be expressed in terms of elementary functions.

```
complexIntegrate(log(1 + sqrt(a * x + b)) / x, x)
```

$$\int^{x} \frac{\log\left(\sqrt{b + \%W\ a} + 1\right)}{\%W}\, d\%W \tag{7}$$

Type: Expression Integer

Sometimes **integrate** can involve symbolic algebraic numbers such as those returned by **rootOf**. To see how to work with these strange generated symbols (such as `%%a0`), see Section 8.3.2 on page 240.

Definite integration is the process of computing the area between the `x`-axis and the curve of a function `f(x)`. The fundamental theorem of calculus states that if `f` is continuous on an interval `a..b` and if there exists a function `g` that is differentiable on `a..b` and such that `D(g, x)` is equal to `f`, then the definite integral of `f` for `x` in the interval `a..b` is equal to `g(b) - g(a)`.

The package RationalFunctionDefiniteIntegration provides the top-level definite integration operation, **integrate**, for integrating real-valued rational functions.

```
integrate((x**4 - 3*x**2 + 6)/(x**6-5*x**4+5*x**2+4), x =
  1..2)
```

$$\frac{2\arctan(8) + 2\arctan(5) + 2\arctan(2) - 2\arctan\left(-\frac{1}{2}\right) - \%pi}{2} \tag{8}$$

Type: Union(f1: OrderedCompletion Expression Integer, ...)

AXIOM checks beforehand that the function you are integrating is defined on the interval `a..b`, and prints an error message if it finds that this is not case, as in the following example:

```
integrate(1/(x**2-2), x = 1..2)

    >> Error detected within library code:
       Pole in path of integration
       You are being returned to the top level
       of the interpreter.
```

When parameters are present in the function, the function may or may not be defined on the interval of integration.

If this is the case, AXIOM issues a warning that a pole might lie in the path of integration, and does not compute the integral.

```
integrate(1/(x**2-a), x = 1..2)
```

$$potentialPole \tag{9}$$

Type: Union(pole: potentialPole, ...)

If you know that you are using values of the parameter for which the function has no pole in the interval of integration, use the string `"noPole"` as a third argument to **integrate**:

The value here is, of course, incorrect if `sqrt(a)` is between `1` and `2`.

```
integrate(1/(x**2-a), x = 1..2, "noPole")
```

$$\left[\frac{\left(\begin{array}{l}-\log\left(\left(-4\,a^2-4\,a\right)\sqrt{a}+a^3+6\,a^2+a\right)+\\ \log\left(\left(-8\,a^2-32\,a\right)\sqrt{a}+a^3+24\,a^2+16\,a\right)+\\ \log\left(a^2-2\,a+1\right)-\log\left(a^2-8\,a+16\right)\end{array}\right)}{4\,\sqrt{a}},\right. \tag{10}$$

$$\left.\frac{\arctan\left(\frac{2}{\sqrt{-a}}\right)-\arctan\left(\frac{1}{\sqrt{-a}}\right)}{\sqrt{-a}}\right]$$

Type: Union(f2: List OrderedCompletion Expression Integer, ...)

8.9 Working with Power Series

AXIOM has very sophisticated facilities for working with power series. Infinite series are represented by a list of the coefficients that have already been determined, together with a function for computing the additional coefficients if needed. The system command that determines how many terms of a series is displayed is `)set streams calculate`. For the purposes of this book, we have used this system command to display fewer than ten terms. Series can be created from expressions, from functions for the series coefficients, and from applications of operations on existing series. The most general function for creating a series is called **series**, although you can also use **taylor**, **laurent** and **puiseux** in situations where you know what kind of exponents are involved.

For information about solving differential equations in terms of power series, see Section 8.10.3 on page 275.

8.9.1 Creation of Power Series

This is the easiest way to create a power series. This tells AXIOM that `x` is to be treated as a power series, so functions of `x` are again power series.

```
x := series 'x
```

$$x \tag{1}$$

Type: UnivariatePuiseuxSeries(Expression Integer, x, 0)

We didn't say anything about the coefficients of the power series, so the coefficients are general expressions over the integers. This allows us to introduce denominators, symbolic constants, and other variables as needed.

Here the coefficients are integers (note that the coefficients are the Fibonacci numbers).

```
1/(1 - x - x**2)
```

$$1 + x + 2\,x^2 + 3\,x^3 + 5\,x^4 + 8\,x^5 + 13\,x^6 + 21\,x^7 + O\left(x^8\right) \tag{2}$$

Type: UnivariatePuiseuxSeries(Expression Integer, x, 0)

This series has coefficients that are rational numbers.

```
sin(x)
```

$$x - \frac{1}{6}\,x^3 + \frac{1}{120}\,x^5 - \frac{1}{5040}\,x^7 + O\left(x^9\right) \tag{3}$$

Type: UnivariatePuiseuxSeries(Expression Integer, x, 0)

When you enter this expression you introduce the symbolic constants `sin(1)` and `cos(1)`.

```
sin(1 + x)
```

$$\sin(1) + \cos(1)\,x - \frac{\sin(1)}{2}\,x^2 - \frac{\cos(1)}{6}\,x^3 + \frac{\sin(1)}{24}\,x^4 + \frac{\cos(1)}{120}\,x^5 - \frac{\sin(1)}{720}\,x^6 - \frac{\cos(1)}{5040}\,x^7 + O\left(x^8\right) \tag{4}$$

Type: UnivariatePuiseuxSeries(Expression Integer, x, 0)

When you enter the expression the variable `a` appears in the resulting series expansion.

```
sin(a * x)
```

$$a\,x - \frac{a^3}{6}\,x^3 + \frac{a^5}{120}\,x^5 - \frac{a^7}{5040}\,x^7 + O\left(x^9\right) \tag{5}$$

Type: UnivariatePuiseuxSeries(Expression Integer, x, 0)

You can also convert an expression into a series expansion. This expression creates the series expansion of `1/log(y)` about `y = 1`. For details and more examples, see Section 8.9.5 on page 261.

```
series(1/log(y),y = 1)
```

$$(y-1)^{(-1)} + \frac{1}{2} - \frac{1}{12}\,(y-1) + \frac{1}{24}\,(y-1)^2 - \frac{19}{720}\,(y-1)^3 + \frac{3}{160}\,(y-1)^4 - \frac{863}{60480}\,(y-1)^5 + \frac{275}{24192}\,(y-1)^6 + O\left((y-1)^7\right) \tag{6}$$

Type: UnivariatePuiseuxSeries(Expression Integer, y, 1)

You can create power series with more general coefficients. You normally accomplish this via a type declaration (see Section 2.3 on page 69). See

Section 8.9.4 on page 259 for some warnings about working with declared series.

We declare that y is a one-variable Taylor series (UTS is the abbreviation for UnivariateTaylorSeries) in the variable z with FLOAT (that is, floating-point) coefficients, centered about 0. Then, by assignment, we obtain the Taylor expansion of exp(z) with floating-point coefficients.

```
y : UTS(FLOAT,'z,0) := exp(z)
```

$$1.0 + z + 0.5\ z^2 + 0.16666666666666666667\ z^3 + 0.041666666666666666667\ z^4 + 0.0083333333333333333334\ z^5 + 0.0013888888888888888889\ z^6 + 0.00019841269841269841271\ z^7 + O\left(z^8\right) \tag{7}$$

Type: UnivariateTaylorSeries(Float, z, 0.0)

You can also create a power series by giving an explicit formula for its n th coefficient. For details and more examples, see Section 8.9.6 on page 263.

To create a series about w = 0 whose n th Taylor coefficient is 1/n!, you can evaluate this expression. This is the Taylor expansion of exp(w) at w = 0.

```
series(1/factorial(n),n,w = 0)
```

$$1 + w + \frac{1}{2}\ w^2 + \frac{1}{6}\ w^3 + \frac{1}{24}\ w^4 + \frac{1}{120}\ w^5 + \frac{1}{720}\ w^6 + \frac{1}{5040}\ w^7 + O\left(w^8\right) \tag{8}$$

Type: UnivariatePuiseuxSeries(Expression Integer, w, 0)

8.9.2 Coefficients of Power Series

You can extract any coefficient from a power series—even one that hasn't been computed yet. This is possible because in AXIOM, infinite series are represented by a list of the coefficients that have already been determined, together with a function for computing the additional coefficients. (This is known as *lazy evaluation.*) When you ask for a coefficient that hasn't yet been computed, AXIOM computes whatever additional coefficients it needs and then stores them in the representation of the power series.

Here's an example of how to extract the coefficients of a power series.

```
x := series(x)
```

$$x \tag{1}$$

Type: UnivariatePuiseuxSeries(Expression Integer, x, 0)

```
y := exp(x) * sin(x)
```

$$x + x^2 + \frac{1}{3}\ x^3 - \frac{1}{30}\ x^5 - \frac{1}{90}\ x^6 - \frac{1}{630}\ x^7 + O\left(x^9\right) \tag{2}$$

Type: UnivariatePuiseuxSeries(Expression Integer, x, 0)

This coefficient is readily available.

```
coefficient(y,6)
```

$$-\frac{1}{90} \tag{3}$$

Type: Expression Integer

But let's get the fifteenth coefficient of y.

```
coefficient(y,15)
```

$$-\frac{1}{10216206000} \tag{4}$$

Type: Expression Integer

If you look at y then you see that the coefficients up to order 15 have all been computed.

```
y
```

$$x+x^2+\frac{1}{3}\,x^3-\frac{1}{30}\,x^5-\frac{1}{90}\,x^6-\frac{1}{630}\,x^7+\frac{1}{22680}\,x^9+ \frac{1}{113400}\,x^{10}+\frac{1}{1247400}\,x^{11}-\frac{1}{97297200}\,x^{13}-\frac{1}{681080400}\,x^{14}- \frac{1}{10216206000}\,x^{15}+O\left(x^{16}\right) \tag{5}$$

Type: UnivariatePuiseuxSeries(Expression Integer, x, 0)

8.9.3 Power Series Arithmetic

You can manipulate power series using the usual arithmetic operations "+", "-", "*", and "/".

The results of these operations are also power series.

```
x := series x
```

$$x \tag{1}$$

Type: UnivariatePuiseuxSeries(Expression Integer, x, 0)

```
(3 + x) / (1 + 7*x)
```

$$3-20\,x+140\,x^2-980\,x^3+6860\,x^4-48020\,x^5+336140\,x^6- 2352980\,x^7+O\left(x^8\right) \tag{2}$$

Type: UnivariatePuiseuxSeries(Expression Integer, x, 0)

You can also compute f(x) ** g(x), where f(x) and g(x) are two power series.

```
base := 1 / (1 - x)
```

$$1+x+x^2+x^3+x^4+x^5+x^6+x^7+O\left(x^8\right) \tag{3}$$

Type: UnivariatePuiseuxSeries(Expression Integer, x, 0)

```
expon := x * base
```

$$x+x^2+x^3+x^4+x^5+x^6+x^7+x^8+O\left(x^9\right) \tag{4}$$

Type: UnivariatePuiseuxSeries(Expression Integer, x, 0)

```
base ** expon
```

$$1 + x^2 + \frac{3}{2}\,x^3 + \frac{7}{3}\,x^4 + \frac{43}{12}\,x^5 + \frac{649}{120}\,x^6 + \frac{241}{30}\,x^7 + O\left(x^8\right) \tag{5}$$

Type: UnivariatePuiseuxSeries(Expression Integer, x, 0)

8.9.4 Functions on Power Series

Once you have created a power series, you can apply transcendental functions (for example, **exp**, **log**, **sin**, **tan**, **cosh**, etc.) to it.

To demonstrate this, we first create the power series expansion of the rational function $\frac{x^2}{1-6x+x^2}$ about `x = 0`.

```
x := series 'x
```

$$x \tag{1}$$

Type: UnivariatePuiseuxSeries(Expression Integer, x, 0)

```
rat := x**2 / (1 - 6*x + x**2)
```

$$x^2 + 6\,x^3 + 35\,x^4 + 204\,x^5 + 1189\,x^6 + 6930\,x^7 + 40391\,x^8 + 235416\,x^9 + O\left(x^{10}\right) \tag{2}$$

Type: UnivariatePuiseuxSeries(Expression Integer, x, 0)

If you want to compute the series expansion of $\sin\left(\frac{x^2}{1-6x+x^2}\right)$ you simply compute the sine of `rat`.

```
sin(rat)
```

$$x^2 + 6\,x^3 + 35\,x^4 + 204\,x^5 + \frac{7133}{6}\,x^6 + 6927\,x^7 + \frac{80711}{2}\,x^8 + 235068\,x^9 + O\left(x^{10}\right) \tag{3}$$

Type: UnivariatePuiseuxSeries(Expression Integer, x, 0)

> **Warning:** the type of the coefficients of a power series may affect the kind of computations that you can do with that series. This can only happen when you have made a declaration to specify a series domain with a certain type of coefficient.

If you evaluate then you have declared that y is a one variable Taylor series (UTS is the abbreviation for UnivariateTaylorSeries) in the variable y with FRAC INT (that is, fractions of integer) coefficients, centered about 0.

```
y : UTS(FRAC INT,y,0) := y
```

$$y \tag{4}$$

Type: UnivariateTaylorSeries(Fraction Integer, y, 0)

You can now compute certain power series in y, *provided* that these series have rational coefficients.

```
exp(y)
```

$$1+y+\frac{1}{2}\,y^2+\frac{1}{6}\,y^3+\frac{1}{24}\,y^4+\frac{1}{120}\,y^5+\frac{1}{720}\,y^6+\frac{1}{5040}\,y^7+O\left(y^8\right) \tag{5}$$

Type: UnivariateTaylorSeries(Fraction Integer, y, 0)

You can get examples of such series by applying transcendental functions to series in y that have no constant terms.

```
tan(y**2)
```

$$y^2+\frac{1}{3}\,y^6+O\left(y^8\right) \tag{6}$$

Type: UnivariateTaylorSeries(Fraction Integer, y, 0)

```
cos(y + y**5)
```

$$1-\frac{1}{2}\,y^2+\frac{1}{24}\,y^4-\frac{721}{720}\,y^6+O\left(y^8\right) \tag{7}$$

Type: UnivariateTaylorSeries(Fraction Integer, y, 0)

Similarly, you can compute the logarithm of a power series with rational coefficients if the constant coefficient is 1.

```
log(1 + sin(y))
```

$$y-\frac{1}{2}\,y^2+\frac{1}{6}\,y^3-\frac{1}{12}\,y^4+\frac{1}{24}\,y^5-\frac{1}{45}\,y^6+\frac{61}{5040}\,y^7+O\left(y^8\right) \tag{8}$$

Type: UnivariateTaylorSeries(Fraction Integer, y, 0)

If you wanted to apply, say, the operation **exp** to a power series with a nonzero constant coefficient a_0, then the constant coefficient of the result would be e^{a_0}, which is *not* a rational number. Therefore, evaluating `exp(2 + tan(y))` would generate an error message.

If you want to compute the Taylor expansion of `exp(2 + tan(y))`, you must ensure that the coefficient domain has an operation **exp** defined for it. An example of such a domain is Expression Integer, the type of formal functional expressions over the integers.

When working with coefficients of this type,

```
z : UTS(EXPR INT,z,0) := z
```

$$z \tag{9}$$

Type: UnivariateTaylorSeries(Expression Integer, z, 0)

this presents no problems.

```
exp(2 + tan(z))
```

$$\%e^2+\%e^2\,z+\frac{\%e^2}{2}\,z^2+\frac{\%e^2}{2}\,z^3+\frac{3\ \%e^2}{8}\,z^4+\frac{37\ \%e^2}{120}\,z^5+\frac{59\ \%e^2}{240}\,z^6+\frac{137\ \%e^2}{720}\,z^7+O\left(z^8\right) \tag{10}$$

Type: UnivariateTaylorSeries(Expression Integer, z, 0)

Another way to create Taylor series whose coefficients are expressions over the integers is to use **taylor** which works similarly to **series**.

This is equivalent to the previous computation, except that now we are using the variable `w` instead of `z`.

```
w := taylor 'w
```

$$w \tag{11}$$

Type: UnivariateTaylorSeries(Expression Integer, w, 0)

```
exp(2 + tan(w))
```

$$\%e^2 + \%e^2\, w + \frac{\%e^2}{2}\, w^2 + \frac{\%e^2}{2}\, w^3 + \frac{3\ \%e^2}{8}\, w^4 + \frac{37\ \%e^2}{120}\, w^5 + \frac{59\ \%e^2}{240}\, w^6 + \frac{137\ \%e^2}{720}\, w^7 + O\left(w^8\right) \tag{12}$$

Type: UnivariateTaylorSeries(Expression Integer, w, 0)

8.9.5 Converting to Power Series

The ExpressionToUnivariatePowerSeries package provides operations for computing series expansions of functions.

Evaluate this to compute the Taylor expansion of `sin x` about `x = 0`. The first argument, `sin(x)`, specifies the function whose series expansion is to be computed and the second argument, `x = 0`, specifies that the series is to be expanded in power of `(x - 0)`, that is, in power of `x`.

```
taylor(sin(x),x = 0)
```

$$x - \frac{1}{6}\, x^3 + \frac{1}{120}\, x^5 - \frac{1}{5040}\, x^7 + O\left(x^8\right) \tag{1}$$

Type: UnivariateTaylorSeries(Expression Integer, x, 0)

Here is the Taylor expansion of `sin x` about $x = \frac{\pi}{6}$:

```
taylor(sin(x),x = %pi/6)
```

$$\frac{1}{2} + \frac{\sqrt{3}}{2}\left(x - \frac{\%pi}{6}\right) - \frac{1}{4}\left(x - \frac{\%pi}{6}\right)^2 - \frac{\sqrt{3}}{12}\left(x - \frac{\%pi}{6}\right)^3 + \frac{1}{48}\left(x - \frac{\%pi}{6}\right)^4 + \frac{\sqrt{3}}{240}\left(x - \frac{\%pi}{6}\right)^5 - \frac{1}{1440}\left(x - \frac{\%pi}{6}\right)^6 - \frac{\sqrt{3}}{10080}\left(x - \frac{\%pi}{6}\right)^7 + O\left(\left(x - \frac{\%pi}{6}\right)^8\right) \tag{2}$$

Type: UnivariateTaylorSeries(Expression Integer, x, pi/6)

The function to be expanded into a series may have variables other than the series variable.

For example, we may expand `tan(x*y)` as a Taylor series in `x`

```
taylor(tan(x*y),x = 0)
```

$$y\ x+\frac{y^3}{3}\ x^3+\frac{2\ y^5}{15}\ x^5+\frac{17\ y^7}{315}\ x^7+O\left(x^8\right) \tag{3}$$

Type: UnivariateTaylorSeries(Expression Integer, x, 0)

or as a Taylor series in `y`.

```
taylor(tan(x*y),y = 0)
```

$$x\ y+\frac{x^3}{3}\ y^3+\frac{2\ x^5}{15}\ y^5+\frac{17\ x^7}{315}\ y^7+O\left(y^8\right) \tag{4}$$

Type: UnivariateTaylorSeries(Expression Integer, y, 0)

A more interesting function is $\frac{te^{xt}}{e^t-1}$. When we expand this function as a Taylor series in `t` the `n` th order coefficient is the `n` th Bernoulli polynomial divided by `n!`.

```
bern := taylor(t*exp(x*t)/(exp(t) - 1),t = 0)
```

$$\begin{aligned}
&1+\frac{2\ x-1}{2}\ t+\frac{6\ x^2-6\ x+1}{12}\ t^2+\frac{2\ x^3-3\ x^2+x}{12}\ t^3+\\
&\frac{30\ x^4-60\ x^3+30\ x^2-1}{720}\ t^4+\frac{6\ x^5-15\ x^4+10\ x^3-x}{720}\ t^5+\\
&\frac{42\ x^6-126\ x^5+105\ x^4-21\ x^2+1}{30240}\ t^6+\\
&\frac{6\ x^7-21\ x^6+21\ x^5-7\ x^3+x}{30240}\ t^7+O\left(t^8\right)
\end{aligned} \tag{5}$$

Type: UnivariateTaylorSeries(Expression Integer, t, 0)

Therefore, this and the next expression produce the same result.

```
factorial(6) * coefficient(bern,6)
```

$$\frac{42\ x^6-126\ x^5+105\ x^4-21\ x^2+1}{42} \tag{6}$$

Type: Expression Integer

```
bernoulliB(6,x)
```

$$x^6-3\ x^5+\frac{5}{2}\ x^4-\frac{1}{2}\ x^2+\frac{1}{42} \tag{7}$$

Type: Polynomial Fraction Integer

Technically, a series with terms of negative degree is not considered to be a Taylor series, but, rather, a *Laurent series*. If you try to compute a Taylor series expansion of $\frac{x}{\log x}$ at `x = 1` via `taylor(x/log(x),x = 1)` you get an error message. The reason is that the function has a *pole* at `x = 1`, meaning that its series expansion about this point has terms of negative degree. A series with finitely many terms of negative degree is called a Laurent series.

You get the desired series expansion by issuing this.

```
laurent(x/log(x),x = 1)
```

$$(x-1)^{(-1)} + \frac{3}{2} + \frac{5}{12}(x-1) - \frac{1}{24}(x-1)^2 + \frac{11}{720}(x-1)^3 - \frac{11}{1440}(x-1)^4 + \frac{271}{60480}(x-1)^5 - \frac{13}{4480}(x-1)^6 + O\left((x-1)^7\right) \quad (8)$$

Type: UnivariateLaurentSeries(Expression Integer, x, 1)

Similarly, a series with terms of fractional degree is neither a Taylor series nor a Laurent series. Such a series is called a *Puiseux series*. The expression `laurent(sqrt(sec(x)),x = 3 * %pi/2)` results in an error message because the series expansion about this point has terms of fractional degree.

However, this command produces what you want.

```
puiseux(sqrt(sec(x)),x = 3 * %pi/2)
```

$$\left(x - \frac{3\,\%pi}{2}\right)^{\left(-\frac{1}{2}\right)} + \frac{1}{12}\left(x - \frac{3\,\%pi}{2}\right)^{\frac{3}{2}} + O\left(\left(x - \frac{3\,\%pi}{2}\right)^{\frac{7}{2}}\right) \quad (9)$$

Type: UnivariatePuiseuxSeries(Expression Integer, x, (3*pi)/2)

Finally, consider the case of functions that do not have Puiseux expansions about certain points. An example of this is x^x about `x = 0`. `puiseux(x**x,x=0)` produces an error message because of the type of singularity of the function at `x = 0`.

The general function **series** can be used in this case. Notice that the series returned is not, strictly speaking, a power series because of the `log(x)` in the expansion.

```
series(x**x,x=0)
```

$$1 + \log(x)\,x + \frac{\log(x)^2}{2}x^2 + \frac{\log(x)^3}{6}x^3 + \frac{\log(x)^4}{24}x^4 + \frac{\log(x)^5}{120}x^5 + \frac{\log(x)^6}{720}x^6 + \frac{\log(x)^7}{5040}x^7 + O\left(x^8\right) \quad (10)$$

Type: UnivariatePuiseuxSeries(Expression Integer, x, 0)

The operation **series** returns the most general type of infinite series. The user who is not interested in distinguishing between various types of infinite series may wish to use this operation exclusively.

8.9.6 Power Series from Formulas

The GenerateUnivariatePowerSeries package enables you to create power series from explicit formulas for their `n` th coefficients. In what follows, we construct series expansions for certain transcendental functions by giving formulas for their coefficients. You can also compute such series expansions directly simply by specifying the function and the point about which the series is to be expanded. See Section 8.9.5 on page 261 for more information.

Consider the Taylor expansion of e^x about `x = 0`:

$$e^x = 1 + x + \frac{x^2}{2} + \frac{x^3}{6} + \cdots$$

$$= \sum_{n=0}^{\infty} \frac{x^n}{n!}$$

The n th Taylor coefficient is `1/n!`.

This is how you create this series in AXIOM.

```
series(n +-> 1/factorial(n),x = 0)
```

$$1 + x + \frac{1}{2}x^2 + \frac{1}{6}x^3 + \frac{1}{24}x^4 + \frac{1}{120}x^5 + \frac{1}{720}x^6 + \frac{1}{5040}x^7 + O\left(x^8\right) \tag{1}$$

Type: UnivariatePuiseuxSeries(Expression Integer, x, 0)

The first argument specifies a formula for the n th coefficient by giving a function that maps `n` to `1/n!`. The second argument specifies that the series is to be expanded in powers of `(x - 0)`, that is, in powers of `x`. Since we did not specify an initial degree, the first term in the series was the term of degree 0 (the constant term). Note that the formula was given as an anonymous function. These are discussed in Section 6.17 on page 165.

Consider the Taylor expansion of `log x` about `x = 1`:

$$\log(x) = (x-1) - \frac{(x-1)^2}{2} + \frac{(x-1)^3}{3} - \cdots$$

$$= \sum_{n=1}^{\infty} (-1)^{n-1} \frac{(x-1)^n}{n}$$

If you were to evaluate the expression `series(n +-> (-1)**(n-1) / n, x = 1)` you would get an error message because AXIOM would try to calculate a term of degree `0` and therefore divide by `0`.

Instead, evaluate this. The third argument, `1..`, indicates that only terms of degree `n = 1, ...` are to be computed.

```
series(n +-> (-1)**(n-1)/n,x = 1,1..)
```

$$(x-1) - \frac{1}{2}(x-1)^2 + \frac{1}{3}(x-1)^3 - \frac{1}{4}(x-1)^4 + \frac{1}{5}(x-1)^5 - \frac{1}{6}(x-1)^6 + \frac{1}{7}(x-1)^7 - \frac{1}{8}(x-1)^8 + O\left((x-1)^9\right) \tag{2}$$

Type: UnivariatePuiseuxSeries(Expression Integer, x, 1)

Next consider the Taylor expansion of an odd function, say, `sin(x)`:

$$\sin(x) = x - \frac{x^3}{3!} + \frac{x^5}{5!} - \cdots$$

Here every other coefficient is zero and we would like to give an explicit formula only for the odd Taylor coefficients.

This is one way to do it. The third argument, `1..`, specifies that the first term to be computed is the term of degree 1. The fourth argument, `2`, specifies that we increment by 2 to find the degrees of subsequent terms, that is, the next term is of degree `1 + 2`, the next of degree `1 + 2 + 2`, etc.

```
series(n +-> (-1)**((n-1)/2)/factorial(n),x = 0,1..,2)
```

$$x - \frac{1}{6}\, x^3 + \frac{1}{120}\, x^5 - \frac{1}{5040}\, x^7 + O\left(x^9\right) \tag{3}$$

Type: UnivariatePuiseuxSeries(Expression Integer, x, 0)

The initial degree and the increment do not have to be integers. For example, this expression produces a series expansion of $\sin(x^{\frac{1}{3}})$.

```
series(n +-> (-1)**((3*n-1)/2)/factorial(3*n),x =
  0,1/3..,2/3)
```

$$x^{\frac{1}{3}} - \frac{1}{6}\, x + \frac{1}{120}\, x^{\frac{5}{3}} - \frac{1}{5040}\, x^{\frac{7}{3}} + O\left(x^3\right) \tag{4}$$

Type: UnivariatePuiseuxSeries(Expression Integer, x, 0)

While the increment must be positive, the initial degree may be negative. This yields the Laurent expansion of `csc(x)` at `x = 0`.

```
cscx := series(n +-> (-1)**((n-1)/2) * 2 * (2**n-1) *
  bernoulli(numer(n+1)) / factorial(n+1), x=0, -1..,2)
```

$$x^{(-1)} + \frac{1}{6}\, x + \frac{7}{360}\, x^3 + \frac{31}{15120}\, x^5 + O\left(x^7\right) \tag{5}$$

Type: UnivariatePuiseuxSeries(Expression Integer, x, 0)

Of course, the reciprocal of this power series is the Taylor expansion of `sin(x)`.

```
1/cscx
```

$$x - \frac{1}{6}\, x^3 + \frac{1}{120}\, x^5 - \frac{1}{5040}\, x^7 + O\left(x^9\right) \tag{6}$$

Type: UnivariatePuiseuxSeries(Expression Integer, x, 0)

As a final example, here is the Taylor expansion of `asin(x)` about `x = 0`.

```
asinx := series(n +-> binomial(n-1,(n-1)/2)/(n*2**(n-
  1)),x=0,1..,2)
```

$$x + \frac{1}{6}\, x^3 + \frac{3}{40}\, x^5 + \frac{5}{112}\, x^7 + O\left(x^9\right) \tag{7}$$

Type: UnivariatePuiseuxSeries(Expression Integer, x, 0)

When we compute the `sin` of this series, we get `x` (in the sense that all higher terms computed so far are zero).

```
sin(asinx)
```

$$x + O\left(x^9\right) \tag{8}$$

Type: UnivariatePuiseuxSeries(Expression Integer, x, 0)

As we discussed in Section 8.9.5 on page 261, you can also use the operations **taylor**, **laurent** and **puiseux** instead of **series** if you know ahead of time what kind of exponents a series has. You can't go wrong using **series**, though.

8.9.7 Substituting Numerical Values in Power Series

Use **eval** to substitute a numerical value for a variable in a power series. For example, here's a way to obtain numerical approximations of `%e` from the Taylor series expansion of `exp(x)`.

First you create the desired Taylor expansion.

```
f := taylor(exp(x))
```

$$1 + x + \frac{1}{2}x^2 + \frac{1}{6}x^3 + \frac{1}{24}x^4 + \frac{1}{120}x^5 + \frac{1}{720}x^6 + \frac{1}{5040}x^7 + O\left(x^8\right) \qquad (1)$$

Type: UnivariateTaylorSeries(Expression Integer, x, 0)

Then you evaluate the series at the value `1.0`. The result is a sequence of the partial sums.

```
eval(f,1.0)
```

$$[1.0,\ 2.0,\ 2.5,\ 2.6666666666666666667,\ 2.7083333333333333333,\ 2.7166666666666666667,\ 2.7180555555555555556,\ \ldots] \qquad (2)$$

Type: Stream Expression Float

8.9.8 Example: Bernoulli Polynomials and Sums of Powers

AXIOM provides operations for computing definite and indefinite sums.

You can compute the sum of the first ten fourth powers by evaluating this. This creates a list whose entries are m^4 as m ranges from 1 to 10, and then computes the sum of the entries of that list.

```
reduce(+,[m**4 for m in 1..10])
```

$$25333 \qquad (1)$$

Type: PositiveInteger

You can also compute a formula for the sum of the first k fourth powers, where k is an unspecified positive integer.

```
sum4 := sum(m**4, m = 1..k)
```

$$\frac{6\,k^5 + 15\,k^4 + 10\,k^3 - k}{30} \qquad (2)$$

Type: Fraction Polynomial Integer

This formula is valid for any positive integer k. For instance, if we replace k by 10, we obtain the number we computed earlier.

```
eval(sum4, k = 10)
```

$$25333 \qquad (3)$$

Type: Fraction Polynomial Integer

You can compute a formula for the sum of the first k $\mathtt{n}^{\text{th}}$ powers in a similar fashion. Just replace the `4` in the definition of **sum4** by any expression not involving k. AXIOM computes these formulas using Bernoulli polynomials; we use the rest of this section to describe this method.

First consider this function of t and x.

```
f := t*exp(x*t) / (exp(t) - 1)
```

$$\frac{t\ \%e^{(t\ x)}}{\%e^{t}-1} \tag{4}$$

Type: Expression Integer

Since the expressions involved get quite large, we tell AXIOM to show us only terms of degree up to 5.

```
)set streams calculate 5
```

If we look at the Taylor expansion of f(x, t) about t = 0, we see that the coefficients of the powers of t are polynomials in x.

```
ff := taylor(f,t = 0)
```

$$1+\frac{2\ x-1}{2}\ t+\frac{6\ x^2-6\ x+1}{12}\ t^2+\frac{2\ x^3-3\ x^2+x}{12}\ t^3+$$
$$\frac{30\ x^4-60\ x^3+30\ x^2-1}{720}\ t^4+\frac{6\ x^5-15\ x^4+10\ x^3-x}{720}\ t^5+ \tag{5}$$
$$O\left(t^6\right)$$

Type: UnivariateTaylorSeries(Expression Integer, t, 0)

In fact, the n th coefficient in this series is essentially the n th Bernoulli polynomial: the n th coefficient of the series is $\frac{1}{n!}B_n(x)$, where $B_n(x)$ is the n th Bernoulli polynomial. Thus, to obtain the n th Bernoulli polynomial, we multiply the n th coefficient of the series ff by n!.

For example, the sixth Bernoulli polynomial is this.

```
factorial(6) * coefficient(ff,6)
```

$$\frac{42\ x^6-126\ x^5+105\ x^4-21\ x^2+1}{42} \tag{6}$$

Type: Expression Integer

We derive some properties of the function f(x,t). First we compute f(x + 1,t) f(x,t).

```
g := eval(f, x = x + 1) - f
```

$$\frac{t\ \%e^{(t\ x+t)}-t\ \%e^{(t\ x)}}{\%e^{t}-1} \tag{7}$$

Type: Expression Integer

If we normalize g, we see that it has a particularly simple form.

```
normalize(g)
```

$$t\ \%e^{(t\ x)} \tag{8}$$

Type: Expression Integer

From this it follows that the n th coefficient in the Taylor expansion of g(x,t) at t = 0 is $\frac{1}{(n-1)!}\ x^{n-1}$.

If you want to check this, evaluate the next expression.

```
taylor(g,t = 0)
```

$$t+x\,t^2+\frac{x^2}{2}\,t^3+\frac{x^3}{6}\,t^4+\frac{x^4}{24}\,t^5+O\left(t^6\right) \tag{9}$$

Type: UnivariateTaylorSeries(Expression Integer, t, 0)

However, since `g(x,t) = f(x+1,t)-f(x,t)`, it follows that the `n` th coefficient is $\frac{1}{n!}\left(B_n(x+1)-B_n(x)\right)$. Equating coefficients, we see that $\frac{1}{(n-1)!}\,x^{n-1}=\frac{1}{n!}\left(B_n(x+1)-B_n(x)\right)$ and, therefore, $x^{n-1}=\frac{1}{n}\left(B_n(x+1)-B_n(x)\right)$. Let's apply this formula repeatedly, letting `x` vary between two integers `a` and `b`, with `a < b`:

$$\begin{array}{lcl}
a^{n-1} & = & \frac{1}{n}(B_n(a+1)-B_n(a)) \\
(a+1)^{n-1} & = & \frac{1}{n}(B_n(a+2)-B_n(a+1)) \\
(a+2)^{n-1} & = & \frac{1}{n}(B_n(a+3)-B_n(a+2)) \\
 & \vdots & \\
(b-1)^{n-1} & = & \frac{1}{n}(B_n(b)-B_n(b-1)) \\
b^{n-1} & = & \frac{1}{n}(B_n(b+1)-B_n(b))
\end{array}$$

When we add these equations we find that the sum of the left-hand sides is $\sum_{m=a}^{b} m^{n-1}$,the sum of the $(n-1)^{\text{st}}$ powers from `a` to `b`. The sum of the right-hand sides is a "telescoping series." After cancellation, the sum is simply $\frac{1}{n}\left(B_n(b+1)-B_n(a)\right)$.

Replacing `n` by `n + 1`, we have shown that

$$\sum_{m=a}^{b} m^n=\frac{1}{n+1}\left(B_{n+1}(b+1)-B_{n+1}(a)\right).$$

Let's use this to obtain the formula for the sum of fourth powers.

First we obtain the Bernoulli polynomial B_5.

```
B5 := factorial(5) * coefficient(ff,5)
```

$$\frac{6\,x^5-15\,x^4+10\,x^3-x}{6} \tag{10}$$

Type: Expression Integer

To find the sum of the first k 4th powers, we multiply `1/5` by $B_5(k+1)-B_5(1)$.

```
1/5 * (eval(B5, x = k + 1) - eval(B5, x = 1))
```

$$\frac{6\,k^5+15\,k^4+10\,k^3-k}{30} \tag{11}$$

Type: Expression Integer

This is the same formula that we obtained via `sum(m**4, m = 1..k)`.

```
sum4
```

$$\frac{6\,k^5+15\,k^4+10\,k^3-k}{30} \tag{12}$$

Type: Fraction Polynomial Integer

At this point you may want to do the same computation, but with an exponent other than `4`. For example, you might try to find a formula for the sum of the first k 20th powers.

8.10 Solution of Differential Equations

In this section we discuss AXIOM's facilities for solving differential equations in closed-form and in series.

AXIOM provides facilities for closed-form solution of single differential equations of the following kinds:

- linear ordinary differential equations, and
- non-linear first order ordinary differential equations when integrating factors can be found just by integration.

For a discussion of the solution of systems of linear and polynomial equations, see Section 8.5 on page 244.

8.10.1 Closed-Form Solutions of Linear Differential Equations

A *differential equation* is an equation involving an unknown *function* and one or more of its derivatives. The equation is called *ordinary* if derivatives with respect to only one dependent variable appear in the equation (it is called *partial* otherwise). The package ElementaryFunctionODESolver provides the top-level operation **solve** for finding closed-form solutions of ordinary differential equations.

To solve a differential equation, you must first create an operator for the unknown function.

We let `y` be the unknown function in terms of `x`.

```
y := operator 'y
```

$$y \qquad (1)$$

Type: BasicOperator

You then type the equation using **D** to create the derivatives of the unknown function `y(x)` where `x` is any symbol you choose (the so-called *dependent variable*).

This is how you enter the equation `y'' + y' + y = 0`.

```
deq := D(y x, x, 2) + D(y x, x) + y x = 0
```

$$y''(x) + y'(x) + y(x) = 0 \qquad (2)$$

Type: Equation Expression Integer

The simplest way to invoke the **solve** command is with three arguments.

- the differential equation,
- the operator representing the unknown function,
- the dependent variable.

So, to solve the above equation, we enter this.

```
solve(deq, y, x)
```

$$\left[particular = 0, \right.$$

$$\left. basis = \left[\cos\left(\frac{x\sqrt{3}}{2}\right) \%e^{\left(-\frac{x}{2}\right)}, \%e^{\left(-\frac{x}{2}\right)} \sin\left(\frac{x\sqrt{3}}{2}\right)\right]\right] \quad (3)$$

Type: Union(Record(particular: Expression Integer, basis: List Expression Integer), ...)

Since linear ordinary differential equations have infinitely many solutions, **solve** returns a *particular solution* f_p and a basis $f_1, \ldots, f_n$ for the solutions of the corresponding homogenuous equation. Any expression of the form $f_p + c_1 f_1 + \ldots c_n f_n$ where the c_i do not involve the dependent variable is also a solution. This is similar to what you get when you solve systems of linear algebraic equations.

A way to select a unique solution is to specify *initial conditions*: choose a value a for the dependent variable and specify the values of the unknown function and its derivatives at a. If the number of initial conditions is equal to the order of the equation, then the solution is unique (if it exists in closed form!) and **solve** tries to find it. To specify initial conditions to **solve**, use an Equation of the form `x = a` for the third parameter instead of the dependent variable, and add a fourth parameter consisting of the list of values `y(a), y'(a), ...`.

To find the solution of `y'' + y = 0` satisfying `y(0) = y'(0) = 1`, do this.

```
deq := D(y x, x, 2) + y x
```

$$y''(x) + y(x) \quad (4)$$

Type: Expression Integer

You can omit the `= 0` when you enter the equation to be solved.

```
solve(deq, y, x = 0, [1, 1])
```

$$\sin(x) + \cos(x) \quad (5)$$

Type: Union(Expression Integer, ...)

AXIOM is not limited to linear differential equations with constant coefficients. It can also find solutions when the coefficients are rational or algebraic functions of the dependent variable. Furthermore, AXIOM is not limited by the order of the equation.

AXIOM can solve the following third order equations with polynomial coefficients.

```
deq := x**3 * D(y x, x, 3) + x**2 * D(y x, x, 2) - 2 * x *
  D(y x, x) + 2 * y x = 2 * x**4
```

$$x^3\, y'''(x) + x^2\, y''(x) - 2\, x\, y'(x) + 2\, y(x) = 2\, x^4 \quad (6)$$

Type: Equation Expression Integer

```
solve(deq, y, x)
```

$$\left[particular = \frac{x^4}{15}, \ basis = \left[\frac{1}{x}, \ x, \ x^2\right]\right] \tag{7}$$

Type: Union(Record(particular: Expression Integer, basis: List Expression Integer), ...)

Here we are solving a homogeneous equation.

```
deq := (x**9+x**3) * D(y x, x, 3) + 18 * x**8 * D(y x, x,
  2) - 90 * x * D(y x, x) - 30 * (11 * x**6 - 3) * y x
```

$$\left(x^9 + x^3\right) \ y''' (x) + 18 \ x^8 \ y'' (x) - 90 \ x \ y' (x) + \left(-330 \ x^6 + 90\right) \ y (x) \tag{8}$$

Type: Expression Integer

```
solve(deq, y, x)
```

$$\left[particular = 0, \ basis = \left[\frac{x}{x^6 + 1}, \ \frac{x \ \%e^{\left(-\sqrt{91} \ \log(x)\right)}}{x^6 + 1}, \ \frac{x \ \%e^{\left(\sqrt{91} \ \log(x)\right)}}{x^6 + 1}\right]\right] \tag{9}$$

Type: Union(Record(particular: Expression Integer, basis: List Expression Integer), ...)

On the other hand, and in contrast with the operation **integrate**, it can happen that AXIOM finds no solution and that some closed-form solution still exists. While it is mathematically complicated to describe exactly when the solutions are guaranteed to be found, the following statements are correct and form good guidelines for linear ordinary differential equations:

- If the coefficients are constants, AXIOM finds a complete basis of solutions (i,e, all solutions).
- If the coefficients are rational functions in the dependent variable, AXIOM at least finds all solutions that do not involve algebraic functions.

Note that this last statement does not mean that AXIOM does not find the solutions that are algebraic functions. It means that it is not guaranteed that the algebraic function solutions will be found.

This is an example where all the algebraic solutions are found.

```
deq := (x**2 + 1) * D(y x, x, 2) + 3 * x * D(y x, x) + y x
  = 0
```

$$\left(x^2 + 1\right) \ y'' (x) + 3 \ x \ y' (x) + y (x) = 0 \tag{10}$$

Type: Equation Expression Integer

```
solve(deq, y, x)
```

$$\left[particular = 0,\ basis = \left[\frac{1}{\sqrt{x^2+1}}, \frac{\log\left(\sqrt{x^2+1} - x\right)}{\sqrt{x^2+1}} \right] \right] \tag{11}$$

Type: Union(Record(particular: Expression Integer, basis: List Expression Integer), ...)

8.10.2 Closed-Form Solutions of Non-Linear Differential Equations

This is an example that shows how to solve a non-linear first order ordinary differential equation manually when an integrating factor can be found just by integration. At the end, we show you how to solve it directly.

Let's solve the differential equation `y' = y / (x + y log y)`.

Using the notation `m(x, y) + n(x, y) y' = 0`, we have `m = -y` and `n = x + y log y`.

```
m := -y
```

$$-y \tag{1}$$

Type: Polynomial Integer

```
n := x + y * log y
```

$$y \log(y) + x \tag{2}$$

Type: Expression Integer

We first check for exactness, that is, does `dm/dy = dn/dx`?

```
D(m, y) - D(n, x)
```

$$-2 \tag{3}$$

Type: Expression Integer

This is not zero, so the equation is not exact. Therefore we must look for an integrating factor: a function `mu(x,y)` such that `d(mu m)/dy = d(mu n)/dx`. Normally, we first search for `mu(x,y)` depending only on `x` or only on `y`.

Let's search for such a `mu(x)` first.

```
mu := operator 'mu
```

$$mu \tag{4}$$

Type: BasicOperator

```
a := D(mu(x) * m, y) - D(mu(x) * n, x)
```

$$\left(-y \log(y) - x\right) mu'(x) - 2\, mu(x) \tag{5}$$

Type: Expression Integer

If the above is zero for a function mu that does *not* depend on y, then mu(x) is an integrating factor.

```
solve(a = 0, mu, x)
```

$$\left[particular = 0,\ basis = \left[\frac{1}{y^2\ \log(y)^2 + 2\ x\ y\ \log(y) + x^2}\right]\right] \quad (6)$$

Type: Union(Record(particular: Expression Integer, basis: List Expression Integer), ...)

The solution depends on y, so there is no integrating factor that depends on x only.

Let's look for one that depends on y only.

```
b := D(mu(y) * m, y) - D(mu(y) * n, x)
```

$$-y\ mu'(y) - 2\ mu(y) \quad (7)$$

Type: Expression Integer

```
sb := solve(b = 0, mu, y)
```

$$\left[particular = 0,\ basis = \left[\frac{1}{y^2}\right]\right] \quad (8)$$

Type: Union(Record(particular: Expression Integer, basis: List Expression Integer), ...)

We've found one!

The above mu(y) is an integrating factor. We must multiply our initial equation (that is, m and n) by the integrating factor.

```
intFactor := sb.basis.1
```

$$\frac{1}{y^2} \quad (9)$$

Type: Expression Integer

```
m := intFactor * m
```

$$-\frac{1}{y} \quad (10)$$

Type: Expression Integer

```
n := intFactor * n
```

$$\frac{y\ \log(y) + x}{y^2} \quad (11)$$

Type: Expression Integer

Let's check for exactness.

```
D(m, y) - D(n, x)
```

$$0 \quad (12)$$

Type: Expression Integer

We must solve the exact equation, that is, find a function s(x,y) such that ds/dx = m and ds/dy = n.

We start by writing `s(x, y) = h(y) + integrate(m, x)` where `h(y)` is an unknown function of `y`. This guarantees that `ds/dx = m`.

```
h := operator 'h
```

$$h \tag{13}$$

Type: BasicOperator

```
sol := h y + integrate(m, x)
```

$$\frac{y\ h(y) - x}{y} \tag{14}$$

Type: Expression Integer

All we want is to find `h(y)` such that `ds/dy = n`.

```
dsol := D(sol, y)
```

$$\frac{y^2\ h'(y) + x}{y^2} \tag{15}$$

Type: Expression Integer

```
nsol := solve(dsol = n, h, y)
```

$$\left[particular = \frac{\log(y)^2}{2},\ basis = [1]\right] \tag{16}$$

Type: Union(Record(particular: Expression Integer, basis: List Expression Integer), ...)

The above particular solution is the `h(y)` we want, so we just replace `h(y)` by it in the implicit solution.

```
eval(sol, h y = nsol.particular)
```

$$\frac{y\ \log(y)^2 - 2\ x}{2\ y} \tag{17}$$

Type: Expression Integer

A first integral of the initial equation is obtained by setting this result equal to an arbitrary constant.

Now that we've seen how to solve the equation "by hand," we show you how to do it with the **solve** operation.

First define `y` to be an operator.

```
y := operator 'y
```

$$y \tag{18}$$

Type: BasicOperator

Next we create the differential equation.

```
deq := D(y x, x) = y(x) / (x + y(x) * log y x)
```

$$y'(x) = \frac{y(x)}{y(x)\log(y(x)) + x} \tag{19}$$

Type: Equation Expression Integer

Finally, we solve it.

```
solve(deq, y, x)
```

$$\frac{y(x)\log(y(x))^2 - 2\ x}{2\ y(x)} \tag{20}$$

Type: Union(Expression Integer, ...)

8.10.3 Power Series Solutions of Differential Equations

The command to solve differential equations in power series around a particular initial point with specific initial conditions is called **seriesSolve**. It can take a variety of parameters, so we illustrate its use with some examples.

Since the coefficients of some solutions are quite large, we reset the default to compute only seven terms.

```
)set streams calculate 7
```

You can solve a single nonlinear equation of any order. For example, we solve `y''' = sin(y'') * exp(y) + cos(x)` subject to `y(0) = 1, y'(0) = 0, y''(0) = 0`.

We first tell AXIOM that the symbol `'y` denotes a new operator.

```
y := operator 'y
```

$$y \tag{1}$$

Type: BasicOperator

Enter the differential equation using `y` like any system function.

```
eq := D(y(x), x, 3) - sin(D(y(x), x, 2))*exp(y(x)) =
  cos(x)
```

$$y''' (x) - \%e^{y(x)} \sin\left(y'' (x)\right) = \cos(x) \tag{2}$$

Type: Equation Expression Integer

Solve it around `x = 0` with the initial conditions `y(0) = 1, y'(0) = y''(0) = 0`.

```
seriesSolve(eq, y, x = 0, [1, 0, 0])
```

$$1 + \frac{1}{6} x^3 + \frac{\%e}{24} x^4 + \frac{\%e^2 - 1}{120} x^5 + \frac{\%e^3 - 2\ \%e}{720} x^6 + \frac{\%e^4 - 8\ \%e^2 + 4\ \%e + 1}{5040} x^7 + O\left(x^8\right) \tag{3}$$

Type: UnivariateTaylorSeries(Expression Integer, x, 0)

You can also solve a system of nonlinear first order equations. For example, we solve a system that has `tan(t)` and `sec(t)` as solutions.

We tell AXIOM that `x` is also an operator.

```
x := operator 'x
```

$$x \tag{4}$$

Type: BasicOperator

Enter the two equations forming our system.

```
eq1 := D(x(t), t) = 1 + x(t)**2
```

$$x' (t) = x(t)^2 + 1 \tag{5}$$

Type: Equation Expression Integer

```
eq2 := D(y(t), t) = x(t) * y(t)
```

$$y' (t) = x(t)\, y(t) \tag{6}$$

Type: Equation Expression Integer

Solve the system around `t = 0` with the initial conditions `x(0) = 0` and `y(0) = 1`. Notice that since we give the unknowns in the order `[x, y]`, the answer is a list of two series in the order `[series for x(t), series for y(t)]`.

```
seriesSolve([eq2, eq1], [x, y], t = 0, [y(0) = 1, x(0) =
  0])
```

$$\left[t + \frac{1}{3}\,t^3 + \frac{2}{15}\,t^5 + \frac{17}{315}\,t^7 + O\left(t^8\right),\right. \qquad (7)$$
$$\left. 1 + \frac{1}{2}\,t^2 + \frac{5}{24}\,t^4 + \frac{61}{720}\,t^6 + O\left(t^8\right)\right]$$

Type: List UnivariateTaylorSeries(Expression Integer, t, 0)

The order in which we give the equations and the initial conditions has no effect on the order of the solution.

8.11 Finite Fields

A *finite field* (also called a *Galois field*) is a finite algebraic structure where one can add, multiply and divide under the same laws (for example, commutativity, associativity or distributivity) as apply to the rational, real or complex numbers. Unlike those three fields, for any finite field there exists a positive prime integer p, called the **characteristic**, such that $p\,x = 0$ for any element x in the finite field. In fact, the number of elements in a finite field is a power of the characteristic and for each prime p and positive integer n there exists exactly one finite field with p^n elements, up to isomorphism.[1]

When `n = 1`, the field has p elements and is called a *prime field*, discussed in the next section. There are several ways of implementing extensions of finite fields, and AXIOM provides quite a bit of freedom to allow you to choose the one that is best for your application. Moreover, we provide operations for converting among the different representations of extensions and different extensions of a single field. Finally, note that you usually need to package-call operations from finite fields if the operations do not take as an argument an object of the field. See Section 2.9 on page 83 for more information on package-calling.

8.11.1 Modular Arithmetic and Prime Fields

Let n be a positive integer. It is well known that you can get the same result if you perform addition, subtraction or multiplication of integers and then take the remainder on dividing by n as if you had first done such remaindering on the operands, performed the arithmetic and then (if necessary) done remaindering again. This allows us to speak of arithmetic *modulo* n or, more simply *mod* n.

[1]For more information about the algebraic structure and properties of finite fields, see, for example, S. Lang, *Algebra*, Second Edition, New York: Addison-Wesley Publishing Company, Inc., 1984, ISBN 0 201 05487 6; or R. Lidl, H. Niederreiter, *Finite Fields*, Encyclopedia of Mathematics and Its Applications, Vol. 20, Cambridge: Cambridge Univ. Press, 1983, ISBN 0 521 30240 4.

In AXIOM, you use IntegerMod to do such arithmetic.

```
(a,b) : IntegerMod 12
```

Type: Void

```
(a, b) := (16, 7)
```

7 (2)

Type: IntegerMod 12

```
[a - b, a * b]
```

[9, 4] (3)

Type: List IntegerMod 12

If n is not prime, there is only a limited notion of reciprocals and division.

```
a / b
```

```
Cannot find a definition or library operation named /
with argument types
                         IntegerMod 12
                         IntegerMod 12
```

```
recip a
```

"failed" (4)

Type: Union("failed", ...)

Here 7 and 12 are relatively prime, so 7 has a multiplicative inverse modulo 12.

```
recip b
```

7 (5)

Type: Union(IntegerMod 12, ...)

If we take n to be a prime number p, then taking inverses and, therefore, division are generally defined.

Use PrimeField instead of IntegerMod for n prime.

```
c : PrimeField 11 := 8
```

8 (6)

Type: PrimeField 11

```
inv c
```

7 (7)

Type: PrimeField 11

You can also use `1/c` and `c**(-1)` for the inverse of c.

```
9/c
```

8 (8)

Type: PrimeField 11

PrimeField (abbreviation PF) checks if its argument is prime when you try to use an operation from it. If you know the argument is prime (particularly if it is large), InnerPrimeField (abbreviation IPF) assumes the argument has already been verified to be prime. If you do use a number that is not prime, you will eventually get an error message, most likely a division by zero message. For computer science applications, the most important finite fields are PrimeField 2 and its extensions.

In the following examples, we work with the finite field with $p = 101$ elements.

```
GF101 := PF 101
```

PrimeField 101 (9)

Type: Domain

Like many domains in AXIOM, finite fields provide an operation for returning a random element of the domain.

```
x := random()$GF101
```

2 (10)

Type: PrimeField 101

```
y : GF101 := 37
```

37 (11)

Type: PrimeField 101

```
z := x/y
```

41 (12)

Type: PrimeField 101

```
z * y - x
```

0 (13)

Type: PrimeField 101

The element 2 is a *primitive element* of this field,

```
pe := primitiveElement()$GF101
```

2 (14)

Type: PrimeField 101

in the sense that its powers enumerate all nonzero elements.

```
[pe**i for i in 0..99]
```

$$[1, 2, 4, 8, 16, 32, 64, 27, 54, 7, 14, 28, 56, 11, 22, 44, 88, 75,$$
$$49, 98, 95, 89, 77, 53, 5, 10, 20, 40, 80, 59, 17, 34, 68, 35, 70,$$
$$39, 78, 55, 9, 18, 36, 72, 43, 86, 71, 41, 82, 63, 25, 50, 100, 99,$$
$$97, 93, 85, 69, 37, 74, 47, 94, 87, 73, 45, 90, 79, 57, 13, 26, 52,$$
$$3, 6, 12, 24, 48, 96, 91, 81, 61, 21, 42, 84, 67, 33, 66, 31, 62,$$
$$23, 46, 92, 83, 65, 29, 58, 15, 30, 60, 19, 38, 76, 51] \tag{15}$$

Type: List PrimeField 101

If every nonzero element is a power of a primitive element, how do you determine what the exponent is? Use **discreteLog**.

```
ex := discreteLog(y)
```

56 (16)

Type: PositiveInteger

```
pe ** ex
```

37 (17)

Type: PrimeField 101

The **order** of a nonzero element x is the smallest positive integer t such $x^t = 1$.

```
order y
```

25 (18)

Type: PositiveInteger

The order of a primitive element is the defining $p - 1$.

```
order pe
```

100 (19)

Type: PositiveInteger

8.11.2 Extensions of Finite Fields

When you want to work with an extension of a finite field in AXIOM, you have three choices to make:

1. Do you want to generate an extension of the prime field (for example, PrimeField 2) or an extension of a given field?
2. Do you want to use a representation that is particularly efficient for multiplication, exponentiation and addition but uses a lot of computer memory (a representation that models the cyclic group structure of the multiplicative group of the field extension and uses a Zech logarithm table), one that uses a normal basis for the vector space structure of the field extension, or one that performs arithmetic modulo an irreducible polynomial? The cyclic group representation is only usable up to "medium" (relative to your machine's performance) sized fields. If the field is large and the normal basis is relatively simple, the normal basis representation is more efficient for exponentiation than the irreducible polynomial representation.
3. Do you want to provide a polynomial explicitly, a root of which "generates" the extension in one of the three senses in (2), or do you wish to have the polynomial generated for you?

This illustrates one of the most important features of AXIOM: you can choose exactly the right data-type and representation to suit your application best.

We first tell you what domain constructors to use for each case above, and then give some examples.

Constructors that automatically generate extensions of the prime field:

FiniteField
FiniteFieldCyclicGroup
FiniteFieldNormalBasis

Constructors that generate extensions of an arbitrary field:

FiniteFieldExtension
FiniteFieldExtensionByPolynomial
FiniteFieldCyclicGroupExtension
FiniteFieldCyclicGroupExtensionByPolynomial
FiniteFieldNormalBasisExtension
FiniteFieldNormalBasisExtensionByPolynomial

Constructors that use a cyclic group representation:

FiniteFieldCyclicGroup
FiniteFieldCyclicGroupExtension
FiniteFieldCyclicGroupExtensionByPolynomial

Constructors that use a normal basis representation:

FiniteFieldNormalBasis
FiniteFieldNormalBasisExtension
FiniteFieldNormalBasisExtensionByPolynomial

Constructors that use an irreducible modulus polynomial representation:

FiniteField
FiniteFieldExtension
FiniteFieldExtensionByPolynomial

Constructors that generate a polynomial for you:

FiniteField
FiniteFieldExtension
FiniteFieldCyclicGroup
FiniteFieldCyclicGroupExtension
FiniteFieldNormalBasis
FiniteFieldNormalBasisExtension

Constructors for which you provide a polynomial:

FiniteFieldExtensionByPolynomial
FiniteFieldCyclicGroupExtensionByPolynomial
FiniteFieldNormalBasisExtensionByPolynomial

These constructors are discussed in the following sections where we collect together descriptions of extension fields that have the same underlying representation.[2]

---

[2]For For more information on the implementation aspects of finite fields, see J. Grabmeier, A. Scheerhorn, *Finite Fields in Axiom,* Technical Report, IBM Heidelberg Scientific Center, 1992.

If you don't really care about all this detail, just use FiniteField. As your knowledge of your application and its AXIOM implementation grows, you can come back and choose an alternative constructor that may improve the efficiency of your code. Note that the exported operations are almost the same for all constructors of finite field extensions and include the operations exported by PrimeField.

8.11.3 Irreducible Modulus Polynomial Representations

All finite field extension constructors discussed in this section use a representation that performs arithmetic with univariate (one-variable) polynomials modulo an irreducible polynomial. This polynomial may be given explicitly by you or automatically generated. The ground field may be the prime field or one you specify. See Section 8.11.2 on page 279 for general information about finite field extensions.

For FiniteField (abbreviation FF) you provide a prime number p and an extension degree n. This degree can be 1.

AXIOM uses the prime field PrimeField(p), here PrimeField 2, and it chooses an irreducible polynomial of degree n, here 12, over the ground field.

```
GF4096 := FF(2,12);
```

(1)

Type: Domain

The objects in the generated field extension are polynomials of degree at most $n-1$ with coefficients in the prime field. The polynomial indeterminate is automatically chosen by AXIOM and is typically something like %A or %D. These (strange) variables are *only* for output display; there are several ways to construct elements of this field.

The operation **index** enumerates the elements of the field extension and accepts as argument the integers from 1 to p^n.

The expression `index(p)` always gives the indeterminate.

```
a := index(2)$GF4096
```

$$\%A \tag{2}$$

Type: FiniteField(2, 12)

You can build polynomials in a and calculate in `GF4096`.

```
b := a**12 - a**5 + a
```

$$\%A^5 + \%A^3 + \%A + 1 \tag{3}$$

Type: FiniteField(2, 12)

```
b ** 1000
```

$$\%A^{10} + \%A^9 + \%A^7 + \%A^5 + \%A^4 + \%A^3 + \%A \tag{4}$$

Type: FiniteField(2, 12)

```
c := a/b
```

$$\%A^{11} + \%A^{8} + \%A^{7} + \%A^{5} + \%A^{4} + \%A^{3} + \%A^{2} \quad (5)$$

Type: FiniteField(2, 12)

Among the available operations are **norm** and **trace**.

```
norm c
```

$$1 \quad (6)$$

Type: PrimeField 2

```
trace c
```

$$0 \quad (7)$$

Type: PrimeField 2

Since any nonzero element is a power of a primitive element, how do we discover what the exponent is?

The operation **discreteLog** calculates the exponent and, if it is called with only one argument, always refers to the primitive element returned by **primitiveElement**.

```
dL := discreteLog a
```

$$1729 \quad (8)$$

Type: PositiveInteger

```
g ** dL
```

$$g^{1729} \quad (9)$$

Type: Polynomial Integer

FiniteFieldExtension (abbreviation FFX) is similar to FiniteField except that the ground-field for FiniteFieldExtension is arbitrary and chosen by you.

In case you select the prime field as ground field, there is essentially no difference between the constructed two finite field extensions.

```
GF16 := FF(2,4);
```

(10)

Type: Domain

```
GF4096 := FFX(GF16,3);
```

(11)

Type: Domain

```
r := (random()$GF4096) ** 20
```

$$\left(\%B^{2} + \%B\right) \%C^{2} + \left(\%B^{3} + \%B^{2}\right) \%C \quad (12)$$

Type: FiniteFieldExtension(FiniteField(2, 4), 3)

```
norm(r)
```

$$\%B^{2} + \%B + 1 \quad (13)$$

Type: FiniteField(2, 4)

FiniteFieldExtensionByPolynomial (abbreviation FFP) is similar to FiniteField and

FiniteFieldExtension but is more general.

```
GF4 := FF(2,2);
```

(14)

Type: Domain

```
f :=
  nextIrreduciblePoly(random(6)$FFPOLY(GF4))$FFPOLY(GF4)
```

$?^6 + ?^5 + (\%D + 1)\ ?^4 + \%D\ ?^2 + \%D + 1$ (15)

Type: Union(SparseUnivariatePolynomial FiniteField(2, 2), ...)

For FFP you choose both the ground field and the irreducible polynomial used in the representation. The degree of the extension is the degree of the polynomial.

```
GF4096 := FFP(GF4,f);
```

(16)

Type: Domain

```
discreteLog random()$GF4096
```

2225 (17)

Type: PositiveInteger

8.11.4 Cyclic Group Representations

In every finite field there exist elements whose powers are all the nonzero elements of the field. Such an element is called a *primitive element*.

In FiniteFieldCyclicGroup (abbreviation FFCG) the nonzero elements are represented by the powers of a fixed primitive element of the field (that is, a generator of its cyclic multiplicative group). Multiplication (and hence exponentiation) using this representation is easy. To do addition, we consider our primitive element as the root of a primitive polynomial (an irreducible polynomial whose roots are all primitive). See Section 8.11.7 on page 289 for examples of how to compute such a polynomial.

To use FiniteFieldCyclicGroup you provide a prime number and an extension degree.

```
GF81 := FFCG(3,4);
```

(1)

Type: Domain

AXIOM uses the prime field, here PrimeField 3, as the ground field and it chooses a primitive polynomial of degree n, here 4, over the prime field.

```
a := primitiveElement()$GF81
```

$\%F^1$ (2)

Type: FiniteFieldCyclicGroup(3, 4)

You can calculate in GF81.

```
b := a**12 - a**5 + a
```

$\%F^{72}$ (3)

Type: FiniteFieldCyclicGroup(3, 4)

In this representation of finite fields the discrete logarithm of an element can be seen directly in its output form.

```
b
```

$$\%F^{72} \tag{4}$$

Type: FiniteFieldCyclicGroup(3, 4)

```
discreteLog b
```

$$72 \tag{5}$$

Type: PositiveInteger

FiniteFieldCyclicGroupExtension (abbreviation FFCGX) is similar to FiniteField-CyclicGroup except that the ground field for FiniteFieldCyclicGroupExtension is arbitrary and chosen by you. In case you select the prime field as ground field, there is essentially no difference between the constructed two finite field extensions.

```
GF9 := FF(3,2);
```

(6)

Type: Domain

```
GF729 := FFCGX(GF9,3);
```

(7)

Type: Domain

```
r := (random()$GF729) ** 20
```

$$\%H^{276} \tag{8}$$

Type: FiniteFieldCyclicGroupExtension(FiniteField(3, 2), 3)

```
trace(r)
```

$$0 \tag{9}$$

Type: FiniteField(3, 2)

FiniteFieldCyclicGroupExtensionByPolynomial (abbreviation FFCGP) is similar to FiniteFieldCyclicGroup and FiniteFieldCyclicGroupExtension but is more general. For FiniteFieldCyclicGroupExtensionByPolynomial you choose both the ground field and the irreducible polynomial used in the representation. The degree of the extension is the degree of the polynomial.

```
GF3 := PrimeField 3;
```

(10)

Type: Domain

We use a utility operation to generate an irreducible primitive polynomial (see Section 8.11.7 on page 289). The polynomial has one variable that is "anonymous": it displays as a question mark.

```
f := createPrimitivePoly(4)$FFPOLY(GF3)
```

$$?^4+?+2 \tag{11}$$

Type: SparseUnivariatePolynomial PrimeField 3

```
GF81 := FFCGP(GF3,f);
```

(12)

Type: Domain

Let's look at a random element from this field.

```
random()$GF81
```

$$\%F^6 \tag{13}$$

Type: FiniteFieldCyclicGroupExtensionByPolynomial(PrimeField 3, ?**4+?+2)

8.11.5 Normal Basis Representations

Let K be a finite extension of degree n of the finite field F and let F have q elements. An element x of K is said to be *normal* over F if the elements

$$1, x^q, x^{q^2}, \ldots, x^{q^{n-1}}$$

form a basis of K as a vector space over F. Such a basis is called a *normal basis*.[3]

If x is normal over F, its minimal polynomial is also said to be *normal* over F. There exist normal bases for all finite extensions of arbitrary finite fields.

In FiniteFieldNormalBasis (abbreviation FFNB), the elements of the finite field are represented by coordinate vectors with respect to a normal basis.

You provide a prime p and an extension degree n.

```
K := FFNB(3,8)
```

$$\text{FiniteFieldNormalBasis }(3,8) \tag{1}$$

Type: Domain

AXIOM uses the prime field PrimeField(p), here PrimeField 3, and it chooses a normal polynomial of degree n, here 8, over the ground field. The remainder class of the indeterminate is used as the normal element. The polynomial indeterminate is automatically chosen by AXIOM and is typically something like %A or %D. These (strange) variables are only for output display; there are several ways to construct elements of this field. The output of the basis elements is something like $\%A^{q^i}$.

[3]This agrees with the general definition of a normal basis because the n distinct powers of the automorphism $x \mapsto x^q$ constitute the Galois group of K/F.

```
a := normalElement()$K
```

$$\%I \quad (2)$$

Type: FiniteFieldNormalBasis(3, 8)

You can calculate in K using a.

```
b := a**12 - a**5 + a
```

$$\%I^{q^7} + 2\ \%I^{q^6} + 2\ \%I^{q^4} + 2\ \%I^{q^3} + 2\ \%I^{q^2} + 2\ \%I^{q} + \%I \quad (3)$$

Type: FiniteFieldNormalBasis(3, 8)

FiniteFieldNormalBasisExtension (abbreviation FFNBX) is similar to FiniteFieldNormalBasis except that the groundfield for FiniteFieldNormalBasisExtension is arbitrary and chosen by you. In case you select the prime field as ground field, there is essentially no difference between the constructed two finite field extensions.

```
GF9 := FFNB(3,2);
```

(4)

Type: Domain

```
GF729 := FFNBX(GF9,3);
```

(5)

Type: Domain

```
r := random()$GF729
```

$$\left(\%J^{q} + 2\ \%J\right)\ \%K^{q^2} + \left(2\ \%J^{q} + 2\ \%J\right)\ \%K^{q} + \%J^{q}\ \%K \quad (6)$$

Type: FiniteFieldNormalBasisExtension(FiniteFieldNormalBasis(3, 2), 3)

```
r + r**3 + r**9 + r**27
```

$$2\ \%J^{q}\ \%K^{q^2} + \left(\%J^{q} + 2\ \%J\right)\ \%K^{q} + \left(\%J^{q} + 2\ \%J\right)\ \%K \quad (7)$$

Type: FiniteFieldNormalBasisExtension(FiniteFieldNormalBasis(3, 2), 3)

FiniteFieldNormalBasisExtensionByPolynomial (abbreviation FFNBP) is similar to FiniteFieldNormalBasis and FiniteFieldNormalBasisExtension but is more general. For FiniteFieldNormalBasisExtensionByPolynomial you choose both the ground field and the irreducible polynomial used in the representation. The degree of the extension is the degree of the polynomial.

```
GF3 := PrimeField 3;
```

(8)

Type: Domain

We use a utility operation to generate an irreducible normal polynomial (see Section 8.11.7 on page 289). The polynomial has one variable that is "anonymous": it displays as a question mark.

```
f := createNormalPoly(4)$FFPOLY(GF3)
```

$$?^4 + 2\,?^3 + 2 \tag{9}$$

Type: SparseUnivariatePolynomial PrimeField 3

```
GF81 := FFNBP(GF3,f);
```

(10)

Type: Domain

Let's look at a random element from this field.

```
r := random()$GF81
```

$$2\,\%L^q \tag{11}$$

Type: FiniteFieldNormalBasisExtensionByPolynomial(PrimeField 3, ?**4+2*?**3+2)

```
r * r**3 * r**9 * r**27
```

$$2\,\%L^{q^3} + 2\,\%L^{q^2} + 2\,\%L^q + 2\,\%L \tag{12}$$

Type: FiniteFieldNormalBasisExtensionByPolynomial(PrimeField 3, ?**4+2*?**3+2)

```
norm r
```

$$2 \tag{13}$$

Type: PrimeField 3

8.11.6 Conversion Operations for Finite Fields

Let K be a finite field.

```
K := PrimeField 3
```

PrimeField 3 (1)

Type: Domain

An extension field K_m of degree m over K is a subfield of an extension field K_n of degree n over K if and only if m divides n.

$$\begin{array}{c} K_n \\ | \\ K_m \quad \Longleftrightarrow \quad m|n \\ | \\ \mathrm{K} \end{array}$$

FiniteFieldHomomorphisms provides conversion operations between different extensions of one fixed finite ground field and between different representations of these finite fields.

Let's choose m and n,

```
(m,n) := (4,8)
```

$$8 \tag{2}$$

Type: PositiveInteger

build the field extensions,

```
Km := FiniteFieldExtension(K,m)
```

$$\text{FiniteFieldExtension (PrimeField } 3, 4) \tag{3}$$

Type: Domain

and pick two random elements from the smaller field.

```
Kn := FiniteFieldExtension(K,n)
```

$$\text{FiniteFieldExtension (PrimeField } 3, 8) \tag{4}$$

Type: Domain

```
a1 := random()$Km
```

$$\%M^3 + 2\,\%M^2 + 2\,\%M + 2 \tag{5}$$

Type: FiniteFieldExtension(PrimeField 3, 4)

```
b1 := random()$Km
```

$$2\,\%M^3 + 2\,\%M + 2 \tag{6}$$

Type: FiniteFieldExtension(PrimeField 3, 4)

Since m divides n, K_m is a subfield of K_n.

```
a2 := a1 :: Kn
```

$$2\,\%N^6 + 2\,\%N^2 + 2 \tag{7}$$

Type: FiniteFieldExtension(PrimeField 3, 8)

Therefore we can convert the elements of K_m into elements of K_n.

```
b2 := b1 :: Kn
```

$$2\,\%N^6 + \%N^4 + \%N^2 + 2 \tag{8}$$

Type: FiniteFieldExtension(PrimeField 3, 8)

To check this, let's do some arithmetic.

```
a1+b1 - ((a2+b2) :: Km)
```

$$0 \tag{9}$$

Type: FiniteFieldExtension(PrimeField 3, 4)

```
a1*b1 - ((a2*b2) :: Km)
```

$$0 \tag{10}$$

Type: FiniteFieldExtension(PrimeField 3, 4)

There are also conversions available for the situation, when K_m and K_n are represented in different ways (see Section 8.11.2 on page 279). For example let's choose K_m where the representation is 0 plus the cyclic multiplicative group and K_n with a normal basis representation.

```
Km := FFCGX(K,m)
```

FiniteFieldCyclicGroupExtension (PrimeField 3, 4) (11)

Type: Domain

```
Kn := FFNBX(K,n)
```

FiniteFieldNormalBasisExtension (PrimeField 3, 8) (12)

Type: Domain

```
(a1,b1) := (random()$Km,random()$Km)
```

$$\%F^{39} \tag{13}$$

Type: FiniteFieldCyclicGroupExtension(PrimeField 3, 4)

```
a2 := a1 :: Kn
```

$$2\ \%I^{q^7} + 2\ \%I^{q^6} + \%I^{q^5} + \%I^{q^4} + 2\ \%I^{q^3} + 2\ \%I^{q^2} + \%I^{q} + \%I \tag{14}$$

Type: FiniteFieldNormalBasisExtension(PrimeField 3, 8)

```
b2 := b1 :: Kn
```

$$\%I^{q^7} + 2\ \%I^{q^6} + \%I^{q^5} + \%I^{q^4} + \%I^{q^3} + 2\ \%I^{q^2} + \%I^{q} + \%I \tag{15}$$

Type: FiniteFieldNormalBasisExtension(PrimeField 3, 8)

Check the arithmetic again.

```
a1+b1 - ((a2+b2) :: Km)
```

$$0 \tag{16}$$

Type: FiniteFieldCyclicGroupExtension(PrimeField 3, 4)

```
a1*b1 - ((a2*b2) :: Km)
```

$$0 \tag{17}$$

Type: FiniteFieldCyclicGroupExtension(PrimeField 3, 4)

8.11.7 Utility Operations for Finite Fields

FiniteFieldPolynomialPackage (abbreviation FFPOLY) provides operations for generating, counting and testing polynomials over finite fields. Let's start with a couple of definitions:

- A polynomial is *primitive* if its roots are primitive elements in an extension of the coefficient field of degree equal to the degree of the polynomial.
- A polynomial is *normal* over its coefficient field if its roots are linearly independent elements in an extension of the coefficient field of degree equal to the degree of the polynomial.

In what follows, many of the generated polynomials have one "anonymous" variable. This indeterminate is displayed as a question mark ("?").

To fix ideas, let's use the field with five elements for the first few examples.

```
GF5 := PF 5;
```

(1)

Type: Domain

You can generate irreducible polynomials of any (positive) degree (within the storage capabilities of the computer and your ability to wait) by using **createIrreduciblePoly**.

```
f := createIrreduciblePoly(8)$FFPOLY(GF5)
```

$?^8 + ?^4 + 2$ (2)

Type: SparseUnivariatePolynomial PrimeField 5

Does this polynomial have other important properties? Use **primitive?** to test whether it is a primitive polynomial.

```
primitive?(f)$FFPOLY(GF5)
```

false (3)

Type: Boolean

Use **normal?** to test whether it is a normal polynomial.

```
normal?(f)$FFPOLY(GF5)
```

false (4)

Type: Boolean

Note that this is actually a trivial case, because a normal polynomial of degree n must have a nonzero term of degree $n - 1$. We will refer back to this later.

To get a primitive polynomial of degree 8 just issue this.

```
p := createPrimitivePoly(8)$FFPOLY(GF5)
```

$?^8 + ?^3 + ?^2 + ? + 2$ (5)

Type: SparseUnivariatePolynomial PrimeField 5

```
primitive?(p)$FFPOLY(GF5)
```

true (6)

Type: Boolean

This polynomial is not normal,

```
normal?(p)$FFPOLY(GF5)
```

false (7)

Type: Boolean

but if you want a normal one simply write this.

```
n := createNormalPoly(8)$FFPOLY(GF5)
```

$?^8 + 4\,?^7 + ?^3 + 1$ (8)

Type: SparseUnivariatePolynomial PrimeField 5

This polynomial is not primitive!

```
primitive?(n)$FFPOLY(GF5)
```

false (9)

Type: Boolean

This could have been seen directly, as the constant term is 1 here, which is not a primitive element up to the factor (-1) raised to the degree of the polynomial.[4]

What about polynomials that are both primitive and normal? The existence of such a polynomial is by no means obvious. [5]

If you really need one use either **createPrimitiveNormalPoly** or **createNormalPrimitivePoly**.

```
createPrimitiveNormalPoly(8)$FFPOLY(GF5)
```

$$?^8 + 4\,?^7 + 2\,?^5 + 2 \tag{10}$$

Type: SparseUnivariatePolynomial PrimeField 5

If you want to obtain additional polynomials of the various types above as given by the **create...** operations above, you can use the **next...** operations. For instance, **nextIrreduciblePoly** yields the next monic irreducible polynomial with the same degree as the input polynomial. By "next" we mean "next in a natural order using the terms and coefficients." This will become more clear in the following examples.

This is the field with five elements.

```
GF5 := PF 5;
```

(11)

Type: Domain

Our first example irreducible polynomial, say of degree 3, must be "greater" than this.

```
h := monomial(1,8)$SUP(GF5)
```

$$?^8 \tag{12}$$

Type: SparseUnivariatePolynomial PrimeField 5

You can generate it by doing this.

```
nh := nextIrreduciblePoly(h)$FFPOLY(GF5)
```

$$?^8 + 2 \tag{13}$$

Type: Union(SparseUnivariatePolynomial PrimeField 5, ...)

Notice that this polynomial is not the same as the one **createIrreduciblePoly**.

```
createIrreduciblePoly(3)$FFPOLY(GF5)
```

$$?^3 + ? + 1 \tag{14}$$

Type: SparseUnivariatePolynomial PrimeField 5

You can step through all irreducible polynomials of degree 8 over the field with 5 elements by repeatedly issuing this.

```
nh := nextIrreduciblePoly(nh)$FFPOLY(GF5)
```

$$?^8 + 3 \tag{15}$$

Type: Union(SparseUnivariatePolynomial PrimeField 5, ...)

You could also ask for the total number of these.

```
numberOfIrreduciblePoly(5)$FFPOLY(GF5)
```

$$624 \tag{16}$$

Type: PositiveInteger

We hope that "natural order" on polynomials is now clear: first we com-

---

[4]Cf. Lidl, R. & Niederreiter, H., *Finite Fields,* Encycl. of Math. 20, (Addison-Wesley, 1983), p.90, Th. 3.18.

[5]The existence of such polynomials is proved in Lenstra, H. W. & Schoof, R. J., *Primitive Normal Bases for Finite Fields,* Math. Comp. 48, 1987, pp. 217-231.

pare the number of monomials of two polynomials ("more" is "greater"); then, if necessary, the degrees of these monomials (lexicographically), and lastly their coefficients (also lexicographically, and using the operation **lookup** if our field is not a prime field). Also note that we make both polynomials monic before looking at the coefficients: multiplying either polynomial by a nonzero constant produces the same result.

The package FiniteFieldPolynomialPackage also provides similar operations for primitive and normal polynomials. With the exception of the number of primitive normal polynomials; we're not aware of any known formula for this.

```
numberOfPrimitivePoly(3)$FFPOLY(GF5)
```

$$20 \tag{17}$$

Type: PositiveInteger

Take these,

```
m := monomial(1,1)$SUP(GF5)
```

$$? \tag{18}$$

Type: SparseUnivariatePolynomial PrimeField 5

```
f := m**3 + 4*m**2 + m + 2
```

$$?^3 + 4\,?^2 + ? + 2 \tag{19}$$

Type: SparseUnivariatePolynomial PrimeField 5

and then we have:

```
f1 := nextPrimitivePoly(f)$FFPOLY(GF5)
```

$$?^3 + 4\,?^2 + 4\,? + 2 \tag{20}$$

Type: Union(SparseUnivariatePolynomial PrimeField 5, ...)

What happened?

```
nextPrimitivePoly(f1)$FFPOLY(GF5)
```

$$?^3 + 2\,?^2 + 3 \tag{21}$$

Type: Union(SparseUnivariatePolynomial PrimeField 5, ...)

Well, for the ordering used in **nextPrimitivePoly** we use as first criterion a comparison of the constant terms of the polynomials. Analogously, in **nextNormalPoly** we first compare the monomials of degree 1 less than the degree of the polynomials (which is nonzero, by an earlier remark).

```
f := m**3 + m**2 + 4*m + 1
```

$$?^3 + ?^2 + 4\,? + 1 \tag{22}$$

Type: SparseUnivariatePolynomial PrimeField 5

```
f1 := nextNormalPoly(f)$FFPOLY(GF5)
```

$$?^3 + ?^2 + 4\,? + 3 \tag{23}$$

Type: Union(SparseUnivariatePolynomial PrimeField 5, ...)

```
nextNormalPoly(f1)$FFPOLY(GF5)
```

$$?^3 + 2\,?^2 + 1 \tag{24}$$

Type: Union(SparseUnivariatePolynomial PrimeField 5, ...)

We don't have to restrict ourselves to prime fields.

Let's consider, say, a field with 16 elements.

```
GF16 := FFX(FFX(PF 2,2),2);
```

(25)

Type: Domain

We can apply any of the operations described above.

```
createIrreduciblePoly(5)$FFPOLY(GF16)
```

$$?^5 + \%P \tag{26}$$

Type: SparseUnivariatePolynomial FiniteFieldExtension(FiniteFieldExtension(PrimeField 2, 2), 2)

AXIOM also provides operations for producing random polynomials of a given degree

```
random(5)$FFPOLY(GF16)
```

$$?^5 + \%D\ ?^4 + \%P\ ?^3 + (\%D\ \%P + 1)\ ?^2 + (\%D + 1)\ ? + \%D\ \%P \tag{27}$$

Type: SparseUnivariatePolynomial FiniteFieldExtension(FiniteFieldExtension(PrimeField 2, 2), 2)

or with degree between two given bounds.

```
random(3,9)$FFPOLY(GF16)
```

$$?^4 + ((\%D + 1)\ \%P + 1)\ ?^3 + (\%D + 1)\ \%P\ ?^2 + (\%D + 1)\ ? + \%P + \%D + 1 \tag{28}$$

Type: SparseUnivariatePolynomial FiniteFieldExtension(FiniteFieldExtension(PrimeField 2, 2), 2)

`FiniteFieldPolynomialPackage2` (abbreviation FFPOLY2) exports an operation **rootOfIrreduciblePoly** for finding one root of an irreducible polynomial `f` in an extension field of the coefficient field. The degree of the extension has to be a multiple of the degree of `f`. It is not checked whether `f` actually is irreducible.

To illustrate this operation, we fix a ground field `GF`

```
GF2 := PrimeField 2;
```

(29)

Type: Domain

and then an extension field.

```
F := FFX(GF2,12)
```

FiniteFieldExtension (PrimeField 2, 12) (30)

Type: Domain

We construct an irreducible polynomial over `GF2`.

```
f := createIrreduciblePoly(6)$FFPOLY(GF2)
```

$$?^6 + ? + 1 \tag{31}$$

Type: SparseUnivariatePolynomial PrimeField 2

We compute a root of f.

```
root := rootOfIrreduciblePoly(f)$FFPOLY2(F,GF2)
```

$$\%A^{11} + \%A^8 + \%A^7 + \%A^5 + \%A + 1 \quad (32)$$

Type: FiniteFieldExtension(PrimeField 2, 12)

8.12 Primary Decomposition of Ideals

AXIOM provides a facility for the primary decomposition of polynomial ideals over fields of characteristic zero. The algorithm works in essentially two steps:

1. the problem is solved for 0-dimensional ideals by "generic" projection on the last coordinate
2. a "reduction process" uses localization and ideal quotients to reduce the general case to the 0-dimensional one.

The AXIOM constructor PolynomialIdeals represents ideals with coefficients in any field and supports the basic ideal operations, including intersection, sum and quotient. IdealDecompositionPackage contains the specific operations for the primary decomposition and the computation of the radical of an ideal with polynomial coefficients in a field of characteristic 0 with an effective algorithm for factoring polynomials.

The following examples illustrate the capabilities of this facility.

First consider the ideal generated by $x^2 + y^2 - 1$ (which defines a circle in the (x,y)-plane) and the ideal generated by $x^2 - y^2$ (corresponding to the straight lines x = y and x = -y.

```
(n,m) : List DMP([x,y],FRAC INT)
```

Type: Void

```
m := [x**2+y**2-1]
```

$$\left[x^2 + y^2 - 1\right] \quad (2)$$

Type: List DistributedMultivariatePolynomial([x, y], Fraction Integer)

```
n := [x**2-y**2]
```

$$\left[x^2 - y^2\right] \quad (3)$$

Type: List DistributedMultivariatePolynomial([x, y], Fraction Integer)

We find the equations defining the intersection of the two loci. This correspond to the sum of the associated ideals.

```
id := ideal m + ideal n
```

$$\left[x^2 - \frac{1}{2},\, y^2 - \frac{1}{2}\right] \quad (4)$$

Type: PolynomialIdeals(Fraction Integer, DirectProduct(2, NonNegativeInteger), [x, y], DistributedMultivariatePolynomial([x, y], Fraction Integer))

We can check if the locus contains only a finite number of points, that is, if the ideal is zero-dimensional.

```
zeroDim? id
```

true (5)

Type: Boolean

```
zeroDim?(ideal m)
```

false (6)

Type: Boolean

```
dimension ideal m
```

1 (7)

Type: PositiveInteger

We can find polynomial relations among the generators (f and g are the parametric equations of the knot).

```
(f,g):DMP([x,y],FRAC INT)
```

Type: Void

```
f := x**2-1
```

$x^2 - 1$ (9)

Type: DistributedMultivariatePolynomial([x, y], Fraction Integer)

```
g := x*(x**2-1)
```

$x^3 - x$ (10)

Type: DistributedMultivariatePolynomial([x, y], Fraction Integer)

```
relationsIdeal [f,g]
```

$\left[-\%V2^2 + \%U2^3 + \%U2^2\right] \mid \left[\%U2 = x^2 - 1,\ \%V2 = x^3 - x\right]$ (11)

Type: SuchThat(List Polynomial Fraction Integer, List Equation Polynomial Fraction Integer)

We can compute the primary decomposition of an ideal.

```
l: List DMP([x,y,z],FRAC INT)
```

Type: Void

```
l:=[x**2+2*y**2,x*z**2-y*z,z**2-4]
```

$\left[x^2 + 2\ y^2,\ x\ z^2 - y\ z,\ z^2 - 4\right]$ (13)

Type: List DistributedMultivariatePolynomial([x, y, z], Fraction Integer)

```
ld:=primaryDecomp ideal 1
```

$$\left[\left[x + \frac{1}{2}\, y,\, y^2,\, z + 2\right],\, \left[x - \frac{1}{2}\, y,\, y^2,\, z - 2\right]\right] \tag{14}$$

Type: List PolynomialIdeals(Fraction Integer, DirectProduct(3, NonNegativeInteger), [x, y, z], DistributedMultivariatePolynomial([x, y, z], Fraction Integer))

We can intersect back.

```
reduce(intersect,ld)
```

$$\left[x - \frac{1}{4}\, y\, z,\, y^2,\, z^2 - 4\right] \tag{15}$$

Type: PolynomialIdeals(Fraction Integer, DirectProduct(3, NonNegativeInteger), [x, y, z], DistributedMultivariatePolynomial([x, y, z], Fraction Integer))

We can compute the radical of every primary component.

```
reduce(intersect,[radical ld.i for i in 1..2])
```

$$\left[x,\, y,\, z^2 - 4\right] \tag{16}$$

Type: PolynomialIdeals(Fraction Integer, DirectProduct(3, NonNegativeInteger), [x, y, z], DistributedMultivariatePolynomial([x, y, z], Fraction Integer))

Their intersection is equal to the radical of the ideal of 1.

```
radical ideal 1
```

$$\left[x,\, y,\, z^2 - 4\right] \tag{17}$$

Type: PolynomialIdeals(Fraction Integer, DirectProduct(3, NonNegativeInteger), [x, y, z], DistributedMultivariatePolynomial([x, y, z], Fraction Integer))

8.13 Computation of Galois Groups

As a sample use of AXIOM's algebraic number facilities, we compute the Galois group of the polynomial $p(x) = x^5 - 5x + 12$.

```
p := x**5 - 5*x + 12
```

$$x^5 - 5\, x + 12 \tag{1}$$

Type: Polynomial Integer

We would like to construct a polynomial $f(x)$ such that the splitting field of $p(x)$ is generated by one root of $f(x)$. First we construct a polynomial $r = r(x)$ such that one root of $r(x)$ generates the field generated by two roots of the polynomial $p(x)$. (As it will turn out, the field generated by two roots of $p(x)$ is, in fact, the splitting field of $p(x)$.)

From the proof of the primitive element theorem we know that if a and b are algebraic numbers, then the field $\mathbf{Q}(a, b)$ is equal to $\mathbf{Q}(a + kb)$ for an appropriately chosen integer k. In our case, we construct the minimal polynomial of $a_i - a_j$, where a_i and a_j are two roots of $p(x)$. We construct this polynomial using **resultant**. The main result we need is the following: If $f(x)$ is a polynomial with roots $a_i \ldots a_m$ and $g(x)$ is a polynomial with roots $b_i \ldots b_n$, then the polynomial `h(x) = resultant(f(y),` g(x-y), y) is a polynomial of degree $m * n$ with roots $a_i + b_j, i = 1 \ldots m, j = 1 \ldots n$.

For $f(x)$ we use the polynomial $p(x)$. For $g(x)$ we use the polynomial $-p(-x)$. Thus, the polynomial we first construct is resultant(p(y), -p(y-x), y).

```
q := resultant(eval(p,x,y),-eval(p,x,y-x),y)
```

$$x^{25} - 50\,x^{21} - 2375\,x^{17} + 90000\,x^{15} - 5000\,x^{13} + 2700000\,x^{11} + 250000\,x^9 + 18000000\,x^7 + 64000000\,x^5 \quad (2)$$

Type: Polynomial Integer

The roots of $q(x)$ are $a_i - a_j, i \le 1, j \le 5$. Of course, there are five pairs (i, j) with $i = j$, so 0 is a 5-fold root of $q(x)$.

Let's get rid of this factor.

```
q1 := exquo(q, x**5)
```

$$x^{20} - 50\,x^{16} - 2375\,x^{12} + 90000\,x^{10} - 5000\,x^8 + 2700000\,x^6 + 250000\,x^4 + 18000000\,x^2 + 64000000 \quad (3)$$

Type: Union(Polynomial Integer, ...)

Factor the polynomial q1.

```
factoredQ := factor q1
```

$$\left(x^{10} - 10\,x^8 - 75\,x^6 + 1500\,x^4 - 5500\,x^2 + 16000\right)\left(x^{10} + 10\,x^8 + 125\,x^6 + 500\,x^4 + 2500\,x^2 + 4000\right) \quad (4)$$

Type: Factored Polynomial Integer

We see that q1 has two irreducible factors, each of degree 10. (The fact that the polynomial q1 has two factors of degree 10 is enough to show that the Galois group of $p(x)$ is the dihedral group of order 10.[6] Note that the type of factoredQ is FR POLY INT, that is, Factored Polynomial Integer. This is a special data type for recording factorizations of polynomials with integer coefficients (see 'Factored' on page 356).

We can access the individual factors using the operation **nthFactor**.

```
r := nthFactor(factoredQ,1)
```

$$x^{10} - 10\,x^8 - 75\,x^6 + 1500\,x^4 - 5500\,x^2 + 16000 \quad (5)$$

Type: Polynomial Integer

Consider the polynomial $r = r(x)$. This is the minimal polynomial of the difference of two roots of $p(x)$. Thus, the splitting field of $p(x)$ contains a subfield of degree 10. We show that this subfield is, in fact, the splitting field of $p(x)$ by showing that $p(x)$ factors completely over this field.

[6] See McKay, Soicher, Computing Galois Groups over the Rationals, Journal of Number Theory 20, 273-281 (1983). We do not assume the results of this paper, however, and we continue with the computation.

First we create a symbolic root of the polynomial $r(x)$. (We replaced `x` by `b` in the polynomial `r` so that our symbolic root would be printed as `b`.)

```
beta:AN := rootOf(eval(r,x,b))
```

$$b \tag{6}$$

Type: AlgebraicNumber

We next tell AXIOM to view $p(x)$ as a univariate polynomial in `x` with algebraic number coefficients. This is accomplished with this type declaration.

```
p := p::UP(x,INT)::UP(x,AN)
```

$$x^5 - 5\,x + 12 \tag{7}$$

Type: UnivariatePolynomial(x, AlgebraicNumber)

Factor $p(x)$ over the field $\mathbf{Q}(\beta)$. (This computation will take some time!)

```
algFactors := factor(p,[beta])
```

$$\left(x + \frac{\left(\begin{array}{l} -85\,b^9 - 116\,b^8 + 780\,b^7 + 2640\,b^6 + 14895\,b^5 - \\ 8820\,b^4 - 127050\,b^3 - 327000\,b^2 - 405200\,b + \\ 2062400 \end{array}\right)}{1339200}\right) \cdot$$
$$\left(x + \frac{-17\,b^8 + 156\,b^6 + 2979\,b^4 - 25410\,b^2 - 14080}{66960}\right) \cdot$$
$$\left(x + \frac{\left(\begin{array}{l} 143\,b^8 - 2100\,b^6 - 10485\,b^4 + 290550\,b^2 - \\ 334800\,b - 960800 \end{array}\right)}{669600}\right) \cdot \tag{8}$$
$$\left(x + \frac{\left(\begin{array}{l} 143\,b^8 - 2100\,b^6 - 10485\,b^4 + 290550\,b^2 + \\ 334800\,b - 960800 \end{array}\right)}{669600}\right) \cdot$$
$$\left(x + \frac{\left(\begin{array}{l} 85\,b^9 - 116\,b^8 - 780\,b^7 + 2640\,b^6 - 14895\,b^5 - \\ 8820\,b^4 + 127050\,b^3 - 327000\,b^2 + 405200\,b + \\ 2062400 \end{array}\right)}{1339200}\right)$$

Type: Factored UnivariatePolynomial(x, AlgebraicNumber)

When factoring over number fields, it is important to specify the field over which the polynomial is to be factored, as polynomials have different factorizations over different fields. When you use the operation **factor**, the field over which the polynomial is factored is the field generated by

1. the algebraic numbers that appear in the coefficients of the polynomial, and
2. the algebraic numbers that appear in a list passed as an optional second argument of the operation.

In our case, the coefficients of `p` are all rational integers and only `beta` appears in the list, so the field is simply $\mathbf{Q}(\beta)$.

It was necessary to give the list `[beta]` as a second argument of the operation because otherwise the polynomial would have been factored over the field generated by its coefficients, namely the rational numbers.

```
factor(p)
```

$$x^5 - 5\ x + 12 \tag{9}$$

Type: Factored UnivariatePolynomial(x, AlgebraicNumber)

We have shown that the splitting field of $p(x)$ has degree `10`. Since the symmetric group of degree 5 has only one transitive subgroup of order `10`, we know that the Galois group of $p(x)$ must be this group, the dihedral group of order `10`. Rather than stop here, we explicitly compute the action of the Galois group on the roots of $p(x)$.

First we assign the roots of $p(x)$ as the values of five variables.

We can obtain an individual root by negating the constant coefficient of one of the factors of $p(x)$.

```
factor1 := nthFactor(algFactors,1)
```

$$x + \frac{\left(\begin{array}{l} -85\ b^9 - 116\ b^8 + 780\ b^7 + 2640\ b^6 + 14895\ b^5 - \\ 8820\ b^4 - 127050\ b^3 - 327000\ b^2 - 405200\ b + 2062400 \end{array}\right)}{1339200} \tag{10}$$

Type: UnivariatePolynomial(x, AlgebraicNumber)

```
root1 := -coefficient(factor1,0)
```

$$\frac{\left(\begin{array}{l} 85\ b^9 + 116\ b^8 - 780\ b^7 - 2640\ b^6 - 14895\ b^5 + 8820\ b^4 + \\ 127050\ b^3 + 327000\ b^2 + 405200\ b - 2062400 \end{array}\right)}{1339200} \tag{11}$$

Type: AlgebraicNumber

We can obtain a list of all the roots in this way.

```
roots := [-coefficient(nthFactor(algFactors,i),0) for i in
  1..5]
```

$$\left[\frac{\left(\begin{array}{l}85\, b^9+116\, b^8-780\, b^7-2640\, b^6-14895\, b^5+8820\, b^4+\\127050\, b^3+327000\, b^2+405200\, b-2062400\end{array}\right)}{1339200},\right.$$

$$\frac{17\, b^8-156\, b^6-2979\, b^4+25410\, b^2+14080}{66960},$$

$$\frac{-143\, b^8+2100\, b^6+10485\, b^4-290550\, b^2+334800\, b+960800}{669600}, \quad (12)$$

$$\frac{-143\, b^8+2100\, b^6+10485\, b^4-290550\, b^2-334800\, b+960800}{669600},$$

$$\left.\frac{\left(\begin{array}{l}-85\, b^9+116\, b^8+780\, b^7-2640\, b^6+14895\, b^5+8820\, b^4-\\127050\, b^3+327000\, b^2-405200\, b-2062400\end{array}\right)}{1339200}\right]$$

Type: List AlgebraicNumber

The expression

```
- coefficient(nthFactor(algFactors, i), 0)}
```

is the `i` th root of $p(x)$ and the elements of `roots` are the `i` th roots of $p(x)$ as `i` ranges from `1` to `5`.

Assign the roots as the values of the variables `a1,...,a5`.

```
(a1,a2,a3,a4,a5) :=
  (roots.1,roots.2,roots.3,roots.4,roots.5)
```

$$\frac{\left(\begin{array}{l}-85\, b^9+116\, b^8+780\, b^7-2640\, b^6+14895\, b^5+8820\, b^4-\\127050\, b^3+327000\, b^2-405200\, b-2062400\end{array}\right)}{1339200} \quad (13)$$

Type: AlgebraicNumber

Next we express the roots of $r(x)$ as polynomials in `beta`. We could obtain these roots by calling the operation **factor**: `factor(r, [beta])` factors `r(x)` over $\mathbf{Q}(\beta)$. However, this is a lengthy computation and we can obtain the roots of $r(x)$ as differences of the roots `a1,...,a5` of $p(x)$. Only ten of these differences are roots of $r(x)$ and the other ten are roots of the other irreducible factor of `q1`. We can determine if a given value is a root of $r(x)$ by evaluating $r(x)$ at that particular value. (Of course, the order in which factors are returned by the operation **factor** is unimportant and may change with different implementations of the operation. Therefore, we cannot predict in advance which differences are roots of $r(x)$ and which are not.)

Let's look at four examples (two are roots of $r(x)$ and two are not).

```
eval(r,x,a1 - a2)
```

$$0 \tag{14}$$

Type: Polynomial AlgebraicNumber

```
eval(r,x,a1 - a3)
```

$$\frac{\left(\begin{array}{l} 47905\ b^9 + 66920\ b^8 - 536100\ b^7 - 980400\ b^6 - 3345075\ b^5 - \\ 5787000\ b^4 + 75572250\ b^3 + 161688000\ b^2 - 184600000\ b - \\ 710912000 \end{array}\right)}{4464} \tag{15}$$

Type: Polynomial AlgebraicNumber

```
eval(r,x,a1 - a4)
```

$$0 \tag{16}$$

Type: Polynomial AlgebraicNumber

```
eval(r,x,a1 - a5)
```

$$\frac{405\ b^8 + 3450\ b^6 - 19875\ b^4 - 198000\ b^2 - 588000}{31} \tag{17}$$

Type: Polynomial AlgebraicNumber

Take one of the differences that was a root of $r(x)$ and assign it to the variable `bb`.

For example, if `eval(r,x,a1 - a4)` returned `0`, you would enter this.

```
bb := a1 - a4
```

$$\frac{\left(\begin{array}{l} 85\ b^9 + 402\ b^8 - 780\ b^7 - 6840\ b^6 - 14895\ b^5 - 12150\ b^4 + \\ 127050\ b^3 + 908100\ b^2 + 1074800\ b - 3984000 \end{array}\right)}{1339200} \tag{18}$$

Type: AlgebraicNumber

Of course, if the difference is, in fact, equal to the root `beta`, you should choose another root of $r(x)$.

Automorphisms of the splitting field are given by mapping a generator of the field, namely beta, to other roots of its minimal polynomial. Let's see what happens when beta is mapped to bb.

We compute the images of the roots a1,...,a5 under this automorphism:

```
aa1 := subst(a1,beta = bb)
```

$$\frac{-143\ b^8 + 2100\ b^6 + 10485\ b^4 - 290550\ b^2 + 334800\ b + 960800}{669600} \tag{19}$$

Type: AlgebraicNumber

```
aa2 := subst(a2,beta = bb)
```

$$\frac{\left(\begin{array}{l}-85\ b^9 + 116\ b^8 + 780\ b^7 - 2640\ b^6 + 14895\ b^5 + 8820\ b^4 - \\ 127050\ b^3 + 327000\ b^2 - 405200\ b - 2062400\end{array}\right)}{1339200} \tag{20}$$

Type: AlgebraicNumber

```
aa3 := subst(a3,beta = bb)
```

$$\frac{\left(\begin{array}{l}85\ b^9 + 116\ b^8 - 780\ b^7 - 2640\ b^6 - 14895\ b^5 + 8820\ b^4 + \\ 127050\ b^3 + 327000\ b^2 + 405200\ b - 2062400\end{array}\right)}{1339200} \tag{21}$$

Type: AlgebraicNumber

```
aa4 := subst(a4,beta = bb)
```

$$\frac{-143\ b^8 + 2100\ b^6 + 10485\ b^4 - 290550\ b^2 - 334800\ b + 960800}{669600} \tag{22}$$

Type: AlgebraicNumber

```
aa5 := subst(a5,beta = bb)
```

$$\frac{17\ b^8 - 156\ b^6 - 2979\ b^4 + 25410\ b^2 + 14080}{66960} \tag{23}$$

Type: AlgebraicNumber

Of course, the values aa1,...,aa5 are simply a permutation of the values a1,...,a5.

Let's find the value of aa1 (execute as many of the following five commands as necessary).

```
(aa1 = a1) :: Boolean
```

false (24)

Type: Boolean

```
(aa1 = a2) :: Boolean
```

false (25)

Type: Boolean

```
(aa1 = a3) :: Boolean
```

true (26)

Type: Boolean

```
(aa1 = a4) :: Boolean
```

false (27)

Type: Boolean

```
(aa1 = a5) :: Boolean
```

false (28)

Type: Boolean

Proceeding in this fashion, you can find the values of `aa2,...aa5`.[7] You have represented the automorphism `beta -> bb` as a permutation of the roots `a1,...,a5`. If you wish, you can repeat this computation for all the roots of $r(x)$ and represent the Galois group of $p(x)$ as a subgroup of the symmetric group on five letters.

Here are two other problems that you may attack in a similar fashion:

1. Show that the Galois group of $p(x) = x^4 + 2x^3 - 2x^2 - 3x + 1$ is the dihedral group of order eight. (The splitting field of this polynomial is the Hilbert class field of the quadratic field $\mathbf{Q}(\sqrt{145})$.)
2. Show that the Galois group of $p(x) = x^6 + 108$ has order 6 and is isomorphic to S_3, the symmetric group on three letters. (The splitting field of this polynomial is the splitting field of $x^3 - 2$.)

8.14 Non-Associative Algebras and Modelling Genetic Laws

Many algebraic structures of mathematics and AXIOM have a multiplication operation "`*`" that satisfies the associativity law $a * (b * c) = (a * b) * c$ for all a, b and c. The octonions (see 'Octonion' on page 423) are a well known exception. There are many other interesting non-associative structures, such as the class of Lie algebras.[8] Lie algebras can be used, for example, to analyse Lie symmetry algebras of partial differential equations. In this section we show a different application of non-associative algebras, the modelling of genetic laws.

The AXIOM library contains several constructors for creating non-associative structures, ranging from the categories Monad, NonAssociativeRng, and FramedNonAssociativeAlgebra, to the domains AlgebraGivenByStructuralConstants and GenericNonAssociativeAlgebra. Furthermore, the package Algebra-

[7] Here you should use the Clef line editor. See Section 1.1.1 on page 21 for more information about Clef.

[8] Two AXIOM implementations of Lie algebras are LieSquareMatrix and FreeNilpotentLie.

Package provides operations for analysing the structure of such algebras.[9]

Mendel's genetic laws are often written in a form like

$$Aa \times Aa = \frac{1}{4}AA + \frac{1}{2}Aa + \frac{1}{4}aa.$$

The implementation of general algebras in AXIOM allows us to use this as the definition for multiplication in an algebra. Hence, it is possible to study questions of genetic inheritance using AXIOM. To demonstrate this more precisely, we discuss one example from a monograph of A. Wörz-Busekros, where you can also find a general setting of this theory.[10]

We assume that there is an infinitely large random mating population. Random mating of two gametes a_i and a_j gives zygotes $a_i a_j$, which produce new gametes. In classical Mendelian segregation we have $a_i a_j = \frac{1}{2}a_i + \frac{1}{2}a_j$. In general, we have

$$a_i a_j = \sum_{k=1}^{n} \gamma_{i,j}^{k} \; a_k.$$

The segregation rates $\gamma_{i,j}$ are the structural constants of an n-dimensional algebra. This is provided in AXIOM by the constructor AlgebraGivenByStructuralConstants (abbreviation ALGSC).

Consider two coupled autosomal loci with alleles A,a, B, and b, building four different gametes $a_1 = AB, a_2 = Ab, a_3 = aB$, and $a_4 = ab$. The zygotes $a_i a_j$ produce gametes a_i and a_j with classical Mendelian segregation. Zygote $a_1 a_4$ undergoes transition to $a_2 a_3$ and vice versa with probability $0 \leq \theta \leq \frac{1}{2}$.

---

[9]The interested reader can learn more about these aspects of the AXIOM library from the paper "Computations in Algebras of Finite Rank," by Johannes Grabmeier and Robert Wisbauer, Technical Report, IBM Heidelberg Scientific Center, 1992.

[10]Wörz-Busekros, A., *Algebras in Genetics*, Springer Lectures Notes in Biomathematics 36, Berlin e.a. (1980). In particular, see example 1.3.

Define a list $[(\gamma_{i,j}^k)$ for $1 \le k \le 4]$ of four four-by-four matrices giving the segregation rates. We use the value $1/10$ for θ.

```
segregationRates : List SquareMatrix(4,FRAC INT) :=
  [matrix [ [1, 1/2, 1/2, 9/20], [1/2, 0, 1/20, 0], [1/2,
  1/20, 0, 0], [9/20, 0, 0, 0] ], matrix [ [0, 1/2, 0,
  1/20], [1/2, 1, 9/20, 1/2], [0, 9/20, 0, 0], [1/20,
  1/2, 0, 0] ], matrix [ [0, 0, 1/2, 1/20], [0, 0, 9/20,
  0], [1/2, 9/20, 1, 1/2], [1/20, 0, 1/2, 0] ], matrix [
  [0, 0, 0, 9/20], [0, 0, 1/20, 1/2], [0, 1/20, 0, 1/2],
  [9/20, 1/2, 1/2, 1] ] ]
```

$$\left[\begin{bmatrix} 1 & \frac{1}{2} & \frac{1}{2} & \frac{9}{20} \\ \frac{1}{2} & 0 & \frac{1}{20} & 0 \\ \frac{1}{2} & \frac{1}{20} & 0 & 0 \\ \frac{9}{20} & 0 & 0 & 0 \end{bmatrix}, \begin{bmatrix} 0 & \frac{1}{2} & 0 & \frac{1}{20} \\ \frac{1}{2} & 1 & \frac{9}{20} & \frac{1}{2} \\ 0 & \frac{9}{20} & 0 & 0 \\ \frac{1}{20} & \frac{1}{2} & 0 & 0 \end{bmatrix}, \begin{bmatrix} 0 & 0 & \frac{1}{2} & \frac{1}{20} \\ 0 & 0 & \frac{9}{20} & 0 \\ \frac{1}{2} & \frac{9}{20} & 1 & \frac{1}{2} \\ \frac{1}{20} & 0 & \frac{1}{2} & 0 \end{bmatrix}, \begin{bmatrix} 0 & 0 & 0 & \frac{9}{20} \\ 0 & 0 & \frac{1}{20} & \frac{1}{2} \\ 0 & \frac{1}{20} & 0 & \frac{1}{2} \\ \frac{9}{20} & \frac{1}{2} & \frac{1}{2} & 1 \end{bmatrix}\right] \quad (1)$$

Type: List SquareMatrix(4, Fraction Integer)

Choose the appropriate symbols for the basis of gametes,

```
gametes := ['AB,'Ab,'aB,'ab]
```

$$[AB,\ Ab,\ aB,\ ab] \quad (2)$$

Type: List Symbol

Define the algebra.

```
A := ALGSC(FRAC INT, 4, gametes, segregationRates);
```

(3)

Type: Domain

What are the probabilities for zygote $a_1 a_4$ to produce the different gametes?

```
a := basis()$A; a.1*a.4
```

$$\frac{9}{20}\ ab + \frac{1}{20}\ aB + \frac{1}{20}\ Ab + \frac{9}{20}\ AB \quad (4)$$

Type: AlgebraGivenByStructuralConstants(Fraction Integer, 4, [AB, Ab, aB, ab], [MATRIX, MATRIX, MATRIX, MATRIX])

Elements in this algebra whose coefficients sum to one play a distinguished role. They represent a population with the distribution of gametes reflected by the coefficients with respect to the basis of gametes.

Random mating of different populations x and y is described by their product $x * y$.

This product is commutative only if the gametes are not sex-dependent, as in our example.

```
commutative?()$A
```

true (5)

Type: Boolean

In general, it is not associative.

```
associative?()$A
```

$$\text{false} \tag{6}$$

Type: Boolean

Random mating within a population x is described by $x * x$. The next generation is $(x * x) * (x * x)$.

Use decimal numbers to compare the distributions more easily.

```
x : ALGSC(DECIMAL, 4, gametes, segregationRates) :=
  convert [3/10, 1/5, 1/10, 2/5]
```

$$0.4\ ab + 0.1\ aB + 0.2\ Ab + 0.3\ AB \tag{7}$$

Type: AlgebraGivenByStructuralConstants(DecimalExpansion, 4, [AB, Ab, aB, ab], [MATRIX, MATRIX, MATRIX, MATRIX])

To compute directly the gametic distribution in the fifth generation, we use **plenaryPower**.

```
plenaryPower(x,5)
```

$$0.36561\ ab + 0.13439\ aB + 0.23439\ Ab + 0.26561\ AB \tag{8}$$

Type: AlgebraGivenByStructuralConstants(DecimalExpansion, 4, [AB, Ab, aB, ab], [MATRIX, MATRIX, MATRIX, MATRIX])

We now ask two questions: Does this distribution converge to an equilibrium state? What are the distributions that are stable?

This is an invariant of the algebra and it is used to answer the first question. The new indeterminates describe a symbolic distribution.

```
q := leftRankPolynomial()$GCNAALG(FRAC INT, 4, gametes,
  segregationRates) :: UP(Y, POLY FRAC INT)
```

$$Y^3 + \left(-\frac{29}{20}\ \%x4 - \frac{29}{20}\ \%x3 - \frac{29}{20}\ \%x2 - \frac{29}{20}\ \%x1\right) Y^2 + \left(\begin{array}{l} \frac{9}{20}\ \%x4^2 + \left(\frac{9}{10}\ \%x3 + \frac{9}{10}\ \%x2 + \frac{9}{10}\ \%x1\right) \%x4 + \\ \frac{9}{20}\ \%x3^2 + \left(\frac{9}{10}\ \%x2 + \frac{9}{10}\ \%x1\right) \%x3 + \frac{9}{20}\ \%x2^2 + \\ \frac{9}{10}\ \%x1\ \%x2 + \frac{9}{20}\ \%x1^2 \end{array}\right) Y \tag{9}$$

Type: UnivariatePolynomial(Y, Polynomial Fraction Integer)

Because the coefficient $\frac{9}{20}$ has absolute value less than 1, all distributions do converge, by a theorem of this theory.

```
factor(q :: POLY FRAC INT)
```

$$(Y - \%x4 - \%x3 - \%x2 - \%x1) \cdot \left(Y - \frac{9}{20}\ \%x4 - \frac{9}{20}\ \%x3 - \frac{9}{20}\ \%x2 - \frac{9}{20}\ \%x1\right) Y \tag{10}$$

Type: Factored Polynomial Fraction Integer

The second question is answered by searching for idempotents in the algebra.

```
cI := conditionsForIdempotents()$GCNAALG(FRAC INT, 4,
  gametes, segregationRates)
```

$$\left[\frac{9}{10}\ \%x1\ \%x4 + \left(\frac{1}{10}\ \%x2 + \%x1\right)\ \%x3 + \%x1\ \%x2 + \%x1^2 - \%x1,\right.$$
$$\left(\%x2 + \frac{1}{10}\ \%x1\right)\ \%x4 + \frac{9}{10}\ \%x2\ \%x3 + \%x2^2 + (\%x1 - 1)\ \%x2,$$
$$\left(\%x3 + \frac{1}{10}\ \%x1\right)\ \%x4 + \%x3^2 + \left(\frac{9}{10}\ \%x2 + \%x1 - 1\right)\ \%x3,$$
$$\left.\%x4^2 + \left(\%x3 + \%x2 + \frac{9}{10}\ \%x1 - 1\right)\ \%x4 + \frac{1}{10}\ \%x2\ \%x3\right] \quad (11)$$

Type: List Polynomial Fraction Integer

Solve these equations and look at the first solution.

```
gbs:= groebnerFactorize cI; gbs.1
```

$$[\%x4 + \%x3 + \%x2 + \%x1 - 1,$$
$$(\%x2 + \%x1)\ \%x3 + \%x1\ \%x2 + \%x1^2 - \%x1] \quad (12)$$

Type: List Polynomial Fraction Integer

Further analysis using the package PolynomialIdeals shows that there is a two-dimensional variety of equilibrium states and all other solutions are contained in it.

Choose one equilibrium state by setting two indeterminates to concrete values.

```
sol := solve concat(gbs.1,[%x1-1/10,%x2-1/10])
```

$$\left[\left[\%x4 = \frac{2}{5},\ \%x3 = \frac{2}{5},\ \%x2 = \frac{1}{10},\ \%x1 = \frac{1}{10}\right]\right] \quad (13)$$

Type: List List Equation Fraction Polynomial Integer

```
e : A := represents reverse (map(rhs, sol.1) :: List FRAC
  INT)
```

$$\frac{2}{5}\ ab + \frac{2}{5}\ aB + \frac{1}{10}\ Ab + \frac{1}{10}\ AB \quad (14)$$

Type: AlgebraGivenByStructuralConstants(Fraction Integer, 4, [AB, Ab, aB, ab], [MATRIX, MATRIX, MATRIX, MATRIX])

Verify the result.

```
e*e-e
```

$$0 \quad (15)$$

Type: AlgebraGivenByStructuralConstants(Fraction Integer, 4, [AB, Ab, aB, ab], [MATRIX, MATRIX, MATRIX, MATRIX])

CHAPTER 9

Some Examples of Domains and Packages

In this chapter we show examples of many of the most commonly used AXIOM domains and packages. The sections are organized by constructor names.

9.1 AssociationList

The AssociationList constructor provides a general structure for associative storage. This type provides association lists in which data objects can be saved according to keys of any type. For a given association list, specific types must be chosen for the keys and entries. You can think of the representation of an association list as a list of records with key and entry fields.

Association lists are a form of table and so most of the operations available for Table are also available for AssociationList. They can also be viewed as lists and can be manipulated accordingly.

This is a Record type with age and gender fields.

```
Data := Record(monthsOld : Integer, gender : String)
```

Record ($monthsOld$: Integer , $gender$: String) (1)

Type: Domain

In this expression, al is declared to be an association list whose keys are strings and whose entries are the above records.

```
al : AssociationList(String,Data)
```

Type: Void

The **table** operation is used to create an empty association list.

```
al := table()
```

$$\text{table}() \tag{3}$$

Type: AssociationList(String, Record(monthsOld: Integer, gender: String))

You can use assignment syntax to add things to the association list.

```
al."bob" := [407,"male"]$Data
```

$$[monthsOld = 407,\ gender = \texttt{"male"}] \tag{4}$$

Type: Record(monthsOld: Integer, gender: String)

```
al."judith" := [366,"female"]$Data
```

$$[monthsOld = 366,\ gender = \texttt{"female"}] \tag{5}$$

Type: Record(monthsOld: Integer, gender: String)

```
al."katie" := [24,"female"]$Data
```

$$[monthsOld = 24,\ gender = \texttt{"female"}] \tag{6}$$

Type: Record(monthsOld: Integer, gender: String)

Perhaps we should have included a species field.

```
al."smokie" := [200,"female"]$Data
```

$$[monthsOld = 200,\ gender = \texttt{"female"}] \tag{7}$$

Type: Record(monthsOld: Integer, gender: String)

Now look at what is in the association list. Note that the last-added (key, entry) pair is at the beginning of the list.

```
al
```

$$\begin{array}{l}\text{table}\left(\texttt{"smokie"} = [monthsOld = 200,\ gender = \texttt{"female"}],\right.\\ \texttt{"katie"} = [monthsOld = 24,\ gender = \texttt{"female"}],\\ \texttt{"judith"} = [monthsOld = 366,\ gender = \texttt{"female"}],\\ \left.\texttt{"bob"} = [monthsOld = 407,\ gender = \texttt{"male"}]\right)\end{array} \tag{8}$$

Type: AssociationList(String, Record(monthsOld: Integer, gender: String))

You can reset the entry for an existing key.

```
al."katie" := [23,"female"]$Data
```

$$[monthsOld = 23,\ gender = \texttt{"female"}] \tag{9}$$

Type: Record(monthsOld: Integer, gender: String)

Use **delete!** to destructively remove an element of the association list. Use **delete** to return a copy of the association list with the element deleted. The second argument is the index of the element to delete.

```
delete!(al,1)
```

$$\begin{array}{l}\text{table}\left(\texttt{"katie"} = [monthsOld = 23,\ gender = \texttt{"female"}],\right.\\ \texttt{"judith"} = [monthsOld = 366,\ gender = \texttt{"female"}],\\ \left.\texttt{"bob"} = [monthsOld = 407,\ gender = \texttt{"male"}]\right)\end{array} \tag{10}$$

Type: AssociationList(String, Record(monthsOld: Integer, gender: String))

For more information about tables, see 'Table' on page 465. For more information about lists, see 'List' on page 404. Issue the system command

)show AssociationList to display the full list of operations defined by AssociationList.

9.2 BalancedBinaryTree

BalancedBinaryTrees(S) is the domain of balanced binary trees with elements of type S at the nodes. A binary tree is either **empty** or else consists of a **node** having a **value** and two branches, each branch a binary tree. A balanced binary tree is one that is balanced with respect its leaves. One with 2^k leaves is perfectly "balanced": the tree has minimum depth, and the **left** and **right** branch of every interior node is identical in shape.

Balanced binary trees are useful in algebraic computation for so-called "divide-and-conquer" algorithms. Conceptually, the data for a problem is initially placed at the root of the tree. The original data is then split into two subproblems, one for each subtree. And so on. Eventually, the problem is solved at the leaves of the tree. A solution to the original problem is obtained by some mechanism that can reassemble the pieces. In fact, an implementation of the Chinese Remainder Algorithm using balanced binary trees was first proposed by David Y. Y. Yun at the IBM T. J. Watson Research Center in Yorktown Heights, New York, in 1978. It served as the prototype for polymorphic algorithms in AXIOM.

In what follows, rather than perform a series of computations with a single expression, the expression is reduced modulo a number of integer primes, a computation is done with modular arithmetic for each prime, and the Chinese Remainder Algorithm is used to obtain the answer to the original problem. We illustrate this principle with the computation of $12^2 = 144$.

A list of moduli.

```
lm := [3,5,7,11]
```

$[3, 5, 7, 11]$ (1)

Type: List PositiveInteger

The expression modTree(n, lm) creates a balanced binary tree with leaf values n mod m for each modulus m in lm.

```
modTree(12,lm)
```

$[0, 2, 5, 1]$ (2)

Type: List Integer

Operation modTree does this using operations on balanced binary trees. We trace its steps. Create a balanced binary tree t of zeros with four leaves.

```
t := balancedBinaryTree(#lm, 0)
```

$[[0, 0, 0], 0, [0, 0, 0]]$ (3)

Type: BalancedBinaryTree NonNegativeInteger

The leaves of the tree are set to the individual moduli.

```
setleaves!(t,lm)
```

$[[3, 0, 5], 0, [7, 0, 11]]$ (4)

Type: BalancedBinaryTree NonNegativeInteger

Use **mapUp!** to do a bottom-up traversal of t, setting each interior node to the product of the values at the nodes of its children.

```
mapUp!(t,_*)
```

1155 (5)

Type: PositiveInteger

The value at the node of every subtree is the product of the moduli of the leaves of the subtree.

```
t
```

[[3, 15, 5], 1155, [7, 77, 11]] (6)

Type: BalancedBinaryTree NonNegativeInteger

Operation **mapDown!**(t,a,fn) replaces the value v at each node of t by fn(a,v).

```
mapDown!(t,12,_rem)
```

[[0, 12, 2], 12, [5, 12, 1]] (7)

Type: BalancedBinaryTree NonNegativeInteger

The operation **leaves** returns the leaves of the resulting tree. In this case, it returns the list of 12 mod m for each modulus m.

```
leaves %
```

[0, 2, 5, 1] (8)

Type: List NonNegativeInteger

Compute the square of the images of 12 modulo each m.

```
squares := [x**2 rem m for x in % for m in lm]
```

[0, 4, 4, 1] (9)

Type: List NonNegativeInteger

Call the Chinese Remainder Algorithm to get the answer for 12^2.

```
chineseRemainder(%,lm)
```

144 (10)

Type: PositiveInteger

9.3 BinaryExpansion

All rational numbers have repeating binary expansions. Operations to access the individual bits of a binary expansion can be obtained by converting the value to RadixExpansion(2). More examples of expansions are available in 'DecimalExpansion' on page 350, 'HexadecimalExpansion' on page 379, and 'RadixExpansion' on page 444.

The expansion (of type BinaryExpansion) of a rational number is returned by the **binary** operation.

```
r := binary(22/7)
```

$11.\overline{001}$ (1)

Type: BinaryExpansion

Arithmetic is exact.

```
r + binary(6/7)
```

100 (2)

Type: BinaryExpansion

The period of the expansion can be short or long . . .

```
[binary(1/i) for i in 102..106]
```

$$\left[0.0\overline{00000101},\right.$$

$$0.\overline{000000100111110001000101100101111001110010010101001},$$

$$0.000\overline{000100111011},\ 0.\overline{000000100111},$$

$$\left.0.0\overline{00000100110101001000011100111110110010101101111100011}\right] \qquad (3)$$

Type: List BinaryExpansion

or very long.

```
binary(1/1007)
```

$$\begin{array}{l}
0.\overline{000000000100000010001010010010111100000111111000010111110} \\
\overline{010110001111101000100111001001100110001100100101010111110110} \\
\overline{100110000000011000011001111011100011010001011110100100011111} \\
\overline{011000010101110111001110101011100110010100101110000000011100} \\
\overline{011110010000001001001001101110010101001110100011011110110101} \\
\overline{110001001000001100101101100000010110010111110001010000001010} \\
\overline{101011010110000011011011101001010111111101011101010011001000} \\
\overline{0010100110110001001100010001000010000110001110100111110001}
\end{array} \qquad (4)$$

Type: BinaryExpansion

These numbers are bona fide algebraic objects.

```
p := binary(1/4)*x**2 + binary(2/3)*x + binary(4/9)
```

$$0.01\ x^2 + 0.\overline{10}\ x + 0.\overline{011100} \qquad (5)$$

Type: Polynomial BinaryExpansion

```
q := D(p, x)
```

$$0.1\ x + 0.\overline{10} \qquad (6)$$

Type: Polynomial BinaryExpansion

```
g := gcd(p, q)
```

$$x + 1.\overline{01} \qquad (7)$$

Type: Polynomial BinaryExpansion

9.4 BinarySearchTree

BinarySearchTree(R) is the domain of binary trees with elements of type R, ordered across the nodes of the tree. A non-empty binary search tree has a value of type R, and **right** and **left** binary search subtrees. If a subtree is empty, it is displayed as a period (".").

Define a list of values to be placed across the tree. The resulting tree has 8 at the root; all other elements are in the left subtree.

```
lv := [8,3,5,4,6,2,1,5,7]
```

$$[8,\ 3,\ 5,\ 4,\ 6,\ 2,\ 1,\ 5,\ 7] \qquad (1)$$

Type: List PositiveInteger

A convenient way to create a binary search tree is to apply the operation **binarySearchTree** to a list of elements.

```
t := binarySearchTree lv
```

$$[[[1, 2, .], 3, [4, 5, [5, 6, 7]]], 8, .] \tag{2}$$

Type: BinarySearchTree PositiveInteger

Another approach is to first create an empty binary search tree of integers.

```
emptybst := empty()$BSTREE(INT)
```

$$[] \tag{3}$$

Type: BinarySearchTree Integer

Insert the value 8. This establishes 8 as the root of the binary search tree. Values inserted later that are less than 8 get stored in the **left** subtree, others in the **right** subtree.

```
t1 := insert!(8,emptybst)
```

$$8 \tag{4}$$

Type: BinarySearchTree Integer

Insert the value 3. This number becomes the root of the **left** subtree of `t1`. For optimal retrieval, it is thus important to insert the middle elements first.

```
insert!(3,t1)
```

$$[3, 8, .] \tag{5}$$

Type: BinarySearchTree Integer

We go back to the original tree `t`. The leaves of the binary search tree are those which have empty **left** and **right** subtrees.

```
leaves t
```

$$[1, 4, 5, 7] \tag{6}$$

Type: List PositiveInteger

The operation **split**`(k,t)` returns a *record* containing the two subtrees: one with all elements "less" than `k`, another with elements "greater" than `k`.

```
split(3,t)
```

$$[less = [1, 2, .], greater = [[., 3, [4, 5, [5, 6, 7]]], 8, .]] \tag{7}$$

Type: Record(less: BinarySearchTree PositiveInteger, greater: BinarySearchTree PositiveInteger)

Define **insertRoot** to insert new elements by creating a new node.

```
insertRoot: (INT,BSTREE INT) -> BSTREE INT
```

Type: Void

The new node puts the inserted value between its "less" tree and "greater" tree.

```
insertRoot(x, t) ==
    a := split(x, t)
    node(a.less, x, a.greater)
```

Type: Void

Function **buildFromRoot** builds a binary search tree from a list of elements `ls` and the empty tree `emptybst`.

```
buildFromRoot ls == reduce(insertRoot,ls,emptybst)
```

Type: Void

Apply this to the reverse of the list `lv`.

```
rt := buildFromRoot reverse lv
```

$$[[[1, 2, .], 3, [4, 5, [5, 6, 7]]], 8, .] \tag{11}$$

Type: BinarySearchTree Integer

Have AXIOM check that these are equal.

```
(t = rt)@Boolean
```

true (12)

Type: Boolean

9.5 CardinalNumber

The CardinalNumber domain can be used for values indicating the cardinality of sets, both finite and infinite. For example, the **dimension** operation in the category VectorSpace returns a cardinal number.

The non-negative integers have a natural construction as cardinals

```
0 = #{ }, 1 = {0}, 2 = {0, 1}, ..., n = {i | 0 <= i < n}.
```

The fact that 0 acts as a zero for the multiplication of cardinals is equivalent to the axiom of choice.

Cardinal numbers can be created by conversion from non-negative integers.

```
c0 := 0 :: CardinalNumber
```

0 (1)

Type: CardinalNumber

```
c1 := 1 :: CardinalNumber
```

1 (2)

Type: CardinalNumber

```
c2 := 2 :: CardinalNumber
```

2 (3)

Type: CardinalNumber

```
c3 := 3 :: CardinalNumber
```

3 (4)

Type: CardinalNumber

They can also be obtained as the named cardinal `Aleph(n)`.

```
A0 := Aleph 0
```

Aleph (0) (5)

Type: CardinalNumber

```
A1 := Aleph 1
```

Aleph (1) (6)

Type: CardinalNumber

The **finite?** operation tests whether a value is a finite cardinal, that is, a non-negative integer.

```
finite? c2
```

true (7)

Type: Boolean

```
finite? A0
```

false (8)

Type: Boolean

Similarly, the **countable?** operation determines whether a value is a countable cardinal, that is, finite or Aleph(0).

```
countable? c2
```

true (9)

Type: Boolean

```
countable? A0
```

true (10)

Type: Boolean

```
countable? A1
```

false (11)

Type: Boolean

Arithmetic operations are defined on cardinal numbers as follows: If x = #X and y = #Y then

x+y	=	#(X+Y)	cardinality of the disjoint union
x-y	=	#(X-Y)	cardinality of the relative complement
x*y	=	#(X*Y)	cardinality of the Cartesian product
x**y	=	#(X**Y)	cardinality of the set of maps from Y to X

Here are some arithmetic examples.

```
[c2 + c2, c2 + A1]
```

[4, Aleph(1)] (12)

Type: List CardinalNumber

```
[c0*c2, c1*c2, c2*c2, c0*A1, c1*A1, c2*A1, A0*A1]
```

[0, 2, 4, 0, Aleph(1), Aleph(1), Aleph(1)] (13)

Type: List CardinalNumber

```
[c2**c0, c2**c1, c2**c2, A1**c0, A1**c1, A1**c2]
```

[1, 2, 4, 1, Aleph(1), Aleph(1)] (14)

Type: List CardinalNumber

Subtraction is a partial operation: it is not defined when subtracting a larger cardinal from a smaller one, nor when subtracting two equal infinite cardinals.

```
[c2-c1, c2-c2, c2-c3, A1-c2, A1-A0, A1-A1]
```

[1, 0, "failed", Aleph(1), Aleph(1), "failed"] (15)

Type: List Union(CardinalNumber, "failed")

The generalized continuum hypothesis asserts that

```
2**Aleph i = Aleph(i+1)
```

and is independent of the axioms of set theory.[1]

The CardinalNumber domain provides an operation to assert whether the hypothesis is to be assumed.

```
generalizedContinuumHypothesisAssumed true
```

true (16)

Type: Boolean

When the generalized continuum hypothesis is assumed, exponentiation to a transfinite power is allowed.

```
[c0**A0, c1**A0, c2**A0, A0**A0, A0**A1, A1**A0, A1**A1]
```

[0, 1, Aleph(1), Aleph(1), Aleph(2), Aleph(1), Aleph(2)] (17)

Type: List CardinalNumber

Three commonly encountered cardinal numbers are

$a = \#\mathbf{Z}$	countable infinity
$c = \#\mathbf{R}$	the continuum
$f = \#\{g \mid g : [0,1] \to \mathbf{R}\}$	

In this domain, these values are obtained under the generalized continuum hypothesis in this way.

```
a := Aleph 0
```

Aleph(0) (18)

Type: CardinalNumber

```
c := 2**a
```

Aleph(1) (19)

Type: CardinalNumber

```
f := 2**c
```

Aleph(2) (20)

Type: CardinalNumber

9.6 CartesianTensor

CartesianTensor(i0,dim,R) provides Cartesian tensors with components belonging to a commutative ring R. Tensors can be described as a generalization of vectors and matrices. This gives a concise *tensor algebra* for multilinear objects supported by the CartesianTensor domain. You can form the inner or outer product of any two tensors and you can add or subtract tensors with the same number of components. Additionally, various forms of traces and transpositions are useful.

The CartesianTensor constructor allows you to specify the minimum index for subscripting. In what follows we discuss in detail how to manipulate tensors.

[1]Goedel, *The consistency of the continuum hypothesis,* Ann. Math. Studies, Princeton Univ. Press, 1940.

Here we construct the domain of Cartesian tensors of dimension 2 over the integers, with indices starting at 1.

```
CT := CARTEN(i0 := 1, 2, Integer)
```

CartesianTensor (1, 2, Integer) (1)

Type: Domain

Forming tensors

Scalars can be converted to tensors of rank zero.

```
t0: CT := 8
```

$$8 \tag{2}$$

Type: CartesianTensor(1, 2, Integer)

```
rank t0
```

$$0 \tag{3}$$

Type: NonNegativeInteger

Vectors (mathematical direct products, rather than one dimensional array structures) can be converted to tensors of rank one.

```
v: DirectProduct(2, Integer) := directProduct [3,4]
```

$$[3, 4] \tag{4}$$

Type: DirectProduct(2, Integer)

```
Tv: CT := v
```

$$[3, 4] \tag{5}$$

Type: CartesianTensor(1, 2, Integer)

Matrices can be converted to tensors of rank two.

```
m: SquareMatrix(2, Integer) := matrix [[1,2],[4,5]]
```

$$\begin{bmatrix} 1 & 2 \\ 4 & 5 \end{bmatrix} \tag{6}$$

Type: SquareMatrix(2, Integer)

```
Tm: CT := m
```

$$\begin{bmatrix} 1 & 2 \\ 4 & 5 \end{bmatrix} \tag{7}$$

Type: CartesianTensor(1, 2, Integer)

```
n: SquareMatrix(2, Integer) := matrix [[2,3],[0,1]]
```

$$\begin{bmatrix} 2 & 3 \\ 0 & 1 \end{bmatrix} \tag{8}$$

Type: SquareMatrix(2, Integer)

```
Tn: CT := n
```

$$\begin{bmatrix} 2 & 3 \\ 0 & 1 \end{bmatrix} \tag{9}$$

Type: CartesianTensor(1, 2, Integer)

In general, a tensor of rank k can be formed by making a list of rank k-1 tensors or, alternatively, a k-deep nested list of lists.

```
t1: CT := [2, 3]
```

$$[2, 3] \qquad (10)$$

Type: CartesianTensor(1, 2, Integer)

```
rank t1
```

$$1 \qquad (11)$$

Type: PositiveInteger

```
t2: CT := [t1, t1]
```

$$\begin{bmatrix} 2 & 3 \\ 2 & 3 \end{bmatrix} \qquad (12)$$

Type: CartesianTensor(1, 2, Integer)

```
t3: CT := [t2, t2]
```

$$\left[\begin{bmatrix} 2 & 3 \\ 2 & 3 \end{bmatrix}, \begin{bmatrix} 2 & 3 \\ 2 & 3 \end{bmatrix}\right] \qquad (13)$$

Type: CartesianTensor(1, 2, Integer)

```
tt: CT := [t3, t3]; tt := [tt, tt]
```

$$\left[\begin{bmatrix} \begin{bmatrix} 2 & 3 \\ 2 & 3 \end{bmatrix} & \begin{bmatrix} 2 & 3 \\ 2 & 3 \end{bmatrix} \\ \begin{bmatrix} 2 & 3 \\ 2 & 3 \end{bmatrix} & \begin{bmatrix} 2 & 3 \\ 2 & 3 \end{bmatrix} \end{bmatrix}, \begin{bmatrix} \begin{bmatrix} 2 & 3 \\ 2 & 3 \end{bmatrix} & \begin{bmatrix} 2 & 3 \\ 2 & 3 \end{bmatrix} \\ \begin{bmatrix} 2 & 3 \\ 2 & 3 \end{bmatrix} & \begin{bmatrix} 2 & 3 \\ 2 & 3 \end{bmatrix} \end{bmatrix}\right] \qquad (14)$$

Type: CartesianTensor(1, 2, Integer)

```
rank tt
```

$$5 \qquad (15)$$

Type: PositiveInteger

Multiplication

Given two tensors of rank k1 and k2, the outer **product** forms a new tensor of rank k1+k2.

Here $T_{mn}(i,j,k,l) = T_m(i,j)\ T_n(k,l)$.

```
Tmn := product(Tm, Tn)
```

$$\begin{bmatrix} \begin{bmatrix} 2 & 3 \\ 0 & 1 \end{bmatrix} & \begin{bmatrix} 4 & 6 \\ 0 & 2 \end{bmatrix} \\ \begin{bmatrix} 8 & 12 \\ 0 & 4 \end{bmatrix} & \begin{bmatrix} 10 & 15 \\ 0 & 5 \end{bmatrix} \end{bmatrix} \qquad (16)$$

Type: CartesianTensor(1, 2, Integer)

The inner product (**contract**) forms a tensor of rank k1+k2-2. This product generalizes the vector dot product and matrix-vector product by summing component products along two indices.

Here we sum along the second index of T_m and the first index of T_v. Here $T_{mv} = \sum_{j=1}^{\dim} T_m(i,j)\ T_v(j)$

```
Tmv := contract(Tm,2,Tv,1)
```

$$[11,\ 32] \tag{17}$$

Type: CartesianTensor(1, 2, Integer)

The multiplication operator "*" is scalar multiplication or an inner product depending on the ranks of the arguments.

If either argument is rank zero it is treated as scalar multiplication. Otherwise, `a*b` is the inner product summing the last index of `a` with the first index of `b`.

```
Tm*Tv
```

$$[11,\ 32] \tag{18}$$

Type: CartesianTensor(1, 2, Integer)

This definition is consistent with the inner product on matrices and vectors.

```
Tmv = m * v
```

$$[11,\ 32] = [11,\ 32] \tag{19}$$

Type: Equation CartesianTensor(1, 2, Integer)

Selecting Components

For tensors of low rank (that is, four or less), components can be selected by applying the tensor to its indices.

```
t0()
```

$$8 \tag{20}$$

Type: PositiveInteger

```
t1(1+1)
```

$$3 \tag{21}$$

Type: PositiveInteger

```
t2(2,1)
```

$$2 \tag{22}$$

Type: PositiveInteger

```
t3(2,1,2)
```

$$3 \tag{23}$$

Type: PositiveInteger

```
Tmn(2,1,2,1)
```

$$0 \tag{24}$$

Type: NonNegativeInteger

A general indexing mechanism is provided for a list of indices.

```
t0[]
```

$$8 \tag{25}$$

Type: PositiveInteger

```
t1[2]
```

3 (26)

Type: PositiveInteger

```
t2[2,1]
```

2 (27)

Type: PositiveInteger

The general mechanism works for tensors of arbitrary rank, but is somewhat less efficient since the intermediate index list must be created.

```
t3[2,1,2]
```

3 (28)

Type: PositiveInteger

```
Tmn[2,1,2,1]
```

0 (29)

Type: NonNegativeInteger

Contraction

A "contraction" between two tensors is an inner product, as we have seen above. You can also contract a pair of indices of a single tensor. This corresponds to a "trace" in linear algebra. The expression `contract(t,k1,k2)` forms a new tensor by summing the diagonal given by indices in position `k1` and `k2`.

This is the tensor given by $xT_{mn} = \sum_{k=1}^{\text{dim}} T_{mn}(k,k,i,j)$.

```
cTmn := contract(Tmn,1,2)
```

$$\begin{bmatrix} 12 & 18 \\ 0 & 6 \end{bmatrix} \tag{30}$$

Type: CartesianTensor(1, 2, Integer)

Since `Tmn` is the outer product of matrix `m` and matrix `n`, the above is equivalent to this.

```
trace(m) * n
```

$$\begin{bmatrix} 12 & 18 \\ 0 & 6 \end{bmatrix} \tag{31}$$

Type: SquareMatrix(2, Integer)

In this and the next few examples, we show all possible contractions of `Tmn` and their matrix algebra equivalents.

```
contract(Tmn,1,2) = trace(m) * n
```

$$\begin{bmatrix} 12 & 18 \\ 0 & 6 \end{bmatrix} = \begin{bmatrix} 12 & 18 \\ 0 & 6 \end{bmatrix} \tag{32}$$

Type: Equation CartesianTensor(1, 2, Integer)

```
contract(Tmn,1,3) = transpose(m) * n
```

$$\begin{bmatrix} 2 & 7 \\ 4 & 11 \end{bmatrix} = \begin{bmatrix} 2 & 7 \\ 4 & 11 \end{bmatrix} \tag{33}$$

Type: Equation CartesianTensor(1, 2, Integer)

```
contract(Tmn,1,4) = transpose(m) * transpose(n)
```

$$\begin{bmatrix} 14 & 4 \\ 19 & 5 \end{bmatrix} = \begin{bmatrix} 14 & 4 \\ 19 & 5 \end{bmatrix} \quad (34)$$

Type: Equation CartesianTensor(1, 2, Integer)

```
contract(Tmn,2,3) = m * n
```

$$\begin{bmatrix} 2 & 5 \\ 8 & 17 \end{bmatrix} = \begin{bmatrix} 2 & 5 \\ 8 & 17 \end{bmatrix} \quad (35)$$

Type: Equation CartesianTensor(1, 2, Integer)

```
contract(Tmn,2,4) = m * transpose(n)
```

$$\begin{bmatrix} 8 & 2 \\ 23 & 5 \end{bmatrix} = \begin{bmatrix} 8 & 2 \\ 23 & 5 \end{bmatrix} \quad (36)$$

Type: Equation CartesianTensor(1, 2, Integer)

```
contract(Tmn,3,4) = trace(n) * m
```

$$\begin{bmatrix} 3 & 6 \\ 12 & 15 \end{bmatrix} = \begin{bmatrix} 3 & 6 \\ 12 & 15 \end{bmatrix} \quad (37)$$

Type: Equation CartesianTensor(1, 2, Integer)

Transpositions

You can exchange any desired pair of indices using the **transpose** operation.

Here the indices in positions one and three are exchanged, that is, $tT_{mn}(i,j,k,l) = T_{mn}(k,j,i,l)$.

```
tTmn := transpose(Tmn,1,3)
```

$$\begin{bmatrix} \begin{bmatrix} 2 & 3 \\ 8 & 12 \end{bmatrix} & \begin{bmatrix} 4 & 6 \\ 10 & 15 \end{bmatrix} \\ \begin{bmatrix} 0 & 1 \\ 0 & 4 \end{bmatrix} & \begin{bmatrix} 0 & 2 \\ 0 & 5 \end{bmatrix} \end{bmatrix} \quad (38)$$

Type: CartesianTensor(1, 2, Integer)

If no indices are specified, the first and last index are exchanged.

```
transpose Tmn
```

$$\begin{bmatrix} \begin{bmatrix} 2 & 8 \\ 0 & 0 \end{bmatrix} & \begin{bmatrix} 4 & 10 \\ 0 & 0 \end{bmatrix} \\ \begin{bmatrix} 3 & 12 \\ 1 & 4 \end{bmatrix} & \begin{bmatrix} 6 & 15 \\ 2 & 5 \end{bmatrix} \end{bmatrix} \quad (39)$$

Type: CartesianTensor(1, 2, Integer)

This is consistent with the matrix transpose.

```
transpose Tm = transpose m
```

$$\begin{bmatrix} 1 & 4 \\ 2 & 5 \end{bmatrix} = \begin{bmatrix} 1 & 4 \\ 2 & 5 \end{bmatrix} \quad (40)$$

Type: Equation CartesianTensor(1, 2, Integer)

If a more complicated reordering of the indices is required, then the **reindex** operation can be used. This operation allows the indices to be arbitrarily permuted.

This defines $rT_{mn}(i,j,k,l) = T_{mn}(i,l,j,k)$.

```
rTmn := reindex(Tmn, [1,4,2,3])
```

$$\left[\begin{array}{cc} \left[\begin{array}{cc} 2 & 0 \\ 4 & 0 \end{array}\right] & \left[\begin{array}{cc} 3 & 1 \\ 6 & 2 \end{array}\right] \\ \left[\begin{array}{cc} 8 & 0 \\ 10 & 0 \end{array}\right] & \left[\begin{array}{cc} 12 & 4 \\ 15 & 5 \end{array}\right] \end{array}\right] \tag{41}$$

Type: CartesianTensor(1, 2, Integer)

Arithmetic

Tensors of equal rank can be added or subtracted so arithmetic expressions can be used to produce new tensors.

```
tt := transpose(Tm)*Tn - Tn*transpose(Tm)
```

$$\left[\begin{array}{cc} -6 & -16 \\ 2 & 6 \end{array}\right] \tag{42}$$

Type: CartesianTensor(1, 2, Integer)

```
Tv*(tt+Tn)
```

$$[-4,\ -11] \tag{43}$$

Type: CartesianTensor(1, 2, Integer)

```
reindex(product(Tn,Tn),[4,3,2,1])+3*Tn*product(Tm,Tm)
```

$$\left[\begin{array}{cc} \left[\begin{array}{cc} 46 & 84 \\ 174 & 212 \end{array}\right] & \left[\begin{array}{cc} 57 & 114 \\ 228 & 285 \end{array}\right] \\ \left[\begin{array}{cc} 18 & 24 \\ 57 & 63 \end{array}\right] & \left[\begin{array}{cc} 17 & 30 \\ 63 & 76 \end{array}\right] \end{array}\right] \tag{44}$$

Type: CartesianTensor(1, 2, Integer)

Specific Tensors

Two specific tensors have properties which depend only on the dimension.

The Kronecker delta satisfies

$$\text{delta}(i,j) = \begin{cases} 1 & \text{if } i = j \\ 0 & \text{if } i \neq j \end{cases}$$

```
delta: CT := kroneckerDelta()
```

$$\left[\begin{array}{cc} 1 & 0 \\ 0 & 1 \end{array}\right] \tag{45}$$

Type: CartesianTensor(1, 2, Integer)

This can be used to reindex via contraction.

```
contract(Tmn, 2, delta, 1) = reindex(Tmn, [1,3,4,2])
```

$$\left[\begin{array}{cc} \left[\begin{array}{cc} 2 & 4 \\ 3 & 6 \end{array}\right] & \left[\begin{array}{cc} 0 & 0 \\ 1 & 2 \end{array}\right] \\ \left[\begin{array}{cc} 8 & 10 \\ 12 & 15 \end{array}\right] & \left[\begin{array}{cc} 0 & 0 \\ 4 & 5 \end{array}\right] \end{array}\right] = \left[\begin{array}{cc} \left[\begin{array}{cc} 2 & 4 \\ 3 & 6 \end{array}\right] & \left[\begin{array}{cc} 0 & 0 \\ 1 & 2 \end{array}\right] \\ \left[\begin{array}{cc} 8 & 10 \\ 12 & 15 \end{array}\right] & \left[\begin{array}{cc} 0 & 0 \\ 4 & 5 \end{array}\right] \end{array}\right] \tag{46}$$

Type: Equation CartesianTensor(1, 2, Integer)

The Levi Civita symbol determines the sign of a permutation of indices.

```
epsilon:CT := leviCivitaSymbol()
```

$$\begin{bmatrix} 0 & 1 \\ -1 & 0 \end{bmatrix} \tag{47}$$

Type: CartesianTensor(1, 2, Integer)

Here we have:

$$\text{epsilon}(i_1, \ldots, i_{\text{dim}}) = \begin{cases} +1 & \text{if } i_1, \ldots, i_{\text{dim}} \text{ is an even permutation of } i_0, \ldots, i_0 + \text{dim} - 1 \\ -1 & \text{if } i_1, \ldots, i_{\text{dim}} \text{ is an odd permutation of } i_0, \ldots, i_0 + \text{dim} - 1 \\ 0 & \text{if } i_1, \ldots, i_{\text{dim}} \text{ is } \textit{not} \text{ a permutation of } i_0, \ldots, i_0 + \text{dim} - 1 \end{cases}$$

This property can be used to form determinants.

```
contract(epsilon*Tm*epsilon, 1,2) = 2 * determinant m
```

$$-6 = -6 \tag{48}$$

Type: Equation CartesianTensor(1, 2, Integer)

Properties of the CartesianTensor domain

GradedModule(R,E) denotes "E-graded R-module", that is, a collection of R-modules indexed by an abelian monoid E. An element g of G[s] for some specific s in E is said to be an element of G with **degree** s. Sums are defined in each module G[s] so two elements of G can be added if they have the same degree. Morphisms can be defined and composed by degree to give the mathematical category of graded modules.

GradedAlgebra(R,E) denotes "E-graded R-algebra." A graded algebra is a graded module together with a degree preserving R-bilinear map, called the **product**.

```
degree(product(a,b))= degree(a) + degree(b)

product(r*a,b) = product(a,r*b) = r*product(a,b)
product(a1+a2,b) = product(a1,b) + product(a2,b)
product(a,b1+b2) = product(a,b1) + product(a,b2)
product(a,product(b,c)) = product(product(a,b),c)
```

The domain CartesianTensor(i0, dim, R) belongs to the category GradedAlgebra(R, NonNegativeInteger). The non-negative integer **degree** is the tensor rank and the graded algebra **product** is the tensor outer product. The graded module addition captures the notion that only tensors of equal rank can be added.

If V is a vector space of dimension dim over R, then the tensor module T[k](V) is defined as

```
T[0](V) = R
T[k](V) = T[k-1](V) * V
```

where "*" denotes the R-module tensor **product**. CartesianTensor(i0,dim,R) is the graded algebra in which the degree k module is T[k](V).

Tensor Calculus

It should be noted here that often tensors are used in the context of tensor-valued manifold maps. This leads to the notion of covariant and contravariant bases with tensor component functions transforming in specific ways under a change of coordinates on the manifold. This is no more directly supported by the CartesianTensor domain than it is by the Vector domain. However, it is possible to have the components implicitly represent component maps by choosing a polynomial or expression type for the components. In this case, it is up to the user to satisfy any constraints which arise on the basis of this interpretation.

9.7 Character

The members of the domain Character are values representing letters, numerals and other text elements. For more information on related topics, see 'CharacterClass' on page 326 and 'String' on page 458.

Characters can be obtained using String notation.

```
chars := [char "a", char "A", char "X", char "8", char
  "+"]
```

$$[a, A, X, 8, +] \tag{1}$$

Type: List Character

Certain characters are available by name. This is the blank character.

```
space()
```

(2)

Type: Character

This is the quote that is used in strings.

```
quote()
```

"   (3)

Type: Character

This is the escape character that allows quotes and other characters within strings.

```
escape()
```

_   (4)

Type: Character

Characters are represented as integers in a machine-dependent way. The integer value can be obtained using the **ord** operation. It is always true that `char(ord c) = c` and `ord(char i) = i`, provided that `i` is in the range `0..size()$Character-1`.

```
[ord c for c in chars]
```

$$[97, 65, 88, 56, 43] \tag{5}$$

Type: List Integer

The **lowerCase** operation converts an upper case letter to the corresponding lower case letter. If the argument is not an upper case letter, then it is returned unchanged.

```
[upperCase c for c in chars]
```

$$[A, A, X, 8, +] \tag{6}$$

Type: List Character

Likewise, the **upperCase** operation converts lower case letters to upper case.

```
[lowerCase c for c in chars]
```

$[a, a, x, 8, +]$ (7)

Type: List Character

A number of tests are available to determine whether characters belong to certain families.

```
[alphabetic? c for c in chars]
```

[true, true, true, false, false] (8)

Type: List Boolean

```
[upperCase? c for c in chars]
```

[false, true, true, false, false] (9)

Type: List Boolean

```
[lowerCase? c for c in chars]
```

[true, false, false, false, false] (10)

Type: List Boolean

```
[digit? c for c in chars]
```

[false, false, false, true, false] (11)

Type: List Boolean

```
[hexDigit? c for c in chars]
```

[true, true, false, true, false] (12)

Type: List Boolean

```
[alphanumeric? c for c in chars]
```

[true, true, true, true, false] (13)

Type: List Boolean

9.8 CharacterClass

The CharacterClass domain allows classes of characters to be defined and manipulated efficiently.

Character classes can be created by giving either a string or a list of characters.

```
cl1 := charClass [char "a", char "e", char "i", char "o",
  char "u", char "y"]
```

"aeiouy" (1)

Type: CharacterClass

```
cl2 := charClass "bcdfghjklmnpqrstvwxyz"
```

"bcdfghjklmnpqrstvwxyz" (2)

Type: CharacterClass

A number of character classes are predefined for convenience.

```
digit()
```

"0123456789" (3)

Type: CharacterClass

```
hexDigit()
```

"0123456789ABCDEFabcdef" (4)

Type: CharacterClass

```
upperCase()
```

"ABCDEFGHIJKLMNOPQRSTUVWXYZ" (5)

Type: CharacterClass

```
lowerCase()
```

"abcdefghijklmnopqrstuvwxyz" (6)

Type: CharacterClass

```
alphabetic()
```

"ABCDEFGHIJKLMNOPQRSTUVWXYZabcdefghijklmnopqrs_
tuvwxyz" (7)

Type: CharacterClass

```
alphanumeric()
```

"0123456789ABCDEFGHIJKLMNOPQRSTUVWXYZabcdefghi_
jklmnopqrstuvwxyz" (8)

Type: CharacterClass

You can quickly test whether a character belongs to a class.

```
member?(char "a", cl1)
```

true (9)

Type: Boolean

```
member?(char "a", cl2)
```

false (10)

Type: Boolean

Classes have the usual set operations because the CharacterClass domain belongs to the category FiniteSetAggregate(Character).

```
intersect(cl1, cl2)
```

"y" (11)

Type: CharacterClass

```
union(cl1,cl2)
```

"abcdefghijklmnopqrstuvwxyz" (12)

Type: CharacterClass

```
difference(cl1,cl2)
```

"aeiou" (13)

Type: CharacterClass

```
intersect(complement(cl1),cl2)
```

"bcdfghjklmnpqrstvwxz" (14)

Type: CharacterClass

You can modify character classes by adding or removing characters.

```
insert!(char "a", cl2)
```

"abcdfghjklmnpqrstvwxyz" (15)

Type: CharacterClass

```
remove!(char "b", cl2)
```

"acdfghjklmnpqrstvwxyz" (16)

Type: CharacterClass

For more information on related topics, see 'Character' on page 325 and 'String' on page 458.

Issue the system command `)show CharacterClass` to display the full list of operations defined by CharacterClass.

9.9 CliffordAlgebra

CliffordAlgebra(n,K,Q) defines a vector space of dimension 2^n over the field K, given a quadratic form Q on If $\{e_1, \ldots, e_n\}$ is a basis for K^n then

$$\begin{array}{lll} \{ & 1 & \\ & e_i & \text{for } 1 \le i \le n \\ & e_{i_1}\, e_{i_2} & \text{for } 1 \le i_1 < i_2 \le n \\ & \ldots & \\ & e_1\, e_2\, \cdots\, e_n & \} \end{array}$$

is a basis for the Clifford algebra. The algebra is defined by the relations

$$\begin{array}{rcll} e_i\, e_i & = & Q(e_i) & \\ e_i\, e_j & = & -e_j\, e_i & \text{for } i \ne j \end{array}$$

Examples of Clifford Algebras are gaussians (complex numbers), quaternions, exterior algebras and spin algebras.

9.9.1 The Complex Numbers as a Clifford Algebra

This is the field over which we will work, rational functions with integer coefficients.

```
K := Fraction Polynomial Integer
```

Fraction Polynomial Integer (1)

Type: Domain

We use this matrix for the quadratic form.

```
m := matrix [[-1]]
```

$$\begin{bmatrix} -1 \end{bmatrix} \tag{2}$$

Type: Matrix Integer

We get complex arithmetic by using this domain.

```
C := CliffordAlgebra(1, K, quadraticForm m)
```

CliffordAlgebra (1, Fraction Polynomial Integer , MATRIX) (3)

Type: Domain

Here is i, the usual square root of -1.

```
i: C := e(1)
```

$$e_1 \tag{4}$$

Type: CliffordAlgebra(1, Fraction Polynomial Integer, MATRIX)

Here are some examples of the arithmetic.

```
x := a + b * i
```

$$a + b\ e_1 \tag{5}$$

Type: CliffordAlgebra(1, Fraction Polynomial Integer, MATRIX)

```
y := c + d * i
```

$$c + d\ e_1 \tag{6}$$

Type: CliffordAlgebra(1, Fraction Polynomial Integer, MATRIX)

See 'Complex' on page 333 for examples of AXIOM's constructor implementing complex numbers.

```
x * y
```

$$-b\ d + a\ c + (a\ d + b\ c)\ e_1 \tag{7}$$

Type: CliffordAlgebra(1, Fraction Polynomial Integer, MATRIX)

9.9.2 The Quaternion Numbers as a Clifford Algebra

```
K := Fraction Polynomial Integer
```

Fraction Polynomial Integer (1)

Type: Domain

We use this matrix for the quadratic form.

```
m := matrix [[-1,0],[0,-1]]
```

$$\begin{bmatrix} -1 & 0 \\ 0 & -1 \end{bmatrix} \tag{2}$$

Type: Matrix Integer

The resulting domain is the quaternions.

```
H := CliffordAlgebra(2, K, quadraticForm m)
```

CliffordAlgebra (2, Fraction Polynomial Integer , MATRIX) (3)

Type: Domain

We use Hamilton's notation for i,j,k.

```
i: H := e(1)
```

$$e_1 \tag{4}$$

Type: CliffordAlgebra(2, Fraction Polynomial Integer, MATRIX)

```
j: H := e(2)
```

$$e_2 \tag{5}$$

Type: CliffordAlgebra(2, Fraction Polynomial Integer, MATRIX)

```
k: H := i * j
```

$$e_1\ e_2 \tag{6}$$

Type: CliffordAlgebra(2, Fraction Polynomial Integer, MATRIX)

```
x := a + b * i + c * j + d * k
```

$$a + b\ e_1 + c\ e_2 + d\ e_1\ e_2 \tag{7}$$

Type: CliffordAlgebra(2, Fraction Polynomial Integer, MATRIX)

```
y := e + f * i + g * j + h * k
```

$$e + f\ e_1 + g\ e_2 + h\ e_1\ e_2 \tag{8}$$

Type: CliffordAlgebra(2, Fraction Polynomial Integer, MATRIX)

```
x + y
```

$$e + a + (f + b)\ e_1 + (g + c)\ e_2 + (h + d)\ e_1\ e_2 \tag{9}$$

Type: CliffordAlgebra(2, Fraction Polynomial Integer, MATRIX)

```
x * y
```

$$\begin{gathered} -d\ h - c\ g - b\ f + a\ e + (c\ h - d\ g + a\ f + b\ e)\ e_1 + \\ (-b\ h + a\ g + d\ f + c\ e)\ e_2 + (a\ h + b\ g - c\ f + d\ e)\ e_1\ e_2 \end{gathered} \tag{10}$$

Type: CliffordAlgebra(2, Fraction Polynomial Integer, MATRIX)

See 'Quaternion' on page 442 for examples of AXIOM's constructor implementing quaternions.

```
y * x
```

$$-d\ h - c\ g - b\ f + a\ e + (-c\ h + d\ g + a\ f + b\ e)\ e_1 + (b\ h + a\ g - d\ f + c\ e)\ e_2 + (a\ h - b\ g + c\ f + d\ e)\ e_1\ e_2 \qquad (11)$$

Type: CliffordAlgebra(2, Fraction Polynomial Integer, MATRIX)

9.9.3 The Exterior Algebra on a Three Space

This is the field over which we will work, rational functions with integer coefficients.

```
K := Fraction Polynomial Integer
```

Fraction Polynomial Integer (1)

Type: Domain

If we chose the three by three zero quadratic form, we obtain the exterior algebra on `e(1),e(2),e(3)`.

```
Ext := CliffordAlgebra(3, K, quadraticForm 0)
```

CliffordAlgebra (3, Fraction Polynomial Integer , MATRIX) (2)

Type: Domain

This is a three dimensional vector algebra. We define `i`, `j`, `k` as the unit vectors.

```
i: Ext := e(1)
```

$$e_1 \qquad (3)$$

Type: CliffordAlgebra(3, Fraction Polynomial Integer, MATRIX)

```
j: Ext := e(2)
```

$$e_2 \qquad (4)$$

Type: CliffordAlgebra(3, Fraction Polynomial Integer, MATRIX)

```
k: Ext := e(3)
```

$$e_3 \qquad (5)$$

Type: CliffordAlgebra(3, Fraction Polynomial Integer, MATRIX)

Now it is possible to do arithmetic.

```
x := x1*i + x2*j + x3*k
```

$$x1\ e_1 + x2\ e_2 + x3\ e_3 \qquad (6)$$

Type: CliffordAlgebra(3, Fraction Polynomial Integer, MATRIX)

```
y := y1*i + y2*j + y3*k
```

$$y1\ e_1 + y2\ e_2 + y3\ e_3 \qquad (7)$$

Type: CliffordAlgebra(3, Fraction Polynomial Integer, MATRIX)

```
x + y
```

$$(y1 + x1)\ e_1 + (y2 + x2)\ e_2 + (y3 + x3)\ e_3 \qquad (8)$$

Type: CliffordAlgebra(3, Fraction Polynomial Integer, MATRIX)

```
x * y + y * x
```

$$0 \tag{9}$$

Type: CliffordAlgebra(3, Fraction Polynomial Integer, MATRIX)

On an n space, a grade p form has a dual n-p form. In particular, in three space the dual of a grade two element identifies e1*e2->e3, e2*e3->e1, e3*e1->e2.

```
dual2 a == coefficient(a,[2,3]) * i + coefficient(a,[3,1]) *
  j + coefficient(a,[1,2]) * k
```

Type: Void

The vector cross product is then given by this.

```
dual2(x*y)
```

$$(x2\ y3 - x3\ y2)\ e_1 + (-x1\ y3 + x3\ y1)\ e_2 + (x1\ y2 - x2\ y1)\ e_3 \tag{11}$$

Type: CliffordAlgebra(3, Fraction Polynomial Integer, MATRIX)

9.9.4 The Dirac Spin Algebra

In this section we will work over the field of rational numbers.

```
K := Fraction Integer
```

Fraction Integer (1)

Type: Domain

We define the quadratic form to be the Minkowski space-time metric.

```
g := matrix [[1,0,0,0], [0,-1,0,0], [0,0,-1,0], [0,0,0,-
  1]]
```

$$\begin{bmatrix} 1 & 0 & 0 & 0 \\ 0 & -1 & 0 & 0 \\ 0 & 0 & -1 & 0 \\ 0 & 0 & 0 & -1 \end{bmatrix} \tag{2}$$

Type: Matrix Integer

We obtain the Dirac spin algebra used in Relativistic Quantum Field Theory.

```
D := CliffordAlgebra(4,K, quadraticForm g)
```

CliffordAlgebra (4, Fraction Integer , MATRIX) (3)

Type: Domain

The usual notation for the basis is γ with a superscript. For AXIOM input we will use gam(i):

```
gam := [e(i)$D for i in 1..4]
```

$$[e_1,\ e_2,\ e_3,\ e_4] \tag{4}$$

Type: List CliffordAlgebra(4, Fraction Integer, MATRIX)

There are various contraction identities of the form

```
g(l,t)*gam(l)*gam(m)*gam(n)*gam(r)*gam(s)*gam(t) =
      2*(gam(s)gam(m)gam(n)gam(r) +
         gam(r)*gam(n)*gam(m)*gam(s))
```

where a sum over l and t is implied.

Verify this identity for particular values of m,n,r,s.

```
m := 1; n:= 2; r := 3; s := 4;
```

(5)

Type: PositiveInteger

```
lhs := reduce(+, [reduce(+, [g(l, t) * gam(l) * gam(m) *
  gam(n) * gam(r) * gam(s) * gam(t) for l in 1..4]) for t
  in 1..4])
```

$$-4\ e_1\ e_2\ e_3\ e_4 \tag{6}$$

Type: CliffordAlgebra(4, Fraction Integer, MATRIX)

```
rhs := 2*(gam s * gam m*gam n*gam r + gam r*gam n*gam
  m*gam s)
```

$$-4\ e_1\ e_2\ e_3\ e_4 \tag{7}$$

Type: CliffordAlgebra(4, Fraction Integer, MATRIX)

9.10 Complex

The Complex constructor implements complex objects over a commutative ring R. Typically, the ring R is Integer, Fraction Integer, Float or SmallFloat. R can also be a symbolic type, like Polynomial Integer. For more information about the numerical and graphical aspects of complex numbers, see Section 8.1 on page 227.

Complex objects are created by the **complex** operation.

```
a := complex(4/3,5/2)
```

$$\frac{4}{3}+\frac{5}{2}\ \%i \tag{1}$$

Type: Complex Fraction Integer

```
b := complex(4/3,-5/2)
```

$$\frac{4}{3}-\frac{5}{2}\ \%i \tag{2}$$

Type: Complex Fraction Integer

The standard arithmetic operations are available.

```
a + b
```

$$\frac{8}{3} \tag{3}$$

Type: Complex Fraction Integer

```
a - b
```

$$5\ \%i \tag{4}$$

Type: Complex Fraction Integer

```
a * b
```

$$\frac{289}{36} \tag{5}$$

Type: Complex Fraction Integer

If R is a field, you can also divide the complex objects.

```
a / b
```

$$-\frac{161}{289}+\frac{240}{289}\ \%i \tag{6}$$

Type: Complex Fraction Integer

Use a conversion (Section 2.7 on page 78) to view the last object as a fraction of complex integers.

```
% :: Fraction Complex Integer
```

$$\frac{-15+8\ \%i}{15+8\ \%i} \tag{7}$$

Type: Fraction Complex Integer

The predefined macro %i is defined to be complex(0,1).

```
3.4 + 6.7 * %i
```

$$3.4+6.7\ \%i \tag{8}$$

Type: Complex Float

You can also compute the **conjugate** and **norm** of a complex number.

```
conjugate a
```

$$\frac{4}{3}-\frac{5}{2}\ \%i \tag{9}$$

Type: Complex Fraction Integer

```
norm a
```

$$\frac{289}{36} \tag{10}$$

Type: Fraction Integer

The **real** and **imag** operations are provided to extract the real and imaginary parts, respectively.

```
real a
```

$$\frac{4}{3} \tag{11}$$

Type: Fraction Integer

```
imag a
```

$$\frac{5}{2} \tag{12}$$

Type: Fraction Integer

The domain Complex Integer is also called the Gaussian integers. If R is the integers (or, more generally, a EuclideanDomain), you can compute greatest common divisors.

```
gcd(13 - 13*%i,31 + 27*%i)
```

$$5+\%i \tag{13}$$

Type: Complex Integer

You can also compute least common multiples.

```
lcm(13 - 13*%i,31 + 27*%i)
```

$$143-39\ \%i \tag{14}$$

Type: Complex Integer

You can **factor** Gaussian integers.

```
factor(13 - 13*%i)
```

$$-(1+\%i)\,(2+3\,\%i)\,(3+2\,\%i) \tag{15}$$

Type: Factored Complex Integer

```
factor complex(2,0)
```

$$-\%i\,(1+\%i)^2 \tag{16}$$

Type: Factored Complex Integer

9.11 ContinuedFraction

Continued fractions have been a fascinating and useful tool in mathematics for well over three hundred years. AXIOM implements continued fractions for fractions of any Euclidean domain. In practice, this usually means rational numbers. In this section we demonstrate some of the operations available for manipulating both finite and infinite continued fractions. It may be helpful if you review 'Stream' on page 457 to remind yourself of some of the operations with streams.

The ContinuedFraction domain is a field and therefore you can add, subtract, multiply and divide the fractions.

The **continuedFraction** operation converts its fractional argument to a continued fraction.

```
c := continuedFraction(314159/100000)
```

$$3+\frac{1|}{|7}+\frac{1|}{|15}+\frac{1|}{|1}+\frac{1|}{|25}+\frac{1|}{|1}+\frac{1|}{|7}+\frac{1|}{|4} \tag{1}$$

Type: ContinuedFraction Integer

This display is a compact form of the bulkier

$$3+\cfrac{1}{7+\cfrac{1}{15+\cfrac{1}{1+\cfrac{1}{25+\cfrac{1}{1+\cfrac{1}{7+\cfrac{1}{4}}}}}}}$$

You can write any rational number in a similar form. The fraction will be finite and you can always take the "numerators" to be 1. That is, any rational number can be written as a simple, finite continued fraction of the form

$$a_1+\cfrac{1}{a_2+\cfrac{1}{a_3+\cfrac{1}{\ddots\, a_{n-1}+\cfrac{1}{a_n}}}}$$

The a_i are called partial quotients and the operation **partialQuotients** creates a stream of them.

```
partialQuotients c
```

$$[3, 7, 15, 1, 25, 1, 7, \ldots] \quad (2)$$

Type: Stream Integer

By considering more and more of the fraction, you get the **convergents**. For example, the first convergent is a_1, the second is $a_1 + 1/a_2$ and so on.

```
convergents c
```

$$\left[3, \frac{22}{7}, \frac{333}{106}, \frac{355}{113}, \frac{9208}{2931}, \frac{9563}{3044}, \frac{76149}{24239}, \ldots\right] \quad (3)$$

Type: Stream Fraction Integer

Since this is a finite continued fraction, the last convergent is the original rational number, in reduced form. The result of **approximants** is always an infinite stream, though it may just repeat the "last" value.

```
approximants c
```

$$\left[3, \frac{22}{7}, \frac{333}{106}, \frac{355}{113}, \frac{9208}{2931}, \frac{9563}{3044}, \frac{76149}{24239}, \ldots\right] \quad (4)$$

Type: Stream Fraction Integer

Inverting `c` only changes the partial quotients of its fraction by inserting a `0` at the beginning of the list.

```
pq := partialQuotients(1/c)
```

$$[0, 3, 7, 15, 1, 25, 1, \ldots] \quad (5)$$

Type: Stream Integer

Do this to recover the original continued fraction from this list of partial quotients. The three-argument form of the **continuedFraction** operation takes an element which is the whole part of the fraction, a stream of elements which are the numerators of the fraction, and a stream of elements which are the denominators of the fraction.

```
continuedFraction(first pq,repeating [1],rest pq)
```

$$\frac{1|}{|3} + \frac{1|}{|7} + \frac{1|}{|15} + \frac{1|}{|1} + \frac{1|}{|25} + \frac{1|}{|1} + \frac{1|}{|7} + \ldots \quad (6)$$

Type: ContinuedFraction Integer

The streams need not be finite for **continuedFraction**. Can you guess which irrational number has the following continued fraction? See the end of this section for the answer.

```
z:=continuedFraction(3,repeating [1],repeating [3,6])
```

$$3 + \frac{1|}{|3} + \frac{1|}{|6} + \frac{1|}{|3} + \frac{1|}{|6} + \frac{1|}{|3} + \frac{1|}{|6} + \frac{1|}{|3} + \ldots \quad (7)$$

Type: ContinuedFraction Integer

In 1737 Euler discovered the infinite continued fraction expansion

$$\frac{e-1}{2} = \cfrac{1}{1 + \cfrac{1}{6 + \cfrac{1}{10 + \cfrac{1}{14 + \cdots}}}}$$

We use this expansion to compute rational and floating point approximations of `e`.[2]

[2]For this and other interesting expansions, see C. D. Olds, *Continued Fractions,* New Mathematical Library, (New York: Random House, 1963), pp. 134–139.

By looking at the above expansion, we see that the whole part is 0 and the numerators are all equal to 1. This constructs the stream of denominators.

```
dens:Stream Integer := cons(1,generate((x+->x+4),6))
```

$$[1, 6, 10, 14, 18, 22, 26, \ldots] \tag{8}$$

Type: Stream Integer

Therefore this is the continued fraction expansion for $(e-1)/2$.

```
cf := continuedFraction(0,repeating [1],dens)
```

$$\frac{1|}{|1} + \frac{1|}{|6} + \frac{1|}{|10} + \frac{1|}{|14} + \frac{1|}{|18} + \frac{1|}{|22} + \frac{1|}{|26} + \ldots \tag{9}$$

Type: ContinuedFraction Integer

These are the rational number convergents.

```
ccf := convergents cf
```

$$\left[0, 1, \frac{6}{7}, \frac{61}{71}, \frac{860}{1001}, \frac{15541}{18089}, \frac{342762}{398959}, \ldots\right] \tag{10}$$

Type: Stream Fraction Integer

You can get rational convergents for e by multiplying by 2 and adding 1.

```
eConvergents := [2*e + 1 for e in ccf]
```

$$\left[1, 3, \frac{19}{7}, \frac{193}{71}, \frac{2721}{1001}, \frac{49171}{18089}, \frac{1084483}{398959}, \ldots\right] \tag{11}$$

Type: Stream Fraction Integer

You can also compute the floating point approximations to these convergents.

```
eConvergents :: Stream Float
```

$$[1.0, 3.0, 2.7142857142857142857, 2.7183098591549295775, 2.7182817182817182817, 2.7182818287356957267, 2.7182818284585634113, \ldots] \tag{12}$$

Type: Stream Float

Compare this to the value of e computed by the **exp** operation in Float.

```
exp 1.0
```

$$2.7182818284590452354 \tag{13}$$

Type: Float

In about 1658, Lord Brouncker established the following expansion for $4/\pi$.

$$1 + \cfrac{1}{2 + \cfrac{9}{2 + \cfrac{25}{2 + \cfrac{49}{2 + \cfrac{81}{2 + \cdots}}}}}$$

Let's use this expansion to compute rational and floating point approximations for π.

```
cf := continuedFraction(1,[(2*i+1)**2 for i in
  0..],repeating [2])
```

$$1+\frac{1|}{|2}+\frac{9|}{|2}+\frac{25|}{|2}+\frac{49|}{|2}+\frac{81|}{|2}+\frac{121|}{|2}+\frac{169|}{|2}+\ldots \tag{14}$$

Type: ContinuedFraction Integer

```
ccf := convergents cf
```

$$\left[1, \frac{3}{2}, \frac{15}{13}, \frac{105}{76}, \frac{315}{263}, \frac{3465}{2578}, \frac{45045}{36979}, \ldots\right] \tag{15}$$

Type: Stream Fraction Integer

```
piConvergents := [4/p for p in ccf]
```

$$\left[4, \frac{8}{3}, \frac{52}{15}, \frac{304}{105}, \frac{1052}{315}, \frac{10312}{3465}, \frac{147916}{45045}, \ldots\right] \tag{16}$$

Type: Stream Fraction Integer

As you can see, the values are converging to π = 3.14159265358979323846..., but not very quickly.

```
piConvergents :: Stream Float
```

$$[4.0,\ 2.6666666666666666667,\ 3.4666666666666666667,\ 2.8952380952380952381,\ 3.3396825396825396825,\ 2.9760461760461760462,\ 3.2837384837384837385,\ \ldots] \tag{17}$$

Type: Stream Float

You need not restrict yourself to continued fractions of integers. Here is an expansion for a quotient of Gaussian integers.

```
continuedFraction((- 122 + 597*%i)/(4 - 4*%i))
```

$$-90+59\ \%i+\frac{1|}{|1-2\ \%i}+\frac{1|}{|-1+2\ \%i} \tag{18}$$

Type: ContinuedFraction Complex Integer

This is an expansion for a quotient of polynomials in one variable with rational number coefficients.

```
r : Fraction UnivariatePolynomial(x,Fraction Integer)
```

Type: Void

```
r := ((x - 1) * (x - 2)) / ((x-3) * (x-4))
```

$$\frac{x^2-3\ x+2}{x^2-7\ x+12} \tag{20}$$

Type: Fraction UnivariatePolynomial(x, Fraction Integer)

```
continuedFraction r
```

$$1+\frac{1|}{\left|\frac{1}{4}\ x-\frac{9}{8}\right.}+\frac{1|}{\left|\frac{16}{3}\ x-\frac{40}{3}\right.} \tag{21}$$

Type: ContinuedFraction UnivariatePolynomial(x, Fraction Integer)

To conclude this section, we give you evidence that

$$z = 3 + \frac{1|}{|3} + \frac{1|}{|6} + \frac{1|}{|3} + \frac{1|}{|6} + \frac{1|}{|3} + \frac{1|}{|6} + \frac{1|}{|3} + \frac{1|}{|6} + \frac{1|}{|3} + \frac{1|}{|6} + \ldots$$

is the expansion of $\sqrt{11}$.

```
[i*i for i in convergents(z) :: Stream Float]
```

$$[9.0,\ 11.1111111111111111,\ 10.99445983379501385,$$

$$11.00027777777777778,\ 10.99998607639879978 6, \qquad (22)$$

$$11.00000069792973103 9,\ 10.99999996501583444 6,\ \ldots]$$

Type: Stream Float

9.12 CycleIndicators

This section is based upon the paper J. H. Redfield, "The Theory of Group-Reduced Distributions", American J. Math.,49 (1927) 433-455, and is an application of group theory to enumeration problems. It is a development of the work by P. A. MacMahon on the application of symmetric functions and Hammond operators to combinatorial theory.

The theory is based upon the power sum symmetric functions s_i which are the sum of the ith powers of the variables. The cycle index of a permutation is an expression that specifies the sizes of the cycles of a permutation, and may be represented as a partition. A partition of a non-negative integer n is a collection of positive integers called its parts whose sum is n. For example, the partition $(3^2\ 2\ 1^2)$ will be used to represent $s_3^2 s_2 s_1^2$ and will indicate that the permutation has two cycles of length 3, one of length 2 and two of length 1. The cycle index of a permutation group is the sum of the cycle indices of its permutations divided by the number of permutations. The cycle indices of certain groups are provided.

We first load what we need from the library.

```
)load cycles evalcyc
library CYCLES has been loaded.
CycleIndicators is now explicitly exposed in frame
G1077
```

The operation **complete** returns the cycle index of the symmetric group of order n for argument n.

```
complete 1
```

$$(1) \qquad (1)$$

Type: SymmetricPolynomial Fraction Integer

Alternatively, it is the nth complete homogeneous symmetric function expressed in terms of power sum symmetric functions.

```
complete 2
```

$$\frac{1}{2}(2) + \frac{1}{2}\left(1^2\right) \qquad (2)$$

Type: SymmetricPolynomial Fraction Integer

```
complete 3
```

$$\frac{1}{3}(3)+\frac{1}{2}(2\ \ 1)+\frac{1}{6}\left(1^3\right) \tag{3}$$

Type: SymmetricPolynomial Fraction Integer

```
complete 7
```

$$\frac{1}{7}(7)+\frac{1}{6}(6\ \ 1)+\frac{1}{10}(5\ \ 2)+\frac{1}{10}\left(5\ \ 1^2\right)+\frac{1}{12}(4\ \ 3)+$$
$$\frac{1}{8}(4\ \ 2\ \ 1)+\frac{1}{24}\left(4\ \ 1^3\right)+\frac{1}{18}\left(3^2\ 1\right)+\frac{1}{24}\left(3\ \ 2^2\right)+$$
$$\frac{1}{12}\left(3\ \ 2\ \ 1^2\right)+\frac{1}{72}\left(3\ \ 1^4\right)+\frac{1}{48}\left(2^3\ 1\right)+\frac{1}{48}\left(2^2\ 1^3\right)+ \tag{4}$$
$$\frac{1}{240}\left(2\ \ 1^5\right)+\frac{1}{5040}\left(1^7\right)$$

Type: SymmetricPolynomial Fraction Integer

The operation **elementary** computes the nth elementary symmetric function for argument n.

```
elementary 7
```

$$\frac{1}{7}(7)-\frac{1}{6}(6\ \ 1)-\frac{1}{10}(5\ \ 2)+\frac{1}{10}\left(5\ \ 1^2\right)-\frac{1}{12}(4\ \ 3)+$$
$$\frac{1}{8}(4\ \ 2\ \ 1)-\frac{1}{24}\left(4\ \ 1^3\right)+\frac{1}{18}\left(3^2\ 1\right)+\frac{1}{24}\left(3\ \ 2^2\right)-$$
$$\frac{1}{12}\left(3\ \ 2\ \ 1^2\right)+\frac{1}{72}\left(3\ \ 1^4\right)-\frac{1}{48}\left(2^3\ 1\right)+\frac{1}{48}\left(2^2\ 1^3\right)- \tag{5}$$
$$\frac{1}{240}\left(2\ \ 1^5\right)+\frac{1}{5040}\left(1^7\right)$$

Type: SymmetricPolynomial Fraction Integer

The operation **alternating** returns the cycle index of the alternating group having an even number of even parts in each cycle partition.

```
alternating 7
```

$$\frac{2}{7}(7)+\frac{1}{5}\left(5\ \ 1^2\right)+\frac{1}{4}(4\ \ 2\ \ 1)+\frac{1}{9}\left(3^2\ 1\right)+\frac{1}{12}\left(3\ \ 2^2\right)+$$
$$\frac{1}{36}\left(3\ \ 1^4\right)+\frac{1}{24}\left(2^2\ 1^3\right)+\frac{1}{2520}\left(1^7\right) \tag{6}$$

Type: SymmetricPolynomial Fraction Integer

The operation **cyclic** returns the cycle index of the cyclic group.

```
cyclic 7
```

$$\frac{6}{7}(7)+\frac{1}{7}\left(1^7\right) \tag{7}$$

Type: SymmetricPolynomial Fraction Integer

The operation **dihedral** is the cycle index of the dihedral group.

```
dihedral 7
```

$$\frac{3}{7}(7)+\frac{1}{2}\left(2^3\ 1\right)+\frac{1}{14}\left(1^7\right) \tag{8}$$

Type: SymmetricPolynomial Fraction Integer

The operation **graphs** for argument n returns the cycle index of the group of permutations on the edges of the complete graph with n nodes induced by applying the symmetric group to the nodes.

```
graphs 5
```

$$\frac{1}{6}(6\ \ 3\ \ 1)+\frac{1}{5}\left(5^2\right)+\frac{1}{4}\left(4^2\ 2\right)+\frac{1}{6}\left(3^3\ 1\right)+\frac{1}{8}\left(2^4\ 1^2\right)+ \frac{1}{12}\left(2^3\ 1^4\right)+\frac{1}{120}\left(1^{10}\right) \tag{9}$$

Type: SymmetricPolynomial Fraction Integer

The cycle index of a direct product of two groups is the product of the cycle indices of the groups. Redfield provided two operations on two cycle indices which will be called "cup" and "cap" here. The **cup** of two cycle indices is a kind of scalar product that combines monomials for permutations with the same cycles. The **cap** operation provides the sum of the coefficients of the result of the **cup** operation which will be an integer that enumerates what Redfield called group-reduced distributions.

We can, for example, represent `complete 2 * complete 2` as the set of objects `a a b b` and `complete 2 * complete 1 * complete 1` as `c c d e`.

This integer is the number of different sets of four pairs.

```
cap(complete 2**2, complete 2*complete 1**2)
```

$$4 \tag{10}$$

Type: Fraction Integer

For example,

```
a a b b     a a b b     a a b b     a a b b
c c d e     c d c e     c e c d     d e c c
```

This integer is the number of different sets of four pairs no two pairs being equal.

```
cap(elementary 2**2, complete 2*complete 1**2)
```

$$2 \tag{11}$$

Type: Fraction Integer

For example,

```
a a b b     a a b b
c d c e     c e c d
```

In this case the configurations enumerated are easily constructed, however the theory merely enumerates them providing little help in actually constructing them.

Here are the number of 6-pairs, first from a a a b b c, second from d d e e f g.

```
cap(complete 3*complete 2*complete 1,complete
  2**2*complete 1**2)
```

$$24 \tag{12}$$

Type: Fraction Integer

Here it is again, but with no equal pairs.

```
cap(elementary 3*elementary 2*elementary 1,complete
  2**2*complete 1**2)
```

$$8 \tag{13}$$

Type: Fraction Integer

```
cap(complete 3*complete 2*complete 1,elementary
  2**2*elementary 1**2)
```

$$8 \tag{14}$$

Type: Fraction Integer

The number of 6-triples, first from a a a b b c, second from d d e e f g, third from h h i i j j.

```
eval(cup(complete 3*complete 2*complete 1, cup(complete
  2**2*complete 1**2,complete 2**3)))
```

$$1500 \tag{15}$$

Type: Fraction Integer

The cycle index of vertices of a square is dihedral 4.

```
square:=dihedral 4
```

$$\frac{1}{4}\,(4) + \frac{3}{8}\,\left(2^2\right) + \frac{1}{4}\,\left(2\ \ 1^2\right) + \frac{1}{8}\,\left(1^4\right) \tag{16}$$

Type: SymmetricPolynomial Fraction Integer

The number of different squares with 2 red vertices and 2 blue vertices.

```
cap(complete 2**2,square)
```

$$2 \tag{17}$$

Type: Fraction Integer

The number of necklaces with 3 red beads, 2 blue beads and 2 green beads.

```
cap(complete 3*complete 2**2,dihedral 7)
```

$$18 \tag{18}$$

Type: Fraction Integer

The number of graphs with 5 nodes and 7 edges.

```
cap(graphs 5,complete 7*complete 3)
```

$$4 \tag{19}$$

Type: Fraction Integer

The cycle index of rotations of vertices of a cube.

```
macro s == powerSum
```

Type: Void

```
cube:=(1/24)*(s 1**8+9*s 2**4 + 8*s 3**2*s 1**2+6*s 4**2)
```

$$\frac{1}{4}\,\left(4^2\right) + \frac{1}{3}\,\left(3^2\ 1^2\right) + \frac{3}{8}\,\left(2^4\right) + \frac{1}{24}\,\left(1^8\right) \tag{21}$$

Type: SymmetricPolynomial Fraction Integer

The number of cubes with 4 red vertices and 4 blue vertices.

```
cap(complete 4**2,cube)
```

$$7 \tag{22}$$

Type: Fraction Integer

The number of labeled graphs with degree sequence 2 2 2 1 1 with no loops or multiple edges.

```
cap(complete 2**3*complete 1**2,wreath(elementary
  4,elementary 2))
```

$$7 \tag{23}$$

Type: Fraction Integer

Again, but with loops allowed but not multiple edges.

```
cap(complete 2**3*complete 1**2,wreath(elementary
  4,complete 2))
```

$$17 \tag{24}$$

Type: Fraction Integer

Again, but with multiple edges allowed, but not loops

```
cap(complete 2**3*complete 1**2,wreath(complete
  4,elementary 2))
```

$$10 \tag{25}$$

Type: Fraction Integer

Again, but with both multiple edges and loops allowed

```
cap(complete 2**3*complete 1**2,wreath(complete 4,complete
  2))
```

$$23 \tag{26}$$

Type: Fraction Integer

Having constructed a cycle index for a configuration we are at liberty to evaluate the s_i components any way we please. For example we can produce enumerating generating functions. This is done by providing a function `f` on an integer `i` to the value required of s_i, and then evaluating `eval(f, cycleindex)`.

```
x: ULS(FRAC INT,'x,0) := 'x
```

$$x \tag{27}$$

Type: UnivariateLaurentSeries(Fraction Integer, x, 0)

```
ZeroOrOne: INT -> ULS(FRAC INT, 'x, 0)
```

Type: Void

```
Integers: INT -> ULS(FRAC INT, 'x, 0)
```

Type: Void

For the integers 0 and 1, or two colors.

```
ZeroOrOne n == 1+x**n
```

Type: Void

```
ZeroOrOne 5
```

$$1 + x^5 \tag{31}$$

Type: UnivariateLaurentSeries(Fraction Integer, x, 0)

For the integers `0, 1, 2, ...` we have this.

```
Integers n == 1/(1-x**n)
```

Type: Void

```
Integers 5
```

$$1 + x^5 + O\left(x^8\right) \tag{33}$$

Type: UnivariateLaurentSeries(Fraction Integer, x, 0)

The coefficient of x^n is the number of graphs with 5 nodes and `n` edges.

```
eval(ZeroOrOne, graphs 5)
```

$$1 + x + 2\,x^2 + 4\,x^3 + 6\,x^4 + 6\,x^5 + 6\,x^6 + 4\,x^7 + O\left(x^8\right) \tag{34}$$

Type: UnivariateLaurentSeries(Fraction Integer, x, 0)

The coefficient of x^n is the number of necklaces with `n` red beads and `n-8` green beads.

```
eval(ZeroOrOne,dihedral 8)
```

$$1 + x + 4\,x^2 + 5\,x^3 + 8\,x^4 + 5\,x^5 + 4\,x^6 + x^7 + O\left(x^8\right) \tag{35}$$

Type: UnivariateLaurentSeries(Fraction Integer, x, 0)

The coefficient of x^n is the number of partitions of `n` into 4 or fewer parts.

```
eval(Integers,complete 4)
```

$$1 + x + 2\,x^2 + 3\,x^3 + 5\,x^4 + 6\,x^5 + 9\,x^6 + 11\,x^7 + O\left(x^8\right) \tag{36}$$

Type: UnivariateLaurentSeries(Fraction Integer, x, 0)

The coefficient of x^n is the number of partitions of `n` into 4 boxes containing ordered distinct parts.

```
eval(Integers,elementary 4)
```

$$x^6 + x^7 + 2\,x^8 + 3\,x^9 + 5\,x^{10} + 6\,x^{11} + 9\,x^{12} + 11\,x^{13} + O\left(x^{14}\right) \tag{37}$$

Type: UnivariateLaurentSeries(Fraction Integer, x, 0)

The coefficient of x^n is the number of different cubes with `n` red vertices and `8-n` green ones.

```
eval(ZeroOrOne,cube)
```

$$1 + x + 3\,x^2 + 3\,x^3 + 7\,x^4 + 3\,x^5 + 3\,x^6 + x^7 + O\left(x^8\right) \tag{38}$$

Type: UnivariateLaurentSeries(Fraction Integer, x, 0)

The coefficient of x^n is the number of different cubes with integers on the vertices whose sum is `n`.

```
eval(Integers,cube)
```

$$1 + x + 4\,x^2 + 7\,x^3 + 21\,x^4 + 37\,x^5 + 85\,x^6 + 151\,x^7 + O\left(x^8\right) \tag{39}$$

Type: UnivariateLaurentSeries(Fraction Integer, x, 0)

The coefficient of x^n is the number of graphs with 5 nodes and with integers on the edges whose sum is n. In other words, the enumeration is of multigraphs with 5 nodes and n edges.

```
eval(Integers,graphs 5)
```

$$1 + x + 3\,x^2 + 7\,x^3 + 17\,x^4 + 35\,x^5 + 76\,x^6 + 149\,x^7 + O\left(x^8\right) \tag{40}$$

Type: UnivariateLaurentSeries(Fraction Integer, x, 0)

Graphs with 15 nodes enumerated with respect to number of edges.

```
eval(ZeroOrOne ,graphs 15)
```

$$1 + x + 2\,x^2 + 5\,x^3 + 11\,x^4 + 26\,x^5 + 68\,x^6 + 177\,x^7 + O\left(x^8\right) \tag{41}$$

Type: UnivariateLaurentSeries(Fraction Integer, x, 0)

Necklaces with 7 green beads, 8 white beads, 5 yellow beads and 10 red beads.

```
cap(dihedral 30,complete 7*complete 8*complete 5*complete
  10)
```

$$49958972383320 \tag{42}$$

Type: Fraction Integer

The operation **SFunction** is the S-function or Schur function of a partition written as a descending list of integers expressed in terms of power sum symmetric functions.

In this case the argument partition represents a tableau shape. For example 3,2,2,1 represents a tableau with three boxes in the first row, two boxes in the second and third rows, and one box in the fourth row. SFunction [3,2,2,1] counts the number of different tableaux of shape 3, 2, 2, 1 filled with objects with an ascending order in the columns and a non-descending order in the rows.

```
sf3221:= SFunction [3,2,2,1]
```

$$\begin{aligned}
&\frac{1}{12}\,(6\ \ 2) - \frac{1}{12}\left(6\ \ 1^2\right) - \frac{1}{16}\left(4^2\right) + \frac{1}{12}\,(4\ \ 3\ \ 1) + \frac{1}{24}\left(4\ \ 1^4\right) - \\
&\frac{1}{36}\left(3^2\ 2\right) + \frac{1}{36}\left(3^2\ 1^2\right) - \frac{1}{24}\left(3\ \ 2^2\ 1\right) - \frac{1}{36}\left(3\ \ 2\ \ 1^3\right) - \\
&\frac{1}{72}\left(3\ \ 1^5\right) - \frac{1}{192}\left(2^4\right) + \frac{1}{48}\left(2^3\ 1^2\right) + \frac{1}{96}\left(2^2\ 1^4\right) - \frac{1}{144}\left(2\ \ 1^6\right) + \\
&\frac{1}{576}\left(1^8\right)
\end{aligned} \tag{43}$$

Type: SymmetricPolynomial Fraction Integer

This is the number filled with a b b c c d d.

```
cap(sf3221,complete 2**4)
```

$$3 \tag{44}$$

Type: Fraction Integer

The configurations enumerated above are:

```
a a b     a a c     a a d
b c       b b       b b
c d       c d       c c
d         d         d
```

This is the number of tableaux filled with 1..8.

```
cap(sf3221, powerSum 1**8)
```

$$70 \tag{45}$$

Type: Fraction Integer

The coefficient of x^n is the number of column strict reverse plane partitions of n of shape 3 2 2 1.

```
eval(Integers, sf3221)
```

$$x^9 + 3\,x^{10} + 7\,x^{11} + 14\,x^{12} + 27\,x^{13} + 47\,x^{14} + O\left(x^{15}\right) \quad (46)$$

Type: UnivariateLaurentSeries(Fraction Integer, x, 0)

The smallest is

```
0 0 0
1 1
2 2
3
```

9.13 DeRhamComplex

The domain constructor DeRhamComplex creates the class of differential forms of arbitrary degree over a coefficient ring. The De Rham complex constructor takes two arguments: a ring, `coefRing`, and a list of coordinate variables.

This is the ring of coefficients.

```
macro coefRing == Integer
```

Type: Void

These are the coordinate variables.

```
lv : List Symbol := [x,y,z]
```

$$[x,\ y,\ z] \quad (2)$$

Type: List Symbol

This is the De Rham complex of Euclidean three-space using coordinates `x`, `y` and `z`.

```
der := DERHAM(coefRing,lv)
```

DeRhamComplex (Integer , $[x, y, z]$) (3)

Type: Domain

This complex allows us to describe differential forms having expressions of integers as coefficients. These coefficients can involve any number of variables, for example, `f(x,t,r,y,u,z)`. As we've chosen to work with ordinary Euclidean three-space, expressions involving these forms are treated as functions of `x`, `y` and `z` with the additional arguments `t`, `r` and `u` regarded as symbolic constants.

Here are some examples of coefficients.

```
R := Expression coefRing
```

Expression Integer (4)

Type: Domain

```
f : R := x**2*y*z-5*x**3*y**2*z**5
```

$$-5\,x^3\,y^2\,z^5 + x^2\,y\,z \quad (5)$$

Type: Expression Integer

```
g : R := z**2*y*cos(z)-7*sin(x**3*y**2)*z**2
```

$$-7\ z^2\ \sin\left(x^3\ y^2\right)+y\ z^2\ \cos(z) \tag{6}$$

Type: Expression Integer

```
h : R :=x*y*z-2*x**3*y*z**2
```

$$-2\ x^3\ y\ z^2+x\ y\ z \tag{7}$$

Type: Expression Integer

We now define the multiplicative basis elements for the exterior algebra over R.

```
dx : der := generator(1)
```

$$dx \tag{8}$$

Type: DeRhamComplex(Integer, [x, y, z])

```
dy : der := generator(2)
```

$$dy \tag{9}$$

Type: DeRhamComplex(Integer, [x, y, z])

```
dz : der := generator(3)
```

$$dz \tag{10}$$

Type: DeRhamComplex(Integer, [x, y, z])

This is an alternative way to give the above assignments.

```
[dx,dy,dz] := [generator(i)$der for i in 1..3]
```

$$[dx,\ dy,\ dz] \tag{11}$$

Type: List DeRhamComplex(Integer, [x, y, z])

Now we define some one-forms.

```
alpha : der := f*dx + g*dy + h*dz
```

$$\begin{array}{l}\left(-2\ x^3\ y\ z^2+x\ y\ z\right)\ dz+ \\ \left(-7\ z^2\ \sin\left(x^3\ y^2\right)+y\ z^2\ \cos(z)\right)\ dy+ \\ \left(-5\ x^3\ y^2\ z^5+x^2\ y\ z\right)\ dx\end{array} \tag{12}$$

Type: DeRhamComplex(Integer, [x, y, z])

```
beta : der := cos(tan(x*y*z)+x*y*z)*dx + x*dy
```

$$x\ dy+\cos(\tan(x\ y\ z)+x\ y\ z)\ dx \tag{13}$$

Type: DeRhamComplex(Integer, [x, y, z])

A well-known theorem states that the composition of **exteriorDifferential** with itself is the zero map for continuous forms. Let's verify this theorem for `alpha`.

```
exteriorDifferential alpha;
```

(14)

Type: DeRhamComplex(Integer, [x, y, z])

We suppressed the lengthy output of the last expression, but nevertheless, the composition is zero.

```
exteriorDifferential %
```

$$0 \tag{15}$$

Type: DeRhamComplex(Integer, [x, y, z])

Now we check that **exteriorDifferential** is a "graded derivation" `D`, that is, `D` satisfies:

$$D(ab) = D(a)b + (-1)^{\deg(a)} aD(b)$$

```
gamma := alpha * beta
```

$$\left(2\ x^4\ y\ z^2 - x^2\ y\ z\right)\ dy\ dz + \left(2\ x^3\ y\ z^2 - x\ y\ z\right)\ \cos\left(\tan\left(x\ y\ z\right) + x\ y\ z\right)\ dx\ dz + \left(\begin{array}{l}\left(7\ z^2\ \sin\left(x^3\ y^2\right) - y\ z^2\ \cos\left(z\right)\right)\ \cos\left(\tan\left(x\ y\ z\right) + x\ y\ z\right) - \\ 5\ x^4\ y^2\ z^5 + x^3\ y\ z\end{array}\right) \cdot dx\ dy \tag{16}$$

Type: DeRhamComplex(Integer, [x, y, z])

We try this for the one-forms `alpha` and `beta`.

```
exteriorDifferential(gamma)
  - (exteriorDifferential(alpha)*beta - alpha *
  exteriorDifferential(beta))
```

$$0 \tag{17}$$

Type: DeRhamComplex(Integer, [x, y, z])

Now we define some "basic operators" (see 'Operator' on page 426).

```
a : BOP := operator('a)
```

$$a \tag{18}$$

Type: BasicOperator

```
b : BOP := operator('b)
```

$$b \tag{19}$$

Type: BasicOperator

```
c : BOP := operator('c)
```

$$c \tag{20}$$

Type: BasicOperator

We also define some indeterminate one- and two-forms using these operators.

```
sigma := a(x,y,z) * dx + b(x,y,z) * dy + c(x,y,z) * dz
```

$$c\left(x,\ y,\ z\right)\ dz + b\left(x,\ y,\ z\right)\ dy + a\left(x,\ y,\ z\right)\ dx \tag{21}$$

Type: DeRhamComplex(Integer, [x, y, z])

```
theta := a(x,y,z) * dx * dy + b(x,y,z) * dx * dz +
  c(x,y,z) * dy * dz
```

$$c\left(x,\ y,\ z\right)\ dy\ dz + b\left(x,\ y,\ z\right)\ dx\ dz + a\left(x,\ y,\ z\right)\ dx\ dy \tag{22}$$

Type: DeRhamComplex(Integer, [x, y, z])

This allows us to get formal definitions for the "gradient" ...

```
totalDifferential(a(x,y,z))$der
```

$$a_{,3}(x, y, z)\ dz + a_{,2}(x, y, z)\ dy + a_{,1}(x, y, z)\ dx \tag{23}$$

Type: DeRhamComplex(Integer, [x, y, z])

the "curl" ...

```
exteriorDifferential sigma
```

$$\begin{aligned}&\left(c_{,2}(x, y, z) - b_{,3}(x, y, z)\right)\ dy\ dz+\\&\left(c_{,1}(x, y, z) - a_{,3}(x, y, z)\right)\ dx\ dz+\\&\left(b_{,1}(x, y, z) - a_{,2}(x, y, z)\right)\ dx\ dy\end{aligned} \tag{24}$$

Type: DeRhamComplex(Integer, [x, y, z])

and the "divergence."

```
exteriorDifferential theta
```

$$\left(c_{,1}(x, y, z) - b_{,2}(x, y, z) + a_{,3}(x, y, z)\right)\ dx\ dy\ dz \tag{25}$$

Type: DeRhamComplex(Integer, [x, y, z])

Note that the De Rham complex is an algebra with unity. This element 1 is the basis for elements for zero-forms, that is, functions in our space.

```
one : der := 1
```

$$1 \tag{26}$$

Type: DeRhamComplex(Integer, [x, y, z])

To convert a function to a function lying in the De Rham complex, multiply the function by "one."

```
g1 : der := a([x,t,y,u,v,z,e]) * one
```

$$a(x, t, y, u, v, z, e) \tag{27}$$

Type: DeRhamComplex(Integer, [x, y, z])

A current limitation of AXIOM forces you to write functions with more than four arguments using square brackets in this way.

```
h1 : der := a([x,y,x,t,x,z,y,r,u,x]) * one
```

$$a(x, y, x, t, x, z, y, r, u, x) \tag{28}$$

Type: DeRhamComplex(Integer, [x, y, z])

Now note how the system keeps track of where your coordinate functions are located in expressions.

```
exteriorDifferential g1
```

$$\begin{aligned}&a_{,6}(x, t, y, u, v, z, e)\ dz + a_{,3}(x, t, y, u, v, z, e)\ dy+\\&a_{,1}(x, t, y, u, v, z, e)\ dx\end{aligned} \tag{29}$$

Type: DeRhamComplex(Integer, [x, y, z])

```
exteriorDifferential h1
```

$$a_{,6}(x,\ y,\ x,\ t,\ x,\ z,\ y,\ r,\ u,\ x)\ dz+ \left(\begin{array}{l} a_{,7}(x,\ y,\ x,\ t,\ x,\ z,\ y,\ r,\ u,\ x)+ \\ a_{,2}(x,\ y,\ x,\ t,\ x,\ z,\ y,\ r,\ u,\ x) \end{array}\right) dy+ \left(\begin{array}{l} a_{,10}(x,\ y,\ x,\ t,\ x,\ z,\ y,\ r,\ u,\ x)+ \\ a_{,5}(x,\ y,\ x,\ t,\ x,\ z,\ y,\ r,\ u,\ x)+ \\ a_{,3}(x,\ y,\ x,\ t,\ x,\ z,\ y,\ r,\ u,\ x)+ \\ a_{,1}(x,\ y,\ x,\ t,\ x,\ z,\ y,\ r,\ u,\ x) \end{array}\right) dx \qquad (30)$$

Type: DeRhamComplex(Integer, [x, y, z])

In this example of Euclidean three-space, the basis for the De Rham complex consists of the eight forms: `1`, `dx`, `dy`, `dz`, `dx*dy`, `dx*dz`, `dy*dz`, and `dx*dy*dz`.

```
coefficient(gamma, dx*dy)
```

$$\left(7\ z^2 \sin\left(x^3\ y^2\right) - y\ z^2 \cos(z)\right) \cos(\tan(x\ y\ z) + x\ y\ z) - 5\ x^4\ y^2\ z^5 + x^3\ y\ z \qquad (31)$$

Type: Expression Integer

```
coefficient(gamma, one)
```

$$0 \qquad (32)$$

Type: Expression Integer

```
coefficient(g1,one)
```

$$a(x,\ t,\ y,\ u,\ v,\ z,\ e) \qquad (33)$$

Type: Expression Integer

9.14 DecimalExpansion

All rationals have repeating decimal expansions. Operations to access the individual digits of a decimal expansion can be obtained by converting the value to RadixExpansion(10). More examples of expansions are available in 'BinaryExpansion' on page 312, 'HexadecimalExpansion' on page 379, and 'RadixExpansion' on page 444. Issue the system command `)show DecimalExpansion` to display the full list of operations defined by DecimalExpansion.

The operation **decimal** is used to create this expansion of type DecimalExpansion.

```
r := decimal(22/7)
```

$$3.\overline{142857} \qquad (1)$$

Type: DecimalExpansion

Arithmetic is exact.

```
r + decimal(6/7)
```

$$4 \tag{2}$$

Type: DecimalExpansion

The period of the expansion can be short or long . . .

```
[decimal(1/i) for i in 350..354]
```

$$\left[0.00\overline{285714},\ 0.\overline{002849},\ 0.00284\overline{09},\right.$$

$$0.\overline{00283286118980169971671388101983}, \tag{3}$$

$$\left.0.0\overline{028248587570621468926553672316384180790960451977401129943 5}\right]$$

Type: List DecimalExpansion

or very long.

```
decimal(1/2049)
```

$$0.\overline{000488042947779404587603709126403123474865788189360663}$$
$$\overline{738408979990239141044411908247925817471937530502684236 21}$$
$$\overline{278672523182040019521717911176183504148365056124938994 63}$$
$$\overline{152757442654953635919960956564177647632991703269887750 12} \tag{4}$$
$$\overline{201073694485114690092728160078086871644704734016593460 22}$$
$$\overline{449975597852611029770619814543679843826256710590531966 81}$$
$$\overline{3079551}$$

Type: DecimalExpansion

These numbers are bona fide algebraic objects.

```
p := decimal(1/4)*x**2 + decimal(2/3)*x + decimal(4/9)
```

$$0.25\ x^2 + 0.\overline{6}\ x + 0.\overline{4} \tag{5}$$

Type: Polynomial DecimalExpansion

```
q := D(p, x)
```

$$0.5\ x + 0.\overline{6} \tag{6}$$

Type: Polynomial DecimalExpansion

```
g := gcd(p, q)
```

$$x + 1.\overline{3} \tag{7}$$

Type: Polynomial DecimalExpansion

9.15 DistributedMultivariatePolynomial

DistributedMultivariatePolynomial and HomogeneousDistributedMultivariatePolynomial, abbreviated DMP and HDMP, respectively, are very similar to MultivariatePolynomial except that they are represented and displayed in a non-recursive manner.

```
(d1,d2,d3) : DMP([z,y,x],FRAC INT)
```

Type: Void

The constructor DMP orders its monomials lexicographically while HDMP orders them by total order refined by reverse lexicographic order.

```
d1 := -4*z + 4*y**2*x + 16*x**2 + 1
```

$$-4\ z + 4\ y^2\ x + 16\ x^2 + 1 \tag{2}$$

Type: DistributedMultivariatePolynomial([z, y, x], Fraction Integer)

```
d2 := 2*z*y**2 + 4*x + 1
```

$$2\ z\ y^2 + 4\ x + 1 \tag{3}$$

Type: DistributedMultivariatePolynomial([z, y, x], Fraction Integer)

```
d3 := 2*z*x**2 - 2*y**2 - x
```

$$2\ z\ x^2 - 2\ y^2 - x \tag{4}$$

Type: DistributedMultivariatePolynomial([z, y, x], Fraction Integer)

These constructors are mostly used in Gröbner basis calculations.

```
groebner [d1,d2,d3]
```

$$\left[z - \frac{1568}{2745}\ x^6 - \frac{1264}{305}\ x^5 + \frac{6}{305}\ x^4 + \frac{182}{549}\ x^3 - \frac{2047}{610}\ x^2 - \frac{103}{2745}\ x - \frac{2857}{10980}, \right.$$
$$y^2 + \frac{112}{2745}\ x^6 - \frac{84}{305}\ x^5 - \frac{1264}{305}\ x^4 - \frac{13}{549}\ x^3 + \frac{84}{305}\ x^2 + \frac{1772}{2745}\ x + \frac{2}{2745}, \tag{5}$$
$$\left. x^7 + \frac{29}{4}\ x^6 - \frac{17}{16}\ x^4 - \frac{11}{8}\ x^3 + \frac{1}{32}\ x^2 + \frac{15}{16}\ x + \frac{1}{4} \right]$$

Type: List DistributedMultivariatePolynomial([z, y, x], Fraction Integer)

```
(n1,n2,n3) : HDMP([z,y,x],FRAC INT)
```

Type: Void

```
(n1,n2,n3) := (d1,d2,d3)
```

$$2\ z\ x^2 - 2\ y^2 - x \tag{7}$$

Type: HomogeneousDistributedMultivariatePolynomial([z, y, x], Fraction Integer)

Note that we get a different Gröbner basis when we use the HDMP polynomials, as expected.

```
groebner [n1,n2,n3]
```

$$\left[y^4 + 2\ x^3 - \frac{3}{2}\ x^2 + \frac{1}{2}\ z - \frac{1}{8},\right.$$
$$x^4 + \frac{29}{4}\ x^3 - \frac{1}{8}\ y^2 - \frac{7}{4}\ z\ x - \frac{9}{16}\ x - \frac{1}{4},\ z\ y^2 + 2\ x + \frac{1}{2},$$
$$y^2\ x + 4\ x^2 - z + \frac{1}{4},\ z\ x^2 - y^2 - \frac{1}{2}\ x, \tag{8}$$
$$\left.z^2 - 4\ y^2 + 2\ x^2 - \frac{1}{4}\ z - \frac{3}{2}\ x\right]$$

Type: List HomogeneousDistributedMultivariatePolynomial([z, y, x], Fraction Integer)

GeneralDistributedMultivariatePolynomial is somewhat more flexible in the sense that as well as accepting a list of variables to specify the variable ordering, it also takes a predicate on exponent vectors to specify the term ordering. With this polynomial type the user can experiment with the effect of using completely arbitrary term orderings. This flexibility is mostly important for algorithms such as Gröbner basis calculations which can be very sensitive to term ordering.

For more information on related topics, see Section 1.9 on page 43, Section 2.7 on page 78, 'Polynomial' on page 436, 'UnivariatePolynomial' on page 472, and 'MultivariatePolynomial' on page 421. Issue the system command `)show DistributedMultivariatePolynomial` to display the full list of operations defined by DistributedMultivariatePolynomial.

9.16 EqTable

The EqTable domain provides tables where the keys are compared using **eq?**. Keys are considered equal only if they are the same instance of a structure. This is useful if the keys are themselves updatable structures. Otherwise, all operations are the same as for type Table. See 'Table' on page 465 for general information about tables. Issue the system command `)show EqTable` to display the full list of operations defined by EqTable.

The operation **table** is here used to create a table where the keys are lists of integers.

```
e: EqTable(List Integer, Integer) := table()
```

table() (1)

Type: EqTable(List Integer, Integer)

These two lists are equal according to "=", but not according to **eq?**.

```
l1 := [1,2,3]
```

$[1, 2, 3]$ (2)

Type: List PositiveInteger

```
l2 := [1,2,3]
```

$[1, 2, 3]$ (3)

Type: List PositiveInteger

Because the two lists are not **eq?**, separate values can be stored under each.

```
e.l1 := 111
```

111 (4)

Type: PositiveInteger

```
e.l2 := 222
```

222 (5)

Type: PositiveInteger

```
e.l1
```

111 (6)

Type: PositiveInteger

9.17 Equation

The Equation domain provides equations as mathematical objects. These are used, for example, as the input to various **solve** operations.

Equations are created using the equals symbol, "=".

```
eq1 := 3*x + 4*y = 5
```

$4\ y + 3\ x = 5$ (1)

Type: Equation Polynomial Integer

```
eq2 := 2*x + 2*y = 3
```

$2\ y + 2\ x = 3$ (2)

Type: Equation Polynomial Integer

The left- and right-hand sides of an equation are accessible using the operations **lhs** and **rhs**.

```
lhs eq1
```

$4\ y + 3\ x$ (3)

Type: Polynomial Integer

```
rhs eq1
```

5 (4)

Type: Polynomial Integer

Arithmetic operations are supported and operate on both sides of the equation.

```
eq1 + eq2
```

$$6\ y+5\ x=8 \tag{5}$$

Type: Equation Polynomial Integer

```
eq1 * eq2
```

$$8\ y^2+14\ x\ y+6\ x^2=15 \tag{6}$$

Type: Equation Polynomial Integer

```
2*eq2 - eq1
```

$$x=1 \tag{7}$$

Type: Equation Polynomial Integer

Equations may be created for any type so the arithmetic operations will be defined only when they make sense. For example, exponentiation is not defined for equations involving non-square matrices.

```
eq1**2
```

$$16\ y^2+24\ x\ y+9\ x^2=25 \tag{8}$$

Type: Equation Polynomial Integer

Note that an equals symbol is also used to *test* for equality of values in certain contexts. For example, `x+1` and `y` are unequal as polynomials.

```
if x+1 = y then "equal" else "unequal"
```

"unequal" (9)

Type: String

```
eqpol := x+1 = y
```

$$x+1=y \tag{10}$$

Type: Equation Polynomial Integer

If an equation is used where a Boolean value is required, then it is evaluated using the equality test from the operand type.

```
if eqpol then "equal" else "unequal"
```

"unequal" (11)

Type: String

If one wants a Boolean value rather than an equation, all one has to do is ask!

```
eqpol::Boolean
```

false (12)

Type: Boolean

9.18 Exit

A function that does not return directly to its caller has Exit as its return type. The operation **error** is an example of one which does not return to its caller. Instead, it causes a return to top-level.

```
n := 0
```

0 (1)

Type: NonNegativeInteger

The function **gasp** is given return type Exit since it is guaranteed never to return a value to its caller.

```
gasp(): Exit ==
    free n
    n := n + 1
    error "Oh no!"
```

```
Function declaration gasp : () -> Exit has been
added to workspace.
```

Type: Void

The return type of **half** is determined by resolving the types of the two branches of the `if`.

```
half(k) ==
  if odd? k then gasp()
  else k quo 2
```

Type: Void

Because **gasp** has the return type Exit, the type of `if` in **half** is resolved to be Integer.

```
half 4
```

```
Compiling function gasp with type () -> Exit
Compiling function half with type PositiveInteger ->
Integer
```

2 (4)

Type: PositiveInteger

```
half 3
```

```
Error signalled from user code in function gasp:
Oh no!
```

```
n
```

1 (5)

Type: NonNegativeInteger

For functions which return no value at all, use Void. See Section 6 on page 135 and 'Void' on page 480 for more information. Issue the system command `)show Exit` to display the full list of operations defined by Exit.

9.19 Factored

Factored creates a domain whose objects are kept in factored form as long as possible. Thus certain operations like "`*`" (multiplication) and **gcd** are relatively easy to do. Others, such as addition, require somewhat more work, and the result may not be completely factored unless the argument domain `R` provides a **factor** operation. Each object consists of a unit and a list of factors, where each factor consists of a member of `R` (the *base*), an exponent, and a flag indicating what is known about the base. A flag may be one of `"nil"`, `"sqfr"`, `"irred"` or `"prime"`, which mean that nothing is known about the base, it is square-free, it is irreducible, or it is prime, respectively. The current restriction to factored objects of integral domains allows simplification to be performed without worrying about multiplication order.

9.19.1 Decomposing Factored Objects

In this section we will work with a factored integer.

```
g := factor(4312)
```

$2^3\ 7^2\ 11$ (1)

Type: Factored Integer

Let's begin by decomposing g into pieces. The only possible units for integers are 1 and -1.

```
unit(g)
```

1 (2)

Type: PositiveInteger

There are three factors.

```
numberOfFactors(g)
```

3 (3)

Type: PositiveInteger

We can make a list of the bases, ...

```
[nthFactor(g,i) for i in 1..numberOfFactors(g)]
```

$[2,\ 7,\ 11]$ (4)

Type: List Integer

and the exponents, ...

```
[nthExponent(g,i) for i in 1..numberOfFactors(g)]
```

$[3,\ 2,\ 1]$ (5)

Type: List Integer

and the flags. You can see that all the bases (factors) are prime.

```
[nthFlag(g,i) for i in 1..numberOfFactors(g)]
```

["prime", "prime", "prime"] (6)

Type: List Union("nil", "sqfr", "irred", "prime")

A useful operation for pulling apart a factored object into a list of records of the components is **factorList**.

```
factorList(g)
```

$[[flg = \texttt{"prime"},\ fctr = 2,\ xpnt = 3],$

$[flg = \texttt{"prime"},\ fctr = 7,\ xpnt = 2],$ (7)

$[flg = \texttt{"prime"},\ fctr = 11,\ xpnt = 1]]$

Type: List Record(flg: Union("nil", "sqfr", "irred", "prime"), fctr: Integer, xpnt: Integer)

If you don't care about the flags, use **factors**.

```
factors(g)
```

$$[[factor = 2,\ exponent = 3],\ [factor = 7,\ exponent = 2],\ [factor = 11,\ exponent = 1]] \quad (8)$$

Type: List Record(factor: Integer, exponent: Integer)

Neither of these operations returns the unit.

```
first(%).factor
```

$$2 \quad (9)$$

Type: PositiveInteger

9.19.2 Expanding Factored Objects

Recall that we are working with this factored integer.

```
g := factor(4312)
```

$$2^3\ 7^2\ 11 \quad (1)$$

Type: Factored Integer

To multiply out the factors with their multiplicities, use **expand**.

```
expand(g)
```

$$4312 \quad (2)$$

Type: PositiveInteger

If you would like, say, the distinct factors multiplied together but with multiplicity one, you could do it this way.

```
reduce(*,[t.factor for t in factors(g)])
```

$$154 \quad (3)$$

Type: PositiveInteger

9.19.3 Arithmetic with Factored Objects

We're still working with this factored integer.

```
g := factor(4312)
```

$$2^3\ 7^2\ 11 \quad (1)$$

Type: Factored Integer

We'll also define this factored integer.

```
f := factor(246960)
```

$$2^4\ 3^2\ 5\ 7^3 \quad (2)$$

Type: Factored Integer

Operations involving multiplication and division are particularly easy with factored objects.

```
f * g
```

$2^7\ 3^2\ 5\ 7^5\ 11$ (3)

Type: Factored Integer

```
f**500
```

$2^{2000}\ 3^{1000}\ 5^{500}\ 7^{1500}$ (4)

Type: Factored Integer

```
gcd(f,g)
```

$2^3\ 7^2$ (5)

Type: Factored Integer

```
lcm(f,g)
```

$2^4\ 3^2\ 5\ 7^3\ 11$ (6)

Type: Factored Integer

If we use addition and subtraction things can slow down because we may need to compute greatest common divisors.

```
f + g
```

$2^3\ 7^2\ 641$ (7)

Type: Factored Integer

```
f - g
```

$2^3\ 7^2\ 619$ (8)

Type: Factored Integer

Test for equality with `0` and `1` by using **zero?** and **one?**, respectively.

```
zero?(factor(0))
```

true (9)

Type: Boolean

```
zero?(g)
```

false (10)

Type: Boolean

```
one?(factor(1))
```

true (11)

Type: Boolean

```
one?(f)
```

false (12)

Type: Boolean

Another way to get the zero and one factored objects is to use package calling (see Section 2.9 on page 83).

```
0$Factored(Integer)
```

0 (13)

Type: Factored Integer

```
1$Factored(Integer)
```

1 (14)

Type: Factored Integer

9.19.4 Creating New Factored Objects

The **map** operation is used to iterate across the unit and bases of a factored object. See 'FactoredFunctions2' on page 361 for a discussion of **map**.

The following four operations take a base and an exponent and create a factored object. They differ in handling the flag component.

```
nilFactor(24,2)
```

24^2 (1)

Type: Factored Integer

This factor has no associated information.

```
nthFlag(%,1)
```

"nil" (2)

Type: Union("nil", ...)

This factor is asserted to be square-free.

```
sqfrFactor(12,2)
```

12^2 (3)

Type: Factored Integer

This factor is asserted to be irreducible.

```
irreducibleFactor(13,10)
```

13^{10} (4)

Type: Factored Integer

This factor is asserted to be prime.

```
primeFactor(11,5)
```

11^5 (5)

Type: Factored Integer

A partial inverse to **factorList** is **makeFR**.

```
h := factor(-720)
```

$-2^4\ 3^2\ 5$ (6)

Type: Factored Integer

The first argument is the unit and the second is a list of records as returned by **factorList**.

```
h - makeFR(unit(h),factorList(h))
```

0 (7)

Type: Factored Integer

9.19.5 Factored Objects with Variables

Some of the operations available for polynomials are also available for factored polynomials.

```
p := (4*x*x-12*x+9)*y*y + (4*x*x-12*x+9)*y + 28*x*x - 84*x
  + 63
```

$$\left(4\ x^2 - 12\ x + 9\right)\ y^2 + \left(4\ x^2 - 12\ x + 9\right)\ y + 28\ x^2 - 84\ x + 63 \qquad (1)$$

Type: Polynomial Integer

```
fp := factor(p)
```

$$(2\ x - 3)^2\ \left(y^2 + y + 7\right) \qquad (2)$$

Type: Factored Polynomial Integer

You can differentiate with respect to a variable.

```
D(p,x)
```

$$(8\ x - 12)\ y^2 + (8\ x - 12)\ y + 56\ x - 84 \qquad (3)$$

Type: Polynomial Integer

```
D(fp,x)
```

$$4\ (2\ x - 3)\ \left(y^2 + y + 7\right) \qquad (4)$$

Type: Factored Polynomial Integer

```
numberOfFactors(%)
```

$$3 \qquad (5)$$

Type: PositiveInteger

9.20 FactoredFunctions2

The FactoredFunctions2 package implements one operation, **map**, for applying an operation to every base in a factored object and to the unit.

```
double(x) == x + x
```

Type: Void

```
f := factor(720)
```

$$2^4\ 3^2\ 5 \qquad (2)$$

Type: Factored Integer

Actually, the **map** operation used in this example comes from Factored itself, since **double** takes an integer argument and returns an integer result.

```
map(double,f)
```

$$2\ 4^4\ 6^2\ 10 \qquad (3)$$

Type: Factored Integer

If we want to use an operation that returns an object that has a type different from the operation's argument, the **map** in Factored cannot be used and we use the one in FactoredFunctions2.

```
makePoly(b) == x + b
```

Type: Void

In fact, the "2" in the name of the package means that we might be using factored objects of two different types.

```
g := map(makePoly,f)
```

$$(x+1)\,(x+2)^4\,(x+3)^2\,(x+5) \qquad (5)$$

Type: Factored Polynomial Integer

It is important to note that both versions of **map** destroy any information known about the bases (the fact that they are prime, for instance).

The flags for each base are set to "nil" in the object returned by **map**.

```
nthFlag(g,1)
```

"nil" (6)

Type: Union("nil", ...)

For more information about factored objects and their use, see 'Factored' on page 356 and Section 8.13 on page 296.

9.21 File

The File(S) domain provides a basic interface to read and write values of type S in files.

Before working with a file, it must be made accessible to AXIOM with the **open** operation.

```
ifile:File List Integer:=open("/tmp/jazz1","output")
```

"/tmp/jazz1" (1)

Type: File List Integer

The **open** function arguments are a FileName and a String specifying the mode. If a full pathname is not specified, the current default directory is assumed. The mode must be one of `"input"` or `"output"`. If it is not specified, `"input"` is assumed. Once the file has been opened, you can read or write data.

The operations **read!** and **write!** are provided.

```
write!(ifile, [-1,2,3])
```

$$[-1,\ 2,\ 3] \qquad (2)$$

Type: List Integer

```
write!(ifile, [10,-10,0,111])
```

$$[10,\ -10,\ 0,\ 111] \qquad (3)$$

Type: List Integer

```
write!(ifile, [7])
```

$[7]$ (4)

Type: List Integer

You can change from writing to reading (or vice versa) by reopening a file.

```
reopen!(ifile, "input")
```

"/tmp/jazz1" (5)

Type: File List Integer

```
read! ifile
```

$[-1, 2, 3]$ (6)

Type: List Integer

```
read! ifile
```

$[10, -10, 0, 111]$ (7)

Type: List Integer

The **read!** operation can cause an error if one tries to read more data than is in the file. To guard against this possibility the **readIfCan!** operation should be used.

```
readIfCan! ifile
```

$[7]$ (8)

Type: Union(List Integer, ...)

```
readIfCan! ifile
```

"failed" (9)

Type: Union("failed", ...)

You can find the current mode of the file, and the file's name.

```
iomode ifile
```

"input" (10)

Type: String

```
name ifile
```

"/tmp/jazz1" (11)

Type: FileName

When you are finished with a file, you should close it.

```
close! ifile
```

"/tmp/jazz1" (12)

Type: File List Integer

```
)system rm /tmp/jazz1
```

A limitation of the underlying LISP system is that not all values can be represented in a file. In particular, delayed values containing compiled functions cannot be saved.

For more information on related topics, see 'TextFile' on page 468, 'KeyedAccessFile' on page 390, 'Library' on page 393, and 'FileName' on page 364. Issue the system command `)show File` to display the full list of operations defined by File.

9.22 FileName

The FileName domain provides an interface to the computer's file system. Functions are provided to manipulate file names and to test properties of files.

The simplest way to use file names in the AXIOM interpreter is to rely on conversion to and from strings. The syntax of these strings depends on the operating system.

```
fn: FileName
```

Type: Void

On AIX, this is a proper file syntax:

```
fn := "/spad/src/input/fname.input"
```

"/spad/src/input/fname.input" (2)

Type: FileName

Although it is very convenient to be able to use string notation for file names in the interpreter, it is desirable to have a portable way of creating and manipulating file names from within programs.

A measure of portability is obtained by considering a file name to consist of three parts: the *directory*, the *name*, and the *extension*.

```
directory fn
```

"/spad/src/input" (3)

Type: String

```
name fn
```

"fname" (4)

Type: String

```
extension fn
```

"input" (5)

Type: String

The meaning of these three parts depends on the operating system. For example, on CMS the file "SPADPROF INPUT M" would have directory "M", name "SPADPROF" and extension "INPUT".

It is possible to create a filename from its parts.

```
fn := filename("/u/smwatt/work", "fname", "input")
```

"/u/smwatt/work/fname.input" (6)

Type: FileName

When writing programs, it is helpful to refer to directories via variables.

```
objdir := "/tmp"
```

"/tmp" (7)

Type: String

```
fn := filename(objdir, "table", "spad")
```

"/tmp/table.spad" (8)

Type: FileName

If the directory or the extension is given as an empty string, then a default is used. On AIX, the defaults are the current directory and no extension.

```
fn := filename("", "letter", "")
```

"letter" (9)

Type: FileName

Three tests provide information about names in the file system.

The **exists?** operation tests whether the named file exists.

```
exists? "/etc/passwd"
```

true (10)

Type: Boolean

The operation **readable?** tells whether the named file can be read. If the file does not exist, then it cannot be read.

```
readable? "/etc/passwd"
```

true (11)

Type: Boolean

```
readable? "/etc/security/passwd"
```

false (12)

Type: Boolean

```
readable? "/ect/passwd"
```

false (13)

Type: Boolean

Likewise, the operation **writable?** tells whether the named file can be written. If the file does not exist, the test is determined by the properties of the directory.

```
writable? "/etc/passwd"
```

false (14)

Type: Boolean

```
writable? "/dev/null"
```

true (15)

Type: Boolean

```
writable? "/etc/DoesNotExist"
```

false (16)

Type: Boolean

```
writable? "/tmp/DoesNotExist"
```

true (17)

Type: Boolean

The **new** operation constructs the name of a new writable file. The argument sequence is the same as for **filename**, except that the name part is actually a prefix for a constructed unique name.

The resulting file is in the specified directory with the given extension, and the same defaults are used.

```
fn := new(objdir, "xxx", "yy")
```

"/tmp/xxx00007.yy" (18)

Type: FileName

9.23 FlexibleArray

The FlexibleArray domain constructor creates one-dimensional arrays of elements of the same type. Flexible arrays are an attempt to provide a data type that has the best features of both one-dimensional arrays (fast, random access to elements) and lists (flexibility). They are implemented by a fixed block of storage. When necessary for expansion, a new, larger block of storage is allocated and the elements from the old storage area are copied into the new block.

Flexible arrays have available most of the operations provided by OneDimensionalArray (see 'OneDimensionalArray' on page 425 and 'Vector' on page 478). Since flexible arrays are also of category ExtensibleLinearAggregate, they have operations **concat!**, **delete!**, **insert!**, **merge!**, **remove!**, **removeDuplicates!**, and **select!**. In addition, the operations **physicalLength** and **physicalLength!** provide user-control over expansion and contraction.

A convenient way to create a flexible array is to apply the operation **flexibleArray** to a list of values.

```
flexibleArray [i for i in 1..6]
```

$[1, 2, 3, 4, 5, 6]$ (1)

Type: FlexibleArray PositiveInteger

Create a flexible array of six zeroes.

```
f : FARRAY INT := new(6,0)
```

$[0, 0, 0, 0, 0, 0]$ (2)

Type: FlexibleArray Integer

For $i = 1 \ldots 6$, set the ith element to i. Display `f`.

```
for i in 1..6 repeat f.i := i; f
```

$[1, 2, 3, 4, 5, 6]$ (3)

Type: FlexibleArray Integer

Initially, the physical length is the same as the number of elements.

```
physicalLength f
```

6 (4)

Type: PositiveInteger

Add an element to the end of f.

```
concat!(f,11)
```

[1, 2, 3, 4, 5, 6, 11] (5)

Type: FlexibleArray Integer

See that its physical length has grown.

```
physicalLength f
```

10 (6)

Type: PositiveInteger

Make f grow to have room for 15 elements.

```
physicalLength!(f,15)
```

[1, 2, 3, 4, 5, 6, 11] (7)

Type: FlexibleArray Integer

Concatenate the elements of f to itself. The physical length allows room for three more values at the end.

```
concat!(f,f)
```

[1, 2, 3, 4, 5, 6, 11, 1, 2, 3, 4, 5, 6, 11] (8)

Type: FlexibleArray Integer

Use **insert!** to add an element to the front of a flexible array.

```
insert!(22,f,1)
```

[22, 1, 2, 3, 4, 5, 6, 11, 1, 2, 3, 4, 5, 6, 11] (9)

Type: FlexibleArray Integer

Create a second flexible array from f consisting of the elements from index 10 forward.

```
g := f(10..)
```

[2, 3, 4, 5, 6, 11] (10)

Type: FlexibleArray Integer

Insert this array at the front of f.

```
insert!(g,f,1)
```

[2, 3, 4, 5, 6, 11, 22, 1, 2, 3, 4, 5, 6, 11, 1, 2, 3, 4, 5, 6, 11] (11)

Type: FlexibleArray Integer

Merge the flexible array f into g after sorting each in place.

```
merge!(sort! f, sort! g)
```

[1, 1, 2, 2, 2, 2, 3, 3, 3, 3, 4, 4, 4, 4, 5, 5, 5, 5, 6, 6, 6, 6, 11, 11, 11, 11, 22] (12)

Type: FlexibleArray Integer

Remove duplicates in place.

```
removeDuplicates! f
```

[1, 2, 3, 4, 5, 6, 11, 22] (13)

Type: FlexibleArray Integer

Remove all odd integers.

```
select!(i +-> even? i,f)
```

[2, 4, 6, 22] (14)

Type: FlexibleArray Integer

All these operations have shrunk the physical length of `f`.

```
physicalLength f
```

8 (15)

Type: PositiveInteger

To force AXIOM not to shrink flexible arrays call the **shrinkable** operation with the argument `false`. You must package call this operation. The previous value is returned.

```
shrinkable(false)$FlexibleArray(Integer)
```

true (16)

Type: Boolean

9.24 Float

AXIOM provides two kinds of floating point numbers. The domain Float (abbreviation FLOAT) implements a model of arbitrary precision floating point numbers. The domain SmallFloat (abbreviation SF) is intended to make available hardware floating point arithmetic in AXIOM. The actual model of floating point that SmallFloat provides is system-dependent. For example, on the IBM system 370 AXIOM uses IBM double precision which has fourteen hexadecimal digits of precision or roughly sixteen decimal digits. Arbitrary precision floats allow the user to specify the precision at which arithmetic operations are computed. Although this is an attractive facility, it comes at a cost. Arbitrary-precision floating-point arithmetic typically takes twenty to two hundred times more time than hardware floating point.

For more information about AXIOM's numeric and graphic facilities, see Section 7 on page 179, Section 8.1 on page 227, and 'SmallFloat' on page 452.

9.24.1 Introduction to Float

Scientific notation is supported for input and output of floating point numbers. A floating point number is written as a string of digits containing a decimal point optionally followed by the letter "`E`", and then the exponent.

We begin by doing some calculations using arbitrary precision floats. The default precision is twenty decimal digits.

```
1.234
```

1.234 (1)

Type: Float

A decimal base for the exponent is assumed, so the number `1.234E2` denotes $1.234 \cdot 10^2$.

```
1.234E2
```

123.4 (2)

Type: Float

The normal arithmetic operations are available for floating point numbers.

```
sqrt(1.2 + 2.3 / 3.4 ** 4.5)
```

1.0996972790671286226 (3)

Type: Float

9.24.2 Conversion Functions

You can use conversion (Section 2.7 on page 78) to go back and forth between Integer, Fraction Integer and Float, as appropriate.

```
i := 3 :: Float
```

3.0 (1)

Type: Float

```
i :: Integer
```

3 (2)

Type: Integer

```
i :: Fraction Integer
```

3 (3)

Type: Fraction Integer

Since you are explicitly asking for a conversion, you must take responsibility for any loss of exactness.

```
r := 3/7 :: Float
```

0.42857142857142857143 (4)

Type: Float

```
r :: Fraction Integer
```

$\frac{3}{7}$ (5)

Type: Fraction Integer

This conversion cannot be performed: use **truncate** or **round** if that is what you intend.

```
r :: Integer
```

```
Cannot convert from type Float to Integer for value
0.42857 14285 71428 57143
```

The operations **truncate** and **round** truncate ...

```
truncate 3.6
```

3.0 (6)

Type: Float

and round to the nearest integral Float respectively.

```
round 3.6
```

4.0 (7)

Type: Float

```
truncate(-3.6)
```

−3.0 (8)

Type: Float

```
round(-3.6)
```

−4.0 (9)

Type: Float

The operation **fractionPart** computes the fractional part of `x`, that is, `x - truncate x`.

```
fractionPart 3.6
```

0.6 (10)

Type: Float

The operation **digits** allows the user to set the precision. It returns the previous value it was using.

```
digits 40
```

20 (11)

Type: PositiveInteger

```
sqrt 0.2
```

0.4472135954999579392818347337462552470881 (12)

Type: Float

```
pi()$Float
```

3.141592653589793238462643383279502884197 (13)

Type: Float

The precision is only limited by the computer memory available. Calculations at 500 or more digits of precision are not difficult.

```
digits 500
```

40 (14)

Type: PositiveInteger

```
pi()$Float
```

3.1415926535897932384626433832795028841971693993751058209749
4459230781640628620899862803482534211706798214808651328230
6647093844609550582231725359408128481117450284102701938521
1055596446229489549303819644288109756659334461284756482337
8678316527120190914564856692346034861045432664821339360726
0249141273724587006606315588174881520920962829254091715364
3678925903600113305305488204665213841469519415116094330572
7036575959195309218611738193261179310511854807446237996274
9567351885752724891227938183011949 1 (15)

Type: Float

Reset **digits** to its default value.

```
digits 20
```

500 (16)

Type: PositiveInteger

Numbers of type Float are represented as a record of two integers, namely, the mantissa and the exponent where the base of the exponent is binary. That is, the floating point number `(m,e)` represents the number $m \cdot 2^e$. A consequence of using a binary base is that decimal numbers can not, in general, be represented exactly.

9.24.3 Output Functions

A number of operations exist for specifying how numbers of type Float are to be displayed. By default, spaces are inserted every ten digits in the output for readability.[3]

Output spacing can be modified with the **outputSpacing** operation. This inserts no spaces and then displays the value of `x`.

```
outputSpacing 0; x := sqrt 0.2
```

0.44721359549995793928 (1)

Type: Float

Issue this to have the spaces inserted every 5 digits.

```
outputSpacing 5; x
```

0.44721 35954 99957 93928 (2)

Type: Float

By default, the system displays floats in either fixed format or scientific format, depending on the magnitude of the number.

```
y := x/10**10
```

0.44721 35954 99957 93928 E -10 (3)

Type: Float

A particular format may be requested with the operations **outputFloating** and **outputFixed**.

```
outputFloating(); x
```

0.44721 35954 99957 93928 E 0 (4)

Type: Float

```
outputFixed(); y
```

0. 00000 0000 0 447 21 35 954 9 9957 93928 (5)

Type: Float

Additionally, you can ask for `n` digits to be displayed after the decimal point.

```
outputFloating 2; y
```

0.45 E -10 (6)

Type: Float

[3]Note that you cannot include spaces in the input form of a floating point number, though you can use underscores.

```
outputFixed 2; x
```

$$0.45 \tag{7}$$

Type: Float

This resets the output printing to the default behavior.

```
outputGeneral()
```

Type: Void

9.24.4 An Example: Determinant of a Hilbert Matrix

Consider the problem of computing the determinant of a `10` by `10` Hilbert matrix. The `(i,j)` th entry of a Hilbert matrix is given by `1/(i+j+1)`.

First do the computation using rational numbers to obtain the exact result.

```
a: Matrix Fraction Integer := matrix [[1/(i+j+1) for j in
  0..9] for i in 0..9]
```

$$\begin{bmatrix} 1 & \frac{1}{2} & \frac{1}{3} & \frac{1}{4} & \frac{1}{5} & \frac{1}{6} & \frac{1}{7} & \frac{1}{8} & \frac{1}{9} & \frac{1}{10} \\ \frac{1}{2} & \frac{1}{3} & \frac{1}{4} & \frac{1}{5} & \frac{1}{6} & \frac{1}{7} & \frac{1}{8} & \frac{1}{9} & \frac{1}{10} & \frac{1}{11} \\ \frac{1}{3} & \frac{1}{4} & \frac{1}{5} & \frac{1}{6} & \frac{1}{7} & \frac{1}{8} & \frac{1}{9} & \frac{1}{10} & \frac{1}{11} & \frac{1}{12} \\ \frac{1}{4} & \frac{1}{5} & \frac{1}{6} & \frac{1}{7} & \frac{1}{8} & \frac{1}{9} & \frac{1}{10} & \frac{1}{11} & \frac{1}{12} & \frac{1}{13} \\ \frac{1}{5} & \frac{1}{6} & \frac{1}{7} & \frac{1}{8} & \frac{1}{9} & \frac{1}{10} & \frac{1}{11} & \frac{1}{12} & \frac{1}{13} & \frac{1}{14} \\ \frac{1}{6} & \frac{1}{7} & \frac{1}{8} & \frac{1}{9} & \frac{1}{10} & \frac{1}{11} & \frac{1}{12} & \frac{1}{13} & \frac{1}{14} & \frac{1}{15} \\ \frac{1}{7} & \frac{1}{8} & \frac{1}{9} & \frac{1}{10} & \frac{1}{11} & \frac{1}{12} & \frac{1}{13} & \frac{1}{14} & \frac{1}{15} & \frac{1}{16} \\ \frac{1}{8} & \frac{1}{9} & \frac{1}{10} & \frac{1}{11} & \frac{1}{12} & \frac{1}{13} & \frac{1}{14} & \frac{1}{15} & \frac{1}{16} & \frac{1}{17} \\ \frac{1}{9} & \frac{1}{10} & \frac{1}{11} & \frac{1}{12} & \frac{1}{13} & \frac{1}{14} & \frac{1}{15} & \frac{1}{16} & \frac{1}{17} & \frac{1}{18} \\ \frac{1}{10} & \frac{1}{11} & \frac{1}{12} & \frac{1}{13} & \frac{1}{14} & \frac{1}{15} & \frac{1}{16} & \frac{1}{17} & \frac{1}{18} & \frac{1}{19} \end{bmatrix} \tag{1}$$

Type: Matrix Fraction Integer

This version of **determinant** uses Gaussian elimination.

```
d:= determinant a
```

$$\frac{1}{46206893947914691316295628839036278726983680000000000} \tag{2}$$

Type: Fraction Integer

```
d :: Float
```

$$0.21641792264314918691E - 52 \tag{3}$$

Type: Float

Now use hardware floats. Note that a semicolon (;) is used to prevent the display of the matrix.

```
b: Matrix SmallFloat := matrix [[1/(i+j+1$SmallFloat) for
  j in 0..9] for i in 0..9];
```

$$\tag{4}$$

Type: Matrix SmallFloat

The result given by hardware floats is correct only to four significant digits of precision. In the jargon of numerical analysis, the Hilbert matrix is said to be "ill-conditioned."

```
determinant b
```

$$2.1643677945721411E - 53 \tag{5}$$

Type: SmallFloat

Now repeat the computation at a higher precision using Float.

```
digits 40
```

$$20 \tag{6}$$

Type: PositiveInteger

```
c: Matrix Float := matrix [[1/(i+j+1$Float) for j in 0..9]
  for i in 0..9];
```

(7)

Type: Matrix Float

```
determinant c
```

$$0.2164179226431491869060594983622617436159E - 52 \tag{8}$$

Type: Float

Reset **digits** to its default value.

```
digits 20
```

$$40 \tag{9}$$

Type: PositiveInteger

9.25 Fraction

The Fraction domain implements quotients. The elements must belong to a domain of category IntegralDomain: multiplication must be commutative and the product of two non-zero elements must not be zero. This allows you to make fractions of most things you would think of, but don't expect to create a fraction of two matrices! The abbreviation for Fraction is FRAC.

Use "/" to create a fraction.

```
a := 11/12
```

$$\frac{11}{12} \tag{1}$$

Type: Fraction Integer

```
b := 23/24
```

$$\frac{23}{24} \tag{2}$$

Type: Fraction Integer

The standard arithmetic operations are available.

```
3 - a*b**2 + a + b/a
```

$$\frac{313271}{76032} \tag{3}$$

Type: Fraction Integer

Extract the numerator and denominator by using **numer** and **denom**, respectively.

```
numer(a)
```

$$11 \tag{4}$$

Type: PositiveInteger

```
denom(b)
```

$$24 \tag{5}$$

Type: PositiveInteger

Operations like **max**, **min**, **negative?**, **positive?** and **zero?** are all available if they are provided for the numerators and denominators. See 'Integer' on page 380 for examples.

Don't expect a useful answer from **factor**, **gcd** or **lcm** if you apply them to fractions.

```
r := (x**2 + 2*x + 1)/(x**2 - 2*x + 1)
```

$$\frac{x^2 + 2\,x + 1}{x^2 - 2\,x + 1} \tag{6}$$

Type: Fraction Polynomial Integer

Since all non-zero fractions are invertible, these operations have trivial definitions.

```
factor(r)
```

$$\frac{x^2 + 2\,x + 1}{x^2 - 2\,x + 1} \tag{7}$$

Type: Factored Fraction Polynomial Integer

Use **map** to apply **factor** to the numerator and denominator, which is probably what you mean.

```
map(factor,r)
```

$$\frac{(x+1)^2}{(x-1)^2} \tag{8}$$

Type: Fraction Factored Polynomial Integer

Other forms of fractions are available. Use **continuedFraction** to create a continued fraction.

```
continuedFraction(7/12)
```

$$\frac{1|}{|1} + \frac{1|}{|1} + \frac{1|}{|2} + \frac{1|}{|2} \tag{9}$$

Type: ContinuedFraction Integer

Use **partialFraction** to create a partial fraction. See 'ContinuedFraction' on page 335 and 'PartialFraction' on page 433 for additional information and examples.

```
partialFraction(7,12)
```

$$1 - \frac{3}{2^2} + \frac{1}{3} \tag{10}$$

Type: PartialFraction Integer

Use conversion to create alternative views of fractions with objects moved in and out of the numerator and denominator.

```
g := 2/3 + 4/5*%i
```

$$\frac{2}{3} + \frac{4}{5}\,\%i \tag{11}$$

Type: Complex Fraction Integer

Conversion is discussed in detail in Section 2.7 on page 78.

```
g :: FRAC COMPLEX INT
```

$$\frac{10 + 12\,\%i}{15} \qquad (12)$$

Type: Fraction Complex Integer

9.26 GeneralSparseTable

Sometimes when working with tables there is a natural value to use as the entry in all but a few cases. The GeneralSparseTable constructor can be used to provide any table type with a default value for entries. See 'Table' on page 465 for general information about tables. Issue the system command `)show GeneralSparseTable` to display the full list of operations defined by GeneralSparseTable.

Suppose we launch a fund-raising campaign to raise fifty thousand dollars. To record the contributions, we want a table with strings as keys (for the names) and integer entries (for the amount). In a data base of cash contributions, unless someone has been explicitly entered, it is reasonable to assume they have made a zero dollar contribution.

This creates a keyed access file with default entry 0.

```
patrons: GeneralSparseTable(String, Integer,
  KeyedAccessFile(Integer), 0) := table()
```

"kaf00056.sdata" (1)

Type: GeneralSparseTable(String, Integer, KeyedAccessFile Integer, 0)

Now `patrons` can be used just as any other table. Here we record two gifts.

```
patrons."Smith" := 10500
```

10500 (2)

Type: PositiveInteger

```
patrons."Jones" := 22000
```

22000 (3)

Type: PositiveInteger

Now let us look up the size of the contributions from Jones and Stingy.

```
patrons."Jones"
```

22000 (4)

Type: PositiveInteger

```
patrons."Stingy"
```

0 (5)

Type: NonNegativeInteger

Have we met our seventy thousand dollar goal?

```
reduce(+, entries patrons)
```

32500 (6)

Type: PositiveInteger

So the project is cancelled and we can delete the data base:

```
)system rm -r kaf*.sdata
```

9.27 Groebner-Factorization-Package

Solving systems of polynomial equations with the Gröbner basis algorithm can often be very time consuming because, in general, the algorithm has exponential run-time. These systems, which often come from concrete applications, frequently have symmetries which are not taken advantage of by the algorithm. However, it often happens in this case that the polynomials which occur during the Gröbner calculations are reducible. Since AXIOM has an excellent polynomial factorization algorithm, it is very natural to combine the Gröbner and factorization algorithms.

GroebnerFactorizationPackage exports the **groebnerFactorize** operation which implements a modified Gröbner basis algorithm. In this algorithm, each polynomial that is to be put into the partial list of the basis is first factored. The remaining calculation is split into as many parts as there are irreducible factors. Call these factors $p_1, \ldots, p_n$. In the branches corresponding to $p_2, \ldots, p_n$, the factor p_1 can be divided out, and so on. This package also contains operations that allow you to specify the polynomials that are not zero on the common roots of the final Gröbner basis.

Here is an example from chemistry. In a theoretical model of the cyclohexan C_6H_{12}, the six carbon atoms each sit in the center of gravity of a tetrahedron that has two hydrogen atoms and two carbon atoms at its corners. We first normalize and set the length of each edge to 1. Hence, the distances of one fixed carbon atom to each of its immediate neighbours is 1. We will denote the distances to the other three carbon atoms by x, y and z.

A. Dress developed a theory to decide whether a set of points and distances between them can be realized in an n-dimensional space. Here, of course, we have $n = 3$.

```
mfzn : SQMATRIX(6,DMP([x,y,z],Fraction INT)) :=
  [[0,1,1,1,1,1], [1,0,1,8/3,x,8/3], [1,1,0,1,8/3,y],
  [1,8/3,1,0,1,8/3], [1,x,8/3,1,0,1], [1,8/3,y,8/3,1,0]]
```

$$\begin{bmatrix} 0 & 1 & 1 & 1 & 1 & 1 \\ 1 & 0 & 1 & \frac{8}{3} & x & \frac{8}{3} \\ 1 & 1 & 0 & 1 & \frac{8}{3} & y \\ 1 & \frac{8}{3} & 1 & 0 & 1 & \frac{8}{3} \\ 1 & x & \frac{8}{3} & 1 & 0 & 1 \\ 1 & \frac{8}{3} & y & \frac{8}{3} & 1 & 0 \end{bmatrix} \quad (1)$$

Type: SquareMatrix(6, DistributedMultivariatePolynomial([x, y, z], Fraction Integer))

For the cyclohexan, the distances have to satisfy this equation.

```
eq := determinant mfzn
```

$$-x^2\,y^2+\frac{22}{3}\,x^2\,y-\frac{25}{9}\,x^2+\frac{22}{3}\,x\,y^2-\frac{388}{9}\,x\,y- \frac{250}{27}\,x-\frac{25}{9}\,y^2-\frac{250}{27}\,y+\frac{14575}{81} \tag{2}$$

Type: DistributedMultivariatePolynomial([x, y, z], Fraction Integer)

They also must satisfy the equations given by cyclic shifts of the indeterminates.

```
groebnerFactorize [eq, eval(eq, [x,y,z], [y,z,x]),
  eval(eq, [x,y,z], [z,x,y])]
```

$$\left[\left[\left(y+x-\frac{22}{3}\right)z+\left(x-\frac{22}{3}\right)y-\frac{22}{3}\,x+\frac{121}{3},\right.\right.$$
$$\left(x^2-\frac{22}{3}\,x+\frac{25}{9}\right)z+\left(x^2-\frac{22}{3}\,x+\frac{25}{9}\right)y-\frac{22}{3}\,x^2+ \frac{388}{9}\,x+\frac{250}{27},$$
$$\left.\left(x^2-\frac{22}{3}\,x+\frac{25}{9}\right)y^2+\left(-\frac{22}{3}\,x^2+\frac{388}{9}\,x+\frac{250}{27}\right)y+ \frac{25}{9}\,x^2+\frac{250}{27}\,x-\frac{14575}{81}\right], \tag{3}$$

$$\left[z+y-\frac{21994}{5625},\ y^2-\frac{21994}{5625}\,y+\frac{4427}{675},\ x-\frac{463}{87}\right],$$
$$\left[z^2+\left(-\frac{1}{2}\,x-\frac{11}{2}\right)z-\frac{5}{6}\,x+\frac{265}{18},\ y-x,\ x^2-\frac{38}{3}\,x+\frac{265}{9}\right],$$
$$\left[z-\frac{25}{9},\ y-\frac{11}{3},\ x-\frac{11}{3}\right],\ \left[z-\frac{11}{3},\ y-\frac{11}{3},\ x-\frac{11}{3}\right],$$
$$\left.\left[z+\frac{5}{3},\ y+\frac{5}{3},\ x+\frac{5}{3}\right],\ \left[z-\frac{19}{3},\ y+\frac{5}{3},\ x+\frac{5}{3}\right]\right]$$

Type: List List Polynomial Fraction Integer

The union of the solutions of this list is the solution of our original problem. If we impose positivity conditions, we get two relevant ideals. One ideal is zero-dimensional, namely $x = y = z = 11/3$, and this determines the "boat" form of the cyclohexan. The other ideal is one-dimensional, which means that we have a solution space given by one parameter. This gives the "chair" form of the cyclohexan. The parameter describes the angle of the "back of the chair."

groebnerFactorize has an optional Boolean-valued second argument. When it is `true` partial results are displayed, since it may happen that the calculation does not terminate in a reasonable time. See the source code for GroebnerFactorizationPackage in **groebf.spad** for more details about the algorithms used.

9.28 Heap

The domain Heap(S) implements a priority queue of objects of type S such that the operation **extract!** removes and returns the maximum element. The implementation represents heaps as flexible arrays (see 'FlexibleArray' on page 366). The representation and algorithms give complexity of $O(\log(n))$ for insertion and extractions, and $O(n)$ for construction.

Create a heap of six elements.

```
h := heap [-4,9,11,2,7,-7]
```

$$[11, 7, 9, -4, 2, -7] \tag{1}$$

Type: Heap Integer

Use **insert!** to add an element.

```
insert!(3,h)
```

$$[11, 7, 9, -4, 2, -7, 3] \tag{2}$$

Type: Heap Integer

The operation **extract!** removes and returns the maximum element.

```
extract! h
```

$$11 \tag{3}$$

Type: PositiveInteger

The internal structure of h has been appropriately adjusted.

```
h
```

$$[9, 7, 3, -4, 2, -7] \tag{4}$$

Type: Heap Integer

Now **extract!** elements repeatedly until none are left, collecting the elements in a list.

```
[extract!(h) while not empty?(h)]
```

$$[9, 7, 3, 2, -4, -7] \tag{5}$$

Type: List Integer

Another way to produce the same result is by defining a **heapsort** function.

```
heapsort(x) == (empty? x => []; cons(extract!(x),heapsort
  x))
```

Type: Void

Create another sample heap.

```
h1 := heap [17,-4,9,-11,2,7,-7]
```

$$[17, 2, 9, -11, -4, 7, -7] \tag{7}$$

Type: Heap Integer

Apply **heapsort** to present elements in order.

```
heapsort h1
```

$$[17, 9, 7, 2, -4, -7, -11] \qquad (8)$$

Type: List Integer

9.29 HexadecimalExpansion

All rationals have repeating hexadecimal expansions. The operation **hex** returns these expansions of type HexadecimalExpansion. Operations to access the individual numerals of a hexadecimal expansion can be obtained by converting the value to RadixExpansion(16). More examples of expansions are available in the 'DecimalExpansion' on page 350, 'BinaryExpansion' on page 312, and 'RadixExpansion' on page 444.

Issue the system command `)show HexadecimalExpansion` to display the full list of operations defined by HexadecimalExpansion.

This is a hexadecimal expansion of a rational number.

```
r := hex(22/7)
```

$$3.\overline{249} \qquad (1)$$

Type: HexadecimalExpansion

Arithmetic is exact.

```
r + hex(6/7)
```

$$4 \qquad (2)$$

Type: HexadecimalExpansion

The period of the expansion can be short or long . . .

```
[hex(1/i) for i in 350..354]
```

$$[0.0\overline{0BB3EE721A54D88},\ 0.\overline{00\ BAB\ 6561},\ 0.00\overline{BA2E8},$$

$$0.\overline{00B9A7862A0FF465879D5F}, \qquad (3)$$

$$0.0\overline{0B92143FA36F5E02E4850FE8\ DBD78}]$$

Type: List HexadecimalExpansion

or very long!

```
hex(1/1007)
```

$$0.\overline{0041149783F0BF2C7D13933192AF6980619EE345E91E C2BB9D5CCA5C071E40926E54E8DDAE24196C0B2F8A0AAD 60DBA57F5D4C8536262210C74F1} \qquad (4)$$

Type: HexadecimalExpansion

These numbers are bona fide algebraic objects.

```
p := hex(1/4)*x**2 + hex(2/3)*x + hex(4/9)
```

$$0.4\ x^2 + 0.\overline{10}\ x + 0.\overline{71C} \qquad (5)$$

Type: Polynomial HexadecimalExpansion

```
q := D(p, x)
```

$$0.8\,x + 0.\overline{10} \tag{6}$$

Type: Polynomial HexadecimalExpansion

```
g := gcd(p, q)
```

$$x + 1.\overline{5} \tag{7}$$

Type: Polynomial HexadecimalExpansion

9.30 Integer

AXIOM provides many operations for manipulating arbitrary precision integers. In this section we will show some of those that come from Integer itself plus some that are implemented in other packages. More examples of using integers are in the following sections: 'Some Numbers' in Section 1.5 on page 29, 'IntegerNumberTheoryFunctions' on page 387, 'DecimalExpansion' on page 350, 'BinaryExpansion' on page 312, 'HexadecimalExpansion' on page 379, and 'RadixExpansion' on page 444.

9.30.1 Basic Functions

The size of an integer in AXIOM is only limited by the amount of computer storage you have available. The usual arithmetic operations are available.

```
2**(5678 - 4856 + 2 * 17)
```

$$\begin{array}{l}480481077043500814718154092512592439123952613987168226347 38\\ 556100880842000763082930863425270914120837430745722782114 96\\ 076276922026433435687527334980249539302425425230458177649 49\\ 544214392905306388478705146745768073877141698859815495632 93\\ 5288783334250628775936\end{array} \tag{1}$$

Type: PositiveInteger

There are a number of ways of working with the sign of an integer. Let's use this **x** as an example.

```
x := -101
```

$$-101 \tag{2}$$

Type: Integer

First of all, there is the absolute value function.

```
abs(x)
```

$$101 \tag{3}$$

Type: PositiveInteger

The **sign** operation returns -1 if its argument is negative, 0 if zero and 1 if positive.

```
sign(x)
```

$$-1 \tag{4}$$

Type: Integer

You can determine if an integer is negative in several other ways.

```
x < 0
```

true (5)

Type: Boolean

```
x <= -1
```

true (6)

Type: Boolean

```
negative?(x)
```

true (7)

Type: Boolean

Similarly, you can find out if it is positive.

```
x > 0
```

false (8)

Type: Boolean

```
x >= 1
```

false (9)

Type: Boolean

```
positive?(x)
```

false (10)

Type: Boolean

This is the recommended way of determining whether an integer is zero.

```
zero?(x)
```

false (11)

Type: Boolean

> Use the **zero?** operation whenever you are testing any mathematical object for equality with zero. This is usually more efficient that using "=" (think of matrices: it is easier to tell if a matrix is zero by just checking term by term than constructing another "zero" matrix and comparing the two matrices term by term) and also avoids the problem that "=" is usually used for creating equations.

This is the recommended way of determining whether an integer is equal to one.

```
one?(x)
```

false (12)

Type: Boolean

This syntax is used to test equality using "=". It says that you want a Boolean (`true` or `false`) answer rather than an equation.

```
(x = -101)@Boolean
```

true (13)

Type: Boolean

The operations **odd?** and **even?** determine whether an integer is odd or even, respectively. They each return a Boolean object.

```
odd?(x)
```

true (14)

Type: Boolean

```
even?(x)
```

false (15)

Type: Boolean

The operation **gcd** computes the greatest common divisor of two integers.

```
gcd(56788,43688)
```

4 (16)

Type: PositiveInteger

The operation **lcm** computes their least common multiple.

```
lcm(56788,43688)
```

620238536 (17)

Type: PositiveInteger

To determine the maximum of two integers, use **max**.

```
max(678,567)
```

678 (18)

Type: PositiveInteger

To determine the minimum, use **min**.

```
min(678,567)
```

567 (19)

Type: PositiveInteger

The **reduce** operation is used to extend binary operations to more than two arguments. For example, you can use **reduce** to find the maximum integer in a list or compute the least common multiple of all integers in the list.

```
reduce(max,[2,45,-89,78,100,-45])
```

100 (20)

Type: PositiveInteger

```
reduce(min,[2,45,-89,78,100,-45])
```

-89 (21)

Type: Integer

```
reduce(gcd,[2,45,-89,78,100,-45])
```

1 (22)

Type: PositiveInteger

```
reduce(lcm,[2,45,-89,78,100,-45])
```

$$1041300 \tag{23}$$

Type: PositiveInteger

The infix operator "/" is *not* used to compute the quotient of integers. Rather, it is used to create rational numbers as described in 'Fraction' on page 373.

```
13 / 4
```

$$\frac{13}{4} \tag{24}$$

Type: Fraction Integer

The infix operation **quo** computes the integer quotient.

```
13 quo 4
```

$$3 \tag{25}$$

Type: PositiveInteger

The infix operation **rem** computes the integer remainder.

```
13 rem 4
```

$$1 \tag{26}$$

Type: PositiveInteger

One integer is evenly divisible by another if the remainder is zero. The operation **exquo** can also be used. See Section 2.5 on page 73 for an example.

```
zero?(167604736446952 rem 2003644)
```

$$\text{true} \tag{27}$$

Type: Boolean

The operation **divide** returns a record of the quotient and remainder and thus is more efficient when both are needed.

```
d := divide(13,4)
```

$$[quotient = 3,\ remainder = 1] \tag{28}$$

Type: Record(quotient: Integer, remainder: Integer)

```
d.quotient
```

$$3 \tag{29}$$

Type: PositiveInteger

Records are discussed in detail in Section 2.4 on page 71.

```
d.remainder
```

$$1 \tag{30}$$

Type: PositiveInteger

9.30.2 Primes and Factorization

Use the operation **factor** to factor integers. It returns an object of type Factored Integer. See 'Factored' on page 356 for a discussion of the manipulation of factored objects.

```
factor 102400
```

$$2^{12}\ 5^{2} \tag{1}$$

Type: Factored Integer

The operation **prime?** returns `true` or `false` depending on whether its argument is a prime.

```
prime? 7
```

true (2)

Type: Boolean

```
prime? 8
```

false (3)

Type: Boolean

The operation **nextPrime** returns the least prime number greater than its argument.

```
nextPrime 100
```

101 (4)

Type: PositiveInteger

The operation **prevPrime** returns the greatest prime number less than its argument.

```
prevPrime 100
```

97 (5)

Type: PositiveInteger

To compute all primes between two integers (inclusively), use the operation **primes**.

```
primes(100,175)
```

$$[173, 167, 163, 157, 151, 149, 139, 137, 131, 127, 113, 109, 107, 103, 101] \quad (6)$$

Type: List Integer

You might sometimes want to see the factorization of an integer when it is considered a *Gaussian integer*. See 'Complex' on page 333 for more details.

```
factor(2 :: Complex Integer)
```

$$-\%i\ (1 + \%i)^2 \quad (7)$$

Type: Factored Complex Integer

9.30.3 Some Number Theoretic Functions

AXIOM provides several number theoretic operations for integers. More examples are in 'IntegerNumberTheoryFunctions' on page 387.

The operation **fibonacci** computes the Fibonacci numbers. The algorithm has running time $O(\log^3(n))$ for argument `n`.

```
[fibonacci(k) for k in 0..]
```

$$[0, 1, 1, 2, 3, 5, 8, \ldots] \quad (1)$$

Type: Stream Integer

The operation **legendre** computes the Legendre symbol for its two integer arguments where the second one is prime. If you know the second argument to be prime, use **jacobi** instead where no check is made.

```
[legendre(i,11) for i in 0..10]
```

$$[0, 1, -1, 1, 1, 1, -1, -1, -1, 1, -1] \quad (2)$$

Type: List Integer

The operation **jacobi** computes the Jacobi symbol for its two integer arguments. By convention, 0 is returned if the greatest common divisor of the numerator and denominator is not 1.

```
[jacobi(i,15) for i in 0..9]
```

$$[0,\ 1,\ 1,\ 0,\ 1,\ 0,\ 0,\ -1,\ 1,\ 0] \tag{3}$$

Type: List Integer

The operation **eulerPhi** computes the values of Euler's ϕ-function where $\phi(n)$ equals the number of positive integers less than or equal to n that are relatively prime to the positive integer n.

```
[eulerPhi i for i in 1..]
```

$$[1,\ 1,\ 2,\ 2,\ 4,\ 2,\ 6,\ \ldots] \tag{4}$$

Type: Stream Integer

The operation **moebiusMu** computes the Möbius μ function.

```
[moebiusMu i for i in 1..]
```

$$[1,\ -1,\ -1,\ 0,\ -1,\ 1,\ -1,\ \ldots] \tag{5}$$

Type: Stream Integer

Although they have somewhat limited utility, AXIOM provides Roman numerals.

```
a := roman(78)
```

LXXVIII (6)

Type: RomanNumeral

```
b := roman(87)
```

LXXXVII (7)

Type: RomanNumeral

```
a + b
```

CLXV (8)

Type: RomanNumeral

```
a * b
```

MMMMMMDCCLXXXVI (9)

Type: RomanNumeral

```
b rem a
```

IX (10)

Type: RomanNumeral

9.31 IntegerLinearDependence

The elements $v_1, \ldots, v_n$ of a module M over a ring R are said to be *linearly dependent over* R if there exist $c_1, \ldots, c_n$ in R, not all 0, such that $c_1 v_1 + \ldots c_n v_n = 0$. If such c_i's exist, they form what is called a *linear dependence relation over* R for the v_i's.

The package IntegerLinearDependence provides functions for testing whether some elements of a module over the integers are linearly dependent over the integers, and to find the linear dependence relations, if any.

Consider the domain of two by two square matrices with integer entries.

```
M := SQMATRIX(2,INT)
```

$$\text{SquareMatrix}(2, \text{Integer}) \quad (1)$$

Type: Domain

Now create three such matrices.

```
m1: M := squareMatrix matrix [[1, 2], [0, -1]]
```

$$\begin{bmatrix} 1 & 2 \\ 0 & -1 \end{bmatrix} \quad (2)$$

Type: SquareMatrix(2, Integer)

```
m2: M := squareMatrix matrix [[2, 3], [1, -2]]
```

$$\begin{bmatrix} 2 & 3 \\ 1 & -2 \end{bmatrix} \quad (3)$$

Type: SquareMatrix(2, Integer)

```
m3: M := squareMatrix matrix [[3, 4], [2, -3]]
```

$$\begin{bmatrix} 3 & 4 \\ 2 & -3 \end{bmatrix} \quad (4)$$

Type: SquareMatrix(2, Integer)

This tells you whether `m1`, `m2` and `m3` are linearly dependent over the integers.

```
linearlyDependentOverZ? vector [m1, m2, m3]
```

$$\text{true} \quad (5)$$

Type: Boolean

Since they are linearly dependent, you can ask for the dependence relation.

```
c := linearDependenceOverZ vector [m1, m2, m3]
```

$$[1, -2, 1] \quad (6)$$

Type: Union(Vector Integer, ...)

This means that the following linear combination should be `0`.

```
c.1 * m1 + c.2 * m2 + c.3 * m3
```

$$\begin{bmatrix} 0 & 0 \\ 0 & 0 \end{bmatrix} \quad (7)$$

Type: SquareMatrix(2, Integer)

When a given set of elements are linearly dependent over `R`, this also means that at least one of them can be rewritten as a linear combination of the others with coefficients in the quotient field of `R`.

To express a given element in terms of other elements, use the operation **solveLinearlyOverQ**.

```
solveLinearlyOverQ(vector [m1, m3], m2)
```

$$\left[\frac{1}{2}, \frac{1}{2}\right] \quad (8)$$

Type: Union(Vector Fraction Integer, ...)

9.32 IntegerNumberTheoryFunctions

The IntegerNumberTheoryFunctions package contains a variety of operations of interest to number theorists. Many of these operations deal with divisibility properties of integers. (Recall that an integer a divides an integer b if there is an integer c such that b = a * c.)

The operation **divisors** returns a list of the divisors of an integer.

```
div144 := divisors(144)
```

$$[1, 2, 3, 4, 6, 8, 9, 12, 16, 18, 24, 36, 48, 72, 144] \quad (1)$$

Type: List Integer

You can now compute the number of divisors of 144 and the sum of the divisors of 144 by counting and summing the elements of the list we just created.

```
#(div144)
```

$$15 \quad (2)$$

Type: PositiveInteger

```
reduce(+,div144)
```

$$403 \quad (3)$$

Type: PositiveInteger

Of course, you can compute the number of divisors of an integer n, usually denoted d(n), and the sum of the divisors of an integer n, usually denoted σ(n), without ever listing the divisors of n.

In AXIOM, you can simply call the operations **numberOfDivisors** and **sumOfDivisors**.

```
numberOfDivisors(144)
```

$$15 \quad (4)$$

Type: PositiveInteger

```
sumOfDivisors(144)
```

$$403 \quad (5)$$

Type: PositiveInteger

The key is that d(n) and σ(n) are "multiplicative functions." This means that when n and m are relatively prime, that is, when n and m have no prime factor in common, then d(nm) = d(n) d(m) and σ(nm) = σ(n) σ(m). Note that these functions are trivial to compute when n is a prime power and are computed for general n from the prime factorization of n. Other examples of multiplicative functions are σ_k(n), the sum of the k th powers of the divisors of n and $\varphi(n)$, the number of integers between 1 and n which are prime to n. The corresponding AXIOM operations are called **sumOfKthPowerDivisors** and **eulerPhi**.

An interesting function is μ(n), the Möbius μ function, defined as follows: μ(1) = 1, μ(n) = 0, when n is divisible by a square, and $\mu = (-1)^k$, when n is the product of k distinct primes. The corresponding AXIOM operation is **moebiusMu**. This function occurs in the following theorem:

Theorem (Möbius Inversion Formula):
Let f(n) be a function on the positive integers and let F(n) be defined by

$$F(n) = \sum_{d|n} f(n)$$

where the sum is taken over the positive divisors of n. Then the values of f(n) can be recovered from the values of F(n):

$$f(n) = \sum_{d|n} \mu(n) F(\frac{n}{d})$$

where again the sum is taken over the positive divisors of n.

When f(n) = 1, then F(n) = d(n). Thus, if you sum μ(d) · d(n/d) over the positive divisors d of n, you should always get 1.

```
f1(n)
  == reduce(+,[moebiusMu(d) * numberOfDivisors(quo(n,d))
  for d in divisors(n)])
```

Type: Void

```
f1(200)
```

1 (7)

Type: PositiveInteger

```
f1(846)
```

1 (8)

Type: PositiveInteger

Similarly, when f(n) = n, then F(n) = σ(n). Thus, if you sum μ(d) · σ(n/d) over the positive divisors d of n, you should always get n.

```
f2(n) == reduce(+,[moebiusMu(d) * sumOfDivisors(quo(n,d))
  for d in divisors(n)])
```

Type: Void

```
f2(200)
```

200 (10)

Type: PositiveInteger

```
f2(846)
```

846 (11)

Type: PositiveInteger

The Möbius inversion formula is derived from the multiplication of formal Dirichlet series. A Dirichlet series is an infinite series of the form

$$\sum_{n=1}^{\infty} a(n) n^{-s}$$

When

$$\sum_{n=1}^{\infty} a(n)n^{-s} \cdot \sum_{n=1}^{\infty} b(n)n^{-s} = \sum_{n=1}^{\infty} c(n)n^{-s}$$

then $c(n) = \sum_{d|n} a(d)b(n/d)$. Recall that the Riemann ζ function is defined by

$$\zeta(s) = \prod_p (1-p^{-s})^{-1} = \sigma_{n=1}^{\infty} n^{-s}$$

where the product is taken over the set of (positive) primes. Thus,

$$\zeta(s)^{-1} = \prod_p (1-p^{-s}) = \sigma_{n=1}^{\infty} \mu(n) n^{-s}$$

Now if $F(n) = \sum_{d|n)f(d)}$, then

$$\sum f(n)n^{-s} \cdot \zeta(s) = \sum F(n)n^{-s}$$

Thus,

$$\zeta(s)^{-1} \cdot \sum F(n)n^{-s} = \sum f(n)n^{-s}$$

and $f(n) = \sum_{d|n} \mu(d)F(n/d)$.

The Fibonacci numbers are defined by `F(1) = F(2) = 1` and `F(n) = F(n-1) + F(n-2)` for `n = 3,4, ....`

The operation **fibonacci** computes the n th Fibonacci number.

```
fibonacci(25)
```

75025 (12)

Type: PositiveInteger

```
[fibonacci(n) for n in 1..15]
```

[1, 1, 2, 3, 5, 8, 13, 21, 34, 55, 89, 144, 233, 377, 610] (13)

Type: List Integer

Fibonacci numbers can also be expressed as sums of binomial coefficients.

```
fib(n) == reduce(+,[binomial(n-1-k,k) for k in 0..quo(n-
  1,2)])
```

Type: Void

```
fib(25)
```

75025 (15)

Type: PositiveInteger

```
[fib(n) for n in 1..15]
```

[1, 1, 2, 3, 5, 8, 13, 21, 34, 55, 89, 144, 233, 377, 610] (16)

Type: List Integer

Quadratic symbols can be computed with the operations **legendre** and **jacobi**. The Legendre symbol $\left(\frac{a}{p}\right)$ is defined for integers a and p with p an odd prime number. By definition, $\left(\frac{a}{p}\right)$, when a is a square (mod p), $\left(\frac{a}{p}\right)$, when a is not a square (mod p), and $\left(\frac{a}{p}\right)$, when a is divisible by p.

You compute $\left(\frac{a}{p}\right)$ via the command legendre(a,p).

```
legendre(3,5)
```

$$-1 \tag{17}$$

Type: Integer

```
legendre(23,691)
```

$$-1 \tag{18}$$

Type: Integer

The Jacobi symbol $\left(\frac{a}{n}\right)$ is the usual extension of the Legendre symbol, where n is an arbitrary integer. The most important property of the Jacobi symbol is the following: if K is a quadratic field with discriminant d and quadratic character χ, then χ(n) = (d/n). Thus, you can use the Jacobi symbol to compute, say, the class numbers of imaginary quadratic fields from a standard class number formula.

This function computes the class number of the imaginary quadratic field with discriminant d.

```
h(d) == quo(reduce(+, [jacobi(d,k) for k in 1..quo(-d,
  2)]), 2 - jacobi(d,2))
```

Type: Void

```
h(-163)
```

$$1 \tag{20}$$

Type: PositiveInteger

```
h(-499)
```

$$3 \tag{21}$$

Type: PositiveInteger

```
h(-1832)
```

$$26 \tag{22}$$

Type: PositiveInteger

9.33 KeyedAccessFile

The domain KeyedAccessFile(S) provides files which can be used as associative tables. Data values are stored in these files and can be retrieved according to their keys. The keys must be strings so this type behaves very much like the StringTable(S) domain. The difference is that keyed access files reside in secondary storage while string tables are kept in memory. For more information on table-oriented operations, see the description of Table.

Before a keyed access file can be used, it must first be opened. A new file can be created by opening it for output.

```
ey: KeyedAccessFile(Integer) := open("/tmp/editor.year",
  "output")
```

"/tmp/editor.year" (1)

Type: KeyedAccessFile Integer

Just as for vectors, tables or lists, values are saved in a keyed access file by setting elements.

```
ey."Char" := 1986
```

1986 (2)

Type: PositiveInteger

```
ey."Caviness" := 1985
```

1985 (3)

Type: PositiveInteger

```
ey."Fitch" := 1984
```

1984 (4)

Type: PositiveInteger

Values are retrieved using application, in any of its syntactic forms.

```
ey."Char"
```

1986 (5)

Type: PositiveInteger

```
ey("Char")
```

1986 (6)

Type: PositiveInteger

```
ey "Char"
```

1986 (7)

Type: PositiveInteger

Attempting to retrieve a non-existent element in this way causes an error. If it is not known whether a key exists, you should use the **search** operation.

```
search("Char", ey)
```

1986 (8)

Type: Union(Integer, ...)

```
search("Smith", ey)
```

"failed" (9)

Type: Union("failed", ...)

When an entry is no longer needed, it can be removed from the file.

```
remove!("Char", ey)
```

1986 (10)

Type: Union(Integer, ...)

The **keys** operation returns a list of all the keys for a given file.

```
keys ey
```

["Fitch", "Caviness"] (11)

Type: List String

The **#** operation gives the number of entries.

```
#ey
```

2 (12)

Type: PositiveInteger

The table view of keyed access files provides safe operations. That is, if the AXIOM program is terminated between file operations, the file is left in a consistent, current state. This means, however, that the operations are somewhat costly. For example, after each update the file is closed.

Here we add several more items to the file, then check its contents.

```
KE := Record(key: String, entry: Integer)
```

Record (*key* : String , *entry* : Integer) (13)

Type: Domain

```
reopen!(ey, "output")
```

"/tmp/editor.year" (14)

Type: KeyedAccessFile Integer

If many items are to be added to a file at the same time, then it is more efficient to use the **write!** operation.

```
write!(ey, ["van Hulzen", 1983]$KE)
```

[*key* = "van Hulzen", *entry* = 1983] (15)

Type: Record(key: String, entry: Integer)

```
write!(ey, ["Calmet", 1982]$KE)
```

[*key* = "Calmet", *entry* = 1982] (16)

Type: Record(key: String, entry: Integer)

```
write!(ey, ["Wang", 1981]$KE)
```

[*key* = "Wang", *entry* = 1981] (17)

Type: Record(key: String, entry: Integer)

```
close! ey
```

"/tmp/editor.year" (18)

Type: KeyedAccessFile Integer

The **read!** operation is also available from the file view, but it returns elements in a random order. It is generally clearer and more efficient to use the **keys** operation and to extract elements by key.

```
keys ey
```

["Wang", "Calmet", "van Hulzen", "Fitch", "Caviness"] (19)

Type: List String

```
members ey
```

[1981, 1982, 1983, 1984, 1985] (20)

Type: List Integer

```
)system rm -r /tmp/editor.year
```

For more information on related topics, see 'File' on page 362, 'TextFile' on page 468, and 'Library' on page 393. Issue the system command `)show KeyedAccessFile` to display the full list of operations defined by KeyedAccessFile.

9.34 Library

The Library domain provides a simple way to store AXIOM values in a file. This domain is similar to KeyedAccessFile but fewer declarations are needed and items of different types can be saved together in the same file.

To create a library, you supply a file name.

```
stuff := library "/tmp/Neat.stuff"
```

"/tmp/Neat.stuff" (1)

Type: Library

Now values can be saved by key in the file. The keys should be mnemonic, just as the field names are for records. They can be given either as strings or symbols.

```
stuff.int := 32**2
```

1024 (2)

Type: PositiveInteger

```
stuff."poly" := x**2 + 1
```

$x^2 + 1$ (3)

Type: Polynomial Integer

```
stuff.str := "Hello"
```

"Hello" (4)

Type: String

You obtain the set of available keys using the **keys** operation.

```
keys stuff
```

["str", "poly", "int"] (5)

Type: List String

You extract values by giving the desired key in this way.

```
stuff.poly
```

$x^2 + 1$ (6)

Type: Polynomial Integer

```
stuff("poly")
```

$x^2 + 1$ (7)

Type: Polynomial Integer

When the file is no longer needed, you should remove it from the file system.

```
)system rm -rf /tmp/Neat.stuff
```

For more information on related topics, see 'File' on page 362, 'TextFile' on page 468, and 'KeyedAccessFile' on page 390. Issue the system command `)show Library` to display the full list of operations defined by Library.

9.35 LinearOrdinary-Differential-Operator

LinearOrdinaryDifferentialOperator(A, M) is the domain of linear ordinary differential operators with coefficients in the differential ring A and operating on M, an A-module. This includes the cases of operators which are polynomials in D acting upon scalar or vector expressions of a single variable. The coefficients of the operator polynomials can be integers, rational functions, matrices or elements of other domains. Issue the system command `)show LinearOrdinaryDifferentialOperator` to display the full list of operations defined by LinearOrdinaryDifferentialOperator.

9.35.1 Differential Operators with Constant Coefficients

This example shows differential operators with rational number coefficients operating on univariate polynomials.

We begin by making type assignments so we can conveniently refer to univariate polynomials in **x** over the rationals.

```
Q := Fraction Integer
```

Fraction Integer (1)

Type: Domain

```
PQ := UnivariatePolynomial('x, Q)
```

UnivariatePolynomial (x, Fraction Integer) (2)

Type: Domain

```
x: PQ := 'x
```

x (3)

Type: UnivariatePolynomial(x, Fraction Integer)

Now we assign Dx to be the differential operator **D** corresponding to d/dx.

```
Dx: LODO(Q, PQ) := D()
```

$$D \tag{4}$$

Type: LinearOrdinaryDifferentialOperator(Fraction Integer, UnivariatePolynomial(x, Fraction Integer))

New operators are created as polynomials in D().

```
a := Dx + 1
```

$$D + 1 \tag{5}$$

Type: LinearOrdinaryDifferentialOperator(Fraction Integer, UnivariatePolynomial(x, Fraction Integer))

```
b := a + 1/2*Dx**2 - 1/2
```

$$\frac{1}{2}\, D^2 + D + \frac{1}{2} \tag{6}$$

Type: LinearOrdinaryDifferentialOperator(Fraction Integer, UnivariatePolynomial(x, Fraction Integer))

To apply the operator a to the value p the usual function call syntax is used.

```
p := 4*x**2 + 2/3
```

$$4\, x^2 + \frac{2}{3} \tag{7}$$

Type: UnivariatePolynomial(x, Fraction Integer)

```
a p
```

$$4\, x^2 + 8\, x + \frac{2}{3} \tag{8}$$

Type: UnivariatePolynomial(x, Fraction Integer)

Operator multiplication is defined by the identity (a*b) p = a(b(p))

```
(a * b) p = a b p
```

$$2\, x^2 + 12\, x + \frac{37}{3} = 2\, x^2 + 12\, x + \frac{37}{3} \tag{9}$$

Type: Equation UnivariatePolynomial(x, Fraction Integer)

Exponentiation follows from multiplication.

```
c := (1/9)*b*(a + b)**2
```

$$\frac{1}{72}\, D^6 + \frac{5}{36}\, D^5 + \frac{13}{24}\, D^4 + \frac{19}{18}\, D^3 + \frac{79}{72}\, D^2 + \frac{7}{12}\, D + \frac{1}{8} \tag{10}$$

Type: LinearOrdinaryDifferentialOperator(Fraction Integer, UnivariatePolynomial(x, Fraction Integer))

Finally, note that operator expressions may be applied directly.

```
(a**2 - 3/4*b + c) (p + 1)
```

$$3\, x^2 + \frac{44}{3}\, x + \frac{541}{36} \tag{11}$$

Type: UnivariatePolynomial(x, Fraction Integer)

9.35.2 Differential Operators with Rational Function Coefficients

This example shows differential operators with rational function coefficients. In this case operator multiplication is non-commutative and, since the coefficients form a field, an operator division algorithm exists.

We begin by defining RFZ to be the rational functions in x with integer coefficients and Dx to be the differential operator for d/dx.

```
RFZ := Fraction UnivariatePolynomial('x, Integer)
```

$$\text{Fraction UnivariatePolynomial } (x, \text{Integer }) \tag{1}$$

Type: Domain

```
x : RFZ := 'x
```

$$x \tag{2}$$

Type: Fraction UnivariatePolynomial(x, Integer)

```
Dx: LODO(RFZ, RFZ) := D()
```

$$D \tag{3}$$

Type: LinearOrdinaryDifferentialOperator(Fraction UnivariatePolynomial(x, Integer), Fraction UnivariatePolynomial(x, Integer))

Operators are created using the usual arithmetic operations.

```
b := 3*x**2*Dx**2 + 2*Dx + 1/x
```

$$3\ x^2\ D^2 + 2\ D + \frac{1}{x} \tag{4}$$

Type: LinearOrdinaryDifferentialOperator(Fraction UnivariatePolynomial(x, Integer), Fraction UnivariatePolynomial(x, Integer))

```
a := b*(5*x*Dx + 7)
```

$$15\ x^3\ D^3 + \left(51\ x^2 + 10\ x\right)\ D^2 + 29\ D + \frac{7}{x} \tag{5}$$

Type: LinearOrdinaryDifferentialOperator(Fraction UnivariatePolynomial(x, Integer), Fraction UnivariatePolynomial(x, Integer))

Operator multiplication corresponds to functional composition.

```
p := x**2 + 1/x**2
```

$$\frac{x^4 + 1}{x^2} \tag{6}$$

Type: Fraction UnivariatePolynomial(x, Integer)

Since operator coefficients depend on x, the multiplication is not commutative.

```
(a*b) p = a(b(p))
```

$$\frac{612\ x^6 + 510\ x^5 + 180\ x^4 - 972\ x^2 + 1026\ x - 120}{x^4} = \frac{612\ x^6 + 510\ x^5 + 180\ x^4 - 972\ x^2 + 1026\ x - 120}{x^4} \tag{7}$$

Type: Equation Fraction UnivariatePolynomial(x, Integer)

```
(b*a) p = b(a(p))
```

$$\frac{612\ x^6 + 510\ x^5 + 255\ x^4 - 972\ x^2 + 486\ x - 45}{x^4} = \frac{612\ x^6 + 510\ x^5 + 255\ x^4 - 972\ x^2 + 486\ x - 45}{x^4} \tag{8}$$

Type: Equation Fraction UnivariatePolynomial(x, Integer)

When the coefficients of operator polynomials come from a field, as in this case, it is possible to define operator division. Division on the left and division on the right yield different results when the multiplication is non-commutative.

The results of **leftDivide** and **rightDivide** are quotient-remainder pairs satisfying:
`leftDivide(a,b) = [q, r]` such that `a = b*q + r`
`rightDivide(a,b) = [q, r]` such that `a = q*b + r`

In both cases, the **degree** of the remainder, `r`, is less than the degree of `b`.

```
ld := leftDivide(a,b)
```

$$\left[quotient = 5\ x\ D + 7,\ remainder = 0\right] \tag{9}$$

Type: Record(quotient: LinearOrdinaryDifferentialOperator(Fraction UnivariatePolynomial(x, Integer), Fraction UnivariatePolynomial(x, Integer)), remainder: LinearOrdinaryDifferentialOperator(Fraction UnivariatePolynomial(x, Integer), Fraction UnivariatePolynomial(x, Integer)))

```
a = b * ld.quotient + ld.remainder
```

$$15\ x^3\ D^3 + \left(51\ x^2 + 10\ x\right)\ D^2 + 29\ D + \frac{7}{x} = 15\ x^3\ D^3 + \left(51\ x^2 + 10\ x\right)\ D^2 + 29\ D + \frac{7}{x} \tag{10}$$

Type: Equation LinearOrdinaryDifferentialOperator(Fraction UnivariatePolynomial(x, Integer), Fraction UnivariatePolynomial(x, Integer))

The operations of left and right division are so-called because the quotient is obtained by dividing `a` on that side by `b`.

```
rd := rightDivide(a,b)
```

$$\left[quotient = 5\ x\ D + 7,\ remainder = 10\ D + \frac{5}{x}\right] \tag{11}$$

Type: Record(quotient: LinearOrdinaryDifferentialOperator(Fraction UnivariatePolynomial(x, Integer), Fraction UnivariatePolynomial(x, Integer)), remainder: LinearOrdinaryDifferentialOperator(Fraction UnivariatePolynomial(x, Integer), Fraction UnivariatePolynomial(x, Integer)))

```
a = rd.quotient * b + rd.remainder
```

$$15\ x^3\ D^3 + \left(51\ x^2 + 10\ x\right)\ D^2 + 29\ D + \frac{7}{x} = 15\ x^3\ D^3 + \left(51\ x^2 + 10\ x\right)\ D^2 + 29\ D + \frac{7}{x} \tag{12}$$

Type: Equation LinearOrdinaryDifferentialOperator(Fraction UnivariatePolynomial(x, Integer), Fraction UnivariatePolynomial(x, Integer))

Operations **rightQuotient** and **rightRemainder** are available if only one of the quotient or remainder are of interest to you. This is the quotient from right division.

```
rightQuotient(a,b)
```

$$5\ x\ D + 7 \tag{13}$$

Type: LinearOrdinaryDifferentialOperator(Fraction UnivariatePolynomial(x, Integer), Fraction UnivariatePolynomial(x, Integer))

This is the remainder from right division. The corresponding "left" functions **leftQuotient** and **leftRemainder** are also available.

```
rightRemainder(a,b)
```

$$10\ D + \frac{5}{x} \tag{14}$$

Type: LinearOrdinaryDifferentialOperator(Fraction UnivariatePolynomial(x, Integer), Fraction UnivariatePolynomial(x, Integer))

For exact division, the operations **leftExactQuotient** and **rightExactQuotient** are supplied. These return the quotient but only if the remainder is zero. The call `rightExactQuotient(a,b)` would yield an error.

```
leftExactQuotient(a,b)
```

$$5\ x\ D + 7 \tag{15}$$

Type: Union(LinearOrdinaryDifferentialOperator(Fraction UnivariatePolynomial(x, Integer), Fraction UnivariatePolynomial(x, Integer)), ...)

The division operations allow the computation of left and right greatest common divisors (**leftGcd** and **rightGcd**) via remainder sequences, and consequently the computation of left and right least common multiples (**rightLcm** and **leftLcm**).

```
e := leftGcd(a,b)
```

$$3\ x^2\ D^2 + 2\ D + \frac{1}{x} \tag{16}$$

Type: LinearOrdinaryDifferentialOperator(Fraction UnivariatePolynomial(x, Integer), Fraction UnivariatePolynomial(x, Integer))

Note that a greatest common divisor doesn't necessarily divide a and b on both sides. Here the left greatest common divisor does not divide a on the right.

```
leftRemainder(a, e)
```

$$0 \tag{17}$$

Type: LinearOrdinaryDifferentialOperator(Fraction UnivariatePolynomial(x, Integer), Fraction UnivariatePolynomial(x, Integer))

```
rightRemainder(a, e)
```

$$10\ D + \frac{5}{x} \tag{18}$$

Type: LinearOrdinaryDifferentialOperator(Fraction UnivariatePolynomial(x, Integer), Fraction UnivariatePolynomial(x, Integer))

Similarly, a least common multiple is not necessarily divisible from both sides.

```
f := rightLcm(a,b)
```

$$20\ x^5\ D^5 + \frac{684\ x^4 + 80\ x^3}{3}\ D^4 + \frac{5832\ x^3 + 1656\ x^2 + 80\ x}{9}\ D^3 + \frac{3672\ x^2 + 2040\ x + 352}{9}\ D^2 + \frac{172}{9\ x}\ D - \frac{28}{9\ x^2} \tag{19}$$

Type: LinearOrdinaryDifferentialOperator(Fraction UnivariatePolynomial(x, Integer), Fraction UnivariatePolynomial(x, Integer))

```
rightRemainder(f, b)
```

$$0 \tag{20}$$

Type: LinearOrdinaryDifferentialOperator(Fraction UnivariatePolynomial(x, Integer), Fraction UnivariatePolynomial(x, Integer))

```
leftRemainder(f, b)
```

$$\frac{-1176\ x + 160}{9\ x}\ D + \frac{312\ x - 80}{9\ x^2} \tag{21}$$

Type: LinearOrdinaryDifferentialOperator(Fraction UnivariatePolynomial(x, Integer), Fraction UnivariatePolynomial(x, Integer))

9.35.3 Differential Operators with Series Coefficients

Problem: Find the first few coefficients in `x` of `L3 phi` where

```
L3  = (d/dx)**3 + G*x**2 * d/dx + H*x**3 - exp(x)
phi = sum s[i]*x**i for i = 0..
```

Solution:

We work with Taylor series in `x`.

```
T := UnivariateTaylorSeries(Expression Integer,'x,0)
```

$$\text{UnivariateTaylorSeries (Expression Integer}, x, 0) \tag{1}$$

Type: Domain

```
x: T := 'x
```

$$x \tag{2}$$

Type: UnivariateTaylorSeries(Expression Integer, x, 0)

Define the operator L3 and the series phi with undetermined coefficients.

```
Dx: LODO(T,T) := D()
```

$$D \tag{3}$$

Type: LinearOrdinaryDifferentialOperator(UnivariateTaylorSeries(Expression Integer, x, 0), UnivariateTaylorSeries(Expression Integer, x, 0))

```
L3 := Dx**3 + G * x**2 * Dx + x**3 * H - exp(x)
```

$$D^3 + G\,x^2\,D - 1 - x - \frac{1}{2}\,x^2 + \frac{6\,H - 1}{6}\,x^3 - \frac{1}{24}\,x^4 - \frac{1}{120}\,x^5 - \frac{1}{720}\,x^6 - \frac{1}{5040}\,x^7 + O\left(x^8\right) \quad (4)$$

Type: LinearOrdinaryDifferentialOperator(UnivariateTaylorSeries(Expression Integer, x, 0), UnivariateTaylorSeries(Expression Integer, x, 0))

```
s: Symbol := 's
```

$$s \quad (5)$$

Type: Symbol

```
phi: T := series([s[i] for i in 0..])
```

$$s_0 + s_1\,x + s_2\,x^2 + s_3\,x^3 + s_4\,x^4 + s_5\,x^5 + s_6\,x^6 + s_7\,x^7 + O\left(x^8\right) \quad (6)$$

Type: UnivariateTaylorSeries(Expression Integer, x, 0)

Apply the operator to get the solution.

```
L3 phi
```

$$
\begin{aligned}
&6\ s_3 - s_0 + \left(24\ s_4 - s_1 - s_0\right)\ x+ \\
&\frac{120\ s_5 - 2\ s_2 + (2\ G - 2)\ s_1 - s_0}{2}\ x^2+ \\
&\frac{720\ s_6 - 6\ s_3 + \left(12\ G - 6\right)\ s_2 - 3\ s_1 + \left(6\ H - 1\right)\ s_0}{6}\ x^3+ \\
&\frac{\left(\begin{array}{l}5040\ s_7 - 24\ s_4 + \left(72\ G - 24\right)\ s_3 - 12\ s_2+ \\ \left(24\ H - 4\right)\ s_1 - s_0\end{array}\right)}{24}\ x^4+ \\
&\frac{\left(\begin{array}{l}40320\ s_8 - 120\ s_5 + \left(480\ G - 120\right)\ s_4 - 60\ s_3+ \\ (120\ H - 20)\ s_2 - 5\ s_1 - s_0\end{array}\right)}{120}\ x^5+ \\
&\frac{\left(\begin{array}{l}362880\ s_9 - 720\ s_6 + \left(3600\ G - 720\right)\ s_5 - 360\ s_4+ \\ (720\ H - 120)\ s_3 - 30\ s_2 - 6\ s_1 - s_0\end{array}\right)}{720}\ x^6+ \\
&\frac{\left(\begin{array}{l}3628800\ s_{10} - 5040\ s_7 + \left(30240\ G - 5040\right)\ s_6- \\ 2520\ s_5 + \left(5040\ H - 840\right)\ s_4 - 210\ s_3 - 42\ s_2 - 7\ s_1- \\ s_0\end{array}\right)}{5040}\ x^7+ \\
&O\left(x^8\right)
\end{aligned}
\tag{7}
$$

Type: UnivariateTaylorSeries(Expression Integer, x, 0)

9.35.4 Differential Operators with Matrix Coefficients Operating on Vectors

This is another example of linear ordinary differential operators with non-commutative multiplication. Unlike the rational function case, the differential ring of square matrices (of a given dimension) with univariate polynomial entries does not form a field. Thus the number of operations available is more limited.

In this section, the operators have three by three matrix coefficients with polynomial entries.

```
PZ := UnivariatePolynomial(x,Integer)
```

UnivariatePolynomial $(x, \text{Integer})$ (1)

Type: Domain

```
x:PZ := 'x
```

x (2)

Type: UnivariatePolynomial(x, Integer)

```
Mat := SquareMatrix(3,PZ)
```

SquareMatrix (3, UnivariatePolynomial $(x, \text{Integer})$) (3)

Type: Domain

The operators act on the vectors considered as a Mat-module.

```
Vect := DPMM(3, PZ, Mat, PZ);
```

(4)

Type: Domain

```
Modo := LODO(Mat, Vect);
```

(5)

Type: Domain

The matrix m is used as a coefficient and the vectors p and q are operated upon.

```
m:Mat := matrix [[x**2,1,0],[1,x**4,0],[0,0,4*x**2]]
```

$$\begin{bmatrix} x^2 & 1 & 0 \\ 1 & x^4 & 0 \\ 0 & 0 & 4\,x^2 \end{bmatrix} \quad (6)$$

Type: SquareMatrix(3, UnivariatePolynomial(x, Integer))

```
p:Vect := directProduct [3*x**2+1,2*x,7*x**3+2*x]
```

$$\left[3\,x^2+1,\ 2\,x,\ 7\,x^3+2\,x\right] \quad (7)$$

Type: DirectProductMatrixModule(3, UnivariatePolynomial(x, Integer), SquareMatrix(3, UnivariatePolynomial(x, Integer)), UnivariatePolynomial(x, Integer))

```
q: Vect := m * p
```

$$\left[3\,x^4+x^2+2\,x,\ 2\,x^5+3\,x^2+1,\ 28\,x^5+8\,x^3\right] \quad (8)$$

Type: DirectProductMatrixModule(3, UnivariatePolynomial(x, Integer), SquareMatrix(3, UnivariatePolynomial(x, Integer)), UnivariatePolynomial(x, Integer))

Now form a few operators.

```
Dx: Modo := D()
```

$$D \tag{9}$$

Type: LinearOrdinaryDifferentialOperator(SquareMatrix(3, UnivariatePolynomial(x, Integer)), DirectProductMatrixModule(3, UnivariatePolynomial(x, Integer), SquareMatrix(3, UnivariatePolynomial(x, Integer)), UnivariatePolynomial(x, Integer)))

```
a := Dx + m
```

$$D + \begin{bmatrix} x^2 & 1 & 0 \\ 1 & x^4 & 0 \\ 0 & 0 & 4\,x^2 \end{bmatrix} \tag{10}$$

Type: LinearOrdinaryDifferentialOperator(SquareMatrix(3, UnivariatePolynomial(x, Integer)), DirectProductMatrixModule(3, UnivariatePolynomial(x, Integer), SquareMatrix(3, UnivariatePolynomial(x, Integer)), UnivariatePolynomial(x, Integer)))

```
b := m*Dx + 1
```

$$\begin{bmatrix} x^2 & 1 & 0 \\ 1 & x^4 & 0 \\ 0 & 0 & 4\,x^2 \end{bmatrix} D + \begin{bmatrix} 1 & 0 & 0 \\ 0 & 1 & 0 \\ 0 & 0 & 1 \end{bmatrix} \tag{11}$$

Type: LinearOrdinaryDifferentialOperator(SquareMatrix(3, UnivariatePolynomial(x, Integer)), DirectProductMatrixModule(3, UnivariatePolynomial(x, Integer), SquareMatrix(3, UnivariatePolynomial(x, Integer)), UnivariatePolynomial(x, Integer)))

```
c := a*b
```

$$\begin{bmatrix} x^2 & 1 & 0 \\ 1 & x^4 & 0 \\ 0 & 0 & 4\,x^2 \end{bmatrix} D^2 + \begin{bmatrix} x^4+2\,x+2 & x^4+x^2 & 0 \\ x^4+x^2 & x^8+4\,x^3+2 & 0 \\ 0 & 0 & 16\,x^4+8\,x+1 \end{bmatrix} D + \begin{bmatrix} x^2 & 1 & 0 \\ 1 & x^4 & 0 \\ 0 & 0 & 4\,x^2 \end{bmatrix} \tag{12}$$

Type: LinearOrdinaryDifferentialOperator(SquareMatrix(3, UnivariatePolynomial(x, Integer)), DirectProductMatrixModule(3, UnivariatePolynomial(x, Integer), SquareMatrix(3, UnivariatePolynomial(x, Integer)), UnivariatePolynomial(x, Integer)))

These operators can be applied to vector values.

```
a p
```

$$\left[3\ x^4+x^2+8\ x,\ 2\ x^5+3\ x^2+3,\ 28\ x^5+8\ x^3+21\ x^2+2\right] \quad (13)$$

Type: DirectProductMatrixModule(3, UnivariatePolynomial(x, Integer), SquareMatrix(3, UnivariatePolynomial(x, Integer)), UnivariatePolynomial(x, Integer))

```
b p
```

$$\left[6\ x^3+3\ x^2+3,\ 2\ x^4+8\ x,\ 84\ x^4+7\ x^3+8\ x^2+2\ x\right] \quad (14)$$

Type: DirectProductMatrixModule(3, UnivariatePolynomial(x, Integer), SquareMatrix(3, UnivariatePolynomial(x, Integer)), UnivariatePolynomial(x, Integer))

```
(a + b + c) (p + q)
```

$$\left[\begin{array}{l}10\ x^8+12\ x^7+16\ x^6+30\ x^5+85\ x^4+94\ x^3+40\ x^2+ \\ 40\ x+17\end{array},\right.$$

$$\begin{array}{l}10\ x^{12}+10\ x^9+12\ x^8+92\ x^7+6\ x^6+32\ x^5+72\ x^4+28\ x^3+ \\ 49\ x^2+32\ x+19\end{array}, \quad (15)$$

$$\left.\begin{array}{l}2240\ x^8+224\ x^7+1280\ x^6+3508\ x^5+492\ x^4+751\ x^3+ \\ 98\ x^2+18\ x+4\end{array}\right]$$

Type: DirectProductMatrixModule(3, UnivariatePolynomial(x, Integer), SquareMatrix(3, UnivariatePolynomial(x, Integer)), UnivariatePolynomial(x, Integer))

9.36 List

A *list* is a finite collection of elements in a specified order that can contain duplicates. A list is a convenient structure to work with because it is easy to add or remove elements and the length need not be constant. There are many different kinds of lists in AXIOM, but the default types (and those used most often) are created by the List constructor. For example, there are objects of type List Integer, List Float and List Polynomial Fraction Integer. Indeed, you can even have List List List Boolean (that is, lists of lists of lists of Boolean values). You can have lists of any type of AXIOM object.

9.36.1 Creating Lists

The easiest way to create a list with, for example, the elements `2, 4, 5, 6` is to enclose the elements with square brackets and separate the elements with commas.

The spaces after the commas are optional, but they do improve the readability.

```
[2, 4, 5, 6]
```

[2, 4, 5, 6] (1)

Type: List PositiveInteger

To create a list with the single element `1`, you can use either `[1]` or the operation **list**.

```
[1]
```

[1] (2)

Type: List PositiveInteger

```
list(1)
```

[1] (3)

Type: List PositiveInteger

Once created, two lists `k` and `m` can be concatenated by issuing `append(k,m)`. **append** does *not* physically join the lists, but rather produces a new list with the elements coming from the two arguments.

```
append([1,2,3],[5,6,7])
```

[1, 2, 3, 5, 6, 7] (4)

Type: List PositiveInteger

Use **cons** to append an element onto the front of a list.

```
cons(10,[9,8,7])
```

[10, 9, 8, 7] (5)

Type: List PositiveInteger

9.36.2 Accessing List Elements

To determine whether a list has any elements, use the operation **empty?**.

```
empty? [x+1]
```

false (1)

Type: Boolean

Alternatively, equality with the list constant **nil** can be tested.

```
([] = nil)@Boolean
```

true (2)

Type: Boolean

We'll use this in some of the following examples.

```
k := [4,3,7,3,8,5,9,2]
```

[4, 3, 7, 3, 8, 5, 9, 2] (3)

Type: List PositiveInteger

Each of the next four expressions extracts the **first** element of k.

```
first k
```

4 (4)

Type: PositiveInteger

```
k.first
```

4 (5)

Type: PositiveInteger

```
k.1
```

4 (6)

Type: PositiveInteger

```
k(1)
```

4 (7)

Type: PositiveInteger

The last two forms generalize to k.i and k(i), respectively, where $1 \leq i \leq n$ and n equals the length of k.

This length is calculated by "#".

```
n := #k
```

8 (8)

Type: PositiveInteger

Performing an operation such as k.i is sometimes referred to as *indexing into k* or *elting into k*. The latter phrase comes about because the name of the operation that extracts elements is called **elt**. That is, k.3 is just alternative syntax for elt(k,3). It is important to remember that list indices begin with 1. If we issue k := [1,3,2,9,5] then k.4 returns 9. It is an error to use an index that is not in the range from 1 to the length of the list.

The last element of a list is extracted by any of the following three expressions.

```
last k
```

2 (9)

Type: PositiveInteger

```
k.last
```

2 (10)

Type: PositiveInteger

This form computes the index of the last element and then extracts the element from the list.

```
k.(#k)
```

2 (11)

Type: PositiveInteger

9.36.3 Changing List Elements

We'll use this in some of the following examples.

```
k := [4,3,7,3,8,5,9,2]
```

$$[4, 3, 7, 3, 8, 5, 9, 2] \qquad (1)$$

Type: List PositiveInteger

List elements are reset by using the `k.i` form on the left-hand side of an assignment. This expression resets the first element of `k` to `999`.

```
k.1 := 999
```

$$999 \qquad (2)$$

Type: PositiveInteger

As with indexing into a list, it is an error to use an index that is not within the proper bounds. Here you see that `k` was modified.

```
k
```

$$[999, 3, 7, 3, 8, 5, 9, 2] \qquad (3)$$

Type: List PositiveInteger

The operation that performs the assignment of an element to a particular position in a list is called **setelt**. This operation is *destructive* in that it changes the list. In the above example, the assignment returned the value `999` and `k` was modified. For this reason, lists are called *mutable* objects: it is possible to change part of a list (mutate it) rather than always returning a new list reflecting the intended modifications.

Moreover, since lists can share structure, changes to one list can sometimes affect others.

```
k := [1,2]
```

$$[1, 2] \qquad (4)$$

Type: List PositiveInteger

```
m := cons(0,k)
```

$$[0, 1, 2] \qquad (5)$$

Type: List Integer

Change the second element of `m`.

```
m.2 := 99
```

$$99 \qquad (6)$$

Type: PositiveInteger

See, `m` was altered.

```
m
```

$$[0, 99, 2] \qquad (7)$$

Type: List Integer

But what about `k`? It changed too!

```
k
```

$$[99, 2] \qquad (8)$$

Type: List PositiveInteger

9.36.4 Other Functions

An operation that is used frequently in list processing is that which returns all elements in a list after the first element.

```
k := [1,2,3]
```

[1, 2, 3] (1)

Type: List PositiveInteger

Use the **rest** operation to do this.

```
rest k
```

[2, 3] (2)

Type: List PositiveInteger

To remove duplicate elements in a list `k`, use **removeDuplicates**.

```
removeDuplicates [4,3,4,3,5,3,4]
```

[4, 3, 5] (3)

Type: List PositiveInteger

To get a list with elements in the order opposite to those in a list `k`, use **reverse**.

```
reverse [1,2,3,4,5,6]
```

[6, 5, 4, 3, 2, 1] (4)

Type: List PositiveInteger

To test whether an element is in a list, use **member?**: `member?(a,k)` returns `true` or `false` depending on whether `a` is in `k` or not.

```
member?(1/2,[3/4,5/6,1/2])
```

true (5)

Type: Boolean

```
member?(1/12,[3/4,5/6,1/2])
```

false (6)

Type: Boolean

As an exercise, the reader should determine how to get a list containing all but the last of the elements in a given non-empty list `k`.[4]

9.36.5 Dot, Dot

Certain lists are used so often that AXIOM provides an easy way of constructing them. If `n` and `m` are integers, then `expand [n..m]` creates a list containing `n, n+1, ... m`. If `n > m` then the list is empty. It is actually permissible to leave off the `m` in the dot-dot construction (see below).

The dot-dot notation can be used more than once in a list construction and with specific elements being given. Items separated by dots are called *segments*.

```
[1..3,10,20..23]
```

[1..3, 10..10, 20..23] (1)

Type: List Segment PositiveInteger

[4] `reverse(rest(reverse(k)))` works.

Segments can be expanded into the range of items between the endpoints by using **expand**.

```
expand [1..3,10,20..23]
```

$$[1, 2, 3, 10, 20, 21, 22, 23] \tag{2}$$

Type: List Integer

What happens if we leave off a number on the right-hand side of "..."?

```
expand [1..]
```

$$[1, 2, 3, 4, 5, 6, 7, \ldots] \tag{3}$$

Type: Stream Integer

What is created in this case is a Stream which is a generalization of a list. See 'Stream' on page 457 for more information.

9.37 MakeFunction

It is sometimes useful to be able to define a function given by the result of a calculation.

Suppose that you have obtained the following expression after several computations and that you now want to tabulate the numerical values of f for x between -1 and +1 with increment 0.1.

```
expr := (x - exp x + 1)**2 * (sin(x**2) * x + 1)**3
```

$$\begin{array}{l}\left(x^3\ \%e^{x^2} + \left(-2\ x^4 - 2\ x^3\right)\ \%e^{x} + x^5 + 2\ x^4 + x^3\right)\ \sin\left(x^2\right)^3 + \\ \left(3\ x^2\ \%e^{x^2} + \left(-6\ x^3 - 6\ x^2\right)\ \%e^{x} + 3\ x^4 + 6\ x^3 + 3\ x^2\right)\ \sin\left(x^2\right)^2 + \\ \left(3\ x\ \%e^{x^2} + \left(-6\ x^2 - 6\ x\right)\ \%e^{x} + 3\ x^3 + 6\ x^2 + 3\ x\right)\ \sin\left(x^2\right) + \\ \%e^{x^2} + (-2\ x - 2)\ \%e^{x} + x^2 + 2\ x + 1\end{array} \tag{1}$$

Type: Expression Integer

You could, of course, use the function **eval** within a loop and evaluate expr twenty-one times, but this would be quite slow. A better way is to create a numerical function f such that f(x) is defined by the expression expr above, but without retyping expr! The package MakeFunction provides the operation **function** which does exactly this.

Issue this to create the function f(x) given by expr.

```
function(expr, f, x)
```

$$f \tag{2}$$

Type: Symbol

To tabulate expr, we can now quickly evaluate f 21 times.

```
tbl := [f(0.1 * i - 1) for i in 0..20];
Compiling function f with type Float -> Float
```

(2)

Type: List Float

Use the list [x1,...,xn] as the third argument to **function** to create a multivariate function f(x1,...,xn).

```
e := (x - y + 1)**2 * (x**2 * y + 1)**2
```

$$x^4\,y^4+\left(-2\,x^5-2\,x^4+2\,x^2\right)\,y^3+\left(x^6+2\,x^5+x^4-4\,x^3-4\,x^2+1\right)\,y^2+\left(2\,x^4+4\,x^3+2\,x^2-2\,x-2\right)\,y+x^2+2\,x+1 \quad (4)$$

Type: Polynomial Integer

```
function(e, g, [x, y])
```

$$g \quad (5)$$

Type: Symbol

In the case of just two variables, they can be given as arguments without making them into a list.

```
function(e, h, x, y)
```

$$h \quad (6)$$

Type: Symbol

Note that the functions created by **function** are not limited to floating point numbers, but can be applied to any type for which they are defined.

```
m1 := squareMatrix [[1, 2], [3, 4]]
```

$$\begin{bmatrix} 1 & 2 \\ 3 & 4 \end{bmatrix} \quad (7)$$

Type: SquareMatrix(2, Integer)

```
m2 := squareMatrix [[1, 0], [-1, 1]]
```

$$\begin{bmatrix} 1 & 0 \\ -1 & 1 \end{bmatrix} \quad (8)$$

Type: SquareMatrix(2, Integer)

```
h(m1, m2)
```

```
Compiling function h with type
(SquareMatrix(2,Integer),
SquareMatrix(2,Integer)) -> SquareMatrix(2,Integer)
```

$$\begin{bmatrix} -7836 & 8960 \\ -17132 & 19588 \end{bmatrix} \quad (9)$$

Type: SquareMatrix(2, Integer)

For more information, see Section 6.14 on page 157. Issue the system command `)show MakeFunction` to display the full list of operations defined by MakeFunction.

9.38 MappingPackage1

Function are objects of type Mapping. In this section we demonstrate some library operations from the packages MappingPackage1, MappingPackage2, and MappingPackage3 that manipulate and create functions. Some terminology: a *nullary* function takes no arguments, a *unary* function takes one argument, and a *binary* function takes two arguments.

We begin by creating an example function that raises a rational number to an integer exponent.

```
power(q: FRAC INT, n: INT): FRAC INT == q**n
```

```
Function declaration power : (Fraction Integer,
Integer) -> Fraction Integer has been added
to workspace.
```

Type: Void

```
power(2,3)
```

```
Compiling function power with type (Fraction Integer,
Integer) -> Fraction Integer
```

8 (2)

Type: Fraction Integer

The **twist** operation transposes the arguments of a binary function. Here `rewop(a, b)` is `power(b, a)`.

```
rewop := twist power
```

theMap (...) (3)

Type: ((Integer, Fraction Integer) → Fraction Integer)

This is 2^3.

```
rewop(3, 2)
```

8 (4)

Type: Fraction Integer

Now we define **square** in terms of **power**.

```
square: FRAC INT -> FRAC INT
```

Type: Void

The **curryRight** operation creates a unary function from a binary one by providing a constant argument on the right.

```
square:= curryRight(power, 2)
```

theMap (...) (6)

Type: (Fraction Integer → Fraction Integer)

Likewise, the **curryLeft** operation provides a constant argument on the left.

```
square 4
```

16 (7)

Type: Fraction Integer

The **constantRight** operation creates (in a trivial way) a binary function from a unary one: `constantRight(f)` is the function g such that `g(a,b)= f(a)`.

```
squirrel:= constantRight(square)$MAPPKG3(FRAC INT,FRAC
  INT,FRAC INT)
```

theMap (...) (8)

Type: ((Fraction Integer, Fraction Integer) → Fraction Integer)

Likewise, `constantLeft(f)` is the function g such that `g(a,b)= f(b)`.

```
squirrel(1/2, 1/3)
```

$$\frac{1}{4} \tag{9}$$

Type: Fraction Integer

The **curry** operation makes a unary function nullary.

```
sixteen := curry(square, 4/1)
```

theMap (...) (10)

Type: (() → Fraction Integer)

```
sixteen()
```

16 (11)

Type: Fraction Integer

The "*" operation constructs composed functions.

```
square2:=square*square
```

theMap (...) (12)

Type: (Fraction Integer → Fraction Integer)

```
square2 3
```

81 (13)

Type: Fraction Integer

Use the "**" operation to create functions that are n-fold iterations of other functions.

```
sc(x: FRAC INT): FRAC INT == x + 1
```

```
Function declaration sc : Fraction Integer -> Fraction
Integer has been added to workspace.
```

Type: Void

This is a list of Mapping objects.

```
incfns := [sc**i for i in 0..10]
```

```
Compiling function sc with type Fraction Integer ->
Fraction Integer
```

[theMap (...), theMap (...), theMap (...), theMap (...),

theMap (...), theMap (...), theMap (...), theMap (...), (15)

theMap (...), theMap (...), theMap (...)]

Type: List (Fraction Integer → Fraction Integer)

This is a list of applications of those functions.

```
[f 4 for f in incfns]
```

[4, 5, 6, 7, 8, 9, 10, 11, 12, 13, 14] (16)

Type: List Fraction Integer

Use the **recur** operation for recursion: `g := recur f` means `g(n,x) == f(n,f(n-1,...f(1,x)))`.

```
times(n:NNI, i:INT):INT == n*i
```

```
Function declaration times :
(NonNegativeInteger,Integer)
-> Integer has been added to workspace.
```

Type: Void

```
r := recur(times)
```

```
Compiling function times with type
(NonNegativeInteger,
Integer) -> Integer
```

theMap (...) (18)

Type: ((NonNegativeInteger, Integer) → Integer)

This is a factorial function.

```
fact := curryRight(r, 1)
```

theMap (...) (19)

Type: (NonNegativeInteger → Integer)

```
fact 4
```

24 (20)

Type: PositiveInteger

Constructed functions can be used within other functions.

```
mto2ton(m, n) ==
  raiser := square**n
  raiser m
```

Type: Void

This is 3^{2^3}.

```
mto2ton(3, 3)
```

```
Compiling function mto2ton with type (PositiveInteger,
PositiveInteger) -> Fraction Integer
```

6561 (22)

Type: Fraction Integer

Here **shiftfib** is a unary function that modifies its argument.

```
shiftfib(r: List INT) : INT ==
  t := r.1
  r.1 := r.2
  r.2 := r.2 + t
  t
```

```
Function declaration shiftfib : List Integer -> Integer

has been added to workspace.
```

Type: Void

By currying over the argument we get a function with private state.

```
fibinit: List INT := [0, 1]
```

[0, 1] (24)

Type: List Integer

```
fibs := curry(shiftfib, fibinit)
```

```
Compiling function shiftfib with type List Integer ->
Integer
```

theMap (...) (25)

Type: (() → Integer)

```
[fibs() for i in 0..30]
```

$$[0, 1, 1, 2, 3, 5, 8, 13, 21, 34, 55, 89, 144, 233, 377, 610, 987, 1597, 2584, 4181, 6765, 10946, 17711, 28657, 46368, 75025, 121393, 196418, 317811, 514229, 832040] \quad (26)$$

Type: List Integer

9.39 Matrix

The Matrix domain provides arithmetic operations on matrices and standard functions from linear algebra. This domain is similar to the TwoDimensionalArray domain, except that the entries for Matrix must belong to a Ring.

9.39.1 Creating Matrices

There are many ways to create a matrix from a collection of values or from existing matrices.

If the matrix has almost all items equal to the same value, use **new** to create a matrix filled with that value and then reset the entries that are different.

```
m : Matrix(Integer) := new(3,3,0)
```

$$\begin{bmatrix} 0 & 0 & 0 \\ 0 & 0 & 0 \\ 0 & 0 & 0 \end{bmatrix} \quad (1)$$

Type: Matrix Integer

To change the entry in the second row, third column to 5, use **setelt**.

```
setelt(m,2,3,5)
```

5 (2)

Type: PositiveInteger

An alternative syntax is to use assignment.

```
m(1,2) := 10
```

10 (3)

Type: PositiveInteger

The matrix was *destructively modified.*

```
m
```

$$\begin{bmatrix} 0 & 10 & 0 \\ 0 & 0 & 5 \\ 0 & 0 & 0 \end{bmatrix} \quad (4)$$

Type: Matrix Integer

If you already have the matrix entries as a list of lists, use **matrix**.

```
matrix [[1,2,3,4],[0,9,8,7]]
```

$$\begin{bmatrix} 1 & 2 & 3 & 4 \\ 0 & 9 & 8 & 7 \end{bmatrix} \tag{5}$$

Type: Matrix Integer

If the matrix is diagonal, use **diagonalMatrix**.

```
dm := diagonalMatrix [1,x**2,x**3,x**4,x**5]
```

$$\begin{bmatrix} 1 & 0 & 0 & 0 & 0 \\ 0 & x^2 & 0 & 0 & 0 \\ 0 & 0 & x^3 & 0 & 0 \\ 0 & 0 & 0 & x^4 & 0 \\ 0 & 0 & 0 & 0 & x^5 \end{bmatrix} \tag{6}$$

Type: Matrix Polynomial Integer

Use **setRow!** and **setColumn!** to change a row or column of a matrix.

```
setRow!(dm,5,vector [1,1,1,1,1])
```

$$\begin{bmatrix} 1 & 0 & 0 & 0 & 0 \\ 0 & x^2 & 0 & 0 & 0 \\ 0 & 0 & x^3 & 0 & 0 \\ 0 & 0 & 0 & x^4 & 0 \\ 1 & 1 & 1 & 1 & 1 \end{bmatrix} \tag{7}$$

Type: Matrix Polynomial Integer

```
setColumn!(dm,2,vector [y,y,y,y,y])
```

$$\begin{bmatrix} 1 & y & 0 & 0 & 0 \\ 0 & y & 0 & 0 & 0 \\ 0 & y & x^3 & 0 & 0 \\ 0 & y & 0 & x^4 & 0 \\ 1 & y & 1 & 1 & 1 \end{bmatrix} \tag{8}$$

Type: Matrix Polynomial Integer

Use **copy** to make a copy of a matrix.

```
cdm := copy(dm)
```

$$\begin{bmatrix} 1 & y & 0 & 0 & 0 \\ 0 & y & 0 & 0 & 0 \\ 0 & y & x^3 & 0 & 0 \\ 0 & y & 0 & x^4 & 0 \\ 1 & y & 1 & 1 & 1 \end{bmatrix} \tag{9}$$

Type: Matrix Polynomial Integer

This is useful if you intend to modify a matrix destructively but want a copy of the original.

```
setelt(dm,4,1,1-x**7)
```

$$-x^7 + 1 \tag{10}$$

Type: Polynomial Integer

```
[dm,cdm]
```

$$\left[\begin{bmatrix} 1 & y & 0 & 0 & 0 \\ 0 & y & 0 & 0 & 0 \\ 0 & y & x^3 & 0 & 0 \\ -x^7+1 & y & 0 & x^4 & 0 \\ 1 & y & 1 & 1 & 1 \end{bmatrix}, \begin{bmatrix} 1 & y & 0 & 0 & 0 \\ 0 & y & 0 & 0 & 0 \\ 0 & y & x^3 & 0 & 0 \\ 0 & y & 0 & x^4 & 0 \\ 1 & y & 1 & 1 & 1 \end{bmatrix}\right] \quad (11)$$

Type: List Matrix Polynomial Integer

Use **subMatrix** to extract part of an existing matrix. The syntax is `subMatrix(`*m, firstrow, lastrow, firstcol, lastcol*`)`.

```
subMatrix(dm,2,3,2,4)
```

$$\begin{bmatrix} y & 0 & 0 \\ y & x^3 & 0 \end{bmatrix} \quad (12)$$

Type: Matrix Polynomial Integer

To change a submatrix, use **setsubMatrix!**.

```
d := diagonalMatrix [1.2,-1.3,1.4,-1.5]
```

$$\begin{bmatrix} 1.2 & 0.0 & 0.0 & 0.0 \\ 0.0 & -1.3 & 0.0 & 0.0 \\ 0.0 & 0.0 & 1.4 & 0.0 \\ 0.0 & 0.0 & 0.0 & -1.5 \end{bmatrix} \quad (13)$$

Type: Matrix Float

If `e` is too big to fit where you specify, an error message is displayed. Use **subMatrix** to extract part of `e`, if necessary.

```
e := matrix [[6.7,9.11],[-31.33,67.19]]
```

$$\begin{bmatrix} 6.7 & 9.11 \\ -31.33 & 67.19 \end{bmatrix} \quad (14)$$

Type: Matrix Float

This changes the submatrix of `d` whose upper left corner is at the first row and second column and whose size is that of `e`.

```
setsubMatrix!(d,1,2,e)
```

$$\begin{bmatrix} 1.2 & 6.7 & 9.11 & 0.0 \\ 0.0 & -31.33 & 67.19 & 0.0 \\ 0.0 & 0.0 & 1.4 & 0.0 \\ 0.0 & 0.0 & 0.0 & -1.5 \end{bmatrix} \quad (15)$$

Type: Matrix Float

```
d
```

$$\begin{bmatrix} 1.2 & 6.7 & 9.11 & 0.0 \\ 0.0 & -31.33 & 67.19 & 0.0 \\ 0.0 & 0.0 & 1.4 & 0.0 \\ 0.0 & 0.0 & 0.0 & -1.5 \end{bmatrix} \quad (16)$$

Type: Matrix Float

Matrices can be joined either horizontally or vertically to make new matrices.

```
a := matrix [[1/2,1/3,1/4],[1/5,1/6,1/7]]
```

$$\begin{bmatrix} \frac{1}{2} & \frac{1}{3} & \frac{1}{4} \\ \frac{1}{5} & \frac{1}{6} & \frac{1}{7} \end{bmatrix} \quad (17)$$

Type: Matrix Fraction Integer

```
b := matrix [[3/5,3/7,3/11],[3/13,3/17,3/19]]
```

$$\begin{bmatrix} \frac{3}{5} & \frac{3}{7} & \frac{3}{11} \\ \frac{3}{13} & \frac{3}{17} & \frac{3}{19} \end{bmatrix} \tag{18}$$

Type: Matrix Fraction Integer

Use **horizConcat** to append them side to side. The two matrices must have the same number of rows.

```
horizConcat(a,b)
```

$$\begin{bmatrix} \frac{1}{2} & \frac{1}{3} & \frac{1}{4} & \frac{3}{5} & \frac{3}{7} & \frac{3}{11} \\ \frac{1}{5} & \frac{1}{6} & \frac{1}{7} & \frac{3}{13} & \frac{3}{17} & \frac{3}{19} \end{bmatrix} \tag{19}$$

Type: Matrix Fraction Integer

Use **vertConcat** to stack one upon the other. The two matrices must have the same number of columns.

```
vab := vertConcat(a,b)
```

$$\begin{bmatrix} \frac{1}{2} & \frac{1}{3} & \frac{1}{4} \\ \frac{1}{5} & \frac{1}{6} & \frac{1}{7} \\ \frac{3}{5} & \frac{3}{7} & \frac{3}{11} \\ \frac{3}{13} & \frac{3}{17} & \frac{3}{19} \end{bmatrix} \tag{20}$$

Type: Matrix Fraction Integer

The operation **transpose** is used to create a new matrix by reflection across the main diagonal.

```
transpose vab
```

$$\begin{bmatrix} \frac{1}{2} & \frac{1}{5} & \frac{3}{5} & \frac{3}{13} \\ \frac{1}{3} & \frac{1}{6} & \frac{3}{7} & \frac{3}{17} \\ \frac{1}{4} & \frac{1}{7} & \frac{3}{11} & \frac{3}{19} \end{bmatrix} \tag{21}$$

Type: Matrix Fraction Integer

9.39.2 Operations on Matrices

AXIOM provides both left and right scalar multiplication.

```
m := matrix [[1,2],[3,4]]
```

$$\begin{bmatrix} 1 & 2 \\ 3 & 4 \end{bmatrix} \tag{1}$$

Type: Matrix Integer

```
4 * m * (-5)
```

$$\begin{bmatrix} -20 & -40 \\ -60 & -80 \end{bmatrix} \tag{2}$$

Type: Matrix Integer

You can add, subtract, and multiply matrices provided, of course, that the matrices have compatible dimensions. If not, an error message is displayed.

```
n := matrix([[1,0,-2],[-3,5,1]])
```

$$\begin{bmatrix} 1 & 0 & -2 \\ -3 & 5 & 1 \end{bmatrix} \tag{3}$$

Type: Matrix Integer

This following product is defined but `n * m` is not.

```
m * n
```

$$\begin{bmatrix} -5 & 10 & 0 \\ -9 & 20 & -2 \end{bmatrix} \qquad (4)$$

Type: Matrix Integer

The operations **nrows** and **ncols** return the number of rows and columns of a matrix. You can extract a row or a column of a matrix using the operations **row** and **column**. The object returned is a Vector.

Here is the third column of the matrix n.

```
vec := column(n,3)
```

$$[-2, 1] \qquad (5)$$

Type: Vector Integer

You can multiply a matrix on the left by a "row vector" and on the right by a "column vector."

```
vec * m
```

$$[1, 0] \qquad (6)$$

Type: Vector Integer

Of course, the dimensions of the vector and the matrix must be compatible or an error message is returned.

```
m * vec
```

$$[0, -2] \qquad (7)$$

Type: Vector Integer

The operation **inverse** computes the inverse of a matrix if the matrix is invertible, and returns `"failed"` if not.

This Hilbert matrix is invertible.

```
hilb := matrix([[1/(i + j) for i in 1..3] for j in 1..3])
```

$$\begin{bmatrix} \frac{1}{2} & \frac{1}{3} & \frac{1}{4} \\ \frac{1}{3} & \frac{1}{4} & \frac{1}{5} \\ \frac{1}{4} & \frac{1}{5} & \frac{1}{6} \end{bmatrix} \qquad (8)$$

Type: Matrix Fraction Integer

```
inverse(hilb)
```

$$\begin{bmatrix} 72 & -240 & 180 \\ -240 & 900 & -720 \\ 180 & -720 & 600 \end{bmatrix} \qquad (9)$$

Type: Union(Matrix Fraction Integer, ...)

This matrix is not invertible.

```
mm := matrix([[1,2,3,4], [5,6,7,8], [9,10,11,12],
  [13,14,15,16]])
```

$$\begin{bmatrix} 1 & 2 & 3 & 4 \\ 5 & 6 & 7 & 8 \\ 9 & 10 & 11 & 12 \\ 13 & 14 & 15 & 16 \end{bmatrix} \qquad (10)$$

Type: Matrix Integer

```
inverse(mm)
```

"failed" (11)

Type: Union("failed", ...)

The operation **determinant** computes the determinant of a matrix provided that the entries of the matrix belong to a CommutativeRing.

The above matrix mm is not invertible and, hence, must have determinant 0.

```
determinant(mm)
```

0 (12)

Type: NonNegativeInteger

The operation **trace** computes the trace of a *square* matrix.

```
trace(mm)
```

34 (13)

Type: PositiveInteger

The operation **rank** computes the *rank* of a matrix: the maximal number of linearly independent rows or columns.

```
rank(mm)
```

2 (14)

Type: PositiveInteger

The operation **nullity** computes the *nullity* of a matrix: the dimension of its null space.

```
nullity(mm)
```

2 (15)

Type: PositiveInteger

The operation **nullSpace** returns a list containing a basis for the null space of a matrix. Note that the nullity is the number of elements in a basis for the null space.

```
nullSpace(mm)
```

$[[1,\ -2,\ 1,\ 0],\ [2,\ -3,\ 0,\ 1]]$ (16)

Type: List Vector Integer

The operation **rowEchelon** returns the row echelon form of a matrix. It is easy to see that the rank of this matrix is two and that its nullity is also two.

```
rowEchelon(mm)
```

$$\begin{bmatrix} 1 & 2 & 3 & 4 \\ 0 & 4 & 8 & 12 \\ 0 & 0 & 0 & 0 \\ 0 & 0 & 0 & 0 \end{bmatrix} \quad (17)$$

Type: Matrix Integer

For more information on related topics, see Section 1.7 on page 38, Section 8.4 on page 241, Section 9.24.4 on page 372, 'Permanent' on page 436, 'Vector' on page 478, 'OneDimensionalArray' on page 425, and 'TwoDimensionalArray' on page 469. Issue the system command `)show Matrix` to display the full list of operations defined by Matrix.

9.40 MultiSet

The domain Multiset(R) is similar to Set(R) except that multiplicities (counts of duplications) are maintained and displayed. Use the operation **multiset** to create multisets from lists. All the standard operations from sets are available for multisets. An element with multiplicity greater than one has the multiplicity displayed first, then a colon, and then the element.

Create a multiset of integers.

```
s := multiset [1,2,3,4,5,4,3,2,3,4,5,6,7,4,10]
```

$$\{1,\ 2:2,\ 3:3,\ 4:4,\ 2:5,\ 6,\ 7,\ 10\} \quad (1)$$

Type: Multiset PositiveInteger

The operation **insert!** adds an element to a multiset.

```
insert!(3,s)
```

$$\{1,\ 2:2,\ 4:3,\ 4:4,\ 2:5,\ 6,\ 7,\ 10\} \quad (2)$$

Type: Multiset PositiveInteger

Use **remove!** to remove an element. Display the resulting multiset.

```
remove!(5,s); s
```

$$\{1,\ 2:2,\ 4:3,\ 4:4,\ 5,\ 6,\ 7,\ 10\} \quad (3)$$

Type: Multiset PositiveInteger

The operation **count** returns the number of copies of a given value.

```
count(5,s)
```

$$1 \quad (4)$$

Type: PositiveInteger

A second multiset.

```
t := multiset [2,2,2,-9]
```

$$\{3:2,\ -9\} \quad (5)$$

Type: Multiset Integer

The **union** of two multisets is additive.

```
U := union(s,t)
```

$$\{1,\ 5:2,\ 4:3,\ 4:4,\ 5,\ 6,\ 7,\ 10,\ -9\} \quad (6)$$

Type: Multiset Integer

The **intersect** operation gives the elements that are in common, with additive multiplicity.

```
I := intersect(s,t)
```

$$\{5:2\} \quad (7)$$

Type: Multiset Integer

The **difference** of `s` and `t` consists of the elements that `s` has but `t` does not. Elements are regarded as indistinguishable, so that if `s` and `t` have any element in common, the **difference** does not contain that element.

```
difference(s,t)
```

$$\{1,\ 4:3,\ 4:4,\ 5,\ 6,\ 7,\ 10\} \quad (8)$$

Type: Multiset Integer

The **symmetricDifference** is the **union** of `difference(s, t)` and `difference(t, s)`.

```
S := symmetricDifference(s,t)
```

$$\{1,\ 4:3,\ 4:4,\ 5,\ 6,\ 7,\ 10,\ -9\} \quad (9)$$

Type: Multiset Integer

Check that the **union** of the **symmetricDifference** and the **intersect** equals the **union** of the elements.

```
(U = union(S,I))@Boolean
```

true (10)

Type: Boolean

Check some inclusion relations.

```
t1 := multiset [1,2,2,3]; [t1 < t, t1 < s, t < s, t1 <= s]
```

[false, true, false, true] (11)

Type: List Boolean

9.41 Multivariate-Polynomial

The domain constructor MultivariatePolynomial is similar to Polynomial except that it specifies the variables to be used. Most functions available for Polynomial are available for MultivariatePolynomial. The abbreviation for MultivariatePolynomial is MPOLY. The type expressions

MultivariatePolynomial([x,y],Integer) and MPOLY([x,y],INT)

refer to the domain of multivariate polynomials in the variables **x** and **y** where the coefficients are restricted to be integers. The first variable specified is the main variable and the display of the polynomial reflects this.

This polynomial appears with terms in descending powers of the variable **x**.

```
m : MPOLY([x,y],INT) := (x**2 - x*y**3 +3*y)**2
```

$$x^4 - 2\,y^3\,x^3 + \left(y^6 + 6\,y\right)\,x^2 - 6\,y^4\,x + 9\,y^2 \quad (1)$$

Type: MultivariatePolynomial([x, y], Integer)

It is easy to see a different variable ordering by doing a conversion.

```
m :: MPOLY([y,x],INT)
```

$$x^2\,y^6 - 6\,x\,y^4 - 2\,x^3\,y^3 + 9\,y^2 + 6\,x^2\,y + x^4 \quad (2)$$

Type: MultivariatePolynomial([y, x], Integer)

You can use other, unspecified variables, by using Polynomial in the coefficient type of MPOLY.

```
p : MPOLY([x,y],POLY INT)
```

Type: Void

```
p := (a**2*x - b*y**2 + 1)**2
```

$$a^4\,x^2 + \left(-2\,a^2\,b\,y^2 + 2\,a^2\right)\,x + b^2\,y^4 - 2\,b\,y^2 + 1 \quad (4)$$

Type: MultivariatePolynomial([x, y], Polynomial Integer)

Conversions can be used to re-express such polynomials in terms of the other variables. For example, you can first push all the variables into a polynomial with integer coefficients.

```
p :: POLY INT
```

$$b^2\,y^4 + \left(-2\,a^2\,b\,x - 2\,b\right)\,y^2 + a^4\,x^2 + 2\,a^2\,x + 1 \quad (5)$$

Type: Polynomial Integer

Now pull out the variables of interest.

```
% :: MPOLY([a,b],POLY INT)
```

$$x^2\ a^4 + \left(-2\ x\ y^2\ b + 2\ x\right)\ a^2 + y^4\ b^2 - 2\ y^2\ b + 1 \tag{6}$$

Type: MultivariatePolynomial([a, b], Polynomial Integer)

Restriction:
AXIOM does not allow you to create types where MultivariatePolynomial is contained in the coefficient type of Polynomial. Therefore, `MPOLY([x,y],POLY INT)` is legal but `POLY MPOLY([x,y],INT)` is not.

Multivariate polynomials may be combined with univariate polynomials to create types with special structures.

```
q : UP(x, FRAC MPOLY([y,z],INT))
```

Type: Void

This is a polynomial in x whose coefficients are quotients of polynomials in y and z.

```
q := (x**2 - x*(z+1)/y +2)**2
```

$$x^4 + \frac{-2\ z - 2}{y}\ x^3 + \frac{4\ y^2 + z^2 + 2\ z + 1}{y^2}\ x^2 + \frac{-4\ z - 4}{y}\ x + 4 \tag{8}$$

Type: UnivariatePolynomial(x, Fraction MultivariatePolynomial([y, z], Integer))

Use conversions for structural rearrangements. z does not appear in a denominator and so it can be made the main variable.

```
q :: UP(z, FRAC MPOLY([x,y],INT))
```

$$\frac{x^2}{y^2}\ z^2 + \frac{-2\ y\ x^3 + 2\ x^2 - 4\ y\ x}{y^2}\ z + \frac{y^2\ x^4 - 2\ y\ x^3 + \left(4\ y^2 + 1\right)\ x^2 - 4\ y\ x + 4\ y^2}{y^2} \tag{9}$$

Type: UnivariatePolynomial(z, Fraction MultivariatePolynomial([x, y], Integer))

Or you can make a multivariate polynomial in x and z whose coefficients are fractions in polynomials in y.

```
q :: MPOLY([x,z], FRAC UP(y,INT))
```

$$x^4 + \left(-\frac{2}{y}\ z - \frac{2}{y}\right)\ x^3 + \left(\frac{1}{y^2}\ z^2 + \frac{2}{y^2}\ z + \frac{4\ y^2 + 1}{y^2}\right)\ x^2 + \left(-\frac{4}{y}\ z - \frac{4}{y}\right)\ x + 4 \tag{10}$$

Type: MultivariatePolynomial([x, z], Fraction UnivariatePolynomial(y, Integer))

A conversion like `q :: MPOLY([x,y], FRAC UP(z,INT))` is not possible in this example because y appears in the denominator of a fraction. As you can see, AXIOM provides extraordinary flexibility in the manipulation and display of expressions via its conversion facility.

For more information on related topics, see 'Polynomial' on page 436,

'UnivariatePolynomial' on page 472, and 'DistributedMultivariatePolynomial' on page 352.

Issue the system command `)show MultivariatePolynomial` to display the full list of operations defined by MultivariatePolynomial.

9.42 None

The None domain is not very useful for interactive work but it is provided nevertheless for completeness of the AXIOM type system.

Probably the only place you will ever see it is if you enter an empty list with no type information.

```
[]
```

$$[] \quad (1)$$

Type: List None

Such an empty list can be converted into an empty list of any other type.

```
[] :: List Float
```

$$[] \quad (2)$$

Type: List Float

If you wish to produce an empty list of a particular type directly, such as List NonNegativeInteger, do it this way.

```
[]$List(NonNegativeInteger)
```

$$[] \quad (3)$$

Type: List NonNegativeInteger

9.43 Octonion

The Octonions, also called the Cayley-Dixon algebra, defined over a commutative ring are an eight-dimensional non-associative algebra. Their construction from quaternions is similar to the construction of quaternions from complex numbers (see 'Quaternion' on page 442).

As Octonion creates an eight-dimensional algebra, you have to give eight components to construct an octonion.

```
oci1 := octon(1,2,3,4,5,6,7,8)
```

$$1+2\,i+3\,j+4\,k+5\,E+6\,I+7\,J+8\,K \quad (1)$$

Type: Octonion Integer

```
oci2 := octon(7,2,3,-4,5,6,-7,0)
```

$$7+2\,i+3\,j-4\,k+5\,E+6\,I-7\,J \quad (2)$$

Type: Octonion Integer

Or you can use two quaternions to create an octonion.

```
oci3 := octon(quatern(-7,-12,3,-10), quatern(5,6,9,0))
```

$$-7-12\,i+3\,j-10\,k+5\,E+6\,I+9\,J \quad (3)$$

Type: Octonion Integer

You can easily demonstrate the non-associativity of multiplication.

```
(oci1 * oci2) * oci3 - oci1 * (oci2 * oci3)
```

$$2696\,i-2928\,j-4072\,k+16\,E-1192\,I+832\,J+2616\,K \quad (4)$$

Type: Octonion Integer

As with the quaternions, we have a real part, the imaginary parts i, j, k, and four additional imaginary parts E, I, J and K. These parts correspond to the canonical basis (1,i,j,k,E,I,J,K).

For each basis element there is a component operation to extract the coefficient of the basis element for a given octonion.

```
[real ocil, imagi ocil, imagj ocil, imagk ocil, imagE
  ocil, imagI ocil, imagJ ocil, imagK ocil]
```

$$[1, 2, 3, 4, 5, 6, 7, 8] \tag{5}$$

Type: List PositiveInteger

A basis with respect to the quaternions is given by (1,E). However, you might ask, what then are the commuting rules? To answer this, we create some generic elements.

We do this in AXIOM by simply changing the ground ring from Integer to Polynomial Integer.

```
q : Quaternion Polynomial Integer := quatern(q1, qi, qj,
  qk)
```

$$q1 + qi\ i + qj\ j + qk\ k \tag{6}$$

Type: Quaternion Polynomial Integer

```
E : Octonion Polynomial Integer:= octon(0,0,0,0,1,0,0,0)
```

$$E \tag{7}$$

Type: Octonion Polynomial Integer

Note that quaternions are automatically converted to octonions in the obvious way.

```
q * E
```

$$q1\ E + qi\ I + qj\ J + qk\ K \tag{8}$$

Type: Octonion Polynomial Integer

```
E * q
```

$$q1\ E - qi\ I - qj\ J - qk\ K \tag{9}$$

Type: Octonion Polynomial Integer

```
q * 1$(Octonion Polynomial Integer)
```

$$q1 + qi\ i + qj\ j + qk\ k \tag{10}$$

Type: Octonion Polynomial Integer

```
1$(Octonion Polynomial Integer) * q
```

$$q1 + qi\ i + qj\ j + qk\ k \tag{11}$$

Type: Octonion Polynomial Integer

Finally, we check that the **norm**, defined as the sum of the squares of the coefficients, is a multiplicative map.

```
o : Octonion Polynomial Integer := octon(o1, oi, oj, ok,
  oE, oI, oJ, oK)
```

$$o1 + oi\ i + oj\ j + ok\ k + oE\ E + oI\ I + oJ\ J + oK\ K \tag{12}$$

Type: Octonion Polynomial Integer

```
norm o
```

$$ok^2 + oj^2 + oi^2 + oK^2 + oJ^2 + oI^2 + oE^2 + o1^2 \quad (13)$$

Type: Polynomial Integer

```
p : Octonion Polynomial Integer := octon(p1, pi, pj, pk,
  pE, pI, pJ, pK)
```

$$p1 + pi\ i + pj\ j + pk\ k + pE\ E + pI\ I + pJ\ J + pK\ K \quad (14)$$

Type: Octonion Polynomial Integer

Since the result is 0, the norm is multiplicative.

```
norm(o*p)-norm(p*o)
```

$$0 \quad (15)$$

Type: Polynomial Integer

Issue the system command `)show Octonion` to display the full list of operations defined by Octonion.

9.44 OneDimensionalArray

The OneDimensionalArray domain is used for storing data in a one-dimensional indexed data structure. Such an array is a homogeneous data structure in that all the entries of the array must belong to the same AXIOM domain. Each array has a fixed length specified by the user and arrays are not extensible. The indexing of one-dimensional arrays is one-based. This means that the "first" element of an array is given the index `1`. See also 'Vector' on page 478 and 'FlexibleArray' on page 366.

To create a one-dimensional array, apply the operation **oneDimensionalArray** to a list.

```
oneDimensionalArray [i**2 for i in 1..10]
```

$$[1, 4, 9, 16, 25, 36, 49, 64, 81, 100] \quad (1)$$

Type: OneDimensionalArray PositiveInteger

Another approach is to first create `a`, a one-dimensional array of 10 `0`'s. OneDimensionalArray has the convenient abbreviation ARRAY1.

```
a : ARRAY1 INT := new(10,0)
```

$$[0, 0, 0, 0, 0, 0, 0, 0, 0, 0] \quad (2)$$

Type: OneDimensionalArray Integer

Set each `i`th element to i, then display the result.

```
for i in 1..10 repeat a.i := i; a
```

$$[1, 2, 3, 4, 5, 6, 7, 8, 9, 10] \quad (3)$$

Type: OneDimensionalArray Integer

Square each element by mapping the function $i \mapsto i^2$ onto each element.

```
map!(i +-> i ** 2,a); a
```

$$[1, 4, 9, 16, 25, 36, 49, 64, 81, 100] \quad (4)$$

Type: OneDimensionalArray Integer

Reverse the elements in place.

```
reverse! a
```

$$[100, 81, 64, 49, 36, 25, 16, 9, 4, 1] \tag{5}$$

Type: OneDimensionalArray Integer

Swap the 4th and 5th element.

```
swap!(a,4,5); a
```

$$[100, 81, 64, 36, 49, 25, 16, 9, 4, 1] \tag{6}$$

Type: OneDimensionalArray Integer

Sort the elements in place.

```
sort! a
```

$$[1, 4, 9, 16, 25, 36, 49, 64, 81, 100] \tag{7}$$

Type: OneDimensionalArray Integer

Create a new one-dimensional array b containing the last 5 elements of a.

```
b := a(6..10)
```

$$[36, 49, 64, 81, 100] \tag{8}$$

Type: OneDimensionalArray Integer

Replace the first 5 elements of a with those of b.

```
copyInto!(a,b,1)
```

$$[36, 49, 64, 81, 100, 36, 49, 64, 81, 100] \tag{9}$$

Type: OneDimensionalArray Integer

9.45 Operator

Given any ring R, the ring of the Integer-linear operators over R is called Operator(R). To create an operator over R, first create a basic operator using the operation **operator**, and then convert it to Operator(R) for the R you want.

We choose R to be the two by two matrices over the integers.

```
R := SQMATRIX(2, INT)
```

$$\text{SquareMatrix}\ (2, \text{Integer}\ ) \tag{1}$$

Type: Domain

Create the operator tilde on R.

```
t := operator("tilde") :: OP(R)
```

$$\text{tilde} \tag{2}$$

Type: Operator SquareMatrix(2, Integer)

To attach an evaluation function (from R to R) to an operator over R, use `evaluate(op, f)` where op is an operator over R and f is a function `R -> R`. This needs to be done only once when the operator is defined. Note that f must be Integer-linear (that is, `f(ax+y) = a f(x) + f(y)` for any integer a, and any x and y in R).

We now attach the transpose map to the above operator t.

```
evaluate(t, m +-> transpose m)
```

$$\text{tilde} \tag{3}$$

Type: Operator SquareMatrix(2, Integer)

Operators can be manipulated formally as in any ring: "+" is the pointwise addition and "*" is composition. Any element `x` of `R` can be converted to an operator $\mathtt{op_x}$ over `R`, and the evaluation function of $\mathtt{op_x}$ is left-multiplication by `x`.

Multiplying on the left by this matrix swaps the two rows.

```
s : R := matrix [[0, 1], [1, 0]]
```

$$\begin{bmatrix} 0 & 1 \\ 1 & 0 \end{bmatrix} \tag{4}$$

Type: SquareMatrix(2, Integer)

Can you guess what is the action of the following operator?

```
rho := t * s
```

$$\text{tilde} \begin{bmatrix} 0 & 1 \\ 1 & 0 \end{bmatrix} \tag{5}$$

Type: Operator SquareMatrix(2, Integer)

Hint: applying `rho` four times gives the identity, so `rho**4-1` should return 0 when applied to any two by two matrix.

```
z := rho**4 - 1
```

$$-1 + \text{tilde} \begin{bmatrix} 0 & 1 \\ 1 & 0 \end{bmatrix} \text{tilde} \begin{bmatrix} 0 & 1 \\ 1 & 0 \end{bmatrix} \text{tilde} \begin{bmatrix} 0 & 1 \\ 1 & 0 \end{bmatrix} \text{tilde} \begin{bmatrix} 0 & 1 \\ 1 & 0 \end{bmatrix} \tag{6}$$

Type: Operator SquareMatrix(2, Integer)

Now check with this matrix.

```
m:R := matrix [[1, 2], [3, 4]]
```

$$\begin{bmatrix} 1 & 2 \\ 3 & 4 \end{bmatrix} \tag{7}$$

Type: SquareMatrix(2, Integer)

```
z m
```

$$\begin{bmatrix} 0 & 0 \\ 0 & 0 \end{bmatrix} \tag{8}$$

Type: SquareMatrix(2, Integer)

As you have probably guessed by now, `rho` acts on matrices by rotating the elements clockwise.

```
rho m
```

$$\begin{bmatrix} 3 & 1 \\ 4 & 2 \end{bmatrix} \tag{9}$$

Type: SquareMatrix(2, Integer)

```
rho rho m
```

$$\begin{bmatrix} 4 & 3 \\ 2 & 1 \end{bmatrix} \tag{10}$$

Type: SquareMatrix(2, Integer)

```
(rho**3) m
```

$$\begin{bmatrix} 2 & 4 \\ 1 & 3 \end{bmatrix} \tag{11}$$

Type: SquareMatrix(2, Integer)

Do the swapping of rows and transposition commute? We can check by computing their bracket.

```
b := t * s - s * t
```

$$-\begin{bmatrix} 0 & 1 \\ 1 & 0 \end{bmatrix} \text{tilde} + \text{tilde} \begin{bmatrix} 0 & 1 \\ 1 & 0 \end{bmatrix} \tag{12}$$

Type: Operator SquareMatrix(2, Integer)

Now apply it to m.

```
b m
```

$$\begin{bmatrix} 1 & -3 \\ 3 & -1 \end{bmatrix} \tag{13}$$

Type: SquareMatrix(2, Integer)

Next we demonstrate how to define a differential operator on a polynomial ring.

This is the recursive definition of the n-th Legendre polynomial.

```
L n ==
  n = 0 => 1
  n = 1 => x
  (2*n-1)/n * x * L(n-1) - (n-1)/n * L(n-2)
```

Type: Void

Create the differential operator $\frac{d}{dx}$ on polynomials in x over the rational numbers.

```
dx := operator("D") :: OP(POLY FRAC INT)
```

$$D \tag{15}$$

Type: Operator Polynomial Fraction Integer

Now attach the map to it.

```
evaluate(dx, p +-> D(p, 'x))
```

$$D \tag{16}$$

Type: Operator Polynomial Fraction Integer

This is the differential equation satisfied by the n-th Legendre polynomial.

```
E n == (1 - x**2) * dx**2 - 2 * x * dx + n*(n+1)
```

Type: Void

Now we verify this for n = 15. Here is the polynomial.

```
L 15
```

$$\frac{9694845}{2048} x^{15} - \frac{35102025}{2048} x^{13} + \frac{50702925}{2048} x^{11} - \frac{37182145}{2048} x^9 + \frac{14549535}{2048} x^7 - \frac{2909907}{2048} x^5 + \frac{255255}{2048} x^3 - \frac{6435}{2048} x \tag{18}$$

Type: Polynomial Fraction Integer

Here is the operator.

```
E 15
```

$$240 - 2\,x\,D + \left(-x^2 + 1\right) D^2 \tag{19}$$

Type: Operator Polynomial Fraction Integer

Here is the evaluation.

```
(E 15)(L 15)
```

0 (20)

Type: Polynomial Fraction Integer

9.46 OrderlyDifferentialPolynomial

Many systems of differential equations may be transformed to equivalent systems of ordinary differential equations where the equations are expressed polynomially in terms of the unknown functions. In AXIOM, the domain constructors OrderlyDifferentialPolynomial (abbreviated ODPOL) and SequentialDifferentialPolynomial (abbreviation SDPOL) implement two domains of ordinary differential polynomials over any differential ring. In the simplest case, this differential ring is usually either the ring of integers, or the field of rational numbers. However, AXIOM can handle ordinary differential polynomials over a field of rational functions in a single indeterminate.

The two domains ODPOL and SDPOL are almost identical, the only difference being the choice of a different ranking, which is an ordering of the derivatives of the indeterminates. The first domain uses an orderly ranking, that is, derivatives of higher order are ranked higher, and derivatives of the same order are ranked alphabetically. The second domain uses a sequential ranking, where derivatives are ordered first alphabetically by the differential indeterminates, and then by order. A more general domain constructor, DifferentialSparseMultivariatePolynomial (abbreviation DSMP) allows both a user-provided list of differential indeterminates as well as a user-defined ranking. We shall illustrate ODPOL(FRAC INT), which constructs a domain of ordinary differential polynomials in an arbitrary number of differential indeterminates with rational numbers as coefficients.

```
dpol:= ODPOL(FRAC INT)
```

OrderlyDifferentialPolynomial Fraction Integer (1)

Type: Domain

A differential indeterminate w may be viewed as an infinite sequence of algebraic indeterminates, which are the derivatives of w. To facilitate referencing these, AXIOM provides the operation **makeVariable** to convert an element of type Symbol to a map from the natural numbers to the differential polynomial ring.

```
w := makeVariable('w)$dpol
```

theMap (...) (2)

Type: (NonNegativeInteger → OrderlyDifferentialPolynomial Fraction Integer)

```
z := makeVariable('z)$dpol
```

theMap (...) (3)

Type: (NonNegativeInteger → OrderlyDifferentialPolynomial Fraction Integer)

The fifth derivative of w can be obtained by applying the map w to the number 5. Note that the order of differentiation is given as a subscript (except when the order is 0).

```
w.5
```

$$w_5 \tag{4}$$

Type: OrderlyDifferentialPolynomial Fraction Integer

```
w 0
```

$$w \tag{5}$$

Type: OrderlyDifferentialPolynomial Fraction Integer

The first five derivatives of z can be generated by a list.

```
[z.i for i in 1..5]
```

$$[z_1, z_2, z_3, z_4, z_5] \tag{6}$$

Type: List OrderlyDifferentialPolynomial Fraction Integer

The usual arithmetic can be used to form a differential polynomial from the derivatives.

```
f:= w.4 - w.1 * w.1 * z.3
```

$$w_4 - {w_1}^2 z_3 \tag{7}$$

Type: OrderlyDifferentialPolynomial Fraction Integer

```
g:=(z.1)**3 * (z.2)**2 - w.2
```

$${z_1}^3 {z_2}^2 - w_2 \tag{8}$$

Type: OrderlyDifferentialPolynomial Fraction Integer

The operation **D** computes the derivative of any differential polynomial.

```
D(f)
```

$$w_5 - {w_1}^2 z_4 - 2 w_1 w_2 z_3 \tag{9}$$

Type: OrderlyDifferentialPolynomial Fraction Integer

The same operation can compute higher derivatives, like the fourth derivative.

```
D(f,4)
```

$$w_8 - {w_1}^2 z_7 - 8 w_1 w_2 z_6 + \left(-12 w_1 w_3 - 12 {w_2}^2\right) z_5 - 2 w_1 z_3 w_5 + \left(-8 w_1 w_4 - 24 w_2 w_3\right) z_4 - 8 w_2 z_3 w_4 - 6 {w_3}^2 z_3 \tag{10}$$

Type: OrderlyDifferentialPolynomial Fraction Integer

The operation **makeVariable** creates a map to facilitate referencing the derivatives of f, similar to the map w.

```
df:=makeVariable(f)$dpol
```

theMap (...) (11)

Type: (NonNegativeInteger → OrderlyDifferentialPolynomial Fraction Integer)

The fourth derivative of f may be referenced easily.

```
df.4
```

$$w_8 - {w_1}^2 \; z_7 - 8 \; w_1 \; w_2 \; z_6 + \left(-12 \; w_1 \; w_3 - 12 \; {w_2}^2\right) \; z_5 - 2 \; w_1 \; z_3 \; w_5 + (-8 \; w_1 \; w_4 - 24 \; w_2 \; w_3) \; z_4 - 8 \; w_2 \; z_3 \; w_4 - 6 \; {w_3}^2 \; z_3 \qquad (12)$$

Type: OrderlyDifferentialPolynomial Fraction Integer

The operation **order** returns the order of a differential polynomial, or the order in a specified differential indeterminate.

```
order(g)
```

$$2 \qquad (13)$$

Type: PositiveInteger

```
order(g, 'w)
```

$$2 \qquad (14)$$

Type: PositiveInteger

The operation **differentialVariables** returns a list of differential indeterminates occurring in a differential polynomial.

```
differentialVariables(g)
```

$$[z, \; w] \qquad (15)$$

Type: List Symbol

The operation **degree** returns the degree, or the degree in the differential indeterminate specified.

```
degree(g)
```

$${z_2}^2 \; {z_1}^3 \qquad (16)$$

Type: IndexedExponents OrderlyDifferentialVariable Symbol

```
degree(g, 'w)
```

$$1 \qquad (17)$$

Type: PositiveInteger

The operation **weights** returns a list of weights of differential monomials appearing in differential polynomial, or a list of weights in a specified differential indeterminate.

```
weights(g)
```

$$[7, \; 2] \qquad (18)$$

Type: List NonNegativeInteger

```
weights(g,'w)
```

$$[2] \qquad (19)$$

Type: List NonNegativeInteger

The operation **weight** returns the maximum weight of all differential monomials appearing in the differential polynomial.

```
weight(g)
```

$$7 \qquad (20)$$

Type: PositiveInteger

A differential polynomial is *isobaric* if the weights of all differential monomials appearing in it are equal.

```
isobaric?(g)
```

$$\text{false} \tag{21}$$

Type: Boolean

To substitute *differentially*, use **eval**. Note that we must coerce `'w` to Symbol, since in ODPOL, differential indeterminates belong to the domain Symbol. Compare this result to the next, which substitutes *algebraically* (no substitution is done since `w.0` does not appear in `g`).

```
eval(g,['w::Symbol],[f])
```

$$-w_6 + {w_1}^2\ z_5 + 4\ w_1\ w_2\ z_4 + \left(2\ w_1\ w_3 + 2\ {w_2}^2\right)\ z_3 + {z_1}^3\ {z_2}^2 \tag{22}$$

Type: OrderlyDifferentialPolynomial Fraction Integer

```
eval(g,['w],[f])
```

$${z_1}^3\ {z_2}^2 - w_2 \tag{23}$$

Type: OrderlyDifferentialPolynomial Fraction Integer

Since OrderlyDifferentialPolynomial belongs to PolynomialCategory, all the operations defined in the latter category, or in packages for the latter category, are available.

```
monomials(g)
```

$$\left[{z_1}^3\ {z_2}^2,\ -w_2\right] \tag{24}$$

Type: List OrderlyDifferentialPolynomial Fraction Integer

```
variables(g)
```

$$\left[z_2,\ w_2,\ z_1\right] \tag{25}$$

Type: List OrderlyDifferentialVariable Symbol

```
gcd(f,g)
```

$$1 \tag{26}$$

Type: OrderlyDifferentialPolynomial Fraction Integer

```
groebner([f,g])
```

$$\left[w_4 - {w_1}^2\ z_3,\ {z_1}^3\ {z_2}^2 - w_2\right] \tag{27}$$

Type: List OrderlyDifferentialPolynomial Fraction Integer

The next three operations are essential for elimination procedures in differential polynomial rings. The operation **leader** returns the leader of a differential polynomial, which is the highest ranked derivative of the differential indeterminates that occurs.

```
lg:=leader(g)
```

$$z_2 \tag{28}$$

Type: OrderlyDifferentialVariable Symbol

The operation **separant** returns the separant of a differential polynomial, which is the partial derivative with respect to the leader.

```
sg:=separant(g)
```

$$2\ z_1^{\ 3}\ z_2 \tag{29}$$

Type: OrderlyDifferentialPolynomial Fraction Integer

The operation **initial** returns the initial, which is the leading coefficient when the given differential polynomial is expressed as a polynomial in the leader.

```
ig:=initial(g)
```

$$z_1^{\ 3} \tag{30}$$

Type: OrderlyDifferentialPolynomial Fraction Integer

Using these three operations, it is possible to reduce `f` modulo the differential ideal generated by `g`. The general scheme is to first reduce the order, then reduce the degree in the leader. First, eliminate `z.3` using the derivative of `g`.

```
g1 := D g
```

$$2\ z_1^{\ 3}\ z_2\ z_3 - w_3 + 3\ z_1^{\ 2}\ z_2^{\ 3} \tag{31}$$

Type: OrderlyDifferentialPolynomial Fraction Integer

Find its leader.

```
lg1:= leader g1
```

$$z_3 \tag{32}$$

Type: OrderlyDifferentialVariable Symbol

Differentiate `f` partially with respect to this leader.

```
pdf:=D(f, lg1)
```

$$-w_1^{\ 2} \tag{33}$$

Type: OrderlyDifferentialPolynomial Fraction Integer

Compute the partial remainder of `f` with respect to `g`.

```
prf:=sg * f- pdf * g1
```

$$2\ z_1^{\ 3}\ z_2\ w_4 - w_1^{\ 2}\ w_3 + 3\ w_1^{\ 2}\ z_1^{\ 2}\ z_2^{\ 3} \tag{34}$$

Type: OrderlyDifferentialPolynomial Fraction Integer

Note that high powers of `lg` still appear in `prf`. Compute the leading coefficient of `prf` as a polynomial in the leader of `g`.

```
lcf:=leadingCoefficient univariate(prf, lg)
```

$$3\ w_1^{\ 2}\ z_1^{\ 2} \tag{35}$$

Type: OrderlyDifferentialPolynomial Fraction Integer

Finally, continue eliminating the high powers of `lg` appearing in `prf` to obtain the (pseudo) remainder of `f` modulo `g` and its derivatives.

```
ig * prf - lcf * g * lg
```

$$2\ z_1^{\ 6}\ z_2\ w_4 - w_1^{\ 2}\ z_1^{\ 3}\ w_3 + 3\ w_1^{\ 2}\ z_1^{\ 2}\ w_2\ z_2 \tag{36}$$

Type: OrderlyDifferentialPolynomial Fraction Integer

9.47 PartialFraction

A *partial fraction* is a decomposition of a quotient into a sum of quotients where the denominators of the summands are powers of primes.[5] For example, the rational number `1/6` is decomposed into `1/2 -1/3`.

[5]Most people first encounter partial fractions when they are learning integral calculus. For a technical discussion of partial fractions, see, for example, Lang's *Algebra*.

You can compute partial fractions of quotients of objects from domains belonging to the category EuclideanDomain. For example, Integer, Complex Integer, and UnivariatePolynomial(x, Fraction Integer) all belong to EuclideanDomain. In the examples following, we demonstrate how to decompose quotients of each of these kinds of object into partial fractions. Issue the system command `)show PartialFraction` to display the full list of operations defined by PartialFraction.

It is necessary that we know how to factor the denominator when we want to compute a partial fraction. Although the interpreter can often do this automatically, it may be necessary for you to include a call to **factor**. In these examples, it is not necessary to factor the denominators explicitly.

The main operation for computing partial fractions is called **partialFraction** and we use this to compute a decomposition of `1 / 10!`. The first argument to **partialFraction** is the numerator of the quotient and the second argument is the factored denominator.

```
partialFraction(1,factorial 10)
```

$$\frac{159}{2^8} - \frac{23}{3^4} - \frac{12}{5^2} + \frac{1}{7} \qquad (1)$$

Type: PartialFraction Integer

Since the denominators are powers of primes, it may be possible to expand the numerators further with respect to those primes. Use the operation **padicFraction** to do this.

```
f := padicFraction(%)
```

$$\frac{1}{2} + \frac{1}{2^4} + \frac{1}{2^5} + \frac{1}{2^6} + \frac{1}{2^7} + \frac{1}{2^8} - \frac{2}{3^2} - \frac{1}{3^3} - \frac{2}{3^4} - \frac{2}{5} - \frac{2}{5^2} + \frac{1}{7} \qquad (2)$$

Type: PartialFraction Integer

The operation **compactFraction** returns an expanded fraction into the usual form. The compacted version is used internally for computational efficiency.

```
compactFraction(f)
```

$$\frac{159}{2^8} - \frac{23}{3^4} - \frac{12}{5^2} + \frac{1}{7} \qquad (3)$$

Type: PartialFraction Integer

You can add, subtract, multiply and divide partial fractions. In addition, you can extract the parts of the decomposition. **numberOfFractionalTerms** computes the number of terms in the fractional part. This does not include the whole part of the fraction, which you get by calling **wholePart**. In this example, the whole part is just `0`.

```
numberOfFractionalTerms(f)
```

$$12 \qquad (4)$$

Type: PositiveInteger

The operation **nthFractionalTerm** returns the individual terms in the decomposition. Notice that the object returned is a partial fraction itself. **firstNumer** and **firstDenom** extract the numerator and denominator of the first term of the fraction.

```
nthFractionalTerm(f,3)
```

$$\frac{1}{2^5} \tag{5}$$

Type: PartialFraction Integer

Given two gaussian integers (see 'Complex' on page 333), you can decompose their quotient into a partial fraction.

```
partialFraction(1,- 13 + 14 * %i)
```

$$-\frac{1}{1+2\,\%i}+\frac{4}{3+8\,\%i} \tag{6}$$

Type: PartialFraction Complex Integer

To convert back to a quotient, simply use a conversion.

```
% :: Fraction Complex Integer
```

$$-\frac{\%i}{14+13\,\%i} \tag{7}$$

Type: Fraction Complex Integer

To conclude this section, we compute the decomposition of

$$\frac{1}{(x+1)(x+2)^2(x+3)^3(x+4)^4}$$

The polynomials in this object have type UnivariatePolynomial(x, Fraction Integer).

We use the **primeFactor** operation (see 'Factored' on page 356) to create the denominator in factored form directly.

```
u : FR UP(x, FRAC INT) := reduce(*,[primeFactor(x+i,i) for
  i in 1..4])
```

$$(x+1)\;(x+2)^2\;(x+3)^3\;(x+4)^4 \tag{8}$$

Type: Factored UnivariatePolynomial(x, Fraction Integer)

These are the compact and expanded partial fractions for the quotient.

```
partialFraction(1,u)
```

$$\frac{\frac{1}{648}}{x+1}+\frac{\frac{1}{4}\,x+\frac{7}{16}}{(x+2)^2}+\frac{-\frac{17}{8}\,x^2-12\,x-\frac{139}{8}}{(x+3)^3}+ \frac{\frac{607}{324}\,x^3+\frac{10115}{432}\,x^2+\frac{391}{4}\,x+\frac{44179}{324}}{(x+4)^4} \tag{9}$$

Type: PartialFraction UnivariatePolynomial(x, Fraction Integer)

```
padicFraction %
```

$$\frac{\frac{1}{648}}{x+1}+\frac{\frac{1}{4}}{x+2}-\frac{\frac{1}{16}}{(x+2)^2}-\frac{\frac{17}{8}}{x+3}+\frac{\frac{3}{4}}{(x+3)^2}-\frac{\frac{1}{2}}{(x+3)^3}+\frac{\frac{607}{324}}{x+4}+\frac{\frac{403}{432}}{(x+4)^2}+\frac{\frac{13}{36}}{(x+4)^3}+\frac{\frac{1}{12}}{(x+4)^4} \quad (10)$$

Type: PartialFraction UnivariatePolynomial(x, Fraction Integer)

9.48 Permanent

The package Permanent provides the function **permanent** for square matrices. The **permanent** of a square matrix can be computed in the same way as the determinant by expansion of minors except that for the permanent the sign for each element is 1, rather than being 1 if the row plus column indices is positive and -1 otherwise. This function is much more difficult to compute efficiently than the **determinant**. An example of the use of **permanent** is the calculation of the n th derangement number, defined to be the number of different possibilities for n couples to dance but never with their own spouse.

Consider an n by n matrix with entries 0 on the diagonal and 1 elsewhere. Think of the rows as one-half of each couple (for example, the males) and the columns the other half. The permanent of such a matrix gives the desired derangement number.

```
kn n ==
  r : MATRIX INT := new(n,n,1)
  for i in 1..n repeat
    r.i.i := 0
  r
```

Type: Void

Here are some derangement numbers, which you see grow quite fast.

```
permanent(kn(5) :: SQMATRIX(5,INT))
```

$$44 \quad (2)$$

Type: PositiveInteger

```
[permanent(kn(n) :: SQMATRIX(n,INT)) for n in 1..13]
```

$$[0, 1, 2, 9, 44, 265, 1854, 14833, 133496, 1334961, 14684570, 176214841, 2290792932] \quad (3)$$

Type: List NonNegativeInteger

9.49 Polynomial

The domain constructor Polynomial (abbreviation: POLY) provides polynomials with an arbitrary number of unspecified variables.

It is used to create the default polynomial domains in AXIOM. Here the coefficients are integers.

```
x + 1
```

$$x + 1 \qquad (1)$$

Type: Polynomial Integer

Here the coefficients have type Float.

```
z - 2.3
```

$$z - 2.3 \qquad (2)$$

Type: Polynomial Float

And here we have a polynomial in two variables with coefficients which have type Fraction Integer.

```
y**2 - z + 3/4
```

$$-z + y^2 + \frac{3}{4} \qquad (3)$$

Type: Polynomial Fraction Integer

The representation of objects of domains created by Polynomial is that of recursive univariate polynomials.[6]

This recursive structure is sometimes obvious from the display of a polynomial.

```
y **2 + x*y + y
```

$$y^2 + (x + 1)\, y \qquad (4)$$

Type: Polynomial Integer

In this example, you see that the polynomial is stored as a polynomial in **y** with coefficients that are polynomials in **x** with integer coefficients. In fact, you really don't need to worry about the representation unless you are working on an advanced application where it is critical. The polynomial types created from DistributedMultivariatePolynomial and HomogeneousDistributedMultivariatePolynomial (discussed in 'DistributedMultivariatePolynomial' on page 352) are stored and displayed in a non-recursive manner.

You see a "flat" display of the above polynomial by converting to one of those types.

```
% :: DMP([y,x],INT)
```

$$y^2 + y\,x + y \qquad (5)$$

Type: DistributedMultivariatePolynomial([y, x], Integer)

We will demonstrate many of the polynomial facilities by using two polynomials with integer coefficients.

By default, the interpreter expands polynomial expressions, even if they are written in a factored format.

```
p := (y-1)**2 * x * z
```

$$\left(x\,y^2 - 2\,x\,y + x\right)\,z \qquad (6)$$

Type: Polynomial Integer

See 'Factored' on page 356 to see how to create objects in factored form directly.

```
q := (y-1) * x * (z+5)
```

$$(x\,y - x)\,z + 5\,x\,y - 5\,x \qquad (7)$$

Type: Polynomial Integer

[6]The term `univariate` means "one variable." `multivariate` means "possibly more than one variable."

The fully factored form can be recovered by using **factor**.

```
factor(q)
```

$$x\ (y-1)\ (z+5) \tag{8}$$

Type: Factored Polynomial Integer

This is the same name used for the operation to factor integers. Such reuse of names is called *overloading* and makes it much easier to think of solving problems in general ways. AXIOM facilities for factoring polynomials created with Polynomial are currently restricted to the integer and rational number coefficient cases. There are more complete facilities for factoring univariate polynomials: see Section 8.2 on page 236.

The standard arithmetic operations are available for polynomials.

```
p - q**2
```

$$\left(-x^2\ y^2+2\ x^2\ y-x^2\right)\ z^2+$$
$$\left(\left(-10\ x^2+x\right)\ y^2+\left(20\ x^2-2\ x\right)\ y-10\ x^2+x\right)\ z-25\ x^2\ y^2+ \tag{9}$$
$$50\ x^2\ y-25\ x^2$$

Type: Polynomial Integer

The operation **gcd** is used to compute the greatest common divisor of two polynomials.

```
gcd(p,q)
```

$$x\ y-x \tag{10}$$

Type: Polynomial Integer

In the case of `p` and `q`, the gcd is obvious from their definitions. We factor the gcd to show this relationship better.

```
factor %
```

$$x\ (y-1) \tag{11}$$

Type: Factored Polynomial Integer

The least common multiple is computed by using **lcm**.

```
lcm(p,q)
```

$$\left(x\ y^2-2\ x\ y+x\right)\ z^2+\left(5\ x\ y^2-10\ x\ y+5\ x\right)\ z \tag{12}$$

Type: Polynomial Integer

Use **content** to compute the greatest common divisor of the coefficients of the polynomial.

```
content p
```

$$1 \tag{13}$$

Type: PositiveInteger

Many of the operations on polynomials require you to specify a variable. For example, **resultant** requires you to give the variable in which the polynomials should be expressed.

This computes the resultant of the values of p and q, considering them as polynomials in the variable z. They do not share a root when thought of as polynomials in z.

```
resultant(p,q,z)
```

$$5\,x^2\,y^3 - 15\,x^2\,y^2 + 15\,x^2\,y - 5\,x^2 \quad (14)$$

Type: Polynomial Integer

This value is 0 because as polynomials in x the polynomials have a common root.

```
resultant(p,q,x)
```

$$0 \quad (15)$$

Type: Polynomial Integer

The data type used for the variables created by Polynomial is Symbol. As mentioned above, the representation used by Polynomial is recursive and so there is a main variable for nonconstant polynomials.

The operation **mainVariable** returns this variable. The return type is actually a union of Symbol and "failed".

```
mainVariable p
```

$$z \quad (16)$$

Type: Union(Symbol, ...)

The latter branch of the union is be used if the polynomial has no variables, that is, is a constant.

```
mainVariable(1 :: POLY INT)
```

"failed" (17)

Type: Union("failed", ...)

You can also use the predicate **ground?** to test whether a polynomial is in fact a member of its ground ring.

```
ground? p
```

false (18)

Type: Boolean

```
ground?(1 :: POLY INT)
```

true (19)

Type: Boolean

The complete list of variables actually used in a particular polynomial is returned by **variables**. For constant polynomials, this list is empty.

```
variables p
```

$$[z,\, y,\, x] \quad (20)$$

Type: List Symbol

The **degree** operation returns the degree of a polynomial in a specific variable.

```
degree(p,x)
```

$$1 \quad (21)$$

Type: PositiveInteger

```
degree(p,y)
```

$$2 \quad (22)$$

Type: PositiveInteger

```
degree(p,z)
```

$$1 \tag{23}$$

Type: PositiveInteger

If you give a list of variables for the second argument, a list of the degrees in those variables is returned.

```
degree(p,[x,y,z])
```

$$[1,\ 2,\ 1] \tag{24}$$

Type: List NonNegativeInteger

The minimum degree of a variable in a polynomial is computed using **minimumDegree**.

```
minimumDegree(p,z)
```

$$1 \tag{25}$$

Type: PositiveInteger

The total degree of a polynomial is returned by **totalDegree**.

```
totalDegree p
```

$$4 \tag{26}$$

Type: PositiveInteger

It is often convenient to think of a polynomial as a leading monomial plus the remaining terms.

```
leadingMonomial p
```

$$x\ y^2\ z \tag{27}$$

Type: Polynomial Integer

The **reductum** operation returns a polynomial consisting of the sum of the monomials after the first.

```
reductum p
```

$$(-2\ x\ y + x)\ z \tag{28}$$

Type: Polynomial Integer

These have the obvious relationship that the original polynomial is equal to the leading monomial plus the reductum.

```
p - leadingMonomial p - reductum p
```

$$0 \tag{29}$$

Type: Polynomial Integer

The value returned by **leadingMonomial** includes the coefficient of that term. This is extracted by using **leadingCoefficient** on the original polynomial.

```
leadingCoefficient p
```

$$1 \tag{30}$$

Type: PositiveInteger

The operation **eval** is used to substitute a value for a variable in a polynomial.

```
p
```

$$\left(x\ y^2 - 2\ x\ y + x\right)\ z \tag{31}$$

Type: Polynomial Integer

This value may be another variable, a constant or a polynomial.

```
eval(p,x,w)
```

$$\left(w\ y^2 - 2\ w\ y + w\right)\ z \tag{32}$$

Type: Polynomial Integer

```
eval(p,x,1)
```

$$\left(y^2 - 2\ y + 1\right)\ z \tag{33}$$

Type: Polynomial Integer

Actually, all the things being substituted are just polynomials, some more trivial than others.

```
eval(p,x,y**2 - 1)
```

$$\left(y^4 - 2\ y^3 + 2\ y - 1\right)\ z \tag{34}$$

Type: Polynomial Integer

Derivatives are computed using the **D** operation.

```
D(p,x)
```

$$\left(y^2 - 2\ y + 1\right)\ z \tag{35}$$

Type: Polynomial Integer

The first argument is the polynomial and the second is the variable.

```
D(p,y)
```

$$(2\ x\ y - 2\ x)\ z \tag{36}$$

Type: Polynomial Integer

Even if the polynomial has only one variable, you must specify it.

```
D(p,z)
```

$$x\ y^2 - 2\ x\ y + x \tag{37}$$

Type: Polynomial Integer

Integration of polynomials is similar and the **integrate** operation is used.

Integration requires that the coefficients support division. Consequently, AXIOM converts polynomials over the integers to polynomials over the rational numbers before integrating them.

```
integrate(p,y)
```

$$\left(\frac{1}{3}\ x\ y^3 - x\ y^2 + x\ y\right)\ z \tag{38}$$

Type: Polynomial Fraction Integer

It is not possible, in general, to divide two polynomials. In our example using polynomials over the integers, the operation **monicDivide** divides a polynomial by a monic polynomial (that is, a polynomial with leading coefficient equal to 1). The result is a record of the quotient and remainder of the division.

You must specify the variable in which to express the polynomial.

```
qr := monicDivide(p,x+1,x)
```

$$\left[quotient = \left(y^2 - 2\ y + 1\right)\ z,\ remainder = \left(-y^2 + 2\ y - 1\right)\ z\right] \tag{39}$$

Type: Record(quotient: Polynomial Integer, remainder: Polynomial Integer)

The selectors of the components of the record are `quotient` and `remainder`. Issue this to extract the remainder.

```
qr.remainder
```

$$\left(-y^2 + 2\ y - 1\right)\ z \tag{40}$$

Type: Polynomial Integer

Now that we can extract the components, we can demonstrate the relationship among them and the arguments to our original expression `qr := monicDivide(p,x+1,x)`.

```
p - ((x+1) * qr.quotient + qr.remainder)
```

$$0 \tag{41}$$

Type: Polynomial Integer

If the "/" operator is used with polynomials, a fraction object is created. In this example, the result is an object of type Fraction Polynomial Integer.

```
p/q
```

$$\frac{(y-1)\, z}{z+5} \tag{42}$$

Type: Fraction Polynomial Integer

If you use rational numbers as polynomial coefficients, the resulting object is of type Polynomial Fraction Integer.

```
(2/3) * x**2 - y + 4/5
```

$$-y+\frac{2}{3}\, x^2+\frac{4}{5} \tag{43}$$

Type: Polynomial Fraction Integer

This can be converted to a fraction of polynomials and back again, if required.

```
% :: FRAC POLY INT
```

$$\frac{-15\, y+10\, x^2+12}{15} \tag{44}$$

Type: Fraction Polynomial Integer

```
% :: POLY FRAC INT
```

$$-y+\frac{2}{3}\, x^2+\frac{4}{5} \tag{45}$$

Type: Polynomial Fraction Integer

To convert the coefficients to floating point, map the **numeric** operation on the coefficients of the polynomial.

```
map(numeric,%)
```

$$-1.0\, y+0.66666666666666666667\, x^2+0.8 \tag{46}$$

Type: Polynomial Float

For more information on related topics, see 'UnivariatePolynomial' on page 472, 'MultivariatePolynomial' on page 421, and 'DistributedMultivariatePolynomial' on page 352. You can also issue the system command `)show Polynomial` to display the full list of operations defined by Polynomial.

9.50 Quaternion

The domain constructor Quaternion implements quaternions over commutative rings. For information on related topics, see 'Complex' on page 333 and 'Octonion' on page 423. You can also issue the system command `)show Quaternion` to display the full list of operations defined by Quaternion.

The basic function for creating quaternions is **quatern**. This is a quaternion over the rational numbers.

```
q := quatern(2/11,-8,3/4,1)
```

$$\frac{2}{11}-8\, i+\frac{3}{4}\, j+k \tag{1}$$

Type: Quaternion Fraction Integer

The four arguments are the real part, the i imaginary part, the j imaginary part and the k imaginary part, respectively.

```
[real q, imagI q, imagJ q, imagK q]
```

$$\left[\frac{2}{11}, -8, \frac{3}{4}, 1\right] \tag{2}$$

Type: List Fraction Integer

Because q is over the rationals (and nonzero), you can invert it.

```
inv q
```

$$\frac{352}{126993} + \frac{15488}{126993}\, i - \frac{484}{42331}\, j - \frac{1936}{126993}\, k \tag{3}$$

Type: Quaternion Fraction Integer

The usual arithmetic (ring) operations are available

```
q**6
```

$$-\frac{2029490709319345}{7256313856} - \frac{48251690851}{1288408}\, i + \frac{144755072553}{41229056}\, j + \frac{48251690851}{10307264}\, k \tag{4}$$

Type: Quaternion Fraction Integer

```
r := quatern(-2,3,23/9,-89); q + r
```

$$-\frac{20}{11} - 5\, i + \frac{119}{36}\, j - 88\, k \tag{5}$$

Type: Quaternion Fraction Integer

In general, multiplication is not commutative.

```
q * r - r * q
```

$$-\frac{2495}{18}\, i - 1418\, j - \frac{817}{18}\, k \tag{6}$$

Type: Quaternion Fraction Integer

There are no predefined constants for the imaginary i, j and k parts but you can easily define them.

```
i := quatern(0,1,0,0); j := quatern(0,0,1,0); k :=
  quatern(0,0,0,1)
```

$$k \tag{7}$$

Type: Quaternion Integer

These satisfy the normal identities.

```
[i*i, j*j, k*k, i*j, j*k, k*i, q*i]
```

$$\left[-1, -1, -1, k, i, j, 8 + \frac{2}{11}\, i + j - \frac{3}{4}\, k\right] \tag{8}$$

Type: List Quaternion Fraction Integer

The norm is the quaternion times its conjugate.

```
norm q
```

$$\frac{126993}{1936} \tag{9}$$

Type: Fraction Integer

```
conjugate q
```

$$\frac{2}{11}+8\,i-\frac{3}{4}\,j-k \quad (10)$$

Type: Quaternion Fraction Integer

```
q * %
```

$$\frac{126993}{1936} \quad (11)$$

Type: Quaternion Fraction Integer

9.51 RadixExpansion

It possible to expand numbers in general bases.

Here we expand 111 in base 5. This means $10^2+10^1+10^0 = 4\cdot 5^2+2\cdot 5^1+5^0$.

```
111::RadixExpansion(5)
```

$$421 \quad (1)$$

Type: RadixExpansion 5

You can expand fractions to form repeating expansions.

```
(5/24)::RadixExpansion(2)
```

$$0.001\overline{10} \quad (2)$$

Type: RadixExpansion 2

```
(5/24)::RadixExpansion(3)
```

$$0.0\overline{12} \quad (3)$$

Type: RadixExpansion 3

```
(5/24)::RadixExpansion(8)
```

$$0.1\overline{52} \quad (4)$$

Type: RadixExpansion 8

```
(5/24)::RadixExpansion(10)
```

$$0.208\overline{3} \quad (5)$$

Type: RadixExpansion 10

For bases from 11 to 36 the letters A through Z are used.

```
(5/24)::RadixExpansion(12)
```

$$0.26 \quad (6)$$

Type: RadixExpansion 12

```
(5/24)::RadixExpansion(16)
```

$$0.3\overline{5} \quad (7)$$

Type: RadixExpansion 16

```
(5/24)::RadixExpansion(36)
```

$$0.7I \tag{8}$$

Type: RadixExpansion 36

For bases greater than 36, the ragits are separated by blanks.

```
(5/24)::RadixExpansion(38)
```

$$0\ .\ 7\ 34\ 31\ \overline{25\ 12} \tag{9}$$

Type: RadixExpansion 38

The RadixExpansion type provides operations to obtain the individual ragits. Here is a rational number in base 8.

```
a := (76543/210)::RadixExpansion(8)
```

$$554.3\overline{7307} \tag{10}$$

Type: RadixExpansion 8

The operation **wholeRagits** returns a list of the ragits for the integral part of the number.

```
w := wholeRagits a
```

$$[5, 5, 4] \tag{11}$$

Type: List Integer

The operations **prefixRagits** and **cycleRagits** return lists of the initial and repeating ragits in the fractional part of the number.

```
f0 := prefixRagits a
```

$$[3] \tag{12}$$

Type: List Integer

```
f1 := cycleRagits a
```

$$[7, 3, 0, 7] \tag{13}$$

Type: List Integer

You can construct any radix expansion by giving the whole, prefix and cycle parts. The declaration is necessary to let AXIOM know the base of the ragits.

```
u:RadixExpansion(8):=wholeRadix(w)+fractRadix(f0,f1)
```

$$554.3\overline{7307} \tag{14}$$

Type: RadixExpansion 8

If there is no repeating part, then the list `[0]` should be used.

```
v: RadixExpansion(12) := fractRadix([1,2,3,11], [0])
```

$$0.123B\overline{0} \tag{15}$$

Type: RadixExpansion 12

If you are not interested in the repeating nature of the expansion, an infinite stream of ragits can be obtained using **fractRagits**.

```
fractRagits(u)
```

$$[3, 7, \overline{3, 0, 7, 7}] \tag{16}$$

Type: Stream Integer

Of course, it's possible to recover the fraction representation:

```
a :: Fraction(Integer)
```

$$\frac{76543}{210} \tag{17}$$

Type: Fraction Integer

Issue the system command `)show RadixExpansion` to display the full list of operations defined by RadixExpansion. More examples of expansions are available in 'DecimalExpansion' on page 350, 'BinaryExpansion' on page 312, and 'HexadecimalExpansion' on page 379.

9.52 RomanNumeral

The Roman numeral package was added to AXIOM in MCMLXXXVI for use in denoting higher order derivatives.

For example, let f be a symbolic operator.

```
f := operator 'f
```

$$f \tag{1}$$

Type: BasicOperator

This is the seventh derivative of f with respect to x.

```
D(f x,x,7)
```

$$f^{(vii)}(x) \tag{2}$$

Type: Expression Integer

You can have integers printed as Roman numerals by declaring variables to be of type RomanNumeral (abbreviation ROMAN).

```
a := roman(1978 - 1965)
```

$$\mathrm{XIII} \tag{3}$$

Type: RomanNumeral

This package now has a small but devoted group of followers that claim this domain has shown its efficacy in many other contexts. They claim that Roman numerals are every bit as useful as ordinary integers.

In a sense, they are correct, because Roman numerals form a ring and you can therefore construct polynomials with Roman numeral coefficients, matrices over Roman numerals, etc..

```
x : UTS(ROMAN,'x,0) := x
```

$$x \tag{4}$$

Type: UnivariateTaylorSeries(RomanNumeral, x, 0)

Was Fibonacci Italian or ROMAN?

```
recip(1 - x - x**2)
```

$$I + x + II\ x^2 + \mathrm{III}\ \ x^3 + V\ x^4 + \mathrm{VIII}\ x^5 + \mathrm{XIII}\ x^6 + \mathrm{XXI}\ x^7 + O\left(x^8\right) \tag{5}$$

Type: Union(UnivariateTaylorSeries(RomanNumeral, x, 0), ...)

You can also construct fractions with Roman numeral numerators and denominators, as this matrix Hilberticus illustrates.

```
m : MATRIX FRAC ROMAN
```

Type: Void

```
m := matrix [[1/(i + j) for i in 1..3] for j in 1..3]
```

$$\begin{bmatrix} \frac{I}{II} & \frac{I}{III} & \frac{I}{IV} \\ \frac{I}{III} & \frac{I}{IV} & \frac{I}{V} \\ \frac{I}{IV} & \frac{I}{V} & \frac{I}{VI} \end{bmatrix} \tag{7}$$

Type: Matrix Fraction RomanNumeral

Note that the inverse of the matrix has integral ROMAN entries.

```
inverse m
```

$$\begin{bmatrix} \mathrm{LXXII} & -\mathrm{CCXL} & \mathrm{CLXXX} \\ -\mathrm{CCXL} & CM & -\mathrm{DCCXX} \\ \mathrm{CLXXX} & -\mathrm{DCCXX} & DC \end{bmatrix} \tag{8}$$

Type: Union(Matrix Fraction RomanNumeral, ...)

Unfortunately, the spoil-sports say that the fun stops when the numbers get big—mostly because the Romans didn't establish conventions about representing very large numbers.

```
y := factorial 10
```

3628800 (9)

Type: PositiveInteger

You work it out!

```
roman y
```

$((((I))))((((I))))((((I))))(((I)))(((I)))(((I)))(((I)))$
$(((I)))(((I)))((I))((I))$MMMMMMMMDCCC (10)

Type: RomanNumeral

Issue the system command `)show RomanNumeral` to display the full list of operations defined by RomanNumeral.

9.53 Segment

The Segment domain provides a generalized interval type.

Segments are created using the "`..`" construct by indicating the (included) end points.

```
s := 3..10
```

3..10 (1)

Type: Segment PositiveInteger

The first end point is called the **lo** and the second is called **hi**.

```
lo s
```

3 (2)

Type: PositiveInteger

These names are used even though the end points might belong to an unordered set.

```
hi s
```

10 (3)

Type: PositiveInteger

In addition to the end points, each segment has an integer "increment." An increment can be specified using the "`by`" construct.

```
t := 10..3 by -2
```

10..3 by -2 (4)

Type: Segment PositiveInteger

This part can be obtained using the **incr** function.

```
incr s
```

1 (5)

Type: PositiveInteger

Unless otherwise specified, the increment is `1`.

```
incr t
```

-2 (6)

Type: Integer

A single value can be converted to a segment with equal end points. This happens if segments and single values are mixed in a list.

```
l := [1..3, 5, 9, 15..11 by -1]
```

[1..3, 5..5, 9..9, 15..11 by -1] (7)

Type: List Segment PositiveInteger

If the underlying type is an ordered ring, it is possible to perform additional operations. The **expand** operation creates a list of points in a segment.

```
expand s
```

[3, 4, 5, 6, 7, 8, 9, 10] (8)

Type: List Integer

If `k > 0`, then `expand(l..h by k)` creates the list `[l, l+k, ..., lN]` where `lN <= h < lN+k`. If `k < 0`, then `lN >= h > lN+k`.

```
expand t
```

[10, 8, 6, 4] (9)

Type: List Integer

It is also possible to expand a list of segments. This is equivalent to appending lists obtained by expanding each segment individually.

```
expand l
```

[1, 2, 3, 5, 9, 15, 14, 13, 12, 11] (10)

Type: List Integer

For more information on related topics, see 'SegmentBinding' on page 448 and 'UniversalSegment' on page 477. Issue the system command `)show Segment` to display the full list of operations defined by Segment.

9.54 SegmentBinding

The SegmentBinding type is used to indicate a range for a named symbol.

First give the symbol, then an "=" and finally a segment of values.

```
x = a..b
```

$x = a..b$ (1)

Type: SegmentBinding Symbol

This is used to provide a convenient syntax for arguments to certain operations.

```
sum(i**2, i = 0..n)
```

$$\frac{2\,n^3+3\,n^2+n}{6} \tag{2}$$

Type: Fraction Polynomial Integer

The **draw** operation uses a SegmentBinding argument as a range of coordinates. This is an example of a two-dimensional parametrized plot; other **draw** options use more than one SegmentBinding argument.

```
draw(x**2, x = -2..2)
```

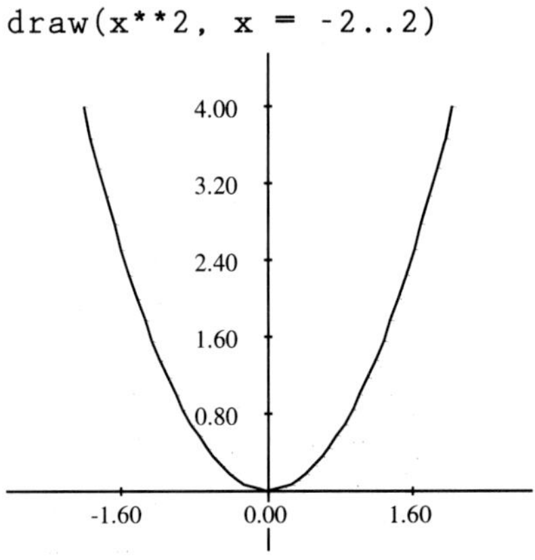

The left-hand side must be of type Symbol but the right-hand side can be a segment over any type.

```
sb := y = 1/2..3/2
```

$$y = \left(\frac{1}{2}\right)..\left(\frac{3}{2}\right) \tag{3}$$

Type: SegmentBinding Fraction Integer

The left- and right-hand sides can be obtained using the **variable** and **segment** operations.

```
variable(sb)
```

$$y \tag{4}$$

Type: Symbol

```
segment(sb)
```

$$\left(\frac{1}{2}\right)..\left(\frac{3}{2}\right) \tag{5}$$

Type: Segment Fraction Integer

Issue the system command `)show SegmentBinding` to display the full list of operations defined by SegmentBinding. For more information on related topics, see 'Segment' on page 447 and 'UniversalSegment' on page 477.

9.55 Set

Sets can be created by giving a fixed set of values ...

The Set domain allows one to represent explicit finite sets of values. These are similar to lists, but duplicate elements are not allowed.

```
s := {x**2-1, y**2-1, z**2-1}
```

$$\left\{x^2-1,\ y^2-1,\ z^2-1\right\} \tag{1}$$

Type: Set Polynomial Integer

or by using a collect form, just as for lists. In either case, the set is formed from a finite collection of values.

```
t := {x**i - i+1 for i in 2..10 | prime? i}
```

$$\left\{x^2-1,\ x^3-2,\ x^5-4,\ x^7-6\right\} \tag{2}$$

Type: Set Polynomial Integer

The basic operations on sets are **intersect**, **union**, **difference**, and **symmetricDifference**.

```
i := intersect(s,t)
```

$$\left\{x^2-1\right\} \tag{3}$$

Type: Set Polynomial Integer

```
u := union(s,t)
```

$$\left\{x^2-1,\ x^3-2,\ x^5-4,\ x^7-6,\ y^2-1,\ z^2-1\right\} \tag{4}$$

Type: Set Polynomial Integer

The set `difference(s,t)` contains those members of `s` which are not in `t`.

```
difference(s,t)
```

$$\left\{y^2-1,\ z^2-1\right\} \tag{5}$$

Type: Set Polynomial Integer

The set `symmetricDifference(s,t)` contains those elements which are in `s` or `t` but not in both.

```
symmetricDifference(s,t)
```

$$\left\{x^3-2,\ x^5-4,\ x^7-6,\ y^2-1,\ z^2-1\right\} \tag{6}$$

Type: Set Polynomial Integer

Set membership is tested using the **member?** operation.

```
member?(y, s)
```

false (7)

Type: Boolean

```
member?((y+1)*(y-1), s)
```

true (8)

Type: Boolean

The **subset?** function determines whether one set is a subset of another.

```
subset?(i, s)
```

true (9)

Type: Boolean

```
subset?(u, s)
```

false (10)

Type: Boolean

When the base type is finite, the absolute complement of a set is defined. This finds the set of all multiplicative generators of PrimeField 11—the integers mod `11`.

```
gs := {g for i in 1..11 | primitive?(g := i::PF 11)}
```

$$\{2,\ 6,\ 7,\ 8\} \tag{11}$$

Type: Set PrimeField 11

The following values are not generators.

```
complement gs
```

{1, 3, 4, 5, 9, 10, 0} (12)

Type: Set PrimeField 11

Often the members of a set are computed individually; in addition, values can be inserted or removed from a set over the course of a computation.

There are two ways to do this:

```
a := {i**2 for i in 1..5}
```

{1, 4, 9, 16, 25} (13)

Type: Set PositiveInteger

One is to view a set as a data structure and to apply updating operations.

```
insert!(32, a)
```

{1, 4, 9, 16, 25, 32} (14)

Type: Set PositiveInteger

```
remove!(25, a)
```

{1, 4, 9, 16, 32} (15)

Type: Set PositiveInteger

```
a
```

{1, 4, 9, 16, 32} (16)

Type: Set PositiveInteger

The other way is to view a set as a mathematical entity and to create new sets from old.

```
b := b0 := {i**2 for i in 1..5}
```

{1, 4, 9, 16, 25} (17)

Type: Set PositiveInteger

```
b := union(b, {32})
```

{1, 4, 9, 16, 25, 32} (18)

Type: Set PositiveInteger

```
b := difference(b, {25})
```

{1, 4, 9, 16, 32} (19)

Type: Set PositiveInteger

```
b0
```

{1, 4, 9, 16, 25} (20)

Type: Set PositiveInteger

For more information about lists, see 'List' on page 404. Issue the system command `)show Set` to display the full list of operations defined by Set.

9.56 SmallFloat

AXIOM provides two kinds of floating point numbers. The domain Float (abbreviation FLOAT) implements a model of arbitrary precision floating point numbers. The domain SmallFloat (abbreviation SF) is intended to make available hardware floating point arithmetic in AXIOM. The actual model of floating point SmallFloat that provides is system-dependent. For example, on the IBM system 370 AXIOM uses IBM double precision which has fourteen hexadecimal digits of precision or roughly sixteen decimal digits. Arbitrary precision floats allow the user to specify the precision at which arithmetic operations are computed. Although this is an attractive facility, it comes at a cost. Arbitrary-precision floating-point arithmetic typically takes twenty to two hundred times more time than hardware floating point.

The usual arithmetic and elementary functions are available for SmallFloat. Use `)show SmallFloat` to get a list of operations or the HyperDoc Browse facility to get more extensive documentation about SmallFloat.

By default, floating point numbers that you enter into AXIOM are of type Float.

```
2.71828
```

2.71828 (1)

Type: Float

You must therefore tell AXIOM that you want to use SmallFloat values and operations. The following are some conservative guidelines for getting AXIOM to use SmallFloat.

To get a value of type SmallFloat, use a target with "@", . . .

```
2.71828@SmallFloat
```

2.71828 (2)

Type: SmallFloat

a conversion, . . .

```
2.71828 :: SmallFloat
```

2.71828 (3)

Type: SmallFloat

or an assignment to a declared variable. It is more efficient if you use a target rather than an explicit or implicit conversion.

```
eApprox : SmallFloat := 2.71828
```

2.71828 (4)

Type: SmallFloat

You also need to declare functions that work with SmallFloat.

```
avg : List SmallFloat -> SmallFloat
```

Type: Void

```
avg l ==
  empty? l => 0 :: SmallFloat
  reduce(_+,l) / #l
```

Type: Void

```
avg []
```

```
Compiling function avg with type List SmallFloat ->
SmallFloat
```

0.0 (7)

Type: SmallFloat

```
avg [3.4,9.7,-6.8]
```

2.1000000000000001 (8)

Type: SmallFloat

Use package-calling for operations from SmallFloat unless the arguments themselves are already of type SmallFloat.

```
cos(3.1415926)$SmallFloat
```

-0.99999999999999856 (9)

Type: SmallFloat

```
cos(3.1415926 :: SmallFloat)
```

-0.99999999999999856 (10)

Type: SmallFloat

By far, the most common usage of SmallFloat is for functions to be graphed. For more information about AXIOM's numerical and graphical facilities, see Section 7 on page 179, Section 8.1 on page 227, and 'Float' on page 368.

9.57 SmallInteger

The SmallInteger domain is intended to provide support in AXIOM for machine integer arithmetic. It is generally much faster than (bignum) Integer arithmetic but suffers from a limited range of values. Since AXIOM can be implemented on top of various dialects of Lisp, the actual representation of small integers may not correspond exactly to the host machines integer representation.

Under *AKCL* on the IBM Risc System/6000, small integers are restricted to the range -2^{26} to $2^{26}-1$, allowing 1 bit for overflow detection.

You can discover the minimum and maximum values in your implementation by using **min** and **max**.

```
min()$SmallInteger
```

-2147483648 (1)

Type: SmallInteger

```
max()$SmallInteger
```

2147483647 (2)

Type: SmallInteger

To avoid confusion with Integer, which is the default type for integers, you usually need to work with declared variables (Section 2.3 on page 69) . . .

```
a := 1234 :: SmallInteger
```

1234 (3)

Type: SmallInteger

or use package calling (Section 2.9 on page 83).

```
b := 124$SmallInteger
```

124 (4)

Type: SmallInteger

You can add, multiply and subtract SmallInteger objects, and ask for the greatest common divisor (**gcd**).

```
gcd(a,b)
```

2 (5)

Type: SmallInteger

The least command multiple (**lcm**) is also available.

```
lcm(a,b)
```

76508 (6)

Type: SmallInteger

Operations **mulmod**, **addmod**, **submod**, and **invmod** are similar—they provide arithmetic modulo a given small integer. Here is `5 * 6 mod 13`.

```
mulmod(5,6,13)$SmallInteger
```

4 (7)

Type: SmallInteger

To reduce a small integer modulo a prime, use **positiveRemainder**.

```
positiveRemainder(37,13)$SmallInteger
```

11 (8)

Type: SmallInteger

Operations **And**, **Or**, **xor**, and **Not** provide bit level operations on small integers.

```
And(3,4)$SmallInteger
```

0 (9)

Type: SmallInteger

Use `shift(int,numToShift)` to shift bits, where `i` is shifted left if `numToShift` is positive, right if negative.

```
shift(1,4)$SmallInteger
```

16 (10)

Type: SmallInteger

```
shift(31,-1)$SmallInteger
```

15 (11)

Type: SmallInteger

Many other operations are available for small integers, including many of those provided for Integer. To see the other operations, use the Browse

HyperDoc facility (Section 14 on page 549). Issue the system command `)show SmallInteger` to display the full list of operations defined by SmallInteger..

9.58 SparseTable

The SparseTable domain provides a general purpose table type with default entries.

Here we create a table to save strings under integer keys. The value `"Try again!"` is returned if no other value has been stored for a key.

```
t: SparseTable(Integer, String, "Try again!") := table()
```

$$table() \tag{1}$$

Type: SparseTable(Integer, String, Try again!)

Entries can be stored in the table.

```
t.3 := "Number three"
```

"Number three" (2)

Type: String

```
t.4 := "Number four"
```

"Number four" (3)

Type: String

These values can be retrieved as usual, but if a look up fails the default entry will be returned.

```
t.3
```

"Number three" (4)

Type: String

```
t.2
```

"Try again!" (5)

Type: String

To see which values are explicitly stored, the **keys** and **entries** functions can be used.

```
keys t
```

$$[4, 3] \tag{6}$$

Type: List Integer

```
entries t
```

["Number four", "Number three"] (7)

Type: List String

If a specific table representation is required, the GeneralSparseTable constructor should be used. The domain SparseTable(K, E, dflt) is equivalent to GeneralSparseTable(K,E, Table(K,E), dflt). For more information, see 'Table' on page 465 and 'GeneralSparseTable' on page 375. Issue the system command `)show SparseTable` to display the full list of operations defined by SparseTable.

9.59 SquareMatrix

The top level matrix type in AXIOM is Matrix (see 'Matrix' on page 414), which provides basic arithmetic and linear algebra functions. However, since the matrices can be of any size it is not true that any pair can be added or multiplied. Thus Matrix has little algebraic structure.

Sometimes you want to use matrices as coefficients for polynomials or in other algebraic contexts. In this case, SquareMatrix should be used. The domain SquareMatrix(n,R) gives the ring of n by n square matrices over R.

Since SquareMatrix is not normally exposed at the top level, you must expose it before it can be used.

```
)set expose add constructor SquareMatrix
```

```
SquareMatrix is now explicitly exposed in frame G1077
```

Once SQMATRIX has been exposed, values can be created using the **squareMatrix** function.

```
m := squareMatrix [[1,-%i],[%i,4]]
```

$$\begin{bmatrix} 1 & -\%i \\ \%i & 4 \end{bmatrix} \tag{1}$$

Type: SquareMatrix(2, Complex Integer)

The usual arithmetic operations are available.

```
m*m - m
```

$$\begin{bmatrix} 1 & -4\ \%i \\ 4\ \%i & 13 \end{bmatrix} \tag{2}$$

Type: SquareMatrix(2, Complex Integer)

Square matrices can be used where ring elements are required. For example, here is a matrix with matrix entries.

```
mm := squareMatrix [[m, 1], [1-m, m**2]]
```

$$\begin{bmatrix} \begin{bmatrix} 1 & -\%i \\ \%i & 4 \end{bmatrix} & \begin{bmatrix} 1 & 0 \\ 0 & 1 \end{bmatrix} \\ \begin{bmatrix} 0 & \%i \\ -\%i & -3 \end{bmatrix} & \begin{bmatrix} 2 & -5\ \%i \\ 5\ \%i & 17 \end{bmatrix} \end{bmatrix} \tag{3}$$

Type: SquareMatrix(2, SquareMatrix(2, Complex Integer))

Or you can construct a polynomial with square matrix coefficients.

```
p := (x + m)**2
```

$$x^2 + \begin{bmatrix} 2 & -2\ \%i \\ 2\ \%i & 8 \end{bmatrix} x + \begin{bmatrix} 2 & -5\ \%i \\ 5\ \%i & 17 \end{bmatrix} \tag{4}$$

Type: Polynomial SquareMatrix(2, Complex Integer)

This value can be converted to a square matrix with polynomial coefficients.

```
p::SquareMatrix(2, ?)
```

$$\begin{bmatrix} x^2+2\ x+2 & -2\ \%i\ x-5\ \%i \\ 2\ \%i\ x+5\ \%i & x^2+8\ x+17 \end{bmatrix} \tag{5}$$

Type: SquareMatrix(2, Polynomial Complex Integer)

For more information on related topics, see Section 2.2.4 on page 68, Section 2.11 on page 87, and 'Matrix' on page 414. Issue the system command `)show SquareMatrix` to display the full list of operations defined by SquareMatrix.

9.60 Stream

A Stream object is represented as a list whose last element contains the wherewithal to create the next element, should it ever be required.

Let `ints` be the infinite stream of non-negative integers.

```
ints := [i for i in 0..]
```

$[0, 1, 2, 3, 4, 5, 6, \ldots]$ (1)

Type: Stream NonNegativeInteger

By default, ten stream elements are calculated. This number may be changed to something else by the system command `)set streams calculate`. For the display purposes of this book, we have chosen a smaller value.

More generally, you can construct a stream by specifying its initial value and a function which, when given an element, creates the next element.

```
f : List INT -> List INT
```

Type: Void

```
f x == [x.1 + x.2, x.1]
```

Type: Void

```
fibs := [i.2 for i in [generate(f,[1,1])]]
```

$[1, 1, 2, 3, 5, 8, 13, \ldots]$ (4)

Type: Stream Integer

You can create the stream of odd non-negative integers by either filtering them from the integers, or by evaluating an expression for each integer.

```
[i for i in ints | odd? i]
```

$[1, 3, 5, 7, 9, 11, 13, \ldots]$ (5)

Type: Stream NonNegativeInteger

```
odds := [2*i+1 for i in ints]
```

$[1, 3, 5, 7, 9, 11, 13, \ldots]$ (6)

Type: Stream NonNegativeInteger

You can accumulate the initial segments of a stream using the **scan** operation.

```
scan(0,+,odds)
```

$[1, 4, 9, 16, 25, 36, 49, \ldots]$ (7)

Type: Stream NonNegativeInteger

The corresponding elements of two or more streams can be combined in this way.

```
[i*j for i in ints for j in odds]
```

$[0, 3, 10, 21, 36, 55, 78, \ldots]$ (8)

Type: Stream NonNegativeInteger

```
map(*,ints,odds)
```

[0, 3, 10, 21, 36, 55, 78, ...] (9)

Type: Stream NonNegativeInteger

Many operations similar to those applicable to lists are available for streams.

```
first ints
```

0 (10)

Type: NonNegativeInteger

```
rest ints
```

[1, 2, 3, 4, 5, 6, 7, ...] (11)

Type: Stream NonNegativeInteger

```
fibs 20
```

6765 (12)

Type: PositiveInteger

The packages StreamFunctions1, StreamFunctions2 and StreamFunctions3 export some useful stream manipulation operations. For more information, see Section 5.5 on page 130, Section 8.9 on page 255, 'ContinuedFraction' on page 335, and 'List' on page 404. Issue the system command `)show Stream` to display the full list of operations defined by Stream.

9.61 String

The type String provides character strings. Character strings provide all the operations for a one-dimensional array of characters, plus additional operations for manipulating text. For more information on related topics, see 'Character' on page 325 and 'CharacterClass' on page 326. You can also issue the system command `)show String` to display the full list of operations defined by String.

String values can be created using double quotes.

```
hello := "Hello, I'm AXIOM!"
```

"Hello, I'm AXIOM!" (1)

Type: String

Note, however, that double quotes and underscores must be preceded by an extra underscore.

```
said := "Jane said, _"Look!_""
```

"Jane said, " Look ! "" (2)

Type: String

```
saw := "She saw exactly one underscore: __."
```

"She saw exactly one underscore: _." (3)

Type: String

It is also possible to use **new** to create a string of any size filled with a given character. Since there are many **new** functions it is necessary to indicate the desired type.

```
gasp: String := new(32, char "x")
```

"xxxxxxxxxxxxxxxxxxxxxxxxxxxxxxxx" (4)

Type: String

The length of a string is given by "#".

```
#gasp
```

32 (5)

Type: PositiveInteger

Indexing operations allow characters to be extracted or replaced in strings. For any string `s`, indices lie in the range `1..#s`.

```
hello.2
```

e (6)

Type: Character

Indexing is really just the application of a string to a subscript, so any application syntax works.

```
hello 2
```

e (7)

Type: Character

```
hello(2)
```

e (8)

Type: Character

If it is important not to modify a given string, it should be copied before any updating operations are used.

```
hullo := copy hello
```

"Hello, I'm AXIOM!" (9)

Type: String

```
hullo.2 := char "u"; [hello, hullo]
```

["Hello, I'm AXIOM!", "Hullo, I'm AXIOM!"] (10)

Type: List String

Operations are provided to split and join strings. The **concat** operation allows several strings to be joined together.

```
saidsaw := concat ["alpha","---","omega"]
```

"alpha---omega" (11)

Type: String

There is a version of **concat** that works with two strings.

```
concat("hello ","goodbye")
```

"hello goodbye" (12)

Type: String

Juxtaposition can also be used to concatenate strings.

```
"This " "is " "several " "strings " "concatenated."
```

"This is several strings concatenated." (13)

Type: String

Substrings are obtained by giving an index range.

```
hello(1..5)
```

"Hello" (14)

Type: String

```
hello(8..)
```

"I'm AXIOM!" (15)

Type: String

A string can be split into several substrings by giving a separation character or character class.

```
split(hello, char " ")
```

["Hello,", "I'm", "AXIOM!"] (16)

Type: List String

```
other := complement alphanumeric();
```

(17)

Type: CharacterClass

```
split(saidsaw, other)
```

["alpha", "omega"] (18)

Type: List String

Unwanted characters can be trimmed from the beginning or end of a string using the operations **trim**, **leftTrim** and **rightTrim**.

```
trim ("## ++ relax ++ ##", char "#")
```

" ++ relax ++ " (19)

Type: String

Each of these functions takes a string and a second argument to specify the characters to be discarded.

```
trim ("## ++ relax ++ ##", other)
```

"relax" (20)

Type: String

The second argument can be given either as a single character or as a character class.

```
leftTrim ("## ++ relax ++ ##", other)
```

"relax ++ ##" (21)

Type: String

```
rightTrim("## ++ relax ++ ##", other)
```

"## ++ relax" (22)

Type: String

Strings can be changed to upper case or lower case using the operations **upperCase**, **upperCase!**, **lowerCase** and **lowerCase!**.

```
upperCase hello
```

"HELLO, I'M AXIOM!" (23)

Type: String

The versions with the exclamation mark change the original string, while the others produce a copy.

```
lowerCase hello
```

"hello, i'm axiom!" (24)

Type: String

Some basic string matching is provided. The function **prefix?** tests whether one string is an initial prefix of another.

```
prefix?("He", "Hello")
```

true (25)

Type: Boolean

```
prefix?("Her", "Hello")
```

false (26)

Type: Boolean

A similar function, **suffix?**, tests for suffixes.

```
suffix?("", "Hello")
```

true (27)

Type: Boolean

```
suffix?("LO", "Hello")
```

false (28)

Type: Boolean

The function **substring?** tests for a substring given a starting position.

```
substring?("ll", "Hello", 3)
```

true (29)

Type: Boolean

```
substring?("ll", "Hello", 4)
```

false (30)

Type: Boolean

A number of **position** functions locate things in strings. If the first argument to position is a string, then `position(s,t,i)` finds the location of `s` as a substring of `t` starting the search at position `i`.

```
n := position("nd", "underground", 1)
```

2 (31)

Type: PositiveInteger

```
n := position("nd", "underground", n+1)
```

10 (32)

Type: PositiveInteger

If `s` is not found, then `0` is returned (`minIndex(s)-1` in IndexedString).

```
n := position("nd", "underground", n+1)
```

0 (33)

Type: NonNegativeInteger

To search for a specific character or a member of a character class, a different first argument is used.

```
position(char "d", "underground", 1)
```

3 (34)

Type: PositiveInteger

```
position(hexDigit(), "underground", 1)
```

3 (35)

Type: PositiveInteger

9.62 StringTable

This domain provides a table type in which the keys are known to be strings so special techniques can be used. Other than performance, the type StringTable(S) should behave exactly the same way as Table(String,S). See 'Table' on page 465 for general information about tables. Issue the system command `)show StringTable` to display the full list of operations defined by StringTable.

This creates a new table whose keys are strings.

```
t: StringTable(Integer) := table()
```

table() (1)

Type: StringTable Integer

The value associated with each string key is the number of characters in the string.

```
for s in split("My name is Ian Watt.",char " ")
  repeat
    t.s := #s
```

Type: Void

```
for key in keys t repeat output [key, t.key]
```

```
["Watt.",5]
["Ian",3]
["is",2]
["name",4]
["My",2]
```

Type: Void

9.63 Symbol

Symbols are one of the basic types manipulated by AXIOM. The Symbol domain provides ways to create symbols of many varieties. Issue the system command `)show Symbol` to display the full list of operations defined by Symbol.

The simplest way to create a symbol is to "single quote" an identifier.

```
X: Symbol := 'x
```

x (1)

Type: Symbol

This gives the symbol even if **x** has been assigned a value. If **x** has not been assigned a value, then it is possible to omit the quote.

```
XX: Symbol := x
```

x (2)

Type: Symbol

Declarations must be used when working with symbols, because otherwise the interpreter tries to place values in a more specialized type Variable.

```
A := 'a
```

a (3)

Type: Variable a

```
B := b
```

b (4)

Type: Variable b

The normal way of entering polynomials uses this fact.

```
x**2 + 1
```

$x^2 + 1$ (5)

Type: Polynomial Integer

Another convenient way to create symbols is to convert a string. This is useful when the name is to be constructed by a program.

```
"Hello"::Symbol
```

$Hello$ (6)

Type: Symbol

Sometimes it is necessary to generate new unique symbols, for example, to name constants of integration. The expression `new()` generates a symbol starting with %.

```
new()$Symbol
```

$\%A$ (7)

Type: Symbol

Successive calls to **new** produce different symbols.

```
new()$Symbol
```

$\%B$ (8)

Type: Symbol

The expression `new("s")` produces a symbol starting with `%s`.

```
new("xyz")$Symbol
```

$\%xyz0$ (9)

Type: Symbol

A symbol can be adorned in various ways. The most basic thing is applying a symbol to a list of subscripts.

```
X[i,j]
```

$x_{i,\,j}$ (10)

Type: Symbol

Somewhat less pretty is to attach subscripts, superscripts or arguments.

```
U := subscript(u, [1,2,1,2])
```

$u_{1,\,2,\,1,\,2}$ (11)

Type: Symbol

```
V := superscript(v, [n])
```

$$v^n \tag{12}$$

Type: Symbol

```
P := argscript(p, [t])
```

$$p\,(t) \tag{13}$$

Type: Symbol

It is possible to test whether a symbol has scripts using the **scripted?** test.

```
scripted? U
```

true (14)

Type: Boolean

```
scripted? X
```

false (15)

Type: Boolean

If a symbol is not scripted, then it may be converted to a string.

```
string X
```

"x" (16)

Type: String

The basic parts can always be extracted using the **name** and **scripts** operations.

```
name U
```

$$u \tag{17}$$

Type: Symbol

```
scripts U
```

$$\left[sub = \left[1,\ 2,\ 1,\ 2\right],\ sup = [],\ presup = [],\ presub = [],\ args = []\right] \tag{18}$$

Type: Record(sub: List OutputForm, sup: List OutputForm, presup: List OutputForm, presub: List OutputForm, args: List OutputForm)

```
name X
```

$$x \tag{19}$$

Type: Symbol

```
scripts X
```

$$\left[sub = [],\ sup = [],\ presup = [],\ presub = [],\ args = []\right] \tag{20}$$

Type: Record(sub: List OutputForm, sup: List OutputForm, presup: List OutputForm, presub: List OutputForm, args: List OutputForm)

The most general form is obtained using the **script** operation. This operation takes an argument which is a list containing, in this order, lists of subscripts, superscripts, presuperscripts, presubscripts and arguments to a symbol.

```
M := script(Mammoth, [[i,j],[k,l],[0,1],[2],[u,v,w]])
```

$${}^{0,\,1}_{2}\mathrm{Mammoth}^{k,\,l}_{i,\,j}(u,\ v,\ w) \tag{21}$$

Type: Symbol

```
scripts M
```

$$\left[sub = \left[i,\ j\right],\ sup = \left[k,\ l\right],\ presup = \left[0,\ 1\right],\ presub = [2],\ args = \left[u,\ v,\ w\right]\right] \tag{22}$$

Type: Record(sub: List OutputForm, sup: List OutputForm, presup: List OutputForm, presub: List OutputForm, args: List OutputForm)

If trailing lists of scripts are omitted, they are assumed to be empty.

```
N := script(Nut, [[i,j],[k,l],[0,1]])
```

$${}^{0,\,1}\mathrm{Nut}^{k,\,l}_{i,\,j} \tag{23}$$

Type: Symbol

```
scripts N
```

$$\left[sub = \left[i,\ j\right],\ sup = \left[k,\ l\right],\ presup = \left[0,\ 1\right],\ presub = [],\ args = []\right] \tag{24}$$

Type: Record(sub: List OutputForm, sup: List OutputForm, presup: List OutputForm, presub: List OutputForm, args: List OutputForm)

9.64 Table

The Table constructor provides a general structure for associative storage. This type provides hash tables in which data objects can be saved according to keys of any type. For a given table, specific types must be chosen for the keys and entries.

In this example the keys to the table are polynomials with integer coefficients. The entries in the table are strings.

```
t: Table(Polynomial Integer, String) := table()
```

table() (1)

Type: Table(Polynomial Integer, String)

To save an entry in the table, the **setelt** operation is used. This can be called directly, giving the table a key and an entry.

```
setelt(t, x**2 - 1, "Easy to factor")
```

"Easy to factor" (2)

Type: String

Alternatively, you can use assignment syntax.

```
t(x**3 + 1) := "Harder to factor"
```

"Harder to factor" (3)

Type: String

```
t(x) := "The easiest to factor"
```

"The easiest to factor" (4)

Type: String

Entries are retrieved from the table by calling the **elt** operation.

```
elt(t, x)
```

"The easiest to factor" (5)

Type: String

This operation is called when a table is "applied" to a key using this or the following syntax.

```
t.x
```

"The easiest to factor" (6)

Type: String

```
t x
```

"The easiest to factor" (7)

Type: String

Parentheses are used only for grouping. They are needed if the key is an infixed expression.

```
t.(x**2 - 1)
```

"Easy to factor" (8)

Type: String

Note that the **elt** operation is used only when the key is known to be in the table—otherwise an error is generated.

```
t (x**3 + 1)
```

"Harder to factor" (9)

Type: String

You can get a list of all the keys to a table using the **keys** operation.

```
keys t
```

$$\left[x,\ x^3+1,\ x^2-1\right] \quad (10)$$

Type: List Polynomial Integer

If you wish to test whether a key is in a table, the **search** operation is used. This operation returns either an entry or `"failed"`.

```
search(x, t)
```

"The easiest to factor" (11)

Type: Union(String, ...)

```
search(x**2, t)
```

"failed" (12)

Type: Union("failed", ...)

The return type is a union so the success of the search can be tested using `case`.

```
search(x**2, t) case "failed"
```

true (13)

Type: Boolean

The **remove!** operation is used to delete values from a table.

```
remove!(x**2-1, t)
```

"Easy to factor" (14)

Type: Union(String, ...)

If an entry exists under the key, then it is returned. Otherwise **remove!** returns "failed".

```
remove!(x-1, t)
```

"failed" (15)

Type: Union("failed", ...)

The number of key-entry pairs can be found using the **#** operation.

```
#t
```

2 (16)

Type: PositiveInteger

Just as **keys** returns a list of keys to the table, a list of all the entries can be obtained using the **members** operation.

```
members t
```

["The easiest to factor", "Harder to factor"] (17)

Type: List String

A number of useful operations take functions and map them on to the table to compute the result. Here we count the entries which have "Hard" as a prefix.

```
count(s: String +-> prefix?("Hard", s), t)
```

1 (18)

Type: PositiveInteger

Other table types are provided to support various needs.

- AssociationList gives a list with a table view. This allows new entries to be appended onto the front of the list to cover up old entries. This is useful when table entries need to be stacked or when frequent list traversals are required. See 'AssociationList' on page 309 for more information.
- EqTable gives tables in which keys are considered equal only when they are in fact the same instance of a structure. See 'EqTable' on page 353 for more information.
- StringTable should be used when the keys are known to be strings. See 'StringTable' on page 462 for more information.
- SparseTable provides tables with default entries, so lookup never fails. The GeneralSparseTable constructor can be used to make any table type behave this way. See 'SparseTable' on page 455 for more information.
- KeyedAccessFile allows values to be saved in a file, accessed as a table. See 'KeyedAccessFile' on page 390 for more information.

Issue the system command `)show Table` to display the full list of operations defined by Table.

9.65 TextFile

The domain TextFile allows AXIOM to read and write character data and exchange text with other programs. This type behaves in AXIOM much like a File of strings, with additional operations to cause new lines. We give an example of how to produce an upper case copy of a file.

This is the file from which we read the text.

```
f1: TextFile := open("/etc/motd", "input")
```

"/etc/motd" (1)

Type: TextFile

This is the file to which we read the text.

```
f2: TextFile := open("/tmp/MOTD", "output")
```

"/tmp/MOTD" (2)

Type: TextFile

Entire lines are handled using the **readLine!** and **writeLine!** operations.

```
l := readLine! f1
```

"Risc System/6000 Model 320H: pascal" (3)

Type: String

```
writeLine!(f2, upperCase l)
```

"RISC SYSTEM/6000 MODEL 320H: PASCAL" (4)

Type: String

Use the **endOfFile?** operation to check if you have reached the end of the file.

```
while not endOfFile? f1 repeat
    s := readLine! f1
    writeLine!(f2, upperCase s)
```

Type: Void

The file `f1` is exhausted and should be closed.

```
close! f1
```

"/etc/motd" (6)

Type: TextFile

It is sometimes useful to write lines a bit at a time. The **write!** operation allows this.

```
write!(f2, "-The-")
```

"-The-" (7)

Type: String

```
write!(f2, "-End-")
```

"-End-" (8)

Type: String

This ends the line. This is done in a machine-dependent manner.

```
writeLine! f2
```

"" (9)

Type: String

```
close! f2
```

"/tmp/MOTD" (10)

Type: TextFile

Finally, clean up.

```
)system rm /tmp/MOTD
```

For more information on related topics, see 'File' on page 362, 'KeyedAccessFile' on page 390, and 'Library' on page 393. Issue the system command `)show TextFile` to display the full list of operations defined by TextFile.

9.66 TwoDimensionalArray

The TwoDimensionalArray domain is used for storing data in a two-dimensional data structure indexed by row and by column. Such an array is a homogeneous data structure in that all the entries of the array must belong to the same AXIOM domain (although see Section 2.6 on page 77). Each array has a fixed number of rows and columns specified by the user and arrays are not extensible. In AXIOM, the indexing of two-dimensional arrays is one-based. This means that both the "first" row of an array and the "first" column of an array are given the index `1`. Thus, the entry in the upper left corner of an array is in position `(1,1)`.

The operation **new** creates an array with a specified number of rows and columns and fills the components of that array with a specified entry. The arguments of this operation specify the number of rows, the number of columns, and the entry.

This creates a five-by-four array of integers, all of whose entries are zero.

```
arr : ARRAY2 INT := new(5,4,0)
```

$$\begin{bmatrix} 0 & 0 & 0 & 0 \\ 0 & 0 & 0 & 0 \\ 0 & 0 & 0 & 0 \\ 0 & 0 & 0 & 0 \\ 0 & 0 & 0 & 0 \end{bmatrix} \quad (1)$$

Type: TwoDimensionalArray Integer

The entries of this array can be set to other integers using the operation **setelt**.

Issue this to set the element in the upper left corner of this array to 17.

```
setelt(arr,1,1,17)
```

17 (2)

Type: PositiveInteger

Now the first element of the array is 17.

```
arr
```

$$\begin{bmatrix} 17 & 0 & 0 & 0 \\ 0 & 0 & 0 & 0 \\ 0 & 0 & 0 & 0 \\ 0 & 0 & 0 & 0 \\ 0 & 0 & 0 & 0 \end{bmatrix} \qquad (3)$$

Type: TwoDimensionalArray Integer

Likewise, elements of an array are extracted using the operation **elt**.

```
elt(arr,1,1)
```

$$17 \qquad (4)$$

Type: PositiveInteger

Another way to use these two operations is as follows. This sets the element in position (3,2) of the array to 15.

```
arr(3,2) := 15
```

$$15 \qquad (5)$$

Type: PositiveInteger

This extracts the element in position (3,2) of the array.

```
arr(3,2)
```

$$15 \qquad (6)$$

Type: PositiveInteger

The operations **elt** and **setelt** come equipped with an error check which verifies that the indices are in the proper ranges. For example, the above array has five rows and four columns, so if you ask for the entry in position (6,2) with arr(6,2) AXIOM displays an error message. If there is no need for an error check, you can call the operations **qelt** and **qsetelt!** which provide the same functionality but without the error check. Typically, these operations are called in well-tested programs.

The operations **row** and **column** extract rows and columns, respectively, and return objects of OneDimensionalArray with the same underlying element type.

```
row(arr,1)
```

$$[17, 0, 0, 0] \qquad (7)$$

Type: OneDimensionalArray Integer

```
column(arr,1)
```

$$[17, 0, 0, 0, 0] \qquad (8)$$

Type: OneDimensionalArray Integer

You can determine the dimensions of an array by calling the operations **nrows** and **ncols**, which return the number of rows and columns, respectively.

```
nrows(arr)
```

$$5 \qquad (9)$$

Type: PositiveInteger

```
ncols(arr)
```

$$4 \tag{10}$$

Type: PositiveInteger

To apply an operation to every element of an array, use **map**. This creates a new array. This expression negates every element.

```
map(-,arr)
```

$$\begin{bmatrix} -17 & 0 & 0 & 0 \\ 0 & 0 & 0 & 0 \\ 0 & -15 & 0 & 0 \\ 0 & 0 & 0 & 0 \\ 0 & 0 & 0 & 0 \end{bmatrix} \tag{11}$$

Type: TwoDimensionalArray Integer

This creates an array where all the elements are doubled.

```
map((x +-> x + x),arr)
```

$$\begin{bmatrix} 34 & 0 & 0 & 0 \\ 0 & 0 & 0 & 0 \\ 0 & 30 & 0 & 0 \\ 0 & 0 & 0 & 0 \\ 0 & 0 & 0 & 0 \end{bmatrix} \tag{12}$$

Type: TwoDimensionalArray Integer

To change the array destructively, use **map!** instead of **map**. If you need to make a copy of any array, use **copy**.

```
arrc := copy(arr)
```

$$\begin{bmatrix} 17 & 0 & 0 & 0 \\ 0 & 0 & 0 & 0 \\ 0 & 15 & 0 & 0 \\ 0 & 0 & 0 & 0 \\ 0 & 0 & 0 & 0 \end{bmatrix} \tag{13}$$

Type: TwoDimensionalArray Integer

```
map!(-,arrc)
```

$$\begin{bmatrix} -17 & 0 & 0 & 0 \\ 0 & 0 & 0 & 0 \\ 0 & -15 & 0 & 0 \\ 0 & 0 & 0 & 0 \\ 0 & 0 & 0 & 0 \end{bmatrix} \tag{14}$$

Type: TwoDimensionalArray Integer

```
arrc
```

$$\begin{bmatrix} -17 & 0 & 0 & 0 \\ 0 & 0 & 0 & 0 \\ 0 & -15 & 0 & 0 \\ 0 & 0 & 0 & 0 \\ 0 & 0 & 0 & 0 \end{bmatrix} \tag{15}$$

Type: TwoDimensionalArray Integer

```
arr
```

$$\begin{bmatrix} 17 & 0 & 0 & 0 \\ 0 & 0 & 0 & 0 \\ 0 & 15 & 0 & 0 \\ 0 & 0 & 0 & 0 \\ 0 & 0 & 0 & 0 \end{bmatrix} \quad (16)$$

Type: TwoDimensionalArray Integer

Use **member?** to see if a given element is in an array.

```
member?(17,arr)
```

true (17)

Type: Boolean

```
member?(10317,arr)
```

false (18)

Type: Boolean

To see how many times an element appears in an array, use **count**.

```
count(17,arr)
```

1 (19)

Type: PositiveInteger

```
count(0,arr)
```

18 (20)

Type: PositiveInteger

For more information about the operations available for TwoDimensionalArray, issue `)show TwoDimensionalArray`. For information on related topics, see 'Matrix' on page 414 and 'OneDimensionalArray' on page 425.

9.67 UnivariatePolynomial

The domain constructor UnivariatePolynomial (abbreviated UP) creates domains of univariate polynomials in a specified variable. For example, the domain UP(a1,POLY FRAC INT) provides polynomials in the single variable `a1` whose coefficients are general polynomials with rational number coefficients.

Restriction:
AXIOM does not allow you to create types where UnivariatePolynomial is contained in the coefficient type of Polynomial. Therefore, UP(x,POLY INT) is legal but POLY UP(x,INT) is not.

UP(x,INT) is the domain of polynomials in the single variable **x** with integer coefficients.

```
(p,q) : UP(x,INT)
```

Type: Void

```
p := (3*x-1)**2 * (2*x + 8)
```

$$18\ x^3 + 60\ x^2 - 46\ x + 8 \quad (2)$$

Type: UnivariatePolynomial(x, Integer)

```
q := (1 - 6*x + 9*x**2)**2
```

$$81\ x^4 - 108\ x^3 + 54\ x^2 - 12\ x + 1 \quad (3)$$

Type: UnivariatePolynomial(x, Integer)

The usual arithmetic operations are available for univariate polynomials.

```
p**2 + p*q
```

$$1458\ x^7 + 3240\ x^6 - 7074\ x^5 + 10584\ x^4 - 9282\ x^3 + 4120\ x^2 - 878\ x + 72 \quad (4)$$

Type: UnivariatePolynomial(x, Integer)

The operation **leadingCoefficient** extracts the coefficient of the term of highest degree.

```
leadingCoefficient p
```

$$18 \quad (5)$$

Type: PositiveInteger

The operation **degree** returns the degree of the polynomial. Since the polynomial has only one variable, the variable is not supplied to operations like **degree**.

```
degree p
```

$$3 \quad (6)$$

Type: PositiveInteger

The reductum of the polynomial, the polynomial obtained by subtracting the term of highest order, is returned by **reductum**.

```
reductum p
```

$$60\ x^2 - 46\ x + 8 \quad (7)$$

Type: UnivariatePolynomial(x, Integer)

The operation **gcd** computes the greatest common divisor of two polynomials.

```
gcd(p,q)
```

$$9\ x^2 - 6\ x + 1 \quad (8)$$

Type: UnivariatePolynomial(x, Integer)

The operation **lcm** computes the least common multiple.

```
lcm(p,q)
```

$$162\ x^5 + 432\ x^4 - 756\ x^3 + 408\ x^2 - 94\ x + 8 \quad (9)$$

Type: UnivariatePolynomial(x, Integer)

The operation **resultant** computes the resultant of two univariate polynomials. In the case of `p` and `q`, the resultant is `0` because they share a common root.

```
resultant(p,q)
```

$$0 \quad (10)$$

Type: NonNegativeInteger

To compute the derivative of a univariate polynomial with resepect to its variable, use **D**.

```
D p
```

$$54\ x^2 + 120\ x - 46 \tag{11}$$

Type: UnivariatePolynomial(x, Integer)

Univariate polynomials can also be used as if they were functions. To evaluate a univariate polynomial at some point, apply the polynomial to the point.

```
p(2)
```

$$300 \tag{12}$$

Type: PositiveInteger

The same syntax is used for composing two univariate polynomials, i.e. substituting one polynomial for the variable in another. This substitutes q for the variable in p.

```
p(q)
```

$$9565938\ x^{12} - 38263752\ x^{11} + 70150212\ x^{10} - 77944680\ x^9 + 58852170\ x^8 - 32227632\ x^7 + 13349448\ x^6 - 4280688\ x^5 + 1058184\ x^4 - 192672\ x^3 + 23328\ x^2 - 1536\ x + 40 \tag{13}$$

Type: UnivariatePolynomial(x, Integer)

This substitutes p for the variable in q.

```
q(p)
```

$$8503056\ x^{12} + 113374080\ x^{11} + 479950272\ x^{10} + 404997408\ x^9 - 1369516896\ x^8 - 626146848\ x^7 + 2939858712\ x^6 - 2780728704\ x^5 + 1364312160\ x^4 - 396838872\ x^3 + 69205896\ x^2 - 6716184\ x + 279841 \tag{14}$$

Type: UnivariatePolynomial(x, Integer)

To obtain a list of coefficients of the polynomial, use **coefficients**.

```
l := coefficients p
```

$$[18,\ 60,\ -46,\ 8] \tag{15}$$

Type: List Integer

From this you can use **gcd** and **reduce** to compute the content of the polynomial.

```
reduce(gcd,l)
```

$$2 \tag{16}$$

Type: PositiveInteger

Alternatively (and more easily), you can just call **content**.

```
content p
```

$$2 \tag{17}$$

Type: PositiveInteger

Note that the operation **coefficients** omits the zero coefficients from the list. Sometimes it is useful to convert a univariate polynomial to a vector whose `i` th position contains the degree `i-1` coefficient of the polynomial.

```
ux := (x**4+2*x+3)::UP(x,INT)
```

$$x^4 + 2\,x + 3 \tag{18}$$

Type: UnivariatePolynomial(x, Integer)

To get a complete vector of coefficients, use the operation **vectorise**, which takes a univariate polynomial and an integer denoting the length of the desired vector.

```
vectorise(ux,5)
```

$$[3,\ 2,\ 0,\ 0,\ 1] \tag{19}$$

Type: Vector Integer

It is common to want to do something to every term of a polynomial, creating a new polynomial in the process.

This is a function for iterating across the terms of a polynomial, squaring each term.

```
squareTerms(p) ==
  reduce(+,[t**2 for t in monomials p])
```

Type: Void

Recall what `p` looked like.

```
p
```

$$18\,x^3 + 60\,x^2 - 46\,x + 8 \tag{21}$$

Type: UnivariatePolynomial(x, Integer)

We can demonstrate **squareTerms** on `p`.

```
squareTerms p
```

```
Compiling function squareTerms with type
UnivariatePolynomial(x,Integer) ->
UnivariatePolynomial(x,Integer)
```

$$324\,x^6 + 3600\,x^4 + 2116\,x^2 + 64 \tag{22}$$

Type: UnivariatePolynomial(x, Integer)

When the coefficients of the univariate polynomial belong to a field,[7] it is possible to compute quotients and remainders.

```
(r,s) : UP(a1,FRAC INT)
```

Type: Void

```
r := a1**2 - 2/3
```

$$a1^2 - \frac{2}{3} \tag{24}$$

Type: UnivariatePolynomial(a1, Fraction Integer)

[7]For example, when the coefficients are rational numbers, as opposed to integers. The important property of a field is that non-zero elements can be divided and produce another element. The quotient of the integers 2 and 3 is not another integer.

```
s := a1 + 4
```

$$a1 + 4 \tag{25}$$

Type: UnivariatePolynomial(a1, Fraction Integer)

When the coefficients are rational numbers or rational expressions, the operation **quo** computes the quotient of two polynomials.

```
r quo s
```

$$a1 - 4 \tag{26}$$

Type: UnivariatePolynomial(a1, Fraction Integer)

The operation **rem** computes the remainder.

```
r rem s
```

$$\frac{46}{3} \tag{27}$$

Type: UnivariatePolynomial(a1, Fraction Integer)

The operation **divide** can be used to return a record of both components.

```
d := divide(r, s)
```

$$\left[quotient = a1 - 4,\ remainder = \frac{46}{3}\right] \tag{28}$$

Type: Record(quotient: UnivariatePolynomial(a1, Fraction Integer), remainder: UnivariatePolynomial(a1, Fraction Integer))

Now we check the arithmetic!

```
r - (d.quotient * s + d.remainder)
```

$$0 \tag{29}$$

Type: UnivariatePolynomial(a1, Fraction Integer)

It is also possible to integrate univariate polynomials when the coefficients belong to a field.

```
integrate r
```

$$\frac{1}{3}\ a1^3 - \frac{2}{3}\ a1 \tag{30}$$

Type: UnivariatePolynomial(a1, Fraction Integer)

```
integrate s
```

$$\frac{1}{2}\ a1^2 + 4\ a1 \tag{31}$$

Type: UnivariatePolynomial(a1, Fraction Integer)

One application of univariate polynomials is to see expressions in terms of a specific variable.

We start with a polynomial in `a1` whose coefficients are quotients of polynomials in `b1` and `b2`.

```
t : UP(a1,FRAC POLY INT)
```

Type: Void

Since in this case we are not talking about using multivariate polynomials in only two variables, we use Polynomial. We also use Fraction because we want fractions.

```
t := a1**2 - a1/b2 + (b1**2-b1)/(b2+3)
```

$$a1^2 - \frac{1}{b2}\ a1 + \frac{b1^2 - b1}{b2 + 3} \tag{33}$$

Type: UnivariatePolynomial(a1, Fraction Polynomial Integer)

We push all the variables into a single quotient of polynomials.

```
u : FRAC POLY INT := t
```

$$\frac{a1^2\ b2^2 + \left(b1^2 - b1 + 3\ a1^2 - a1\right)\ b2 - 3\ a1}{b2^2 + 3\ b2} \tag{34}$$

Type: Fraction Polynomial Integer

Alternatively, we can view this as a polynomial in the variable This is a *mode-directed* conversion: you indicate as much of the structure as you care about and let AXIOM decide on the full type and how to do the transformation.

```
u :: UP(b1,?)
```

$$\frac{1}{b2+3}\ b1^2 - \frac{1}{b2+3}\ b1 + \frac{a1^2\ b2 - a1}{b2} \tag{35}$$

Type: UnivariatePolynomial(b1, Fraction Polynomial Integer)

See Section 8.2 on page 236 for a discussion of the factorization facilities in AXIOM for univariate polynomials. For more information on related topics, see Section 1.9 on page 43, Section 2.7 on page 78, 'Polynomial' on page 436, 'MultivariatePolynomial' on page 421, and 'DistributedMultivariatePolynomial' on page 352. Issue the system command `)show UnivariatePolynomial` to display the full list of operations defined by UnivariatePolynomial.

9.68 UniversalSegment

The UniversalSegment domain generalizes Segment by allowing segments without a "hi" end point.

```
pints := 1..
```

1.. (1)

Type: UniversalSegment PositiveInteger

```
nevens := (0..) by -2
```

0.. by -2 (2)

Type: UniversalSegment NonNegativeInteger

Values of type Segment are automatically converted to type UniversalSegment when appropriate.

```
useg: UniversalSegment(Integer) := 3..10
```

3..10 (3)

Type: UniversalSegment Integer

The operation **hasHi** is used to test whether a segment has a `hi` end point.

```
hasHi pints
```

false (4)

Type: Boolean

```
hasHi nevens
```

false (5)

Type: Boolean

```
hasHi useg
```

true (6)

Type: Boolean

All operations available on type Segment apply to UniversalSegment, with the proviso that expansions produce streams rather than lists. This is to accommodate infinite expansions.

```
expand pints
```

[1, 2, 3, 4, 5, 6, 7, ...] (7)

Type: Stream Integer

```
expand nevens
```

[0, −2, −4, −6, −8, −10, −12, ...] (8)

Type: Stream Integer

```
expand [1, 3, 10..15, 100..]
```

[1, 3, 10, 11, 12, 13, 14, ...] (9)

Type: Stream Integer

For more information on related topics, see 'Segment' on page 447, 'SegmentBinding' on page 448, 'List' on page 404, and 'Stream' on page 457. Issue the system command `)show UniversalSegment` to display the full list of operations defined by UniversalSegment.

9.69 Vector

The Vector domain is used for storing data in a one-dimensional indexed data structure. A vector is a homogeneous data structure in that all the components of the vector must belong to the same AXIOM domain. Each vector has a fixed length specified by the user; vectors are not extensible. This domain is similar to the OneDimensionalArray domain, except that when the components of a Vector belong to a Ring, arithmetic operations are provided. For more examples of operations that are defined for both Vector and OneDimensionalArray, see 'OneDimensionalArray' on page 425.

As with the OneDimensionalArray domain, a Vector can be created by calling the operation **new**, its components can be accessed by calling the operations **elt** and **qelt**, and its components can be reset by calling the operations **setelt** and **qsetelt!**.

This creates a vector of integers of length 5 all of whose components are 12.

```
u : VECTOR INT := new(5,12)
```

[12, 12, 12, 12, 12] (1)

Type: Vector Integer

This is how you create a vector from a list of its components.

```
v : VECTOR INT := vector([1,2,3,4,5])
```

$$[1, 2, 3, 4, 5] \tag{2}$$

Type: Vector Integer

Indexing for vectors begins at `1`. The last element has index equal to the length of the vector, which is computed by "#".

```
#(v)
```

$$5 \tag{3}$$

Type: PositiveInteger

This is the standard way to use **elt** to extract an element. Functionally, it is the same as if you had typed `elt(v,2)`.

```
v.2
```

$$2 \tag{4}$$

Type: PositiveInteger

This is the standard way to use **setelt** to change an element. It is the same as if you had typed `setelt(v,3,99)`.

```
v.3 := 99
```

$$99 \tag{5}$$

Type: PositiveInteger

Now look at `v` to see the change. You can use **qelt** and **qsetelt!** (instead of **elt** and **setelt**, respectively) but *only* when you know that the index is within the valid range.

```
v
```

$$[1, 2, 99, 4, 5] \tag{6}$$

Type: Vector Integer

When the components belong to a Ring, AXIOM provides arithmetic operations for Vector. These include left and right scalar multiplication.

```
5 * v
```

$$[5, 10, 495, 20, 25] \tag{7}$$

Type: Vector Integer

```
v * 7
```

$$[7, 14, 693, 28, 35] \tag{8}$$

Type: Vector Integer

```
w : VECTOR INT := vector([2,3,4,5,6])
```

$$[2, 3, 4, 5, 6] \tag{9}$$

Type: Vector Integer

Addition and subtraction are also available.

```
v + w
```

$$[3, 5, 103, 9, 11] \tag{10}$$

Type: Vector Integer

Of course, when adding or subtracting, the two vectors must have the same length or an error message is displayed.

```
v - w
```

$$[-1, -1, 95, -1, -1] \tag{11}$$

Type: Vector Integer

For more information about other aggregate domains, see the following: 'List' on page 404, 'Matrix' on page 414, 'OneDimensionalArray' on page 425,

'Set' on page 449, 'Table' on page 465, and 'TwoDimensionalArray' on page 469. Issue the system command `)show Vector` to display the full list of operations defined by Vector.

9.70 Void

When an expression is not in a value context, it is given type Void. For example, in the expression

```
r := (a; b; if c then d else e; f)
```

values are used only from the subexpressions c and f: all others are thrown away. The subexpressions a, b, d and e are evaluated for side-effects only and have type Void. There is a unique value of type Void.

You will most often see results of type Void when you declare a variable.

```
a : Integer
```

Type: Void

Usually no output is displayed for Void results. You can force the display of a rather ugly object by issuing `)set message void on`.

```
)set message void on
```

```
b : Fraction Integer
```

```
"()"
```

Type: Void

```
)set message void off
```

All values can be converted to type Void.

```
3::Void
```

Type: Void

Once a value has been converted to Void, it cannot be recovered.

```
% :: PositiveInteger
```

```
Cannot convert from type Void to PositiveInteger for
value
"()"
```

PART III

Advanced Programming in AXIOM

CHAPTER 10

Interactive Programming

Programming in the interpreter is easy. So is the use of AXIOM's graphics facility. Both are rather flexible and allow you to use them for many interesting applications. However, both require learning some basic ideas and skills.

All graphics examples in the AXIOM Images section are either produced directly by interactive commands or by interpreter programs. Four of these programs are introduced here. By the end of this chapter you will know enough about graphics and programming in the interpreter to not only understand all these examples, but to tackle interesting and difficult problems on your own. Appendix F lists all the remaining commands and programs used to create these images.

10.1 Drawing Ribbons Interactively

We begin our discussion of interactive graphics with the creation of a useful facility: plotting ribbons of two-graphs in three-space. Suppose you want to draw the two-dimensional graphs of n functions $f_i(x), 1 \leq i \leq n$, all over some fixed range of x. One approach is to create a two-dimensional graph for each one, then superpose one on top of the other. What you will more than likely get is a jumbled mess. Even if you make each function a different color, the result is likely to be confusing.

A better approach is to display each of the $f_i(x)$ in three dimensions as a "ribbon" of some appropriate width along the y-direction, laying down each ribbon next to the previous one. A ribbon is simply a function of x and y depending only on x.

We illustrate this for $f_i(x)$ defined as simple powers of x for x ranging between -1 and 1.

Draw the ribbon for $z = x^2$.

```
draw(x**2,x=-1..1,y=0..1)
```

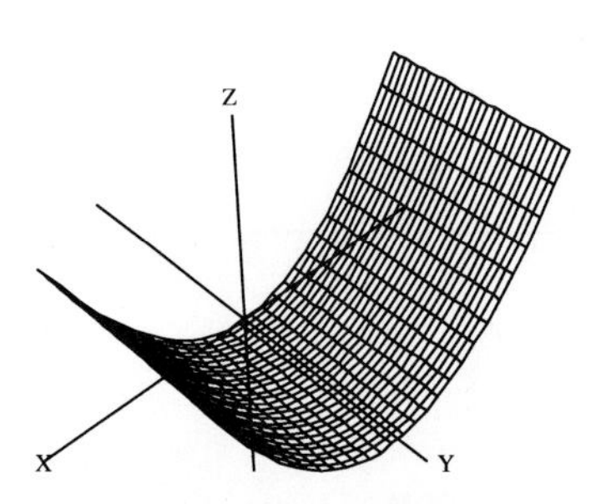

Now that was easy! What you get is a "wire-mesh" rendition of the ribbon. That's fine for now. Notice that the mesh-size is small in both the x and the y directions. AXIOM normally computes points in both these directions. This is unnecessary. One step is all we need in the y-direction. To have AXIOM economize on y-points, we re-draw the ribbon with option `var2Steps == 1`.

Re-draw the ribbon, but with option `var2Steps == 1` so that only 1 step is computed in the y direction.

```
vp := draw(x**2,x=-1..1,y=0..1,var2Steps==1)
```

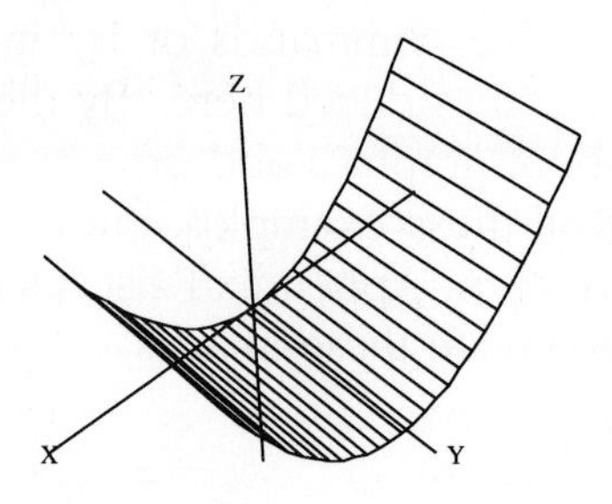

The operation has created a viewport, that is, a graphics window on your screen. We assigned the viewport to `vp` and now we manipulate its contents.

10.1.1 Graphs as Objects

Graphs are objects, like numbers and algebraic expressions. You may want to do some experimenting with graphs. For example, say

```
showRegion(vp, "on")
```

to put a bounding box around the ribbon. Try it! Issue `rotate(vp,`

-45, 90) to rotate the figure −45 longitudinal degrees and 90 latitudinal degrees.

Here is a different rotation. This turns the graph so you can view it along the y-axis.

```
rotate(vp, 0, -90); vp
```

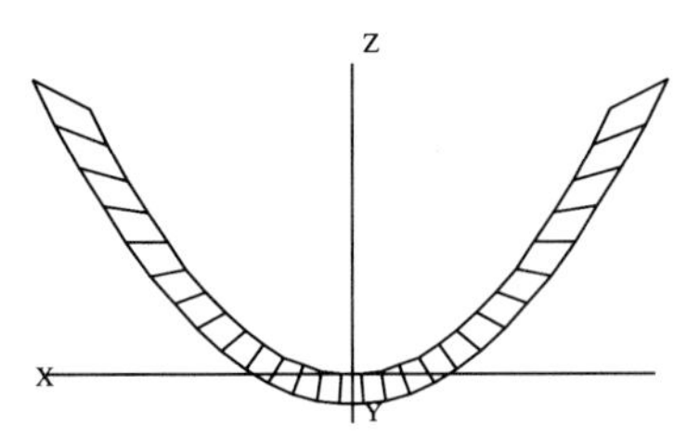

There are many other things you can do. In fact, most everything you can do interactively using the three-dimensional control panel (such as translating, zooming, resizing, coloring, perspective and lighting selections) can also be done directly by operations (see Chapter 7 for more details).

When you are done experimenting, say `reset(vp)` to restore the picture to its original position and settings.

10.1.2 Adding Ribbons

Let's add another ribbon to our picture—one for x^3. Since y ranges from 0 to 1 for the first ribbon, now let y range from 1 to 2. This puts the second ribbon next to the first one.

How do you add a second ribbon to the viewport? One method is to extract the "space" component from the viewport using the operation **subspace**. You can think of the space component as the object inside the window (here, the ribbon). Let's call it `sp`. To add the second ribbon, you draw the second ribbon using the option `space == sp`.

Extract the space component of `vp`.

```
sp := subspace(vp)
```

Add the ribbon for x^3 alongside that for x^2.

```
vp := draw(x**3,x=-1..1,y=1..2,var2Steps==1, space==sp)
```

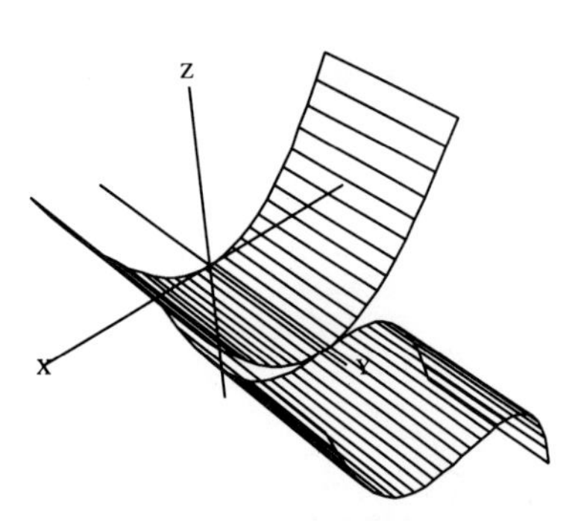

Unless you moved the original viewport, the new viewport covers the old one. You might want to check that the old object is still there by moving the top window.

Let's show quadrilateral polygon outlines on the ribbons and then enclose the ribbons in a box.

Show quadrilateral polygon outlines.

```
drawStyle(vp,"shade");outlineRender(vp,"on")
```

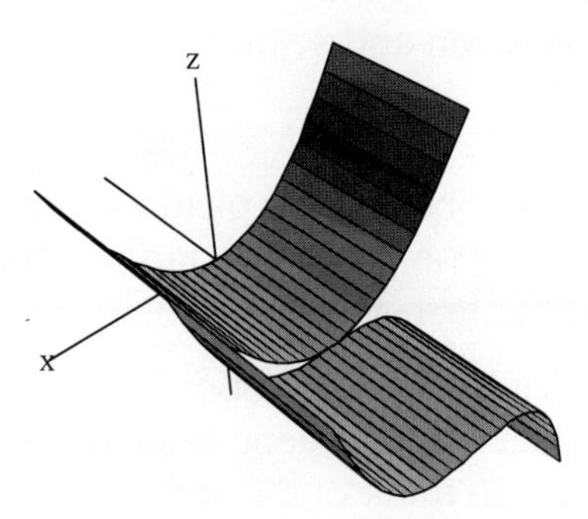

Enclose the ribbons in a box.

```
rotate(vp,20, 60); showRegion(vp,"on")
```

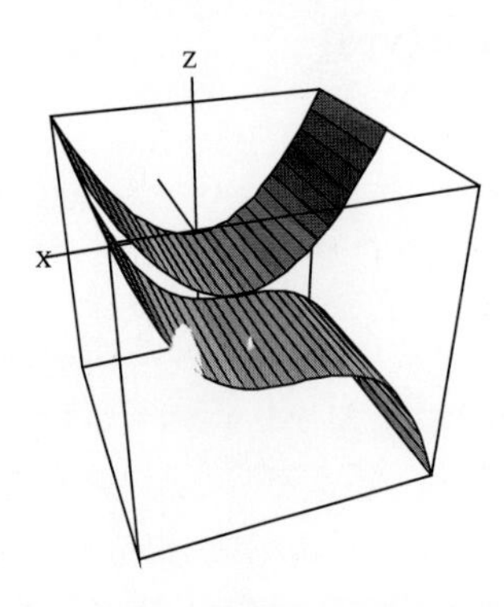

This process has become tedious! If we had to add two or three more ribbons, we would have to repeat the above steps several more times. It is time to write an interpreter program to help us take care of the details.

10.2 A Ribbon Program

The above approach creates a new viewport for each additional ribbon. A better approach is to build one object composed of all ribbons before creating a viewport. To do this, use **makeObject** rather than **draw**. The operations have similar formats, but **draw** returns a viewport and **makeObject** returns a space object.

We now create a function **drawRibbons** of two arguments: `flist`, a list of formulas for the ribbons you want to draw, and `xrange`, the range over which you want them drawn. Using this function, you can just say

```
drawRibbons([x**2, x**3], x=-1..1)
```

to do all of the work required in the last section. Here is the **drawRibbons** program. Invoke your favorite editor and create a file called **ribbon.input** containing the following program.

Comment	Code	Line
	`drawRibbons(flist, xrange) ==`	1
Create empty space `sp`.	`  sp := createThreeSpace()`	2
The initial ribbon position.	`  y0 := 0`	3
For each function `f`,	`  for f in flist repeat`	4
create and add a ribbon	`    makeObject(f, xrange, y=y0..y0+1,`	5
for `f` to the space `sp`.	`       space==sp, var2Steps == 1)`	6
The next ribbon position.	`    y0 := y0 + 1`	7
Create viewport.	`  vp := makeViewport3D(sp, "Ribbons")`	8
Select shading style.	`  drawStyle(vp, "shade")`	9
Show polygon outlines.	`  outlineRender(vp, "on")`	10
Enclose in a box.	`  showRegion(vp,"on")`	11
The number of ribbons	`  n := # flist`	12
Zoom in x- and z-directions.	`  zoom(vp,n,1,n)`	13
Change the angle of view.	`  rotate(vp,0,75)`	14
Return the viewport.	`  vp`	15

Figure 10.1: The first **drawRibbons** function.

Here are some remarks on the syntax used in the **drawRibbons** function (consult Chapter 6 for more details). Unlike most other programming languages which use semicolons, parentheses, or *begin–end* brackets to delineate the structure of programs, the structure of an AXIOM program is determined by indentation. The first line of the function definition always begins in column 1. All other lines of the function are indented with respect to the first line and form a *pile* (see Section 5.2 on page 112).

The definition of **drawRibbons** consists of a pile of expressions to be executed one after another. Each expression of the pile is indented at the same level. Lines 4-7 designate one single expression: since lines 5-7 are indented with respect to the others, these lines are treated as a continuation of line 4. Also since lines 5 and 7 have the same indentation level, these lines designate a pile within the outer pile.

The last line of a pile usually gives the value returned by the pile. Here it is also the value returned by the function. AXIOM knows this is the last line of the function because it is the last line of the file. In other cases, a new expression beginning in column one signals the end of a function.

The line `drawStyle(vp,"shade")` is given after the viewport has been created to select the draw style. We have also used the **zoom** option. Without the zoom, the viewport region would be scaled equally in all three coordinate directions.

Let's try the function **drawRibbons**. First you must read the file to give AXIOM the function definition.

Read the input file.

```
)read ribbons
```

Draw ribbons for $x, x^2, \ldots, x^5$ for $-1 \leq x \leq 1$

```
drawRibbons([x**i for i in 1..5],x=-1..1)
```

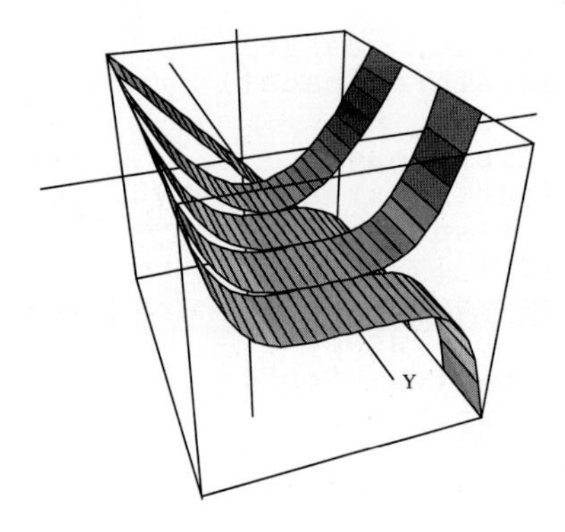

10.3 Coloring and Positioning Ribbons

Before leaving the ribbon example, we make two improvements. Normally, the color given to each point in the space is a function of its height within a bounding box. The points at the bottom of the box are red, those at the top are purple.

To change the normal coloring, you can give an option `colorFunction == ` *function*. When AXIOM goes about displaying the data, it determines the range of colors used for all points within the box. AXIOM then distributes these numbers uniformly over the number of hues. Here we use the simple color function $(x, y) \mapsto i$ for the ith ribbon.

Also, we add an argument `yrange` so you can give the range of `y` occupied by the ribbons. For example, if the `yrange` is given as `y=0..1`

```
drawRibbons(flist, xrange, yrange) ==                                         1
  sp := createThreeSpace()                                                    2
  num := # flist                                                              3
  yVar := variable yrange                                                     4
  y0:Float    := lo segment yrange                                            5
  width:Float := (hi segment yrange - y0)/num                                 6
  for f in flist for color in 1..num repeat                                   7
    makeObject(f, xrange, yVar = y0..y0+width,                                8
      var2Steps == 1, colorFunction == (x,y) +-> color,                       9
      space == sp)                                                           10
    y0 := y0 + width                                                         11
  vp := makeViewport3D(sp, "Ribbons")                                        12
  drawStyle(vp, "shade")                                                     13
  outlineRender(vp, "on")                                                    14
  showRegion(vp, "on")                                                       15
  vp                                                                         16
```

Line annotations: Create empty space sp (line 2). The number of ribbons (line 3). The ribbon variable (line 4). The first ribbon coordinate (line 5). The width of a ribbon (line 6). For each function f, create and add ribbon to sp of a different color (lines 7–10). The next ribbon coordinate (line 11). Create viewport (line 12). Select shading style (line 13). Show polygon outlines (line 14). Enclose in a box (line 15). Return the viewport (line 16).

Figure 10.2: The final **drawRibbons** function.

and there are 5 ribbons to be displayed, each ribbon would have width 0.2 and would appear in the range $0 \leq y \leq 1$.

Refer to lines 4-9. Line 4 assigns to `yVar` the variable part of the `yrange` (after all, it need not be `y`). Suppose that `yrange` is given as `t = a..b` where `a` and `b` have numerical values. Then line 5 assigns the value of `a` to the variable `y0`. Line 6 computes the width of the ribbon by dividing the difference of `a` and `b` by the number, `num`, of ribbons. The result is assigned to the variable `width`. Note that in the for-loop in line 7, we are iterating in parallel; it is not a nested loop.

10.4 Points, Lines, and Curves

What you have seen so far is a high-level program using the graphics facility. We now turn to the more basic notions of points, lines, and curves in three-dimensional graphs. These facilities use small floats (objects of type SmallFloat) for data. Let us first give names to the small float values 0 and 1.

The small float 0.

```
zero := 0.0@SF
```

0.0 (1)

Type: SmallFloat

The small float 1.

```
one := 1.0@SF
```

1.0 (2)

Type: SmallFloat

The “`@`” sign means “of the type.” Thus `zero` is 0.0 of the type SmallFloat.

You can also say `0.0::SF`.

Points can have four small float components: x, y, z coordinates and an optional color. A "curve" is simply a list of points connected by straight line segments.

Create the point `origin` with color zero, that is, the lowest color on the color map.

```
origin := point [zero,zero,zero,zero]
```

$$[0.0, 0.0, 0.0, 0.0] \tag{3}$$

Type: Point SmallFloat

Create the point `unit` with color zero.

```
unit := point [one,one,one,zero]
```

$$[1.0, 1.0, 1.0, 0.0] \tag{4}$$

Type: Point SmallFloat

Create the curve (well, here, a line) from `origin` to `unit`.

```
curve := [origin, unit]
```

$$[[0.0, 0.0, 0.0, 0.0], [1.0, 1.0, 1.0, 0.0]] \tag{5}$$

Type: List Point SmallFloat

10.4.1 Drawing Arrows

We make this line segment into an arrow by adding an arrowhead. The arrowhead extends to, say, `p3` on the left, and to, say, `p4` on the right. To describe an arrow, you tell AXIOM to draw the two curves `[p1, p2, p3]` and `[p2, p4]`. We also decide through experimentation on values for `arrowScale`, the ratio of the size of the arrowhead to the stem of the arrow, and `arrowAngle`, the angle between the arrowhead and the arrow.

Invoke your favorite editor and create an input file called **arrows.input**. This input file first defines the values of `arrowAngle` and `arrowScale`, then defines the function **makeArrow**(p_1, p_2) to draw an arrow from point p_1 to p_2.

The angle of the arrowhead.
The size of the arrowhead relative to the stem.

The arrow.
The length of the arrowhead.
The angle from the x-axis
The x-coord of left endpoint.
The y-coord of left endpoint.
The x-coord of right endpoint.
The y-coord of right endpoint.
The z-coord of both endpoints.
The left endpoint of head.
The right endpoint of head.
The arrow as a list of curves.

```
arrowAngle := %pi-%pi/10.0@SF                              1
arrowScale := 0.2@SF                                       2
                                                           3
makeArrow(p1, p2) ==                                       4
  delta := p2 - p1                                         5
  len := arrowScale * length delta                         6
  theta := atan(delta.1, delta.2)                          7
  c1 := len*cos(theta + arrowAngle)                        8
  s1 := len*sin(theta + arrowAngle)                        9
  c2 := len*cos(theta - arrowAngle)                       10
  s2 := len*sin(theta - arrowAngle)                       11
  z  := p2.3*(1 - arrowScale)                             12
  p3 := point [p2.1 + c1, p2.2 + s1, z, p2.4]             13
  p4 := point [p2.1 + c2, p2.2 + s2, z, p2.4]             14
  [[p1, p2, p3], [p2, p4]]                                15
```

Read the file and then create an arrow from the point `origin` to the point `unit`.

Read the input file defining **makeArrow**.

```
)read arrows
```

Type: Void

Construct the arrow (a list of two curves).

```
arrow := makeArrow(origin,unit)
```

$$\left[\begin{bmatrix}[[0.0,\ 0.0,\ 0.0,\ 0.0],\ [1.0,\ 1.0,\ 1.0,\ 0.0], \\ [0.69134628604607973,\ 0.842733077659504, \\ 0.80000000000000004,\ 0.0]\end{bmatrix}, \begin{bmatrix}[[1.0,\ 1.0,\ 1.0,\ 0.0], \\ [0.842733077659504,\ 0.69134628604607973, \\ 0.80000000000000004,\ 0.0]\end{bmatrix}\right] \quad (8)$$

Type: List List Point SmallFloat

Create an empty object `sp` of type `ThreeSpace`.

```
sp := createThreeSpace()
```

3-Space with 0 components (9)

Type: ThreeSpace SmallFloat

Add each curve of the arrow to the space `sp`.

```
for a in arrow repeat sp := curve(sp,a)
```

Type: Void

Create a three-dimensional viewport containing that space.

```
vp := makeViewport3D(sp,"Arrow")
```

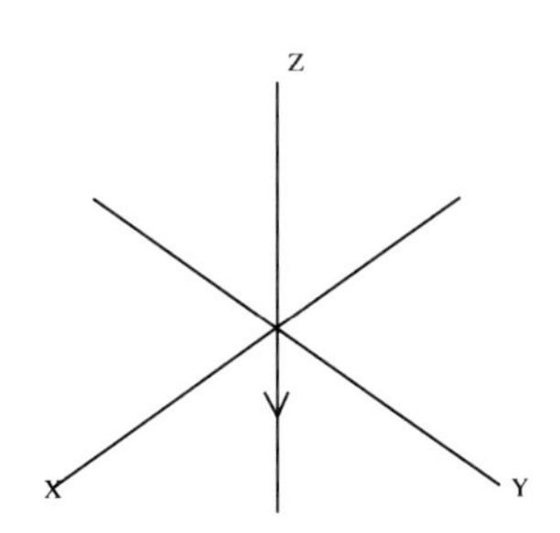

Here is a better viewing angle.

```
rotate(vp,200,-60); vp
```

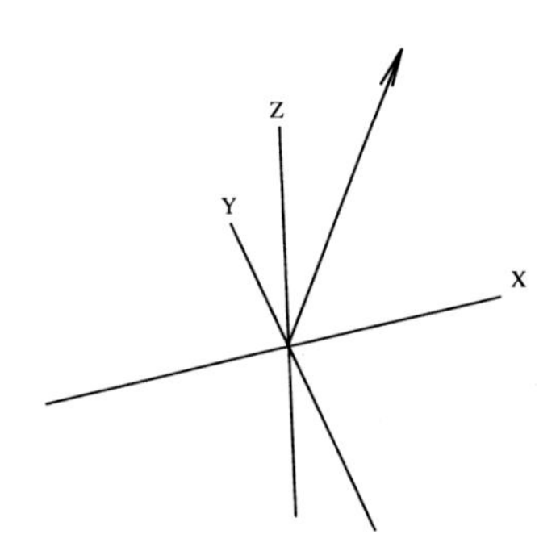

10.5 A Bouquet of Arrows

Let's draw a "bouquet" of arrows. Each arrow is identical. The arrowheads are uniformly placed on a circle parallel to the xy-plane. Thus the position of each arrow differs only by the angle θ, $0 \leq \theta < 2\pi$, between the arrow and the x-axis on the xy-plane.

Our bouquet is rather special: each arrow has a different color (which won't be evident here, unfortunately). This is arranged by letting the color of each successive arrow be denoted by θ. In this way, the color of arrows ranges from red to green to violet. Here is a program to draw a bouquet of n arrows.

The initial angle.
Create empty space sp.
For each index i, create:
—the point at base of arrow;
—the point at tip of arrow;
—the ith arrow.
For each arrow component,
add the component to sp.
The next angle.
Create the viewport from sp.

```
drawBouquet(n,title) ==                                          1
  angle := 0.0@SF                                                2
  sp := createThreeSpace()                                       3
  for i in 0..n-1 repeat                                         4
    start := point [0.0@SF,0.0@SF,0.0@SF,angle]                  5
    end   := point [cos angle, sin angle, 1.0@SF, angle]         6
    arrow := makeArrow(start,end)                                7
    for a in makeArrow(start,end) repeat                         8
      curve(sp,a)                                                9
    angle := angle + 2*%pi/n                                    10
  makeViewport3D(sp,title)                                      11
```

Read the input file.

```
)read bouquet
```

A bouquet of a dozen arrows.

```
drawBouquet(12,"A Dozen Arrows")
```

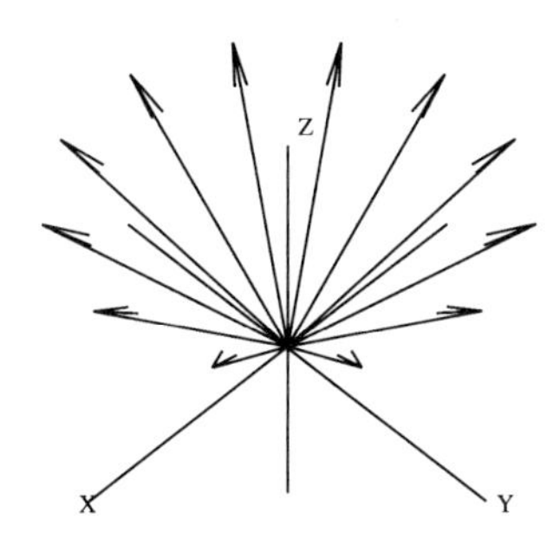

10.6 Drawing Complex Vector Fields

We now put our arrows to good use drawing complex vector fields. These vector fields give a representation of complex-valued functions of complex variables. Consider a Cartesian coordinate grid of points (x, y) in the plane, and some complex-valued function f defined on this grid. At every point on this grid, compute the value of $f(x + iy)$ and call it z. Since z has both a real and imaginary value for a given (x, y) grid point, there are four dimensions to plot. What do we do? We represent the values of z by arrows planted at each grid point. Each arrow represents the value of z in polar coordinates (r, θ). The length of the arrow is proportional to r. Its direction is given by θ.

The code for drawing vector fields is in the file **vectors.input**. We discuss its contents from top to bottom.

Before showing you the code, we have two small matters to take care of. First, what if the function has large spikes, say, ones that go off to infinity? We define a variable `clipValue` for this purpose. When `r` exceeds the value of `clipValue`, then the value of `clipValue` is used instead of that for `r`. For convenience, we define a function `clipFun(x)` which uses `clipValue` to "clip" the value of `x`.

Maximum value allowed.

```
clipValue : SF = 6                                    1
clipFun(x) == min(max(x,-clipValue),clipValue)        2
```

Notice that we identify `clipValue` as a small float but do not declare the type of the function **clipFun**. As it turns out, **clipFun** is called with a small float value. This declaration ensures that **clipFun** never does a conversion when it is called.

The second matter concerns the possible "poles" of a function, the actual points where the spikes have infinite values. AXIOM uses normal Small-

Float arithmetic which does not directly handle infinite values. If your function has poles, you must adjust your step size to avoid landing directly on them (AXIOM calls **error** when asked to divide a value by 0, for example).

We set the variables `realSteps` and `imagSteps` to hold the number of steps taken in the real and imaginary directions, respectively. Most examples will have ranges centered around the origin. To avoid a pole at the origin, the number of points is taken to be odd.

Number of real steps.
Number of imaginary steps.

```
realSteps: INT := 25                                              3
imagSteps: INT := 25                                              4
```

Now define the function **drawComplexVectorField** to draw the arrows. It is good practice to declare the type of the main function in the file. This one declaration is usually sufficient to ensure that other lower-level functions are compiled with the correct types.

```
C := Complex SmallFloat                                           5
S := Segment SmallFloat                                           6
drawComplexVectorField: (C -> C, S, S) -> VIEW3D                  7
```

The first argument is a function mapping complex small floats into complex small floats. The second and third arguments give the range of real and imaginary values as segments like `a..b`. The result is a three-dimensional viewport. Here is the full function definition:

The real step size.
The imaginary step size.
Create empty space `sp`.
The initial real value.
Begin real iteration.
The initial imaginary value.
Begin imaginary iteration.
The value of `f` at the point.
The direction of the arrow.
The length of the arrow.
The base point of the arrow.
The scaled length of the arrow.
The tip point of the arrow.

Create the arrow.
Add arrow to the space `sp`.
The next imaginary value.
The next real value.
Draw it!

```
drawComplexVectorField(f, realRange,imagRange) ==                 8
  delReal := (hi(realRange)-lo(realRange))/realSteps              9
  delImag := (hi(imagRange)-lo(imagRange))/imagSteps              10
  sp := createThreeSpace()                                        11
  real := lo(realRange)                                           12
  for i in 1..realSteps+1 repeat                                  13
    imag := lo(imagRange)                                         14
    for j in 1..imagSteps+1 repeat                                15
      z := f complex(real,imag)                                   16
      arg := argument z                                           17
      len := clipFun sqrt norm z                                  18
      p1 :=  point [real, imag, 0.0@SF, arg]                      19
      scaleLen := delReal * len                                   20
      p2 := point [p1.1 + scaleLen*cos(arg),                      21
                   p1.2 + scaleLen*sin(arg),0.0@SF, arg]          22
      arrow := makeArrow(p1, p2)                                  23
      for a in arrow repeat curve(sp, a)                          24
      imag := imag + delImag                                      25
    real := real + delReal                                        26
  makeViewport3D(sp, "Complex Vector Field")                      27
```

As a first example, let us draw `f(z) == sin(z)`. There is no need to create a user function: just pass the **sin** from Complex SmallFloat.

Read the file.

```
)read vectors
```

Draw the complex vector field of `sin(x)`.

```
drawComplexVectorField(sin,-2..2,-2..2)
```

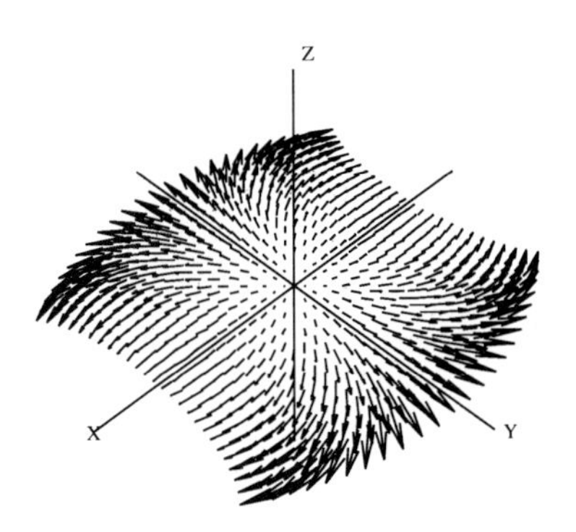

10.7 Drawing Complex Functions

Here is another way to graph a complex function of complex arguments. For each complex value z, compute $f(z)$, again expressing the value in polar coordinates (r, θ). We draw the complex valued function, again considering the (x, y)-plane as the complex plane, using r as the height (or z-coordinate) and θ as the color. This is a standard plot—we learned how to do this in Chapter 7—but here we write a new program to illustrate the creation of polygon meshes, or grids.

Call this function **drawComplex**. It displays the points using the "mesh" of points. The function definition is in three parts.

The first part.
The real step size.
The imaginary step size.
Initial list of list of points `llp`.

```
drawComplex: (C -> C, S, S) -> VIEW3D                              1
drawComplex(f, realRange, imagRange) ==                            2
  delReal := (hi(realRange)-lo(realRange))/realSteps               3
  delImag := (hi(imagRange)-lo(imagRange))/imagSteps               4
  llp:List List Point SF := []                                     5
```

Variables `delReal` and `delImag` give the step sizes along the real and imaginary directions as computed by the values of the global variables `realSteps` and `imagSteps`. The mesh is represented by a list of lists of points `llp`, initially empty. Now `[ ]` alone is ambiguous, so to set this initial value you have to tell AXIOM what type of empty list it is. Next comes the loop which builds `llp`.

The initial real value.
Begin real iteration.
The initial imaginary value.
The initial list of points `lp`.

```
  real := lo(realRange)                                            6
  for i in 1..realSteps+1 repeat                                   7
    imag := lo(imagRange)                                          8
    lp := []$(List Point SF)                                       9
```

Begin imaginary iteration.
The value of f at the point.
Create a point.

Add the point to lp.
The next imaginary value.
The next real value.
Add lp to llp.

```
for j in 1..imagSteps+1 repeat                              10
  z := f complex(real,imag)                                 11
  pt := point [real,imag, clipFun sqrt norm z,              12
               argument z]                                  13
  lp := cons(pt,lp)                                         14
  imag := imag + delImag                                    15
real := real + delReal                                      16
llp := cons(lp, llp)                                        17
```

The code consists of both an inner and outer loop. Each pass through the inner loop adds one list lp of points to the list of lists of points llp. The elements of lp are collected in reverse order.

Create a mesh and display.

```
makeViewport3D(mesh(llp), "Complex Function")               18
```

The operation **mesh** then creates an object of type ThreeSpace(SmallFloat) from the list of lists of points. This is then passed to **makeViewport3D** to display the image.

Now add this function directly to your **vectors.input** file and re-read the file using)read vectors. We try **drawComplex** using a user-defined function f.

This one has a pole at $z = 0$.

```
f(z) == exp(1/z)
```

Type: Void

Draw it with an odd number of steps to avoid the pole.

```
drawComplex(f,-2..2,-2..2)
```

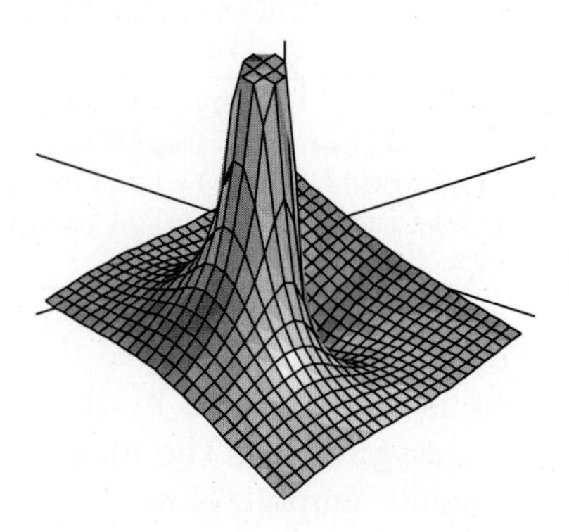

10.8 Functions Producing Functions

In Section 6.14 on page 157, you learned how to use the operation **function** to create a function from symbolic formulas. Here we introduce a similar operation which not only creates functions, but functions from functions.

The facility we need is provided by the package MakeUnaryCompiledFunction(E,S,T). This package produces a unary (one-argument) compiled function from some symbolic data generated by a previous computation.[1] The E tells where the symbolic data comes from; the S and T give AXIOM the source and target type of the function, respectively. The compiled function produced has type S → T. To produce a compiled function with definition `p(x) == expr`, call `compiledFunction(expr, x)` from this package. The function you get has no name. You must to assign the function to the variable p to give it that name.

Do some computation.

```
(x+1/3)**5
```

$$x^5 + \frac{5}{3}\,x^4 + \frac{10}{9}\,x^3 + \frac{10}{27}\,x^2 + \frac{5}{81}\,x + \frac{1}{243} \tag{1}$$

Type: Polynomial Fraction Integer

Convert this to an anonymous function of **x**. Assign it to the variable p to give the function a name.

```
p := compiledFunction(%,x)$MakeUnaryCompiledFunction(POLY
  FRAC INT,SF,SF)
```

theMap (...) (2)

Type: (SmallFloat → SmallFloat)

Apply the function.

```
p(sin(1.3))
```

3.668751115057229 (3)

Type: SmallFloat

For a more sophisticated application, read on.

10.9 Automatic Newton Iteration Formulas

We resume our continuing saga of arrows and complex functions. Suppose we want to investigate the behavior of Newton's iteration function in the complex plane. Given a function f, we want to find the complex values z such that $f(z) = 0$.

The first step is to produce a Newton iteration formula for a given f: $x_{n+1} = x_n - \frac{f(x_n)}{f'(x_n)}$. We represent this formula by a function g that performs the computation on the right-hand side, that is, $x_{n+1} = g(x_n)$.

The type Expression Integer (abbreviated EXPR INT) is used to represent general symbolic expressions in AXIOM. To make our facility as general as possible, we assume f has this type. Given f, we want to produce

[1] MakeBinaryCompiledFunction is available for binary functions.

a Newton iteration function g which, given a complex point x_n, delivers the next Newton iteration point x_{n+1}.

This time we write an input file called **newton.input**. We need to import MakeUnaryCompiledFunction (discussed in the last section), call it with appropriate types, and then define the function `newtonStep` which references it. Here is the function `newtonStep`:

The complex numbers.
Package for making functions.

Newton's iteration function.
Function for f.
Function for f'.
Return the iterator function.

Turn an expression f into a function.

Create an nth derivative function.

Returns the variable in f.
The list of variables.
The number of variables.

Return a dummy variable.

```
C := Complex SmallFloat                                           1
complexFunPack:=MakeUnaryCompiledFunction(EXPR INT,C,C)           2
                                                                  3
newtonStep(f) ==                                                  4
  fun   := complexNumericFunction f                               5
  deriv := complexDerivativeFunction(f,1)                         6
  (x:C):C +->                                                     7
    x - fun(x)/deriv(x)                                           8
                                                                  9
complexNumericFunction f ==                                      10
  v := theVariableIn f                                           11
  compiledFunction(f, v)$complexFunPack                          12
                                                                 13
complexDerivativeFunction(f,n) ==                                14
  v := theVariableIn f                                           15
  df := D(f,v,n)                                                 16
  compiledFunction(df, v)$complexFunPack                         17
                                                                 18
theVariableIn f ==                                               19
  vl := variables f                                              20
  nv := # vl                                                     21
  nv > 1 => error "Expression is not univariate."                22
  nv = 0 => 'x                                                   23
  first vl                                                       24
```

Do you see what is going on here? A formula `f` is passed into the function **newtonStep**. First, the function turns `f` into a compiled program mapping complex numbers into complex numbers. Next, it does the same thing for the derivative of `f`. Finally, it returns a function which computes a single step of Newton's iteration.

The function **complexNumericFunction** extracts the variable from the expression `f` and then turns `f` into a function which maps complex numbers into complex numbers. The function **complexDerivativeFunction** does the same thing for the derivative of `f`. The function **theVariableIn** extracts the variable from the expression `f`, calling the function **error** if `f` has more than one variable. It returns the dummy variable `x` if `f` has no variables.

Let's now apply **newtonStep** to the formula for computing cube roots of two.

Read the input file with the definitions.

```
)read newton.input
```

Type: Void

The cube root of two.

```
f := x**3 - 2
```

$$x^3 - 2 \quad (7)$$

Type: Polynomial Integer

Get Newton's iteration formula.

```
g := newtonStep f
```

theMap (...) (8)

Type: (Complex SmallFloat → Complex SmallFloat)

Let a denote the result of applying Newton's iteration once to the complex number `1 + %i`.

```
a := g(1.0 + %i)
```

$$0.66666666666666674 + 0.33333333333333337\ \%i \quad (9)$$

Type: Complex SmallFloat

Now apply it repeatedly. How fast does it converge?

```
[(a := g(a)) for i in 1..]
```

$$\begin{array}{l} [1.1644444444444444 - 0.73777777777777775\ \%i, \\ 0.92614004697164776 - 0.17463006425584393\ \%i, \\ 1.3164444838140228 + 0.15690694583015852\ \%i, \\ 1.2462991025761463 + 0.015454763610132094\ \%i, \\ 1.2598725296532081 - 3.3827162059311272E - 4\ \%i, \\ 1.259920960928212 + 2.6023534653422681E - 8\ \%i, \\ 1.259921049894879 - 3.6751942591616685E - 15\ \%i, \\ 1.2599210498948732 - 3.3132158019282496E - 29\ \%i, \\ 1.2599210498948732 - 5.6051938572992683E - 45\ \%i, \\ 1.2599210498948732, \ldots] \end{array} \quad (10)$$

Type: Stream Complex SmallFloat

Check the accuracy of the last iterate.

```
a**3
```

$$2.0 \quad (11)$$

Type: Complex SmallFloat

In 'MappingPackage1' on page 411, we show how functions can be manipulated as objects in AXIOM. A useful operation to consider here is "*", which means composition. For example `g*g` causes the Newton iteration formula to be applied twice. Correspondingly, `g**n` means to apply the iteration formula `n` times.

Apply `g` twice to the point `1 + %i`.

```
(g*g) (1.0 + %i)
```

$$1.16444444444444444 - 0.737777777777777775\ \%i \tag{12}$$

Type: Complex SmallFloat

Apply `g` 11 times.

```
(g**11) (1.0 + %i)
```

$$1.2599210498948732 \tag{13}$$

Type: Complex SmallFloat

Look now at the vector field and surface generated after two steps of Newton's formula for the cube root of two. The poles in these pictures represent bad starting values, and the flat areas are the regions of convergence to the three roots.

The vector field.

```
drawComplexVectorField(g**3,-3..3,-3..3)
```

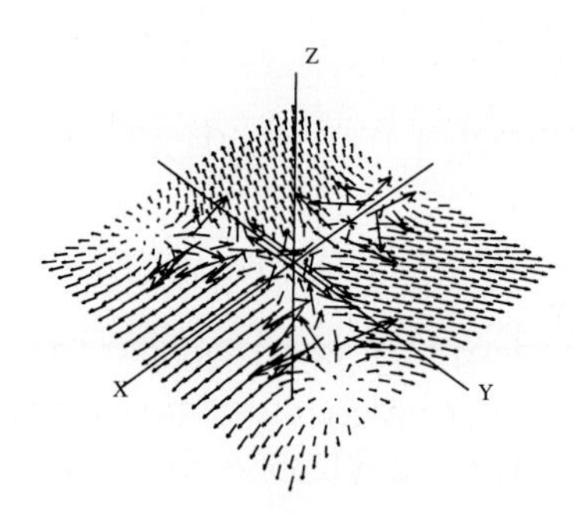

The surface.

```
drawComplex(g**3,-3..3,-3..3)
```

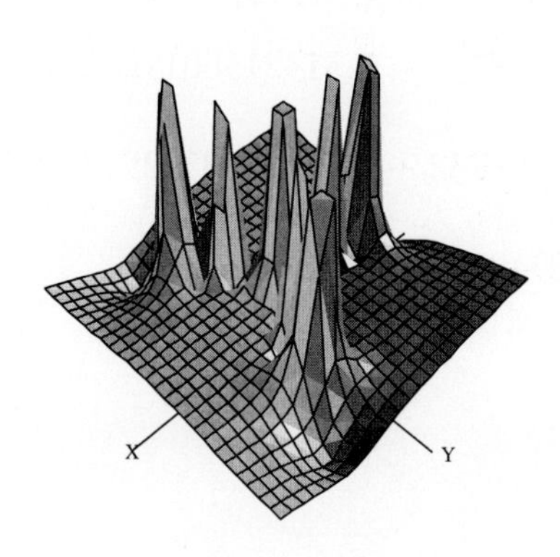

CHAPTER 11

Packages

Packages provide the bulk of AXIOM's algorithmic library, from numeric packages for computing special functions to symbolic facilities for differential equations, symbolic integration, and limits.

In Chapter 10, we developed several useful functions for drawing vector fields and complex functions. We now show you how you can add these functions to the AXIOM library to make them available for general use.

The way we created the functions in Chapter 10 is typical of how you, as an advanced AXIOM user, may interact with AXIOM. You have an application. You go to your editor and create an input file defining some functions for the application. Then you run the file and try the functions. Once you get them all to work, you will often want to extend them, add new features, perhaps write additional functions.

Eventually, when you have a useful set of functions for your application, you may want to add them to your local AXIOM library. To do this, you embed these function definitions in a package and add that package to the library.

To introduce new packages, categories, and domains into the system, you need to use the AXIOM compiler to convert the constructors into executable machine code. An existing compiler in AXIOM is available on an "as-is" basis. A new, faster compiler will be available in version 2.0 of AXIOM.

All constructors used in a file must be spelled out in full unless abbreviated by macros like these at the top of a file.

Identify kinds and abbreviations
Type definition begins here.

Export part begins.
Exported Operations

Implementation part begins.
Local variable 1.
Local variable 2.
Local variable 3.
Local variable 4.
Local variable 5.

Exported function definition 1.
Exported function definition 2.
Exported function definition 3.

Local function definition 1.

Local function definition 2.

Exported function definition 4.

```
C       ==> Complex SmallFloat                                    1
S       ==> Segment SmallFloat                                    2
INT     ==> Integer                                               3
SF      ==> SmallFloat                                            4
VIEW3D  ==> ThreeDimensionalViewport                              5
CURVE   ==> List List Point SF                                    6
                                                                  7
)abbrev package DRAWCX DrawComplex                                8
DrawComplex(): Exports == Implementation where                    9
                                                                  10
  Exports == with                                                 11
    drawComplex: (C -> C,S,S,Boolean) -> VIEW3D                   12
    drawComplexVectorField: (C -> C,S,S) -> VIEW3D                13
    setRealSteps: INT -> INT                                      14
    setImagSteps: INT -> INT                                      15
    setClipValue: SF-> SF                                         16
                                                                  17
  Implementation == add                                           18
    arrowScale : SF := (0.2)::SF --relative size                  19
    arrowAngle : SF := pi()-pi()/(20::SF)                         20
    realSteps  : INT := 11 --# real steps                         21
    imagSteps  : INT := 11 --# imaginary steps                    22
    clipValue  : SF  := 10::SF --maximum vector length            23
                                                                  24
    setRealSteps(n) == realSteps := n                             25
    setImagSteps(n) == imagSteps := n                             26
    setClipValue(c) == clipValue := c                             27
                                                                  28
    clipFun: SF -> SF --Clip large magnitudes.                    29
    clipFun(x) == min(max(x, -clipValue), clipValue)              30
                                                                  31
    makeArrow: (Point SF,Point SF,SF,SF) -> CURVE                 32
    makeArrow(p1, p2, len, arg) == ...                            33
                                                                  34
    drawComplex(f, realRange, imagRange, arrows?) == ...          35
```

Figure 11.1: The DrawComplex package.

11.1 Names, Abbreviations, and File Structure

Each package has a name and an abbreviation. For a package of the complex draw functions from Chapter 10, we choose the name DrawComplex and abbreviation DRAWCX.[1] To be sure that you have not chosen a name or abbreviation already used by the system, issue the system command `)show` for both the name and the abbreviation.

Once you have named the package and its abbreviation, you can choose any new filename you like with extension "**.spad**" to hold the definition of your package. We choose the name **drawpak.spad**. If your application involves more than one package, you can put them all in the same file.

[1] An abbreviation can be any string of between two and seven capital letters and digits, beginning with a letter. See Section 2.2.5 on page 68 for more information.

AXIOM assumes no relationship between the name of a library file, and the name or abbreviation of a package.

Near the top of the "**.spad**" file, list all the abbreviations for the packages using `)abbrev`, each command beginning in column one. Macros giving names to AXIOM expressions can also be placed near the top of the file. The macros are only usable from their point of definition until the end of the file.

Consider the definition of DrawComplex in Figure 11.1. After the macro definition

```
S      ==> Segment SmallFloat
```

the name `S` can be used in the file as a shorthand for Segment SmallFloat.[2] The abbreviation command for the package

```
)abbrev package DRAWCX DrawComplex
```

is given after the macros (although it could precede them).

11.2 Syntax

The definition of a package has the syntax:

PackageForm : *Exports* == *Implementation*

The syntax for defining a package constructor is the same as that for defining any function in AXIOM. In practice, the definition extends over many lines so that this syntax is not practical. Also, the type of a package is expressed by the operator `with` followed by an explicit list of operations. A preferable way to write the definition of a package is with a `where` expression:

The definition of a package usually has the form:
PackageForm : `Exports == Implementation where`
  optional type declarations
  `Exports == with`
    list of exported operations
  `Implementation == add`
    list of function definitions for exported operations

The DrawComplex package takes no parameters and exports five operations, each a separate item of a *pile*. Each operation is described as a

[2]The interpreter also allows `macro` for macro definitions.

declaration: a name, followed by a colon ("`:`"), followed by the type of the operation. All operations have types expressed as *mappings* with the syntax

source `->` *target*

11.3 Abstract Datatypes

A constructor as defined in AXIOM is called an *abstract datatype* in the computer science literature. Abstract datatypes separate "specification" (what operations are provided) from "implementation" (how the operations are implemented). The `Exports` (specification) part of a constructor is said to be "public" (it provides the user interface to the package) whereas the `Implementation` part is "private" (information here is effectively hidden—programs cannot take advantage of it).

The `Exports` part specifies what operations the package provides to users. As an author of a package, you must ensure that the `Implementation` part provides a function for each operation in the `Exports` part.[3]

An important difference between interactive programming and the use of packages is in the handling of global variables such as `realSteps` and `imagSteps`. In interactive programming, you simply change the values of variables by *assignment*. With packages, such variables are local to the package—their values can only be set using functions exported by the package. In our example package, we provide two functions **setRealSteps** and **setImagSteps** for this purpose.

Another local variable is `clipValue` which can be changed using the exported operation **setClipValue**. This value is referenced by the internal function **clipFun** that decides whether to use the computed value of the function at a point or, if the magnitude of that value is too large, the value assigned to `clipValue` (with the appropriate sign).

11.4 Capsules

The part to the right of `add` in the `Implementation` part of the definition is called a *capsule*. The purpose of a capsule is:

- to define a function for each exported operation, and
- to define a *local environment* for these functions to run.

What is a local environment? First, what is an environment? Think of

[3] The DrawComplex package enhances the facility described in Chapter 10.7 by allowing a complex function to have arrows emanating from the surface to indicate the direction of the complex argument.

the capsule as an input file that AXIOM reads from top to bottom. Think of the input file as having a `)clear all` at the top so that initially no variables or functions are defined. When this file is read, variables such as `realSteps` and `arrowSize` in DrawComplex are set to initial values. Also, all the functions defined in the capsule are compiled. These include those that are exported (like `drawComplex`), and those that are not (like `makeArrow`). At the end, you get a set of name-value pairs: variable names (like `realSteps` and `arrowSize`) are paired with assigned values, while operation names (like `drawComplex` and `makeArrow`) are paired with function values.

This set of name-value pairs is called an *environment*. Actually, we call this environment the "initial environment" of a package: it is the environment that exists immediately after the package is first built. Afterwards, functions of this capsule can access or reset a variable in the environment. The environment is called *local* since any changes to the value of a variable in this environment can be seen *only* by these functions.

Only the functions from the package can change the variables in the local environment. When two functions are called successively from a package, any changes caused by the first function called are seen by the second.

Since the environment is local to the package, its names don't get mixed up with others in the system or your workspace. If you happen to have a variable called `realSteps` in your workspace, it does not affect what the DrawComplex functions do in any way.

The functions in a package are compiled into machine code. Unlike function definitions in input files that may be compiled repeatedly as you use them with varying argument types, functions in packages have a unique type (generally parameterized by the argument parameters of a package) and a unique compilation residing on disk.

The capsule itself is turned into a compiled function. This so-called *capsule function* is what builds the initial environment spoken of above. If the package has arguments (see below), then each call to the package constructor with a distinct pair of arguments builds a distinct package, each with its own local environment.

11.5 Input Files vs. Packages

A good question at this point would be "Is writing a package more difficult than writing an input file?"

The programs in input files are designed for flexibility and ease-of-use. AXIOM can usually work out all of your types as it reads your program and does the computations you request. Let's say that you define a one-argument function without giving its type. When you first apply the function to a value, this value is understood by AXIOM as identifying the type for the argument parameter. Most of the time AXIOM goes through the body of your function and figures out the target type that you have in mind. AXIOM sometimes fails to get it right. Then—and only then—do you need a declaration to tell AXIOM what type you want.

Input files are usually written to be read by AXIOM—and by you. Without suitable documentation and declarations, your input files are likely incomprehensible to a colleague—and to you some months later!

Packages are designed for legibility, as well as run-time efficiency. There are few new concepts you need to learn to write packages. Rather, you just have to be explicit about types and type conversions. The types of all functions are pre-declared so that AXIOM—and the reader— knows precisely what types of arguments can be passed to and from the functions (certainly you don't want a colleague to guess or to have to work this out from context!). The types of local variables are also declared. Type conversions are explicit, never automatic.[4]

In summary, packages are more tedious to write than input files. When writing input files, you can casually go ahead, giving some facts now, leaving others for later. Writing packages requires forethought, care and discipline.

11.6 Compiling Packages

Once you have defined the package DrawComplex, you need to compile and test it. To compile the package, issue the system command `)compile drawpak`. AXIOM reads the file **drawpak.spad** and compiles its contents into machine binary. If all goes well, the file **DRAWCX.NRLIB** is created in your local directory for the package. To test the package, you must load the package before trying an operation.

Compile the package.

```
)compile drawpak
```

Load the package.

```
)load DRAWCX
```

```
library DRAWCX has been loaded.
DrawComplex is now explicitly exposed in frame G1077
```

Use an odd step size to avoid a pole at the origin.

```
setRealSteps 51
```

$$51 \tag{1}$$

Type: PositiveInteger

[4]There is one exception to this rule: conversions from a subdomain to a domain are automatic. After all, the objects both have the domain as a common type.

```
setImagSteps 51
```

51 (2)

Type: PositiveInteger

Define **f** to be the Gamma function.

```
f(z) == Gamma(z)
```

Type: Void

Clip values of function with magnitude larger than 7.

```
setClipValue 7
```

7.0 (4)

Type: SmallFloat

Draw the **Gamma** function.

```
drawComplex(f,-%pi..%pi,-%pi..%pi, false)
```

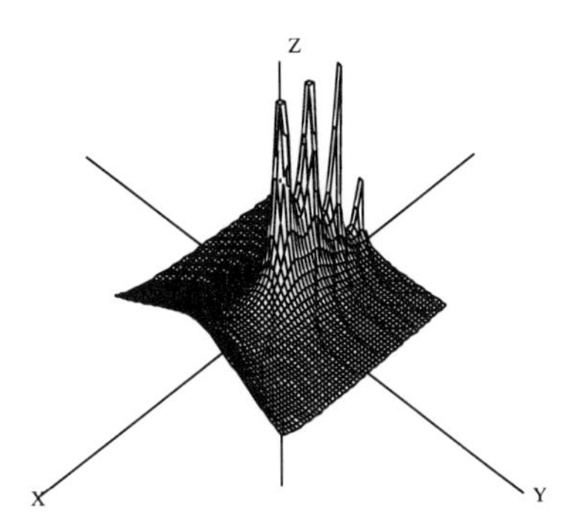

11.7 Parameters

The power of packages becomes evident when packages have parameters. Usually these parameters are domains and the exported operations have types involving these parameters.

In Chapter 2, you learned that categories denote classes of domains. Although we cover this notion in detail in the next chapter, we now give you a sneak preview of its usefulness.

In Section 6.15 on page 159, we defined functions `bubbleSort(m)` and `insertionSort(m)` to sort a list of integers. If you look at the code for these functions, you see that they may be used to sort *any* structure `m` with the right properties. Also, the functions can be used to sort lists of *any* elements—not just integers. Let us now recall the code for `bubbleSort`.

```
bubbleSort(m) ==
  n := #m
  for i in 1..(n-1) repeat
    for j in n..(i+1) by -1 repeat
      if m.j < m.(j-1) then swap!(m,j,j-1)
  m
```

What properties of "lists of integers" are assumed by the sorting algorithm? In the first line, the operation **#** computes the maximum index of the list. The first obvious property is that `m` must have a finite number of elements. In AXIOM, this is done by your telling AXIOM that `m` has the "attribute" `finiteAggregate`. An *attribute* is a property that a domain either has or does not have. As we show later in Section 12.9 on page 522, programs can query domains as to the presence or absence of an attribute.

The operation **swap!** swaps elements of `m`. Using Browse, you find that **swap!** requires its elements to come from a domain of category IndexedAggregate with attribute `shallowlyMutable`. This attribute means that you can change the internal components of `m` without changing its external structure. Shallowly-mutable data structures include lists, streams, one- and two-dimensional arrays, vectors, and matrices.

The category IndexedAggregate designates the class of aggregates whose elements can be accessed by the notation `m.s` for suitable selectors `s`. The category IndexedAggregate takes two arguments: `Index`, a domain of selectors for the aggregate, and `Entry`, a domain of entries for the aggregate. Since the sort functions access elements by integers, we must choose `Index =` Integer. The most general class of domains for which `bubbleSort` and `insertionSort` are defined are those of category IndexedAggregate(Integer,Entry) with the two attributes `shallowlyMutable` and `finiteAggregate`.

Using Browse, you can also discover that AXIOM has many kinds of domains with attribute `shallowlyMutable`. Those of class IndexedAggregate(Integer,Entry) include Bits, FlexibleArray, OneDimensionalArray, List, String, and Vector, and also HashTable and EqTable with integer keys. Although you may never want to sort all such structures, we nonetheless demonstrate AXIOM's ability to do so.

Another requirement is that Entry has an operation "<". One way to get this operation is to assume that Entry has category OrderedSet. By definition, will then export a "<" operation. A more general approach is to allow any comparison function `f` to be used for sorting. This function will be passed as an argument to the sorting functions.

Our sorting package then takes two arguments: a domain `S` of objects of *any* type, and a domain A, an aggregate of type IndexedAggregate(Integer, S) with the above two attributes. Here is its definition using what are close to the original definitions of `bubbleSort` and `insertionSort` for sorting lists of integers. The symbol "`!`" is added to the ends of the operation names. This uniform naming convention is used for AXIOM operation names that destructively change one or more of their arguments.

```
SortPackage(S,A) : Exports == Implementation where          1
  S: Object                                                  2
  A: IndexedAggregate(Integer,S)                             3
    with (finiteAggregate; shallowlyMutable)                 4
                                                             5
  Exports ==                                                 6
    bubbleSort!: (A,(S,S) -> Boolean) -> A                   7
    insertionSort!: (A, (S,S) -> Boolean) -> A               8
                                                             9
  Implementation ==                                          10
    bubbleSort!(m,f) ==                                      11
      n := #m                                                12
      for i in 1..(n-1) repeat                               13
        for j in n..(i+1) by -1 repeat                       14
          if f(m.j,m.(j-1)) then swap!(m,j,j-1)              15
      m                                                      16
    insertionSort!(m,f) ==                                   17
      for i in 2..#m repeat                                  18
        j := i                                               19
        while j > 1 and f(m.j,m.(j-1)) repeat                20
          swap!(m,j,j-1)                                     21
          j := (j - 1) pretend PositiveInteger               22
      m                                                      23
```

11.8 Conditionals

When packages have parameters, you can say that an operation is or is not exported depending on the values of those parameters. When the domain of objects S has an "<" operation, we can supply one-argument versions of bubbleSort and insertionSort which use this operation for sorting. The presence of the operation "<" is guaranteed when S is an ordered set.

```
Exports ==                                                   1
    bubbleSort!: (A,(S,S) -> Boolean) -> A                   2
    insertionSort!: (A, (S,S) -> Boolean) -> A               3
                                                             4
    if S has OrderedSet then                                 5
      bubbleSort!: A -> A                                    6
      insertionSort!: A -> A                                 7
```

In addition to exporting the one-argument sort operations conditionally, we must provide conditional definitions for the operations in the Implementation part. This is easy: just have the one-argument functions call the corresponding two-argument functions with the operation "<" from S.

```
  Implementation ==                                          1
      ...                                                    2
    if S has OrderedSet then                                 3
```

```
    bubbleSort!(m) == bubbleSort!(m,<$S)                    4
    insertionSort!(m) == insertionSort!(m,<$S)              5
```

In Section 6.15 on page 159, we give an alternative definition of **bubbleSort** using **first** and **rest** that is more efficient for a list (for which access to any element requires traversing the list from its first node). To implement a more efficient algorithm for lists, we need the operation **setelt** which allows us to destructively change the **first** and **rest** of a list. Using Browse, you find that these operations come from category UnaryRecursiveAggregate. Several aggregate types are unary recursive aggregates including those of List and AssociationList. We provide two different implementations for **bubbleSort!** and **insertionSort!**: one for list-like structures, another for array-like structures.

```
Implementation == add                                          1
    ...                                                        2
    if A has UnaryRecursiveAggregate(S) then                   3
      bubbleSort!(m,fn) ==                                     4
        null m => m                                            5
        l := m                                                 6
        while not null (r := l.rest) repeat                    7
           r := sort r                                         8
           x := l.first                                        9
           if fm(x,r.first) then                              10
             l.first := r.first                               11
             r.first := x                                     12
           l.rest := r                                        13
           l := l.rest                                        14
         m                                                    15
       insertionSort!(m,fn) ==                                16
          ...                                                 17
```

The ordering of definitions is important. The standard definitions come first and then the predicate

```
A has UnaryRecursiveAggregate(S)
```

is evaluated. If `true`, the special definitions cover up the standard ones.

Another equivalent way to write the capsule is to use an `if-then-else` expression:

```
    if A has UnaryRecursiveAggregate(S) then                   1
       ...                                                     2
    else                                                       3
       ...                                                     4
```

11.9 Testing

Once you have written the package, embed it in a file, for example, **sortpak.spad**. Be sure to include an `)abbrev` command at the top of the file:

```
)abbrev package SORTPAK SortPackage
```

Now compile the file (using `)compile sortpak.spad`).

Load the file. You are then ready to begin testing.

```
)load SORTPAK
```

```
library SORTPAK has been loaded.
SortPackage is now explicitly exposed in frame G1077
```

Define a list.

```
l := [1,7,4,2,11,-7,3,2]
```

$$[1, 7, 4, 2, 11, -7, 3, 2] \quad (1)$$

Type: List Integer

Since the integers are an ordered set, a one-argument operation will do.

```
bubbleSort!(l)
```

$$[-7, 1, 2, 2, 3, 4, 7, 11] \quad (2)$$

Type: List Integer

Re-sort it using "greater than."

```
bubbleSort!(l,(x,y) +-> x > y)
```

$$[11, 7, 4, 3, 2, 2, 1, -7] \quad (3)$$

Type: List Integer

Now sort it again using "<" on integers.

```
bubbleSort!(l, <$Integer)
```

$$[-7, 1, 2, 2, 3, 4, 7, 11] \quad (4)$$

Type: List Integer

A string is an aggregate of characters so we can sort them as well.

```
bubbleSort! "Mathematical Sciences"
```

" MSaaaccceeehiilmnstt" (5)

Type: String

Is "<" defined on booleans?

```
false < true
```

true (6)

Type: Boolean

Good! Create a bit string representing ten consecutive boolean values `true`.

```
u : Bits := new(10,true)
```

"1111111111" (7)

Type: Bits

Set bits 3 through 5 to `false`, then display the result.

```
u(3..5) := false; u
```

"1100011111" (8)

Type: Bits

Now sort these booleans.

```
bubbleSort! u
```

"0001111111" (9)

Type: Bits

Create an "eq-table" (see 'EqTable' on page 353), a table having integers as keys and strings as values.

```
t : EqTable(Integer,String) := table()
```

table() (10)

Type: EqTable(Integer, String)

Give the table a first entry.

```
t.1 := "robert"
```

"robert" (11)

Type: String

And a second.

```
t.2 := "richard"
```

"richard" (12)

Type: String

What does the table look like?

```
t
```

table (2 = "richard", 1 = "robert") (13)

Type: EqTable(Integer, String)

Now sort it.

```
bubbleSort! t
```

table (2 = "robert", 1 = "richard") (14)

Type: EqTable(Integer, String)

11.10 How Packages Work

Recall that packages as abstract datatypes are compiled independently and put into the library. The curious reader may ask: "How is the interpreter able to find an operation such as **bubbleSort!**? Also, how is a single compiled function such as **bubbleSort!** able to sort data of different types?"

After the interpreter loads the package SortPackage, the four operations from the package become known to the interpreter. Each of these operations is expressed as a *modemap* in which the type of the operation is written in terms of symbolic domains.

See the modemaps for **bubbleSort!**.

```
)display op bubbleSort!
```

```
There are 2 exposed functions called bubbleSort! :

   [1] D1 -> D1 from SortPackage(D2,D1)
          if D2 has ORDSET and D2 has OBJECT and D1 has
          IndexedAggregate(Integer, D2) with
              finiteAggregate
              shallowlyMutable
```

```
[2] (D1,((D3,D3) -> Boolean)) -> D1 from
      SortPackage(D3,D1) if D3 has OBJECT
      and D1 has IndexedAggregate(Integer,D3) with
           finiteAggregate
           shallowlyMutable
```

What happens if you ask for `bubbleSort!([1,-5,3])`? There is a unique modemap for an operation named **bubbleSort!** with one argument. Since `[1,-5,3]` is a list of integers, the symbolic domain `D1` is defined as List(Integer). For some operation to apply, it must satisfy the predicate for some `D2`. What `D2`? The third expression of the `and` requires `D1 has IndexedAggregate(Integer, D2) with` two attributes. So the interpreter searches for an IndexedAggregate among the ancestors of List (Integer) (see Section 12.4 on page 519). It finds one: IndexedAggregate(Integer, Integer). The interpreter tries defining `D2` as Integer. After substituting for `D1` and `D2`, the predicate evaluates to `true`. An applicable operation has been found!

Now AXIOM builds the package SortPackage(List(Integer), Integer). According to its definition, this package exports the required operation: **bubbleSort!**: List Integer → List Integer. The interpreter then asks the package for a function implementing this operation. The package gets all the functions it needs (for example, **rest** and **swap!**) from the appropriate domains and then it returns a **bubbleSort!** to the interpreter together with the local environment for **bubbleSort!**. The interpreter applies the function to the argument `[1,-5,3]`. The **bubbleSort!** function is executed in its local environment and produces the result.

CHAPTER 12

Categories

This chapter unravels the mysteries of categories—what they are, how they are related to domains and packages, how they are defined in AXIOM, and how you can extend the system to include new categories of your own.

We assume that you have read the introductory material on domains and categories in Section 2.1.1 on page 61. There you learned that the notion of packages covered in the previous chapter are special cases of domains. While this is in fact the case, it is useful here to regard domains as distinct from packages.

Think of a domain as a datatype, a collection of objects (the objects of the domain). From your "sneak preview" in the previous chapter, you might conclude that categories are simply named clusters of operations exported by domains. As it turns out, categories have a much deeper meaning. Categories are fundamental to the design of AXIOM. They control the interactions between domains and algorithmic packages, and, in fact, between all the components of AXIOM.

Categories form hierarchies as shown on the inside cover pages of this book. The inside front-cover pages illustrate the basic algebraic hierarchy of the AXIOM programming language. The inside back-cover pages show the hierarchy for data structures.

Think of the category structures of AXIOM as a foundation for a city on which superstructures (domains) are built. The algebraic hierarchy, for example, serves as a foundation for constructive mathematical algorithms embedded in the domains of AXIOM. Once in place, domains can be constructed, either independently or from one another.

Superstructures are built for quality—domains are compiled into machine code for run-time efficiency. You can extend the foundation in directions beyond the space directly beneath the superstructures, then extend selected superstructures to cover the space. Because of the compilation strategy, changing components of the foundation generally means that the existing superstructures (domains) built on the changed parts of the foundation (categories) have to be rebuilt—that is, recompiled.

Before delving into some of the interesting facts about categories, let's see how you define them in AXIOM.

12.1 Definitions

A category is defined by a function with exactly the same format as any other function in AXIOM.

> The definition of a category has the syntax:
>
> *CategoryForm* : `Category` == *Extensions* `[ with` *Exports* `]`
>
> The brackets `[ ]` here indicate optionality.

The first example of a category definition is SetCategory, the most basic of the algebraic categories in AXIOM.

```
SetCategory(): Category ==                      1
   Join(Type,CoercibleTo OutputForm) with       2
      "=" : ($, $) -> Boolean                   3
```

The definition starts off with the name of the category (SetCategory); this is always in column one in the source file. All parts of a category definition are then indented with respect to this first line.

In Chapter 2, we talked about Ring as denoting the class of all domains that are rings, in short, the class of all rings. While this is the usual naming convention in AXIOM, it is also common to use the word "Category" at the end of a category name for clarity. The interpretation of the name SetCategory is, then, "the category of all domains that are (mathematical) sets."

The name SetCategory is followed in the definition by its formal parameters enclosed in parentheses "()". Here there are no parameters. As required, the type of the result of this category function is the distinguished name Category.

Then comes the "==". As usual, what appears to the right of the "==" is a definition, here, a category definition. A category definition always has two parts separated by the reserved word `with`.

The first part tells what categories the category extends. Here, the category extends two categories: Type, the category of all domains, and CoercibleTo(OutputForm). The operation `Join` is a system-defined operation that forms a single category from two or more other categories.

Every category other than Type is an extension of some other category. If, for example, SetCategory extended only the category Type, the definition here would read "`Type with ...`". In fact, the `Type` is optional in this line; "`with ...`" suffices.

12.2 Exports

To the right of the `with` is a list of all the *exports* of the category. Each exported operation has a name and a type expressed by a *declaration* of the form "*name*: *type*".

Categories can export symbols, as well as 0 and 1 which denote domain constants.[1] In the current implementation, all other exports are operations with types expressed as *mappings* with the syntax

source -> *target*

The category SetCategory has a single export: the operation "=" whose type is given by the mapping `($, $) -> Boolean`. The "$" in a mapping type always means "the domain." Thus the operation "=" takes two arguments from the domain and returns a value of type Boolean.

The source part of the mapping here is given by a *tuple* consisting of two or more types separated by commas and enclosed in parentheses. If an operation takes only one argument, you can drop the parentheses around the source type. If the mapping has no arguments, the source part of the mapping is either left blank or written as "`()`". Here are examples of formats of various operations with some contrived names.

```
someIntegerConstant   :     $
aZeroArgumentOperation:     () -> Integer
aOneArgumentOperation:      Integer -> $
aTwoArgumentOperation:      (Integer,$) -> Void
aThreeArgumentOperation:    ($,Integer,$) -> Fraction($)
```

[1]The numbers 0 and 1 are operation names in AXIOM.

12.3 Documentation

The definition of SetCategory above is missing an important component: its library documentation. Here is its definition, complete with documentation.

```
++ Description:                                                  1
++ \axiomType{SetCategory} is the basic category                 2
++ for describing a collection of elements with                  3
++ \axiomOp{=} (equality) and a \axiomFun{coerce}                4
++ to \axiomType{OutputForm}.                                    5
                                                                 6
SetCategory(): Category ==                                       7
  Join(Type, CoercibleTo OutputForm) with                        8
    "=": ($, $) -> Boolean                                       9
      ++ \axiom{x = y} tests if \axiom{x} and                   10
      ++ \axiom{y} are equal.                                   11
```

Documentary comments are an important part of constructor definitions. Documentation is given both for the category itself and for each export. A description for the category precedes the code. Each line of the description begins in column one with "++". The description starts with the word `Description:`.[2] All lines of the description following the initial line are indented by the same amount.

Surround the name of any constructor (with or without parameters) with an `\axiomType{}`. Similarly, surround an operator name with `\axiomOp{}`, an AXIOM operation with `\axiomFun{}`, and a variable or AXIOM expression with `\axiom{}`. Library documentation is given in a TeX-like language so that it can be used both for hard-copy and for Browse. These different wrappings cause operations and types to have mouse-active buttons in Browse. For hard-copy output, wrapped expressions appear in a different font. The above documentation appears in hard-copy as:

> SetCategory is the basic category for describing a collection of elements with "=" (equality) and a **coerce** to OutputForm.

and

> `x = y` tests if `x` and `y` are equal.

For our purposes in this chapter, we omit the documentation from further category descriptions.

[2] Other information such as the author's name, date of creation, and so on, can go in this area as well but are currently ignored by AXIOM.

12.4 Hierarchies

A second example of a category is SemiGroup, defined by:

```
SemiGroup(): Category == SetCategory with                1
      "*":   ($,$) -> $                                   2
      "**": ($, PositiveInteger) -> $                     3
```

This definition is as simple as that for SetCategory, except that there are two exported operations. Multiple exported operations are written as a *pile*, that is, they all begin in the same column. Here you see that the category mentions another type, PositiveInteger, in a signature. Any domain can be used in a signature.

Since categories extend one another, they form hierarchies. Each category other than Type has one or more parents given by the one or more categories mentioned before the `with` part of the definition. SemiGroup extends SetCategory and SetCategory extends both Type and CoercibleTo (OutputForm). Since CoercibleTo (OutputForm) also extends Type, the mention of Type in the definition is unnecessary but included for emphasis.

12.5 Membership

We say a category designates a class of domains. What class of domains? That is, how does AXIOM know what domains belong to what categories? The simple answer to this basic question is key to the design of AXIOM:

Domains belong to categories by assertion.

When a domain is defined, it is asserted to belong to one or more categories. Suppose, for example, that an author of domain String wishes to use the binary operator "`*`" to denote concatenation. Thus `"hello " * "there"` would produce the string `"hello there"`[3]. The author of String could then assert that String is a member of SemiGroup. According to our definition of SemiGroup, strings would then also have the operation "`**`" defined automatically. Then `"--" ** 4` would produce a string of eight dashes `"--------"`. Since String is a member of SemiGroup, it also is a member of SetCategory and thus has an operation "=" for testing that two strings are equal.

Now turn to the algebraic category hierarchy inside the front cover of this book. Any domain that is a member of a category extending SemiGroup is a member of SemiGroup (that is, it *is* a semigroup). In particular, any

[3]Actually, concatenation of strings in AXIOM is done by juxtaposition or by using the operation **concat**. The expression `"hello " "there"` produces the string `"hello there"`.

domain asserted to be a Ring is a semigroup since Ring extends Monoid, that, in turn, extends SemiGroup. The definition of Integer in AXIOM asserts that Integer is a member of category IntegerNumberSystem, that, in turn, asserts that it is a member of EuclideanDomain. Now EuclideanDomain extends PrincipalIdealDomain and so on. If you trace up the hierarchy, you see that EuclideanDomain extends Ring, and, therefore, SemiGroup. Thus Integer is a semigroup and also exports the operations "*" and "**".

12.6 Defaults

We actually omitted the last part of the definition of SemiGroup in Section 12.4 on page 519. Here now is its complete AXIOM definition.

```
SemiGroup(): Category == SetCategory with                1
      "*": ($, $) -> $                                   2
      "**": ($, PositiveInteger) -> $                    3
    add                                                  4
      import RepeatedSquaring($)                         5
      x: $ ** n: PositiveInteger == expt(x,n)            6
```

The add part at the end is used to give "default definitions" for exported operations. Once you have a multiplication operation "*", you can define exponentiation for positive integer exponents using repeated multiplication:

$$x^n = \underbrace{x\,x\,x\,\cdots\,x}_{n \text{ times}}$$

This definition for "**" is called a *default* definition. In general, a category can give default definitions for any operation it exports. Since SemiGroup and all its category descendants in the hierarchy export "**", any descendant category may redefine "**" as well.

A domain of category SemiGroup (such as Integer) may or may not choose to define its own "**" operation. If it does not, a default definition that is closest (in a "tree-distance" sense of the hierarchy) to the domain is chosen.

The part of the category definition following an "add" operation is a *capsule*, as discussed in the previous chapter. The line

```
import RepeatedSquaring($)
```

references the package RepeatedSquaring($), that is, the package RepeatedSquaring that takes "this domain" as its parameter. For example, if the semigroup Polynomial (Integer) does not define its own exponentiation operation, the definition used may come from the package RepeatedSquaring (Polynomial (Integer)). The next line gives the definition in terms of **expt** from that package.

The default definitions are collected to form a "default package" for the category. The name of the package is the same as the category but with an ampersand ("&") added at the end. A default package always takes an additional argument relative to the category. Here is the definition of the default package SemiGroup& as automatically generated by AXIOM from the above definition of SemiGroup.

```
SemiGroup_&($): Exports == Implementation where          1
  $: SemiGroup                                           2
  Exports == with                                        3
    "**": ($, PositiveInteger) -> $                      4
  Implementation == add                                  5
    import RepeatedSquaring($)                           6
    x:$ ** n:PositiveInteger == expt(x,n)                7
```

12.7 Axioms

In the previous section you saw the complete AXIOM program defining SemiGroup. According to this definition, semigroups (that is, are sets with the operations "*" and "**".

You might ask: "Aside from the notion of default packages, isn't a category just a *macro*, that is, a shorthand equivalent to the two operations "*" and "**" with their types?" If a category were a macro, every time you saw the word SemiGroup, you would rewrite it by its list of exported operations. Furthermore, every time you saw the exported operations of SemiGroup among the exports of a constructor, you could conclude that the constructor exported SemiGroup.

A category is *not* a macro and here is why. The definition for SemiGroup has documentation that states:

> Category SemiGroup denotes the class of all multiplicative semigroups, that is, a set with an associative operation "*".
>
> Axioms:
> `associative("*" :  ($,$)->$) -- (x*y)*z = x*(y*z)`

According to the author's remarks, the mere exporting of an operation named "*" and "**" is not enough to qualify the domain as a SemiGroup. In fact, a domain can be a semigroup only if it explicitly exports a "**" and a "*" satisfying the associativity axiom.

In general, a category name implies a set of axioms, even mathematical theorems. There are numerous axioms from Ring, for example, that are well-understood from the literature. No attempt is made to list them all. Nonetheless, all such mathematical facts are implicit by the use of the name Ring.

12.8 Correctness

While such statements are only comments, AXIOM can enforce their intention simply by shifting the burden of responsibility onto the author of a domain. A domain belongs to category `Ring` only if the author asserts that the domain belongs to Ring or to a category that extends Ring.

This principle of assertion is important for large user-extendable systems. AXIOM has a large library of operations offering facilities in many areas. Names such as **norm** and **product**, for example, have diverse meanings in diverse contexts. An inescapable hindrance to users would be to force those who wish to extend AXIOM to always invent new names for operations. AXIOM allows you to reuse names, and then use context to disambiguate one from another.

Here is another example of why this is important. Some languages, such as **APL**, denote the Boolean constants `true` and `false` by the integers `1` and `0`. You may want to let infix operators "+" and "`*`" serve as the logical operators **or** and **and**, respectively. But note this: Boolean is not a ring. The *inverse axiom* for Ring states:

> Every element `x` has an additive inverse `y` such that `x + y = 0`.

Boolean is not a ring since `true` has no inverse—there is no inverse element `a` such that `1 + a = 0` (in terms of booleans, `(true or a) = false`). Nonetheless, AXIOM *could* easily and correctly implement Boolean this way. Boolean simply would not assert that it is of category Ring. Thus the "+" for Boolean values is not confused with the one for Ring. Since the Polynomial constructor requires its argument to be a ring, AXIOM would then refuse to build the domain Polynomial(Boolean). Also, AXIOM would refuse to wrongfully apply algorithms to Boolean elements that presume that the ring axioms for "+" hold.

12.9 Attributes

Most axioms are not computationally useful. Those that are can be explicitly expressed by what AXIOM calls an *attribute*. The attribute `commutative("*")`, for example, is used to assert that a domain has commutative multiplication. Its definition is given by its documentation:

> A domain R has `commutative("*")` if it has an operation "*": (R,R) → R such that `x * y = y * x`.

Just as you can test whether a domain has the category Ring, you can test that a domain has a given attribute.

Do polynomials over the integers have commutative multiplication?

```
Polynomial Integer has commutative("*")
```

true (1)

Type: Boolean

Do matrices over the integers have commutative multiplication?

```
Matrix Integer has commutative("*")
```

false (2)

Type: Boolean

Attributes are used to conditionally export and define operations for a domain (see Section 13.3 on page 529). Attributes can also be asserted in a category definition.

After mentioning category Ring many times in this book, it is high time that we show you its definition:

```
Ring(): Category ==                                    1
  Join(Rng,Monoid,LeftModule($: Rng)) with             2
      characteristic: -> NonNegativeInteger            3
      coerce: Integer -> $                             4
      unitsKnown                                       5
    add                                                6
      n:Integer                                        7
      coerce(n) == n * 1$$                             8
```

There are only two new things here. First, look at the "`$$`" on the last line. This is not a typographic error! The first "`$`" says that the `1` is to come from some domain. The second "`$`" says that the domain is "this domain." If "`$`" is Fraction(Integer), this line reads `coerce(n) == n * 1$Fraction(Integer)`.

The second new thing is the presence of attribute "`unitsKnown`". AXIOM can always distinguish an attribute from an operation. An operation has a name and a type. An attribute has no type. The attribute `unitsKnown` asserts a rather subtle mathematical fact that is normally taken for granted when working with rings.[4] Because programs can test for this attribute, AXIOM can correctly handle rather more complicated mathematical structures (ones that are similar to rings but do not have this attribute).

[4]With this axiom, the units of a domain are the set of elements **x** that each have a multiplicative inverse y in the domain. Thus `1` and `-1` are units in domain Integer. Also, for Fraction Integer, the domain of rational numbers, all non-zero elements are units.

12.10 Parameters

Like domain constructors, category constructors can also have parameters. For example, category MatrixCategory is a parameterized category for defining matrices over a ring R so that the matrix domains can have different representations and indexing schemes. Its definition has the form:

```
MatrixCategory(R,Row,Col): Category ==                    1
    TwoDimensionalArrayCategory(R,Row,Col) with ...       2
```

The category extends TwoDimensionalArrayCategory with the same arguments. You cannot find TwoDimensionalArrayCategory in the algebraic hierarchy listing. Rather, it is a member of the data structure hierarchy, given inside the back cover of this book. In particular, TwoDimensionalArrayCategory is an extension of HomogeneousAggregate since its elements are all one type.

The domain Matrix(R), the class of matrices with coefficients from domain R, asserts that it is a member of category MatrixCategory(R, Vector(R), Vector(R)). The parameters of a category must also have types. The first parameter to MatrixCategory R is required to be a ring. The second and third are required to be domains of category FiniteLinearAggregate(R).[5] In practice, examples of categories having parameters other than domains are rare.

Adding the declarations for parameters to the definition for MatrixCategory, we have:

```
R: Ring                                                      1
(Row, Col): FiniteLinearAggregate(R)                         2
                                                             3
MatrixCategory(R, Row, Col): Category ==                     4
    TwoDimensionalArrayCategory(R, Row, Col) with ...        5
```

12.11 Conditionals

As categories have parameters, the actual operations exported by a category can depend on these parameters. As an example, the operation **determinant** from category MatrixCategory is only exported when the underlying domain R has commutative multiplication:

```
if R has commutative("*") then
   determinant: $ -> R
```

Conditionals can also define conditional extensions of a category. Here is a portion of the definition of QuotientFieldCategory:

[5] This is another extension of HomogeneousAggregate that you can see in the data structure hierarchy.

```
QuotientFieldCategory(R) : Category == ... with ...          1
     if R has OrderedSet then OrderedSet                      2
     if R has IntegerNumberSystem then                        3
       ceiling: $ -> R                                        4
         ...                                                  5
```

Think of category QuotientFieldCategory(R) as denoting the domain Fraction(R), the class of all fractions of the form a/b for elements of R. The first conditional means in English: "If the elements of R are totally ordered (R is an OrderedSet), then so are the fractions a/b".

The second conditional is used to conditionally export an operation **ceiling** which returns the smallest integer greater than or equal to its argument. Clearly, "ceiling" makes sense for integers but not for polynomials and other algebraic structures. Because of this conditional, the domain Fraction(Integer) exports an operation **ceiling**: Fraction Integer → Integer, but Fraction Polynomial Integer does not.

Conditionals can also appear in the default definitions for the operations of a category. For example, a default definition for **ceiling** within the part following the "`add`" reads:

```
if R has IntegerNumberSystem then
    ceiling x == ...
```

Here the predicate used is identical to the predicate in the `Exports` part. This need not be the case. See Section 11.8 on page 509 for a more complicated example.

12.12 Anonymous Categories

The part of a category to the right of a `with` is also regarded as a category—an "anonymous category." Thus you have already seen a category definition in Chapter 11. The `Exports` part of the package DrawComplex (Section 11.3 on page 504) is an anonymous category. This is not necessary. We could, instead, give this category a name:

```
DrawComplexCategory(): Category == with                       1
   drawComplex: (C -> C,S,S,Boolean) -> VIEW3D                2
   drawComplexVectorField: (C -> C,S,S) -> VIEW3D             3
   setRealSteps: INT -> INT                                   4
   setImagSteps: INT -> INT                                   5
   setClipValue: SF-> SF                                      6
```

and then define DrawComplex by:

```
DrawComplex(): DrawComplexCategory == Implementation    1
   where                                                2
      ...                                               3
```

There is no reason, however, to give this list of exports a name since no other domain or package exports it. In fact, it is rare for a package to export a named category. As you will see in the next chapter, however, it is very common for the definition of domains to mention one or more category before the `with`.

CHAPTER 13

Domains

We finally come to the *domain constructor.* A few subtle differences between packages and domains turn up some interesting issues. We first discuss these differences then describe the resulting issues by illustrating a program for the QuadraticForm constructor. After a short example of an algebraic constructor, CliffordAlgebra, we show how you use domain constructors to build a database query facility.

13.1 Domains vs. Packages

Packages are special cases of domains. What is the difference between a package and a domain that is not a package? By definition, there is only one difference: a domain that is not a package has the symbol "$" appearing somewhere among the types of its exported operations. The "$" denotes "this domain." If the "$" appears before the "->" in the type of a signature, it means the operation takes an element from the domain as an argument. If it appears after the "->", then the operation returns an element of the domain.

If no exported operations mention "$", then evidently there is nothing of interest to do with the objects of the domain. You might then say that a package is a "boring" domain! But, as you saw in Chapter 11, packages are a very useful notion indeed. The exported operations of a package depend solely on the parameters to the package constructor and other explicit domains.

To summarize, domain constructors are versatile structures that serve two distinct practical purposes: Those like Polynomial and List describe classes of computational objects; others, like SortPackage, describe packages of useful operations. As in the last chapter, we focus here on the first kind.

13.2 Definitions

The syntax for defining a domain constructor is the same as for any function in AXIOM:

DomainForm : *Exports* == *Implementation*

As this definition usually extends over many lines, a `where` expression is generally used instead.

A recommended format for the definition of a domain is:
DomainForm : `Exports == Implementation where`
  optional type declarations
  `Exports == [`*Category Assertions*`] with`
    list of exported operations
  `Implementation == [`*Add Domain*`] add`
    `[Rep :=` *Representation*`]`
    list of function definitions for exported operations

Note: The brackets `[ ]` here denote optionality.

A complete domain constructor definition for QuadraticForm is shown in Figure 13.1. Interestingly, this little domain illustrates all the new concepts you need to learn.

A domain constructor can take any number and type of parameters. QuadraticForm takes a positive integer `n` and a field `K` as arguments. Like a package, a domain has a set of explicit exports and an implementation described by a capsule. Domain constructors are documented in the same way as package constructors.

Domain QuadraticForm(n, K), for a given positive integer `n` and domain `K`, explicitly exports three operations:

- `quadraticForm(A)` creates a quadratic form from a matrix `A`.
- `matrix(q)` returns the matrix `A` used to create the quadratic form `q`.
- `q.v` computes the scalar $v^T A v$ for a given vector `v`.

Compared with the corresponding syntax given for the definition of a package, you see that a domain constructor has three optional parts to its definition: *Category Assertions*, *Add Domain*, and *Representation*.

The exports.
The export **quadraticForm**.
The export **matrix**.
The export **elt**.
The definitions of the exports
The "representation."
The definition of **quadraticForm**.
The definition of **matrix**.
The definition of **elt**.

```
)abbrev domain QFORM QuadraticForm

++ Description:
++   This domain provides modest support for
++   quadratic forms.
QuadraticForm(n, K): Exports == Implementation where
    n: PositiveInteger
    K: Field

    Exports == AbelianGroup with
      quadraticForm: SquareMatrix(n,K) -> $
        ++ \axiom{quadraticForm(m)} creates a quadratic
        ++ quadratic form from a symmetric,
        ++ square matrix \axiom{m}.
      matrix: $ -> SquareMatrix(n,K)
        ++ \axiom{matrix(qf)} creates a square matrix
        ++ from the quadratic form {qf}.
      elt: ($, DirectProduct(n,K)) -> K
        ++ \axiom{qf(v)} evaluates the quadratic form
        ++ \axiom{qf} on the vector \axiom{v},
        ++ producing a scalar.

    Implementation == SquareMatrix(n,K) add
      Rep := SquareMatrix(n,K)
      quadraticForm m ==
        not symmetric? m => error
          "quadraticForm requires a symmetric matrix"
        m :: $
      matrix q == q :: Rep
      elt(q,v) == dot(v, (matrix q * v))
```

Figure 13.1: The QuadraticForm domain.

13.3 Category Assertions

The *Category Assertions* part of your domain constructor definition lists those categories of which all domains created by the constructor are unconditionally members. The word "unconditionally" means that membership in a category does not depend on the values of the parameters to the domain constructor. This part thus defines the link between the domains and the category hierarchies given on the inside covers of this book. As described in Section 12.8 on page 522, it is this link that makes it possible for you to pass objects of the domains as arguments to other operations in AXIOM.

Every QuadraticForm domain is declared to be unconditionally a member of category AbelianGroup. An abelian group is a collection of elements closed under addition. Every object x of an abelian group has an additive inverse y such that $x + y = 0$. The exports of an abelian group include 0, "+", "-", and scalar multiplication by an integer. After asserting that QuadraticForm domains are abelian groups, it is possible to pass quadratic forms to algorithms that only assume arguments to have these abelian group properties.

In Section 12.11 on page 524, you saw that Fraction(R), a member of QuotientFieldCategory(R), is a member of OrderedSet if R is a member of OrderedSet. Likewise, from the Exports part of the definition of ModMonic(R, S),

```
UnivariatePolynomialCategory(R) with
  if R has Finite then Finite
     ...
```

you see that ModMonic(R, S) is a member of Finite is R is.

The Exports part of a domain definition is the same kind of expression that can appear to the right of an "==" in a category definition. If a domain constructor is unconditionally a member of two or more categories, a Join form is used. The Exports part of the definition of FlexibleArray(S) reads, for example:

```
Join(ExtensibleLinearAggregate(S),
     OneDimensionalArrayAggregate(S)) with...
```

13.4 A Demo

Before looking at the *Implementation* part of QuadraticForm, let's try some examples.

Build a domain QF.

```
QF := QuadraticForm(2,Fraction Integer)
```

QuadraticForm (2, Fraction Integer) (1)

Type: Domain

Define a matrix to be used to construct a quadratic form.

```
A := matrix [[-1,1/2],[1/2,1]]
```

$$\begin{bmatrix} -1 & \frac{1}{2} \\ \frac{1}{2} & 1 \end{bmatrix} \quad (2)$$

Type: Matrix Fraction Integer

Construct the quadratic form. A package call $QF is necessary since there are other QuadraticForm domains.

```
q : QF := quadraticForm(A)
```

$$\begin{bmatrix} -1 & \frac{1}{2} \\ \frac{1}{2} & 1 \end{bmatrix} \quad (3)$$

Type: QuadraticForm(2, Fraction Integer)

Looks like a matrix. Try computing the number of rows. AXIOM won't let you.

```
nrows q
```

```
Cannot find a definition or library operation named
nrows with argument types
         QuadraticForm(2,Fraction Integer)
```

Create a direct product element v. A package call is again necessary, but AXIOM understands your list as denoting a vector.

```
v := directProduct([2,-1])$DirectProduct(2,Fraction
  Integer)
```

$$[2, \ -1] \tag{4}$$

Type: DirectProduct(2, Fraction Integer)

Compute the product $v^T Av$.

```
q.v
```

$$-5 \tag{5}$$

Type: Fraction Integer

What is 3 times q minus q plus q?

```
3*q-q+q
```

$$\begin{bmatrix} -3 & \frac{3}{2} \\ \frac{3}{2} & 3 \end{bmatrix} \tag{6}$$

Type: QuadraticForm(2, Fraction Integer)

13.5 Browse

The Browse facility of HyperDoc is useful for investigating the properties of domains, packages, and categories. From the main HyperDoc menu, move your mouse to **Browse** and click on the left mouse button. This brings up the Browse first page. Now, with your mouse pointer somewhere in this window, enter the string "quadraticform" into the input area (all lower case letters will do). Move your mouse to **Constructors** and click. Up comes a page describing QuadraticForm.

From here, click on **Description**. This gives you a page that includes a part labeled by "*Description:*". You also see the types for arguments n and K displayed as well as the fact that QuadraticForm returns an AbelianGroup. You can go and experiment a bit by selecting **Field** with your mouse. Eventually, use [up-arrow button] several times to return to the first page on QuadraticForm.

Select **Operations** to get a list of operations for QuadraticForm. You can select an operation by clicking on it to get an individual page with information about that operation. Or you can select the buttons along the bottom to see alternative views or get additional information on the operations. Then return to the page on QuadraticForm.

Select **Cross Reference** to get another menu. This menu has buttons for **Parents**, **Ancestors**, and others. Clicking on **Parents**, you see that

QuadraticForm has one parent AbelianMonoid.

13.6 Representation

The Implementation part of an AXIOM capsule for a domain constructor uses the special variable Rep to identify the lower level data type used to represent the objects of the domain. The Rep for quadratic forms is SquareMatrix(n, K). This means that all objects of the domain are required to be n by n matrices with elements from K.

The code for **quadraticForm** in Figure 13.1 on page 529 checks that the matrix is symmetric and then converts it to "$", which means, as usual, "this domain." Such explicit conversions are generally required by the compiler. Aside from checking that the matrix is symmetric, the code for this function essentially does nothing. The m :: $ on line 28 coerces m to a quadratic form. In fact, the quadratic form you created in step (3) of Section 13.4 on page 530 is just the matrix you passed it in disguise! Without seeing this definition, you would not know that. Nor can you take advantage of this fact now that you do know! When we try in the next step of Section 13.4 on page 530 to regard q as a matrix by asking for **nrows**, the number of its rows, AXIOM gives you an error message saying, in effect, "Good try, but this won't work!"

The definition for the **matrix** function could hardly be simpler: it just returns its argument after explicitly *coercing* its argument to a matrix. Since the argument is already a matrix, this coercion does no computation.

Within the context of a capsule, an object of "$" is regarded both as a quadratic form *and* as a matrix.[1] This makes the definition of q.v easy—it just calls the **dot** product from DirectProduct to perform the indicated operation.

13.7 Multiple Representations

To write functions that implement the operations of a domain, you want to choose the most computationally efficient data structure to represent the elements of your domain.

A classic problem in computer algebra is the optimal choice for an internal representation of polynomials. If you create a polynomial, say $3x^2+5$, how does AXIOM hold this value internally? There are many ways. AXIOM has nearly a dozen different representations of polynomials, one to suit almost any purpose. Algorithms for solving polynomial equations work most efficiently with polynomials represented one way, whereas those

[1] In case each of "$" and Rep have the same named operation available, the one from $ takes precedence. Thus, if you want the one from "Rep", you must package call it using a "$Rep" suffix.

for factoring polynomials are most efficient using another. One often-used representation is a list of terms, each term consisting of exponent-coefficient records written in the order of decreasing exponents. For example, the polynomial $3x^2+5$ is represented by the list `[[e:2, c:3], [e:0, c:5]]`.

What is the optimal data structure for a matrix? It depends on the application. For large sparse matrices, a linked-list structure of records holding only the non-zero elements may be optimal. If the elements can be defined by a simple formula $f(i, j)$, then a compiled function for `f` may be optimal. Some programmers prefer to represent ordinary matrices as vectors of vectors. Others prefer to represent matrices by one big linear array where elements are accessed with linearly computable indexes.

While all these simultaneous structures tend to be confusing, AXIOM provides a helpful organizational tool for such a purpose: categories. PolynomialCategory, for example, provides a uniform user interface across all polynomial types. Each kind of polynomial implements functions for all these operations, each in its own way. If you use only the top-level operations in PolynomialCategory you usually do not care what kind of polynomial implementation is used.

Within a given domain, however, you define (at most) one representation.[2] If you want to have multiple representations (that is, several domains, each with its own representation), use a category to describe the `Exports`, then define separate domains for each representation.

13.8 Add Domain

The capsule part of `Implementation` defines functions that implement the operations exported by the domain—usually only some of the operations. In our demo in Section 13.4 on page 530, we asked for the value of `3*q-q+q`. Where do the operations "`*`", "+", and "`-`" come from? There is no definition for them in the capsule!

The `Implementation` part of a definition can optionally specify an "add-domain" to the left of an `add` (for QuadraticForm, defines SquareMatrix(n,K) is the add-domain). The meaning of an add-domain is simply this: if the capsule part of the `Implementation` does not supply a function for an operation, AXIOM goes to the add-domain to find the function. So do "`*`", "+" and "`-`" come from SquareMatrix(n,K)?

[2]You can make that representation a Union type, however. See Section 2.5 on page 73 for examples of unions.

13.9 Defaults

In Chapter 11, we saw that categories can provide default implementations for their operations. How and when are they used? When AXIOM finds that QuadraticForm(2, Fraction Integer) does not implement the operations "*", "+", and "-", it goes to SquareMatrix(2,Fraction Integer) to find it. As it turns out, SquareMatrix(2, Fraction Integer) does not implement *any* of these operations!

What does AXIOM do then? Here is its overall strategy. First, AXIOM looks for a function in the capsule for the domain. If it is not there, AXIOM looks in the add-domain for the operation. If that fails, AXIOM searches the add-domain of the add-domain, and so on. If all those fail, it then searches the default packages for the categories of which the domain is a member. In the case of QuadraticForm, it searches AbelianGroup, then its parents, grandparents, and so on. If this fails, it then searches the default packages of the add-domain. Whenever a function is found, the search stops immediately and the function is returned. When all fails, the system calls **error** to report this unfortunate news to you. To find out the actual order of constructors searched for QuadraticForm, consult Browse: from the QuadraticForm, click on **Cross Reference**, then on **Lineage**.

Let's apply this search strategy for our example `3*q-q+q`. The scalar multiplication comes first. AXIOM finds a default implementation in AbelianGroup&. Remember from Section 12.6 on page 520 that SemiGroup provides a default definition for x^n by repeated squaring? AbelianGroup similarly provides a definition for nx by repeated doubling.

But the search of the defaults for QuadraticForm fails to find any "+" or "*" in the default packages for the ancestors of QuadraticForm. So it now searches among those for SquareMatrix. Category MatrixCategory, which provides a uniform interface for all matrix domains, is a grandparent of SquareMatrix and has a capsule defining many functions for matrices, including matrix addition, subtraction, and scalar multiplication. The default package MatrixCategory& is where the functions for "+" and - come from.

You can use Browse to discover where the operations for QuadraticForm are implemented. First, get the page describing QuadraticForm. With your mouse somewhere in this window, type a "2", press the **Tab** key, and then enter "Fraction Integer" to indicate that you want the domain QuadraticForm(2, Fraction Integer). Now click on **Operations** to get a table of operations and on "*" to get a page describing the "*" operation. Finally, click on **implementation** at the bottom.

13.10 Origins

Aside from the notion of where an operation is implemented, a useful notion is the *origin* or "home" of an operation. When an operation (such as **quadraticForm**) is explicitly exported by a domain (such as QuadraticForm), you can say that the origin of that operation is that domain. If an operation is not explicitly exported from a domain, it is inherited from, and has as origin, the (closest) category that explicitly exports it. The operations "+" and "-" of QuadraticForm, for example, are inherited from AbelianMonoid. As it turns out, AbelianMonoid is the origin of virtually every "+" operation in AXIOM!

Again, you can use Browse to discover the origins of operations. From the Browse page on QuadraticForm, click on **Operations**, then on **origins** at the bottom of the page.

The origin of the operation is the *only* place where on-line documentation is given. However, you can re-export an operation to give it special documentation. Suppose you have just invented the world's fastest algorithm for inverting matrices using a particular internal representation for matrices. If your matrix domain just declares that it exports MatrixCategory, it exports the **inverse** operation, but the documentation the user gets from Browse is the standard one from MatrixCategory. To give your version of **inverse** the attention it deserves, simply export the operation explicitly with new documentation. This redundancy gives **inverse** a new origin and tells Browse to present your new documentation.

13.11 Short Forms

In AXIOM, a domain could be defined using only an add-domain and no capsule. Although we talk about rational numbers as quotients of integers, there is no type RationalNumber in AXIOM. To create such a type, you could compile the following "short-form" definition:

```
RationalNumber() == Fraction(Integer)                    1
```

The `Exports` part of this definition is missing and is taken to be equivalent to that of Fraction(Integer). Because of the add-domain philosophy, you get precisely what you want. The effect is to create a little stub of a domain. When a user asks to add two rational numbers, AXIOM would ask RationalNumber for a function implementing this "+". Since the domain has no capsule, the domain then immediately sends its request to Fraction (Integer).

The short form definition for domains is used to define such domains as MultivariatePolynomial:

```
MultivariatePolynomial(vl: List Symbol, R: Ring) ==     1
   SparseMultivariatePolynomial(R,                      2
      OrderedVariableList vl)                           3
```

13.12 Example 1: Clifford Algebra

Now that we have QuadraticForm available, let's put it to use. Given some quadratic form Q described by an n by n matrix over a field K, the domain CliffordAlgebra(n, K, Q) defines a vector space of dimension 2^n over K. This is an interesting domain since complex numbers, quaternions, exterior algebras and spin algebras are all examples of Clifford algebras.

The basic idea is this: the quadratic form Q defines a basis $e_1, e_2 \ldots, e_n$ for the vector space K^n—the direct product of K with itself n times. From this, the Clifford algebra generates a basis of 2^n elements given by all the possible products of the e_i in order without duplicates, that is, 1, e_1, e_2, $e_1 e_2$, e_3, $e_1 e_3$, $e_2 e_3$, $e_1 e_2, e_3$, and so on.

The algebra is defined by the relations

$$
\begin{aligned}
e_i\, e_i &= Q(e_i) \\
e_i\, e_j &= -e_j\, e_i \quad \text{for } i \neq j
\end{aligned}
$$

Now look at the snapshot of its definition given in Figure 13.2. Lines 9-10 show part of the definitions of the `Exports`. A Clifford algebra over a field K is asserted to be a ring, an algebra over K, and a vector space over K. Its explicit exports include `e(n)`, which returns the n th unit element.

The `Implementation` part begins by defining a local variable `Qeelist` to hold the list of all `q.v` where v runs over the unit vectors from 1 to the dimension n. Another local variable `dim` is set to 2^n, computed once and for all. The representation for the domain is PrimitiveArray(K), which is a basic array of elements from domain K. Line 18 defines `New` as shorthand for the more lengthy expression `new(dim, 0$K)$Rep`, which computes a primitive array of length 2^n filled with 0's from domain K.

Lines 19-22 define the sum of two elements `x` and `y` straightforwardly. First, a new array of all 0's is created, then filled with the sum of the corresponding elements. Indexing for primitive arrays starts at 0. The definition of the product of `x` and `y` first requires the definition of a local function **addMonomProd**. AXIOM knows it is local since it is not an exported function. The types of all local functions must be declared.

For a demonstration of CliffordAlgebra, see 'CliffordAlgebra' on page 328.

```
NNI ==> NonNegativeInteger                                            1
PI  ==> PositiveInteger                                               2
                                                                      3
CliffordAlgebra(n,K,q): Exports == Implementation where               4
    n: PI                                                             5
    K: Field                                                          6
    q: QuadraticForm(n, K)                                            7
                                                                      8
    Exports == Join(Ring,Algebra(K),VectorSpace(K)) with              9
      e: PI -> $                                                     10
          ...                                                        11
                                                                     12
    Implementation == add                                            13
      Qeelist :=                                                     14
        [q.unitVector(i::PI) for i in 1..n]                          15
      dim     :=  2**n                                               16
      Rep     := PrimitiveArray K                                    17
      New ==> new(dim, 0$K)$Rep                                      18
      x + y ==                                                       19
        z := New                                                     20
        for i in 0..dim-1 repeat z.i := x.i + y.i                    21
        z                                                            22
      addMonomProd: (K, NNI, K, NNI, $) -> $                         23
      addMonomProd(c1, b1, c2, b2, z) ==  ...                        24
      x * y ==                                                       25
        z := New                                                     26
        for ix in 0..dim-1 repeat                                    27
          if x.ix ~= 0 then for iy in 0..dim-1 repeat                28
            if y.iy ~= 0                                             29
            then addMonomProd(x.ix,ix,y.iy,iy,z)                     30
          z                                                          31
           ...                                                       32
```

Figure 13.2: Part of the CliffordAlgebra domain.

13.13 Example 2: Building A Query Facility

We now turn to an entirely different kind of application, building a query language for a database.

Here is the practical problem to solve. The Browse facility of AXIOM has a database for all operations and constructors which is stored on disk and accessed by HyperDoc. For our purposes here, we regard each line of this file as having eight fields: `class`, `name`, `type`, `nargs`, `exposed`, `kind`, `origin`, and `condition`. Here is an example entry:

```
o`determinant`$->R`1`x`d`Matrix(R)`has(R,commutative("*"))
```

In English, the entry means:

> The operation **determinant**: $ → R with *1* argument, is *exposed* and is exported by *domain* Matrix(R) if `R has commutative("*")`.

Our task is to create a little query language that allows us to get useful information from this database.

13.13.1 A Little Query Language

First we design a simple language for accessing information from the database. We have the following simple model in mind for its design. Think of the database as a box of index cards. There is only one search operation—it takes the name of a field and a predicate (a boolean-valued function) defined on the fields of the index cards. When applied, the search operation goes through the entire box selecting only those index cards for which the predicate is true. The result of a search is a new box of index cards. This process can be repeated again and again.

The predicates all have a particularly simple form: *symbol* = *pattern*, where *symbol* designates one of the fields, and *pattern* is a "search string"—a string that may contain a "`*`" as a wildcard. Wildcards match any substring, including the empty string. Thus the pattern `"*ma*t"` matches `"mat"`, `"doormat"` and `"smart"`.

To illustrate how queries are given, we give you a sneak preview of the facility we are about to create.

Extract the database of all AXIOM operations.

```
ops := getDatabase("o")
```

4129 (1)

Type: Database IndexCard

How many exposed three-argument **map** operations involving streams?

```
ops.(name="map").(nargs="3").(type="*Stream*")
```

3 (2)

Type: Database IndexCard

As usual, the arguments of **elt** ("`.`") associate to the left. The first **elt** produces the set of all operations with name `map`. The second **elt** produces the set of all map operations with three arguments. The third **elt** produces the set of all three-argument map operations having a type mentioning Stream.

Another thing we'd like to do is to extract one field from each of the index cards in the box and look at the result. Here is an example of that kind of request.

What constructors explicitly export a **determinant** operation?

```
ops.(name="determinant").origin.sort.unique
```

```
["InnerMatrixLinearAlgebraFunctions",
 "MatrixCategory", "MatrixLinearAlgebraFunctions",    (3)
 "SquareMatrixCategory"]
```

Type: DataList String

The arguments again associate to the left. The first **elt** produces the set of all index cards with name `determinant`. The second **elt** extracts the `origin` component from each index card. Each origin component is the name of a constructor which directly exports the operation represented by the index card. Extracting a component from each index card produces what we call a *datalist*. The third **elt**, `sort`, causes the datalist of origins to be sorted in alphabetic order. The fourth, `unique`, causes duplicates to be removed.

Before giving you a more extensive demo of this facility, we now build the necessary domains and packages to implement it.

13.13.2 The Database Constructor

We work from the top down. First, we define a database, our box of index cards, as an abstract datatype. For sake of illustration and generality, we assume that an index card is some type `S`, and that a database is a box of objects of type `S`. Here is the AXIOM program defining the Database domain.

Select by an equation.
Select by a field name.
Combine two databases.
Subtract one from another.
A brief database display.
A full database display.
A selective display.
Display a database.

```
PI ==> PositiveInteger                                        1
Database(S): Exports == Implementation where                  2
  S: Object with                                              3
    elt: ($, Symbol) -> String                                4
    display: $ -> Void                                        5
    fullDisplay: $ -> Void                                    6
                                                              7
  Exports == with                                             8
    elt: ($,QueryEquation) -> $                               9
    elt: ($, Symbol) -> DataList String                      10
    "+": ($,$) -> $                                          11
    "-": ($,$) -> $                                          12
    display: $ -> Void                                       13
    fullDisplay: $ -> Void                                   14
    fullDisplay: ($,PI,PI) -> Void                           15
    coerce: $ -> OutputForm                                  16
  Implementation == add                                      17
     ...                                                     18
```

The domain constructor takes a parameter `S`, which stands for the class of index cards. We describe an index card later. Here think of an index card as a string which has the eight fields mentioned above.

First, we tell AXIOM what operations we are going to require from index cards. We need an **elt** to extract the contents of a field (such as `name` and `type`) as a string. For example, `c.name` returns a string that is the content of the `name` field on the index card `c`. We need to display an index card in two ways: **display** shows only the name and type of an operation; **fullDisplay** displays all fields. The display operations return no useful information and thus have return type Void.

Next, we tell AXIOM what operations the user can apply to the database. This part defines our little query language. The most important operation is `db . field = pattern` which returns a new database, consisting of all index cards of `db` such that the `field` part of the index card is matched by the string pattern called `pattern`. The expression `field = pattern` is an object of type QueryEquation (defined in the next section).

Another **elt** is needed to produce a DataList object. Operation "+" is to merge two databases together; "-" is used to subtract away common entries in a second database from an initial database. There are three display functions. The **fullDisplay** function has two versions: one that prints all the records, the other that prints only a fixed number of records. A **coerce** to OutputForm creates a display object.

The `Implementation` part of Database is straightforward.

```
Implementation == add                                          1
  s: Symbol                                                    2
  Rep := List S                                                3
  elt(db,equation) == ...                                      4
  elt(db,key) == [x.key for x in db]::DataList(String)         5
  display(db) ==  for x in db repeat display x                 6
  fullDisplay(db) == for x in db repeat fullDisplay x          7
  fullDisplay(db, n, m) == for x in db for i in 1..m           8
    repeat                                                     9
      if i >= n then fullDisplay x                             10
  x+y == removeDuplicates! merge(x,y)                          11
  x-y == mergeDifference(copy(x::Rep),                         12
                         y::Rep)$MergeThing(S)                 13
  coerce(db): OutputForm == (#db):: OutputForm                 14
```

The database is represented by a list of elements of `S` (index cards). We leave the definition of the first **elt** operation (on line 4) until the next section. The second **elt** collects all the strings with field name *key* into a list. The **display** function and first **fullDisplay** function simply call the corresponding functions from `S`. The second **fullDisplay** function provides an efficient way of printing out a portion of a large list. The "+" is defined by using the existing **merge** operation defined on lists, then removing duplicates from the result. The "-" operation requires writing a corresponding subtraction operation. A package MergeThing (not shown) provides this.

The **coerce** function converts the database to an OutputForm by computing the number of index cards. This is a good example of the independence of the representation of an AXIOM object from how it presents itself to the user. We usually do not want to look at a database—but do care how many "hits" we get for a given query. So we define the output representation of a database to be simply the number of index cards our query finds.

13.13.3 Query Equations

The predicate for our search is given by an object of type QueryEquation. AXIOM does not have such an object yet so we have to invent it.

```
QueryEquation(): Exports == Implementation where                1
  Exports == with                                                2
    equation: (Symbol, String) -> $                              3
    variable: $ -> Symbol                                        4
    value:    $ -> String                                        5
                                                                 6
  Implementation == add                                          7
    Rep := Record(var:Symbol, val:String)                        8
    equation(x, s) == [x, s]                                     9
    variable q == q.var                                         10
    value    q == q.val                                         11
```

AXIOM converts an input expression of the form $a = b$ to `equation(a, b)`. Our equations always have a symbol on the left and a string on the right. The `Exports` part thus specifies an operation **equation** to create a query equation, and **variable** and **value** to select the left- and right-hand sides. The `Implementation` part uses Record for a space-efficient representation of an equation.

Here is the missing definition for the **elt** function of Database in the last section:

```
    elt(db,eq) ==                                               1
      field  := variable eq                                     2
      value  := value eq                                        3
      [x for x in db | matches?(value,x.field)]                 4
```

Recall that a database is represented by a list. Line 4 simply runs over that list collecting all elements such that the pattern (that is, `value`) matches the selected field of the element.

13.13.4 DataLists

Type DataList is a new type invented to hold the result of selecting one field from each of the index cards in the box. It is useful to make datalists extensions of lists—lists that have special **elt** operations defined on them for sorting and removing duplicates.

```
DataList(S:OrderedSet) : Exports == Implementation where   1
  Exports == ListAggregate(S) with                           2
    elt: ($,"unique") -> $                                   3
    elt: ($,"sort") -> $                                     4
    elt: ($,"count") -> NonNegativeInteger                   5
    coerce: List S -> $                                      6
                                                             7
  Implementation ==  List(S) add                             8
    Rep := List S                                            9
    elt(x,"unique") == removeDuplicates(x)                   10
    elt(x,"sort") == sort(x)                                 11
    elt(x,"count") == #x                                     12
    coerce(x:List S) == x :: $                               13
```

The `Exports` part asserts that datalists belong to the category ListAggregate. Therefore, you can use all the usual list operations on datalists, such as **first**, **rest**, and **concat**. In addition, datalists have four explicit operations. Besides the three **elt** operations, there is a **coerce** operation that creates datalists from lists.

The `Implementation` part needs only to define four functions. All the rest are obtained from List(S).

13.13.5 Index Cards

An index card comes from a file as one long string. We define functions that extract substrings from the long string. Each field has a name that is passed as a second argument to **elt**.

```
IndexCard() == Implementation where      1
  Exports == with                        2
    elt: ($, Symbol) -> String           3
    display: $ -> Void                   4
    fullDisplay: $ -> Void               5
    coerce: String -> $                  6
  Implementation == String add ...       7
```

We leave the `Implementation` part to the reader. All operations involve straightforward string manipulations.

13.13.6 Creating a Database

We must not forget one important operation: one that builds the database in the first place! We'll name it **getDatabase** and put it in a package. This function is implemented by calling the Common LISP function `getBrowseDatabase(s)` to get appropriate information from Browse. This operation takes a string indicating which lines you want from the database: `"o"` gives you all operation lines, and `"k"`, all constructor lines. Similarly, `"c"`, `"d"`, and `"p"` give you all category, domain and package lines respectively.

```
OperationsQuery(): Exports == Implementation where          1
  Exports == with                                           2
    getDatabase: String -> Database(IndexCard)              3
                                                            4
  Implementation == add                                     5
    getDatabase(s) == getBrowseDatabase(s)$Lisp             6
```

We do not bother creating a special name for databases of index cards. Database (IndexCard) will do. Notice that we used the package OperationsQuery to create, in effect, a new kind of domain: Database(IndexCard).

13.13.7 Putting It All Together

To create the database facility, you put all these constructors into one file.[3] At the top of the file put `)abbrev` commands, giving the constructor abbreviations you created.

```
)abbrev domain  ICARD   IndexCard                           1
)abbrev domain  QEQUAT  QueryEquation                       2
)abbrev domain  MTHING  MergeThing                          3
)abbrev domain  DLIST   DataList                            4
)abbrev domain  DBASE   Database                            5
)abbrev package OPQUERY OperationsQuery                     6
```

With all this in **alql.spad**, for example, compile it using

```
)compile alql
```

and then load each of the constructors:

```
)load ICARD QEQUAT MTHING DLIST DBASE OPQUERY
```

You are ready to try some sample queries.

13.13.8 Example Queries

Our first set of queries give some statistics on constructors in the current AXIOM system.

How many constructors does AXIOM have?

```
k := getDatabase "k"
```

760 (1)

Type: Database IndexCard

How many of these constructors are exposed?

```
kx := k.(exposed="x")
```

455 (2)

Type: Database IndexCard

[3]You could use separate files, but we are putting them all together because, organizationally, that is the logical thing to do.

Break this down into the number of categories, domains, and packages.

```
[kx.(kind=k) for k in ["c","d","p"]]
```

[170, 109, 176] (3)

Type: List Database IndexCard

What are all the exposed domain constructors that take no parameters?

```
kx.(kind="d").(nargs="0").name
```

["AlgebraicNumber", "Any", "BasicOperator", "BinaryExpansion", "Bits", "Boolean", "CardinalNumber", "CharacterClass", "Character", "Color", "Commutator", "DecimalExpansion", "DrawOption", "Exit", "FileName", "Float", "HexadecimalExpansion", "IndexCard", "Integer", "Library", "NonNegativeInteger", "OrdSetInts", "Palette", "Pi", "PositiveInteger", "QueryEquation", "RomanNumeral", "ScriptFormulaFormat", "SmallFloat", "SmallInteger", "String", "Symbol", "TexFormat", "TextFile", "ThreeDimensionalViewport", "Void"] (4)

Type: DataList String

How many constructors have "Matrix" in their name?

```
mk := k.(name="*Matrix*")
```

21 (5)

Type: Database IndexCard

What are the names of those that are exposed and are domains?

```
mk.(exposed="x").(kind="d").name
```

["DenavitHartenbergMatrix", "LieSquareMatrix", "Matrix"] (6)

Type: DataList String

How many operations are there in the library?

```
o := getDatabase "o"
```

4129 (7)

Type: Database IndexCard

How many of these are exposed operations?

```
x := o.(exposed="x")
```

2632 (8)

Type: Database IndexCard

Break this down into categories, domains, and packages.

```
[x.(kind=k) for k in ["c","d","p"]]
```

[1212, 619, 801] (9)

Type: List Database IndexCard

The query language is helpful in getting information about a particular operation you might like to apply. While this information can be obtained with Browse, the use of the query database gives you data that you can manipulate in the workspace.

How many exposed operations have "eigen" in the name?

```
eigens := x.(name="*eigen*")
```

5 (10)

Type: Database IndexCard

What are their names?

```
eigens.name
```

["eigenMatrix", "eigenvalues", "eigenvectors", "eigenvector", "inteigen"] (11)

Type: DataList String

Where do they come from?

```
eigens.origin.sort.unique
```

["EigenPackage", "RadicalEigenPackage"] (12)

Type: DataList String

The operations "+" and "-" are useful for constructing small databases and combining them. However, remember that the only matching you can do is string matching. Thus a pattern such as `"*Matrix*"` on the type field matches any type containing Matrix, MatrixCategory, SquareMatrix, and so on.

How many exposed operations mention "Matrix" in their type?

```
tm := x.(type="*Matrix*")
```

103 (13)

Type: Database IndexCard

How many operations come from constructors with "Matrix" in their name?

```
fm := x.(origin="*Matrix*")
```

106 (14)

Type: Database IndexCard

How many operations are in `fm` but not in `tm`?

```
fm-tm
```

103 (15)

Type: Database IndexCard

Display the operations that both mention "Matrix" in their type and come from a constructor having "Matrix" in their name.

```
fullDisplay(fm-%)
```

```
clearDenominator : (Matrix(Q))->Matrix(R)
   from MatrixCommonDenominator(R,Q)
commonDenominator : (Matrix(Q))->R
   from MatrixCommonDenominator(R,Q)
splitDenominator :
(Matrix(Q))->Record(num:Matrix(R),den:R)
   from MatrixCommonDenominator(R,Q)
```

Type: Void

How many operations involve matrices?

```
m := tm+fm
```

206 (17)

Type: Database IndexCard

Display four of them.

```
fullDisplay(m, 202, 205)
```

```
translate : (R,R,R)->_$ from
DenavitHartenbergMatrix(R)
transpose : (Row)->_$ from MatrixCategory(R,Row,Col)
transpose : (_$)->_$ from MatrixCategory(R,Row,Col)
vertConcat : (_$,_$)->_$ from MatrixCategory(R,Row,Col)
```

Type: Void

How many distinct names of operations involving matrices are there?

```
m.name.unique.count
```

119 (19)

Type: PositiveInteger

CHAPTER 14

Browse

This chapter discusses the Browse component of HyperDoc. We suggest you invoke AXIOM and work through this chapter, section by section, following our examples to gain some familiarity with Browse.

14.1 The Front Page: Searching the Library

To enter Browse, click on **Browse** on the top level page of HyperDoc to get the *front page* of Browse.

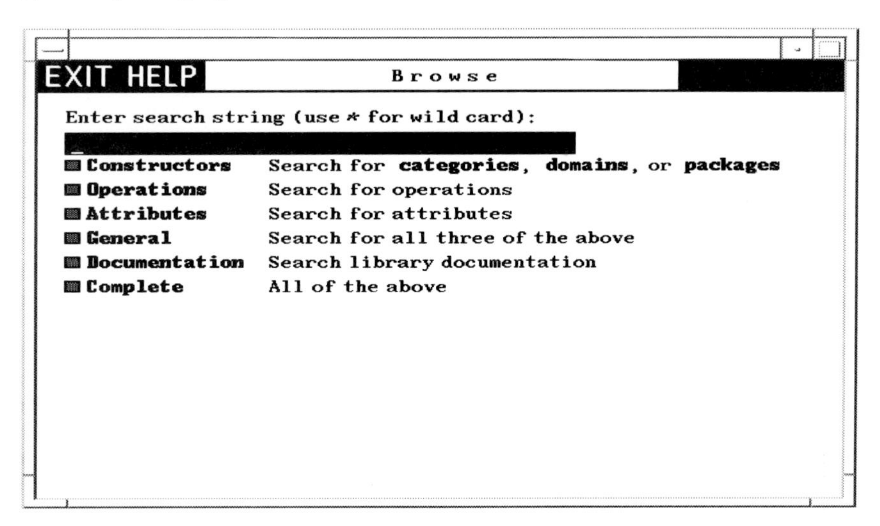

Figure 14.1: The Browse front page.

To use this page, you first enter a *search string* into the input area at the top, then click on one of the buttons below. We show the use of each of the buttons by example.

Constructors

First enter the search string `Matrix` into the input area and click on **Constructors**. What you get is the *constructor page* for Matrix. We show and describe this page in detail in Section 14.2 on page 551. By convention, AXIOM does a case-insensitive search for a match. Thus `matrix` is just as good as `Matrix`, has the same effect as `MaTrix`, and so on. We recommend that you generally use small letters for names however. A search string with only capital letters has a special meaning (see Section 14.3.3 on page 567).

Click on [up-arrow button] to return to the Browse front page.

Use the symbol "`*`" in search strings as a *wild card*. A wild card matches any substring, including the empty string. For example, enter the search string `*matrix*` into the input area and click on **Constructors**.[1] What you get is a table of all constructors whose names contain the string "`matrix`."

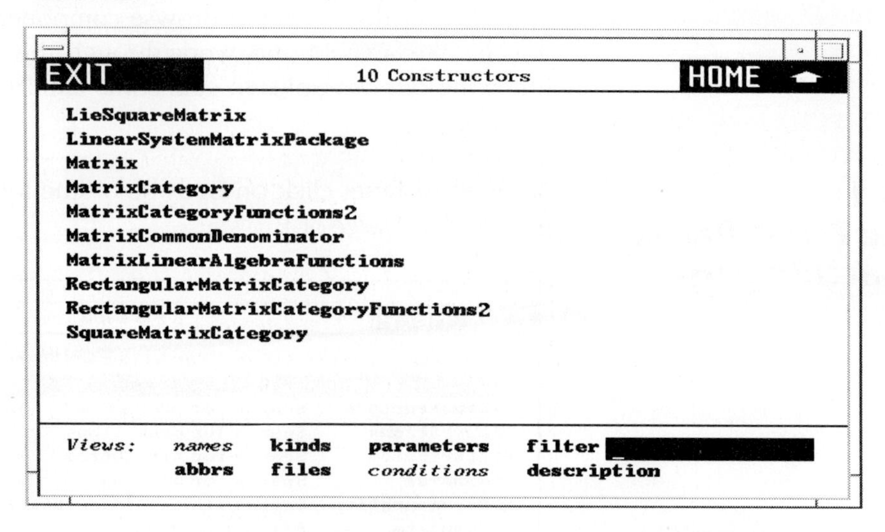

Figure 14.2: Table of exposed constructors matching `*matrix*` .

These are all the *exposed* constructors in AXIOM. To see how to get all exposed and unexposed constructors in AXIOM, skip to the section entitled **Exposure** in Section 14.3.4 on page 567.

One of the names in this table is Matrix. Click on Matrix. What you get is again the constructor page for Matrix. As you see, Browse gives you a large network of information in which there are many ways to reach the same pages.

Again click on the [up-arrow button] to return to the table of constructors whose

[1]To get only categories, domains, or packages, rather than all constructors, you can click on the corresponding button to the right of **Constructors**.

names contain `matrix`. Below the table is a **Views** panel. This panel contains buttons that let you view constructors in different ways. To learn about views of constructors, skip to Section 14.2.3 on page 560.

Click on [up-arrow button] to return to the Browse front page.

Operations

Enter `*matrix` into the input area and click on **Operations**. This time you get a table of *operations* whose names end with `matrix` or `Matrix`.

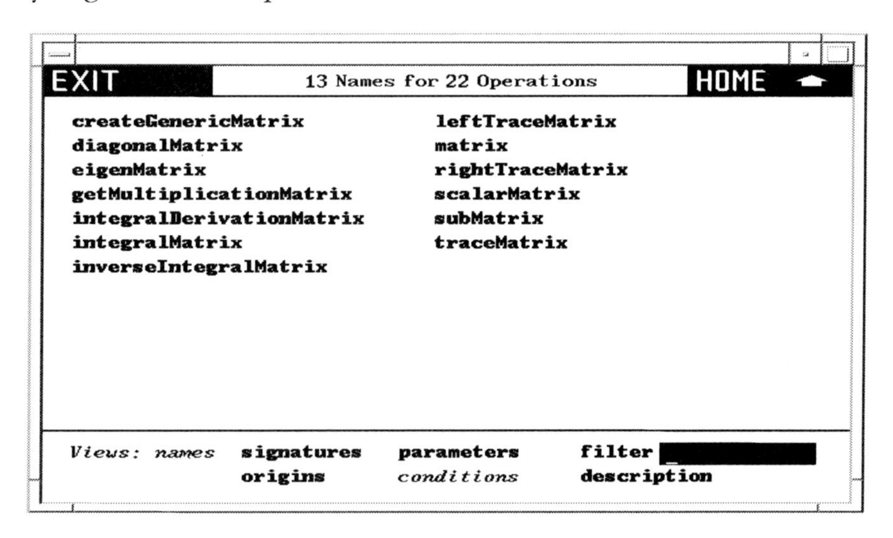

Figure 14.3: Table of operations matching `*matrix` .

If you select an operation name, you go to a page describing all the operations in AXIOM of that name. At the bottom of an operation page is another kind of **Views** panel, one for operation pages. To learn more about these views, skip to Section 14.3.2 on page 563.

Click on [up-arrow button] to return to the Browse front page.

Attributes

This button gives you a table of attribute names that match the search string. Enter the search string `*` and click on **Attributes** to get a list of all system attributes.

Click on [up-arrow button] to return to the Browse front page.

Again there is a **Views** panel at the bottom with buttons that let you view the attributes in different ways.

General

This button does a general search for all constructor, operation, and attribute names matching the search string. Enter the search string `*matrix*` into the input area. Click on **General** to find all constructs that have `matrix` as a part of their name.

The summary gives you all the names under a heading when the number

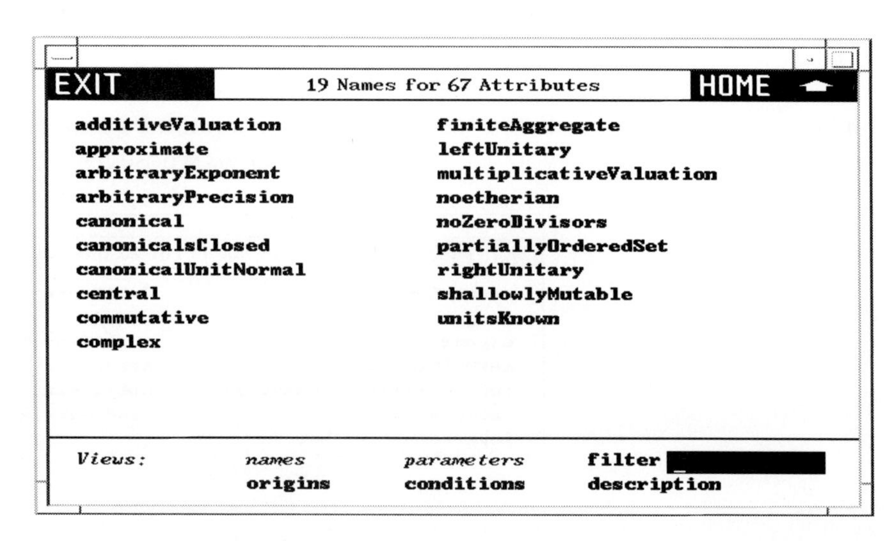

Figure 14.4: Table of AXIOM attributes.

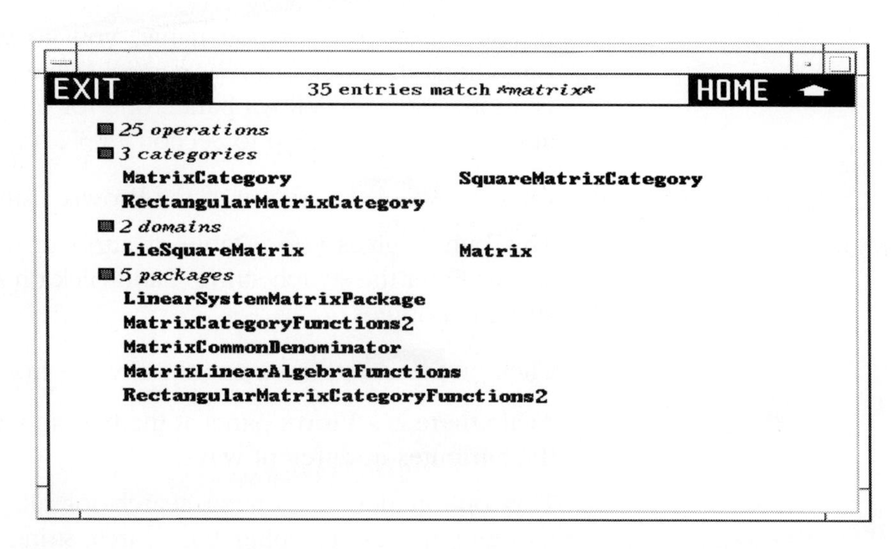

Figure 14.5: Table of all constructs matching `*matrix*` .

of entries is sufficiently small.[2].

Click on [button] to return to the Browse front page.

Documentation

Again enter the search key `*matrix*` and this time click on **Documentation**. This search matches any constructor, operation, or attribute name whose documentation contains a substring matching `matrix`.

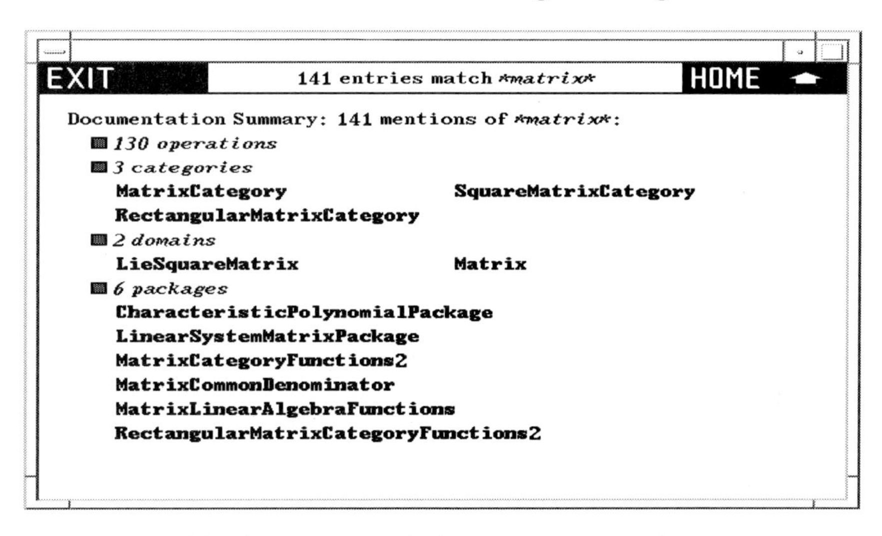

Figure 14.6: Table of constructs with documentation matching `*matrix*` .

Click on [button] to return to the Browse front page.

Complete

This search combines both **General** and **Documentation**.

14.2 The Constructor Page

In this section we look in detail at a constructor page for domain Matrix. Enter `matrix` into the input area on the main Browse page and click on **Constructors**.

The header part tells you that Matrix has abbreviation MATRIX and one argument called `R` that must be a domain of category Ring. Just what domains can be arguments of Matrix? To find this out, click on the `R` on the second line of the heading. What you get is a table of all acceptable domain parameter values of `R`, or a table of *rings* in AXIOM.

Click on [button] to return to the constructor page for Matrix.

[2]See Section 14.3.4 on page 567 to see how you can change this.

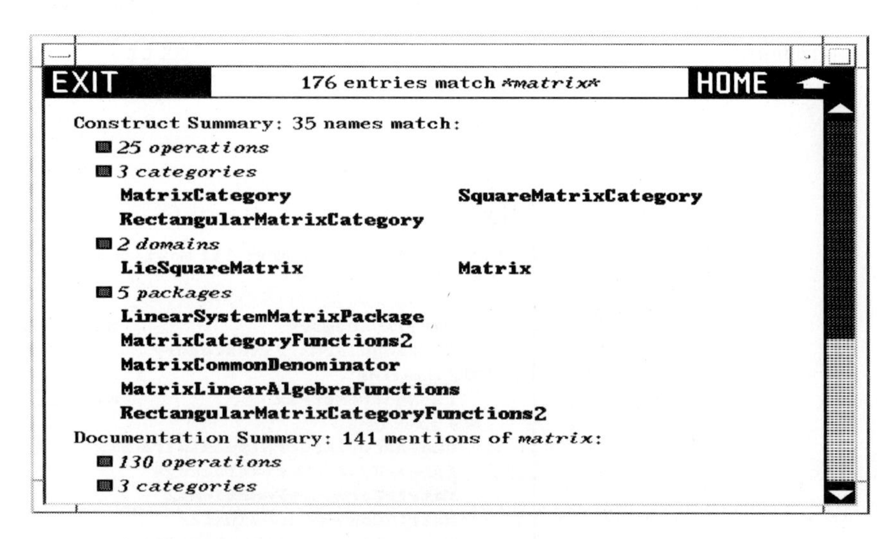

Figure 14.7: Table summarizing complete search for pattern `*matrix*` .

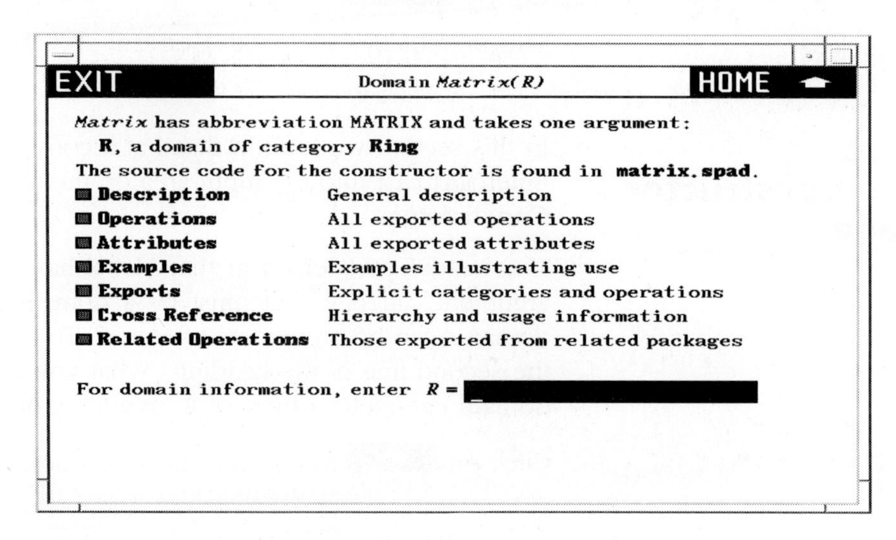

Figure 14.8: Constructor page for Matrix.

EXIT
85 Domain Arguments of Matrix(R)
HOME
AlgebraicFunctionField
AlgebraicNumber
BalancedPAdicInteger
BalancedPAdicRational
BinaryExpansion
CliffordAlgebra
Complex
ContinuedFraction
DecimalExpansion
DifferentialSparseMultivariatePolynomial
DirectProduct
DirectProductMatrixModule
DirectProductModule
Views: names kinds parameters filter
abbrs files conditions description

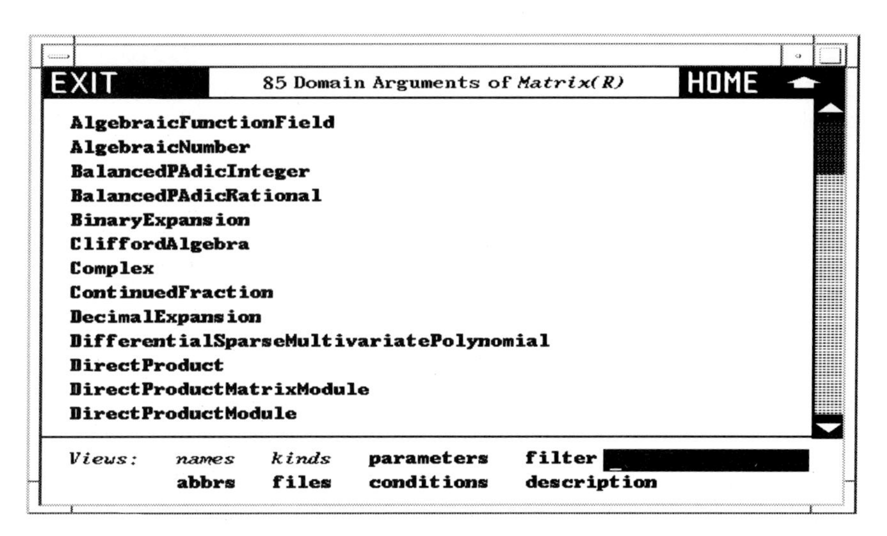

Figure 14.9: Table of acceptable domain parameters to Matrix.

If you have access to the source code of AXIOM, the third line of the heading gives you the name of the source file containing the definition of Matrix. Click on it to pop up an editor window containing the source code of Matrix.

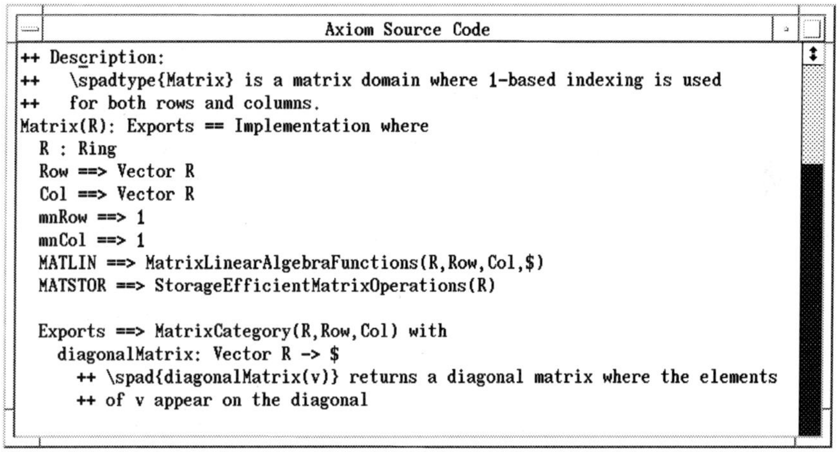

Axiom Source Code
++ Description:
++ \spadtype{Matrix} is a matrix domain where 1-based indexing is used
++ for both rows and columns.
Matrix(R): Exports == Implementation where
R : Ring
Row ==> Vector R
Col ==> Vector R
mnRow ==> 1
mnCol ==> 1
MATLIN ==> MatrixLinearAlgebraFunctions(R,Row,Col,$)
MATSTOR ==> StorageEfficientMatrixOperations(R)
Exports ==> MatrixCategory(R,Row,Col) with
diagonalMatrix: Vector R -> $
++ \spad{diagonalMatrix(v)} returns a diagonal matrix where the elements
++ of v appear on the diagonal

Figure 14.10: Source code for Matrix.

We recommend that you leave the editor window up while working through this chapter as you occasionally may want to refer to it.

14.2.1 Constructor Page Buttons

We examine each button on this page in order.

Description

Click here to bring up a page with a brief description of constructor Matrix. If you have access to system source code, note that these comments can be found directly over the constructor definition.

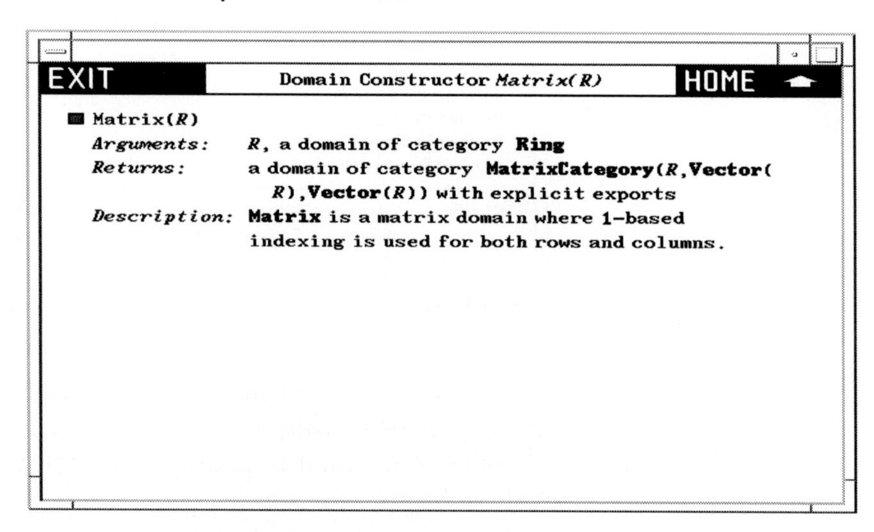

Figure 14.11: Description page for Matrix.

Operations

Click here to get a table of operations exported by Matrix. You may wish to widen the window to have multiple columns as below.

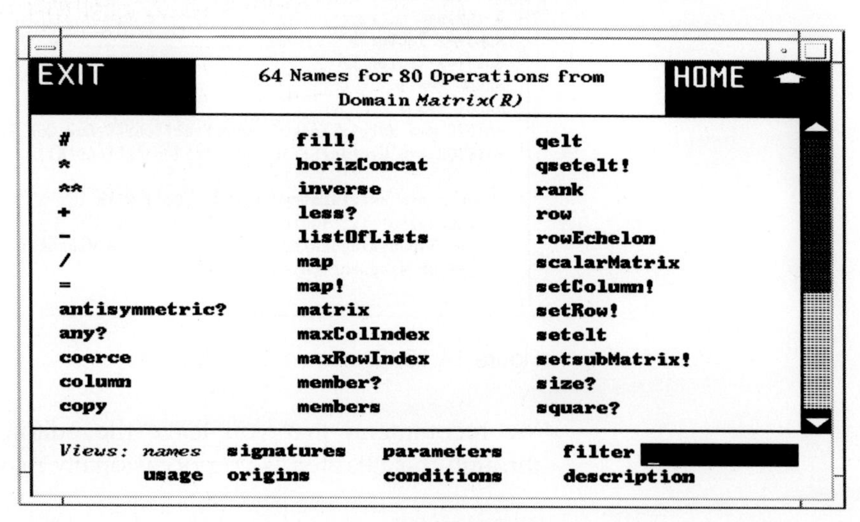

Figure 14.12: Table of operations from Matrix.

If you click on an operation name, you bring up a description page for the operations. For a detailed description of these pages, skip to Section 14.3.2 on page 563.

Attributes

Click here to get a table of the two attributes exported by Matrix: `finiteAggregate` and `shallowlyMutable`. These are two computational properties that result from Matrix being regarded as a data structure.

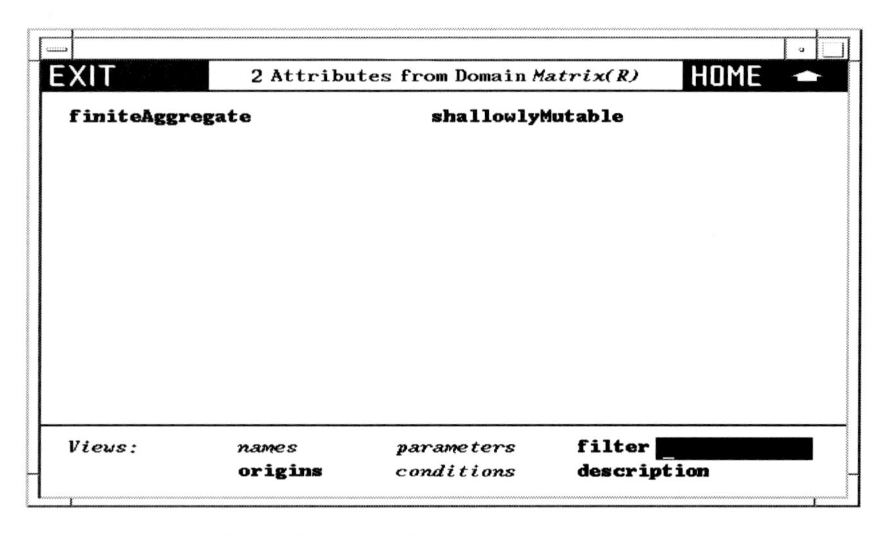

Figure 14.13: Attributes from Matrix.

Examples

Click here to get an *examples page* with examples of operations to create and manipulate matrices.

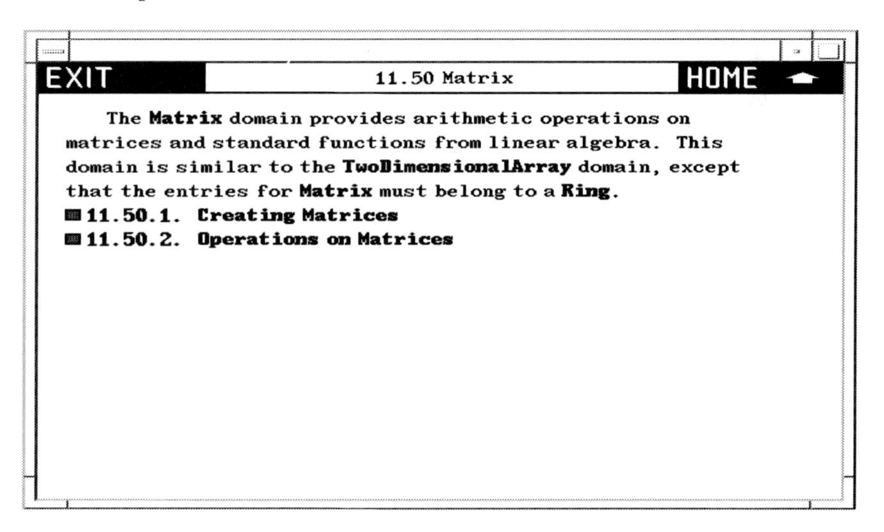

Figure 14.14: Example page for Matrix.

Read through this section. Try selecting the various buttons. Notice that if you click on an operation name, such as **new**, you bring up a description page for that operation from Matrix.

Example pages have several examples of AXIOM commands. Each example has an active button to its left. Click on it! A pre-computed answer is pasted into the page immediately following the command. If you click on the button a second time, the answer disappears. This button thus acts as a toggle: "now you see it; now you don't."

Note also that the AXIOM commands themselves are active. If you want to see AXIOM execute the command, then click on it! A new AXIOM window appears on your screen and the command is executed.

Exports

Click here to see a page describing the exports of Matrix exactly as described by the source code.

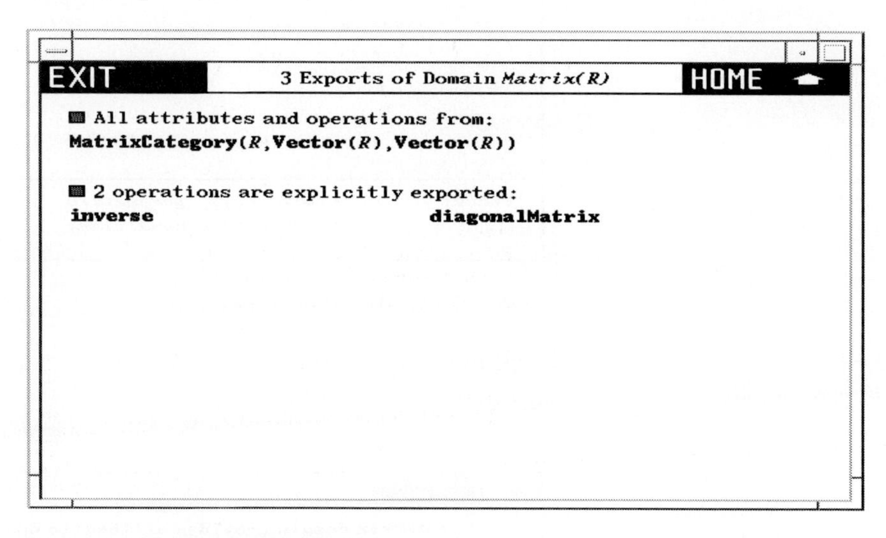

Figure 14.15: Exports of Matrix.

As you see, Matrix declares that it exports all the operations and attributes exported by category MatrixCategory(R, Row, Col). In addition, two operations, **diagonalMatrix** and **inverse**, are explicitly exported.

To learn a little about the structure of AXIOM, we suggest you do the following exercise. Otherwise, click on [button] and go on to the next section.

Matrix explicitly exports only two operations. The other operations are thus exports of MatrixCategory. In general, operations are usually not explicitly exported by a domain. Typically they are *inherited* from several different categories. Let's find out from where the operations of Matrix come.

1. Click on **MatrixCategory**, then on **Exports**. Here you see that

MatrixCategory explicitly exports many matrix operations. Also, it inherits its operations from TwoDimensionalArrayCategory.

2. Click on **TwoDimensionalArrayCategory**, then on **Exports**. Here you see explicit operations dealing with rows and columns. In addition, it inherits operations from HomogeneousAggregate.
3. Click on **Object**, then on **Exports**, where you see there are no exports.
4. Click on [up-arrow button] repeatedly to return to the constructor page for Matrix.

Related Operations

Click here bringing up a table of operations that are exported by *packages* but not by Matrix itself.

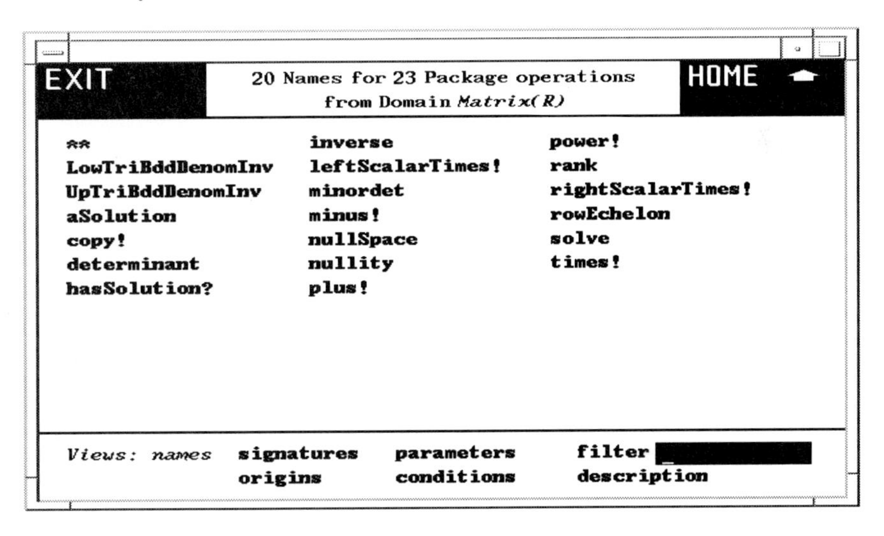

Figure 14.16: Related operations of Matrix.

To see a table of such packages, use the **Relatives** button on the **Cross Reference** page described next.

14.2.2 Cross Reference

Click on the **Cross Reference** button on the main constructor page for Matrix. This gives you a page having various cross reference information stored under the respective buttons.

Parents

The parents of a domain are the same as the categories mentioned under the **Exports** button on the first page. Domain Matrix has only one parent but in general a domain can have any number.

Ancestors

The *ancestors* of a constructor consist of its parents, the parents of its parents, and so on. Did you perform the exercise in the last section under **Exports**? If so, you see here all the categories you found while ascending

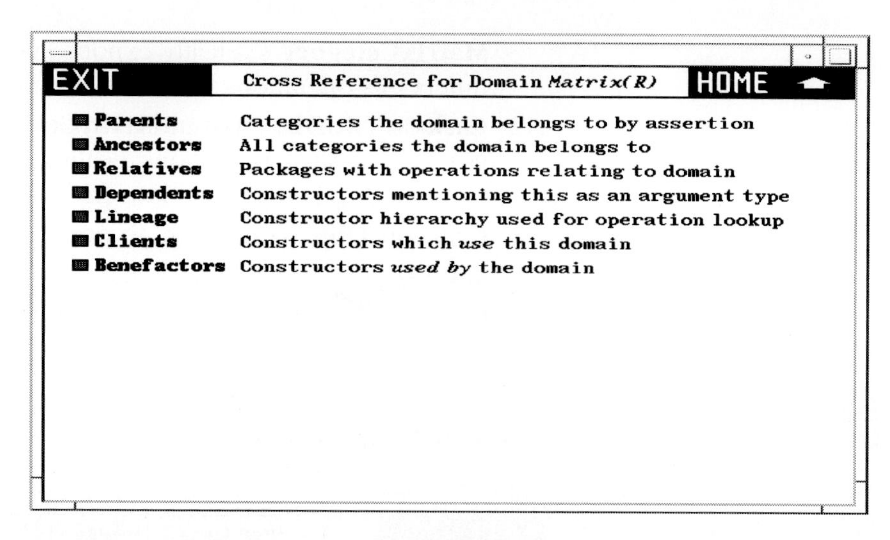

Figure 14.17: Cross-reference page for Matrix.

the **Exports** chain for Matrix.

Relatives

The *relatives* of a domain constructor are package constructors that provide operations in addition to those *exported* by the domain.

Try this exercise.

1. Click on **Relatives**, bringing up a list of *packages*.
2. Click on **LinearSystemMatrixPackage** bringing up its constructor page.[3]
3. Click on **Operations**. Here you see **rank**, an operation also exported by Matrix itself.
4. Click on **rank**. This **rank** has two arguments and thus is different from the **rank** from Matrix.
5. Click on [up-arrow button] to return to the list of operations for the package LinearSystemMatrixPackage.
6. Click on **solve** to bring up a **solve** for linear systems of equations.
7. Click on [up-arrow button] several times to return to the cross reference page for Matrix.

Dependents

The *dependents* of a constructor are those *domains* or *packages* that mention that constructor either as an argument or in its *exports*.

If you click on **Dependents** two entries may surprise you: RectangularMatrix and SquareMatrix. This happens because Matrix, as it turns out, appears in signatures of operations exported by these domains.

[3]You may want to widen your HyperDoc window to make what follows more legible.

Lineage

The term *lineage* refers to the *search order* for functions. If you are an expert user or curious about how the AXIOM system works, try the following exercise. Otherwise, you best skip this button and go on to **Clients**.

Clicking on **Lineage** gives you a list of domain constructors: InnerIndexedTwoDimensionalArray, MatrixCategory&, TwoDimensionalArrayCategory&, HomogeneousAggregate&, Aggregate&. What are these constructors and how are they used?

We explain by an example. Suppose you create a matrix using the interpreter, then ask for its **rank**. AXIOM must then find a function implementing the **rank** operation for matrices. The first place AXIOM looks for **rank** is in the Matrix domain.

If not there, the lineage of Matrix tells AXIOM where else to look. Associated with the matrix domain are five other lineage domains. Their order is important. AXIOM first searches the first one, InnerIndexedTwoDimensionalArray. If not there, it searches the second MatrixCategory&. And so on.

Where do these *lineage constructors* come from? The source code for Matrix contains this syntax for the *function body* of Matrix:[4]

```
InnerIndexedTwoDimensionalArray(R,mnRow,mnCol,Row,Col)
   add ...
```

where the "`...`" denotes all the code that follows. In English, this means: "The functions for matrices are defined as those from InnerIndexedTwoDimensionalArray domain augmented by those defined in '`...`'," where the latter take precedence.

This explains InnerIndexedTwoDimensionalArray. The other names, those with names ending with an ampersand "&" are *default packages* for categories to which Matrix belongs. Default packages are ordered by the notion of "closest ancestor."

Clients

A client of Matrix is any constructor that uses Matrix in its implementation. For example, Complex is a client of Matrix; it exports several operations that take matrices as arguments or return matrices as values.[5]

Benefactors

[4]InnerIndexedTwoDimensionalArray is a special domain implemented for matrix-like domains to provide efficient implementations of two-dimensional arrays. For example, domains of category TwoDimensionalArrayCategory can have any integer as their `minIndex`. Matrices and other members of this special "inner" array have their `minIndex` defined as `1`.

[5]A constructor is a client of Matrix if it handles any matrix. For example, a constructor having internal (unexported) operations dealing with matrices is also a client.

A *benefactor* of Matrix is any constructor that Matrix uses in its implementation. This information, like that for clients, is gathered from run-time structures.[6]

Cross reference pages for categories have some different buttons on them. Starting with the constructor page of Matrix, click on Ring producing its constructor page. Click on **Cross Reference**, producing the cross-reference page for Ring. Here are buttons **Parents** and **Ancestors** similar to the notion for domains, except for categories the relationship between parent and child is defined through *category extension.*

Children

Category hierarchies go both ways. There are children as well as parents. A child can have any number of parents, but always at least one. Every category is therefore a descendant of exactly one category: Object.

Descendants

These are children, children of children, and so on.

Category hierarchies are complicated by the fact that categories take parameters. Where a parameterized category fits into a hierarchy *may* depend on values of its parameters. In general, the set of categories in AXIOM forms a *directed acyclic graph*, that is, a graph with directed arcs and no cycles.

Domains

This produces a table of all domain constructors that can possibly be rings (members of category Ring). Some domains are unconditional rings. Others are rings for some parameters and not for others. To find out which, select the **conditions** button in the views panel. For example, DirectProduct(n, R) is a ring if R is a ring.

14.2.3 Views Of Constructors

Below every constructor table page is a **Views** panel. As an example, click on **Cross Reference** from the constructor page of Matrix, then on **Benefactors** to produce a short table of constructor names.

The **Views** panel is at the bottom of the page. Two items, *names* and *conditions,* are in italics. Others are active buttons. The active buttons are those that give you useful alternative views on this table of constructors. Once you select a view, you notice that the button turns off (becomes italicized) so that you cannot reselect it.

names

This view gives you a table of names. Selecting any of these names brings up the constructor page for that constructor.

[6]The benefactors exclude constructors such as PrimitiveArray whose operations macro-expand and so vanish from sight!

abbrs

This view gives you a table of abbreviations, in the same order as the original constructor names. Abbreviations are in capitals and are limited to 7 characters. They can be used interchangeably with constructor names in input areas.

kinds

This view organizes constructor names into the three kinds: categories, domains and packages.

files

This view gives a table of file names for the source code of the constructors in alphabetic order after removing duplicates.

parameters

This view presents constructors with the arguments. This view of the benefactors of Matrix shows that Matrix uses as many as five different List domains in its implementation.

filter

This button is used to refine the list of names or abbreviations. Starting with the *names* view, enter `m*` into the input area and click on **filter**. You then get a shorter table with only the names beginning with `m`.

documentation

This gives you documentation for each of the constructors.

conditions

This page organizes the constructors according to predicates. The view is not available for your example page since all constructors are unconditional. For a table with conditions, return to the **Cross Reference** page for Matrix, click on **Ancestors**, then on **conditions** in the view panel. This page shows you that CoercibleTo(OutputForm) and SetCategory are ancestors of Matrix(R) only if `R` belongs to category SetCategory.

14.2.4 Giving Parameters to Constructors

Notice the input area at the bottom of the constructor page. If you leave this blank, then the information you get is for the domain constructor Matrix(R), that is, Matrix for an arbitrary underlying domain `R`.

In general, however, the exports and other information *do* usually depend on the actual value of `R`. For example, Matrix exports the **inverse** operation only if the domain `R` is a Field. To see this, try this from the main constructor page:

1. Enter `Integer` into the input area at the bottom of the page.
2. Click on **Operations**, producing a table of operations. Note the number of operation names that appear at the top of the page.
3. Click on [up-arrow button] to return to the constructor page.
4. Use the **Delete** or **Backspace** keys to erase `Integer` from the input area.
5. Click on **Operations** to produce a new table of operations. Look at the number of operations you get. This number is greater than what you had before. Find, for example, the operation **inverse**.

6. Click on **inverse** to produce a page describing the operation **inverse**. At the bottom of the description, you notice that the **Conditions** line says "R has Field." This operation is *not* exported by Matrix(Integer) since Integer is not a *field*.
   Try putting the name of a domain such as Fraction Integer (which is a field) into the input area, then clicking on **Operations**. As you see, the operation **inverse** is exported.

14.3 Miscellaneous Features of Browse

14.3.1 The Description Page for Operations

From the constructor page of Matrix, click on **Operations** to bring up the table of operations for Matrix.

Find the operation **inverse** in the table and click on it. This takes you to a page showing the documentation for this operation.

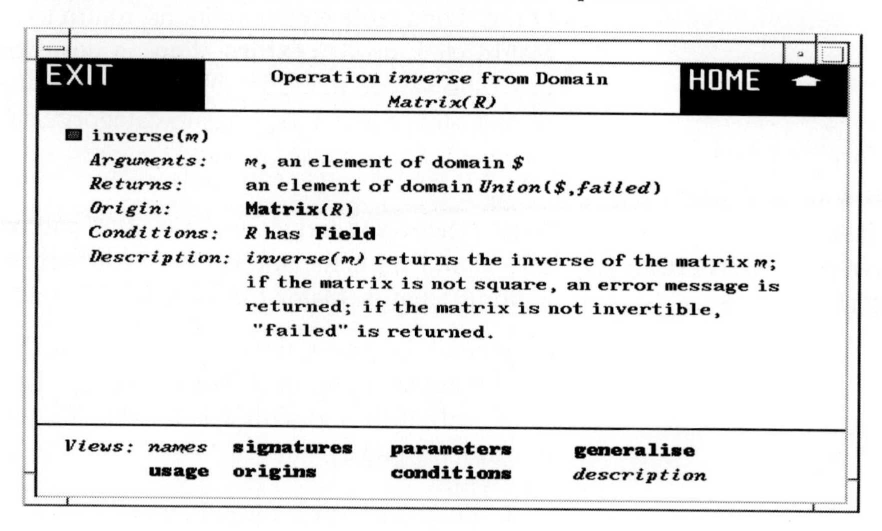

Figure 14.18: Operation **inverse** from Matrix.

Here is the significance of the headings you see.

Arguments

This lists each of the arguments of the operation in turn, paraphrasing the *signature* of the operation. As for signatures, a "$" is used to designate *this domain*, that is, Matrix(R).

Returns

This describes the return value for the operation, analogous to the **Arguments** part.

Origin This tells you which domain or category explicitly exports the operation. In this example, the domain itself is the *Origin.*

Conditions This tells you that the operation is exported by Matrix(R) only if "R has Field," that is, "R is a member of category Field." When no **Conditions** part is given, the operation is exported for all values of R.

Description Here are the "++" comments that appear in the source code of its *Origin*, here Matrix. You find these comments in the source code for Matrix.

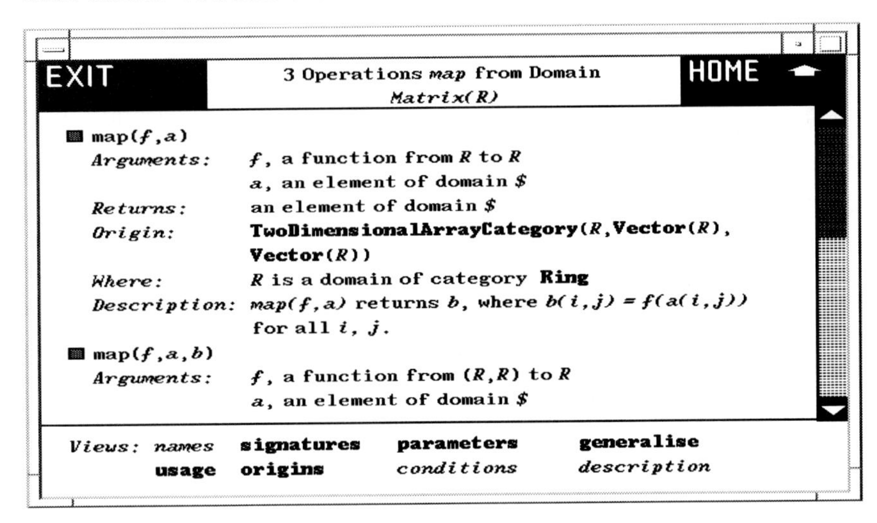

Figure 14.19: Operations **map** from Matrix.

Click on [button] to return to the table of operations. Click on **map**. Here you find three different operations named **map**. This should not surprise you. Operations are identified by name and *signature.* There are three operations named **map**, each with different signatures. What you see is the *descriptions* view of the operations. If you like, select the button in the heading of one of these descriptions to get *only* that operation.

Where This part qualifies domain parameters mentioned in the arguments to the operation.

14.3.2 Views of Operations

We suggest that you go to the constructor page for Matrix and click on **Operations** to bring up a table of operations with a **Views** panel at the bottom.

names This view lists the names of the operations. Unlike constructors, however, there may be several operations with the same name. The heading for the page tells you the number of unique names and the number of distinct operations when these numbers are different.

filter

As for constructors, you can use this button to cut down the list of operations you are looking at. Enter, for example, `m*` into the input area to the right of **filter** then click on **filter**. As usual, any logical expression is permitted. For example, use

```
*! or *?
```

to get a list of destructive operations and predicates.

documentation

This gives you the most information: a detailed description of all the operations in the form you have seen before. Every other button summarizes these operations in some form.

signatures

This views the operations by showing their signatures.

parameters

This views the operations by their distinct syntactic forms with parameters.

origins

This organizes the operations according to the constructor that explicitly exports them.

conditions

This view organizes the operations into conditional and unconditional operations.

usage

This button is only available if your user-level is set to *development*. The **usage** button produces a table of constructors that reference this operation.[7]

implementation

This button is only available if your user-level is set to *development*. If you enter values for all domain parameters on the constructor page, then the **implementation** button appears in place of the **conditions** button. This button tells you what domains or packages actually implement the various operations.[8]

With your user-level set to *development*, we suggest you try this exercise. Return to the main constructor page for Matrix, then enter `Integer` into the input area at the bottom as the value of R. Then click on **Operations** to produce a table of operations. Note that the **conditions** part of the **Views** table is replaced by **implementation**. Click on **implementation**. After some delay, you get a page describing what implements each of the matrix operations, organized by the various domains and packages.

generalize

This button only appears for an operation page of a constructor involving a unique operation name.

From an operations page for Matrix, select any operation name, say **rank**. In the views panel, the **filter** button is replaced by **generalize**. Click on it! What you get is a description of all AXIOM operations named **rank**.[9]

all domains

[7]AXIOM requires an especially long time to produce this table, so anticipate this when

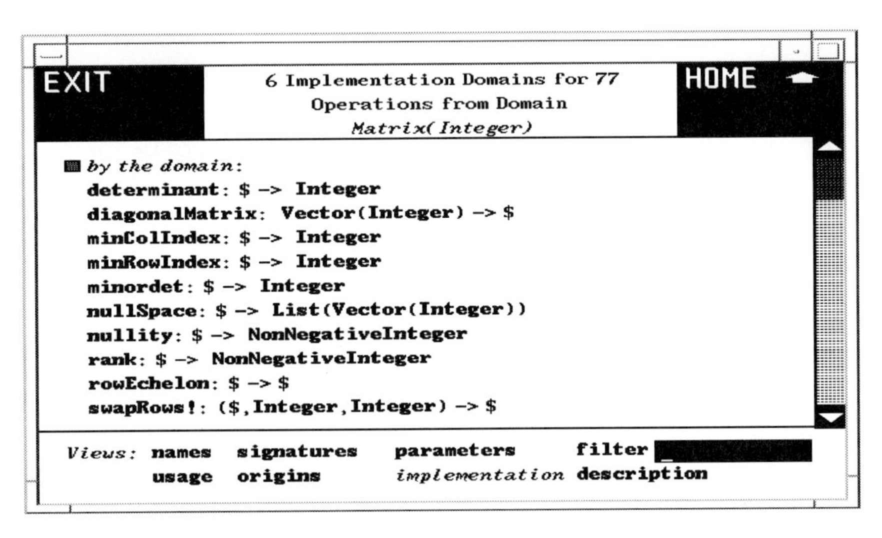

Figure 14.20: Implementation domains for Matrix.

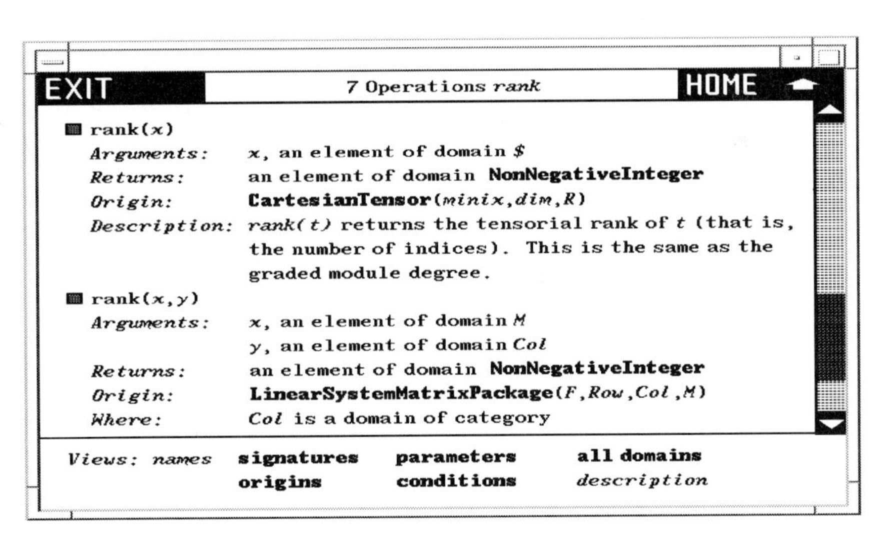

Figure 14.21: All operations named **rank** in AXIOM.

This button only appears on an operation page resulting from a search from the front page of Browse or from selecting **generalize** from an operation page for a constructor.

Note that the **filter** button in the **Views** panel is replaced by **all domains**. Click on it to produce a table of *all* domains or packages that export a **rank** operation.

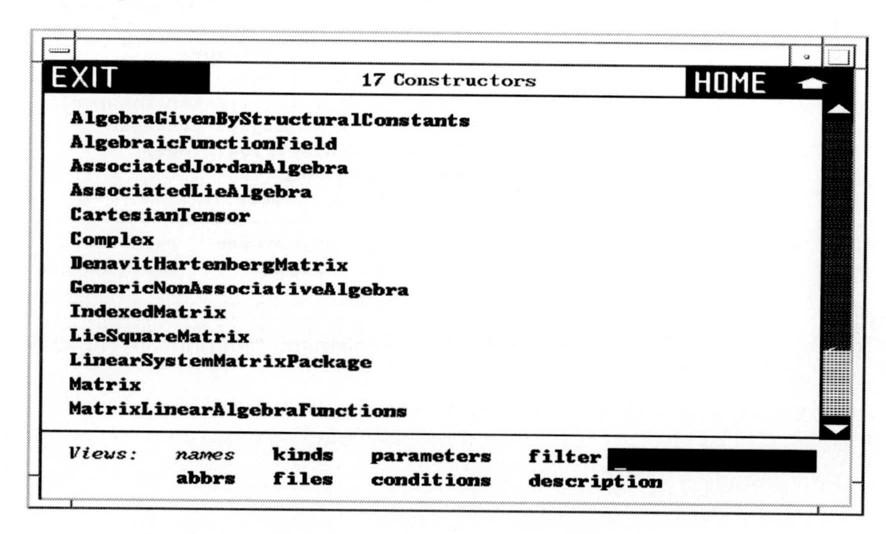

Figure 14.22: Table of all domains that export **rank**.

We note that this table specifically refers to all the **rank** operations shown in the preceding page. Return to the descriptions of all the **rank** operations and select one of them by clicking on the button in its heading. Select **all domains**. As you see, you have a smaller table of constructors. When there is only one constructor, you get the constructor page for that constructor.

---

requesting this information.

[8]This button often takes a long time; expect a delay while you wait for an answer.

[9]If there were more than **threshold** operations of the name, you get instead a page with a **Views** panel at the bottom and the message to **Select a view below**. To get the descriptions of all these operations as mentioned above, select the **description** button. See the discussion of **threshold** in Section 14.3.4 on page 567.

14.3.3 Capitalization Convention

When entering search keys for constructors, you can use capital letters to search for abbreviations. For example, enter `UTS` into the input area and click on **Constructors**. Up comes a page describing UnivariateTaylorSeries whose abbreviation is UTS.

Constructor abbreviations always have three or more capital letters. For short constructor names (six letters or less), abbreviations are not generally helpful as their abbreviation is typically the constructor name in capitals. For example, the abbreviation for Matrix is MATRIX.

Abbreviations can also contain numbers. For example, POLY2 is the abbreviation for constructor PolynomialFunctions2. For default packages, the abbreviation is the same as the abbreviation for the corresponding category with the "&" replaced by "-". For example, for the category default package MatrixCategory& the abbreviation is MATCAT- since the corresponding category MatrixCategory has abbreviation MATCAT.

14.3.4 Browse Options

You can set two options for using Browse: exposure and threshold.

Exposure

By default, the only constructors, operations, and attributes shown by Browse are those from *exposed constructors*. To change this, you can issue

```
)set hyperdoc browse exposure on
```

After you make this setting, you will see both exposed and unexposed constructs. By definition, an operation or attribute is exposed only if it is exported from an exposed constructor. Unexposed items are generally marked by Browse with an asterisk. For more information on exposure, see Section 2.11 on page 87.

With this setting, try the following experiment. Starting with the main Browse page, enter `*matrix*` into the input area and click on **Constructors**. The result is the following table.

Threshold

For General, Documentation or Complete searches, a summary is presented of all matches. When the number of items of a given kind is less than a number called **threshold**, AXIOM presents a table of names with the heading for that kind.

Also, when an operation name is chosen and there are less than **threshold** distinct operations, the operations are initially shown in **description** mode.

The default value of **threshold** is 10. To change its value to say 5, issue

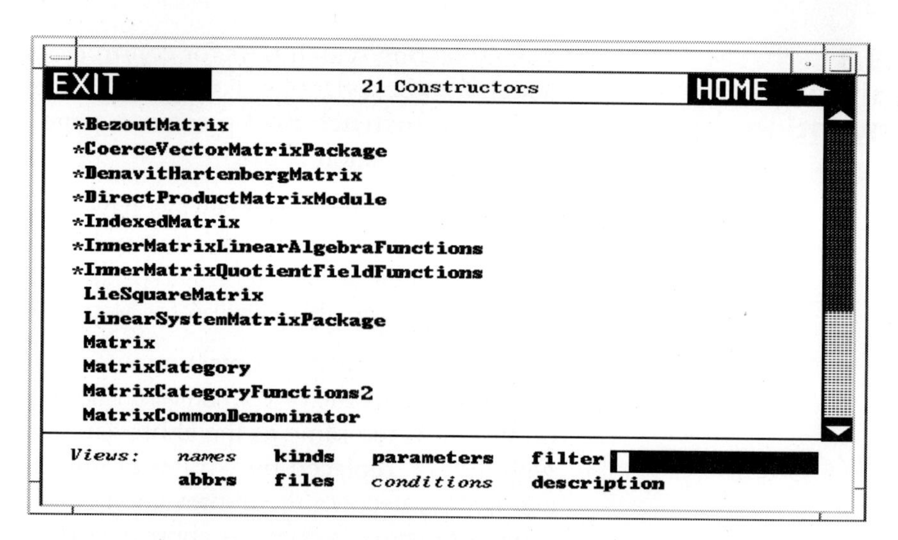

Figure 14.23: Table of all constructors matching `*matrix*` .

```
)set hyperdoc browse threshold 5
```

Notice that the headings in the summary are active. If you click on a heading, you bring up a separate page for those entries.

APPENDICES

APPENDIX A

AXIOM System Commands

This chapter describes system commands, the command-line facilities used to control the AXIOM environment. The first section is an introduction and discusses the common syntax of the commands available.

A.1 Introduction

System commands are used to perform AXIOM environment management. Among the commands are those that display what has been defined or computed, set up multiple logical AXIOM environments (frames), clear definitions, read files of expressions and commands, show what functions are available, and terminate AXIOM.

Some commands are restricted: the commands

```
)set userlevel interpreter
)set userlevel compiler
)set userlevel development
```

set the user-access level to the three possible choices. All commands are available at `development` level and the fewest are available at `interpreter` level. The default user-level is `interpreter`. In addition to the `)set` command (discussed in Section A.19 on page 586) you can use the HyperDoc settings facility to change the *user-level*.

Each command listing begins with one or more syntax pattern descriptions plus examples of related commands. The syntax descriptions are intended to be easy to read and do not necessarily represent the most compact way of specifying all possible arguments and options; the descriptions may occasionally be redundant.

All system commands begin with a right parenthesis which should be in the first available column of the input line (that is, immediately after the input prompt, if any). System commands may be issued directly to AXIOM or be included in **.input** files.

A system command *argument* is a word that directly follows the command name and is not followed or preceded by a right parenthesis. A system command *option* follows the system command and is directly preceded by a right parenthesis. Options may have arguments: they directly follow the option. This example may make it easier to remember what is an option and what is an argument:

`)syscmd` *arg1 arg2* `)opt1` *opt1arg1 opt1arg2* `)opt2` *opt2arg1* `...`

In the system command descriptions, optional arguments and options are enclosed in brackets ("`[`" and "`]`"). If an argument or option name is in italics, it is meant to be a variable and must have some actual value substituted for it when the system command call is made. For example, the syntax pattern description

`)read` *fileName* `[)quietly]`

would imply that you must provide an actual file name for *fileName* but need not use the `)quietly` option. Thus

```
)read matrix.input
```

is a valid instance of the above pattern.

System command names and options may be abbreviated and may be in upper or lower case. The case of actual arguments may be significant, depending on the particular situation (such as in file names). System command names and options may be abbreviated to the minimum number of starting letters so that the name or option is unique. Thus

```
)s Integer
```

is not a valid abbreviation for the `)set` command, because both `)set` and `)show` begin with the letter "s". Typically, two or three letters are sufficient for disambiguating names. In our descriptions of the commands, we have used no abbreviations for either command names or options.

In some syntax descriptions we use a vertical line "|" to indicate that you must specify one of the listed choices. For example, in

```
)set output fortran on | off
```

only `on` and `off` are acceptable words for following `boot`. We also sometimes use "..." to indicate that additional arguments or options of the listed form are allowed. Finally, in the syntax descriptions we may also list the syntax of related commands.

A.2)abbreviation

User Level Required: compiler

Command Syntax:

`)abbreviation query [`*nameOrAbbrev*`]`
`)abbreviation category` *abbrev fullname* `[)quiet]`
`)abbreviation domain` *abbrev fullname* `[)quiet]`
`)abbreviation package` *abbrev fullname* `[)quiet]`
`)abbreviation remove` *nameOrAbbrev*

Command Description:

This command is used to query, set and remove abbreviations for category, domain and package constructors. Every constructor must have a unique abbreviation. This abbreviation is part of the name of the subdirectory under which the components of the compiled constructor are stored. It is this abbreviation that is used to bring compiled code into AXIOM with the `)load` command. Furthermore, by issuing this command you let the system know what file to load automatically if you use a new constructor. Abbreviations must start with a letter and then be followed by up to seven letters or digits. Any letters appearing in the abbreviation must be in uppercase.

When used with the `query` argument, this command may be used to list the name associated with a particular abbreviation or the abbreviation for a constructor. If no abbreviation or name is given, the names and corresponding abbreviations for *all* constructors are listed.

The following shows the abbreviation for the constructor List:

```
)abbreviation query List
```

The following shows the constructor name corresponding to the abbreviation NNI:

```
)abbreviation query NNI
```

The following lists all constructor names and their abbreviations.

```
)abbreviation query
```

To add an abbreviation for a constructor, use this command with `category`, `domain` or `package`. The following add abbreviations to the system for a category, domain and package, respectively:

```
)abbreviation domain   SET Set
)abbreviation category COMPCAT  ComplexCategory
)abbreviation package  LIST2MAP ListToMap
```

If the `)quiet` option is used, no output is displayed from this command. You would normally only define an abbreviation in a library source file. If this command is issued for a constructor that has already been loaded, the constructor will be reloaded next time it is referenced. In particular, you can use this command to force the automatic reloading of constructors.

To remove an abbreviation, the `remove` argument is used. This is usually only used to correct a previous command that set an abbreviation for a constructor name. If, in fact, the abbreviation does exist, you are prompted for confirmation of the removal request. Either of the following commands will remove the abbreviation VECTOR2 and the constructor name VectorFunctions2 from the system:

```
)abbreviation remove VECTOR2
)abbreviation remove VectorFunctions2
```

Also See: '`)compile`' in Section A.6 on page 575 and '`)load`' in Section A.14 on page 583.

A.3)boot

User Level Required: development

Command Syntax:

`)boot` *bootExpression*

Command Description:

This command is used by AXIOM system developers to execute expressions written in the BOOT language. For example,

```
)boot times3(x) == 3*x
```

creates and compiles the Common LISP function "times3" obtained by translating the BOOT code.

Also See: ')fin' in Section A.9 on page 578, ')lisp' in Section A.13 on page 582, ')set' in Section A.19 on page 586, and ')system' in Section A.23 on page 588.

A.4)cd

User Level Required: interpreter

Command Syntax:

)cd *directory*

Command Description:

This command sets the AXIOM working current directory. The current directory is used for looking for input files (for `)read`), AXIOM library source files (for `)compile`), saved history environment files (for `)history )restore`), compiled AXIOM library files (for `)load`), and files to edit (for `)edit`). It is also used for writing spool files (via `)spool`), writing history input files (via `)history )write`) and history environment files (via `)history )save`),and compiled AXIOM library files (via `)compile`).

If issued with no argument, this command sets the AXIOM current directory to your home directory. If an argument is used, it must be a valid directory name. Except for the "`)`" at the beginning of the command, this has the same syntax as the operating system `cd` command.

Also See: ')compile' in Section A.6 on page 575, ')edit' in Section A.8 on page 578, ')history' in Section A.12 on page 580, ')load' in Section A.14 on page 583, ')read' in Section A.18 on page 585, and ')spool' in Section A.21 on page 587.

A.5)clear

User Level Required: interpreter

Command Syntax:

`)clear all`
`)clear completely`
`)clear properties all`
`)clear properties` *obj1* [*obj2* ...]
`)clear value all`
`)clear value` *obj1* [*obj2* ...]
`)clear mode all`
`)clear mode` *obj1* [*obj2* ...]

Command Description:

This command is used to remove function and variable declarations, definitions and values from the workspace. To empty the entire workspace and reset the step counter to 1, issue

```
)clear all
```

To remove everything in the workspace but not reset the step counter, issue

```
)clear properties all
```

To remove everything about the object **x**, issue

```
)clear properties x
```

To remove everything about the objects `x, y` and `f`, issue

```
)clear properties x y f
```

The word `properties` may be abbreviated to the single letter "`p`".

```
)clear p all
)clear p x
)clear p x y f
```

All definitions of functions and values of variables may be removed by either

```
)clear value all
)clear v all
```

This retains whatever declarations the objects had. To remove definitions and values for the specific objects `x, y` and `f`, issue

```
)clear value x y f
)clear v x y f
```

To remove the declarations of everything while leaving the definitions and values, issue

```
)clear mode  all
)clear m all
```

To remove declarations for the specific objects `x, y` and `f`, issue

```
)clear mode x y f
)clear m x y f
```

The `)display names` and `)display properties` commands may be used to see what is currently in the workspace.

The command

```
)clear completely
```

does everything that `)clear all` does, and also clears the internal system function and constructor caches.

Also See: '`)display`' in Section A.7 on page 577, '`)history`' in Section A.12 on page 580, and '`)undo`' in Section A.25 on page 592.

A.6)compile

User Level Required: compiler

Command Syntax:

)compile
)compile *fileName*
)compile *fileName*.spad
)compile *directory/fileName*.spad
)compile *fileName*)break
)compile *fileName*)nobreak
)compile *fileName*)library
)compile *fileName*)nolibrary

)compile *fileName*)constructor *nameOrAbbrev*

)compile *fileName*)constructor *nameOrAbbrev*
)functions *funName1* [*funName2* ...]

Command Description:

This command is used to compile category, domain, and package constructors. You can compile individual constructors in a file or even individual functions from constructors.

A file containing AXIOM constructor definitions must have file extension ".spad." The full filename is remembered between invocations of this command and `)edit` commands. The sequence of commands

```
)compile matrix.spad
)edit
)compile
```

will call the compiler, edit, and then call the compiler again on the file **matrix.spad.** If you do not specify a *directory,* the working current directory (see Section A.4 on page 574) is searched for the file. If the file is not found, the standard system directories are searched.

If no options are given, all constructors within a file are compiled. Each constructor should have an `)abbreviation` command in the file in which it is defined. We suggest that you place the `)abbreviation` commands at the top of the file in the order in which the constructors are defined. The list of commands serves as a table of contents for the file.

The `)library` option causes directories containing the compiled code for each constructor to be created in the working current directory. The name of such a directory consists of the constructor abbreviation and the **.NRLIB** file extension. For example, the directory containing the compiled code for the MATRIX constructor is called **MATRIX.NRLIB.** The `)nolibrary` option says that such files should not be created. The default is `)library.`

The `)constructor` option is used to specify a particular constructor to compile. All other constructors in the file are ignored. The constructor name or abbreviation follows `)constructor.` Thus either

```
)compile matrix.spad )constructor RectangularMatrix
```

or

```
)compile matrix.spad )constructor RMATRIX
```

compiles the RectangularMatrix constructor defined in **matrix.spad.**

The `)functions` option allows you to recompile specific functions within a constructor. If several functions have the same name, all will be recompiled. The `)constructor` option is required if `)functions` is specified. The following will recompile the function **coerce** defined within the RectangularMatrix constructor.

```
)compile matrix.spad )constructor RMATRIX )functions coerce
```

You cannot use `)nolibrary` with this option. The compiled code for the functions will be appended to the library file in the working current directory. You must ensure that a copy of the library exists in the working current directory before invoking the `)compile` command with the `)functions` option. We recommend that you recompile the whole constructor after debugging of individual functions is complete.

The `)break` and `)nobreak` options determine what the compiler does when it encounters an error. `)break` is the default and it indicates that processing should stop at the first error. The value of the `)set break` variable then controls what happens.

It is important for you to realize that it does not suffice to compile a constructor to use the new code in the interpreter. After compilation, the `)load` command with the `)update` option should be used to bring in the new code and update internal system tables with information about the constructor.

Also See: '`)abbreviation`' in Section A.2 on page 572, '`)edit`' in Section A.8 on page 578, and '`)load`' in Section A.14 on page 583.

A.7)display

User Level Required: interpreter

Command Syntax:

```
)display all
)display properties
)display properties all
)display properties [obj1 [obj2 ...]]
)display value all
)display value [obj1 [obj2 ...]]
)display mode all
)display mode [obj1 [obj2 ...]]
)display names
)display operations opName
```

Command Description:

This command is used to display the contents of the workspace and signatures of functions with a given name.[1]

The command

```
)display names
```

lists the names of all user-defined objects in the workspace. This is useful if you do not wish to see everything about the objects and need only be reminded of their names.

The commands

```
)display all
)display properties
)display properties all
```

all do the same thing: show the values and types and declared modes of all variables in the workspace. If you have defined functions, their signatures and definitions will also be displayed.

To show all information about a particular variable or user functions, for example, something named `d`, issue

```
)display properties d
```

To just show the value (and the type) of `d`, issue

```
)display value d
```

To just show the declared mode of `d`, issue

[1] A *signature* gives the argument and return types of a function.

```
)display mode d
```

All modemaps for a given operation may be displayed by using `)display operations`. A *modemap* is a collection of information about a particular reference to an operation. This includes the types of the arguments and the return value, the location of the implementation and any conditions on the types. The modemap may contain patterns. The following displays the modemaps for the operation **complex**:

```
)d op complex
```

Also See: '`)clear`' in Section A.5 on page 574, '`)history`' in Section A.12 on page 580, '`)set`' in Section A.19 on page 586, '`)show`' in Section A.20 on page 586, and '`)what`' in Section A.26 on page 592.

A.8)edit

User Level Required: interpreter

Command Syntax:

`)edit` [*filename*]

Command Description:

This command is used to edit files. It works in conjunction with the `)read` and `)compile` commands to remember the name of the file on which you are working. By specifying the name fully, you can edit any file you wish. Thus

```
)edit /u/julius/matrix.input
```

will place you in an editor looking at the file `/u/julius/matrix.input`. By default, the editor is `vi`, but if you have an EDITOR shell environment variable defined, that editor will be used. When AXIOM is running under the X Window System, it will try to open a separate `xterm` running your editor if it thinks one is necessary. For example, under the Korn shell, if you issue

```
export EDITOR=emacs
```

then the emacs editor will be used by `)edit`.

If you do not specify a file name, the last file you edited, read or compiled will be used. If there is no "last file" you will be placed in the editor editing an empty unnamed file.

It is possible to use the `)system` command to edit a file directly. For example,

```
)system emacs /etc/rc.tcpip
```

calls `emacs` to edit the file.

Also See: '`)system`' in Section A.23 on page 588, '`)compile`' in Section A.6 on page 575, and '`)read`' in Section A.18 on page 585.

A.9)fin

User Level Required: development

Command Syntax:

`)fin`

Command Description:

This command is used by AXIOM developers to leave the AXIOM system and return to the underlying Common LISP system. To return to AXIOM, issue the "`(|spad|)`" function call to Common LISP.

Also See: '`)pquit`' in Section A.16 on page 584 and '`)quit`' in Section A.17 on page 585.

A.10)frame

User Level Required: interpreter

Command Syntax:

`)frame new` *frameName*
`)frame drop` [*frameName*]
`)frame next`
`)frame last`
`)frame names`
`)frame import` *frameName* [*objectName1* [*objectName2* ...]]
`)set message frame on | off`
`)set message prompt frame`

Command Description:

A *frame* can be thought of as a logical session within the physical session that you get when you start the system. You can have as many frames as you want, within the limits of your computer's storage, paging space, and so on. Each frame has its own *step number*, *environment* and *history*. You can have a variable named `a` in one frame and it will have nothing to do with anything that might be called `a` in any other frame.

Some frames are created by the HyperDoc program and these can have pretty strange names, since they are generated automatically. To find out the names of all frames, issue

```
)frame names
```

It will indicate the name of the current frame.

You create a new frame "**quark**" by issuing

```
)frame new quark
```

The history facility can be turned on by issuing either `)set history on` or `)history )on`. If the history facility is on and you are saving history information in a file rather than in the AXIOM environment then a history file with filename **quark.axh** will be created as you enter commands. If you wish to go back to what you were doing in the "**initial**" frame, use

```
)frame next
```

or

```
)frame last
```

to cycle through the ring of available frames to get back to "**initial**".

If you want to throw away a frame (say "**quark**"), issue

```
)frame drop quark
```

If you omit the name, the current frame is dropped.

If you do use frames with the history facility on and writing to a file, you may want to delete some of the older history files. These are directories, so you may want to issue a command like `rm -r quark.axh` to the operating system.

You can bring things from another frame by using `)frame import`. For example, to bring the `f` and `g` from the frame "**quark**" to the current frame, issue

```
)frame import quark f g
```

If you want everything from the frame "**quark**", issue

```
)frame import quark
```

You will be asked to verify that you really want everything.

There are two `)set` flags to make it easier to tell where you are.

```
)set message frame on | off
```

will print more messages about frames when it is set on. By default, it is off.

```
)set message prompt frame
```

will give a prompt that looks like

```
initial (1) ->
```

when you start up. In this case, the frame name and step make up the prompt.

Also See: '`)history`' in Section A.12 on page 580 and '`)set`' in Section A.19 on page 586.

A.11)help

User Level Required: interpreter

Command Syntax:

`)help`
`)help` *commandName*

Command Description:

This command displays help information about system commands. If you issue

```
)help
```

then this very text will be shown. You can also give the name or abbreviation of a system command to display information about it. For example,

```
)help clear
```

will display the description of the `)clear` system command.

All this material is available in the AXIOM User Guide and in HyperDoc. In HyperDoc, choose the **Commands** item from the **Reference** menu.

A.12)history

User Level Required: interpreter

Command Syntax:

`)history )on`

```
)history )off
)history )write historyInputFileName
)history )show [n] [both]
)history )save savedHistoryName
)history )restore [savedHistoryName]
)history )reset
)history )change n
)history )memory
)history )file
%
%%(n)
)set history on | off
```

Command Description:

The *history* facility within AXIOM allows you to restore your environment to that of another session and recall previous computational results. Additional commands allow you to review previous input lines and to create an **.input** file of the lines typed to AXIOM.

AXIOM saves your input and output if the history facility is turned on (which is the default). This information is saved if either of

```
)set history on
)history )on
```

has been issued. Issuing either

```
)set history off
)history )off
```

will discontinue the recording of information.

Whether the facility is disabled or not, the value of "%" in AXIOM always refers to the result of the last computation. If you have not yet entered anything, "%" evaluates to an object of type Variable('%). The function "%%" may be used to refer to other previous results if the history facility is enabled. In that case, `%%(n)` is the output from step `n` if `n > 0`. If `n < 0`, the step is computed relative to the current step. Thus `%%(-1)` is also the previous step, `%%(-2)`, is the step before that, and so on. If an invalid step number is given, AXIOM will signal an error.

The *environment* information can either be saved in a file or entirely in memory (the default). Each frame (Section A.10 on page 579) has its own history database. When it is kept in a file, some of it may also be kept in memory for efficiency. When the information is saved in a file, the name of the file is of the form **FRAME.axh** where "**FRAME**" is the name of the current frame. The history file is placed in the current working directory (see Section A.4 on page 574). Note that these history database files are not text files (in fact, they are directories themselves), and so are not in human-readable format.

The options to the `)history` command are as follows:

`)change` *n* will set the number of steps that are saved in memory to *n*. This option only has effect when the history data is maintained in a file. If you have issued `)history )memory` (or not changed the default) there is no need to use `)history )change`.

`)on` will start the recording of information. If the workspace is not empty, you will be asked to confirm this request. If you do so, the workspace will be cleared and history data will begin being saved. You can also turn the facility on by issuing `)set history on`.

`)off` will stop the recording of information. The `)history )show` command will not work after issuing this command. Note that this command may be issued to save time, as there is some performance penalty paid for saving the environment data. You can also turn the facility off by issuing `)set history off`.

`)file` indicates that history data should be saved in an external file on disk.

`)memory` indicates that all history data should be kept in memory rather than saved in a file. Note that if you are computing with very large objects it may not be practical to kept this data in memory.

`)reset` will flush the internal list of the most recent workspace calculations so that the data structures may be garbage collected by the underlying Common LISP system. Like `)history )change`, this option only has real effect when history data is being saved in a file.

`)restore [`*savedHistoryName*`]` completely clears the environment and restores it to a saved session, if possible. The `)save` option below allows you to save a session to a file with a given name. If you had issued `)history )save jacobi` the command `)history )restore jacobi` would clear the current workspace and load the contents of the named saved session. If no saved session name is specified, the system looks for a file called **last.axh**.

`)save` *savedHistoryName* is used to save a snapshot of the environment in a file. This file is placed in the current working directory (see Section A.4 on page 574). Use `)history )restore` to restore the environment to the state preserved in the file. This option also creates an input file containing all the lines of input since you created the workspace frame (for example, by starting your AXIOM session) or last did a `)clear all` or `)clear completely`.

`)show [`*n*`] [both]` can show previous input lines and output results. `)show` will display up to twenty of the last input lines (fewer if you haven't typed in twenty lines). `)show` *n* will display up to *n* of the last input lines. `)show both` will display up to five of the last input lines and output results. `)show` *n* `both` will display up to *n* of the last input lines and output results.

`)write` *historyInputFile* creates an **.input** file with the input lines typed since the start of the session/frame or the last `)clear all` or `)clear completely`. If *historyInputFileName* does not contain a period (".") in the filename, **.input** is appended to it. For example, `)history )write chaos` and `)history )write chaos.input` both write the input lines to a file called **chaos.input** in your current working directory. If you issued one or more `)undo` commands, `)history )write` eliminates all input lines backtracked over as a result of `)undo`. You can edit this file and then use `)read` to have AXIOM process the contents.

Also See: '`)frame`' in Section A.10 on page 579, '`)read`' in Section A.18 on page 585, '`)set`' in Section A.19 on page 586, and '`)undo`' in Section A.25 on page 592.

A.13)lisp

User Level Required: development

Command Syntax:

`)lisp [`*lispExpression*`]`

Command Description:

This command is used by AXIOM system developers to have single expressions evaluated by the Common LISP system on which AXIOM is built. The *lispExpression* is read by the Common LISP reader and evaluated. If this expression is not complete (unbalanced parentheses, say), the reader will wait until a complete expression is entered.

Since this command is only useful for evaluating single expressions, the)fin command may be used to drop out of AXIOM into Common LISP.

Also See: ')system' in Section A.23 on page 588, ')boot' in Section A.3 on page 573, and ')fin' in Section A.9 on page 578.

A.14)load

User Level Required: interpreter

Command Syntax:

```
)load libName1 [libName2 ...]   [)update]
)load libName1 [libName2 ...]   )cond [)update]
)load libName1 [libName2 ...]   )query
)load libName1 [libName2 ...]   )noexpose
```

Command Description:

The `)load` command is used to bring in the compiled library code for constructors and update internal system tables with information about the constructors. This command is usually only used by AXIOM library developers.

The abbreviation of a constructor serves as part of the name of the directory in which the compiled code is stored (see Section A.2 on page 572 for a discussion of defining and querying abbreviations). The abbreviation is used in the `)load` command. For example, to load the constructors Integer, NonNegativeInteger and List which have abbreviations INT, NNI and LIST, respectively, issue the command

```
)load INT NNI LIST
```

To load constructors only if they have not already been loaded (that is., load *conditionally*), use the `)cond` option:

```
)load INT NNI LIST )cond
```

To query whether particular constructors have been loaded, use the `)query` option:

```
)load I NNI L )query
```

When constructors are loaded from AXIOM system directories, some checks and updates are not performed because it is assumed that the system knows about these constructors. To force these checks and updates to occur, add the `)update` option to the command:

```
)load INT NNI LIST )update
)load INT NNI LIST )cond )update
```

The only time it is really necessary to use the `)load` command is when a new constructor has been compiled or an existing constructor has been modified and then compiled. If an `)abbreviate` command has been issued for a constructor, it will be automatically loaded when needed. In particular, any constructor that comes with the AXIOM system will be automatically loaded.

If you write several interdependent constructors it is important that they all get loaded when needed. To accomplish this, either load them manually or issue `)abbreviate` commands for each of the constructors so that they will be automatically loaded when needed.

Constructors are automatically exposed in the frame in which you load them unless you use the `)noexpose` option.

```
)load MATCAT- )noexpose
```

See Section 2.11 on page 87 for more information about constructor exposure.

Also See: ')abbreviation' in Section A.2 on page 572 and ')compile' in Section A.6 on page 575.

A.15)ltrace

User Level Required: development

Command Syntax:

This command has the same arguments as options as the)trace command.

Command Description:

This command is used by AXIOM system developers to trace Common LISP or BOOT functions. It is not supported for general use.

Also See: ')boot' in Section A.3 on page 573, ')lisp' in Section A.13 on page 582, and ')trace' in Section A.24 on page 589.

A.16)pquit

User Level Required: interpreter

Command Syntax:

```
)pquit
```

Command Description:

This command is used to terminate AXIOM and return to the operating system. Other than by redoing all your computations or by using the)history)restore command to try to restore your working environment, you cannot return to AXIOM in the same state.

)pquit differs from the)quit in that it always asks for confirmation that you want to terminate AXIOM (the "p" is for "protected"). When you enter the)pquit command, AXIOM responds

Please enter **y** or **yes** if you really want to leave the interactive environment and return to the operating system:

If you respond with y or yes, you will see the message

You are now leaving the AXIOM interactive environment.
Issue the command **axiom** to the operating system to start a new session.

and AXIOM will terminate and return you to the operating system (or the environment from which you invoked the system). If you responded with something other than y or yes, then the message

You have chosen to remain in the AXIOM interactive environment.

will be displayed and, indeed, AXIOM would still be running.

Also See: ')fin' in Section A.9 on page 578, ')history' in Section A.12 on page 580, ')quit' in Section A.17 on page 585, and ')system' in Section A.23 on page 588.

A.17)quit

User Level Required: interpreter

Command Syntax:

```
)quit
)set quit protected | unprotected
```

Command Description:

This command is used to terminate AXIOM and return to the operating system. Other than by redoing all your computations or by using the `)history )restore` command to try to restore your working environment, you cannot return to AXIOM in the same state.

`)quit` differs from the `)pquit` in that it asks for confirmation only if the command

```
)set quit protected
```

has been issued. Otherwise, `)quit` will make AXIOM terminate and return you to the operating system (or the environment from which you invoked the system).

The default setting is `)set quit protected` so that `)quit` and `)pquit` behave in the same way. If you do issue

```
)set quit unprotected
```

we suggest that you do not (somehow) assign `)quit` to be executed when you press, say, a function key.

Also See: ')fin' in Section A.9 on page 578, '`)history`' in Section A.12 on page 580, '`)pquit`' in Section A.16 on page 584, and '`)system`' in Section A.23 on page 588.

A.18)read

User Level Required: interpreter

Command Syntax:

)read [*fileName*]
)read [*fileName*] [)quiet] [)ifthere]

Command Description:

This command is used to read **.input** files into AXIOM. The command

```
)read matrix.input
```

will read the contents of the file **matrix.input** into AXIOM. The ".input" file extension is optional. See Section 4.1 on page 99 for more information about **.input** files.

This command remembers the previous file you edited, read or compiled. If you do not specify a file name, the previous file will be read.

The `)ifthere` option checks to see whether the **.input** file exists. If it does not, the `)read` command does nothing. If you do not use this option and the file does not exist, you are asked to give the name of an existing **.input** file.

The `)quiet` option suppresses output while the file is being read.

Also See: '`)compile`' in Section A.6 on page 575, '`)edit`' in Section A.8 on page 578, and '`)history`' in Section A.12 on page 580.

A.19)set

User Level Required: interpreter

Command Syntax:

)set
)set *label1* [... *labelN*]
)set *label1* [... *labelN*] *newValue*

Command Description:

The)set command is used to view or set system variables that control what messages are displayed, the type of output desired, the status of the history facility, the way AXIOM user functions are cached, and so on. Since this collection is very large, we will not discuss them here. Rather, we will show how the facility is used. We urge you to explore the)set options to familiarize yourself with how you can modify your AXIOM working environment. There is a HyperDoc version of this same facility available from the main HyperDoc menu.

The)set command is command-driven with a menu display. It is tree-structured. To see all top-level nodes, issue)set by itself.

```
)set
```

Variables with values have them displayed near the right margin. Subtrees of selections have "..." displayed in the value field. For example, there are many kinds of messages, so issue)set message to see the choices.

```
)set message
```

The current setting for the variable that displays whether computation times are displayed is visible in the menu displayed by the last command. To see more information, issue

```
)set message time
```

This shows that time printing is on now. To turn it off, issue

```
)set message time off
```

As noted above, not all settings have so many qualifiers. For example, to change the)quit command to being unprotected (that is, you will not be prompted for verification), you need only issue

```
)set quit unprotected
```

Also See: ')quit' in Section A.17 on page 585.

A.20)show

User Level Required: interpreter

Command Syntax:

)show *nameOrAbbrev*
)show *nameOrAbbrev*)operations
)show *nameOrAbbrev*)attributes

Command Description: This command displays information about AXIOM domain, package and category *constructors*. If no options are given, the)operations option is assumed. For example,

```
)show POLY
```

```
)show POLY )operations
)show Polynomial
)show Polynomial )operations
```

each display basic information about the Polynomial domain constructor and then provide a listing of operations. Since Polynomial requires a Ring (for example, Integer) as argument, the above commands all refer to a unspecified ring R. In the list of operations, "$" means Polynomial(R).

The basic information displayed includes the *signature* of the constructor (the name and arguments), the constructor *abbreviation*, the *exposure status* of the constructor, and the name of the *library source file* for the constructor.

If operation information about a specific domain is wanted, the full or abbreviated domain name may be used. For example,

```
)show POLY INT
)show POLY INT )operations
)show Polynomial Integer
)show Polynomial Integer )operations
```

are among the combinations that will display the operations exported by the domain Polynomial(Integer) (as opposed to the general *domain constructor* Polynomial). Attributes may be listed by using the `)attributes` option.

Also See: '`)display`' in Section A.7 on page 577, '`)set`' in Section A.19 on page 586, and '`)what`' in Section A.26 on page 592.

A.21)spool

User Level Required: interpreter

Command Syntax:

`)spool` [*fileName*]
`)spool`

Command Description:

This command is used to save *(spool)* all AXIOM input and output into a file, called a *spool file*. You can only have one spool file active at a time. To start spool, issue this command with a filename. For example,

```
)spool integrate.out
```

To stop spooling, issue `)spool` with no filename.

If the filename is qualified with a directory, then the output will be placed in that directory. If no directory information is given, the spool file will be placed in the *current directory*. The current directory is the directory from which you started AXIOM or is the directory you specified using the `)cd` command.

Also See: '`)cd`' in Section A.4 on page 574.

A.22)synonym

User Level Required: interpreter

Command Syntax:

`)synonym`
`)synonym` *synonym fullCommand*

```
)what synonyms
```

Command Description:

This command is used to create short synonyms for system command expressions. For example, the following synonyms might simplify commands you often use.

```
)synonym save         history )save
)synonym restore      history )restore
)synonym mail         system mail
)synonym ls           system ls
)synonym fortran      set output fortran
```

Once defined, synonyms can be used in place of the longer command expressions. Thus

```
)fortran on
```

is the same as the longer

```
)set fortran output on
```

To list all defined synonyms, issue either of

```
)synonyms
)what synonyms
```

To list, say, all synonyms that contain the substring "`ap`", issue

```
)what synonyms ap
```

Also See: '`)set`' in Section A.19 on page 586 and '`)what`' in Section A.26 on page 592.

A.23)system

User Level Required: interpreter

Command Syntax:

`)system` *cmdExpression*

Command Description:

This command may be used to issue commands to the operating system while remaining in AXIOM. The *cmdExpression* is passed to the operating system for execution.

To get an operating system shell, issue, for example, `)system sh`. When you enter the key combination, **Ctrl**-**D** (pressing and holding the **Ctrl** key and then pressing the **D** key) the shell will terminate and you will return to AXIOM. We do not recommend this way of creating a shell because Common LISP may field some interrupts instead of the shell. If possible, use a shell running in another window.

If you execute programs that misbehave you may not be able to return to AXIOM. If this happens, you may have no other choice than to restart AXIOM and restore the environment via `)history )restore`, if possible.

Also See: '`)boot`' in Section A.3 on page 573, '`)fin`' in Section A.9 on page 578, '`)lisp`' in Section A.13 on page 582, '`)pquit`' in Section A.16 on page 584, and '`)quit`' in Section A.17 on page 585.

A.24)trace

User Level Required: interpreter

Command Syntax:

```
)trace
)trace )off
)trace function [options]
)trace constructor [options]
)trace domainOrPackage [options]
```

where options can be one or more of

```
)after S-expression
)before S-expression
)break after
)break before
)cond S-expression
)count
)count n
)depth n
)local op1 [... opN]
)nonquietly
)nt
)off
)only listOfDataToDisplay
)ops
)ops op1 [... opN ]
)restore
)stats
)stats reset
)timer
)varbreak
)varbreak var1 [... varN ]
)vars
)vars var1 [... varN ]
)within executingFunction
```

Command Description:

This command is used to trace the execution of functions that make up the AXIOM system, functions defined by users, and functions from the system library. Almost all options are available for each type of function but exceptions will be noted below.

To list all functions, constructors, domains and packages that are traced, simply issue

```
)trace
```

To untrace everything that is traced, issue

```
)trace )off
```

When a function is traced, the default system action is to display the arguments to the function and the return value when the function is exited. Note that if a function is left via an action such as a `THROW`, no return value will be displayed. Also, optimization of tail recursion may decrease the number of times a function is actually invoked and so may cause less trace information to be displayed. Other information can be displayed or collected when a function is traced and this is controlled by the various options. Most options will be of interest only to AXIOM system developers. If a domain or package is traced, the default action is to trace all functions exported.

Individual interpreter, lisp or boot functions can be traced by listing their names after `)trace`. Any options that are present must follow the functions to be traced.

```
)trace f
```

traces the function `f`. To untrace `f`, issue

```
)trace f )off
```

Note that if a function name contains a special character, it will be necessary to escape the character with an underscore

```
)trace \_/D\_.1
```

To trace all domains or packages that are or will be created from a particular constructor, give the constructor name or abbreviation after `)trace`.

```
)trace MATRIX
)trace List Integer
```

The first command traces all domains currently instantiated with Matrix. If additional domains are instantiated with this constructor (for example, if you have used Matrix(Integer) and Matrix(Float)), they will be automatically traced. The second command traces List(Integer). It is possible to trace individual functions in a domain or package. See the `)ops` option below.

The following are the general options for the `)trace` command.

`)break after` causes a Common LISP break loop to be entered after exiting the traced function.

`)break before` causes a Common LISP break loop to be entered before entering the traced function.

`)break` is the same as `)break before`.

`)count` causes the system to keep a count of the number of times the traced function is entered. The total can be displayed with `)trace )stats` and cleared with `)trace )stats reset`.

`)count` n causes information about the traced function to be displayed for the first n executions. After the nth execution, the function is untraced.

`)depth` n causes trace information to be shown for only n levels of recursion of the traced function. The command

```
)trace fib )depth 10
```

will cause the display of only 10 levels of trace information for the recursive execution of a user function **fib**.

`)math` causes the function arguments and return value to be displayed in the AXIOM monospace two-dimensional math format.

`)nonquietly` causes the display of additional messages when a function is traced.

`)nt` This suppresses all normal trace information. This option is useful if the `)count` or `)timer` options are used and you are interested in the statistics but not the function calling information.

`)off` causes untracing of all or specific functions. Without an argument, all functions, constructors, domains and packages are untraced. Otherwise, the given functions and other objects are untraced. To immediately retrace the untraced functions, issue `)trace )restore`.

`)only` *listOfDataToDisplay* causes only specific trace information to be shown. The items are listed by using the following abbreviations:

a display all arguments

v display return value

1 display first argument

2 display second argument

15 display the 15th argument, and so on

`)restore` causes the last untraced functions to be retraced. If additional options are present, they are added to those previously in effect.

`)stats` causes the display of statistics collected by the use of the `)count` and `)timer` options.

`)stats reset` resets to 0 the statistics collected by the use of the `)count` and `)timer` options.

`)timer` causes the system to keep a count of execution times for the traced function. The total can be displayed with `)trace )stats` and cleared with `)trace )stats reset`.

`)varbreak` *var1* [... *varN*] causes a Common LISP break loop to be entered after the assignment to any of the listed variables in the traced function.

`)vars` *var1* [... *varN*] causes the display of the value of any of the specified variables after they are assigned in the traced function.

`)within` *executingFunction* causes the display of trace information only if the traced function is called when the given *executingFunction* is running.

The following are the options for tracing constructors, domains and packages.

`)local` [*op1* [... *opN*]] causes local functions of the constructor to be traced. Note that to untrace an individual local function, you must use the fully qualified internal name, using the escape character "_" before the semicolon.

```
)trace FRAC )local
)trace FRAC\_;cancelGcd )off
```

`)ops` *op1* [... *opN*] By default, all operations from a domain or package are traced when the domain or package is traced. This option allows you to specify that only particular operations should be traced. The command

```
)trace Integer )ops min max \_+ \_-
```

traces four operations from the domain Integer. Since + and - are special characters, it is necessary to escape them with an underscore.

Also See: '`)boot`' in Section A.3 on page 573, '`)lisp`' in Section A.13 on page 582, and '`)ltrace`' in Section A.15 on page 584.

A.25)undo

User Level Required: interpreter

Command Syntax:

```
)undo
)undo integer
)undo integer [option]
)undo )redo
```

where *option* is one of

```
)after
)before
```

Command Description:

This command is used to restore the state of the user environment to an earlier point in the interactive session. The argument of an `)undo` is an integer which must designate some step number in the interactive session.

```
)undo n
)undo n )after
```

These commands return the state of the interactive environment to that immediately after step `n`. If `n` is a positive number, then `n` refers to step nummber `n`. If `n` is a negative number, it refers to the `n` th previous command (that is, undoes the effects of the last $-n$ commands).

A `)clear all` resets the `)undo` facility. Otherwise, an `)undo` undoes the effect of `)clear` with options `properties`, `value`, and `mode`, and that of a previous `undo`. If any such system commands are given between steps n and $n+1$ ($n > 0$), their effect is undone for `)undo m` for any $0 < m \leq n$.

The command `)undo` is equivalent to `)undo -1` (it undoes the effect of the previous user expression). The command `)undo 0` undoes any of the above system commands issued since the last user expression.

```
)undo n )before
```

This command returns the state of the interactive environment to that immediately before step `n`. Any `)undo` or `)clear` system commands given before step `n` will not be undone.

```
)undo )redo
```

This command reads the file `redo.input`. created by the last `)undo` command. This file consists of all user input lines, excluding those backtracked over due to a previous `)undo`.

Also See: '`)history`' in Section A.12 on page 580. The command `)history )write` will eliminate the "undone" command lines of your program.

A.26)what

User Level Required: interpreter

Command Syntax:

)what categories *pattern1* [*pattern2* ...]
)what commands *pattern1* [*pattern2* ...]
)what domains *pattern1* [*pattern2* ...]

```
)what operations pattern1 [pattern2 ...]
)what packages  pattern1 [pattern2 ...]
)what synonym  pattern1 [pattern2 ...]
)what things  pattern1 [pattern2 ...]
)apropos  pattern1 [pattern2 ...]
```

Command Description:

This command is used to display lists of things in the system. The patterns are all strings and, if present, restrict the contents of the lists. Only those items that contain one or more of the strings as substrings are displayed. For example,

```
)what synonym
```

displays all command synonyms,

```
)what synonym ver
```

displays all command synonyms containing the substring "`ver`",

```
)what synonym ver pr
```

displays all command synonyms containing the substring "`ver`" or the substring "`pr`". Output similar to the following will be displayed

```
---------------- System Command Synonyms -----------------

user-defined synonyms satisfying patterns:
      ver pr

  )apr ........................... )what things
  )apropos ....................... )what things
  )prompt ........................ )set message prompt
  )version ....................... )lisp *yearweek*
```

Several other things can be listed with the `)what` command:

`categories` displays a list of category constructors.

`commands` displays a list of system commands available at your user-level. Your user-level is set via the `)set userlevel` command. To get a description of a particular command, such as "`)what`", issue `)help what`.

`domains` displays a list of domain constructors.

`operations` displays a list of operations in the system library. It is recommended that you qualify this command with one or more patterns, as there are thousands of operations available. For example, say you are looking for functions that involve computation of eigenvalues. To find their names, try `)what operations eig`. A rather large list of operations is loaded into the workspace when this command is first issued. This list will be deleted when you clear the workspace via `)clear all` or `)clear completely`. It will be re-created if it is needed again.

`packages` displays a list of package constructors.

`synonym` lists system command synonyms.

`things` displays all of the above types for items containing the pattern strings as substrings. The command synonym `)apropos` is equivalent to `)what things`.

Also See: '`)display`' in Section A.7 on page 577, '`)set`' in Section A.19 on page 586, and '`)show`' in Section A.20 on page 586.

APPENDIX B

Categories

This is a listing of all categories in the AXIOM library at the time this book was produced. Use the Browse facility (described in Chapter 14) to get more information about these constructors.

This sample entry will help you read the following table:

CategoryName{CategoryAbbreviation}: $\text{Category}_1 \ldots \text{Category}_N$ *with* $\text{operation}_1 \ldots \text{operation}_M$

where

CategoryName	is the full category name, for example, CommutativeRing.
CategoryAbbreviation	is the category abbreviation, for example, COMRING.
Category_i	is a category to which the category belongs.
operation_j	is an operation explicitly exported by the category.

AbelianGroup{ABELGRP}: CancellationAbelianMonoid *with* * -

AbelianMonoidRing{AMR}: Algebra BiModule CharacteristicNonZero CharacteristicZero CommutativeRing IntegralDomain Ring *with* / coefficient degree leadingCoefficient leadingMonomial map monomial monomial? reductum

AbelianMonoid{ABELMON}: AbelianSemiGroup *with* * Zero zero?

AbelianSemiGroup{ABELSG}: SetCategory *with* * +

Aggregate{AGG}: Object *with* # copy empty empty? eq? less? more? size?

AlgebraicallyClosedField{ACF}: Field RadicalCategory *with* rootOf rootsOf zeroOf zerosOf

AlgebraicallyClosedFunctionSpace{ACFS}: AlgebraicallyClosedField FunctionSpace *with* rootOf rootsOf zeroOf zerosOf

Algebra{ALGEBRA}: Module Ring *with* coerce

ArcHyperbolicFunctionCategory{AHYP}: *with* acosh acoth acsch asech asinh atanh

ArcTrigonometricFunctionCategory{ATRIG}: *with* acos acot acsc asec asin atan

AssociationListAggregate{ALAGG}: ListAggregate TableAggregate *with* assoc

AttributeRegistry{ATTREG}: *with*

BagAggregate{BGAGG}: HomogeneousAggregate *with* bag extract! insert! inspect

BiModule{BMODULE}: LeftModule RightModule *with*

BinaryRecursiveAggregate{BRAGG}: RecursiveAggregate *with* elt left right setelt setleft! setright!

BinaryTreeCategory{BTCAT}: BinaryRecursiveAggregate *with* node

BitAggregate{BTAGG}: OneDimensionalArrayAggregate OrderedSet *with* ^ and nand nor not or xor

CachableSet{CACHSET}: OrderedSet *with* position

setPosition

CancellationAbelianMonoid{CABMON}: AbelianMonoid *with* -

CharacteristicNonZero{CHARNZ}: Ring *with* charthRoot

CharacteristicZero{CHARZ}: Ring *with*

CoercibleTo{KOERCE}: *with* coerce

Collection{CLAGG}: ConvertibleTo HomogeneousAggregate *with* construct find reduce remove removeDuplicates select

CombinatorialFunctionCategory{CFCAT}: *with* binomial factorial permutation

CombinatorialOpsCategory{COMBOPC}: CombinatorialFunctionCategory *with* factorials product summation

CommutativeRing{COMRING}: BiModule Ring *with*

ComplexCategory{COMPCAT}: CharacteristicNonZero CharacteristicZero CommutativeRing ConvertibleTo DifferentialExtension EuclideanDomain Field FullyEvalableOver FullyLinearlyExplicitRingOver FullyRetractableTo IntegralDomain MonogenicAlgebra OrderedSet PolynomialFactorizationExplicit RadicalCategory TranscendentalFunctionCategory *with* abs argument complex conjugate exquo imag imaginary norm polarCoordinates rational rational? rationalIfCan real

ConvertibleTo{KONVERT}: *with* convert

DequeueAggregate{DQAGG}: QueueAggregate StackAggregate *with* bottom! dequeue extractBottom! extractTop! height insertBottom! insertTop! reverse! top!

DictionaryOperations{DIOPS}: BagAggregate Collection *with* dictionary remove! select!

Dictionary{DIAGG}: DictionaryOperations *with*

DifferentialExtension{DIFEXT}: DifferentialRing PartialDifferentialRing Ring *with* D differentiate

DifferentialPolynomialCategory{DPOLCAT}: DifferentialExtension Evalable InnerEvalable PolynomialCategory RetractableTo *with* degree differentialVariables initial isobaric? leader makeVariable order separant weight weights

DifferentialRing{DIFRING}: Ring *with* D differentiate

DifferentialVariableCategory{DVARCAT}: OrderedSet RetractableTo *with* D coerce differentiate makeVariable order variable weight

DirectProductCategory{DIRPCAT}: AbelianSemiGroup Algebra BiModule CancellationAbelianMonoid CoercibleTo CommutativeRing DifferentialExtension Finite FullyLinearlyExplicitRingOver FullyRetractableTo IndexedAggregate OrderedAbelianMonoidSup OrderedRing VectorSpace *with* * directProduct dot unitVector

DivisionRing{DIVRING}: Algebra EntireRing *with* ** inv

DoublyLinkedAggregate{DLAGG}: RecursiveAggregate *with* concat! head last next previous setnext! setprevious! tail

ElementaryFunctionCategory{ELEMFUN}: *with* ** exp log

EltableAggregate{ELTAGG}: Eltable *with* elt qelt qsetelt! setelt

Eltable{ELTAB}: *with* elt

EntireRing{ENTIRER}: BiModule Ring *with*

EuclideanDomain{EUCDOM}: PrincipalIdealDomain *with* divide euclideanSize extendedEuclidean multiEuclidean quo rem sizeLess?

Evalable{EVALAB}: *with* eval

ExpressionSpace{ES}: Evalable InnerEvalable OrderedSet RetractableTo *with* belong? box definingPolynomial distribute elt eval freeOf? height is? kernel kernels mainKernel map minPoly operator operators paren subst tower

ExtensibleLinearAggregate{ELAGG}: LinearAggregate *with* concat! delete! insert! merge! remove! removeDuplicates! select!

ExtensionField{XF}: CharacteristicZero Field FieldOfPrimeCharacteristic RetractableTo VectorSpace *with* Frobenius algebraic? degree extensionDegree inGroundField? transcendenceDegree transcendent?

FieldOfPrimeCharacteristic{FPC}: CharacteristicNonZero Field *with* discreteLog order primeFrobenius

Field{FIELD}: DivisionRing EuclideanDomain UniqueFactorizationDomain *with* /

FileCategory{FILECAT}: SetCategory *with* close! iomode name open read! reopen! write!

FileNameCategory{FNCAT}: SetCategory *with* coerce directory exists? extension filename name new readable? writable?

FiniteAbelianMonoidRing{FAMR}: AbelianMonoidRing FullyRetractableTo *with* coefficients content exquo ground ground? mapExponents minimumDegree numberOfMonomials primitivePart

FiniteAlgebraicExtensionField{FAXF}: ExtensionField FiniteFieldCategory RetractableTo *with* basis coordinates createNormalElement definingPolynomial degree extensionDegree generator minimalPolynomial norm normal? normalElement represents trace

FiniteFieldCategory{FFIELDC}: FieldOfPrimeCharacteristic Finite StepThrough *with* charthRoot conditionP createPrimitiveElement discreteLog factorsOfCyclicGroupSize order primitive? primitiveElement representationType tableForDiscreteLogarithm

FiniteLinearAggregate{FLAGG}: LinearAggregate OrderedSet *with* copyInto! merge position reverse reverse!

sort sort! sorted?

FiniteRankAlgebra{FINRALG}: Algebra CharacteristicNonZero CharacteristicZero *with* characteristicPolynomial coordinates discriminant minimalPolynomial norm rank regularRepresentation represents trace traceMatrix

FiniteRankNonAssociativeAlgebra{FINAALG}: NonAssociativeAlgebra *with* JacobiIdentity? JordanAlgebra? alternative? antiAssociative? antiCommutative? associative? associatorDependence commutative? conditionsForIdempotents coordinates flexible? jordanAdmissible? leftAlternative? leftCharacteristicPolynomial leftDiscriminant leftMinimalPolynomial leftNorm leftRecip leftRegularRepresentation leftTrace leftTraceMatrix leftUnit leftUnits lieAdmissible? lieAlgebra? noncommutativeJordanAlgebra? powerAssociative? rank recip represents rightAlternative? rightCharacteristicPolynomial rightDiscriminant rightMinimalPolynomial rightNorm rightRecip rightRegularRepresentation rightTrace rightTraceMatrix rightUnit rightUnits someBasis structuralConstants unit

FiniteSetAggregate{FSAGG}: Dictionary Finite SetAggregate *with* cardinality complement max min universe

Finite{FINITE}: SetCategory *with* index lookup random size

FloatingPointSystem{FPS}: RealNumberSystem *with* base bits decreasePrecision digits exponent float increasePrecision mantissa max order precision

FramedAlgebra{FRAMALG}: FiniteRankAlgebra *with* basis convert coordinates discriminant regularRepresentation represents traceMatrix

FramedNonAssociativeAlgebra{FRNAALG}: FiniteRankNonAssociativeAlgebra *with* apply basis conditionsForIdempotents convert coordinates elt leftDiscriminant leftRankPolynomial leftRegularRepresentation leftTraceMatrix represents rightDiscriminant rightRankPolynomial rightRegularRepresentation rightTraceMatrix structuralConstants

FreeAbelianMonoidCategory{FAMONC}: CancellationAbelianMonoid RetractableTo *with* * + coefficient highCommonTerms mapCoef mapGen nthCoef nthFactor size terms

FullyEvalableOver{FEVALAB}: Eltable Evalable InnerEvalable *with* map

FullyLinearlyExplicitRingOver{FLINEXP}: LinearlyExplicitRingOver *with*

FullyPatternMatchable{FPATMAB}: Object PatternMatchable *with*

FullyRetractableTo{FRETRCT}: RetractableTo *with*

FunctionFieldCategory{FFCAT}: MonogenicAlgebra *with* D absolutelyIrreducible? branchPoint? branchPointAtInfinity? complementaryBasis differentiate elt genus integral? integralAtInfinity? integralBasis integralBasisAtInfinity integralCoordinates integralDerivationMatrix integralMatrix integralMatrixAtInfinity integralRepresents inverseIntegralMatrix inverseIntegralMatrixAtInfinity nonSingularModel normalizeAtInfinity numberOfComponents primitivePart ramified? ramifiedAtInfinity? rationalPoint? rationalPoints reduceBasisAtInfinity represents singular? singularAtInfinity? yCoordinates

FunctionSpace{FS}: AbelianGroup AbelianMonoid Algebra CharacteristicNonZero CharacteristicZero ConvertibleTo ExpressionSpace Field FullyLinearlyExplicitRingOver FullyPatternMatchable FullyRetractableTo Group Monoid PartialDifferentialRing Patternable RetractableTo Ring *with* ** / applyQuote coerce convert denom denominator eval ground ground? isExpt isMult isPlus isPower isTimes numer numerator univariate variables

GcdDomain{GCDDOM}: IntegralDomain *with* gcd lcm

GradedAlgebra{GRALG}: GradedModule *with* One product

GradedModule{GRMOD}: RetractableTo SetCategory *with* * + - Zero degree

Group{GROUP}: Monoid *with* ** / commutator conjugate inv

HomogeneousAggregate{HOAGG}: Aggregate SetCategory *with* any? count every? map map! member? members parts

HyperbolicFunctionCategory{HYPCAT}: *with* cosh coth csch sech sinh tanh

IndexedAggregate{IXAGG}: EltableAggregate HomogeneousAggregate *with* entries entry? fill! first index? indices maxIndex minIndex swap!

IndexedDirectProductCategory{IDPC}: SetCategory *with* leadingCoefficient leadingSupport map monomial reductum

InnerEvalable{IEVALAB}: *with* eval

IntegerNumberSystem{INS}: CharacteristicZero CombinatorialFunctionCategory ConvertibleTo DifferentialRing EuclideanDomain LinearlyExplicitRingOver OrderedRing PatternMatchable RealConstant RetractableTo StepThrough UniqueFactorizationDomain *with* addmod base bit? copy dec even? hash inc invmod length mask mulmod odd? positiveRemainder powmod random rational rational? rationalIfCan shift submod symmetricRemainder

IntegralDomain{INTDOM}: Algebra CommutativeRing EntireRing *with* associates? exquo unit? unitCanonical unitNormal

KeyedDictionary{KDAGG}: Dictionary *with* key? keys remove! search

LazyStreamAggregate{LZSTAGG}: StreamAggregate *with* complete explicitEntries? explicitlyEmpty? extend frst lazy?

lazyEvaluate numberOfComputedEntries remove rst select

LeftAlgebra{LALG}: LeftModule Ring *with* coerce

LeftModule{LMODULE}: AbelianGroup *with* *

LinearAggregate{LNAGG}: Collection IndexedAggregate *with* concat delete elt insert map new setelt

LinearlyExplicitRingOver{LINEXP}: Ring *with* reducedSystem

LiouvillianFunctionCategory{LFCAT}: PrimitiveFunctionCategory TranscendentalFunctionCategory *with* Ci Ei Si dilog erf li

ListAggregate{LSAGG}: ExtensibleLinearAggregate FiniteLinearAggregate StreamAggregate *with* list

MatrixCategory{MATCAT}: TwoDimensionalArrayCategory *with* * ** + - / antisymmetric? coerce determinant diagonal? diagonalMatrix elt exquo horizConcat inverse listOfLists matrix minordet nullSpace nullity rank rowEchelon scalarMatrix setelt setsubMatrix! square? squareTop subMatrix swapColumns! swapRows! symmetric? transpose vertConcat zero

Module{MODULE}: BiModule *with*

MonadWithUnit{MONADWU}: Monad *with* ** One leftPower leftRecip one? recip rightPower rightRecip

Monad{MONAD}: SetCategory *with* * ** leftPower rightPower

MonogenicAlgebra{MONOGEN}: CommutativeRing ConvertibleTo DifferentialExtension Field Finite FiniteFieldCategory FramedAlgebra FullyLinearlyExplicitRingOver FullyRetractableTo *with* convert definingPolynomial derivationCoordinates generator lift reduce

MonogenicLinearOperator{MLO}: Algebra BiModule Ring *with* coefficient degree leadingCoefficient minimumDegree monomial reductum

Monoid{MONOID}: SemiGroup *with* ** One one? recip

MultiDictionary{MDAGG}: DictionaryOperations *with* duplicates insert! removeDuplicates!

MultiSetAggregate{MSAGG}: MultiDictionary SetAggregate *with*

MultivariateTaylorSeriesCategory{MTSCAT}: Evalable InnerEvalable PartialDifferentialRing PowerSeriesCategory RadicalCategory TranscendentalFunctionCategory *with* coefficient extend integrate monomial order polynomial

NonAssociativeAlgebra{NAALG}: Module NonAssociativeRng *with* plenaryPower

NonAssociativeRing{NASRING}: MonadWithUnit NonAssociativeRng *with* characteristic coerce

NonAssociativeRng{NARNG}: AbelianGroup Monad *with* antiCommutator associator commutator

Object{OBJECT}: *with*

OctonionCategory{OC}: Algebra CharacteristicNonZero CharacteristicZero ConvertibleTo Finite FullyEvalableOver FullyRetractableTo OrderedSet *with* abs conjugate imagE imagI imagJ imagK imagi imagj imagk inv norm octon rational rational? rationalIfCan real

OneDimensionalArrayAggregate{A1AGG}: FiniteLinearAggregate *with*

OrderedAbelianGroup{OAGROUP}: AbelianGroup OrderedCancellationAbelianMonoid *with*

OrderedAbelianMonoidSup{OAMONS}: OrderedCancellationAbelianMonoid *with* sup

OrderedAbelianMonoid{OAMON}: AbelianMonoid OrderedAbelianSemiGroup *with*

OrderedAbelianSemiGroup{OASGP}: AbelianMonoid OrderedSet *with*

OrderedCancellationAbelianMonoid{OCAMON}: CancellationAbelianMonoid OrderedAbelianMonoid *with*

OrderedFinite{ORDFIN}: Finite OrderedSet *with*

OrderedMonoid{ORDMON}: Monoid OrderedSet *with*

OrderedMultiSetAggregate{OMAGG}: MultiSetAggregate PriorityQueueAggregate *with* min

OrderedRing{ORDRING}: OrderedAbelianGroup OrderedMonoid Ring *with* abs negative? positive? sign

OrderedSet{ORDSET}: SetCategory *with* < max min

PAdicIntegerCategory{PADICCT}: CharacteristicZero EuclideanDomain *with* approximate complete digits extend moduloP modulus order quotientByP sqrt

PartialDifferentialRing{PDRING}: Ring *with* D differentiate

PartialTranscendentalFunctions{PTRANFN}: *with* acosIfCan acoshIfCan acotIfCan acothIfCan acscIfCan acschIfCan asecIfCan asechIfCan asinIfCan asinhIfCan atanIfCan atanhIfCan cosIfCan coshIfCan cotIfCan cothIfCan cscIfCan cschIfCan expIfCan logIfCan nthRootIfCan secIfCan sechIfCan sinIfCan sinhIfCan tanIfCan tanhIfCan

Patternable{PATAB}: ConvertibleTo Object *with*

PatternMatchable{PATMAB}: SetCategory *with* patternMatch

PermutationCategory{PERMCAT}: Group OrderedSet *with* < cycle cycles elt eval orbit

PlottablePlaneCurveCategory{PPCURVE}: CoercibleTo *with* listBranches xRange yRange

PlottableSpaceCurveCategory{PSCURVE}: CoercibleTo *with* listBranches xRange yRange zRange

PointCategory{PTCAT}: VectorCategory *with* convert cross dimension extend length point

PolynomialCategory{POLYCAT}: ConvertibleTo Evalable FiniteAbelianMonoidRing FullyLinearlyExplicitRingOver GcdDomain InnerEvalable OrderedSet PartialDifferentialRing PatternMatchable PolynomialFactorizationExplicit RetractableTo *with* coefficient content degree discriminant isExpt isPlus isTimes mainVariable minimumDegree monicDivide monomial monomials multivariate primitiveMonomials primitivePart resultant squareFree squareFreePart totalDegree univariate variables

PolynomialFactorizationExplicit{PFECAT}: UniqueFactorizationDomain *with* charthRoot conditionP factorPolynomial factorSquareFreePolynomial gcdPolynomial solveLinearPolynomialEquation squareFreePolynomial

PowerSeriesCategory{PSCAT}: AbelianMonoidRing *with* complete monomial pole? variables

PrimitiveFunctionCategory{PRIMCAT}: *with* integral

PrincipalIdealDomain{PID}: GcdDomain *with* expressIdealMember principalIdeal

PriorityQueueAggregate{PRQAGG}: BagAggregate *with* max merge merge!

QuaternionCategory{QUATCAT}: Algebra CharacteristicNonZero CharacteristicZero ConvertibleTo DifferentialExtension DivisionRing EntireRing FullyEvalableOver FullyLinearlyExplicitRingOver FullyRetractableTo OrderedSet *with* abs conjugate imagI imagJ imagK norm quatern rational rational? rationalIfCan real

QueueAggregate{QUAGG}: BagAggregate *with* back dequeue! enqueue! front length rotate!

QuotientFieldCategory{QFCAT}: Algebra CharacteristicNonZero CharacteristicZero ConvertibleTo DifferentialExtension Field FullyEvalableOver FullyLinearlyExplicitRingOver FullyPatternMatchable OrderedRing OrderedSet Patternable PolynomialFactorizationExplicit RealConstant RetractableTo StepThrough *with* / ceiling denom denominator floor fractionPart numer numerator random wholePart

RadicalCategory{RADCAT}: *with* ** nthRoot sqrt

RealConstant{REAL}: ConvertibleTo *with*

RealNumberSystem{RNS}: CharacteristicZero ConvertibleTo Field OrderedRing PatternMatchable RadicalCategory RealConstant RetractableTo *with* abs ceiling floor fractionPart norm round truncate wholePart

RectangularMatrixCategory{RMATCAT}: BiModule HomogeneousAggregate Module *with* / antisymmetric? column diagonal? elt exquo listOfLists map matrix maxColIndex maxRowIndex minColIndex minRowIndex ncols nrows nullSpace nullity qelt rank row rowEchelon square? symmetric?

RecursiveAggregate{RCAGG}: HomogeneousAggregate *with* children cyclic? elt leaf? leaves node? nodes setchildren! setelt setvalue! value

RetractableTo{RETRACT}: *with* coerce retract retractIfCan

RightModule{RMODULE}: AbelianGroup *with* *

Ring{RING}: LeftModule Monoid Rng *with* characteristic coerce

Rng{RNG}: AbelianGroup SemiGroup *with*

SegmentCategory{SEGCAT}: SetCategory *with* BY SEGMENT convert hi high incr lo low segment

SegmentExpansionCategory{SEGXCAT}: SegmentCategory *with* expand map

SemiGroup{SGROUP}: SetCategory *with* * **

SetAggregate{SETAGG}: Collection SetCategory *with* < brace difference intersect subset? symmetricDifference union

SetCategory{SETCAT}: CoercibleTo Object *with* =

SExpressionCategory{SEXCAT}: SetCategory *with* # atom? car cdr convert destruct elt eq expr float float? integer integer? list? null? pair? string string? symbol symbol? uequal

SpecialFunctionCategory{SPFCAT}: *with* Beta Gamma abs airyAi airyBi besselI besselJ besselK besselY digamma polygamma

SquareMatrixCategory{SMATCAT}: Algebra BiModule DifferentialExtension FullyLinearlyExplicitRingOver FullyRetractableTo Module RectangularMatrixCategory *with* * ** determinant diagonal diagonalMatrix diagonalProduct inverse minordet scalarMatrix trace

StackAggregate{SKAGG}: BagAggregate *with* depth pop! push! top

StepThrough{STEP}: SetCategory *with* init nextItem

StreamAggregate{STAGG}: LinearAggregate UnaryRecursiveAggregate *with* explicitlyFinite? possiblyInfinite?

StringAggregate{SRAGG}: OneDimensionalArrayAggregate *with* coerce elt leftTrim lowerCase lowerCase! match match? position prefix? replace rightTrim split substring? suffix? trim upperCase upperCase!

StringCategory{STRICAT}: StringAggregate *with* string

TableAggregate{TBAGG}: IndexedAggregate KeyedDictionary *with* map setelt table

ThreeSpaceCategory{SPACEC}: SetCategory *with* check closedCurve closedCurve? coerce components composite composites copy create3Space curve curve? enterPointData lllip lllp llprop lp lprop merge mesh mesh? modifyPointData numberOfComponents numberOfComposites objects point point? polygon polygon? subspace

TranscendentalFunctionCategory{TRANFUN}: ArcHyperbolicFunctionCategory

ArcTrigonometricFunctionCategory ElementaryFunctionCategory HyperbolicFunctionCategory TrigonometricFunctionCategory *with* pi

TrigonometricFunctionCategory{TRIGCAT}: *with* cos cot csc sec sin tan

TwoDimensionalArrayCategory{ARR2CAT}: HomogeneousAggregate *with* column elt fill! map map! maxColIndex maxRowIndex minColIndex minRowIndex ncols new nrows parts qelt qsetelt! row setColumn! setRow! setelt

UnaryRecursiveAggregate{URAGG}: RecursiveAggregate *with* concat concat! cycleEntry cycleLength cycleSplit! cycleTail elt first last rest second setelt setfirst! setlast! setrest! split! tail third

UniqueFactorizationDomain{UFD}: GcdDomain *with* factor prime? squareFree squareFreePart

UnivariateLaurentSeriesCategory{ULSCAT}: Field RadicalCategory TranscendentalFunctionCategory UnivariatePowerSeriesCategory *with* integrate multiplyCoefficients rationalFunction

UnivariateLaurentSeriesConstructorCategory{ULSCCAT}: QuotientFieldCategory RetractableTo UnivariateLaurentSeriesCategory *with* coerce degree laurent removeZeroes taylor taylorIfCan taylorRep

UnivariatePolynomialCategory{UPOLYC}: DifferentialExtension DifferentialRing Eltable EuclideanDomain PolynomialCategory StepThrough *with* D composite differentiate discriminant divideExponents elt integrate makeSUP monicDivide multiplyExponents order pseudoDivide pseudoQuotient pseudoRemainder resultant separate subResultantGcd unmakeSUP vectorise

UnivariatePowerSeriesCategory{UPSCAT}: DifferentialRing Eltable PowerSeriesCategory *with* approximate center elt eval extend multiplyExponents order series terms truncate variable

UnivariatePuiseuxSeriesCategory{UPXSCAT}: Field RadicalCategory TranscendentalFunctionCategory UnivariatePowerSeriesCategory *with* integrate multiplyExponents

UnivariatePuiseuxSeriesConstructorCategory{UPXSCCA}: RetractableTo UnivariatePuiseuxSeriesCategory *with* coerce degree laurent laurentIfCan laurentRep puiseux rationalPower

UnivariateTaylorSeriesCategory{UTSCAT}: RadicalCategory TranscendentalFunctionCategory UnivariatePowerSeriesCategory *with* ** coefficients integrate multiplyCoefficients polynomial quoByVar series

VectorCategory{VECTCAT}: OneDimensionalArrayAggregate *with* * + - dot zero

VectorSpace{VSPACE}: Module *with* / dimension

Domains

This is a listing of all domains in the AXIOM library at the time this book was produced. Use the Browse facility (described in Chapter 14) to get more information about these constructors.

This sample entry will help you read the following table:

DomainName{DomainAbbreviation}: $\text{Category}_1 \ldots \text{Category}_N$ *with* $\text{operation}_1 \ldots \text{operation}_M$

where

DomainName	is the full domain name, for example, Integer.
DomainAbbreviation	is the domain abbreviation, for example, INT.
Category_i	is a category to which the domain belongs.
operation_j	is an operation exported by the domain.

AlgebraGivenByStructuralConstants{ALGSC}: FramedNonAssociativeAlgebra LeftModule *with* 0 * ** + - = JacobiIdentity? JordanAlgebra? alternative? antiAssociative? antiCommutative? antiCommutator apply associative? associator associatorDependence basis coerce commutative? commutator conditionsForIdempotents convert coordinates elt flexible? jordanAdmissible? leftAlternative? leftCharacteristicPolynomial leftDiscriminant leftMinimalPolynomial leftNorm leftPower leftRankPolynomial leftRecip leftRegularRepresentation leftTrace leftTraceMatrix leftUnit leftUnits lieAdmissible? lieAlgebra? noncommutativeJordanAlgebra? plenaryPower powerAssociative? rank recip represents rightAlternative? rightCharacteristicPolynomial rightDiscriminant rightMinimalPolynomial rightNorm rightPower rightRankPolynomial rightRecip rightRegularRepresentation rightTrace rightTraceMatrix rightUnit rightUnits someBasis structuralConstants unit zero?

AlgebraicFunctionField{ALGFF}: FunctionFieldCategory *with* 0 1 * ** + - / = D absolutelyIrreducible? associates? basis branchPoint? branchPointAtInfinity? characteristic characteristicPolynomial charthRoot coerce complementaryBasis convert coordinates definingPolynomial derivationCoordinates differentiate discriminant divide elt euclideanSize expressIdealMember exquo extendedEuclidean factor gcd generator genus integral? integralAtInfinity? integralBasis integralBasisAtInfinity integralCoordinates integralDerivationMatrix integralMatrix integralMatrixAtInfinity integralRepresents inv inverseIntegralMatrix inverseIntegralMatrixAtInfinity knownInfBasis lcm lift minimalPolynomial multiEuclidean nonSingularModel norm normalizeAtInfinity numberOfComponents one? prime? primitivePart principalIdeal quo ramified? ramifiedAtInfinity? rank rationalPoint? rationalPoints recip reduce reduceBasisAtInfinity reducedSystem regularRepresentation rem represents retract retractIfCan singular? singularAtInfinity? sizeLess? squareFree squareFreePart trace traceMatrix unit? unitCanonical unitNormal yCoordinates zero?

AlgebraicNumber{AN}: AlgebraicallyClosedField CharacteristicZero ConvertibleTo DifferentialRing

ExpressionSpace LinearlyExplicitRingOver RealConstant RetractableTo *with* 0 1 * ** + - / < = D associates? belong? box characteristic coerce convert definingPolynomial denom differentiate distribute divide elt euclideanSize eval expressIdealMember exquo extendedEuclidean factor freeOf? gcd height inv is? kernel kernels lcm mainKernel map max min minPoly multiEuclidean nthRoot numer one? operator operators paren prime? principalIdeal quo recip reduce reducedSystem rem retract retractIfCan rootOf rootsOf sizeLess? sqrt squareFree squareFreePart subst tower unit? unitCanonical unitNormal zero? zeroOf zerosOf

AnonymousFunction{ANON}: SetCategory *with* = coerce

AntiSymm{ANTISYM}: LeftAlgebra RetractableTo *with* 0 1 * ** + - = characteristic coefficient coerce degree exp generator homogeneous? leadingBasisTerm leadingCoefficient map one? recip reductum retract retractIfCan retractable? zero?

Any{ANY}: SetCategory *with* = any coerce domain domainOf obj objectOf showTypeInOutput

ArrayStack{ASTACK}: StackAggregate *with* # = any? arrayStack bag coerce copy count depth empty empty? eq? every? extract! insert! inspect less? map map! member? members more? parts pop! push! size? top

AssociatedJordanAlgebra{JORDAN}: CoercibleTo FiniteRankNonAssociativeAlgebra FramedNonAssociativeAlgebra NonAssociativeAlgebra *with* 0 * ** + - = JacobiIdentity? JordanAlgebra? alternative? antiAssociative? antiCommutative? antiCommutator apply associative? associator associatorDependence basis coerce commutative? commutator conditionsForIdempotents convert coordinates elt flexible? jordanAdmissible? leftAlternative? leftCharacteristicPolynomial leftDiscriminant leftMinimalPolynomial leftNorm leftPower leftRankPolynomial leftRecip leftRegularRepresentation leftTrace leftTraceMatrix leftUnit leftUnits lieAdmissible? lieAlgebra? noncommutativeJordanAlgebra? plenaryPower powerAssociative? rank recip represents rightAlternative? rightCharacteristicPolynomial rightDiscriminant rightMinimalPolynomial rightNorm rightPower rightRankPolynomial rightRecip rightRegularRepresentation rightTrace rightTraceMatrix rightUnit rightUnits someBasis structuralConstants unit zero?

AssociatedLieAlgebra{LIE}: CoercibleTo FiniteRankNonAssociativeAlgebra FramedNonAssociativeAlgebra NonAssociativeAlgebra *with* 0 * ** + - = JacobiIdentity? JordanAlgebra? alternative? antiAssociative? antiCommutative? antiCommutator apply associative? associator associatorDependence basis coerce commutative? commutator conditionsForIdempotents convert coordinates elt flexible? jordanAdmissible? leftAlternative? leftCharacteristicPolynomial leftDiscriminant leftMinimalPolynomial leftNorm leftPower leftRankPolynomial leftRecip leftRegularRepresentation leftTrace leftTraceMatrix leftUnit leftUnits lieAdmissible? lieAlgebra? noncommutativeJordanAlgebra? plenaryPower powerAssociative? rank recip represents rightAlternative? rightCharacteristicPolynomial rightDiscriminant rightMinimalPolynomial rightNorm rightPower rightRankPolynomial rightRecip rightRegularRepresentation rightTrace rightTraceMatrix rightUnit rightUnits someBasis structuralConstants unit zero?

AssociationList{ALIST}: AssociationListAggregate *with* # = any? assoc bag child? children coerce concat concat! construct copy copyInto! count cycleEntry cycleLength cycleSplit! cycleTail cyclic? delete delete! dictionary distance elt empty empty? entries entry? eq? every? explicitlyFinite? extract! fill! find first index? indices insert insert! inspect key? keys last leaf? less? list map map! maxIndex member? members merge merge! minIndex more? new node? nodes parts position possiblyInfinite? qelt qsetelt! reduce remove remove! removeDuplicates removeDuplicates! rest reverse reverse! search second select select! setchildren! setelt setfirst! setlast! setrest! setvalue! size? sort sort! sorted? split! swap! table tail third value

BalancedBinaryTree{BBTREE}: BinaryTreeCategory *with* # = any? balancedBinaryTree children coerce copy count cyclic? elt empty empty? eq? every? leaf? leaves left less? map map! mapDown! mapUp! member? members more? node node? nodes parts right setchildren! setelt setleaves! setleft! setright! setvalue! size? value

BalancedPAdicInteger{BPADIC}: PAdicIntegerCategory *with* 0 1 * ** + - = approximate associates? characteristic coerce complete digits divide euclideanSize expressIdealMember exquo extend extendedEuclidean gcd lcm moduloP modulus multiEuclidean one? order principalIdeal quo quotientByP recip rem sizeLess? sqrt unit? unitCanonical unitNormal zero?

BalancedPAdicRational{BPADICRT}: QuotientFieldCategory *with* 0 1 * ** + - / = D approximate associates? characteristic coerce continuedFraction denom denominator differentiate divide euclideanSize expressIdealMember exquo extendedEuclidean factor fractionPart gcd inv lcm map multiEuclidean numer numerator one? prime? principalIdeal quo recip reducedSystem rem removeZeroes retract retractIfCan sizeLess? squareFree squareFreePart unit? unitCanonical unitNormal wholePart zero?

BasicOperator{BOP}: OrderedSet *with* < = arity assert coerce comparison copy deleteProperty! display equality has? input is? max min name nary? nullary? operator properties property setProperties setProperty unary? weight

BinaryExpansion{BINARY}: QuotientFieldCategory *with* 0 1 * ** + - / < = D abs associates? binary ceiling characteristic coerce convert denom denominator differentiate divide euclideanSize expressIdealMember exquo extendedEuclidean factor floor fractionPart gcd init inv lcm map max min multiEuclidean negative? nextItem numer numerator one? patternMatch positive? prime? principalIdeal quo random recip reducedSystem rem retract retractIfCan sign sizeLess? squareFree squareFreePart unit? unitCanonical unitNormal wholePart zero?

BinarySearchTree{BSTREE}: BinaryTreeCategory *with* # = any? binarySearchTree children coerce copy count cyclic? elt empty empty? eq? every? insert! insertRoot! leaf? leaves left less? map map! member? members more? node node? nodes parts right setchildren! setelt setleft! setright! setvalue! size? split value

BinaryTournament{BTOURN}: BinaryTreeCategory *with* # = any? binaryTournament children coerce copy count cyclic? elt empty empty? eq? every? insert! leaf? leaves left less? map map! member? members more? node node? nodes parts right setchildren! setelt setleft! setright! setvalue! size? value

BinaryTree{BTREE}: BinaryTreeCategory *with* # = any? binaryTree children coerce copy count cyclic? elt empty empty? eq? every? leaf? leaves left less? map map! member? members more? node node? nodes parts right setchildren! setelt setleft! setright! setvalue! size? value

Bits{BITS}: BitAggregate *with* # < = ^ and any? bits coerce concat construct convert copy copyInto! count delete elt empty empty? entries entry? eq? every? fill! find first index? indices insert less? map map! max maxIndex member? members merge min minIndex more? nand new nor not or parts position qelt qsetelt! reduce remove removeDuplicates reverse reverse! select setelt size? sort sort! sorted? swap! xor

Boolean{BOOLEAN}: ConvertibleTo Finite OrderedSet *with* < = ^ and coerce convert false implies index lookup max min nand nor not or random size true xor

CardinalNumber{CARD}: CancellationAbelianMonoid Monoid OrderedSet RetractableTo *with* 0 1 * ** + - < = Aleph coerce countable? finite? generalizedContinuumHypothesisAssumed generalizedContinuumHypothesisAssumed? max min one? recip retract retractIfCan zero?

CartesianTensor{CARTEN}: GradedAlgebra *with* 0 1 * + - = coerce contract degree elt kroneckerDelta leviCivitaSymbol product rank ravel reindex retract retractIfCan transpose unravel

CharacterClass{CCLASS}: ConvertibleTo FiniteSetAggregate SetCategory *with* # < = alphabetic alphanumeric any? bag brace cardinality charClass coerce complement construct convert copy count dictionary difference digit empty empty? eq? every? extract! find hexDigit index insert! inspect intersect less? lookup lowerCase map map! max member? members min more? parts random reduce remove remove! removeDuplicates select select! size size? subset? symmetricDifference union universe upperCase

Character{CHAR}: OrderedFinite *with* < = alphabetic? alphanumeric? char coerce digit? escape hexDigit? index lookup lowerCase lowerCase? max min ord quote random size space upperCase upperCase?

CliffordAlgebra{CLIF}: Algebra Ring VectorSpace *with* 0 1 * ** + - / = characteristic coefficient coerce dimension e monomial one? recip zero?

Color{COLOR}: AbelianSemiGroup *with* * + = blue coerce color green hue numberOfHues red yellow

Commutator{COMM}: SetCategory *with* = coerce mkcomm

Complex{COMPLEX}: ComplexCategory *with* 0 1 * ** + - / < = D abs acos acosh acot acoth acsc acsch argument asec asech asin asinh associates? atan atanh basis characteristic characteristicPolynomial charthRoot coerce complex conditionP conjugate convert coordinates cos cosh cot coth createPrimitiveElement csc csch definingPolynomial derivationCoordinates differentiate discreteLog discriminant divide elt euclideanSize eval exp expressIdealMember exquo extendedEuclidean factor factorPolynomial factorSquareFreePolynomial factorsOfCyclicGroupSize gcd gcdPolynomial generator imag imaginary index init inv lcm lift log lookup map max min minimalPolynomial multiEuclidean nextItem norm nthRoot one? order pi polarCoordinates prime? primeFrobenius primitive? primitiveElement principalIdeal quo random rank rational rational? rationalIfCan real recip reduce reducedSystem regularRepresentation rem representationType represents retract retractIfCan sec sech sin sinh size sizeLess? solveLinearPolynomialEquation sqrt squareFree squareFreePart squareFreePolynomial tableForDiscreteLogarithm tan tanh trace traceMatrix unit? unitCanonical unitNormal zero?

ContinuedFraction{CONTFRAC}: Algebra Field *with* 0 1 * ** + - / = approximants associates? characteristic coerce complete continuedFraction convergents denominators divide euclideanSize expressIdealMember exquo extend extendedEuclidean factor gcd inv lcm multiEuclidean numerators one? partialDenominators partialNumerators partialQuotients prime? principalIdeal quo recip reducedContinuedFraction reducedForm rem sizeLess? squareFree squareFreePart unit? unitCanonical unitNormal wholePart zero?

Database{DBASE}: SetCategory *with* + - = coerce display elt fullDisplay

DataList{DLIST}: ListAggregate *with* # < = any? children coerce concat concat! construct convert copy copyInto! count cycleEntry cycleLength cycleSplit! cycleTail cyclic? datalist delete delete! elt empty empty? entries entry? eq? every? explicitlyFinite? fill! find first index? indices insert insert! last leaf? leaves less? list map map! max maxIndex member? members merge merge! min minIndex more? new node? nodes parts position possiblyInfinite? qelt qsetelt! reduce remove remove! removeDuplicates removeDuplicates! rest reverse reverse! second select select! setchildren! setelt setfirst! setlast! setrest! setvalue! size? sort sort! sorted? split! swap! tail third value

DecimalExpansion{DECIMAL}: QuotientFieldCategory *with*

0 1 * ** + - / < = D abs associates? ceiling characteristic coerce convert decimal denom denominator differentiate divide euclideanSize expressIdealMember exquo extendedEuclidean factor floor fractionPart gcd init inv lcm map max min multiEuclidean negative? nextItem numer numerator one? patternMatch positive? prime? principalIdeal quo random recip reducedSystem rem retract retractIfCan sign sizeLess? squareFree squareFreePart unit? unitCanonical unitNormal wholePart zero?

DenavitHartenbergMatrix{DHMATRIX}: MatrixCategory *with* # * ** + - / = antisymmetric? any? coerce column copy count determinant diagonal? diagonalMatrix elt empty empty? eq? every? exquo fill! horizConcat identity inverse less? listOfLists map map! matrix maxColIndex maxRowIndex member? members minColIndex minRowIndex minordet more? ncols new nrows nullSpace nullity parts qelt qsetelt! rank rotatex rotatey rotatez row rowEchelon scalarMatrix scale setColumn! setRow! setelt setsubMatrix! size? square? squareTop subMatrix swapColumns! swapRows! symmetric? translate transpose vertConcat zero

Dequeue{DEQUEUE}: DequeueAggregate *with* # = any? back bag bottom! coerce copy count depth dequeue dequeue! empty empty? enqueue! eq? every? extract! extractBottom! extractTop! front height insert! insertBottom! insertTop! inspect length less? map map! member? members more? parts pop! push! reverse! rotate! size? top top!

DeRhamComplex{DERHAM}: LeftAlgebra RetractableTo *with* 0 1 * ** + - = characteristic coefficient coerce degree exteriorDifferential generator homogeneous? leadingBasisTerm leadingCoefficient map one? recip reductum retract retractIfCan retractable? totalDifferential zero?

DifferentialSparseMultivariatePolynomial{DSMP}: DifferentialPolynomialCategory RetractableTo *with* 0 1 * ** + - / < = D associates? characteristic charthRoot coefficient coefficients coerce conditionP content convert degree differentialVariables differentiate discriminant eval exquo factor factorPolynomial factorSquareFreePolynomial gcd gcdPolynomial ground ground? initial isExpt isPlus isTimes isobaric? lcm leader leadingCoefficient leadingMonomial mainVariable makeVariable map mapExponents max min minimumDegree monicDivide monomial monomial? monomials multivariate numberOfMonomials one? order patternMatch prime? primitiveMonomials primitivePart recip reducedSystem reductum resultant retract retractIfCan separant solveLinearPolynomialEquation squareFree squareFreePart squareFreePolynomial totalDegree unit? unitCanonical unitNormal univariate variables weight weights zero?

DirectProductMatrixModule{DPMM}: DirectProductCategory LeftModule *with* 0 1 # * ** + - / < = D abs any? characteristic coerce copy count differentiate dimension directProduct dot elt empty empty? entries entry? eq? every? fill! first index index? indices less? lookup map map! max maxIndex member? members min minIndex more? negative? one? parts positive? qelt qsetelt! random recip reducedSystem retract retractIfCan setelt sign size size? sup swap! unitVector zero?

DirectProductModule{DPMO}: DirectProductCategory LeftModule *with* 0 1 # * ** + - / < = D abs any? characteristic coerce copy count differentiate dimension directProduct dot elt empty empty? entries entry? eq? every? fill! first index index? indices less? lookup map map! max maxIndex member? members min minIndex more? negative? one? parts positive? qelt qsetelt! random recip reducedSystem retract retractIfCan setelt sign size size? sup swap! unitVector zero?

DirectProduct{DIRPROD}: DirectProductCategory *with* 0 1 # * ** + - / < = D abs any? characteristic coerce copy count differentiate dimension directProduct dot elt empty empty? entries entry? eq? every? fill! first index index? indices less? lookup map map! max maxIndex member? members min minIndex more? negative? one? parts positive? qelt qsetelt! random recip reducedSystem retract retractIfCan setelt sign size size? sup swap! unitVector zero?

DistributedMultivariatePolynomial{DMP}: PolynomialCategory *with* 0 1 * ** + - / < = D associates? characteristic charthRoot coefficient coefficients coerce conditionP const content convert degree differentiate discriminant eval exquo factor factorPolynomial factorSquareFreePolynomial gcd gcdPolynomial ground ground? isExpt isPlus isTimes lcm leadingCoefficient leadingMonomial mainVariable map mapExponents max min minimumDegree monicDivide monomial monomial? monomials multivariate numberOfMonomials one? prime? primitiveMonomials primitivePart recip reducedSystem reductum reorder resultant retract retractIfCan solveLinearPolynomialEquation squareFree squareFreePart squareFreePolynomial totalDegree unit? unitCanonical unitNormal univariate variables zero?

DrawOption{DROPT}: SetCategory *with* = adaptive clip coerce colorFunction coordinate coordinates curveColor option option? pointColor range ranges space style title toScale tubePoints tubeRadius unit var1Steps var2Steps

ElementaryFunctionsUnivariateLaurentSeries{EFULS}: PartialTranscendentalFunctions *with* ** acos acosIfCan acosh acoshIfCan acot acotIfCan acoth acothIfCan acsc acscIfCan acsch acschIfCan asec asecIfCan asech asechIfCan asin asinIfCan asinh asinhIfCan atan atanIfCan atanh atanhIfCan cos cosIfCan cosh coshIfCan cot cotIfCan coth cothIfCan csc cscIfCan csch cschIfCan exp expIfCan log logIfCan nthRootIfCan sec secIfCan sech sechIfCan sin sinIfCan sinh sinhIfCan tan tanIfCan tanh tanhIfCan

ElementaryFunctionsUnivariatePuiseuxSeries{EFUPXS}: PartialTranscendentalFunctions *with* ** acos acosIfCan acosh acoshIfCan acot acotIfCan acoth acothIfCan acsc acscIfCan acsch acschIfCan asec asecIfCan asech asechIfCan asin asinIfCan asinh asinhIfCan atan atanIfCan atanh atanhIfCan cos cosIfCan cosh coshIfCan cot cotIfCan coth cothIfCan csc cscIfCan csch cschIfCan exp expIfCan log logIfCan nthRootIfCan sec secIfCan sech sechIfCan sin sinIfCan sinh sinhIfCan tan tanIfCan tanh tanhIfCan

EqTable{EQTBL}: TableAggregate *with* # = any? bag coerce construct copy count dictionary elt empty empty? entries entry? eq? every? extract! fill! find first index? indices insert! inspect key? keys less? map map! maxIndex member? members minIndex more? parts qelt qsetelt! reduce remove remove! removeDuplicates search select select! setelt size? swap! table

Equation{EQ}: CoercibleTo InnerEvalable Object SetCategory *with* * ** + - = coerce equation eval lhs map rhs

EuclideanModularRing{EMR}: EuclideanDomain *with* 0 1 * ** + - = associates? characteristic coerce divide euclideanSize exQuo expressIdealMember exquo extendedEuclidean gcd inv lcm modulus multiEuclidean one? principalIdeal quo recip reduce rem sizeLess? unit? unitCanonical unitNormal zero?

Exit{EXIT}: SetCategory *with* = coerce

Expression{EXPR}: AlgebraicallyClosedFunctionSpace CombinatorialOpsCategory FunctionSpace LiouvillianFunctionCategory RetractableTo SpecialFunctionCategory TranscendentalFunctionCategory *with* 0 1 * ** + - / < = Beta Ci D Ei Gamma Si abs acos acosh acot acoth acsc acsch airyAi airyBi applyQuote asec asech asin asinh associates? atan atanh belong? besselI besselJ besselK besselY binomial box characteristic charthRoot coerce commutator conjugate convert cos cosh cot coth csc csch definingPolynomial denom denominator differentiate digamma dilog distribute divide elt erf euclideanSize eval exp expressIdealMember exquo extendedEuclidean factor factorial factorials freeOf? gcd ground ground? height integral inv is? isExpt isMult isPlus isPower isTimes kernel kernels lcm li log mainKernel map max min minPoly multiEuclidean nthRoot numer numerator one? operator operators paren patternMatch permutation pi polygamma prime? principalIdeal product quo recip reduce reducedSystem rem retract retractIfCan rootOf rootsOf sec sech sin sinh sizeLess? sqrt squareFree squareFreePart subst summation tan tanh tower unit? unitCanonical unitNormal univariate variables zero? zeroOf zerosOf

ExtAlgBasis{EAB}: OrderedSet *with* < = Nul coerce degree exponents max min

Factored{FR}: Algebra DifferentialExtension Eltable Evalable FullyEvalableOver FullyRetractableTo GcdDomain InnerEvalable IntegralDomain RealConstant UniqueFactorizationDomain *with* 0 1 * ** + - = D associates? characteristic coerce convert differentiate elt eval expand exponent exquo factor factorList factors flagFactor gcd irreducibleFactor lcm makeFR map nilFactor nthExponent nthFactor nthFlag numberOfFactors one? prime? primeFactor rational rational? rationalIfCan recip retract retractIfCan sqfrFactor squareFree squareFreePart unit unit? unitCanonical unitNormal unitNormalize zero?

FileName{FNAME}: FileNameCategory *with* = coerce directory exists? extension filename name new readable? writable?

File{FILE}: FileCategory *with* = close! coerce iomode name open read! readIfCan! reopen! write!

FiniteDivisor{FDIV}: AbelianGroup *with* 0 * + - = algsplit coerce divisor finiteBasis generator ideal lSpaceBasis mkBasicDiv principal? reduce zero?

FiniteFieldCyclicGroupExtensionByPolynomial{FFCGP}: FiniteAlgebraicExtensionField *with* 0 1 * ** + - / = Frobenius algebraic? associates? basis characteristic charthRoot coerce conditionP coordinates createNormalElement createPrimitiveElement definingPolynomial degree dimension discreteLog divide euclideanSize expressIdealMember exquo extendedEuclidean extensionDegree factor factorsOfCyclicGroupSize gcd generator getZechTable inGroundField? index init inv lcm lookup minimalPolynomial multiEuclidean nextItem norm normal? normalElement one? order prime? primeFrobenius primitive? primitiveElement principalIdeal quo random recip rem representationType represents retract retractIfCan size sizeLess? squareFree squareFreePart tableForDiscreteLogarithm trace transcendenceDegree transcendent? unit? unitCanonical unitNormal zero?

FiniteFieldCyclicGroupExtension{FFCGX}: FiniteAlgebraicExtensionField *with* 0 1 * ** + - / = Frobenius algebraic? associates? basis characteristic charthRoot coerce conditionP coordinates createNormalElement createPrimitiveElement definingPolynomial degree dimension discreteLog divide euclideanSize expressIdealMember exquo extendedEuclidean extensionDegree factor factorsOfCyclicGroupSize gcd generator getZechTable inGroundField? index init inv lcm lookup minimalPolynomial multiEuclidean nextItem norm normal? normalElement one? order prime? primeFrobenius primitive? primitiveElement principalIdeal quo random recip rem representationType represents retract retractIfCan size sizeLess? squareFree squareFreePart tableForDiscreteLogarithm trace transcendenceDegree transcendent? unit? unitCanonical unitNormal zero?

FiniteFieldCyclicGroup{FFCG}: FiniteAlgebraicExtensionField *with* 0 1 * ** + - / = Frobenius algebraic? associates? basis characteristic charthRoot coerce conditionP coordinates createNormalElement createPrimitiveElement definingPolynomial degree dimension discreteLog divide euclideanSize expressIdealMember exquo extendedEuclidean extensionDegree factor factorsOfCyclicGroupSize gcd generator getZechTable inGroundField? index init inv lcm lookup minimalPolynomial multiEuclidean nextItem norm normal? normalElement one? order prime? primeFrobenius primitive? primitiveElement principalIdeal quo random recip rem representationType represents retract retractIfCan size sizeLess? squareFree squareFreePart tableForDiscreteLogarithm trace transcendenceDegree transcendent? unit? unitCanonical unitNormal zero?

FiniteFieldExtensionByPolynomial{FFP}: FiniteAlgebraicExtensionField *with* 0 1 * ** + - / = Frobenius algebraic? associates? basis characteristic charthRoot coerce conditionP coordinates createNormalElement createPrimitiveElement definingPolynomial degree dimension discreteLog divide euclideanSize expressIdealMember exquo extendedEuclidean extensionDegree factor factorsOfCyclicGroupSize gcd generator inGroundField? index init inv lcm lookup minimalPolynomial multiEuclidean nextItem norm normal? normalElement one? order prime? primeFrobenius primitive? primitiveElement principalIdeal quo random recip rem representationType represents retract retractIfCan size sizeLess? squareFree squareFreePart tableForDiscreteLogarithm trace transcendenceDegree transcendent? unit? unitCanonical unitNormal zero?

FiniteFieldExtension{FFX}: FiniteAlgebraicExtensionField *with* 0 1 * ** + - / = Frobenius algebraic? associates? basis characteristic charthRoot coerce conditionP coordinates createNormalElement createPrimitiveElement definingPolynomial degree dimension discreteLog divide euclideanSize expressIdealMember exquo extendedEuclidean extensionDegree factor factorsOfCyclicGroupSize gcd generator inGroundField? index init inv lcm lookup minimalPolynomial multiEuclidean nextItem norm normal? normalElement one? order prime? primeFrobenius primitive? primitiveElement principalIdeal quo random recip rem representationType represents retract retractIfCan size sizeLess? squareFree squareFreePart tableForDiscreteLogarithm trace transcendenceDegree transcendent? unit? unitCanonical unitNormal zero?

FiniteFieldNormalBasisExtensionByPolynomial{FFNBP}: FiniteAlgebraicExtensionField *with* 0 1 * ** + - / = Frobenius algebraic? associates? basis characteristic charthRoot coerce conditionP coordinates createNormalElement createPrimitiveElement definingPolynomial degree dimension discreteLog divide euclideanSize expressIdealMember exquo extendedEuclidean extensionDegree factor factorsOfCyclicGroupSize gcd generator getMultiplicationMatrix getMultiplicationTable inGroundField? index init inv lcm lookup minimalPolynomial multiEuclidean nextItem norm normal? normalElement one? order prime? primeFrobenius primitive? primitiveElement principalIdeal quo random recip rem representationType represents retract retractIfCan size sizeLess? sizeMultiplication squareFree squareFreePart tableForDiscreteLogarithm trace transcendenceDegree transcendent? unit? unitCanonical unitNormal zero?

FiniteFieldNormalBasisExtension{FFNBX}: FiniteAlgebraicExtensionField *with* 0 1 * ** + - / = Frobenius algebraic? associates? basis characteristic charthRoot coerce conditionP coordinates createNormalElement createPrimitiveElement definingPolynomial degree dimension discreteLog divide euclideanSize expressIdealMember exquo extendedEuclidean extensionDegree factor factorsOfCyclicGroupSize gcd generator getMultiplicationMatrix getMultiplicationTable inGroundField? index init inv lcm lookup minimalPolynomial multiEuclidean nextItem norm normal? normalElement one? order prime? primeFrobenius primitive? primitiveElement principalIdeal quo random recip rem representationType represents retract retractIfCan size sizeLess? sizeMultiplication squareFree squareFreePart tableForDiscreteLogarithm trace transcendenceDegree transcendent? unit? unitCanonical unitNormal zero?

FiniteFieldNormalBasis{FFNB}: FiniteAlgebraicExtensionField *with* 0 1 * ** + - / = Frobenius algebraic? associates? basis characteristic charthRoot coerce conditionP coordinates createNormalElement createPrimitiveElement definingPolynomial degree dimension discreteLog divide euclideanSize expressIdealMember exquo extendedEuclidean extensionDegree factor factorsOfCyclicGroupSize gcd generator getMultiplicationMatrix getMultiplicationTable inGroundField? index init inv lcm lookup minimalPolynomial multiEuclidean nextItem norm normal? normalElement one? order prime? primeFrobenius primitive? primitiveElement principalIdeal quo random recip rem representationType represents retract retractIfCan size sizeLess? sizeMultiplication squareFree squareFreePart tableForDiscreteLogarithm trace transcendenceDegree transcendent? unit? unitCanonical unitNormal zero?

FiniteField{FF}: FiniteAlgebraicExtensionField *with* 0 1 * ** + - / = Frobenius algebraic? associates? basis characteristic charthRoot coerce conditionP coordinates createNormalElement createPrimitiveElement definingPolynomial degree dimension discreteLog divide euclideanSize expressIdealMember exquo extendedEuclidean extensionDegree factor factorsOfCyclicGroupSize gcd generator inGroundField? index init inv lcm lookup minimalPolynomial multiEuclidean nextItem norm normal? normalElement one? order prime? primeFrobenius primitive? primitiveElement principalIdeal quo random recip rem representationType represents retract retractIfCan size sizeLess? squareFree squareFreePart tableForDiscreteLogarithm trace transcendenceDegree transcendent? unit? unitCanonical unitNormal zero?

FlexibleArray{FARRAY}: ExtensibleLinearAggregate OneDimensionalArrayAggregate *with* # < = any? coerce concat concat! construct convert copy copyInto! count delete delete! elt empty empty? entries entry? eq? every? fill! find first flexibleArray index? indices insert insert! less? map map! max maxIndex member? members merge merge! min minIndex more? new parts physicalLength physicalLength! position qelt qsetelt! reduce remove remove! removeDuplicates removeDuplicates! reverse reverse! select

select! setelt shrinkable size? sort sort! sorted? swap!

Float{FLOAT}: CoercibleTo ConvertibleTo DifferentialRing FloatingPointSystem TranscendentalFunctionCategory *with* 0 1 * ** + - / < = D abs acos acosh acot acoth acsc acsch asec asech asin asinh associates? atan atanh base bits ceiling characteristic coerce convert cos cosh cot coth csc csch decreasePrecision differentiate digits divide euclideanSize exp exp1 exponent expressIdealMember exquo extendedEuclidean factor float floor fractionPart gcd increasePrecision inv lcm log log10 log2 mantissa max min multiEuclidean negative? norm normalize nthRoot one? order outputFixed outputFloating outputGeneral outputSpacing patternMatch pi positive? precision prime? principalIdeal quo rationalApproximation recip relerror rem retract retractIfCan round sec sech shift sign sin sinh sizeLess? sqrt squareFree squareFreePart tan tanh truncate unit? unitCanonical unitNormal wholePart zero?

FractionalIdeal{FRIDEAL}: Group *with* 1 * ** / = basis coerce commutator conjugate denom ideal inv minimize norm numer one? randomLC recip

Fraction{FRAC}: QuotientFieldCategory *with* 0 1 * ** + - / < = D abs associates? ceiling characteristic charthRoot coerce conditionP convert denom denominator differentiate divide elt euclideanSize eval expressIdealMember exquo extendedEuclidean factor factorPolynomial factorSquareFreePolynomial floor fractionPart gcd gcdPolynomial init inv lcm map max min multiEuclidean negative? nextItem numer numerator one? patternMatch positive? prime? principalIdeal quo random recip reducedSystem rem retract retractIfCan sign sizeLess? solveLinearPolynomialEquation squareFree squareFreePart squareFreePolynomial unit? unitCanonical unitNormal wholePart zero?

FramedModule{FRMOD}: Monoid *with* 1 * ** = basis coerce module norm one? recip

FreeAbelianGroup{FAGROUP}: AbelianGroup FreeAbelianMonoidCategory Module OrderedSet *with* 0 * + - < = coefficient coerce highCommonTerms mapCoef mapGen max min nthCoef nthFactor retract retractIfCan size terms zero?

FreeAbelianMonoid{FAMONOID}: FreeAbelianMonoidCategory *with* 0 * + - = coefficient coerce highCommonTerms mapCoef mapGen nthCoef nthFactor retract retractIfCan size terms zero?

FreeGroup{FGROUP}: Group RetractableTo *with* 1 * ** / = coerce commutator conjugate factors inv mapExpon mapGen nthExpon nthFactor one? recip retract retractIfCan size

FreeModule{FM}: BiModule IndexedDirectProductCategory Module *with* 0 * + - = coerce leadingCoefficient leadingSupport map monomial reductum zero?

FreeMonoid{FMONOID}: Monoid OrderedSet RetractableTo *with* 1 * ** < = coerce divide factors hclf hcrf lquo mapExpon mapGen max min nthExpon nthFactor one? overlap recip retract retractIfCan rquo size

FreeNilpotentLie{FNLA}: NonAssociativeAlgebra *with* 0 * ** + - = antiCommutator associator coerce commutator deepExpand dimension generator leftPower rightPower shallowExpand zero?

FunctionCalled{FUNCTION}: SetCategory *with* = coerce name

GeneralDistributedMultivariatePolynomial{GDMP}: PolynomialCategory *with* 0 1 * ** + - / < = D associates? characteristic charthRoot coefficient coefficients coerce conditionP const content convert degree differentiate discriminant eval exquo factor factorPolynomial factorSquareFreePolynomial gcd gcdPolynomial ground ground? isExpt isPlus isTimes lcm leadingCoefficient leadingMonomial mainVariable map mapExponents max min minimumDegree monicDivide monomial monomial? monomials multivariate numberOfMonomials one? prime? primitiveMonomials primitivePart recip reducedSystem reductum reorder resultant retract retractIfCan solveLinearPolynomialEquation squareFree squareFreePart squareFreePolynomial totalDegree unit? unitCanonical unitNormal univariate variables zero?

GeneralSparseTable{GSTBL}: TableAggregate *with* # = any? bag coerce construct copy count dictionary elt empty empty? entries entry? eq? every? extract! fill! find first index? indices insert! inspect key? keys less? map map! maxIndex member? members minIndex more? parts qelt qsetelt! reduce remove remove! removeDuplicates search select select! setelt size? swap! table

GenericNonAssociativeAlgebra{GCNAALG}: FramedNonAssociativeAlgebra LeftModule *with* 0 * ** + - = JacobiIdentity? JordanAlgebra? alternative? antiAssociative? antiCommutative? antiCommutator apply associative? associator associatorDependence basis coerce commutative? commutator conditionsForIdempotents convert coordinates elt flexible? generic genericLeftDiscriminant genericLeftMinimalPolynomial genericLeftNorm genericLeftTrace genericLeftTraceForm genericRightDiscriminant genericRightMinimalPolynomial genericRightNorm genericRightTrace genericRightTraceForm jordanAdmissible? leftAlternative? leftCharacteristicPolynomial leftDiscriminant leftMinimalPolynomial leftNorm leftPower leftRankPolynomial leftRecip leftRegularRepresentation leftTrace leftTraceMatrix leftUnit leftUnits lieAdmissible? lieAlgebra? noncommutativeJordanAlgebra? plenaryPower powerAssociative? rank recip represents rightAlternative? rightCharacteristicPolynomial rightDiscriminant rightMinimalPolynomial rightNorm rightPower rightRankPolynomial rightRecip rightRegularRepresentation rightTrace rightTraceMatrix rightUnit rightUnits someBasis structuralConstants unit zero?

GraphImage{GRIMAGE}: SetCategory *with* = appendPoint

coerce component graphImage key makeGraphImage point pointLists putColorInfo ranges units

HashTable{HASHTBL}: TableAggregate *with* # = any? bag coerce construct copy count dictionary elt empty empty? entries entry? eq? every? extract! fill! find first index? indices insert! inspect key? keys less? map map! maxIndex member? members minIndex more? parts qelt qsetelt! reduce remove remove! removeDuplicates search select select! setelt size? swap! table

Heap{HEAP}: PriorityQueueAggregate *with* # = any? bag coerce copy count empty empty? eq? every? extract! heap insert! inspect less? map map! max member? members merge merge! more? parts size?

HexadecimalExpansion{HEXADEC}: QuotientFieldCategory *with* 0 1 * ** + - / < = D abs associates? ceiling characteristic coerce convert denom denominator differentiate divide euclideanSize expressIdealMember exquo extendedEuclidean factor floor fractionPart gcd hex init inv lcm map max min multiEuclidean negative? nextItem numer numerator one? patternMatch positive? prime? principalIdeal quo random recip reducedSystem rem retract retractIfCan sign sizeLess? squareFree squareFreePart unit? unitCanonical unitNormal wholePart zero?

IndexCard{ICARD}: OrderedSet *with* < = coerce display elt fullDisplay max min

IndexedBits{IBITS}: BitAggregate *with* # < = And Not Or ^ and any? coerce concat construct convert copy copyInto! count delete elt empty empty? entries entry? eq? every? fill! find first index? indices insert less? map map! max maxIndex member? members merge min minIndex more? nand new nor not or parts position qelt qsetelt! reduce remove removeDuplicates reverse reverse! select setelt size? sort sort! sorted? swap! xor

IndexedDirectProductAbelianGroup{IDPAG}: AbelianGroup IndexedDirectProductCategory *with* 0 * + - = coerce leadingCoefficient leadingSupport map monomial reductum zero?

IndexedDirectProductAbelianMonoid{IDPAM}: AbelianMonoid IndexedDirectProductCategory *with* 0 * + = coerce leadingCoefficient leadingSupport map monomial reductum zero?

IndexedDirectProductObject{IDPO}: IndexedDirectProductCategory *with* = coerce leadingCoefficient leadingSupport map monomial reductum

IndexedDirectProductOrderedAbelianMonoidSup {IDPOAMS}: IndexedDirectProductCategory OrderedAbelianMonoidSup *with* 0 * + - < = coerce leadingCoefficient leadingSupport map max min monomial reductum sup zero?

IndexedDirectProductOrderedAbelianMonoid{IDPOAM}: IndexedDirectProductCategory OrderedAbelianMonoid *with* 0 * + < = coerce leadingCoefficient leadingSupport map max min monomial reductum zero?

IndexedExponents{INDE}: IndexedDirectProductCategory OrderedAbelianMonoidSup *with* 0 * + - < = coerce leadingCoefficient leadingSupport map max min monomial reductum sup zero?

IndexedFlexibleArray{IFARRAY}: ExtensibleLinearAggregate OneDimensionalArrayAggregate *with* # < = any? coerce concat concat! construct convert copy copyInto! count delete delete! elt empty empty? entries entry? eq? every? fill! find first flexibleArray index? indices insert insert! less? map map! max maxIndex member? members merge merge! min minIndex more? new parts physicalLength physicalLength! position qelt qsetelt! reduce remove remove! removeDuplicates removeDuplicates! reverse reverse! select select! setelt shrinkable size? sort sort! sorted? swap!

IndexedList{ILIST}: ListAggregate *with* # < = any? child? children coerce concat concat! construct convert copy copyInto! count cycleEntry cycleLength cycleSplit! cycleTail cyclic? delete delete! distance elt empty empty? entries entry? eq? every? explicitlyFinite? fill! find first index? indices insert insert! last leaf? less? list map map! max maxIndex member? members merge merge! min minIndex more? new node? nodes parts position possiblyInfinite? qelt qsetelt! reduce remove remove! removeDuplicates removeDuplicates! rest reverse reverse! second select select! setchildren! setelt setfirst! setlast! setrest! setvalue! size? sort sort! sorted? split! swap! tail third value

IndexedMatrix{IMATRIX}: MatrixCategory *with* # * ** + - / = antisymmetric? any? coerce column copy count determinant diagonal? diagonalMatrix elt empty empty? eq? every? exquo fill! horizConcat inverse less? listOfLists map map! matrix maxColIndex maxRowIndex member? members minColIndex minRowIndex minordet more? ncols new nrows nullSpace nullity parts qelt qsetelt! rank row rowEchelon scalarMatrix setColumn! setRow! setelt setsubMatrix! size? square? squareTop subMatrix swapColumns! swapRows! symmetric? transpose vertConcat zero

IndexedOneDimensionalArray{IARRAY1}: OneDimensionalArrayAggregate *with* # < = any? coerce concat construct convert copy copyInto! count delete elt empty empty? entries entry? eq? every? fill! find first index? indices insert less? map map! max maxIndex member? members merge min minIndex more? new parts position qelt qsetelt! reduce remove removeDuplicates reverse reverse! select setelt size? sort sort! sorted? swap!

IndexedString{ISTRING}: StringAggregate *with* # < = any? coerce concat construct copy copyInto! count delete elt empty empty? entries entry? eq? every? fill! find first hash index? indices insert leftTrim less? lowerCase lowerCase! map map! match? max maxIndex member? members merge min minIndex more? new parts position prefix? qelt qsetelt! reduce remove removeDuplicates replace reverse reverse!

rightTrim select setelt size? sort sort! sorted? split substring? suffix? swap! trim upperCase upperCase!

IndexedTwoDimensionalArray{IARRAY2}: TwoDimensionalArrayCategory *with* # = any? coerce column copy count elt empty empty? eq? every? fill! less? map map! maxColIndex maxRowIndex member? members minColIndex minRowIndex more? ncols new nrows parts qelt qsetelt! row setColumn! setRow! setelt size?

IndexedVector{IVECTOR}: VectorCategory *with* # * + - < = any? coerce concat construct convert copy copyInto! count delete dot elt empty empty? entries entry? eq? every? fill! find first index? indices insert less? map map! max maxIndex member? members merge min minIndex more? new parts position qelt qsetelt! reduce remove removeDuplicates reverse reverse! select setelt size? sort sort! sorted? swap! zero

InfiniteTuple{ITUPLE}: CoercibleTo *with* coerce construct filterUntil filterWhile generate map select

InnerFiniteField{IFF}: FiniteAlgebraicExtensionField *with* 0 1 * ** + - / = Frobenius algebraic? associates? basis characteristic charthRoot coerce conditionP coordinates createNormalElement createPrimitiveElement definingPolynomial degree dimension discreteLog divide euclideanSize expressIdealMember exquo extendedEuclidean extensionDegree factor factorsOfCyclicGroupSize gcd generator inGroundField? index init inv lcm lookup minimalPolynomial multiEuclidean nextItem norm normal? normalElement one? order prime? primeFrobenius primitive? primitiveElement principalIdeal quo random recip rem representationType represents retract retractIfCan size sizeLess? squareFree squareFreePart tableForDiscreteLogarithm trace transcendenceDegree transcendent? unit? unitCanonical unitNormal zero?

InnerFreeAbelianMonoid{IFAMON}: FreeAbelianMonoidCategory *with* 0 * + - = coefficient coerce highCommonTerms mapCoef mapGen nthCoef nthFactor retract retractIfCan size terms zero?

InnerIndexedTwoDimensionalArray{IIARRAY2}: TwoDimensionalArrayCategory *with* # = any? coerce column copy count elt empty empty? eq? every? fill! less? map map! maxColIndex maxRowIndex member? members minColIndex minRowIndex more? ncols new nrows parts qelt qsetelt! row setColumn! setRow! setelt size?

InnerPAdicInteger{IPADIC}: PAdicIntegerCategory *with* 0 1 * ** + - = approximate associates? characteristic coerce complete digits divide euclideanSize expressIdealMember exquo extend extendedEuclidean gcd lcm moduloP modulus multiEuclidean one? order principalIdeal quo quotientByP recip rem sizeLess? sqrt unit? unitCanonical unitNormal zero?

InnerPrimeField{IPF}: ConvertibleTo FiniteAlgebraicExtensionField FiniteFieldCategory *with* 0 1 * ** + - / = Frobenius algebraic? associates? basis characteristic charthRoot coerce conditionP convert coordinates createNormalElement createPrimitiveElement definingPolynomial degree dimension discreteLog divide euclideanSize expressIdealMember exquo extendedEuclidean extensionDegree factor factorsOfCyclicGroupSize gcd generator inGroundField? index init inv lcm lookup minimalPolynomial multiEuclidean nextItem norm normal? normalElement one? order prime? primeFrobenius primitive? primitiveElement principalIdeal quo random recip rem representationType represents retract retractIfCan size sizeLess? squareFree squareFreePart tableForDiscreteLogarithm trace transcendenceDegree transcendent? unit? unitCanonical unitNormal zero?

InnerTaylorSeries{ITAYLOR}: IntegralDomain Ring *with* 0 1 * ** + - = associates? characteristic coefficients coerce exquo one? order pole? recip series unit? unitCanonical unitNormal zero?

InputForm{INFORM}: ConvertibleTo SExpressionCategory *with* 0 1 # * ** + / = atom? binary car cdr coerce compile convert declare destruct elt eq expr flatten float float? function integer integer? interpret lambda list? null? pair? string string? symbol symbol? uequal unparse

IntegerMod{ZMOD}: CommutativeRing ConvertibleTo Finite StepThrough *with* 0 1 * ** + - = characteristic coerce convert index init lookup nextItem one? random recip size zero?

Integer{INT}: ConvertibleTo IntegerNumberSystem *with* 0 1 * ** + - < = D abs addmod associates? base binomial bit? characteristic coerce convert copy dec differentiate divide euclideanSize even? expressIdealMember exquo extendedEuclidean factor factorial gcd hash inc init invmod lcm length mask max min mulmod multiEuclidean negative? nextItem odd? one? patternMatch permutation positive? positiveRemainder powmod prime? principalIdeal quo random rational rational? rationalIfCan recip reducedSystem rem retract retractIfCan shift sign sizeLess? squareFree squareFreePart submod symmetricRemainder unit? unitCanonical unitNormal zero?

IntegrationResult{IR}: Module RetractableTo *with* 0 * + - = D coerce differentiate elem? integral logpart mkAnswer notelem ratpart retract retractIfCan zero?

Kernel{KERNEL}: CachableSet ConvertibleTo Patternable *with* < = argument coerce convert height is? kernel max min name operator position setPosition symbolIfCan

KeyedAccessFile{KAFILE}: FileCategory TableAggregate *with* # = any? bag close! coerce construct copy count dictionary elt empty empty? entries entry? eq? every? extract! fill! find first index? indices insert! inspect iomode key? keys less? map map! maxIndex member? members minIndex more? name open pack! parts qelt qsetelt! read! reduce remove remove! removeDuplicates reopen! search select select! setelt size? swap! table write!

LaurentPolynomial{LAUPOL}: CharacteristicNonZero

CharacteristicZero ConvertibleTo DifferentialExtension EuclideanDomain FullyRetractableTo IntegralDomain RetractableTo *with* 0 1 * ** + - = D associates? characteristic charthRoot coefficient coerce convert degree differentiate divide euclideanSize expressIdealMember exquo extendedEuclidean gcd lcm leadingCoefficient monomial monomial? multiEuclidean one? order principalIdeal quo recip reductum rem retract retractIfCan separate sizeLess? trailingCoefficient unit? unitCanonical unitNormal zero?

Library{LIB}: TableAggregate *with* # = any? bag coerce construct copy count dictionary elt empty empty? entries entry? eq? every? extract! fill! find first index? indices insert! inspect key? keys less? library map map! maxIndex member? members minIndex more? pack! parts qelt qsetelt! reduce remove remove! removeDuplicates search select select! setelt size? swap! table

LieSquareMatrix{LSQM}: CoercibleTo FramedNonAssociativeAlgebra SquareMatrixCategory *with* 0 1 # * ** + - / = D JacobiIdentity? JordanAlgebra? alternative? antiAssociative? antiCommutative? antiCommutator antisymmetric? any? apply associative? associator associatorDependence basis characteristic coerce column commutative? commutator conditionsForIdempotents convert coordinates copy count determinant diagonal diagonal? diagonalMatrix diagonalProduct differentiate elt empty empty? eq? every? exquo flexible? inverse jordanAdmissible? leftAlternative? leftCharacteristicPolynomial leftDiscriminant leftMinimalPolynomial leftNorm leftPower leftRankPolynomial leftRecip leftRegularRepresentation leftTrace leftTraceMatrix leftUnit leftUnits less? lieAdmissible? lieAlgebra? listOfLists map map! matrix maxColIndex maxRowIndex member? members minColIndex minRowIndex minordet more? ncols noncommutativeJordanAlgebra? nrows nullSpace nullity one? parts plenaryPower powerAssociative? qelt rank recip reducedSystem represents retract retractIfCan rightAlternative? rightCharacteristicPolynomial rightDiscriminant rightMinimalPolynomial rightNorm rightPower rightRankPolynomial rightRecip rightRegularRepresentation rightTrace rightTraceMatrix rightUnit rightUnits row rowEchelon scalarMatrix size? someBasis square? structuralConstants symmetric? trace unit zero?

LinearOrdinaryDifferentialOperator{LODO}: MonogenicLinearOperator *with* 0 1 * ** + - = D characteristic coefficient coerce degree elt leadingCoefficient leftDivide leftExactQuotient leftGcd leftLcm leftQuotient leftRemainder minimumDegree monomial one? recip reductum rightDivide rightExactQuotient rightGcd rightLcm rightQuotient rightRemainder zero?

ListMonoidOps{LMOPS}: RetractableTo SetCategory *with* = coerce leftMult listOfMonoms makeMulti makeTerm makeUnit mapExpon mapGen nthExpon nthFactor outputForm plus retract retractIfCan reverse reverse! rightMult size

ListMultiDictionary{LMDICT}: MultiDictionary *with* # = any? bag coerce construct convert copy count dictionary duplicates duplicates? empty empty? eq? every? extract! find insert! inspect less? map map! member? members more? parts reduce remove remove! removeDuplicates removeDuplicates! select select! size? substitute

List{LIST}: ListAggregate *with* # < = any? append child? children coerce concat concat! cons construct convert copy copyInto! count cycleEntry cycleLength cycleSplit! cycleTail cyclic? delete delete! distance elt empty empty? entries entry? eq? every? explicitlyFinite? fill! find first index? indices insert insert! last leaf? less? list map map! max maxIndex member? members merge merge! min minIndex more? new nil node? nodes null parts position possiblyInfinite? qelt qsetelt! reduce remove remove! removeDuplicates removeDuplicates! rest reverse reverse! second select select! setDifference setIntersection setUnion setchildren! setelt setfirst! setlast! setrest! setvalue! size? sort sort! sorted? split! swap! tail third value

LocalAlgebra{LA}: Algebra OrderedRing *with* 0 1 * ** + - / < = abs characteristic coerce denom max min negative? numer one? positive? recip sign zero?

Localize{LO}: Module OrderedAbelianGroup *with* 0 * + - / < = coerce denom max min numer zero?

MakeCachableSet{MKCHSET}: CachableSet CoercibleTo *with* < = coerce max min position setPosition

MakeOrdinaryDifferentialRing{MKODRING}: CoercibleTo DifferentialRing *with* 0 1 * ** + - = D characteristic coerce differentiate one? recip zero?

Matrix{MATRIX}: MatrixCategory *with* # * ** + - / = antisymmetric? any? coerce column copy count determinant diagonal? diagonalMatrix elt empty empty? eq? every? exquo fill! horizConcat inverse less? listOfLists map map! matrix maxColIndex maxRowIndex member? members minColIndex minRowIndex minordet more? ncols new nrows nullSpace nullity parts qelt qsetelt! rank row rowEchelon scalarMatrix setColumn! setRow! setelt setsubMatrix! size? square? squareTop subMatrix swapColumns! swapRows! symmetric? transpose vertConcat zero

ModMonic{MODMON}: Finite UnivariatePolynomialCategory *with* 0 1 * ** + - / < = An D UnVectorise Vectorise associates? characteristic charthRoot coefficient coefficients coerce composite computePowers conditionP content degree differentiate discriminant divide divideExponents elt euclideanSize eval expressIdealMember exquo extendedEuclidean factor factorPolynomial factorSquareFreePolynomial gcd gcdPolynomial ground ground? index init integrate isExpt isPlus isTimes lcm leadingCoefficient leadingMonomial lift lookup mainVariable makeSUP map mapExponents max min minimumDegree modulus monicDivide monomial monomial? monomials multiEuclidean multiplyExponents multivariate nextItem numberOfMonomials one? order pow prime? primitiveMonomials primitivePart principalIdeal pseudoDivide pseudoQuotient pseudoRemainder quo random recip reduce reducedSystem reductum rem resultant retract retractIfCan separate setPoly size sizeLess? solveLinearPolynomialEquation squareFree squareFreePart squareFreePolynomial subResultantGcd totalDegree unit? unitCanonical unitNormal univariate unmakeSUP variables vectorise zero?

ModularField{MODFIELD}: Field *with* 0 1 * ** + - / = associates? characteristic coerce divide euclideanSize exQuo expressIdealMember exquo extendedEuclidean factor gcd inv lcm modulus multiEuclidean one? prime? principalIdeal quo recip reduce rem sizeLess? squareFree squareFreePart unit? unitCanonical unitNormal zero?

ModularRing{MODRING}: Ring *with* 0 1 * ** + - = characteristic coerce exQuo inv modulus one? recip reduce zero?

MoebiusTransform{MOEBIUS}: Group *with* 1 * ** / = coerce commutator conjugate eval inv moebius one? recip scale shift

MonoidRing{MRING}: Algebra CharacteristicNonZero CharacteristicZero Finite RetractableTo Ring *with* 0 1 * ** + - = characteristic charthRoot coefficient coefficients coerce index leadingCoefficient leadingMonomial lookup map monomial monomial? monomials numberOfMonomials one? random recip reductum retract retractIfCan size terms zero?

Multiset{MSET}: MultiSetAggregate *with* # < = any? bag brace coerce construct convert copy count dictionary difference duplicates empty empty? eq? every? extract! find insert! inspect intersect less? map map! member? members more? multiset parts reduce remove remove! removeDuplicates removeDuplicates! select select! size? subset? symmetricDifference union

MultivariatePolynomial{MPOLY}: PolynomialCategory *with* 0 1 * ** + - / < = D associates? characteristic charthRoot coefficient coefficients coerce conditionP content convert degree differentiate discriminant eval exquo factor factorPolynomial factorSquareFreePolynomial gcd gcdPolynomial ground ground? isExpt isPlus isTimes lcm leadingCoefficient leadingMonomial mainVariable map mapExponents max min minimumDegree monicDivide monomial monomial? monomials multivariate numberOfMonomials one? prime? primitiveMonomials primitivePart recip reducedSystem reductum resultant retract retractIfCan solveLinearPolynomialEquation squareFree squareFreePart squareFreePolynomial totalDegree unit? unitCanonical unitNormal univariate variables zero?

NewDirectProduct{NDP}: DirectProductCategory *with* 0 1 # * ** + - / < = D abs any? characteristic coerce copy count differentiate dimension directProduct dot elt empty empty? entries entry? eq? every? fill! first index index? indices less? lookup map map! max maxIndex member? members min minIndex more? negative? one? parts positive? qelt qsetelt! random recip reducedSystem retract retractIfCan setelt sign size size? sup swap! unitVector zero?

NewDistributedMultivariatePolynomial{NDMP}: PolynomialCategory *with* 0 1 * ** + - / < = D associates? characteristic charthRoot coefficient coefficients coerce conditionP const content convert degree differentiate discriminant eval exquo factor factorPolynomial factorSquareFreePolynomial gcd gcdPolynomial ground ground? isExpt isPlus isTimes lcm leadingCoefficient leadingMonomial mainVariable map mapExponents max min minimumDegree monicDivide monomial monomial? monomials multivariate numberOfMonomials one? prime? primitiveMonomials primitivePart recip reducedSystem reductum reorder resultant retract retractIfCan solveLinearPolynomialEquation squareFree squareFreePart squareFreePolynomial totalDegree unit? unitCanonical unitNormal univariate variables zero?

None{NONE}: SetCategory *with* = coerce

NonNegativeInteger{NNI}: Monoid OrderedAbelianMonoidSup *with* 0 1 * ** + - < = coerce divide exquo gcd max min one? quo recip rem sup zero?

Octonion{OCT}: FullyRetractableTo OctonionCategory *with* 0 1 * ** + - < = abs characteristic charthRoot coerce conjugate convert elt eval imagE imagI imagJ imagK imagi imagj imagk index inv lookup map max min norm octon one? random rational rational? rationalIfCan real recip retract retractIfCan size zero?

OneDimensionalArray{ARRAY1}: OneDimensionalArrayAggregate *with* # < = any? coerce concat construct convert copy copyInto! count delete elt empty empty? entries entry? eq? every? fill! find first index? indices insert less? map map! max maxIndex member? members merge min minIndex more? new oneDimensionalArray parts position qelt qsetelt! reduce remove removeDuplicates reverse reverse! select setelt size? sort sort! sorted? swap!

OnePointCompletion{ONECOMP}: AbelianGroup FullyRetractableTo OrderedRing SetCategory *with* 0 1 * ** + - < = abs characteristic coerce finite? infinite? infinity max min negative? one? positive? rational rational? rationalIfCan recip retract retractIfCan sign zero?

Operator{OP}: Algebra CharacteristicNonZero CharacteristicZero Eltable RetractableTo Ring *with* 0 1 * ** + - = characteristic charthRoot coerce elt evaluate one? opeval recip retract retractIfCan zero?

OppositeMonogenicLinearOperator{OMLO}: DifferentialRing MonogenicLinearOperator *with* 0 1 * ** + - = D characteristic coefficient coerce degree differentiate leadingCoefficient minimumDegree monomial one? op po recip reductum zero?

OrderedCompletion{ORDCOMP}: AbelianGroup FullyRetractableTo OrderedRing SetCategory *with* 0 1 * ** + - < = abs characteristic coerce finite? infinite? max min minusInfinity negative? one? plusInfinity positive? rational

rational? rationalIfCan recip retract retractIfCan sign whatInfinity zero?

OrderedDirectProduct{ODP}: DirectProductCategory *with* 0 1 # * ** + - / < = D abs any? characteristic coerce copy count differentiate dimension directProduct dot elt empty empty? entries entry? eq? every? fill! first index index? indices less? lookup map map! max maxIndex member? members min minIndex more? negative? one? parts positive? qelt qsetelt! random recip reducedSystem retract retractIfCan setelt sign size size? sup swap! unitVector zero?

OrderedVariableList{OVAR}: ConvertibleTo OrderedFinite *with* < = coerce convert index lookup max min random size variable

OrderlyDifferentialPolynomial{ODPOL}: DifferentialPolynomialCategory RetractableTo *with* 0 1 * ** + - / < = D associates? characteristic charthRoot coefficient coefficients coerce conditionP content degree differentialVariables differentiate discriminant eval exquo factor factorPolynomial factorSquareFreePolynomial gcd gcdPolynomial ground ground? initial isExpt isPlus isTimes isobaric? lcm leader leadingCoefficient leadingMonomial mainVariable makeVariable map mapExponents max min minimumDegree monicDivide monomial monomial? monomials multivariate numberOfMonomials one? order prime? primitiveMonomials primitivePart recip reducedSystem reductum resultant retract retractIfCan separant solveLinearPolynomialEquation squareFree squareFreePart squareFreePolynomial totalDegree unit? unitCanonical unitNormal univariate variables weight weights zero?

OrderlyDifferentialVariable{ODVAR}: DifferentialVariableCategory *with* < = D coerce differentiate makeVariable max min order retract retractIfCan variable weight

OrdinaryDifferentialRing{ODR}: Algebra DifferentialRing Field *with* 0 1 * ** + - / = D associates? characteristic coerce differentiate divide euclideanSize expressIdealMember exquo extendedEuclidean factor gcd inv lcm multiEuclidean one? prime? principalIdeal quo recip rem sizeLess? squareFree squareFreePart unit? unitCanonical unitNormal zero?

OrdSetInts{OSI}: OrderedSet *with* < = coerce max min value

OutputForm{OUTFORM}: SetCategory *with* * ** + - / < <= = > >= D SEGMENT ^= and assign blankSeparate box brace bracket center coerce commaSeparate differentiate div dot elt empty exquo hconcat height hspace infix infix? int label left matrix message messagePrint not or outputForm over overbar paren pile postfix prefix presub presuper prime print prod quo quote rarrow rem right root rspace scripts semicolonSeparate slash string sub subHeight sum super superHeight supersub vconcat vspace width zag

PAdicInteger{PADIC}: PAdicIntegerCategory *with* 0 1 * ** + - = approximate associates? characteristic coerce complete digits divide euclideanSize expressIdealMember exquo extend extendedEuclidean gcd lcm moduloP modulus multiEuclidean one? order principalIdeal quo quotientByP recip rem sizeLess? sqrt unit? unitCanonical unitNormal zero?

PAdicRationalConstructor{PADICRC}: QuotientFieldCategory *with* 0 1 * ** + - / < = D abs approximate associates? ceiling characteristic charthRoot coerce conditionP continuedFraction convert denom denominator differentiate divide elt euclideanSize eval expressIdealMember exquo extendedEuclidean factor factorPolynomial factorSquareFreePolynomial floor fractionPart gcd gcdPolynomial init inv lcm map max min multiEuclidean negative? nextItem numer numerator one? patternMatch positive? prime? principalIdeal quo random recip reducedSystem rem removeZeroes retract retractIfCan sign sizeLess? solveLinearPolynomialEquation squareFree squareFreePart squareFreePolynomial unit? unitCanonical unitNormal wholePart zero?

PAdicRational{PADICRAT}: QuotientFieldCategory *with* 0 1 * ** + - / = D approximate associates? characteristic coerce continuedFraction denom denominator differentiate divide euclideanSize expressIdealMember exquo extendedEuclidean factor fractionPart gcd inv lcm map multiEuclidean numer numerator one? prime? principalIdeal quo recip reducedSystem rem removeZeroes retract retractIfCan sizeLess? squareFree squareFreePart unit? unitCanonical unitNormal wholePart zero?

Palette{PALETTE}: SetCategory *with* = bright coerce dark dim hue light pastel shade

ParametricPlaneCurve{PARPCURV}: *with* coordinate curve

ParametricSpaceCurve{PARSCURV}: *with* coordinate curve

ParametricSurface{PARSURF}: *with* coordinate surface

PartialFraction{PFR}: Algebra Field *with* 0 1 * ** + - / = associates? characteristic coerce compactFraction divide euclideanSize expressIdealMember exquo extendedEuclidean factor firstDenom firstNumer gcd inv lcm multiEuclidean nthFractionalTerm numberOfFractionalTerms one? padicFraction padicallyExpand partialFraction prime? principalIdeal quo recip rem sizeLess? squareFree squareFreePart unit? unitCanonical unitNormal wholePart zero?

Partition{PRTITION}: ConvertibleTo OrderedCancellationAbelianMonoid *with* 0 * + - < = coerce conjugate convert max min partition pdct powers zero?

PatternMatchListResult{PATLRES}: SetCategory *with* = atoms coerce failed failed? lists makeResult new

PatternMatchResult{PATRES}: SetCategory *with* = addMatch addMatchRestricted coerce construct destruct failed failed? getMatch insertMatch new satisfy? union

Pattern{PATTERN}: RetractableTo SetCategory *with* 0 1 * ** + / = addBadValue coerce constant? convert copy depth elt generic? getBadValues hasPredicate? hasTopPredicate? inR?

isExpt isList isOp isPlus isPower isQuotient isTimes multiple? optional? optpair patternVariable predicates quoted? resetBadValues retract retractIfCan setPredicates setTopPredicate symbol? topPredicate variables withPredicates

PendantTree{PENDTREE}: BinaryRecursiveAggregate *with* # = any? children coerce copy count cyclic? elt empty empty? eq? every? leaf? leaves left less? map map! member? members more? node? nodes parts ptree right setchildren! setelt setleft! setright! setvalue! size? value

PermutationGroup{PERMGRP}: SetCategory *with* < <= = base coerce degree elt generators initializeGroupForWordProblem member? movedPoints orbit orbits order permutationGroup random strongGenerators wordInGenerators wordInStrongGenerators wordsForStrongGenerators

Permutation{PERM}: PermutationCategory *with* 1 * ** / < = coerce coerceImages coerceListOfPairs coercePreimagesImages commutator conjugate cycle cyclePartition cycles degree elt eval even? fixedPoints inv listRepresentation max min movedPoints numberOfCycles odd? one? orbit order recip sign sort

Pi{HACKPI}: CharacteristicZero CoercibleTo ConvertibleTo Field RealConstant RetractableTo *with* 0 1 * ** + - / = associates? characteristic coerce convert divide euclideanSize expressIdealMember exquo extendedEuclidean factor gcd inv lcm multiEuclidean one? pi prime? principalIdeal quo recip rem retract retractIfCan sizeLess? squareFree squareFreePart unit? unitCanonical unitNormal zero?

PlaneAlgebraicCurvePlot{ACPLOT}: PlottablePlaneCurveCategory *with* coerce listBranches makeSketch refine xRange yRange

Plot3D{PLOT3D}: PlottableSpaceCurveCategory *with* adaptive3D? coerce debug3D listBranches maxPoints3D minPoints3D numFunEvals3D plot pointPlot refine screenResolution3D setAdaptive3D setMaxPoints3D setMinPoints3D setScreenResolution3D tRange tValues xRange yRange zRange zoom

Plot{PLOT}: PlottablePlaneCurveCategory *with* adaptive? coerce debug listBranches maxPoints minPoints numFunEvals parametric? plot plotPolar pointPlot refine screenResolution setAdaptive setMaxPoints setMinPoints setScreenResolution tRange xRange yRange zoom

Point{POINT}: PointCategory *with* # * + - < = any? coerce concat construct convert copy copyInto! count cross delete dimension dot elt empty empty? entries entry? eq? every? extend fill! find first index? indices insert length less? map map! max maxIndex member? members merge min minIndex more? new parts point position qelt qsetelt! reduce remove removeDuplicates reverse reverse! select setelt size? sort sort! sorted? swap! zero

PolynomialIdeals{IDEAL}: SetCategory *with* * ** + = backOldPos coerce contract dimension element? generalPosition generators groebner groebner? groebnerIdeal ideal in? inRadical? intersect leadingIdeal quotient relationsIdeal saturate zeroDim?

PolynomialRing{PR}: FiniteAbelianMonoidRing *with* 0 1 * ** + - / = associates? characteristic charthRoot coefficient coefficients coerce content degree exquo ground ground? leadingCoefficient leadingMonomial map mapExponents minimumDegree monomial monomial? numberOfMonomials one? primitivePart recip reductum retract retractIfCan unit? unitCanonical unitNormal zero?

Polynomial{POLY}: PolynomialCategory *with* 0 1 * ** + - / < = D associates? characteristic charthRoot coefficient coefficients coerce conditionP content convert degree differentiate discriminant eval exquo factor factorPolynomial factorSquareFreePolynomial gcd gcdPolynomial ground ground? integrate isExpt isPlus isTimes lcm leadingCoefficient leadingMonomial mainVariable map mapExponents max min minimumDegree monicDivide monomial monomial? monomials multivariate numberOfMonomials one? patternMatch prime? primitiveMonomials primitivePart recip reducedSystem reductum resultant retract retractIfCan solveLinearPolynomialEquation squareFree squareFreePart squareFreePolynomial totalDegree unit? unitCanonical unitNormal univariate variables zero?

PositiveInteger{PI}: AbelianSemiGroup Monoid OrderedSet *with* 1 * ** + < = coerce gcd max min one? recip

PrimeField{PF}: ConvertibleTo FiniteAlgebraicExtensionField FiniteFieldCategory *with* 0 1 * ** + - / = Frobenius algebraic? associates? basis characteristic charthRoot coerce conditionP convert coordinates createNormalElement createPrimitiveElement definingPolynomial degree dimension discreteLog divide euclideanSize expressIdealMember exquo extendedEuclidean extensionDegree factor factorsOfCyclicGroupSize gcd generator inGroundField? index init inv lcm lookup minimalPolynomial multiEuclidean nextItem norm normal? normalElement one? order prime? primeFrobenius primitive? primitiveElement principalIdeal quo random recip rem representationType represents retract retractIfCan size sizeLess? squareFree squareFreePart tableForDiscreteLogarithm trace transcendenceDegree transcendent? unit? unitCanonical unitNormal zero?

PrimitiveArray{PRIMARR}: OneDimensionalArrayAggregate *with* # < = any? coerce concat construct convert copy copyInto! count delete elt empty empty? entries entry? eq? every? fill! find first index? indices insert less? map map! max maxIndex member? members merge min minIndex more? new parts position qelt qsetelt! reduce remove removeDuplicates reverse reverse! select setelt size? sort sort! sorted? swap!

Product{PRODUCT}: AbelianGroup AbelianMonoid CancellationAbelianMonoid Finite Group Monoid OrderedAbelianMonoidSup OrderedSet SetCategory *with* 0 1

* ** + - / < = coerce commutator conjugate index inv lookup makeprod max min one? random recip selectfirst selectsecond size sup zero?

QuadraticForm{QFORM}: AbelianGroup *with* 0 * + - = coerce elt matrix quadraticForm zero?

QuasiAlgebraicSet{QALGSET}: CoercibleTo SetCategory *with* = coerce definingEquations definingInequation empty? idealSimplify quasiAlgebraicSet setStatus simplify

Quaternion{QUAT}: QuaternionCategory *with* 0 1 * ** + - < = D abs characteristic charthRoot coerce conjugate convert differentiate elt eval imagI imagJ imagK inv map max min norm one? quatern rational rational? rationalIfCan real recip reducedSystem retract retractIfCan zero?

QueryEquation{QEQUAT}: *with* equation value variable

Queue{QUEUE}: QueueAggregate *with* # = any? back bag coerce copy count dequeue! empty empty? enqueue! eq? every? extract! front insert! inspect length less? map map! member? members more? parts queue rotate! size?

RadicalFunctionField{RADFF}: FunctionFieldCategory *with* 0 1 * ** + - / = D absolutelyIrreducible? associates? basis branchPoint? branchPointAtInfinity? characteristic characteristicPolynomial charthRoot coerce complementaryBasis convert coordinates definingPolynomial derivationCoordinates differentiate discriminant divide elt euclideanSize expressIdealMember exquo extendedEuclidean factor gcd generator genus integral? integralAtInfinity? integralBasis integralBasisAtInfinity integralCoordinates integralDerivationMatrix integralMatrix integralMatrixAtInfinity integralRepresents inv inverseIntegralMatrix inverseIntegralMatrixAtInfinity lcm lift minimalPolynomial multiEuclidean nonSingularModel norm normalizeAtInfinity numberOfComponents one? prime? primitivePart principalIdeal quo ramified? ramifiedAtInfinity? rank rationalPoint? rationalPoints recip reduce reduceBasisAtInfinity reducedSystem regularRepresentation rem represents retract retractIfCan singular? singularAtInfinity? sizeLess? squareFree squareFreePart trace traceMatrix unit? unitCanonical unitNormal yCoordinates zero?

RadixExpansion{RADIX}: QuotientFieldCategory *with* 0 1 * ** + - / < = D abs associates? ceiling characteristic coerce convert cycleRagits denom denominator differentiate divide euclideanSize expressIdealMember exquo extendedEuclidean factor floor fractRadix fractRagits fractionPart gcd init inv lcm map max min multiEuclidean negative? nextItem numer numerator one? patternMatch positive? prefixRagits prime? principalIdeal quo random recip reducedSystem rem retract retractIfCan sign sizeLess? squareFree squareFreePart unit? unitCanonical unitNormal wholePart wholeRadix wholeRagits zero?

RectangularMatrix{RMATRIX}: CoercibleTo RectangularMatrixCategory VectorSpace *with* 0 # * + - / = antisymmetric? any? coerce column copy count diagonal? dimension elt empty empty? eq? every? exquo less? listOfLists map map! matrix maxColIndex maxRowIndex member? members minColIndex minRowIndex more? ncols nrows nullSpace nullity parts qelt rank rectangularMatrix row rowEchelon size? square? symmetric? zero?

Reference{REF}: Object SetCategory *with* = coerce deref elt ref setelt setref

RewriteRule{RULE}: Eltable RetractableTo SetCategory *with* = coerce elt lhs pattern quotedOperators retract retractIfCan rhs rule suchThat

RomanNumeral{ROMAN}: IntegerNumberSystem *with* 0 1 * ** + - < = D abs addmod associates? base binomial bit? characteristic coerce convert copy dec differentiate divide euclideanSize even? expressIdealMember exquo extendedEuclidean factor factorial gcd hash inc init invmod lcm length mask max min mulmod multiEuclidean negative? nextItem odd? one? patternMatch permutation positive? positiveRemainder powmod prime? principalIdeal quo random rational rational? rationalIfCan recip reducedSystem rem retract retractIfCan roman shift sign sizeLess? squareFree squareFreePart submod symmetricRemainder unit? unitCanonical unitNormal zero?

RuleCalled{RULECOLD}: SetCategory *with* = coerce name

Ruleset{RULESET}: Eltable SetCategory *with* = coerce elt rules ruleset

ScriptFormulaFormat1{FORMULA1}: Object *with* coerce

ScriptFormulaFormat{FORMULA}: SetCategory *with* = coerce convert display epilogue formula new prologue setEpilogue! setFormula! setPrologue!

SegmentBinding{SEGBIND}: SetCategory *with* = coerce equation segment variable

Segment{SEG}: SegmentCategory SegmentExpansionCategory *with* = BY SEGMENT coerce convert expand hi high incr lo low map segment

SemiCancelledFraction{SCFRAC}: ConvertibleTo QuotientFieldCategory *with* 0 1 * ** + - / < = D abs associates? ceiling characteristic charthRoot coerce conditionP convert denom denominator differentiate divide elt euclideanSize eval expressIdealMember exquo extendedEuclidean factor factorPolynomial factorSquareFreePolynomial floor fractionPart gcd gcdPolynomial init inv lcm map max min multiEuclidean negative? nextItem normalize numer numerator one? patternMatch positive? prime? principalIdeal quo random recip reducedSystem rem retract retractIfCan sign sizeLess? solveLinearPolynomialEquation squareFree squareFreePart squareFreePolynomial unit? unitCanonical unitNormal wholePart zero?

SequentialDifferentialPolynomial{SDPOL}: DifferentialPolynomialCategory RetractableTo *with* 0 1 * ** + - / < = D associates? characteristic charthRoot coefficient coefficients coerce conditionP content degree differentialVariables differentiate discriminant eval exquo factor factorPolynomial factorSquareFreePolynomial gcd gcdPolynomial ground ground? initial isExpt isPlus isTimes isobaric? lcm leader leadingCoefficient leadingMonomial mainVariable makeVariable map mapExponents max min minimumDegree monicDivide monomial monomial? monomials multivariate numberOfMonomials one? order prime? primitiveMonomials primitivePart recip reducedSystem reductum resultant retract retractIfCan separant solveLinearPolynomialEquation squareFree squareFreePart squareFreePolynomial totalDegree unit? unitCanonical unitNormal univariate variables weight weights zero?

SequentialDifferentialVariable{SDVAR}: DifferentialVariableCategory *with* < = D coerce differentiate makeVariable max min order retract retractIfCan variable weight

Set{SET}: FiniteSetAggregate *with* # < = any? bag brace cardinality coerce complement construct convert copy count dictionary difference empty empty? eq? every? extract! find index insert! inspect intersect less? lookup map map! max member? members min more? parts random reduce remove remove! removeDuplicates select select! size size? subset? symmetricDifference union universe

SExpressionOf{SEXOF}: SExpressionCategory *with* # = atom? car cdr coerce convert destruct elt eq expr float float? integer integer? list? null? pair? string string? symbol symbol? uequal

SExpression{SEX}: SExpressionCategory *with* # = atom? car cdr coerce convert destruct elt eq expr float float? integer integer? list? null? pair? string string? symbol symbol? uequal

SimpleAlgebraicExtension{SAE}: MonogenicAlgebra *with* 0 1 * ** + - / = D associates? basis characteristic characteristicPolynomial charthRoot coerce conditionP convert coordinates createPrimitiveElement definingPolynomial derivationCoordinates differentiate discreteLog discriminant divide euclideanSize expressIdealMember exquo extendedEuclidean factor factorsOfCyclicGroupSize gcd generator index init inv lcm lift lookup minimalPolynomial multiEuclidean nextItem norm one? order prime? primeFrobenius primitive? primitiveElement principalIdeal quo random rank recip reduce reducedSystem regularRepresentation rem representationType represents retract retractIfCan size sizeLess? squareFree squareFreePart tableForDiscreteLogarithm trace traceMatrix unit? unitCanonical unitNormal zero?

SingletonAsOrderedSet{SAOS}: OrderedSet *with* < = coerce create max min

SmallFloat{SF}: ConvertibleTo DifferentialRing FloatingPointSystem TranscendentalFunctionCategory *with* 0 1 * ** + - / < = D abs acos acosh acot acoth acsc acsch asec asech asin asinh associates? atan atanh base bits ceiling characteristic coerce convert cos cosh cot coth csc csch decreasePrecision differentiate digits divide euclideanSize exp exp1 exponent expressIdealMember exquo extendedEuclidean factor float floor fractionPart gcd hash increasePrecision inv lcm log log10 log2 mantissa max min multiEuclidean negative? norm nthRoot one? order patternMatch pi positive? precision prime? principalIdeal quo rationalApproximation recip rem retract retractIfCan round sec sech sign sin sinh sizeLess? sqrt squareFree squareFreePart tan tanh truncate unit? unitCanonical unitNormal wholePart zero?

SmallInteger{SI}: IntegerNumberSystem *with* 0 1 * ** + - < = And D Not Or ^ abs addmod and associates? base binomial bit? characteristic coerce convert copy dec differentiate divide euclideanSize even? expressIdealMember exquo extendedEuclidean factor factorial gcd hash inc init invmod lcm length mask max min mulmod multiEuclidean negative? nextItem not odd? one? or patternMatch permutation positive? positiveRemainder powmod prime? principalIdeal quo random rational rational? rationalIfCan recip reducedSystem rem retract retractIfCan shift sign sizeLess? squareFree squareFreePart submod symmetricRemainder unit? unitCanonical unitNormal xor zero?

SparseMultivariatePolynomial{SMP}: PolynomialCategory *with* 0 1 * ** + - / < = D associates? characteristic charthRoot coefficient coefficients coerce conditionP content convert degree differentiate discriminant eval exquo factor factorPolynomial factorSquareFreePolynomial gcd gcdPolynomial ground ground? isExpt isPlus isTimes lcm leadingCoefficient leadingMonomial mainVariable map mapExponents max min minimumDegree monicDivide monomial monomial? monomials multivariate numberOfMonomials one? patternMatch prime? primitiveMonomials primitivePart recip reducedSystem reductum resultant retract retractIfCan solveLinearPolynomialEquation squareFree squareFreePart squareFreePolynomial totalDegree unit? unitCanonical unitNormal univariate variables zero?

SparseMultivariateTaylorSeries{SMTS}: MultivariateTaylorSeriesCategory *with* 0 1 * ** + - / = D acos acosh acot acoth acsc acsch asec asech asin asinh associates? atan atanh characteristic charthRoot coefficient coerce complete cos cosh cot coth csc csch csubst degree differentiate eval exp exquo extend fintegrate integrate leadingCoefficient leadingMonomial log map monomial monomial? nthRoot one? order pi pole? polynomial recip reductum sec sech sin sinh sqrt tan tanh unit? unitCanonical unitNormal variables zero?

SparseTable{STBL}: TableAggregate *with* # = any? bag coerce construct copy count dictionary elt empty empty? entries entry? eq? every? extract! fill! find first index? indices insert! inspect key? keys less? map map! maxIndex member? members minIndex more? parts qelt qsetelt! reduce remove remove! removeDuplicates search select select! setelt size? swap! table

SparseUnivariatePolynomial{SUP}: UnivariatePolynomialCategory *with* 0 1 * ** + - / < = D associates? characteristic charthRoot coefficient coefficients coerce composite conditionP content degree differentiate discriminant divide divideExponents elt euclideanSize eval expressIdealMember exquo extendedEuclidean factor factorPolynomial factorSquareFreePolynomial gcd gcdPolynomial ground ground? init integrate isExpt isPlus isTimes lcm leadingCoefficient leadingMonomial mainVariable makeSUP map mapExponents max min minimumDegree monicDivide monomial monomial? monomials multiEuclidean multiplyExponents multivariate nextItem numberOfMonomials one? order outputForm prime? primitiveMonomials primitivePart principalIdeal pseudoDivide pseudoQuotient pseudoRemainder quo recip reducedSystem reductum rem resultant retract retractIfCan separate sizeLess? solveLinearPolynomialEquation squareFree squareFreePart squareFreePolynomial subResultantGcd totalDegree unit? unitCanonical unitNormal univariate unmakeSUP variables vectorise zero?

SparseUnivariateTaylorSeries{SUTS}: UnivariateTaylorSeriesCategory *with* 0 1 * ** + - / = D acos acosh acot acoth acsc acsch approximate asec asech asin asinh associates? atan atanh center characteristic charthRoot coefficient coefficients coerce complete cos cosh cot coth csc csch degree differentiate elt eval exp exquo extend integrate leadingCoefficient leadingMonomial log map monomial monomial? multiplyCoefficients multiplyExponents nthRoot one? order pi pole? polynomial quoByVar recip reductum sec sech series sin sinh sqrt tan tanh terms truncate unit? unitCanonical unitNormal variable variables zero?

SquareMatrix{SQMATRIX}: CoercibleTo SquareMatrixCategory *with* 0 1 # * ** + - / = D antisymmetric? any? characteristic coerce column copy count determinant diagonal diagonal? diagonalMatrix diagonalProduct differentiate elt empty empty? eq? every? exquo inverse less? listOfLists map map! matrix maxColIndex maxRowIndex member? members minColIndex minRowIndex minordet more? ncols nrows nullSpace nullity one? parts qelt rank recip reducedSystem retract retractIfCan row rowEchelon scalarMatrix size? square? squareMatrix symmetric? trace transpose zero?

Stack{STACK}: StackAggregate *with* # = any? bag coerce copy count depth empty empty? eq? every? extract! insert! inspect less? map map! member? members more? parts pop! push! size? stack top

Stream{STREAM}: LazyStreamAggregate *with* # = any? child? children coerce complete concat concat! cons construct convert copy count cycleEntry cycleLength cycleSplit! cycleTail cyclic? delay delete distance elt empty empty? entries entry? eq? every? explicitEntries? explicitlyEmpty? explicitlyFinite? extend fill! filterUntil filterWhile find findCycle first frst generate index? indices insert last lazy? lazyEvaluate leaf? less? map map! maxIndex member? members minIndex more? new node? nodes numberOfComputedEntries output parts possiblyInfinite? qelt qsetelt! reduce remove removeDuplicates repeating repeating? rest rst second select setchildren! setelt setfirst! setlast! setrest! setvalue! showAll? showAllElements size? split! swap! tail third value

StringTable{STRTBL}: TableAggregate *with* # = any? bag coerce construct copy count dictionary elt empty empty? entries entry? eq? every? extract! fill! find first index? indices insert! inspect key? keys less? map map! maxIndex member? members minIndex more? parts qelt qsetelt! reduce remove remove! removeDuplicates search select select! setelt size? swap! table

String{STRING}: StringCategory *with* # < = any? coerce concat construct copy copyInto! count delete elt empty empty? entries entry? eq? every? fill! find first index? indices insert leftTrim less? lowerCase lowerCase! map map! match? max maxIndex member? members merge min minIndex more? new parts position prefix? qelt qsetelt! reduce remove removeDuplicates replace reverse reverse! rightTrim select setelt size? sort sort! sorted? split string substring? suffix? swap! trim upperCase upperCase!

SubSpaceComponentProperty{COMPPROP}: SetCategory *with* = close closed? coerce copy new solid solid?

SubSpace{SUBSPACE}: SetCategory *with* = addPoint addPoint2 addPointLast birth child children closeComponent coerce deepCopy defineProperty extractClosed extractIndex extractPoint extractProperty internal? leaf? level merge modifyPoint new numberOfChildren parent pointData root? separate shallowCopy subspace traverse

SuchThat{SUCH}: SetCategory *with* = coerce construct lhs rhs

Symbol{SYMBOL}: ConvertibleTo OrderedSet PatternMatchable *with* < = argscript coerce convert elt list max min name new patternMatch resetNew script scripted? scripts string subscript superscript

SymmetricPolynomial{SYMPOLY}: FiniteAbelianMonoidRing *with* 0 1 * ** + - / = associates? characteristic charthRoot coefficient coefficients coerce content degree exquo ground ground? leadingCoefficient leadingMonomial map mapExponents minimumDegree monomial monomial? numberOfMonomials one? primitivePart recip reductum retract retractIfCan unit? unitCanonical unitNormal zero?

Tableau{TABLEAU}: Object *with* coerce listOfLists tableau

Table{TABLE}: TableAggregate *with* # = any? bag coerce construct copy count dictionary elt empty empty? entries entry? eq? every? extract! fill! find first index? indices insert! inspect key? keys less? map map! maxIndex member? members minIndex more? parts qelt qsetelt! reduce remove remove! removeDuplicates search select select! setelt size? swap! table

TaylorSeries{TS}: MultivariateTaylorSeriesCategory *with* 0 1 * ** + - / = D acos acosh acot acoth acsc acsch asec asech asin asinh associates? atan atanh characteristic charthRoot coefficient coerce complete cos cosh cot coth csc csch degree differentiate eval exp exquo extend fintegrate integrate leadingCoefficient leadingMonomial log map monomial monomial? nthRoot one? order pi pole? polynomial recip reductum sec sech sin sinh sqrt tan tanh unit? unitCanonical unitNormal variables zero?

TexFormat1{TEX1}: Object *with* coerce

TexFormat{TEX}: SetCategory *with* = coerce convert display epilogue new prologue setEpilogue! setPrologue! setTex! tex

TextFile{TEXTFILE}: FileCategory *with* = close! coerce endOfFile? iomode name open read! readIfCan! readLine! readLineIfCan! reopen! write! writeLine!

ThreeDimensionalViewport{VIEW3D}: SetCategory *with* = axes clipSurface close coerce colorDef controlPanel diagonals dimensions drawStyle eyeDistance hitherPlane intensity key lighting makeViewport3D modifyPointData move options outlineRender perspective reset resize rotate showClipRegion showRegion subspace title translate viewDeltaXDefault viewDeltaYDefault viewPhiDefault viewThetaDefault viewZoomDefault viewpoint viewport3D write zoom

ThreeSpace{SPACE3}: ThreeSpaceCategory *with* = check closedCurve closedCurve? coerce components composite composites copy create3Space curve curve? enterPointData lllip lllp llprop lp lprop merge mesh mesh? modifyPointData numberOfComponents numberOfComposites objects point point? polygon polygon? subspace

Tree{TREE}: RecursiveAggregate *with* # = any? children coerce copy count cyclic? elt empty empty? eq? every? leaf? leaves less? map map! member? members more? node? nodes parts setchildren! setelt setvalue! size? tree value

TubePlot{TUBE}: *with* closed? getCurve listLoops open? setClosed tube

Tuple{TUPLE}: CoercibleTo SetCategory *with* = coerce length select

TwoDimensionalArray{ARRAY2}: TwoDimensionalArrayCategory *with* # = any? coerce column copy count elt empty empty? eq? every? fill! less? map map! maxColIndex maxRowIndex member? members minColIndex minRowIndex more? ncols new nrows parts qelt qsetelt! row setColumn! setRow! setelt size?

TwoDimensionalViewport{VIEW2D}: SetCategory *with* = axes close coerce connect controlPanel dimensions getGraph graphState graphStates graphs key makeViewport2D move options points putGraph region reset resize scale show title translate units viewport2D write

UnivariateLaurentSeriesConstructor{ULSCONS}: UnivariateLaurentSeriesConstructorCategory *with* 0 1 * ** + - / < = D abs acos acosh acot acoth acsc acsch approximate asec asech asin asinh associates? atan atanh ceiling center characteristic charthRoot coefficient coerce complete conditionP convert cos cosh cot coth csc csch degree denom denominator differentiate divide elt euclideanSize eval exp expressIdealMember exquo extend extendedEuclidean factor factorPolynomial factorSquareFreePolynomial floor fractionPart gcd gcdPolynomial init integrate inv laurent lcm leadingCoefficient leadingMonomial log map max min monomial monomial? multiEuclidean multiplyCoefficients multiplyExponents negative? nextItem nthRoot numer numerator one? order patternMatch pi pole? positive? prime? principalIdeal quo random rationalFunction recip reducedSystem reductum rem removeZeroes retract retractIfCan sec sech series sign sin sinh sizeLess? solveLinearPolynomialEquation sqrt squareFree squareFreePart squareFreePolynomial tan tanh taylor taylorIfCan taylorRep terms truncate unit? unitCanonical unitNormal variable variables wholePart zero?

UnivariateLaurentSeries{ULS}: UnivariateLaurentSeriesConstructorCategory *with* 0 1 * ** + - / = D acos acosh acot acoth acsc acsch approximate asec asech asin asinh associates? atan atanh center characteristic charthRoot coefficient coerce complete cos cosh cot coth csc csch degree denom denominator differentiate divide elt euclideanSize eval exp expressIdealMember exquo extend extendedEuclidean factor gcd integrate inv laurent lcm leadingCoefficient leadingMonomial log map monomial monomial? multiEuclidean multiplyCoefficients multiplyExponents nthRoot numer numerator one? order pi pole? prime? principalIdeal quo rationalFunction recip reducedSystem reductum rem removeZeroes retract retractIfCan sec sech series sin sinh sizeLess? sqrt squareFree squareFreePart tan tanh taylor taylorIfCan taylorRep terms truncate unit? unitCanonical unitNormal variable variables zero?

UnivariatePolynomial{UP}: UnivariatePolynomialCategory *with* 0 1 * ** + - / < = D associates? characteristic charthRoot coefficient coefficients coerce composite conditionP content degree differentiate discriminant divide divideExponents elt euclideanSize eval expressIdealMember exquo extendedEuclidean factor factorPolynomial factorSquareFreePolynomial gcd gcdPolynomial ground ground? init integrate isExpt isPlus isTimes lcm leadingCoefficient leadingMonomial mainVariable makeSUP map mapExponents max min minimumDegree monicDivide monomial monomial? monomials multiEuclidean multiplyExponents multivariate nextItem numberOfMonomials one? order prime? primitiveMonomials primitivePart principalIdeal pseudoDivide pseudoQuotient pseudoRemainder quo recip reducedSystem reductum rem resultant retract retractIfCan separate sizeLess? solveLinearPolynomialEquation squareFree squareFreePart squareFreePolynomial subResultantGcd totalDegree unit? unitCanonical unitNormal univariate unmakeSUP variables vectorise zero?

UnivariatePuiseuxSeriesConstructor{UPXSCONS}: UnivariatePuiseuxSeriesConstructorCategory *with* 0 1 * ** + - / = D acos acosh acot acoth acsc acsch approximate asec asech asin asinh associates? atan atanh center characteristic charthRoot coefficient coerce complete cos cosh cot coth csc csch degree differentiate divide elt euclideanSize eval exp expressIdealMember exquo extend extendedEuclidean factor gcd integrate inv laurent laurentIfCan laurentRep lcm leadingCoefficient leadingMonomial log map monomial monomial? multiEuclidean multiplyExponents nthRoot one? order pi pole? prime? principalIdeal puiseux quo rationalPower recip reductum rem retract retractIfCan sec sech series sin sinh sizeLess? sqrt squareFree squareFreePart tan tanh terms truncate unit? unitCanonical unitNormal variable variables zero?

UnivariatePuiseuxSeries{UPXS}: UnivariatePuiseuxSeriesConstructorCategory *with* 0 1 * ** + - / = D acos acosh acot acoth acsc acsch approximate asec asech asin asinh associates? atan atanh center characteristic charthRoot coefficient coerce complete cos cosh cot coth csc csch degree differentiate divide elt euclideanSize eval exp expressIdealMember exquo extend extendedEuclidean factor gcd integrate inv laurent laurentIfCan laurentRep lcm leadingCoefficient leadingMonomial log map monomial monomial? multiEuclidean multiplyExponents nthRoot one? order pi pole? prime? principalIdeal puiseux quo rationalPower recip reductum rem retract retractIfCan sec sech series sin sinh sizeLess? sqrt squareFree squareFreePart tan tanh terms truncate unit? unitCanonical unitNormal variable variables zero?

UnivariateTaylorSeries{UTS}: UnivariateTaylorSeriesCategory *with* 0 1 * ** + - / = D acos acosh acot acoth acsc acsch approximate asec asech asin asinh associates? atan atanh center characteristic charthRoot coefficient coefficients coerce complete cos cosh cot coth csc csch degree differentiate elt eval evenlambert exp exquo extend generalLambert integrate invmultisect lagrange lambert leadingCoefficient leadingMonomial log map monomial monomial? multiplyCoefficients multiplyExponents multisect nthRoot oddlambert one? order pi pole? polynomial quoByVar recip reductum revert sec sech series sin sinh sqrt tan tanh terms truncate unit? unitCanonical unitNormal univariatePolynomial variable variables zero?

UniversalSegment{UNISEG}: SegmentCategory SegmentExpansionCategory *with* = BY SEGMENT coerce convert expand hasHi hi high incr lo low map segment

Variable{VARIABLE}: CoercibleTo SetCategory *with* = coerce variable

Vector{VECTOR}: VectorCategory *with* # * + - < = any? coerce concat construct convert copy copyInto! count delete dot elt empty empty? entries entry? eq? every? fill! find first index? indices insert less? map map! max maxIndex member? members merge min minIndex more? new parts position qelt qsetelt! reduce remove removeDuplicates reverse reverse! select setelt size? sort sort! sorted? swap! vector zero

Void{VOID}: *with* coerce void

APPENDIX D

Packages

This is a listing of all packages in the AXIOM library at the time this book was produced. Use the Browse facility (described in Chapter 14) to get more information about these constructors.

This sample entry will help you read the following table:

PackageName{PackageAbbreviation}: $\text{Category}_1 \ldots \text{Category}_N$ *with* $\text{operation}_1 \ldots \text{operation}_M$

where

PackageName	is the full package name, for example, PadeApproximantPackage.
PackageAbbreviation	is the package abbreviation, for example, PADEPAC.
Category_i	is a category to which the package belongs.
operation_j	is an operation exported by the package.

AlgebraicFunction{AF}: *with* ** belong? definingPolynomial inrootof iroot minPoly operator rootOf

AlgebraicHermiteIntegration{INTHERAL}: *with* HermiteIntegrate

AlgebraicIntegrate{INTALG}: *with* algintegrate palginfieldint palgintegrate

AlgebraicIntegration{INTAF}: *with* algint

AlgebraicManipulations{ALGMANIP}: *with* ratDenom ratPoly rootKerSimp rootSimp rootSplit

AlgebraicMultFact{ALGMFACT}: *with* factor

AlgebraPackage{ALGPKG}: *with* basisOfCenter basisOfCentroid basisOfCommutingElements basisOfLeftAnnihilator basisOfLeftNucleus basisOfLeftNucloid basisOfMiddleNucleus basisOfNucleus basisOfRightAnnihilator basisOfRightNucleus basisOfRightNucloid biRank doubleRank leftRank radicalOfLeftTraceForm rightRank weakBiRank

AlgFactor{ALGFACT}: *with* doublyTransitive? factor split

AnyFunctions1{ANY1}: *with* coerce retract retractIfCan retractable?

ApplyRules{APPRULE}: *with* applyRules localUnquote

AttachPredicates{PMPRED}: *with* suchThat

BalancedFactorisation{BALFACT}: *with* balancedFactorisation

BasicOperatorFunctions1{BOP1}: *with* constantOpIfCan constantOperator derivative evaluate

BezoutMatrix{BEZOUT}: *with* bezoutDiscriminant bezoutMatrix bezoutResultant

BoundIntegerRoots{BOUNDZRO}: *with* integerBound

CartesianTensorFunctions2{CARTEN2}: *with* map reshape

ChangeOfVariable{CHVAR}: *with* chvar eval goodPoint mkIntegral radPoly rootPoly

CharacteristicPolynomialPackage{CHARPOL}: *with* characteristicPolynomial

CoerceVectorMatrixPackage{CVMP}: *with* coerce coerceP

CombinatorialFunction{COMBF}: *with* ** belong? binomial factorial factorials iibinom iidprod iidsum iifact iiperm iipow ipow operator permutation product summation

CommonDenominator{CDEN}: *with* clearDenominator commonDenominator splitDenominator

CommonOperators{COMMONOP}: *with* operator

CommuteUnivariatePolynomialCategory{COMMUPC}: *with* swap

ComplexFactorization{COMPFACT}: *with* factor

ComplexFunctions2{COMPLEX2}: *with* map

ComplexIntegerSolveLinearPolynomialEquation {CINTSLPE}: *with* solveLinearPolynomialEquation

ComplexRootFindingPackage{CRFP}: *with* complexZeros divisorCascade factor graeffe norm pleskenSplit reciprocalPolynomial rootRadius schwerpunkt setErrorBound startPolynomial

ComplexRootPackage{CMPLXRT}: *with* complexZeros

ConstantLODE{ODECONST}: *with* constDsolve

CoordinateSystems{COORDSYS}: *with* bipolar bipolarCylindrical cartesian conical cylindrical elliptic ellipticCylindrical oblateSpheroidal parabolic parabolicCylindrical paraboloidal polar prolateSpheroidal spherical toroidal

CRApackage{CRAPACK}: *with* chineseRemainder modTree multiEuclideanTree

CycleIndicators{CYCLES}: *with* SFunction alternating cap complete cup cyclic dihedral elementary eval graphs powerSum skewSFunction wreath

CyclicStreamTools{CSTTOOLS}: *with* computeCycleEntry computeCycleLength cycleElt

CyclotomicPolynomialPackage{CYCLOTOM}: *with* cyclotomic cyclotomicDecomposition cyclotomicFactorization

DegreeReductionPackage{DEGRED}: *with* expand reduce

DiophantineSolutionPackage{DIOSP}: *with* dioSolve

DirectProductFunctions2{DIRPROD2}: *with* map reduce scan

DiscreteLogarithmPackage{DLP}: *with* shanksDiscLogAlgorithm

DisplayPackage{DISPLAY}: *with* bright center copies newLine say sayLength

DistinctDegreeFactorize{DDFACT}: *with* distdfact exptMod factor irreducible? separateDegrees separateFactors tracePowMod

DoubleResultantPackage{DBLRESP}: *with* doubleResultant

DrawNumericHack{DRAWHACK}: *with* coerce

DrawOptionFunctions0{DROPT0}: *with* adaptive clipBoolean coordinate curveColorPalette pointColorPalette ranges space style title toScale tubePoints tubeRadius units var1Steps var2Steps

DrawOptionFunctions1{DROPT1}: *with* option

EigenPackage{EP}: *with* characteristicPolynomial eigenvalues eigenvector eigenvectors inteigen

ElementaryFunctionODESolver{ODEEF}: *with* solve

ElementaryFunctionSign{SIGNEF}: *with* sign

ElementaryFunctionStructurePackage{EFSTRUC}: *with* normalize realElementary rischNormalize validExponential

ElementaryFunctionsUnivariateTaylorSeries{EFUTS}: *with* ** acos acosh acot acoth acsc acsch asec asech asin asinh atan atanh cos cosh cot coth csc csch exp log sec sech sin sincos sinh sinhcosh tan tanh

ElementaryFunction{EF}: *with* acos acosh acot acoth acsc acsch asec asech asin asinh atan atanh belong? cos cosh cot coth csc csch exp iiacos iiacosh iiacot iiacoth iiacsc iiacsch iiasec iiasech iiasin iiasinh iiatan iiatanh iicos iicosh iicot iicoth iicsc iicsch iiexp iilog iisec iisech iisin iisinh iitan iitanh log operator pi sec sech sin sinh specialTrigs tan tanh

ElementaryIntegration{INTEF}: *with* lfextendedint lfextlimint lfinfieldint lfintegrate lflimitedint

ElementaryRischDE{RDEEF}: *with* rischDE

EllipticFunctionsUnivariateTaylorSeries{ELFUTS}: *with* cn dn sn sncndn

EquationFunctions2{EQ2}: *with* map

ErrorFunctions{ERROR}: *with* error

EuclideanGroebnerBasisPackage{GBEUCLID}: *with* euclideanGroebner euclideanNormalForm

EvaluateCycleIndicators{EVALCYC}: *with* eval

ExpressionFunctions2{EXPR2}: *with* map

ExpressionSpaceFunctions1{ES1}: *with* map

ExpressionSpaceFunctions2{ES2}: *with* map

ExpressionSpaceODESolver{EXPRODE}: *with* seriesSolve

ExpressionToUnivariatePowerSeries{EXPR2UPS}: *with* laurent puiseux series taylor

ExpressionTubePlot{EXPRTUBE}: *with* constantToUnaryFunction tubePlot

FactoredFunctions2{FR2}: *with* map

FactoredFunctions{FACTFUNC}: *with* log nthRoot

FactoredFunctionUtilities{FRUTIL}: *with* mergeFactors refine

FactoringUtilities{FACUTIL}: *with* completeEval degree lowerPolynomial normalDeriv raisePolynomial ran variables

FindOrderFinite{FORDER}: *with* order

FiniteDivisorFunctions2{FDIV2}: *with* map

FiniteFieldFunctions{FFF}: *with* createMultiplicationMatrix createMultiplicationTable createZechTable sizeMultiplication

FiniteFieldHomomorphisms{FFHOM}: *with* coerce

FiniteFieldPolynomialPackage2{FFPOLY2}: *with* rootOfIrreduciblePoly

FiniteFieldPolynomialPackage{FFPOLY}: *with* createIrreduciblePoly createNormalPoly createNormalPrimitivePoly createPrimitiveNormalPoly createPrimitivePoly leastAffineMultiple nextIrreduciblePoly nextNormalPoly nextNormalPrimitivePoly nextPrimitiveNormalPoly nextPrimitivePoly normal? numberOfIrreduciblePoly numberOfNormalPoly numberOfPrimitivePoly primitive? random reducedQPowers

FiniteFieldSolveLinearPolynomialEquation{FFSLPE}: *with* solveLinearPolynomialEquation

FiniteLinearAggregateFunctions2{FLAGG2}: *with* map reduce scan

FiniteLinearAggregateSort{FLASORT}: *with* heapSort quickSort shellSort

FiniteSetAggregateFunctions2{FSAGG2}: *with* map reduce scan

FloatingComplexPackage{FLOATCP}: *with* complexRoots complexSolve

FloatingRealPackage{FLOATRP}: *with* realRoots solve

FractionalIdealFunctions2{FRIDEAL2}: *with* map

FractionFunctions2{FRAC2}: *with* map

FunctionalSpecialFunction{FSPECF}: *with* Beta Gamma abs airyAi airyBi belong? besselI besselJ besselK besselY digamma iiGamma iiabs operator polygamma

FunctionFieldCategoryFunctions2{FFCAT2}: *with* map

FunctionFieldIntegralBasis{FFINTBAS}: *with* integralBasis

FunctionSpaceAssertions{PMASSFS}: *with* assert constant multiple optional

FunctionSpaceAttachPredicates{PMPREDFS}: *with* suchThat

FunctionSpaceComplexIntegration{FSCINT}: *with* complexIntegrate internalIntegrate

FunctionSpaceFunctions2{FS2}: *with* map

FunctionSpaceIntegration{FSINT}: *with* integrate

FunctionSpacePrimitiveElement{FSPRMELT}: *with* primitiveElement

FunctionSpaceReduce{FSRED}: *with* bringDown newReduc

FunctionSpaceSum{SUMFS}: *with* sum

FunctionSpaceToUnivariatePowerSeries{FS2UPS}: *with* exprToGenUPS exprToUPS

FunctionSpaceUnivariatePolynomialFactor{FSUPFACT}: *with* ffactor qfactor

GaussianFactorizationPackage{GAUSSFAC}: *with* factor prime? sumSquares

GeneralHenselPackage{GHENSEL}: *with* HenselLift completeHensel

GeneralPolynomialGcdPackage{GENPGCD}: *with* gcdPolynomial randomR

GenerateUnivariatePowerSeries{GENUPS}: *with* laurent puiseux series taylor

GenExEuclid{GENEEZ}: *with* compBound reduction solveid tablePow testModulus

GenUFactorize{GENUFACT}: *with* factor

GenusZeroIntegration{INTG0}: *with* palgLODE0 palgRDE0 palgextint0 palgint0 palglimint0

GosperSummationMethod{GOSPER}: *with* GospersMethod

GraphicsDefaults{GRDEF}: *with* adaptive clipPointsDefault drawToScale maxPoints minPoints screenResolution

GrayCode{GRAY}: *with* firstSubsetGray nextSubsetGray

GroebnerFactorizationPackage{GBF}: *with* factorGroebnerBasis groebnerFactorize

GroebnerInternalPackage{GBINTERN}: *with* credPol critB critBonD critM critMTonD1 critMonD1 critT critpOrder fprindINFO gbasis hMonic lepol makeCrit minGbasis prinb prindINFO prinpolINFO prinshINFO redPo redPol sPol updatD updatF virtualDegree

GroebnerPackage{GB}: *with* groebner normalForm

GroebnerSolve{GROEBSOL}: *with* genericPosition groebSolve testDim

HallBasis{HB}: *with* generate inHallBasis? lfunc

HeuGcd{HEUGCD}: *with* content contprim gcd gcdcofact gcdcofactprim gcdprim lintgcd

IdealDecompositionPackage{IDECOMP}: *with* primaryDecomp prime? radical zeroDimPrimary? zeroDimPrime?

IncrementingMaps{INCRMAPS}: *with* increment incrementBy

InfiniteTupleFunctions2{ITFUN2}: *with* map

InfiniteTupleFunctions3{ITFUN3}: *with* map

Infinity{INFINITY}: *with* infinity minusInfinity plusInfinity

InnerAlgFactor{IALGFACT}: *with* factor

InnerCommonDenominator{ICDEN}: *with* clearDenominator commonDenominator splitDenominator

InnerMatrixLinearAlgebraFunctions{IMATLIN}: *with* determinant inverse nullSpace nullity rank rowEchelon

InnerMatrixQuotientFieldFunctions{IMATQF}: *with* inverse nullSpace nullity rank rowEchelon

InnerModularGcd{INMODGCD}: *with* modularGcd reduction

InnerMultFact{INNMFACT}: *with* factor

InnerNormalBasisFieldFunctions{INBFF}: *with* * ** / basis dAndcExp expPot index inv lookup minimalPolynomial norm normal? normalElement pol qPot random repSq setFieldInfo trace xn

InnerNumericEigenPackage{INEP}: *with* charpol innerEigenvectors

InnerNumericFloatSolvePackage{INFSP}: *with* innerSolve innerSolve1 makeEq

InnerPolySign{INPSIGN}: *with* signAround

InnerPolySum{ISUMP}: *with* sum

InnerTrigonometricManipulations{ITRIGMNP}: *with* F2FG FG2F GF2FG explogs2trigs trigs2explogs

InputFormFunctions1{INFORM1}: *with* interpret packageCall

IntegerCombinatoricFunctions{COMBINAT}: *with* binomial factorial multinomial partition permutation stirling1 stirling2

IntegerFactorizationPackage{INTFACT}: *with* BasicMethod PollardSmallFactor factor squareFree

IntegerLinearDependence{ZLINDEP}: *with* linearDependenceOverZ linearlyDependentOverZ? solveLinearlyOverQ

IntegerNumberTheoryFunctions{INTHEORY}: *with* bernoulli chineseRemainder divisors euler eulerPhi fibonacci harmonic jacobi legendre moebiusMu numberOfDivisors sumOfDivisors sumOfKthPowerDivisors

IntegerPrimesPackage{PRIMES}: *with* nextPrime prevPrime prime? primes

IntegerRetractions{INTRET}: *with* integer integer? integerIfCan

IntegerRoots{IROOT}: *with* approxNthRoot approxSqrt perfectNthPower? perfectNthRoot perfectSqrt perfectSquare?

IntegralBasisTools{IBATOOL}: *with* diagonalProduct idealiser leastPower

IntegrationResultFunctions2{IR2}: *with* map

IntegrationResultRFToFunction{IRRF2F}: *with* complexExpand complexIntegrate expand integrate split

IntegrationResultToFunction{IR2F}: *with* complexExpand expand split

IntegrationTools{INTTOOLS}: *with* kmax ksec mkPrim union vark varselect

InverseLaplaceTransform{INVLAPLA}: *with* inverseLaplace

IrredPolyOverFiniteField{IRREDFFX}: *with* generateIrredPoly

IrrRepSymNatPackage{IRSN}: *with* dimensionOfIrreducibleRepresentation irreducibleRepresentation

KernelFunctions2{KERNEL2}: *with* constantIfCan constantKernel

Kovacic{KOVACIC}: *with* kovacic

LaplaceTransform{LAPLACE}: *with* laplace

LeadingCoefDetermination{LEADCDET}: *with* distFact polCase

LinearDependence{LINDEP}: *with* linearDependence linearlyDependent? solveLinear

LinearPolynomialEquationByFractions{LPEFRAC}: *with* solveLinearPolynomialEquationByFractions

LinearSystemMatrixPackage{LSMP}: *with* aSolution hasSolution? rank solve

LinearSystemPolynomialPackage{LSPP}: *with* linSolve

LinGrobnerPackage{LGROBP}: *with* anticoord choosemon computeBasis coordinate groebgen intcompBasis linGenPos minPol totolex transform

LiouvillianFunction{LF}: *with* Ci Ei Si belong? dilog erf integral li operator

ListFunctions2{LIST2}: *with* map reduce scan

ListFunctions3{LIST3}: *with* map

ListToMap{LIST2MAP}: *with* match

MakeBinaryCompiledFunction{MKBCFUNC}: *with* binaryFunction compiledFunction

MakeFloatCompiledFunction{MKFLCFN}: *with* makeFloatFunction

MakeFunction{MKFUNC}: *with* function

MakeRecord{MKRECORD}: *with* makeRecord

MakeUnaryCompiledFunction{MKUCFUNC}: *with* compiledFunction unaryFunction

MappingPackage1{MAPPKG1}: *with* ** coerce fixedPoint id nullary recur

MappingPackage2{MAPPKG2}: *with* const constant curry diag

MappingPackage3{MAPPKG3}: *with* * constantLeft constantRight curryLeft curryRight twist

MappingPackageInternalHacks1{MAPHACK1}: *with* iter recur

MappingPackageInternalHacks2{MAPHACK2}: *with* arg1 arg2

MappingPackageInternalHacks3{MAPHACK3}: *with* comp

MatrixCategoryFunctions2{MATCAT2}: *with* map reduce

MatrixCommonDenominator{MCDEN}: *with* clearDenominator commonDenominator splitDenominator

MatrixLinearAlgebraFunctions{MATLIN}: *with* determinant inverse minordet nullSpace nullity rank rowEchelon

MergeThing{MTHING}: *with* mergeDifference

MeshCreationRoutinesForThreeDimensions{MESH}: *with* meshFun2Var meshPar1Var meshPar2Var ptFunc

ModularDistinctDegreeFactorizer{MDDFACT}: *with* ddFact exptMod factor gcd separateFactors

ModularHermitianRowReduction{MHROWRED}: *with* rowEch rowEchelon

MonoidRingFunctions2{MRF2}: *with* map

MoreSystemCommands{MSYSCMD}: *with* systemCommand

MPolyCatFunctions2{MPC2}: *with* map reshape

MPolyCatFunctions3{MPC3}: *with* map

MPolyCatRationalFunctionFactorizer{MPRFF}: *with* factor pushdown pushdterm pushucoef pushuconst pushup totalfract

MRationalFactorize{MRATFAC}: *with* factor

MultFiniteFactorize{MFINFACT}: *with* factor

MultipleMap{MMAP}: *with* map

MultivariateFactorize{MULTFACT}: *with* factor

MultivariateLifting{MLIFT}: *with* corrPoly lifting lifting1

MultivariateSquareFree{MULTSQFR}: *with* squareFree squareFreePrim

NonCommutativeOperatorDivision{NCODIV}: *with* leftDivide leftExactQuotient leftGcd leftLcm leftQuotient leftRemainder

NoneFunctions1{NONE1}: *with* coerce

NonLinearFirstOrderODESolver{NODE1}: *with* solve

NonLinearSolvePackage{NLINSOL}: *with* solve solveInField

NPCoef{NPCOEF}: *with* listexp npcoef

NumberFieldIntegralBasis{NFINTBAS}: *with* discriminant integralBasis

NumberFormats{NUMFMT}: *with* FormatArabic FormatRoman ScanArabic ScanRoman

NumberTheoreticPolynomialFunctions{NTPOLFN}: *with* bernoulliB cyclotomic eulerE

NumericalOrdinaryDifferentialEquations{NUMODE}: *with* rk4 rk4a rk4f rk4qc

NumericalQuadrature{NUMQUAD}: *with* aromberg asimpson atrapezoidal romberg rombergo simpson simpsono trapezoidal trapezoidalo

NumericComplexEigenPackage{NCEP}: *with* characteristicPolynomial complexEigenvalues complexEigenvectors

NumericContinuedFraction{NCNTFRAC}: *with* continuedFraction

NumericRealEigenPackage{NREP}: *with* characteristicPolynomial realEigenvalues realEigenvectors

NumericTubePlot{NUMTUBE}: *with* tube

Numeric{NUMERIC}: *with* complexNumeric numeric

OctonionCategoryFunctions2{OCTCT2}: *with* map

ODEIntegration{ODEINT}: *with* expint int

ODETools{ODETOOLS}: *with* particularSolution variationOfParameters wronskianMatrix

OneDimensionalArrayFunctions2{ARRAY12}: *with* map reduce scan

OnePointCompletionFunctions2{ONECOMP2}: *with* map

OperationsQuery{OPQUERY}: *with* getDatabase

OrderedCompletionFunctions2{ORDCOMP2}: *with* map

OrderingFunctions{ORDFUNS}: *with* pureLex reverseLex totalLex

OrthogonalPolynomialFunctions{ORTHPOL}: *with* ChebyshevU chebyshevT hermiteH laguerreL legendreP

OutputPackage{OUT}: *with* output

PadeApproximantPackage{PADEPAC}: *with* pade

PadeApproximants{PADE}: *with* pade padecf

ParadoxicalCombinatorsForStreams{YSTREAM}: *with* Y

PartitionsAndPermutations{PARTPERM}: *with* conjugate conjugates partitions permutations sequences shuffle shufflein

PatternFunctions1{PATTERN1}: *with* addBadValue

badValues predicate satisfy? suchThat

PatternFunctions2{PATTERN2}: *with* map

PatternMatchAssertions{PMASS}: *with* assert constant multiple optional

PatternMatchFunctionSpace{PMFS}: *with* patternMatch

PatternMatchIntegerNumberSystem{PMINS}: *with* patternMatch

PatternMatchKernel{PMKERNEL}: *with* patternMatch

PatternMatchListAggregate{PMLSAGG}: *with* patternMatch

PatternMatchPolynomialCategory{PMPLCAT}: *with* patternMatch

PatternMatchPushDown{PMDOWN}: *with* fixPredicate patternMatch

PatternMatchQuotientFieldCategory{PMQFCAT}: *with* patternMatch

PatternMatchResultFunctions2{PATRES2}: *with* map

PatternMatchSymbol{PMSYM}: *with* patternMatch

PatternMatchTools{PMTOOLS}: *with* patternMatch patternMatchTimes

PatternMatch{PATMATCH}: *with* Is is?

Permanent{PERMAN}: *with* permanent

PermutationGroupExamples{PGE}: *with* abelianGroup alternatingGroup cyclicGroup dihedralGroup janko2 mathieu11 mathieu12 mathieu22 mathieu23 mathieu24 rubiksGroup symmetricGroup youngGroup

PiCoercions{PICOERCE}: *with* coerce

PlotFunctions1{PLOT1}: *with* plot plotPolar

PlotTools{PLOTTOOL}: *with* calcRanges

PointFunctions2{PTFUNC2}: *with* map

PointPackage{PTPACK}: *with* color hue phiCoord rCoord shade thetaCoord xCoord yCoord zCoord

PointsOfFiniteOrderRational{PFOQ}: *with* order torsion? torsionIfCan

PointsOfFiniteOrderTools{PFOTOOLS}: *with* badNum doubleDisc getGoodPrime mix polyred

PointsOfFiniteOrder{PFO}: *with* order torsion? torsionIfCan

PolToPol{POLTOPOL}: *with* dmpToNdmp dmpToP ndmpToDmp ndmpToP pToDmp pToNdmp

PolyGroebner{PGROEB}: *with* lexGroebner totalGroebner

PolynomialAN2Expression{PAN2EXPR}: *with* coerce

PolynomialCategoryLifting{POLYLIFT}: *with* map

PolynomialCategoryQuotientFunctions{POLYCATQ}: *with* isExpt isPlus isPower isTimes mainVariable multivariate univariate variables

PolynomialFactorizationByRecursionUnivariate{PFBRU}: *with* bivariateSLPEBR factorByRecursion factorSFBRlcUnit factorSquareFreeByRecursion randomR solveLinearPolynomialEquationByRecursion

PolynomialFactorizationByRecursion{PFBR}: *with* bivariateSLPEBR factorByRecursion factorSFBRlcUnit factorSquareFreeByRecursion randomR solveLinearPolynomialEquationByRecursion

PolynomialFunctions2{POLY2}: *with* map

PolynomialGcdPackage{PGCD}: *with* gcd gcdPrimitive

PolynomialInterpolationAlgorithms{PINTERPA}: *with* LagrangeInterpolation

PolynomialInterpolation{PINTERP}: *with* interpolate

PolynomialNumberTheoryFunctions{PNTHEORY}: *with* bernoulli chebyshevT chebyshevU cyclotomic euler fixedDivisor hermite laguerre legendre

PolynomialRoots{POLYROOT}: *with* froot qroot rroot

PolynomialSolveByFormulas{SOLVEFOR}: *with* aCubic aLinear aQuadratic aQuartic aSolution cubic linear mapSolve quadratic quartic solve

PolynomialSquareFree{PSQFR}: *with* squareFree

PolynomialToUnivariatePolynomial{POLY2UP}: *with* univariate

PowerSeriesLimitPackage{LIMITPS}: *with* complexLimit limit

PrimitiveArrayFunctions2{PRIMARR2}: *with* map reduce scan

PrimitiveElement{PRIMELT}: *with* primitiveElement

PrimitiveRatDE{ODEPRIM}: *with* denomLODE

PrimitiveRatRicDE{ODEPRRIC}: *with* changevar constantCoefficientRicDE denomRicDE leadingCoefficientRicDE polyRicDE singRicDE

PrintPackage{PRINT}: *with* print

PureAlgebraicIntegration{INTPAF}: *with* palgLODE palgRDE palgextint palgint palglimint

PureAlgebraicLODE{ODEPAL}: *with* algDsolve

QuasiAlgebraicSet2{QALGSET2}: *with* radicalSimplify

QuaternionCategoryFunctions2{QUATCT2}: *with* map

QuotientFieldCategoryFunctions2{QFCAT2}: *with* map

RadicalEigenPackage{REP}: *with* eigenMatrix gramschmidt normalise orthonormalBasis radicalEigenvalues radicalEigenvector radicalEigenvectors

RadicalSolvePackage{SOLVERAD}: *with* contractSolve radicalRoots radicalSolve

RadixUtilities{RADUTIL}: *with* radix

RandomNumberSource{RANDSRC}: *with* randnum reseed size

RationalFactorize{RATFACT}: *with* factor

RationalFunctionDefiniteIntegration{DEFINTRF}: *with* integrate

RationalFunctionFactorizer{RFFACTOR}: *with* factorFraction

RationalFunctionFactor{RFFACT}: *with* factor

RationalFunctionIntegration{INTRF}: *with* extendedIntegrate infieldIntegrate internalIntegrate limitedIntegrate

RationalFunctionLimitPackage{LIMITRF}: *with* complexLimit limit

RationalFunctionSign{SIGNRF}: *with* sign

RationalFunctionSum{SUMRF}: *with* sum

RationalFunction{RF}: *with* coerce eval mainVariable multivariate univariate variables

RationalIntegration{INTRAT}: *with* extendedint infieldint integrate limitedint

RationalLODE{ODERAT}: *with* ratDsolve

RationalRetractions{RATRET}: *with* rational rational? rationalIfCan

RationalRicDE{ODERTRIC}: *with* changevar constantCoefficientRicDE polyRicDE ricDsolve singRicDE

RatODETools{RTODETLS}: *with* genericPolynomial

RealSolvePackage{REALSOLV}: *with* realSolve solve

RealZeroPackageQ{REAL0Q}: *with* realZeros refine

RealZeroPackage{REAL0}: *with* midpoint midpoints realZeros refine

RectangularMatrixCategoryFunctions2{RMCAT2}: *with* map reduce

ReducedDivisor{RDIV}: *with* order

ReduceLODE{ODERED}: *with* reduceLODE

ReductionOfOrder{REDORDER}: *with* ReduceOrder

RepeatedDoubling{REPDB}: *with* double

RepeatedSquaring{REPSQ}: *with* expt

RepresentationPackage1{REP1}: *with* antisymmetricTensors createGenericMatrix permutationRepresentation symmetricTensors tensorProduct

RepresentationPackage2{REP2}: *with* areEquivalent? completeEchelonBasis createRandomElement cyclicSubmodule isAbsolutelyIrreducible? meatAxe scanOneDimSubspaces split standardBasisOfCyclicSubmodule

ResolveLatticeCompletion{RESLATC}: *with* coerce

RetractSolvePackage{RETSOL}: *with* solveRetract

SAERationalFunctionAlgFactor{SAERFFC}: *with* factor

SegmentBindingFunctions2{SEGBIND2}: *with* map

SegmentFunctions2{SEG2}: *with* map

SimpleAlgebraicExtensionAlgFactor{SAEFACT}: *with* factor

SmallFloatSpecialFunctions{SFSFUN}: *with* Beta Gamma airyAi airyBi besselI besselJ besselK besselY digamma hypergeometric0F1 logGamma polygamma

SortedCache{SCACHE}: *with* cache clearCache enterInCache

SparseUnivariatePolynomialFunctions2{SUP2}: *with* map

SpecialOutputPackage{SPECOUT}: *with* outputAsFortran outputAsScript outputAsTex

StorageEfficientMatrixOperations{MATSTOR}: *with* ** copy! leftScalarTimes! minus! plus! power! rightScalarTimes! times!

StreamFunctions1{STREAM1}: *with* concat

StreamFunctions2{STREAM2}: *with* map reduce scan

StreamFunctions3{STREAM3}: *with* map

StreamTaylorSeriesOperations{STTAYLOR}: *with* * + - / addiag coerce compose deriv eval evenlambert gderiv generalLambert int integers integrate invmultisect lagrange lambert lazyGintegrate lazyIntegrate mapdiv mapmult monom multisect nlde oddintegers oddlambert power powern recip revert

StreamTranscendentalFunctions{STTF}: *with* ** acos acosh acot acoth acsc acsch asec asech asin asinh atan atanh cos cosh cot coth csc csch exp log sec sech sin sincos sinh sinhcosh tan tanh

SubResultantPackage{SUBRESP}: *with* primitivePart subresultantVector

SymmetricFunctions{SYMFUNC}: *with* symFunc

SymmetricGroupCombinatoricFunctions{SGCF}: *with* coleman inverseColeman listYoungTableaus makeYoungTableau nextColeman nextLatticePermutation nextPartition numberOfImproperPartitions subSet unrankImproperPartitions0 unrankImproperPartitions1

SystemODESolver{ODESYS}: *with* solveInField triangulate

SystemSolvePackage{SYSSOLP}: *with* solve triangularSystems

TableauxBumpers{TABLBUMP}: *with* bat bat1 bumprow bumptab bumptab1 inverse lex maxrow mr slex tab tab1 untab

TangentExpansions{TANEXP}: *with* tanAn tanNa tanSum

ToolsForSign{TOOLSIGN}: *with* direction nonQsign sign

TopLevelDrawFunctionsForAlgebraicCurves {DRAWCURV}: *with* draw

TopLevelDrawFunctionsForCompiledFunctions {DRAWCFUN}: *with* draw makeObject recolor

TopLevelDrawFunctions{DRAW}: *with* draw makeObject

TopLevelThreeSpace{TOPSP}: *with* createThreeSpace

TranscendentalHermiteIntegration{INTHERTR}: *with* HermiteIntegrate

TranscendentalIntegration{INTTR}: *with* expextendedint expintegrate expintfldpoly explimitedint primextendedint primextintfrac primintegrate primintegratefrac primintfldpoly primlimintfrac primlimitedint

TranscendentalManipulations{TRMANIP}: *with* cos2sec cosh2sech cot2tan cot2trig coth2tanh coth2trigh csc2sin csch2sinh expand expandLog expandPower htrigs removeCosSq removeCoshSq removeSinSq removeSinhSq sec2cos sech2cosh simplify simplifyExp sin2csc sinh2csch tan2cot tan2trig tanh2coth tanh2trigh

TranscendentalRischDE{RDETR}: *with* DSPDE SPDE baseRDE expRDE primRDE

TransSolvePackageService{SOLVESER}: *with* decomposeFunc unvectorise

TransSolvePackage{SOLVETRA}: *with* solve

TriangularMatrixOperations{TRIMAT}: *with* LowTriBddDenomInv UpTriBddDenomInv

TrigonometricManipulations{TRIGMNIP}: *with* complexElementary complexNormalize imag real real? trigs

TubePlotTools{TUBETOOL}: *with* * + - cosSinInfo cross dot loopPoints point unitVector

TwoDimensionalPlotClipping{CLIP}: *with* clip clipParametric clipWithRanges

TwoFactorize{TWOFACT}: *with* generalSqFr generalTwoFactor twoFactor

UnivariateFactorize{UNIFACT}: *with* factor factorSquareFree genFact henselFact henselfact quadratic sqroot trueFactors

UnivariateLaurentSeriesFunctions2{ULS2}: *with* map

UnivariatePolynomialCategoryFunctions2{UPOLYC2}: *with* map

UnivariatePolynomialCommonDenominator{UPCDEN}: *with* clearDenominator commonDenominator splitDenominator

UnivariatePolynomialFunctions2{UP2}: *with* map

UnivariatePolynomialSquareFree{UPSQFREE}: *with* BumInSepFFE squareFree squareFreePart

UnivariatePuiseuxSeriesFunctions2{UPXS2}: *with* map

UnivariateTaylorSeriesFunctions2{UTS2}: *with* map

UnivariateTaylorSeriesODESolver{UTSODE}: *with* mpsode ode ode1 ode2 stFunc1 stFunc2 stFuncN

UniversalSegmentFunctions2{UNISEG2}: *with* map

UserDefinedPartialOrdering{UDPO}: *with* getOrder largest less? more? setOrder userOrdered?

UserDefinedVariableOrdering{UDVO}: *with* getVariableOrder resetVariableOrder setVariableOrder

VectorFunctions2{VECTOR2}: *with* map reduce scan

ViewDefaultsPackage{VIEWDEF}: *with* axesColorDefault lineColorDefault pointColorDefault pointSizeDefault tubePointsDefault tubeRadiusDefault unitsColorDefault var1StepsDefault var2StepsDefault viewDefaults viewPosDefault viewSizeDefault viewWriteAvailable viewWriteDefault

ViewportPackage{VIEW}: *with* coerce drawCurves graphCurves

WeierstrassPreparation{WEIER}: *with* cfirst clikeUniv crest qqq sts2stst weierstrass

WildFunctionFieldIntegralBasis{WFFINTBS}: *with* integralBasis listSquaredFactors

APPENDIX E

Operations

This appendix contains a partial list of AXIOM operations with brief descriptions. For more details, use the Browse facility of HyperDoc: enter the name of the operation for which you want more information in the input area on the main Browse menu and then click on **Operations.**

#aggregate

$\#a$ returns the number of items in a.

xy**

$x**y$ returns x to the power y. Also, this operation returns, if x is:

an equation: a new equation by raising both sides of x to the power y.

a float or small float: **sign** (x)**exp** $(y \log(|x|))$.

See also InputForm and OutputForm.

x*y

The binary operator $*$ denotes multiplication. Its meaning depends on the type of its arguments:

if x and y are members of a ring (more generally, a domain of category SemiGroup), $x * y$ returns the product of x and y.

if r is an integer and x is an element of a ring, or if r is a scalar and x is a vector, matrix, or direct product: $r * x$ returns the left multiplication of r by x. More generally, if r is an integer and x is a member of a domain of category AbelianMonoid, or r is a member of domain R and x is a domain of category Module(R), GradedModule, or GradedAlgebra defined over R, $r * x$ returns the left multiplication of r by x. Here x can be a vector, a matrix, or a direct product. Similarly, $x * n$ returns the right integer multiple of x.

if a and b are monad elements, the product of a and b (see Monad).

if A and B are matrices, returns the product of A and B. If v is a row vector, $v * A$ returns the product of v and A. If v is column vector, $A * v$ returns the product of A with column vector v. In each case, the operation calls **error** if the dimensions are incompatible.

if s is an integer or float and c is a color, $s * c$ returns the weighted shade scaled by s.

if s and t are Cartesian tensors, $s * t$ is the inner product of the tensors s and t. This contracts the last index of s with the first index of t, that is,
$t * s = \texttt{contract}(t, \texttt{rank}\ t, s, 1)$,
$t * s = \sum_{k=1}^{N} t([i_1, .., i_N, k] * s[k, j_1, .., j_M])$.

if eq is an equation, $r * eq$ multiplies both sides of eq by r.

if I and J are ideals, the product of ideals.

See also OutputForm, Monad, LeftModule, RightModule, and FreeAbelianMonoidCategory,

See also InputForm and OutputForm.

x+y

The binary operator $+$ denotes addition. Its meaning depends on the type of its arguments. If x and y are:

members of a ring (more generally, of a domain of category AbelianSemiGroup): the sum of x and y.

matrices: the matrix sum if x and y have the same dimensions, and **error** otherwise.

vectors: the component-wise sum if x and y have the same length, and **error** otherwise.

colors: a color which additively mixes colors x and y.

equations: an equation created by adding the respective left- and right-hand sides of x and y.

elements of graded module or algebra: the sum of x and y

in the module of elements of the same degree as x and y.

ideals: the ideal generated by the union of x and y.

See also FreeAbelianMonoidCategory, InputForm and OutputForm.

[x]−y

$-x$ returns the negative (additive inverse) of x, where x is a member of a ring (more generally, a domain of category AbelianGroup). Also, x may be a matrix, a vector, or a member of a graded module.
$x - y$ returns $x + (-y)$.
See also CancellationAbelianMonoid and OutputForm.

x/y

The binary operator / generally denotes binary division. Its precise meaning, however, depends on the type of its arguments:

x and y are elements of a group: multiplies x by the inverse **inv** (y) of y.

x and y are elements of a field: divides x by y, calling **error** if $y = 0$.

x is a matrix or a vector and y is a scalar: divides each element of x by y.

x and y are floats or small floats: divides x by y.

x and y are fractions: returns the quotient as another fraction.

x and y are polynomials: returns the quotient as a fraction of polynomials.

See also AbelianMonoidRing, InputForm and OutputForm.

0

The additive identity element for a ring (more generally, for an AbelianMonoid). Also, for a graded module or algebra, the zero of degree 0 (see GradedModule). See also InputForm.

1

The multiplicative identity element for a ring (more generally, for a Monoid and MonadWithUnit). or a graded algebra. See also InputForm.

x<y

The binary operator $<$ denotes the boolean-valued "less than" function. Its meaning depends on the type of its arguments. The operation $x < y$ for x and y:

elements of a totally ordered set (such as integer and floating point numbers): tests if x is less than y.

sets: tests if all the elements of x are also elements of y.

permutations: tests if x is less than y; see Permutation for details. Note: this order relation is total if and only if the underlying domain is of category Finite or OrderedSet.

permutation groups: tests if x is a proper subgroup of y.

See also OutputForm.

x=y

The meaning of binary operator $x = y$ depends on the value expected of the operation. If the value is expected to be:

a boolean: $x = y$ tests that x and y are equal.

an equation: $x = y$ creates an equation.

See also OutputForm.

abelianGroup (*listOfPositiveIntegers*)

abelianGroup $([p_1, \ldots, p_k])$ constructs the abelian group that is the direct product of cyclic groups with order p_i.

absolutelyIrreducible? ()

absolutelyIrreducible? ()$F tests if the algebraic function field F remains irreducible over the algebraic closure of the ground field. See FunctionFieldCategory using Browse.

abs (*element*)

abs (x) returns the absolute value of x, an element of an OrderedRing or a Complex, Quaternion, or Octonion value.

acos (*expression*)
acosIfCan (*expression*)

Argument x can be a Complex, Float, SmallFloat, or Expression value or a series.
acos (x) returns the arccosine of x.
acosIfCan (x) returns **acos** (x) if possible, and `"failed"` otherwise.

acosh (*expression*)
acoshIfCan (*expression*)

Argument x can be a Complex, Float, SmallFloat, or Expression value or a series.
acosh (x) returns the hyperbolic arccosine of x.
acoshIfCan (x) returns **acosh** (x) if possible, and `"failed"` otherwise.

acoth (*expression*)
acothIfCan (*expression*)

Argument x can be a Complex, Float, SmallFloat, or Expression value or a series.
acoth (x) returns the hyperbolic arccotangent of x.
acothIfCan (x) returns **acoth** (x) if possible, and `"failed"` otherwise.

acot (*expression*)
acotIfCan (*expression*)

Argument x can be a Complex, Float, SmallFloat, or Expression value or a series.
acot (x) returns the arccotangent of x.
acotIfCan (x) returns **acot** (x) if possible, and `"failed"` otherwise.

acsch (*expression*)
acschIfCan (*expression*)

Argument x can be a Complex, Float, SmallFloat, or

Expression value or a series.
acsch (x) returns the hyperbolic arccosecant of x.
acschIfCan (x) returns **acsch** (x) if possible, and `"failed"` otherwise.

acsc (*expression*)
acscIfCan (*expression*)

Argument x can be a Complex, Float, SmallFloat, or Expression value or a series.
acsc (x) returns the arccosecant of x.
acscIfCan (x) returns **acsc** (x) if possible, and `"failed"` otherwise.

adaptive ([*boolean*])

adaptive () tests whether plotting will be done adaptively.
adaptive $(true)$ turns adaptive plotting on; **adaptive** $(false)$ turns it off. Note: this command can be expressed by the draw option $adaptive == b$.

addmod (*integer, integer, integer*)

addmod (a, b, p), $0 \leq a, b < p > 1$, means $a + b$ **mod** p.

airyAi (*complexSmallFloat*)
airyBi (*complexSmallFloat*)

airyAi (x) is the Airy function $\mathrm{Ai}(x)$ satisfying the differential equation $\mathrm{Ai}''(x) - x\mathrm{Ai}(x) = 0$.
airyBi (x) is the Airy function $\mathrm{Bi}(x)$ satisfying the differential equation $\mathrm{Bi}''(x) - x\mathrm{Bi}(x) = 0$.

Aleph (*nonNegativeInteger*)

Aleph (n) provides the named (infinite) cardinal number.

algebraic? ()

algebraic? (a) tests whether an element a is algebraic with respect to the ground field F.

alphabetic ()
alphabetic? (*character*)

alphabetic () returns the class of all characters ch for which **alphabetic?** (ch) is $true$.
alphabetic? (ch) tests if ch is an alphabetic character a...z, A...B.

alphanumeric ()
alphanumeric? (*character*)

alphanumeric () returns the class of all characters ch for which **alphanumeric?** (ch) is $true$.
alphanumeric? (ch) tests if ch is either an alphabetic character a...z, A...B or digit 0...9.

alternating (*integer*)

alternating (n) is the cycle index of the alternating group of degree n. See CycleIndicators for details.

alternatingGroup (*listOfIntegers*)

alternatingGroup (li) constructs the alternating group acting on the integers in the list li. If n is odd, the generators are in general the $(n-2)$-cycle $(li.3, \ldots, li.n)$ and the 3-cycle $(li.1, li.2, li.3)$. If n is even, the generators are the product of the 2-cycle $(li.1, li.2)$ with $(n-2)$-cycle $(li.3, \ldots, li.n)$ and the 3-cycle $(li.1, li.2, li.3)$. Duplicates in the list will be removed.
alternatingGroup (n) constructs the alternating group A_n acting on the integers $1, \ldots, n$. If n is odd, the generators are in general the $(n-2)$-cycle $(3, \ldots, n)$ and the 3-cycle $(1, 2, 3)$. If n is even, the generators are the product of the 2-cycle $(1, 2)$ with $(n-2)$-cycle $(3, \ldots, n)$ and the 3-cycle $(1, 2, 3)$ if n is even.

alternative? ()

alternative? ()F tests if 2**associator**$(a, a, b) = 0 = 2$**associator**(a, b, b) for all a, b in the algebra F. Note: in general, $2a = 0$ does not necessarily imply $a = 0$.

and (*boolean, boolean*)

x **and** y returns the logical *and* of two BitAggregates x and y.
b_1 **and** b_2 returns the logical *and* of Boolean b_1 and b_2.
si_1 **and** si_2 returns the bit-by-bit logical *and* of the small integers si_1 and si_2.
See also OutputForm.

approximants (*continuedFraction*)

approximants (cf) returns the stream of approximants of the continued fraction cf. If the continued fraction is finite, then the stream will be infinite and periodic with period 1.

approximate (*series, integer*)

approximate (s, r) returns a truncated power series as an expression in the coefficient domain of the power series. For example, if R is Fraction Polynomial Integer and s is a series over R, then approximate(s, r) returns the power series s truncated after the exponent r term.

approximate (*pAdicInteger, integer*)

approximate (x, n), x a p-adic integer, returns an integer y such that $y = x$ **mod** p^n when n is positive, and 0 otherwise.

approxNthRoot (*integer, nonNegativeInteger*)

approxNthRoot (n, p) returns an integer approximation i to $n^{1/p}$ such that $-1 < i - n^{1/p} < 1$.

approxSqrt (*integer*)

approxSqrt (n) returns an integer approximation i to $\sqrt{(n)}$ such that $-1 < i - \sqrt{(n)} < 1$. A variable precision Newton iteration is used with running time $O(\log(n)^2)$.

areEquivalent? (*listOfMatrices, listOfMatrices* [, *randomElements?, numberOfTries*])

areEquivalent? $(lM, lM', b, numberOfTries)$ tests whether

the two lists of matrices, assumed of the same square shape, can be simultaneously conjugated by a non-singular matrix. If these matrices represent the same group generators, the representations are equivalent. The algorithm tries $numberOfTries$ times to create elements in the generated algebras in the same fashion. For details, consult Browse.
areEquivalent? $(aG0, aG1, numberOfTries)$ calls **areEquivalent?** $(aG0, aG1, true, 25)$.
areEquivalent? $(aG0, aG1)$ calls **areEquivalent?** $(aG0, aG1, true, 25)$.

argscript (*symbol, listOfOutputForms*)

argscript $(f, [o_1, \ldots, o_n])$ returns a new symbol with f with scripts $o_1, \ldots, o_n$.

argument (*complexExpression*)

argument (c) returns the angle made by complex expression c with the positive real axis.

arity (*basicOperator*)

arity (*op*) returns n if *op* is n-ary, and `"failed"` if *op* has arbitrary arity.

asec (*expression*)
asecIfCan (*expression*)

Argument x can be a Complex, Float, SmallFloat, or Expression value or a series.
asec (x) returns the arcsecant of x.
asecIfCan (x) returns **asec** (x) if possible, and `"failed"` otherwise.

asech (*expression*)
asechIfCan (*expression*)

Argument x can be a Complex, Float, SmallFloat, or Expression value or a series.
asech (x) returns the hyperbolic arcsecant of x.
asechIfCan (x) returns **asech** (x) if possible, and `"failed"` otherwise.

asin (*expression*)
asinIfCan (*expression*)

Argument x can be a Complex, Float, SmallFloat, or Expression value or a series.
asin (x) returns the arcsine of x.
asinIfCan (x) returns **asin** (x) if possible, and `"failed"` otherwise.

asinh (*expression*)
asinhIfCan (*expression*)

Argument x can be a Complex, Float, SmallFloat, or Expression value or a series.
asinh (x) returns the hyperbolic arcsine of x.
asinhIfCan (x) returns **asinh** (x) if possible, and `"failed"` otherwise.

assign (*outputForm, outputForm*)

assign (f, g) creates an OutputForm object for the assignment $f{:=}g$.

associates? (*element, element*)

associates? (x, y) tests whether x and y are associates, that is, that x and y differ by a unit factor.

associative? ()

associative? $()\$F$ tests if multiplication in F is associative, where F is a FiniteRankNonAssociativeAlgebra.

associatorDependence ()

associatorDependence $()\$F$ computes associator identities for F. Consult FiniteRankNonAssociativeAlgebra using Browse for details..

associator (*element, element, element*)

associator (a, b, c) returns $(ab)c - a(bc)$, where a, b, and c are all members of a domain of category NonAssociateRng.

assoc (*element, associationList*)

assoc (k, al) returns the element x in the AssociationList al stored under key k, or `"failed"` if no such element exists.

atan (*expression* [, *phase*])
atanIfCan (*expression*)

Argument x can be a Complex, Float, SmallFloat, or Expression value or a series.
atan (x) returns the arctangent of x.
atan (x, y) computes the arc tangent from x with phase y.
atanIfCan (x) returns the **atan** (x) if possible, and `"failed"` otherwise.

atanh (*expression*)
atanhIfCan (*expression*)

Argument x can be a Complex, Float, SmallFloat, or Expression value or a series.
atanh (x) returns the hyperbolic arctangent of x.
atanhIfCan (x) returns **atanh** (x) if possible, and `"failed"` otherwise.

atom? (*sExpression*)

atom? (s) tests if x is atomic, where x is an SExpression or OutputForm.

antiCommutator (*element, element*)

antiCommutator (x, y) returns $xy + yx$, where x and y are elements of a non-associative ring, possibly without identity. See NonAssociativeRng using Browse.

antisymmetric? (*matrix*)

antisymmetric? (m) tests if the matrix m is square and antisymmetric, that is, $m_{i,j} = -m_{j,i}$ for all i and j.

antisymmetricTensors (*matrices, positiveInteger*)

antisymmetricTensors (A, n), where A is an m by m matrix, returns a matrix obtained by applying to A the irreducible, polynomial representation of the general linear group GL_m corresponding to the partition $(1, 1, \ldots, 1, 0, 0, \ldots, 0)$ of n. A call to **error** occurs if n is greater than m. Note: this corresponds to the symmetrization of the representation with the sign representation of the symmetric group S_n. The carrier spaces of the representation are the antisymmetric tensors of the n-fold tensor product.
antisymmetricTensors (lA, n), where lA is a list of m by m matrices, similarly applies the representation of GL_m to each matrix A of lA, returning a list of matrices.

any? (*predicate, aggregate*)

any? $(pred, a)$ tests if predicate **pred** (x) is $true$ for any element x of aggregate a. Note: for collections, `any?(p, u) = reduce(or, map(p, u), false, true)`.

any (*type, object*)

any $(type, object)$ is a technical function for creating an $object$ of Any. Argument $type$ is a *LISP* form for the $type$ of $object$.

append (*list, list*)

append (l_1, l_2) appends the elements of list l_1 onto the front of list l_2. See also **concat**.

axesColorDefault ([*palette*])

axesColorDefault (p) sets the default color of the axes in a two-dimensional viewport to the palette p.
axesColorDefault () returns the default color of the axes in a two-dimensional viewport.

back (*queue*)

back (q) returns the element at the back of the queue, or calls **error** if q is empty.

bag ([*bag*])

bag $([x, y, \ldots, z])$ creates a bag with elements x, y, ..., z.

balancedBinaryTree (*nonNegativeInteger, element*)

balancedBinaryTree (n, s) creates a balanced binary tree with n nodes, each with value s.

base (*group*)

base (gp) returns a base for the group *gp*. Consult PermutationGroup using Browse for details.

basis ()

basis ()$\$R$ returns a fixed basis of R or a subspace of R. See FiniteAlgebraicExtensionField, FramedAlgebra, FramedNonAssociativeAlgebra using Browse for details.

basisOfCenter ()

basisOfCenter ()$\$R$ returns a basis of the space of all x in R satisfying **commutator** $(x, a) = 0$ and **associator** (x, a, b) = **associator** (a, x, b) = **associator** (a, b, x) = 0 for all a, b in R. Domain R is a domain of category FramedNonAssociativeAlgebra.

basisOfCentroid ()

basisOfCentroid ()$\$R$ returns a basis of the centroid of R, that is, the endomorphism ring of R considered as (R, R)-bimodule. Domain R is a domain of category FramedNonAssociativeAlgebra.

basisOfCommutingElements ()

basisOfCommutingElements ()$\$R$ returns a basis of the space of all x of R satisfying **commutator** (x, a) = 0 for all a in R. Domain R is a domain of category FramedNonAssociativeAlgebra.

basisOfLeftAnnihilator (*element*)
basisOfRightAnnihilator (*element*)

These operations return a basis of the space of all x in R of category FramedNonAssociativeAlgebra, satisfying

basisOfLeftAnnihilator (a): $0 = xa$.
basisOfRightAnnihilator (a): $0 = ax$.

basisOfNucleus ()
basisOfLeftNucleus ()
basisOfMiddleNucleus ()
basisOfRightNucleus ()

Each operation returns a basis of the space of all x of R, a domain of category FramedNonAssociativeAlgebra, satisfying for all a and b:

basisOfNucleus ()$\$R$: **associator** (x, a, b) = **associator** (a, x, b) = **associator** (a, b, x) = 0;
basisOfLeftNucleus ()$\$R$: **associator** (x, a, b) = 0;
basisOfMiddleNucleus ()$\$R$: **associator** (a, x, b) = 0;
basisOfRightNucleus ()$\$R$: **associator** (a, b, x) = 0.

basisOfLeftNucloid ()
basisOfRightNucloid ()

Each operation returns a basis of the space of endomorphisms of R, a domain of category FramedNonAssociativeAlgebra, considered as:

basisOfLeftNucloid (): a right module.
basisOfRightNucloid (): a left module.

Note: if R has a unit, the left and right nucloid coincide with the left and right nucleus.

belong? (*operator*)

belong? $(op)\$R$ tests if *op* is known as an operator to R. For example, R is an Expression domain or AlgebraicNumber.

bernoulli (*integer*)

bernoulli (n) returns the nth Bernoulli number, that is,

$B(n, 0)$ where $B(n, x)$ is the nth Bernoulli polynomial.

besselI (*complexSmallFloat, complexSmallFloat*)
besselJ (*complexSmallFloat, complexSmallFloat*)
besselK (*complexSmallFloat, complexSmallFloat*)
besselY (*complexSmallFloat, complexSmallFloat*)

besselI (v, x) is the modified Bessel function of the first kind, $I(v, x)$, satisfying the differential equation $x^2w''(x) + xw'(x) - (x^2 + v^2)w(x) = 0$.

besselJ (v, x) is the Bessel function of the second kind, $J(v, x)$, satisfying the differential equation $x^2w''(x) + xw'(x) + (x^2 - v^2)w(x) = 0$.

besselK (v, x) is the modified Bessel function of the first kind, $K(v, x)$, satisfying the differential equation $x^2w''(x) + xw'(x) - (x^2 + v^2)w(x) = 0$. Note: The default implementation uses the relation $K(v, x) = \pi/2(I(-v, x) - I(v, x))/\sin(v\pi)$ so is not valid for integer values of v.

besselY (v, x) is the Bessel function of the second kind, $Y(v, x)$, satisfying the differential equation $x^2w''(x) + xw'(x) + (x^2 - v^2)w(x) = 0$. Note: The default implementation uses the relation $Y(v, x) = (J(v, x)\cos(v\pi) - J(-v, x))/\sin(v\pi)$ so is not valid for integer values of v.

Beta (*complexSmallFloat, complexSmallFloat*)

Beta (x, y) is the Euler beta function, $B(x, y)$, defined by **Beta** (x, y) $\int_0^1 t^{x-1}(1-t)^{y-1}dt$. Note: this function is defined by **Beta** $(x, y) = \frac{(x)(y)}{(x+y)}$.

binaryTournament (*listOfElements*)

binaryTournament (ls) creates a BinaryTournament tree with the elements of ls as values at the nodes.

binaryTree (*value*)

binaryTree (x) creates a binary tree consisting of one node for which the **value** is x and the **left** and **right** subtrees are empty.

binary (*various*)

binary (rn) converts rational number rn to a binary expansion.
binary $(op, [a_1, \ldots, a_n])$ returns the input form corresponding to $a_1 \mathrm{op} \ldots op a_n$, where op and the a_i's are of type InputForm.

binomial (*integerNumber, integerNumber*)

binomial (x, y) returns the binomial coefficient $C(x, y) = x!/(y!(x-y)!)$, where $x \geq y \geq 0$, the number of combinations of x objects taken y at a time. Arguments x and y can come from any Expression or IntegerNumberSystem domain.

bipolar (x)
bipolarCylindrical (x)

bipolar (a) returns a function for transforming bipolar coordinates to Cartesian coordinates; this function maps the point (u, v) to $(x = a\sinh(v)/(\cosh(v) - \cos(u)), y = a\sin(u)/(\cosh(v) - \cos(u)))$.
bipolarCylindrical (a) returns a function for transforming bipolar cylindrical coordinates to Cartesian coordinates; this function maps the point (u, v, z) to $(x = a\sinh(v)/(\cosh(v) - \cos(u)), y = a\sin(u)/(\cosh(v) - \cos(u)), z)$.

biRank (*element*)

biRank $(x)\$R$, where R is a domain of category FramedNonAssociativeAlgebra, returns the number of linearly independent elements among x, xb_i, b_ix, b_ixb_j, $i, j = 1, \ldots, n$, where $b = [b_1, \ldots, b_n]$ is the fixed basis for R. Note: if R has a unit, then **doubleRank**, **weakBiRank** and **biRank** coincide.

bit? (*integer, integer*)

bit? (i, n) tests if the nth bit of i is a 1.

bits ()

bits () returns the precision of floats in bits. Also see **precision**.

blankSeparate (*listOfOutputForms*)

blankSeparate (*lo*), where *lo* is a list of objects of type OutputForm (normally unexposed), returns a single output form consisting of the elements of *lo* separated by blanks.

blue ()

blue () returns the position of the blue hue from total hues.

bottom! (*dequeue*)

bottom! (q) removes then returns the element at the bottom (back) of the dequeue q.

box (*expression*)

box (e), where e is an expression, returns e with a box around it that prevents e from being evaluated when operators are applied to it. For example, **log** (1) returns 0, but **log** (**box**(1)) returns the formal kernel **log** (1).
box $(f_1, \ldots, f_n)$, where the f_i are expressions, returns $(f_1, \ldots, f_n)$ with a box around them that prevents the f_i from being evaluated when operators are applied to them, and makes them applicable to a unary operator. For example, **atan** (**box**$[x, 2]$) returns the formal kernel **atan** $(x, 2)$.
box (o), where o is an object of type OutputForm (normally unexposed), returns an output form enclosing o in a box.

brace (*outputForm*)

brace (o), where o is an object of type OutputForm (normally unexposed), returns an output form enclosing o in braces.

bracket (*outputForm*)

bracket (o), where o is an object of type OutputForm (normally unexposed), returns an output form enclosing o in brackets.

branchPoint (*element*)
branchPointAtInfinity? ()

branchPoint? (a)F tests if $x = a$ is a branch point of the algebraic function field F.
branchPointAtInfinity? ()F tests if the algebraic function field F has a branch point at infinity.

bright (*color*)

bright (c) sets the shade of a hue, c, above dim but below pastel.
bright (ls) sets the font property of a list of strings ls to bold-face type.

cap (*symmetricPolynomial, symmetricPolynomial*)

cap (s_1, s_2), introduced by Redfield, is the scalar product of two cycle indices, where the s_i are SymmetricPolynomials with rational number coefficients. See also **cup**. See CycleIndicators for details.

cardinality (*finiteSetAggregate*)

cardinality (u) returns the number of elements of u. Note: `cardinality(u) = #u`.

car (*sExpression*)

car (se) returns a_1 when se is the SExpression object $(a_1, \ldots, a_n)$.

cdr (*sExpression*)

cdr (se) returns $(a_2, \ldots, a_n)$ when se is the SExpression object $(a_1, \ldots, a_n)$.

ceiling (*floatOrRationalNumber*)

Argument x is a floating point number or fraction of numbers.
ceiling (x) returns the smallest integral element above x.

center (*stringsOrSeries*)

center (s) returns the point about which the series s is expanded.
center (ls, n, s) takes a list of strings ls, and centers them within a list of strings which is n characters long. The remaining spaces are filled with strings composed of as many repetitions as possible of the last string parameter s.
center (s_1, n, s_2) is equivalent to **center** ($[s_1], n, s_2$).

char (*character*)

char (i) returns a Character object with integer code i. Note: **ord(char(*i*))** = *i*.
char (s) returns the unique character of a string s of length one.

characteristic ()

characteristic ()R returns the characteristic of ring R: the smallest positive integer n such that $nx = 0$ for all x in the ring, or zero if no such n exists.

characteristicPolynomial (*matrix*[, *symbol*])

characteristicPolynomial (a) returns the characteristic polynomial of the regular representation of a with respect to any basis.
characteristicPolynomial (m) returns the characteristic polynomial of the matrix m expressed as polynomial with a new symbol as variable.
characteristicPolynomial (m, sy) is similar except that the resulting polynomial has variable sy.
characteristicPolynomial (m, r), where r is a member of the coefficient domain of matrix m, evaluates the characteristic polynomial at r. In particular, if r is the polynomial $'x$, then it returns the characteristic polynomial expressed as a polynomial in $'x$.

charClass (*strings*)

charClass (s) creates a character class containing exactly the characters given in the string s.
charClass (ls) creates a character class which contains exactly the characters given in the list ls of strings.

charthRoot (*element*)

charthRoot (r), where r is an element of domain with **characteristic** $p \neq 0$, returns the pth root of r, or `"failed"` if none exists in the domain.
charthRoot (f)R takes the pth root of finite field element f, where p is the characteristic of the finite field R. Note: such a root is always defined in finite fields.

chebyshevT (*positiveInteger, element*)

chebyshevT (n, x) returns the nth Chebyshev polynomial of the first kind, $T_n(x)$, defined by $(1 - tx)/(1 - 2tx + t^2) = \sum_{n=0}^{\infty} T_n(x)\, t^n$.

children (*recursiveAggregate*)

children (u) returns a list of the children of aggregate u.

chineseRemainder (*listOfElements, listOfModuli*)
chineseRemainder (*integer, modulus, integer, modulus*)

chineseRemainder (lv, lm) where lv is a list of values $[v_1, \ldots, v_n]$ and lm is a list of moduli $[m_1, \ldots, m_n]$, returns m such that $m = n_i \bmod p_i$; the p_i must be relatively prime.
chineseRemainder (n_1, p_1, n_2, p_2) is equivalent to **chineseRemainder** ($[n_1, n_2], [p_1, p_2]$), where all arguments are integers.

clearDenominator (*fraction*)

clearDenominator ($[q_1, \ldots,]$) returns $[p_1, \ldots,]$ such that $q_i = p_i/d$ where d is a common denominator for the q_i's.
clearDenominator (A), where A is a matrix of fractions,

returns matrix B such that $A = B/d$ where d is a common denominator for the elements of A.
clearDenominator (p) returns polynomial q such that $p = q/d$ where d is a common denominator for the coefficients of polynomial p.

clip (*rangeOrBoolean*)

clip (b) turns two-dimensional clipping on if b is $true$, and off if b is $false$. This command may be given as a draw option: `clip == b`.
clip ([*a*..*b*]) defines the range for user-defined clipping. This command may be given as a draw option: `range == [a..b]`.

clipPointsDefault ([*boolean*])

clipPointsDefault () tests if automatic clipping is to be done.

clipPointsDefault (b) turns on automatic clipping for $b = true$, and off if $b = false$. This command may be given as a draw option: `clip == b`.

close (*filename*)

close (v) closes the viewport window of the given two-dimensional or three-dimensional viewport v and terminates the corresponding **Unix** process. Argument v is a member of domain TwoDimensionalViewport or ThreeDimensionalViewport.

close! (*filename*)

close! (fn) returns the file fn closed to input and output.

closedCurve? (*threeSpace*)

closedCurve? (sp) tests if the ThreeSpace object sp contains a single closed curve component.

closedCurve (*listsOfPoints* [, *listOfPoints*])

closedCurve (lpt) returns a ThreeSpace object containing a single closed curve described by the list of points lpt of the form $[p_0, p_1, \ldots, p_n, p_0]$.
closedCurve (sp) returns a closed curve as a list of points, where sp must be a ThreeSpace object containing a single closed curve.
closedCurve (sp, lpt) returns ThreeSpace object with the closed curve denoted by lpt added. Argument lpt is a list of points of the form $[p_0, p_1, \ldots, p_n, p_0]$.

coefficient (*polynomialOrSeries, nonNegativeInteger*)

coefficient (p, n) extracts the coefficient of the monomial with exponent n from polynomial p, or returns zero if exponent is not present.
coefficient (u, x, n) returns the coefficient of variable x to the power n in u, a multivariate polynomial or series.
coefficient $(u, [x_1, \ldots,], [n_1, \ldots,])$ returns the coefficient of $x_1^{n_1} \cdots x_k^{n_k}$ in u, a multivariate series or polynomial.
Also defined for domain CliffordAlgebra and categories AbelianMonoidRing, FreeAbelianCategory, and MonogenicLinearOperator.
coefficient (s, n) returns the terms of total degree n of series s as a polynomial.

coefficients (*polynomialOrStream*)

coefficients (p) returns the list of non-zero coefficients of polynomial p starting with the coefficient of the maximum degree.
coefficients (s) returns a stream of coefficients $[a_0, a_1, a_2, \ldots]$ for the stream s: $a_0 + a_1 x + a_2 x^2 + \cdots$. Note: the entries of the stream may be zero.

coerceImages (*listOfElements*)

coerceImages (ls) coerces the list ls to a permutation whose image is given by ls and whose preimage is fixed to be $[1, \ldots, n]$. Note: **coerceImages** (ls) $= coercePreimagesImages([1, \ldots, n], ls)$.

coerceListOfPairs (*listOfPairsOfElements*)

coerceListOfPairs (lls) coerces a list of pairs lls to a permutation, or calls $error$ if not consistent, that is, the set of the first elements coincides with the set of second elements.

coercePreimagesImages (*listOfListOfElements*)

coercePreimagesImages (lls) coerces the representation lls of a permutation as a list of preimages and images to a permutation.

coleman (*listOfIntegers, listOfIntegers, listOfIntegers*)

coleman $(alpha, beta, pi)$ generates the Coleman-matrix of a certain double coset of the symmetric group given by an representing element pi and $alpha$ and $beta$. The matrix has nonnegative entries, row sums $alpha$ and column sums $beta$. Consult SymmetricGroupCombinatoricFunctions using Browse for details.

color (*integer*)

color (i) returns a color of the indicated hue i.

colorDef (*viewPort, color, color*)

colorDef (v, c_1, c_2) sets the range of colors along the colormap so that the lower end of the colormap is defined by c_1 and the top end of the colormap is defined by $c2$ for the given three-dimensional viewport v.

colorFunction (*smallFloatFunction*)

colorFunction (fn) specifies the color for three-dimensional plots. Function fn can take one to three SmallFloat arguments and always returns a SmallFloat value. If one argument, the color is based upon the z-component of plot. If two arguments, the color is based on two parameter values. If three arguments, the color is based on the x, y, and z components. This command may be given as a draw option: `colorFunction == fn`.

column (*matrix, positiveInteger*)

column (M, j) returns the j th column of the matrix or TwoDimensionalArrayCategory object M, or calls **error** if the index is outside the proper range.

commaSeparate (*listOfOutputForms*)

commaSeparate (lo), where lo is a list of objects of type OutputForm (normally unexposed), returns an output form which separates the elements of lo by commas.

commonDenominator (*fraction*)

commonDenominator $([q_1, \ldots,])$ returns a common denominator for the q_i's.
commonDenominator (A), where A is a matrix of fractions, returns a common denominator for the elements of A.
commonDenominator (p) returns a common denominator for the coefficients of polynomial p.

commutative? ()

commutative? ()$R tests if multiplication in the algebra R is commutative.

commutator (*groupElement, groupElement*)

commutator (p, q) computes **inv** $(p) * $ **inv**$(q) * p * q$ where p and q are members of a Group domain.
commutator (a, b) returns $ab - ba$ where a and b are members of a NonAssociativeRing domain.

compactFraction (*partialFraction*)

compactFraction (u) normalizes the partial fraction u to a compact representation where it has only one fractional term per prime in the denominator.

comparison (*basicOperator, property*)

comparison (op, p) attaches p as the `"%less?"` property to op. If $op1$ and $op2$ have the same name, and one of them has a `"%less?"` property p, then $p(op1, op2)$ is called to decide whether $op1 < op2$.

compile (*symbol, listOfTypes*)

compile $(f, [T_1, \ldots, T_n])$ forces the interpreter to compile the function with name f with signature $(T_1, \ldots, T_n) -> T$, where T is a type determined by type analysis of the function body of f. If the compilation is successful, the operation returns the name f. The operation calls **error** if f is not defined beforehand in the interpreter, or if the T_i's are not valid types, or if the compiler fails. See also **function**, **interpret**, **lambda**, and **compiledFunction**.

compiledFunction (*expression, symbol* [, *symbol*])

Argument *expression* may be of any type that is coercible to type InputForm (most commonly used types). These functions must be package called to define the type of the function produced.
compiledFunction $(expr, x)\$P$, where P is MakeUnaryCompiledFunction(E, S, T), returns an anonymous function of type ST defined by defined by $x \mapsto$ expr. The anonymous function is compiled and directly applicable to objects of type S.
compiledFunction $(expr, x, y)\$P$, where P is MakeBinaryCompiledFunction(E, A, B, T) returns an anonymous function of type (A, B) → T defined by $(x, y) \mapsto$ expr. The anonymous function is compiled and is then directly applicable to objects of type (A, B).
See also **compile**, **function**, and **lambda**.

complement (*finiteSetElement*)

complement (u) returns the complement of the finite set u, that is, the set of all values not in u.

complementaryBasis (*vector*)

complementaryBasis $(b_1, \ldots, b_n)$ returns the complementary basis $(b_1', \ldots, b_n')$ of $(b_1, \ldots, b_n)$ for a domain of category FunctionFieldCategory.

complete (*streamOrInteger*)

complete (u) causes all terms of a stream or continued fraction u to be computed. If not called on a finite stream or continued fraction, this function will compute until interrupted.
complete (n) is the n th complete homogeneous symmetric function expressed in terms of power sums. Alternatively, it is the cycle index of the symmetric group of degree n. See CycleIndicators for details.

completeEchelonBasis (*vectorOfVectors*)

completeEchelonBasis (vv) returns a completed basis from vv, a vector of vectors of domain elements. Consult RepresentationPackage2 using Browse for details.

complex (*element, element*)

complex (x, y) creates the complex expression x + %i*y.

complexEigenvalues (*matrix, precision*)

complexEigenvalues (m, eps) computes the eigenvalues of the matrix m to precision eps, chosen as a float or a rational number so as to agree with the type of the coefficients of the matrix m.

complexEigenvectors (*matrix, precision*)

complexEigenvectors (m, eps) (m, a matrix) returns a list of records, each containing a complex eigenvalue, its algebraic multiplicity, and a list of associated eigenvectors. All results are expressed as complex floats or rationals with precision eps.

complexElementary (*expression*[, *symbol*])

complexElementary (e) rewrites e in terms of the two fundamental complex transcendental elementary functions: log, exp.

complexElementary (e, x) does the same but only rewrites kernels of e involving x.

complexExpand (*integrationResult*)

complexExpand (ir), where ir is an IntegrationResult, returns the expanded complex function corresponding to ir.

complexIntegrate (*expression, variable*)

complexIntegrate (f, x) returns $\int f(x)dx$ where x is viewed as a complex variable.

complexLimit (*expression, equation*)

complexLimit $(f(x), x = a)$ computes the complex limit of f as its argument x approaches a.

complexNormalize (*expression* [, *symbol*])

complexNormalize (e) rewrites e using the least possible number of complex independent kernels.
complexNormalize (e, x) rewrites e using the least possible number of complex independent kernels involving x.

complexNumeric (*expression* [, *positiveInteger*])

complexNumeric (u) returns a complex approximation of u, where u is a polynomial or an expression.
complexNumeric (u, n) does the same but requires accuracy to be up to n decimal places.

complexRoots (*rationalFunctions* [, *options*])

complexRoots (rf, eps) finds all the complex solutions of a univariate rational function with rational number coefficients with precision given by *eps*. The complex solutions are returned either as rational numbers or floats depending on whether *eps* is a rational number or a float.
complexRoots (lrf, lv, eps) similarly finds all the complex solutions of a list of rational functions with rational number coefficients with respect the variables appearing in lv. Solutions are computed to precision *eps* and returned as a list of values corresponding to the order of variables in lv.

complexSolve (*eq, x*)

See **solve** (u, v).

complexZeros (*polynomial, floatOrRationaNumber*)

complexZeros $(poly, eps)$ finds the complex zeros of the univariate polynomial *poly* to precision *eps*. Solutions are returned either as complex floats or rationals depending on the type of *eps*.

components (*threeSpace*)

components (sp) takes the ThreeSpace object sp, and returns a list of ThreeSpace objects, each having a single component.

composite (*polynomial, polynomial*)

composite (p, q), for polynomials p and q, returns f if $p = f(q)$, and `"failed"` if no such f exists.

composite (lsp), where lsp is a list $[sp_1, sp_2, \ldots, sp_n]$ of ThreeSpace objects, returns a single ThreeSpace object containing the union of all objects in the parameter list grouped as a single composite.

composites (*threeSpace*)

composites (sp) takes the ThreeSpace object sp and returns a list of ThreeSpace objects, one for each single composite of sp. If sp has no defined composites (composites need to be explicitly created), the list returned is empty. Note that not all the components need to be part of a composite.

concat (*aggregate, aggregate*)
concat! (*aggregate, aggregate*)

concat (u, x) returns list u with additional element x at the end. Note: equivalent to **concat** $(u, [x])$.
concat (u, v) returns an aggregate consisting of the elements of u followed by the elements of v.
concat (u), where u is a list of aggregates $[a, b, \ldots, c]$, returns a single aggregate consisting of the elements of a followed by those of b followed ... by the elements of c.
concat! (u, x), where u is extensible, destructively adds element x to the end of aggregate u; if u is a stream, it must be finite.
concat! (u, v) destructively appends v to the end of u; if u is a stream, it must be finite.

conditionP (*matrix*)

conditionP (M), given a matrix M representing a homogeneous system of equations over a field F with

characteristic p, returns a non-zero vector whose p^{th} power is a non-trivial solution to these equations, or `"failed"` if no such vector exists.

conditionsForIdempotents ()

conditionsForIdempotents () determines a complete list of polynomial equations for the coefficients of idempotents with respect to the R-module basis. See also FramedNonAssociativeAlgebra for an alternate definition.

conical (*smallFloat, smallFloat*)

conical (a, b) returns a function of two parameters for mapping conical coordinates to Cartesian coordinates. The function maps the point (λ, μ, ν) to $x = \lambda\mu\nu/(ab)$,
$y = \lambda/a\sqrt{((mu^2 - a^2)(\nu^2 - a^2)/(a^2 - b^2))}$,
$z = \lambda/b\sqrt{((mu^2 - b^2)(nu^2 - b^2)/(b^2 - a^2))}$.

conjugate (*element* [, *element*])

conjugate (u) returns the conjugate of a complex, quaternion, or octonian expression u. For example, if u is the complex expression $x + \%iy$, **conjugate** (u) returns $x - \%iy$.
conjugate (pt) returns the conjugate of a partition pt. See PartitionsAndPermutations using Browse.
conjugate (p, q) returns **inv** $(q) * p * q$ for elements p and q of a group. Note: this operation is called *right action by*

conjugation.

conjugates (*streamOfPartitions*)

conjugates (lp) is the stream of conjugates of a stream of partitions lp.

connect (*twoDimensionalViewport, positiveInteger, string*)

connect (v, n, s) displays the lines connecting the graph points in field n of the two-dimensional viewport v if $s =$ "on", and does not display the lines if $s =$ "off".

constant (*variableOrfunction*)
constantLeft (*function, element*)
constantRight (*function, element*)

These operations add an argument to a function and must be package-called from package P as indicated. See also **curry**, **curryLeft**, and **curryRight**.
constant (f)$P returns the function g such that $g(a) = f()$, where function f has type $\rightarrow$ C and a has type A. The function must be package-called from $P =$ MappingPackage2(A, C).
constantRight (f)$P returns the function g such that $g(a, b) = f(a)$, where function f has type A $\rightarrow$ C and b has type B. This function must be package-called from $P =$ MappingPackage3(A, B, C).
constantLeft (f)$P returns the function g such that $g(a, b) = f(b)$, where function f has type B $\rightarrow$ C and a has type A. The function must be package-called from $P =$ MappingPackage3(A, B, C).
constant (x) tells the pattern matcher that x should match the symbol $'x$ and no other quantity, or calls **error** if x is not a symbol.

constantOperator (*property*)
constantOpIfCan (*f*)

constantOperator (f) returns a nullary operator op such that *op*() always evaluate to f.
constantOpIfCan (*op*) returns f if *op* is the constant nullary operator always returning f, and `"failed"` otherwise.

construct (*element, ..*)

construct ($x, y, \ldots, z$)$R returns the collection of elements $x, y, \ldots, z$ from domain R ordered as given. This is equivalently written as $[x, y, \ldots, z]$. The qualification R may be omitted for domains of type List. Infinite tuples such as $[x_i$ `for` i `in` 1..] are converted to a Stream object.

cons (*element, listOrStream*)

cons (x, u), where u is a list or stream, creates a new list or stream whose **first** element is x and whose **rest** is u. Equivalent to **concat** (x, u).

content (*polynomial*[, *symbol*])

content (p) returns the greatest common divisor (**gcd**) of the coefficients of polynomial p.
content (p, v), where p is a multivariate polynomial type, returns the *gcd* of the coefficients of the polynomial p viewed as a univariate polynomial with respect to the variable v. For example, if $p = 7x^2y + 14xy^2$, the *gcd* of the coefficients with respect to x is $7y$.

continuedFraction (*fractionOrFloat*[, *options*])

continuedFraction (f) converts the floating point number f to a reduced continued fraction.
continuedFraction (r) converts the fraction r with components of type R to a continued fraction over R.
continuedFraction (r, s, s'), where s and s' are streams over a domain R, constructs a continued fraction in the following way: if $s = [a1, a2, \ldots]$ and $s' = [b1, b2, \ldots]$ then the result is the continued fraction $r + a1/(b1 + a2/(b2 + \ldots))$.

contract (*idealOrTensors*[, *options*])

contract ($I, lvar$) contracts the ideal I to the polynomial ring $F[lvar]$.
contract (t, i, j) is the contraction of tensor t which sums along the ith and j^{th} indices. For example, if $r = contract(t, 1, 3)$ for a rank 4 tensor t, then r is the rank 2 $(= 4 - 2)$ tensor given by $r(i, j) = \sum_{h=1}^{\text{dim}} t(h, i, h, j)$.
contract (t, i, s, j) is the inner product of tensors s and t which sums along the k_1st index of t and the k_2st index of s. For example, if $r = contract(s, 2, t, 1)$ for rank 3 tensors s and t, then r is the rank 4 $(= 3 + 3 - 2)$ tensor given by $r(i, j, k, l) = \sum_{h=1}^{\text{dim}} s(i, h, j)t(h, k, l)$.

contractSolve (*equation, symbol*)

contractSolve (eq, x) finds the solutions expressed in terms of radicals of the equation of rational functions eq with respect to the symbol x. The result contains new symbols for common subexpressions in order to reduce the size of the output. Alternatively, an expression u may be given for eq in which case the equation eq is defined as $u = 0$

controlPanel (*viewport, string*)

controlPanel (v, s) displays the control panel of the given two-dimensional or three-dimensional viewport v if $s =$ "on", or hides the control panel if $s =$ "off".

convergents (*continuedFraction*)

convergents (cf) returns the stream of the convergents of the continued fraction cf. If the continued fraction is finite, then the stream will be finite.

coordinate (*curveOrSurface, nonNegativeInteger*)

coordinate (u, n) returns the nth coordinate function for the curve or surface u. See ParametericPlaneCurve, ParametricSpaceCurve, and ParametericSurface, using Browse.

coordinates (*pointOrvector*[, *basis*])

coordinates (pt) specifies a change of coordinate systems of point pt. This option is expressed in the form *coordinates* $==$ *pt*.

The following operations return a matrix representation of the coordinates of an argument vector v of the form $[v_1 \ldots v_n]$ with respect to the basis a domain R. The coordinates of v_i are contained in the ith row of the matrix returned.
coordinates (v, b) returns the matrix representation with respect to the basis b for vector v of elements from domain R of category FiniteRankNonAssociativeAlgebra or FiniteRankAlgebra. If a second argument is not given, the basis is taken to be the fixed basis of R.
coordinates $(v)\$R$, returns a matrix representation for v with respect to a fixed basis for domain R of category FiniteAlgebraicExtensionField, FramedNonAssociativeAlgebra, or FramedAlgebra.

copies (*integer, string*)

copies (n, s) returns a string composed of n copies of string s.

copy (*aggregate*)

copy (u) returns a top-level (non-recursive) copy of an aggregate u. Note: for lists, `copy(u) == [x for x in u]`.

copyInto! (*aggregate, aggregate, integer*)

copyInto! (u, v, p) returns linear aggregate u with elements of u replaced by the successive elements of v starting at index p. Arguments u and v can be elements of any FiniteLinearAggregate.

cos (*expression*)
cosIfCan (*expression*)

Argument x can be a Complex, Float, SmallFloat, or Expression value or a series.
cos (x) returns the cosine of x.
cosIfCan (x) returns **cos** (x) if possible, and `"failed"` otherwise.

cos2sec (*expression*)

cos2sec (e) converts every **cos** (u) appearing in e into $1/\sec(u)$.

cosh2sech (*expression*)

cosh2sech (e) converts every **cosh** (u) appearing in e into $1/\text{sech}(u)$.

cosh (*expression*)
coshIfCan (*expression*)

Argument x can be a Complex, Float, SmallFloat, or Expression value or a series.
cosh (x) returns the hyperbolic cosine of x.
coshIfCan (x) returns **cosh** (x) if possible, and `"failed"` otherwise.

cot (*expression*)

Argument x can be a Complex, Float, SmallFloat, or Expression value or a series.
cot (x) returns the cotangent of x.
cotIfCan (x) returns **cot** (x) if possible, and `"failed"` otherwise.

cot2tan (*expression*)

cot2tan $(expression)$ converts every $\cot(u)$ appearing in e into $1/\tan(u)$.

cot2trig (*expression*)

cot2trig $(expression)$ converts every $\cot(u)$ appearing in e into $\cos(u)/\sin(u)$.

coth (*expression*)
cothIfCan (*expression*)

Argument x can be a Complex, Float, SmallFloat, or Expression value or a series.
coth (x) returns the hyperbolic cotangent of x.
cothIfCan (x) returns **coth** (x) if possible, and `"failed"` otherwise.

coth2tanh (*expression*)

coth2tanh $(expression)$ converts every $\coth(u)$ appearing in e into $1/\tanh(u)$.

coth2trigh (*expression*)

coth2trigh $(expression)$ converts every $\coth(u)$ appearing in e into $\cosh(u)/\sinh(u)$.

count (*predicate, aggregate*)

count $(pred, u)$ returns the number of elements x in u such that **pred** (x) is $true$. For collections, `count(p, u) = reduce(+, [1 for x in u | p(x)], 0)`.
count (x, u) returns the number of occurrences of x in u. For collections, `count(x, u) = reduce(+, [x=y for y in u], 0)`.

countable? (*cardinal*)

countable? (u) tests if the cardinal number u is countable, that is, if $u \leq Aleph0$.

createThreeSpace ()

createThreeSpace ()\$ThreeSpace(R) creates a ThreeSpace object capable of holding point, curve, mesh components or any combination of the three. The ring R is usually SmallFloat. If you do not package call this function, SmallFloat is assumed.
createThreeSpace (s) creates a ThreeSpace object containing objects pre-defined within some SubSpace s.

createGenericMatrix (*nonNegativeInteger*)

createGenericMatrix (n) creates a square matrix of dimension n whose entry at the i-th row and j-th column is the indeterminate $x_{i,j}$ (double subscripted). See

RepresentationPackage1 using Browse.

createIrreduciblePoly (*nonNegativeInteger*)

createIrreduciblePoly $(n)$$FFPOLY(GF) generates a monic irreducible polynomial of degree n over the finite field GF.

createNormalElement ()

createNormalElement ()$$F$ computes a normal element over the ground field of a finite algebraic extension field F, that is, an element a such that $a^{q^i}, 0 \leq i <$ **extensionDegree**()$$F$ is an F-basis, where q is the size of the ground field.

createNormalPrimitivePoly (*element*)

createNormalPrimitivePoly $(n)$$FFPOLY(GF) generates a normal and primitive polynomial of degree n over the field GF.

createPrimitiveElement ()

createPrimitiveElement ()$$F$ computes a generator of the (cyclic) multiplicative group of a finite field F.

createRandomElement (*listOfMatrices, matrix*)

createRandomElement (lm, m) creates a random element of the group algebra generated by lm, where lm is a list of matrices and m is a matrix. See RepresentationPackage2 using Browse.

csc2sin (*expression*)

csc2sin $(expression)$ converts every **csc** (u) appearing in f into $1/\mathtt{sin}(u)$.

csch2sinh (*expression*)

csch2sinh $(expression)$ converts every **csch** (u) appearing in f into $1/\sinh(u)$.

csch (*expression*)
cschIfCan (*expression*)

Argument x can be a Complex, Float, SmallFloat, or Expression value or a series.
csch (x) returns the hyperbolic cosecant of x.
cschIfCan (x) returns **csch** (x) if possible, and `"failed"` otherwise.

cscIfCan (*expression*)

Argument x can be a Complex, Float, SmallFloat, or Expression value or a series.
csc (x) returns the cosecant of x.
cscIfCan (x) returns **csc** (x) if possible, and `"failed"` otherwise.

cup (*symmetricPolynomial, symmetricPolynomial*)

cup (s_1, s_2), introduced by Redfield, is the scalar product of two cycle indices, where the s_i are of type SymmetricPolynomial with rational number coefficients. See also **cap**. See CycleIndicators for details.

curry (*function*)
curryLeft (*function, element*)
curryRight (*function, element*)

These functions drop an argument from a function.
curry (f, a) returns the function g such that $g() = f(a)$, where function f has type A → C and element a has type A.
curryRight (f, b) returns the function g such that $g(a) = f(a, b)$, where function f has type (A, B) → C and element b has type B.
curryLeft (f, a) is the function g such that $g(b) = f(a, b)$, where function f has type (A, B) → C and element a has type A.
See also **constant**, **constantLeft**, and **constantRight**.

curve (*listOfPoints*[, *options*])

curve $([p_0, p_1, \ldots, p_n])$ creates a space curve defined by the list of points p_0 through p_n and returns a ThreeSpace object whose component is the curve.
curve (sp) checks to see if the ThreeSpace object sp is composed of a single curve defined by a list of points; if so, the list of points defining the curve is returned. Otherwise, the operation calls **error**.
curve (c_1, c_2) creates a plane curve from two component functions c_1 and c_2. See ComponentFunction using Browse.
$curve(sp, [[p_0], [p_1], \ldots, [p_n]])$ adds a space curve defined by a list of points p_0 through p_n to a ThreeSpace object sp. Each p_i is from a domain **PointDomain** (m, R), where R is the Ring over which the point elements are defined and m is the dimension of the points.
curve $(s, [p_0, p_1, \ldots, p_n])$ adds the space curve component designated by the list of points p_0 through p_n to the ThreeSpace object sp.
curve (c_1, c_2, c_3) creates a space curve from three component functions c_1, c_2, and c_3.

curve? (*threeSpace*)

curve? (sp) tests if the ThreeSpace object sp contains a single curve object.

curveColor (*float*)

curveColor (p) specifies a color index for two-dimensional graph curves from the palette p. This option is expressed in the form $curveColor == p$.

cycle (*listOfPermutations*)

cycle (ls) converts a cycle ls, a list with no repetitions, to the permutation, which maps $ls.i$ to $ls.(i+1)$ (index modulo the length of the list).

cycleEntry (*aggregate*)

cycleEntry (u) returns the head of a top-level cycle contained in aggregate u, or **empty** () if none exists.

cycleLength (*aggregate*)

cycleLength (u) returns the length of a top-level cycle contained in aggregate u, or 0 if u has no such cycle.

cyclePartition (*permutation*)

cyclePartition (p) returns the cycle structure of a permutation p including cycles of length 1. The permutation is assumed to be a member of Permutation(S) where S is a finite set.

cycleRagits (*radixExpansion*)

cycleRagits (rx) returns the cyclic part of the ragits of the fractional part of a radix expansion. For example, if $x = 3/28 = 0.10714285714285\ldots$, then `cycleRagits(x) = [7, 1, 4, 2, 8, 5]`.

cycleSplit! (*aggregate*)

cycleSplit! (u) splits the recursive aggregate (for example, a list) u into two aggregates by dropping off the cycle. The value returned is the cycle entry, or *nil* if none exists. For example, if $w = \mathbf{concat}(u, v)$ is the cyclic list where v is the head of the cycle, **cycleSplit!** (w) will drop v off w. Thus w is destructively changed to u, and v is returned.

cycles (*listOfListOfElements*)

cycles (lls) coerces a list of list of cycles lls to a permutation. Each cycle, represented as a list ls with no repetitions, is coerced to the permutation, which maps $ls.i$ to $ls.(i+1)$ (index modulo the length of the list). These permutations are then multiplied.

cycleTail (*aggregate*)

cycleTail (u) returns the last node in the cycle of a recursive aggregate (for example, a list) u, or empty if none exists.

cyclic (*integer*)

cyclic (n) returns the cycle index of the cyclic group of degree n. CycleIndicators for details.

cyclic? (*aggregate*)

cyclic? (u) tests if recursive aggregate (for example, a list) u has a cycle.

cyclicGroup (*listOfIntegers*)

cyclicGroup ($[i_1, \ldots, i_k]$) constructs the cyclic group of order k acting on the list of integers $i_1, \ldots, i_k$. Note: duplicates in the list will be removed.

cyclicGroup (*positiveInteger*)

cyclicGroup (n) constructs the cyclic group of order n acting on the integers $1, \ldots, n$, $n > 0$.

cyclicSubmodule (*listOfMatrices, vector*)

cyclicSubmodule (lm, v), where lm is a list of n by n square matrices and v is a vector of size n, generates a basis in echelon form. Consult RepresentationPackage2 using Browse for details.

cylindrical (*point*)

cylindrical (pt) transforms pt from polar coordinates to Cartesian coordinates, by mapping the point $(r, theta, z)$ to $x = r\cos(theta)$, $y = r\sin(theta)$, z.

D (*expression*[, *options*])

D (x) returns the derivative of x. This function is a simple differential operator where no variable needs to be specified.
D ($x, [s_1, \ldots s_n]$) computes successive partial derivatives, that is, $D(\ldots \mathbf{D}(x, s_1) \ldots, s_n)$.
D (u, x) computes the partial derivative of u with respect to x.
D ($u, deriv[, n]$) differentiates u n times using a derivation which extends $deriv$ on R. Argument n defaults to 1.
D (p, d, x') extends the R-derivation d to an extension R in $R[x]$ where Dx is given by x', and returns Dp.
$D(x, [s_1, \ldots, s_n], [n_1, \ldots, n_m])$ computes multiple partial derivatives, that is, $\mathbf{D}(\ldots \mathbf{D}(x, s_1, n_1) \ldots, s_n, n_m)$.
D (u, x, n) computes multiple partial derivatives, that is, n th derivative of u with respect to x.
D ($of[, n]$), where of is an object of type OutputForm (normally unexposed), returns an output form for the n th derivative of f, for example, f', f'', f''', f^{iv}, and so on.
D ()$A provides the operator corresponding to the derivation in the differential ring A.

dark (*color*)

dark (*color*) returns the shade of the indicated hue of *color* to its lowest value.

ddFact (*polynomial, primeInteger*)

ddFact (q, p) computes a distinct degree factorization of the polynomial q modulo the prime p, that is, such that each factor is a product of irreducibles of the same degrees.

decimal (*rationalNumber*)

decimal (rn) converts a rational number rn to a decimal expansion.

declare (*listOfInputForms*)

declare (t) returns a name f such that f has been declared to the interpreter to be of type t, but has not been assigned a value yet.

decreasePrecision (*integer*)

decreasePrecision (n)$R decreases the current **precision** by n decimal digits.

definingPolynomial ()

definingPolynomial ()$R returns the minimal polynomial for a MonogenicAlgebra domain R, that is, one which **generator** ()$R satisfies.
definingPolynomial (x) returns an expression p such that $p(x) = 0$, where x is an AlgebraicNumber or an object of type Expression.

degree (*polynomial*[, *symbol*])
The meaning of degree(u[, s]) depends on the type of u.

if u is a polynomial: **degree** (u, x) returns the degree of polynomial u with respect to the variable x. Similarly, **degree** (u, lv), where lv is a list of variables, returns a list of degrees of polynomial u with respect to each of the variables in lv.

if u is an element of an AbelianMonoidRing or GradedModule domain: **degree** (u) returns the maximum of the exponents of the terms of u.

if u is a series: **degree** (u) returns the degree of the leading term of u.

if u is an element of a domain of category ExtensionField: **degree** (u) returns the degree of the minimal polynomial of u if u is algebraic with respect to the ground field F, and `%infinity` otherwise.

if u is a permutation: **degree** (u) returns the number of points moved by the permutation.

if u is a permutation group: **degree** (u) returns the number of points moved by all permutations of the group u. For additional information on **degree**, consult Browse.

delete (*aggregate, integerOrSegment*)

delete (u, i) returns a copy of linear aggregate u with the ith element deleted. Note: for lists, `delete(a, i) == concat(a(0..i-1), a(i + 1, ..))`.
delete $(u, i..j)$ returns a copy of u with the ith through j^{th} element deleted. Note: for lists, `delete(a, i..j) = concat(a(0..i-1), a(j+1..))`.
delete! (u, i) destructively deletes the ith element of u.
delete! $(u, i..j)$ destructively deletes elements $u.i$ through $u.j$ of u.

deleteProperty (*basicOperator, string*)

deleteProperty (*op*, s) destructively removes property s from *op*.

denom (*expression*)
denominator (*expression*)

Argument x can be from domain Fraction(R) for some domain R, or of type Expression if the result is of type R.
denom (x) returns the denominator of x as an object of domain R; if x is of type Expression, it returns an object of domain SMP(R, Kernel(Expression R)).
denominator (x) returns the denominator of x as an element of Fraction(R); if x is of type Expression, it returns an object of domain Expression(R).

denominators (*fractionOrContinuedFraction*)

denominator $(frac)$ is the denominator of the fraction $frac$.
denominators (cf) returns the stream of denominators of the approximants of the continued fraction x. If the continued fraction is finite, then the stream will be finite.

depth (*stack*)
depth (st) returns the number of elements of stack st.

dequeue (*queue*)
dequeue! (*queue*)
dequeue $([x, y, \ldots, z])$ creates a dequeue with first (top or front) element x, second element y, ..., and last (bottom or back) element z.
dequeue! (q) destructively extracts the first (top) element from queue q. The element previously second in the queue becomes the first element. A call to **error** occurs if q is empty.

derivationCoordinates (*vectorOfElements, derivationFunction*)
derivationCoordinates $(v, \ ')$ returns a matrix M such that $v' = Mv$. Argument v is a vector of elements from R, a domain of category MonogenicAlgebra over a ring R. Argument $'$ is a derivation function defined on R.

derivative (*basicOperator*[, *property*])
derivative (*op*) returns the value of the `"%diff"` property of *op* if it has one, and `"failed"` otherwise.
derivative (*op, dprop*) attaches *dprop* as the `"%diff"` property of *op*. Note: if *op* has a `"%diff"` property f, then applying a derivation D to *op*(a) returns $f(a)D(a)$. Argument *op* must be unary.
derivative (*op*, $[f_1, \ldots, f_n]$) attaches $[f_1, \ldots, f_n]$ as the `"%diff"` property of *op*. Note: if *op* has such a `"%diff"` property, then applying a derivation D to *op*$(a_1, \ldots, a_n)$ returns $f_1(a_1, \ldots, a_n)D(a_1) + \cdots + fn(a_1, \ldots, a_n)D(a_n)$. See also D.

destruct (*sExpression*)
destruct (se), where se is the $SExpression$ $(a_1, \ldots, a_n)$, returns the list $[a_1, \ldots, a_n]$.

determinant (*matrix*)
determinant (m) returns the determinant of the matrix m, or calls **error** if the matrix is not square. Note: the underlying coefficient domain of m is assumed to have a commutative "*".

diagonal (*matrix*)
diagonal (m), where m is a square matrix, returns a vector consisting of the diagonal elements of m.
diagonal (f), where f is a function of type (A, A) $\to$ T is the function g such that $g(a) = f(a, a)$. See MappingPackage for related functions.

diagonal? (*matrix*)
diagonal? (m) tests if the matrix m is square and diagonal.

diagonalMatrix (*listOfElements*)
diagonalMatrix (l), where l is a list or vector of elements, returns a (square) diagonal matrix with those elements of l

on the diagonal.
diagonalMatrix $([m_1, \ldots, m_k])$ creates a block diagonal matrix M with block matrices $m_1, \ldots, m_k$ down the diagonal, with 0 block matrices elsewhere.

diagonalProduct (*matrix*)
diagonalProduct (m) returns the product of the elements on the diagonal of the matrix m.

dictionary ()
dictionary ()$R creates an empty dictionary of type R.
dictionary $([x, y, \ldots, z])$ creates a dictionary consisting of entries $x, y, \ldots, z$.

difference (*setAggregate*, *element*)
difference (u, x) returns the set aggregate u with element x removed.
difference (u, v) returns the set aggregate w consisting of elements in set aggregate u but not in set aggregate v.

differentialVariables (*differentialPolynomial*)
differentialVariables (p) returns a list of differential indeterminates occurring in a differential polynomial p.

differentiate (*expression*[, *options*])
See **D**.

digamma (*complexSmallFloat*)
digamma (x) is the function, $\psi(x)$, defined by $\psi(x) = '(x)/(x)$. Argument x is either a small float or a complex small float.

digit ()
digit () returns the class of all characters for which **digit?** is $true$.

digit? (*character*)
digit? (*ch*) tests if character c is a digit character, that is, one of 0..9.

digits ([*positiveInteger*])
digits () returns the current precision of floats in numbers of digits.
digits (n) set the **precision** of floats to n digits.
digits (x) returns a stream of *p*-adic digits of p-adic integer n. See PAdicInteger using Browse.

dihedral (*integer*)
dihedral (n) is the cycle index of the dihedral group of degree n.

dihedralGroup (*listOfIntegers*)
dihedralGroup $([i_1, \ldots, i_k])$ constructs the dihedral group of order $2k$ acting on the integers $i_1, \ldots, i_k$. Note: duplicates in the list will be removed.
dihedralGroup (n) constructs the dihedral group of order $2n$ acting on integers $1, \ldots, n$.

dilog (*expression*)
dilog (x) returns the dilogarithm of x, that is, $\int log(x)/(1-x)dx$.

dim (*color*)
dim (c) sets the shade of a hue c, above dark but below bright.

dimension ([*various*])
dimension ()$R returns the dimensionality of the vector space or rank of Lie algebra R.
dimension (I) gives the dimension of the ideal I.
dimension (s) returns the dimension of the point category s.

dioSolve (*equation*)
dioSolve (*eq*) computes a basis of all minimal solutions for a linear homomogeneous Diophantine equation *eq*, then all minimal solutions of the inhomogeneous equation. Alternatively, an expression u may be given for *eq* in which case the equation *eq* is defined as $u = 0$.

directory (*filename*)
directory (f) returns the directory part of the file name.

directProduct (*vector*)
directProduct (v) converts the vector v to become a direct product

discreteLog (*finiteFieldElement*)
discreteLog (a)\$$F$ computes the discrete logarithm of a with respect to **primitiveElement** ()$F of the field F.

discreteLog (*finiteFieldElement*, *finiteFieldElement*)
discreteLog (b, a) computes s such that $b^s = a$ if such an s exists.

discriminant (*polynomial*[, *symbol*])
discriminant $(p[, x])$ returns the discriminant of the polynomial p with respect to the variable x. If x is univariate, the second argument may be omitted.
discriminant ()$R returns **determinant** (**traceMatrix**()$R) of a FramedAlgebra domain R.
discriminant $([v_1, .., v_n])$ returns **determinant** $(traceMatrix([v_1, .., v_n]))$ where the v_i each have n elements.

display (*text*[, *width*])
display $(t[, w])$, where t is either IBM SCRIPT Formula Format or TeX text, outputs t so that each line has length $\leq w$. The default value of w is that length set by the system command)set output length.
display (*op*, f) attaches f as the `"%display"` property of *op*.
display (*op*) returns the `"%display"` property of *op* if it has

one attached, and "failed" otherwise.
Value f either has type OutputForm → OutputForm or else List(OutputForm) → OutputForm. Argument *op* must be unary. Note: if *op* has a "%display" property f of the former type, then $op(a)$ gets converted to OutputForm as $f(a)$. If f has the latter type, then $op(a_1, \ldots, a_n)$ gets converted to OutputForm as $f(a_1, \ldots, a_n)$.

distance (*aggregate, aggregate*)
distance (u, v), where u and v are recursive aggregates (for example, lists) returns the path length (an integer) from node u to v.

distdfact (*polynomial, boolean*)
distdfact $(p, squareFreeFlag)$ produces the complete factorization of the polynomial p returning an internal data structure. If argument $squareFreeFlag$ is $true$, the polynomial is assumed square free.

distribute (*expression* [, *f*])
distribute $(f[, g])$ expands all the kernels in f that contain g in their arguments and that are formally enclosed by a **box** or a **paren** expression. By default, g is the list of all kernels in f.

divide (*element, element*)
divide (x, y) divides x by y producing a record containing a $quotient$ and $remainder$, where the remainder is smaller (see **sizeLess?**) than the divisor y.

divideExponents (*polynomial, nonNegativeInteger*)
divideExponents (p, n) returns a new polynomial resulting from dividing all exponents of the polynomial p by the non negative integer n, or "failed" if no exponent is exactly divisible by n.

divisors (*integer*)
divisors (i) returns a list of the divisors of integer i.

domain (*typeAnyObject*)
domain (a) returns the type of the original object that was converted to Any as object of type SExpression

domainOf (*typeAnyObject*)
domainOf (a) returns a printable form of the type of the original type of a, an object of type Any.

dot (*vector, vector*)
dot (v_1, v_2) computes the inner product of the vectors v_1 and v_2, or calls **error** if x and y are not of the same length.
dot (of), where of is an object of type OutputForm (normally unexposed), returns an output form with one dot overhead ($\dot{x}$).

doubleRank (*element*)
doubleRank (x), where x is an element of a domain R of category FramedNonAssociativeAlgebra, determines the number of linearly independent elements in $b_1 x, \ldots, b_n x$, where $b = [b_1, \ldots, b_n]$ is the fixed basis for R.

doublyTransitive? ()
doublyTransitive? (p) tests if polynomial p, is irreducible over the field K generated by its coefficients, and if $p(X)/(X - a)$ is irreducible over $K(a)$ where $p(a) = 0$.

draw (*functionOrExpression, range* [, *options*])
f, g, and h below denote user-defined functions which map one or more SmallFloat values to a SmallFloat value.

draw $(f, a..b)$ draws the two-dimensional graph of $y = f(x)$ as x ranges from **min** (a, b) to **max** (a, b).

draw $(curve(f, g), a..b)$ draws the two-dimensional graph of the parametric curve $x = f(t), y = g(t)$ as t ranges from **min** (a, b) to **max** (a, b).

draw $(f, a..b, c..d)$ draws the three-dimensional graph of $z = f(x, y)$ as x ranges from **min** (a, b) to **max** (a, b) and y ranges from **min** (c, d) to **max** (c, d).

draw $(curve(f, g, h), a..b)$ draws a three-dimensional graph of the parametric curve $x = f(t), y = g(t), z = h(t)$ as t ranges from **min** (a, b) to **max** (a, b).

draw $(surface(f, g, h), a..b, c..d)$ draws the three-dimensional graph of the parametric surface $x = f(u, v)$, $y = g(u, v)$, $z = h(u, v)$ as u ranges from **min** (a, b) to **max** (a, b) and v ranges from **min** (c, d) to **max** (c, d).

Arguments f, g, and h below denote an Expression involving the variables indicated as arguments. For example, $f(x, y)$ denotes an expression involving the variables x and y.

draw $(f(x), x = a..b)$ draws the two-dimensional graph of $y = f(x)$ as x ranges from **min** (a, b) to **max** (a, b).

draw $(curve(f(t), g(t)), t = a..b)$ draws the two-dimensional graph of the parametric curve $x = f(t), y = g(t)$ as t ranges from **min** (a, b) to **max** (a, b).

draw $(f(x, y), x = a..b, y = c..d)$ draws the three-dimensional graph of $z = f(x, y)$ as x ranges from **min** (a, b) to **max** (a, b) and y ranges from **min** (c, d) to **max** (c, d).

draw $(curve(f(t), g(t), h(t)), t = a..b)$ draws the three-dimensional graph of the parametric curve $x = f(t)$, $y = g(t)$, $z = h(t)$ as t ranges from **min** (a, b) to **max** (a, b).

$draw(surface(f(u, v), g(u, v), h(u, v)), u = a..b, v = c..d)$ draws the three-dimensional graph of the parametric surface $x = f(u, v)$, $y = g(u, v)$, $z = h(u, v)$ as u ranges from **min** (a, b) to **max** (a, b) and v ranges from **min** (c, d) to **max** (c, d).

Each of the **draw** operations optionally take options given as

extra arguments.
`adaptive== true` turns on adaptive plotting.
`clip== true` turns on two-dimensional clipping.
`colorFunction== f` specifies the color based on a function.
`coordinates== p` specifies a change of coordinate systems of point p: *bipolar*, *bipolarCylindrical*, *conical*, *elliptic*, *ellipticCylindrical*, *oblateSpheroidal*, *parabolic*, *parabolicCylindrical*, *paraboloidal*, *prolateSpheroidal*, *spherical*, and *toroidal*.
`curveColor== p` specifies a color index for two-dimensional graph curves from the pallete p.
`pointColor== p` specifies a color index for two-dimensional graph points from the palette p.
`range== [a..b]` provides a user-specified range for implicit curve plots.
`space== sp` adds the current graph to ThreeSpace object *sp*.
`style== s` specifies the drawing style in which the graph will be plotted: *wire*, *solid*, *shade*, *smooth*.
`title== s` titles the graph with string s.
`toScale== true` causes the graph to be drawn to scale.
`tubePoints== n` specifies the number of points n defining the circle which creates the tube around a three-dimensional curve. The default value is 6.
`tubeRadius== r` specifies a Float radius r for a tube plot around a three-dimensional curve.
`unit== [`f_1, f_2`]` marks off the units of a two-dimensional graph in increments f_1 along the x-axis, f_2 along the y-axis.
`var1Steps== n` indicates the number of subdivisions n of the first range variable.
`var2Steps== n` indicates the number of subdivisions n of the second range variable.

drawToScale ([*boolean*])

drawToScale () tests if plots are currently to be drawn to scale.
drawToScale (*true*) causes plots to be drawn to scale.
drawToScale (*false*) causes plots to be drawn to fill up the viewport window. The default setting is *false*.

duplicates (*dictionary*)

duplicates (d) returns a list of values which have duplicates in d

Ei (*variable*)

Ei (x) returns the exponential integral of x: $\int exp(x)/x \mathrm{d}x$.

eigenMatrix (*matrix*)

eigenMatrix (A) returns the matrix B such that $BA($**inverse** $B)$ is diagonal, or `"failed"` if no such B exists.

eigenvalues (*matrix*)

eigenvalues (A), where A is a matrix with rational function coefficients, returns the eigenvalues of the matrix A which are expressible as rational functions over the rational numbers.

eigenvector (*eigenvalue, matrix*)

eigenvector ($eigval, A$) returns the eigenvectors belonging to the eigenvalue $eigval$ for the matrix A.

eigenvectors (*matrix*)

eigenvectors (A) returns the eigenvalues and eigenvectors for the matrix A. The rational eigenvalues and the corresponding eigenvectors are explicitly computed. The non-rational eigenvalues are defined via their minimal polynomial. Their corresponding eigenvectors are expressed in terms of a "generic" root of this polynomial.

element? (*polynomial, ideal*)

element? (f, I) tests if the polynomial f belongs to the ideal I.

elementary (*integer*)

elementary (n) is the nth elementary symmetric function expressed in terms of power sums. See CycleIndicators for details.

elliptic (*scaleFactor*)

elliptic (r) returns a function for transforming elliptic coordinates to Cartesian coordinates. The function returned will map the point (u, v) to $x = r\cosh(u)\cos(v)$, $y = r\sinh(u)\sin(v)$.

ellipticCylindrical (*scaleFactor*)

ellipticCylindrical (r) returns a function for transforming elliptic cylindrical coordinates to Cartesian coordinates as a function of the scale factor r. The function returned will map the point (u, v, z) to $x = r\cosh(u)\cos(v)$, $y = r\sinh(u)\sin(v)$, z.

elt (*structure, various*[, ...])

elt (u, v), usually written as $u.v$ or $u(v)$, regards the structure u as a function and applies structure u to argument v. Many types export **elt** with multiple arguments; **elt** ($u, v, w\ldots$) is generally written $u(v, w\ldots)$. The interpretation of u depends on its type. If u is:

> an indexed aggregate such as a list, stream, vector, or string: $u.i$, $1 \leq i \leq maxIndex(u)$, is equivalently written $u(i)$ and returns the ith element of u. Also, $u(i, y)$ returns $u(i)$ if i is an appropriate index for u, and y otherwise.
>
> a linear aggregate: $u(i..j)$ returns the aggregate of elements of $u(k)$ for $k = i, i+1, \ldots, j$ in that order.
>
> a basic operator: $u(x)$ applies the unary operator u to x; similarly, $u.[x_1, \ldots, x_n]$ applies the n-ary operator u to $x_1, \ldots, x_n$. Also, $u(x, y)$, $u(x, y, z)$, and $u(x, y, z, w)$ respectively apply the binary, ternary, or 4-ary operator u to arguments.

a univariate polynomial or rational function: $u(y)$ evaluates the rational function or polynomial with the distinguished variable replaced by the value of y; this value may either be another rational function or polynomial or a member of the underlying coefficient domain.

a list: $u.first$ is equivalent to **first** (u) and returns the first element of list u. Also, $u.last$ is equivalent to **last** (u) and returns the last element of list u. Both of these call **error** if u is the empty list. Similarly, $u.rest$ is equivalent to **rest** (u) and returns the list u beginning at its second element, or calls **error** if u has less than two elements.

a library: $u(name)$ returns the entry in the library stored under the key $name$.

a linear ordinary differential operator: $u(x)$ applies the differential operator u to the value x.

a matrix or two-dimensional array: $u(i, j[, x])$, $1 \leq i \leq nrows(u), 1 \leq j \leq ncols(m)$, returns the element in the ith row and j^{th} column of the matrix m. If the indices are out of range and an extra argument x is provided, then x is returned; otherwise, **error** is called. Also, $u([i_1, \ldots, i_m], [j_1, \ldots, j_m])$ returns the m-by-n matrix consisting of elements $u(i_k, j_l)$ of u.

a permutation group: $u(i)$ returns the i-th generator of the group u.

a point: $u.i$ returns the ith component of the point u.

a rewrite rule: $u(f[, n])$ applies rewrite rule u to expression f at most n times, where $n = \infty$ by default. When the left-hand side of u matches a subexpression of f, the subexpression is replaced by the right-hand side of u producing a new f. After n iterations or when no further match occurs, the transformed f is returned.

a ruleset: $u(f[, n])$ applies ruleset u to expression f at most n times, where $n = \infty$ by default. Similar to last case, except that on each iteration, each rule in the ruleset is applied in turn in attempt to find a match.

an SExpression $(a_1, \ldots, a_n \quad . \quad b)$ (where b denotes the **cdr** of the last node): $u.i$ returns a_i; similarly $u.[i_1, \ldots, i_m]$ returns $(a_{i_1}, \ldots, a_{i_m})$.

a univariate series: $u(r)$ returns the coefficient of the term of degree r in u.

a symbol: $u[a_1, \ldots, a_n]$ returns u subscripted by $a_1, \ldots, a_n$.

a cartesian tensor: $u(r)$ gives a component of a rank 1 tensor; $u([i_1, \ldots, l_n])$ gives a component of a rank n tensor; $u()$ gives the component of a rank 0 tensor. Also: $u(i, j)$, $u(i, j, k)$, and $u(i, j, k, l)$ gives a component of a rank 2, 3, and 4 tensors respectively.

See also QuadraticForm, FramedNonAssociativeAlgebra, and FunctionFieldCategory.

empty ()

empty ()$\$R$ creates an aggregate of type R with 0 elements.

empty? (*aggregate*)

empty? (u) tests if aggregate u has 0 elements.

endOfFile? (*file*)

endOfFile? (f) tests whether the file f is positioned after the end of all text. If the file is open for output, then this test always returns $true$.

enqueue! (*value, queue*)

enqueue! (x, q) inserts x into the queue q at the back end.

enterPointData (*space, listOfPoints*)

enterPointData $(s, [p_0, p_1, \ldots, p_n])$ adds a list of points from p_0 through p_n to the ThreeSpace s, and returns the index of the start of the list.

entry? (*value, aggregate*)

entry? (x, u), where u is an indexed aggregate (such as a list, vector, or string), tests if x equals $u.i$ for some index i.

epilogue (*formattedObject*)

epilogue (t) extracts the epilogue section of an IBM SCRIPT Formula Format or TEX formatted object t.

eq (*sExpression, sExpression*)

$eq(s, t)$, for SExpressions s and t returns $true$ if EQ(s, t) is $true$ in Common LISP.

eq? (*aggregate, aggregate*)

eq? (u, v) tests if two aggregates u and v are same objects in the AXIOM store.

equality (*operator, function*)

equality (op, f) attaches f as the `"%equal?"` property to op. Argument f must be a boolean-valued "equality" function defined on BasicOperator objects. If $op1$ and $op2$ have the same name, and one of them has an `"%equal?"` property f, then $f(op1, op2)$ is called to decide whether $op1$ and $op2$ should be considered equal.

equation (*expression, expression*)

equation (a, b) creates the equation $a = b$.
equation $(v, a..b)$, also written: $v = a..b$, creates a segment binding value with variable v and segment $a..b$.

erf (*variable*)

erf (x) returns the error function of x: $\frac{2}{\sqrt{(\pi)}} \int exp^{-x^2} dx$.

error (*string*[, *string*])

error (msg) displays error message msg and terminates. Argument msg is either a string or a list of strings.
error $(name, msg)$ is similar except that the error message is preceded by a message saying that the error occured in a function named $name$.

euclideanGroebner (*ideal*[, *string, string*])

euclideanGroebner (lp[, "$info$", "*redcrit*"]) computes a Gröbner basis for a polynomial ideal over a Euclidean domain generated by the list of polynomials lp. If the string `"info"` is given as a second argument, a summary is given of the critical pairs. If the string `"redcrit"` is given as a third argument, the critical pairs are printed.

euclideanNormalForm (*polynomial, groebnerBasis*)

euclideanNormalForm ($poly, gb$) reduces the polynomial $poly$ modulo the precomputed Gröbner basis gb giving a canonical representative of the residue class.

euclideanSize (*element*)

euclideanSize (x) returns the Euclidean size of the element x, or calls **error** if x is zero.

eulerPhi (*positiveInteger*)

eulerPhi (n) returns the number of integers between 1 and n (including 1) which are relatively prime to n. This is the Euler phi function $\phi(n)$, also called the totient function.

euler (*positiveInteger*)

euler (n) returns the n th Euler number. This is $2^n E(n, 1/2)$, where $E(n, x)$ is the n th Euler polynomial.

eval (*expression*[, *options*])

Many domains have forms of the **eval** defined. Here are some the most common forms.
eval (f) unquotes all the quoted operators in f.
eval ($f, x = v$) replaces symbol or expression x by v in f; if x is an expression, it must be retractable to a single Kernel.
eval ($f, [x_1 = v_1, \ldots, x_n = v_n]$) returns f with symbols or expressions x_i replaced by v_i in parallel; if x_i is an expression, it must be retractable to a single Kernel.
eval ($f, [x_1, \ldots, x_n]$) unquotes all the quoted operations in f whose name is one of the x_i.'s.
eval (f, x) unquotes all quoted operators in f whose name is x.
eval (e, s, f) replaces every subexpression of e of the form $s(a_1, \ldots, a_n)$ by $f(a_1, \ldots, a_n)$. The function f can have type Expression → Expression if s is a unary operator; otherwise f must have signature List(Expression) → Expression.
eval ($e, [s_1, \ldots, s_n], [f_1, \ldots, f_n]$), replaces every subexpression of e of the form $s_i(a_1, \ldots, a_{n_i})$ by $f_i(a_1, \ldots, a_{n_i})$. If all the s_i's are unary operators, the functions f_i can have signature Expression → Expression; otherwise, the f_i must have signature List(Expression) → Expression.
eval (p, el), where p is a permutation, returns the image of element el under p.
eval (s), where `s` is of type SymmetricPolynomial with rational number coefficients, returns the sum of the coefficients of a cycle index. See CycleIndicators for details.
eval (f, s), where `s` is of type SymmetricPolynomial with rational number coefficients and `f` is a function of type Integer → Algebra Fraction Integer, evaluates the cycle index s by applying the function `f` to each integer in a monomial partition, forms their product and sums the results over all monomials. See EvaluateCycleIndicators for details.

evaluate (*operator, function*)

evaluate (*op*) returns the value of the `"%eval"` property of BasicOperator object *op* if it has one, and `"failed"` otherwise.
evaluate (*op*, f) attaches f as the `"%eval"` property of *op*. If *op* has an `"%eval"` property f, then applying *op* to a returns the result of $f(a)$. If f takes a single argument, then applying *op* to a value a returns the result $f(a)$. If f takes a list of arguments, then applying *op* to $a_1, \ldots, a_n$ returns the result of $f(a_1, \ldots, a_n)$.
Argument f may also be an anonymous function of the form $u + - > g(u)$. In this case, g *must* be additive, that is, $g(a + b) = g(a) + g(b)$ for any a and b in R. This implies that $g(na) = ng(a)$ for any a in R and integer $n > 0$.

even? (*integerNumber*)

even? (n) tests if integer n is even.
even? (p) tests if permutation p is an even permutation, that is, that the **sign** (p) = 1.

every? (*predicate, aggregate*)

every? ($pred, u$) tests if *pred(x)* is $true$ for all elements x of u.

exists? (*file*)

exists? (f) tests if the file f exists in the file system.

exp (*expression*)
expIfCan (x)

exp (x) returns `%e` to the power x.
expIfCan (z) returns exp(z) if possible, and `"failed"` otherwise.

exp1 ()

exp1 ()$R returns exp 1: 2.7182818284... either a float or a small float according to whether R = Float or R = SmallFloat.

expand (*expression*)

expand (f) performs the following expansions on Expression f:

Logs of products are expanded into sums of logs.

Trigonometric and hyperbolic trigonometric functions of sums are expanded into sums of products of trigonometric and hyperbolic trigonometric functions.

Formal powers of the form $(a/b)^c$ are expanded into $a^c b^{(-c)}$.

expand (ir), where ir is an IntegrationResult, returns the list of possible real functions corresponding to ir.

expand $(lseg)$, where $lseg$ is a list of segments, returns a list with all segments expanded. For example, `expand [1..4, 7..9] = [1, 2, 3, 4, 7, 8, 9]`.
expand $(l..h$ `by` $k)$ returns a list of explicit elements. For example, `expand(1..5 by 2) = [1, 3, 5]`.
expand (f) returns an unfactored form of factored object f.

expandLog (*expression*)

expandLog (f) converts every **log** (a/b) appearing in Expression f into $\log(a) - \log(b)$.

expandPower (*expression*)

expandPower (f) converts every power $(a/b)^c$ appearing in Expression f into $a^c b^{-c}$.

explicitEntries? (*stream*)

explicitEntries? (s) tests if the stream s has explicitly computed entries.

explicitlyEmpty? (*stream*)

explicitlyEmpty? (s) tests if the stream is an (explicitly) empty stream. Note: this is a null test which will not cause lazy evaluation.

explicitlyFinite? (*stream*)

explicitlyFinite? (s) tests if the stream s has a finite number of elements. Note: for many datatypes, `explicitlyFinite?(s) = not possiblyInfinite?(s)`.

exponent (*floatOrFactored*)

exponent (fl) returns the **exponent** part of a float or small float fl.
exponent (u), where u is a factored object, returns the exponent of the first factor of u, or 0 if the factored object consists solely of a unit.

expressIdealMember (*listOfIdeals, ideal*)

expressIdealMember $([f_1, \ldots, f_n], h)$ returns a representation of ideal h as a linear combination of the ideals f_i or `"failed"` if h is not in the ideal generated by the f_i.

exptMod (*polynomial, nonNegativeInteger, polynomial*[, *prime*])

exptMod $(u, k, v[, p])$ raises the polynomial u to the k th power modulo the polynomial v. If a prime p is given, the power is also computed modulo that prime.

exquo (*element, element*)

exquo (a, b) either returns an element c such that $cb = a$ or `"failed"` if no such element can be found. Values a and b are members of a domain of category IntegralDomain.
exquo (A, r) returns the exact quotient of the elements of matrix A by coefficient r, or calls **error** if this is not possible.

extend (*stream, integer*)

extend (ps, n), where ps is a power series, causes all terms of ps of degree $\leq n$ to be computed.
extend (st, n), where st is a stream, causes entries to be computed so that st has at least n explicit entries, or so that all entries of st are finite with length $\leq n$.

extendedEuclidean (*element, element*[, *element*])

Argments x, y, and z are members of a domain of category EuclideanDomain.
extendedEuclidean (x, y) returns a record rec containing three fields: $coef1$, $coef2$, and $generator$ where $rec.coef1 * x + rec.coef2 * y = rec.generator$ and $rec.generator$ is a gcd of x and y. The gcd is unique only up to associates if `canonicalUnitNormal` is not asserted. Note: See **principalIdeal** for a version of this operation which accepts an arbitrary length list of arguments.
extendedEuclidean (x, y, z) either returns a record rec of two fields $coef1$ and $coef2$ where $rec.coef1 * x + rec.coef2 * y = z$, and `"failed"` if z cannot be expressed as such a linear combination of x and y.

extendedIntegrate (*rationalFnct, symbol, rationalFnct*)

extendedIntegrate (f, x, g) returns fractions $[h, c]$ such that $dc/dx = 0$ and $dh/dx = f - cg$ if (h, c) exist, and `"failed"` otherwise.

extensionDegree ()

extensionDegree ()$\$F$ returns the degree of the field extension F if the extension is algebraic, and `infinity` if it is not.

extension (*filename*)

extension (fn) returns the type part of the file name fn as a string.

extract! (*bag*)

extract! (bg) destructively removes a (random) item from bag bg.

extractBottom! (*dequeue*)

extractBottom! (d) destructively extracts the bottom (back) element from the dequeue d, or calls **error** if d is empty.

extractTop! (*dequeue*)

extractTop! (d) destructively extracts the top (front) element from the dequeue d, or calls **error** if d is empty.

e (*positiveInteger*)

e (n) produces the appropriate unit element of a CliffordAlgebra.

factor (*polynomial*[, *numbers*])

factor (x) returns the factorization of x into irreducibles, where x is a member of any domain of category UniqueFactorizationDomain.

factor (p, lan), where p is a polynomial and lan is a list of algebraic numbers, factors p over the extension generated by the algebraic numbers given by the list lan.
factor $(upoly, prime)$, where $upoly$ is a univariate polynomial and $prime$ is a prime integer, returns the list of factors of $upoly$ modulo the integer prime p, or calls **error** if $upoly$ is not square-free modulo p.

factorFraction (*fraction*)

factorFraction (r) factors the numerator and the denominator of the polynomial fraction r.

factorGroebnerBasis (*listOfPolynomials* [, *boolean*])

factorGroebnerBasis $(basis[, flag])$ checks whether the $basis$ contains reducible polynomials and uses these to split the $basis$. Information about partial results is given if a second argument of $true$ is given.

factorials (*expression* [, *symbol*])

factorials $(f[, x])$ rewrites the permutations and binomials in f in terms of factorials. If a symbol x is given as a second argument, the operation rewrites only those terms involving x.

factorial (*expression*)

factorial (n), where n is an integer, returns the integer value of $n! = \prod_1^n i$.
factorial (n), where n is an expression, returns a formal expression denoting n! Note: $n! = n(n-1)!$ when $n > 0$; also, $0! = 1$.

factorList (*factoredForm*)

factorList (f), for a factored form f, returns list of records. Each record corresponds to a factor of f and has three fields: flg, $fctr$, and $xpnt$. The $fctr$ lists the factor and $xpnt$, the exponent. The flg is one of the strings: `"nil"`, `"sqfr"`, `"irred"`, or `"prime"`.

factorPolynomial (*polynomial*)

factorPolynomial (p) returns the factorization of a sparse univariate polynomial p as a factored form.

factors (*factoredForm*)

factors (u) returns a list of the factors of a factored form u in a form as a list suitable for iteration. Each element in the list is a record containing both a $factor$ and $exponent$ field.

factorsOfCyclicGroupSize ()

factorsOfCyclicGroupSize () returns the factorization of **size** () $- 1$

factorSquareFreePolynomial (*polynomial*)

factorSquareFreePolynomial (p) factors the univariate polynomial p into irreducibles, where p is known to be square free and primitive with respect to its main variable.

fibonacci (*nonNegativeInteger*)

fibonacci (n) returns the nth Fibonacci number. The Fibonacci numbers $F[n]$ are defined by $F[0] = F[1] = 1$ and $F[n] = F[n-1] + F[n-2]$. The algorithm has running time $O(\log(n)^3)$.

filename (*directory, name, extension*)

filename (d, n, e) creates a file name with string d as its directory, string n as its name and string e as its extension.

fill! (*aggregate, value*)

fill! (a, x) replaces each entry in aggregate a by x. The modified a is returned. If a is a domain of category $TwoDimensionalArrayCategory$ such as a matrix, **fill!** (a, x) sets every element of a to x.

filterUntil (*predicate, stream*)

filterUntil (p, s) returns $[x_0, x_1, \ldots, x_n]$, where stream $s = [x_0, x_1, x_2, ..]$ and n is the smallest index such that $p(x_n) = true$.

filterWhile (*predicate, stream*)

filterWhile $(pred, s)$ returns $[x_0, x_1, \ldots, x_{(n-1)}]$ where $s = [x_0, x_1, x_2, ..]$ and n is the smallest index such that $p(x_n) = false$.

find (*predicate, aggregate*)

find $(pred, u)$ returns the first x in u such that **pred** (x) is $true$, and `"failed"` if no such x exists.

findCycle (*nonNegativeInteger, stream*)

findCycle (n, st) determines if stream st is periodic within n terms. The operation returns a record with three fields: *cycle?*, $prefix$, and *period*. If *cycle?* has value true, *period* denotes the period of the cycle, and $prefix$ gives the number of terms in the stream before the cycle begins.

finite? (*cardinalNumber*)

finite? (f) tests if expression f is finite.
finite? (a) tests if cardinal number a is a finite cardinal, that is, an integer.

fintegrate (*taylorSeries, symbol, coefficient*)

fintegrate (s, v, c) integrates the series s with respect to variable v and having c as the constant of integration.

first (*aggregate* [, *nonNegativeInteger*])

first (u) returns the first element x of aggregate u.
first (u, n) returns a copy of the first n elements of u.

fixedPoint (*function* [, *positiveInteger*])

fixedPoint (f), a function of type A $\rightarrow$ A, is the fixed point of function f. That is, **fixedPoint** $(f) = f($**fixedPoint**$(f))$.
fixedPoint (f, n), where f is a function of type List(A) $\rightarrow$ List(A) and n is a positive integer, is the fixed point of function f which is assumed to transform a list of length n.

fixedPoints (*permutation*)

fixedPoints (p) returns the points fixed by the permutation p.

flagFactor (*base, exponent, flag*)

flagFactor $(base, exponent, flag)$ creates a factored object with a single factor whose $base$ is asserted to be properly described by the information $flag$: one of the strings "nil", "sqfr", "irred", and "prime".

flatten (*inputForm*)

flatten (s) returns an input form corresponding to s with all the nested operations flattened to triples using new local variables. This operation is used to optimize compiled code.

flexible? ()

flexible? ()R tests if 2**associator**$(a, b, a) = 0$ for all a, b in a domain R of category FiniteRankNonAssociativeAlgebra. Note: only this can be tested since, in general, it is not known whether $2a = 0$ implies $a = 0$.

flexibleArray (*listOfElements*)

flexibleArray (ls) creates a flexible array from a list of elements ls.

float? (*sExpression*)

float? (s) is $true$ if s is an atom and belongs o Flt.

float (*integer, integer*[, *positiveinteger*])

float (a, e) returns $a\mathtt{base}()^e$ as a float.
float (a, e, b) returns ab^e as a float.

floor (*rationalNumber*)

floor (fr), where fr is a fraction, returns the largest integral element below fr.
floor (fl), where fl is a float, returns the largest integer $<= fl$.

formula (*formulaFormat*)

formula (t) extracts the formula section of an IBM SCRIPT Formula formatted object t.

fractionPart (*fraction*)

fractionPart (x) returns the fractional part of x. Argument x can be a fraction, a radix (binary, decimal, or hexadecimal) expansion, or a float. Note: x = whole(x) + fractionPart(x).

fractRadix (*listOfIntegers, listOfIntegers*)

fractRadix (pre, cyc) creates a fractional radix expansion from a list of prefix ragits and a list of cyclic ragits. For example, **fractRadix** ([1], [6]) will return 0.16666666 ....

fractRagits (*radixExpansion*)

fractRagits (rx) returns the ragits of the fractional part of a radix expansion as a stream of integers.

freeOf? (*expression, kernel*)

freeOf? (x, k) tests if expression x does not contain any operator whose name is the symbol or kernel k.

Frobenius (*element*)

Frobenius $(a)\$F$ returns a^q where q is the **size** ()F of extension field F.

front (*queue*)

front (q) returns the element at the front of the queue, or calls **error** if q is empty.

frst (*stream*)

frst (s) returns the first element of stream s. Warning: this function should only be called after a *empty?* test has been made since there is no error check.

function (*expression, name*[, *options*])

Most domains provide an operation which converts objects to type InputForm. Argument e below denotes an object from such a domain. These operations create user-functions from already computed results.
function (e, f) creates a function $f() == e$.
function $(e, f, [x_1, \ldots, x_n])$ creates a function $f(x_1, \ldots, x_n) == e$.
function (e, f, x) creates a function $f(x) == e$.
function (e, f, x, y) creates a function $f(x, y) == e$.
function $(expr, [x_1, \ldots, x_n], f)$, where $expr$ is an input form and where f and the x_i's are symbols, returns the input form corresponding to $f(x_1, \ldots, x_n) ==$ i. See also **unparse**.

Gamma (*smallFloat*)

Gamma (x) is the Euler gamma function, **Gamma** (x), defined by $(x) = \int_0^\infty t^{(x-1)} * exp(-t)dt$.

gcdPolynomial (*polynomial, polynomial*)

gcdPolynomial (p, q) returns the **gcd** of the univariate polynomials p and q.

gcd (*element*[, *element, element*])

gcd (x, y) returns the greatest common divisor of x and y. Arguments x and y are elements of a domain of category $GcdDomain$.
gcd $([x_1, \ldots, x_n])$ returns the common gcd of the elements of the list of x_i.
gcd $(p_1, p_2, prime)$ computes the gcd of the univariate polynomials p_1 and p_2 modulo the prime integer $prime$.

generalizedContinuumHypothesisAssumed? ([*bool*])

generalizedContinuumHypothesisAssumed? () tests if the hypothesis is currently assumed.
generalizedContinuumHypothesisAssumed (*bool*) dictates that the hypothesis is or is not to be assumed, according to whether *bool* is true or false.

generalPosition (*ideal*, *listOfVariables*)

generalPosition ($I, listvar$) performs a random linear transformation on the variables in $listvar$ and returns the transformed ideal I along with the change of basis matrix.

generate (*function* [, *element*])

generate (f), where f is a function of no arguments, creates an infinite stream all of whose elements are equal to the value of $f()$. Note: **generate** $(f) = [f(), f(), f(), \ldots]$.
generate (f, x), where f is a function of one argument, creates an infinite stream whose first element is x and whose n^{th} element ($n > 1$) is f applied to the previous element. Note: **generate** $(f, x) = [x, f(x), f(f(x)), \ldots]$.
See also HallBasis.

generator ()

generator ()$R returns a root of the defining polynomial of a domain of category FiniteAlgebraicExtensionField R. This element generates the field as an algebra over the ground field.
See also MonogenicAlgebra and FreeNilpotentLie.

generators (*ideal*)

generators (I) returns a list of generators for the ideal I.
generators (gp) returns the generators of a permutation group *gp*.

genus ()

genus ()$R returns the genus of the algebraic function field R. If R has several absolutely irreducible components, then the genus of one of them is returned.

getMultiplicationMatrix ()
getMultiplicationTable ()

getMultiplicationMatrix ()$R returns a matrix multiplication table for domain FiniteFieldNormalBasis(p, n), a finite extension field of degree n over the domain PrimeField(p) with p elements. Each element of the matrix is a member of the underlying prime field.
getMultiplicationTable ()$R is similar except that the multiplication table for the normal basis of the field is represented by a vector of lists of records, each record having two fields: $value$, an element of the prime field over which the domain is built, and $index$, a small integer. This table is used to perform multiplications between field elements.

getVariableOrder ()

getVariableOrder () returns $[[b_1, \ldots, b_m], [a_1, \ldots, a_n]]$ such that the ordering on the variables was given by **setVariableOrder** $([b_1, \ldots, b_m], [a_1, \ldots, a_n])$.

getZechTable ()

getZechTable ()$F returns the Zech logarithm table of the field F where F is some domain FiniteFieldCyclicGroup(p, extdeg). This table is used to perform additions in the field quickly.

gramschmidt (*listOfMatrices*)

Argument lv has the form of a list of matrices of elements of type Expression.
gramschmidt (lv) converts the list of column vectors lv into a set of orthogonal column vectors of Euclidean length 1 using the Gram-Schmidt algorithm.

graphs (*integer*)

graphs (n) is the cycle index of the group induced on the edges of a graph by applying the symmetric function to the n nodes. See CycleIndicators for details.

green ()

green () returns the position of the green hue from total hues.

groebner (*listOfPolynomials*)

groebner (lp) computes a Gröbner basis for a polynomial ideal generated by the list of polynomials lp.
groebner (I) returns a set of generators of ideal I that are a Gröbner basis for I.
groebner ($lp, infoflag$) computes a Gröbner basis for a polynomial ideal generated by the list of polynomials lp. Argument $infoflag$ is used to get information on the computation. If $infoflag$ is `"info"`, then summary information is displayed for each s-polynomial generated. If $infoflag$ is `"redcrit"`, the reduced critical pairs are displayed. To get the display of both kinds of information, use **groebner** (lp, "$info$", "$redcrit$").

groebner? (*ideal*)

groebner? (I) tests if the generators of the ideal I are a Gröbner basis.

groebnerIdeal (*listOfPolynomials*)

groebnerIdeal (lp) constructs the ideal generated by the list of polynomials lp assumed to be a Gröbner basis. Note: this operation avoids a Gröbner basis computation.

groebnerFactorize (*listOfPolynomials* [*options*])

groebnerFactorize (lp[, *bool*]) returns a list of list of polynomials, each inner list denoting a Gröbner basis. The union of the solutions of the bases is the solution of the system of equations given by lp. Information about partial results is printed if a second argument is given with value $true$.
groebnerFactorize ($lp, nonZeroRestrictions$[, *bool*]), where $nonZeroRestrictions$ is a list of polynomials, is similar. Here, however, the solutions to the system of equations are computed under the restriction that the polynomials in the second argument do not vanish. Information about partial results is printed if a third argument with value $true$ is given.

ground (*expression*)
ground? (*expression*)

ground (p) retracts expression polynomial p to the coefficient

ring, or calls **error** if such a retraction is not possible.
ground? (p) tests if an expression or polynomial p is a member of the coefficient ring. See also **ground?**.

harmonic (*positiveInteger*)

harmonic (n) returns the nth harmonic number, defined by $H[n] = \sum_{k=1}^{n} 1/k$.

has (*domain, property*)

has $(R, prop)$ tests if domain R has property $prop$. Argument $prop$ is either a category, operation, an attribute, or a combination of these. For example, `Integer has Ring` and `Integer has commutative("*")`.

has? (*operation, property*)

has? (op, s) tests if property s is attached to op.

hash (*number*)

hash (n) returns the hash code for n, an integer or a float.

hasHi (*segment*)

hasHi (seg) tests whether the segment seg has an upper bound. For example, **hasHi** $(1..) = false$.

hasSolution? (*matrix, vector*)

hasSolution? (A, B) tests if the linear system $AX = B$ has a solution, where A is a matrix and B is a (column) vector.

hconcat (*outputForms*[, *outputForm*])

hconcat (o_1, o_2), where o_1 and o_2 are objects of type OutputForm (normally unexposed), returns an output form for the horizontal concatenation of forms o_1 and o_2.
hconcat (lof), where lof is a list of objects of type OutputForm (normally unexposed), returns an output form for the horizontal concatenation of the elements of lof.

heap (*listOfElements*)

heap (ls) creates a Heap of elements consisting of the elements of ls.

heapSort (*predicate, aggregate*)

heapSort $(pred, agg)$ sorts the aggregate agg with the ordering function $pred$ using the heapsort algorithm.

height (*expression*)

height (f), where f is an expression, returns the highest nesting level appearing in f. Constants have height 0. Symbols have height 1. For any operator op and expressions $f_1, \ldots, f_n$, $op(f_1, \ldots, f_n)$ has height equal to $1 + max(height(f_1), \ldots, height(f_n))$.
height (d) returns the number of elements in dequeue d. Note: **height** $(d) = \#d$.

hermiteH (*nonNegativeInteger, element*)

hermiteH (n, x) is the nth Hermite polynomial, $H[n](x)$, defined by **exp** $(2tx - t^2) = \sum_{n=0}^{\infty} H[n](x)t^n/n!$.

hexDigit ()

hexDigit () returns the class of all characters for which **hexDigit?** is $true$.

hexDigit? (*character*)

hexDigit? (c) tests if c is a hexadecimal numeral, that is, one of 0..9, a..f or A..F.

hex (*rationalNumber*)

hex (r) converts a rational number to a hexadecimal expansion.

hi (*segment*)

hi (s) returns the second endpoint of segment s. For example, **hi** $(l..h) = h$.

horizConcat (*matrix, matrix*)

horizConcat (x, y) horizontally concatenates two matrices with an equal number of rows. The entries of y appear to the right of the entries of x. The operation calls **error** if the matrices do not have the same number of rows.

htrigs (*expression*)

htrigs (f) converts all the exponentials in expression f into hyperbolic sines and cosines.

hue (*palette*)

hue (p) returns the hue field of the indicated palette p.

hue (*color*)

hue (c) returns the hue index of the indicated color c.

hypergeometric0F1 (*complexSF, complexSF*)

hypergeometric0F1 (c, z) is the hypergeometric function $0F1(c; z)$. Arguments c and z are both either small floats or complex small floats.

ideal (*polyList*)

ideal $(polyList)$ constructs the ideal generated by the list of polynomials $polyList$.

imag (*expression*)
imagi (*quaternionOrOctonion*)
imagI (*octonion*)

imag (x) extracts the imaginary part of a complex value or expression x.
imagI (q) extracts the i part of quaternion q. Similarly, operations **imagJ**, and **imagK** are used to extract the j and k parts.
imagi (o) extracts the i part of octonion o. Similarly, **imagj**, **imagk**, **imagE**, **imagI**, **imagJ**, and **imagK** are used to extract other parts.

implies (*boolean, boolean*)

implies (a, b) tests if boolean value a implies boolean value b. The result is $true$ except when a is $true$ and b is $false$.

in? (*ideal, ideal*)

in? (I, J) tests if the ideal I is contained in the ideal J.

inHallBasis (*integer, integer, integer, integer*)

$inHallBasis?(n, leftCandidate, rightCandidate, left)$ tests to see if a new element should be added to the P. Hall basis being constructed. The list $[leftCandidate, wt, rightCandidate]$ is included in the basis if in the unique factorization of $rightCandidate$, we have left factor $leftOfRight$, and $leftOfRight <= leftCandidate$

increasePrecision (*integer*)

increasePrecision (n) increases the current **precision** by n decimal digits.

index (*positiveInteger*)

index (i) takes a positive integer i less than or equal to **size** () and returns the ith element of the set. This operation establishes a bijection between the elements of the finite set and 1..**size**().

index? (*index, aggregate*)

index? (i, u) tests if i is an index of aggregate u. For example, `index?(2, [1, 2, 3])` is $true$ but `index?(4, [1, 2, 3])` is $false$.

infieldIntegrate (*rationalFunction, symbol*)

infieldIntegrate (f, x), where f is a fraction of polynomials, returns a fraction g such that $\frac{dg}{dx} = f$ if g exists, and `"failed"` otherwise.

infinite? (*orderedCompletion*)

infinite? (x) tests if x is infinite, where x is a member of the ordered completion of a domain. See OrderedCompletion using Browse.

infinity ()

infinity () returns $infinity$ denoting $+\infty$ as a one point completion of the integers. See OnePointCompletion using Browse. See also **minusInfinity** and **plusInfinitity**.

infix (*outputForm, outputForms* [, *OutputForm*])

infix (o, lo), where o is an object of type OutputForm (normally unexposed) and lo is a list of objects of type OutputForm, creates a form depicting the $n$ary application of infix operation o to a tuple of arguments lo.
infix (o, a, b), where o, a, and b are objects of type OutputForm (normally unexposed), creates an output form which displays as: a op b.

initial (*differentialPolynomial*)

initial (p) returns the leading coefficient of differential polynomial p expressed as a univariate polynomial in its leader.

initializeGroupForWordProblem (*group* [, *integer, integer*])

initializeGroupForWordProblem $(gp[, n, m])$ initializes the group *gp* for the word problem. Consult PermutationGroup using Browse for details.

input (*operator*[, *function*])

input (op) returns the `"%input"` property of *op* if it has one attached, and `"failed"` otherwise.
input (op, f) attaches f as the `"%input"` property of *op*. If *op* has a `"%input"` property f, then $op(a1, \ldots, an)$ is converted to InputForm using $f(a1, \ldots, an)$. Argument f must be a function with signature List(InputForm) $\rightarrow$ InputForm.

inRadical? (*polynomial, ideal*)

inRadical? (f, I) tests if some power of the polynomial f belongs to the ideal I.

insert (*x, aggregate* [, *integer*])

insert (x, u, i) returns a copy of u having x as its ith element.
insert (v, u, k) returns a copy of u having v inserted beginning at the ith element.
insert! (x, u) destructively inserts item x into bag u.
insert! (x, u) destructively inserts item x as a leaf into binary search tree or binary tournament u.
insert! (x, u, i) destructively inserts x into aggregate u at position i.
insert! (v, u, i) destructively inserts aggregate v into u at position i.
insert! (x, d, n) destructively inserts n copies of x into dictionary d.

insertBottom! (*element, queue*)

insertBottom! (x, d) destructively inserts x into the dequeue d at the bottom (back) of the dequeue.

insertTop! (*element, dequeue*)

insertTop! (x, d) destructively inserts x into the dequeue d at the top (front). The element previously at the top of the dequeue becomes the second in the dequeue, and so on.

integer (*expression*)
integer? (*expression*)
integerIfCan (*expression*)

integer (x) returns x as an integer, or calls **error** if this is not possible.
integer? (x) tests if expression x is an integer.
integerIfCan (x) returns expression x as of type Integer or

else "failed" if it cannot.

integerPart (*float*)

integerPart (fl) returns the integer part of the mantissa of float fl.

integral (*expression, symbol*)
integral (*expression, segmentBinding*)

integral (f, x) returns the formal integral $\int f dx$.
integral ($f, x = a..b$) returns the formal definite integral $\int_a^b f(x)dx$.

integralBasis ()
integralBasisAtInfinity ()

Domain F is the domain of functions on a fixed curve. See FunctionFieldCategory using Browse.
integralBasisAtInfinity ()$F returns the local integral basis at infinity.
integralBasis ()$F returns the integral basis for the curve.

integralCoordinates (*function*)

integralCoordinates (f), where f is a function on a curve defined by domain F, returns the coordinates of f with respect to the **integralBasis** ()$F as polynomials A_i together with a common denominator d. Specifically, the operation returns a record having selector num with value $[A_1, \ldots, A_n]$ and selector den with value d such that $f = (A_1 w_1 + \ldots + A_n w_n)/d$ where $(w_1, \ldots, w_n)$ is the integral basis. See FunctionFieldCategory using Browse.

integralDerivationMatrix (*function*)

integralDerivationMatrix (d) extends the derivation d and returns the coordinates of the derivative of f with respect to the **integralBasis** ()$F as a matrix of polynomials and a common denominator Q. Specifically, the operation returns a record having selector num with value M and selector den with value Q such that the ith row of M divided by Q form the coordinates of f with respect to integral basis $(w1, \ldots, wn)$. See FunctionFieldCategory using Browse.

integralMatrix ()
integralMatrixAtInfinity ()

Domain F is a domain of functions on a fixed curve. These operations return a matrix which transform the natural basis to an integral basis. See FunctionFieldCategory using Browse.

integralMatrix () returns M such that $(w_1, \ldots, w_n) = M(1, y, \ldots, y^{n-1})$, where $(w_1, \ldots, w_n)$ is the integral basis returned by **integralBasis** ()$F.
integralMatrixAtInfinity ()$F returns matrix M which transforms the natural basis such that $(v_1, \ldots, v_n) = M(1, y, \ldots, y^{n-1})$ where $(v_1, \ldots, v_n)$ is the local integral basis at infinity returned by **integralBasisAtInfinity** ()$F.

integralRepresents (*vector, commonDenominator*)

integralRepresents ($[A_1, \ldots, A_n], d$) is the inverse of the operation **integralCoordinates** defined for domain F, a domain of functions on a fixed curve. Given the coordinates as polynomials $[A_1, \ldots, A_n]$ over a common denominator d, this operation returns the function represented as$(A_1 w_1 + \ldots + A_n w_n)/d$ where $(w_1, \ldots, w_n)$ is the integral basisreturned by **integralBasis** ()$F.See FunctionFieldCategory using Browse.

integrate (*expression*)
integrate (*expression, variable* [, *options*])

integrate (f) returns the integral of a univariate polynomial or power series f with respect to its distinguished variable.
integrate (f, x) returns the integral of $f(x)dx$, where x is viewed as a real variable.
integrate ($f, x = a..b$[, "*noPole*"]) returns the integral of $f(x)dx$ from a to b. If it is not possible to check whether f has a pole for x between a and b, then a third argument "noPole" will make this function assume that$f$ has no such pole.This operation calls **error** if f has a pole for x between a and b or if a third argument different from "noPole" is given.

interpret (*inputForm*)

interpret (f) passes f of type InputForm to the interpreter.
interpret (f)$P, where P is the package InputFormFunctions1(R) for some type R, passes f of type InputForm to the interpreter, and transforms the result into an object of type R.

intersect (*elements*[, *element*])

intersect (li), where li is a list of ideals, computes the intersection of the list of ideals li.
intersect (u, v), where u and v are sets, returns the set w consisting of elements common to both sets u and v. See also Multiset.
intersect (I, J), where I and J are ideals, computes the intersection of the ideals I and J.

inv (*element*)

inv (x) returns the multiplicative inverse of x, where x is an element of a domain of category Group or DivisionRing, or calls **error** if x is 0.

inverse (*matrix*)

inverse (A) returns the inverse of the matrix A, or "failed" if the matrix is not invertible, or calls **error** if the matrix is not square.

inverseColeman (*listOfIntegers, listOfIntegers, matrix*)

inverseColeman ($alpha, beta, C$) returns the lexicographically smallest permutation in a double coset of the symmetric group corresponding to a non-negative Coleman-matrix. Consult SymmetricGroupCombinatoricFunctions using Browse for details.

inverseIntegralMatrix ()
inverseIntegralMatrixAtInfinity ()

Domain F is a domain of functions on a fixed curve. These operations return a matrix which transform an integral basis to a natural basis. See FunctionFieldCategory using Browse.
inverseIntegralMatrix ()$F returns M such that $M(w_1, \ldots, w_n) = (1, y, \ldots, y^{n-1})$ where $(w_1, \ldots, w_n)$ is the integral basis returned by **integralBasis** ()$F. See also **integralMatrix**.
inverseIntegralMatrixAtInfinity () returns M such that $M(v_1, \ldots, v_n) = (1, y, \ldots, y^(n-1))$ where $(v_1, \ldots, v_n)$ is the local integral basis at infinity returned by **integralBasisAtInfinity** ()$F. See also **integralMatrixAtInfinity**.

inverseLaplace (*expression, symbol, symbol*)

inverseLaplace (f, s, t) returns the Inverse Laplace transform of $f(s)$ using t as the new variable, or `"failed"` if unable to find a closed form.

invmod (*positiveInteger, positiveInteger*)

invmod (a, b), for relatively prime positive integers a and b such that $a < b$, returns $1/a$ **mod** b.

iomode (*file*)

iomode (f) returns the status of the file f as one of the following strings: `"input"`, `"output"` or `"closed"`.

irreducible? (*polynomial*)

irreducible? (p) tests whether the polynomial p is irreducible.

irreducibleFactor (*element, integer*)

irreducibleFactor $(base, exponent)$ creates a factored object with a single factor whose $base$ is asserted to be irreducible (flag = `"irred"`).

irreducibleRepresentation (*listOfIntegers* [, *permutations*])

irreducibleRepresentation ($lambda$[, *pi*]) returns a matrix giving the irreducible representation corresponding to partition $lambda$, represented as a list of integers, in Young's natural form of the permutation *pi* in the symmetric group whose elements permute $1, 2, \ldots, n$. If a second argument is not given, the permutation is taken to be the following two generators of the symmetric group, namely (12) (2-cycle) and $(12\ldots n)$ ((n)-cycle).

is? (*expression, pattern*)

is? $(expr, pat)$ tests if the expression $expr$ matches the pattern pat.
is? ($expression$, *op*) tests if $expression$ is a kernel and is its operator is op.

isAbsolutelyIrreducible? (*listOfMatrices, integer*)

isAbsolutelyIrreducible? $(aG, numberOfTries)$ uses Norton's irreducibility test to check for absolute irreduciblity. Consult RepresentationPackage2 using Browse for details.

isExpt (*expression* [, *operator*])

isExpt (p[, *op*]) returns a record with two fields: var denoting a kernel x, and $exponent$ denoting an integer n, if expression p has the form $p = x^n$ and $n \neq 0$. If a second argument *op* is given, x must have the form *op*(a) for some a.

isMult (*expression*)

isMult (p) returns a record with two fields: $coef$ denoting an integer n, and var denoting a kernel x, if p has the form $n * x$ and $n \neq 0$, and `"failed"` if this is not possible.

isobaric? (*differentialPolynomial*)

isobaric? (p) tests if every differential monomial appearing in the differential polynomial p has the same weight.

isPlus (*expression*)

isPlus (p) returns $[m_1, \ldots, m_n]$ if p has the form $m_1 + \ldots + m_n$ for $n > 1$ and $m_i \neq 0$, and `"failed"` if this is not possible.

isTimes (*expression*)

isTimes (p) returns $[a_1, \ldots, a_n]$ if p has the form $a_1 * \ldots * a_n$ for $n > 1$ and $m_i \neq 1$, and `"failed"` if this is not possible.

Is (*subject, pattern*)

$Is(expr, pat)$ matches the pattern pat on the expression $expr$ and returns a list of matches $[v_1 = e_1, \ldots, v_n = e_n]$ or `"failed"` if matching fails. An empty list is returned if either $expr$ is exactly equal to pat or if pat does not match $expr$.

jacobi (*integer, integer*)

jacobi (a, b) returns the Jacobi symbol $J(a/b)$. When b is odd, $J(a/b) = \prod_{p \in \mathbf{factors}(b)} L(a/p)$. Note: by convention, 0 is returned if **gcd** $(a, b) \neq 1$.

jacobiIdentity? ()

jacobiIdentity? () tests if $(ab)c + (bc)a + (ca)b = 0$ for all a, b, c in a domain of FiniteRankNonAssociativeAlgebra. For example, this relation holds for crossed products of three-dimensional vectors.

janko2 ([*listOfIntegers*])

janko2 () constructs the janko group acting on the integers $1, \ldots, 100$.
janko2 ([*li*]) constructs the janko group acting on the 100 integers given in the list li. The default value of li is $[1, \ldots, 100]$. This operation removes duplicates in the list and calls **error** if li does not have exactly 100 distinct entries.

jordanAdmissible? ()

jordanAlgebra? ()

jordanAdmissible? ()$F, where F is a member of FiniteRankNonAssociativeAlgebra(R) over a commutative ring R, tests if 2 is invertible in R and if the algebra defined by $\{a, b\}$ defined by $(1/2)(ab + ba)$ is a Jordan algebra, that is, satisfies the Jordan identity.
jordanAlgebra? ()$F tests if the algebra is commutative, that **characteristic** ()$F $\neq 2$, and $(ab)a^2 - a(ba^2) = 0$ for all a and b in the algebra (Jordan identity). Example: for every associative algebra $(A, +, @)$, you can construct a Jordan algebra $(A, +, *)$, where $a * b := (a@b + b@a)/2$.

kernel (*operator, expression*)

kernel (op, x) constructs $op(x)$ without evaluating it.
kernel $(op, [f_1, \ldots, f_n])$ constructs $op(f1, \ldots, fn)$ without evaluating it.

kernels (*expression*)

kernels (f) returns the list of all the top-level kernels appearing in expression f, but not the ones appearing in the arguments of the top-level kernels.

key? (*key, dictionary*)
keys (*dictionary*)

key? (k, d) tests if k is a key in dictionary d. Dictionary d is an element of a domain of category KeyedDictionary(K, E), where K and E denote the domains of keys and entries.
keys (d) returns the list the keys in table d.

kroneckerDelta (*[integer, integer]*)

kroneckerDelta () is the rank 2 tensor defined by **kroneckerDelta** $(i, j) = 1$ if $i = j$, and 0 otherwise.

label (*outputForm, outputForm*)

label (o_1, o_2), where o_1 and o_2 are objects of type OutputForm (normally unexposed), returns an output form displaying equation o_2 with label o_1.

laguerreL (*nonNegativeInteger, x*)
laguerreL (*nonNegativeInteger, nonNegativeInteger, x*)

laguerreL (n, x) is the nth Laguerre polynomial, $L[n](x)$, defined by $exp(\frac{-tx}{1-t})/(1-t) = \sum_{n=0}^{\infty} L[n](x)t^n/n!$.
laguerreL (m, n, x) is the associated Laguerre polynomial, $L_m[n](x)$, defined as the mth derivative of $L[n](x)$.

lambda (*inputForm, listOfSymbols*)

lambda $(i, [x_1, \ldots x_n])$ returns the input form corresponding to $(x_1, \ldots, x_n) \mapsto i$ if $n > 1$. See also **compiledFunction**, **flatten**, and **unparse**.

laplace (*expression, symbol, symbol*)

laplace (f, t, s) returns the Laplace transform of $f(t)$, defined by $\int_{t=0}^{\infty} exp(-st)f(t)\mathrm{d}t$. If the transform cannot be computed, the formal object **laplace** (f, t, s) is returned.

last (*indexedAggregate*[, *nonNegativeInteger*])

last (u) returns the last element of u.
last (u, n) returns a copy of the last n $(n \geq 0)$ elements of u.

laurent (*expression*)
laurentIfCan (*expression*)

laurent (u) converts u to a Laurent series, or calls **error** if this is not possible.
laurentIfCan (u) converts the Puiseux series u to a Laurent series, or returns `"failed"` if this is not possible.
laurent $(f, x = a)$ expands the expression f as a Laurent series in powers of $(x - a)$.
laurent (f, n) expands the expression f as a Laurent series in powers of x; at least n terms are computed.
laurent $(n \mapsto a_n, x = a, n_0..[n_1])$ returns a Laurent series defined by $\sum_{n=n_0}^{n_1} a_n(x-a)^n$, where n_1 is ∞ by default.
laurent $(a_n, n, x = a, n_0..[n_1])$ returns a Laurent series defined by $\sum_{n=n_0}^{n_1} a_n(x-a)^n$, where n_1 is ∞ by default.

laurentRep (*expression*)

laurentRep $(f(x))$ returns $g(x)$ where the Puiseux series $f(x) = g(x^r)$ is represented by $[r, g(x)]$.

lazy? (*stream*)

lazy? (s) tests if the first node of the stream s is a lazy evaluation mechanism which could produce an additional entry to s.

lazyEvaluate (*stream*)

lazyEvaluate (s) causes one lazy evaluation of stream s. Caution: s must be a "lazy node" satisfying **lazy?** $(s) = true$, as there is no error check. A call to this function may or may not produce an explicit first entry.

lcm (*elements*[, *element*])

lcm (x, y) returns the least common multiple of x and y.
lcm (lx) returns the least common multiple of the elements of the list lx.

ldexquo (*lodOperator, lodOperator*)

ldexquo (a, b) returns q such that $a = b * q$, or `"failed"` if no such q exists.

leftDivide (*lodOperator, lodOperator*)
leftQuotient (*lodOperator, lodOperator*)
leftRemainder (*lodOperator, lodOperator*)

leftDivide (a, b) returns a record with two fields: "quotient" q and "remainder" r such that $a = bq + r$ and the degree of r is less than the degree of b. This operation is called "left division." Operation **leftQuotient** (a, b) returns q, and **leftRemainder** (a, b) returns r.

leader (*differentialPolynomial*)

leader (p) returns the derivative of the highest rank appearing in the differential polynomial p, or calls **error** if p

is in the ground ring.

leadingCoefficient (*polynomial*)

leadingCoefficient (p) returns the coefficient of the highest degree term of polynomial p. See also IndexedDirectProductCategory and MonogenicLinearOperator.

leadingIdeal (*ideal*)

leadingIdeal (I) is the ideal generated by the leading terms of the elements of the ideal I.

leadingMonomial (*polynomial*)

leadingMonomial (p) returns the monomial of polynomial p with the highest degree.

leaf? (*aggregate*)
leafValues (*aggregate*)
leaves (*aggregate*)

These operations apply to a recursive aggregate a. See, for example, BinaryTree.
leaf? (a) tests if a is a terminal node.
leaves (a) returns the list of values at the leaf nodes in left-to-right order.

left (*binaryRecursiveAggregate*)

left (a) returns the left child of binary aggregate a.

leftAlternative? ()

leftAlternative? $()\$F$, where F is a domain of FiniteRankNonAssociativeAlgebra, tests if $2 * \texttt{associator}(a, a, b) = 0$ for all a, b in F. Note: in general, you do not know whether $2 * a = 0$ implies $a = 0$.

leftCharacteristicPolynomial (*polynomial*)

leftCharacteristicPolynomial $(p)\$F$ returns the characteristic polynomial of the left regular representation of p of domain F with respect to any basis. Argument p is a member of a domain of category FiniteRankNonAssociativeAlgebra(R) where R is a commutative ring.

leftDiscriminant ([*listOfVectors*])

leftDiscriminant $([v_1, \ldots, v_n])\$F$ where F is a domain of category FramedNonAssociativeAlgebra over a commutative ring R, returns the determinant of the n-by-n matrix whose element at the ith row and j^{th} column is given by the left trace of the product $v_i * v_j$. Same as **determinant(leftTraceMatrix** $([v_1, \ldots, v_n])$). If no argument is given, $v_1, \ldots, v_n$ are taken to elements of the fixed R-basis.

leftGcd (*lodOperator, lodOperator*)

leftGcd (a, b) computes the value g of highest degree such that $a = aa * g$ and $b = bb * g$ for some values aa and bb. The value g is computed using left-division.

leftLcm (*lodOperator, lodOperator*)

leftLcm (a, b) computes the value m of lowest degree such that $m = a * aa = b * bb$ for some values aa and bb. The value m is computed using left-division.

leftMinimalPolynomial (*element*)

leftMinimalPolynomial (a) returns the polynomial determined by the smallest non-trivial linear combination of left powers of a, an element of a domain of category FiniteRankNonAssociativeAlgebra. Note: the polynomial has no a constant term because, in general, the algebra has no unit.

leftNorm (*element*)

leftNorm (a) returns the determinant of the left regular representation of a, an element of a domain of category FiniteRankNonAssociativeAlgebra.

leftPower (*monad, nonNegativeInteger*)

leftPower (a, n) returns the nth left power of monad a, that is, **leftPower** $(a, n) := a$**leftPower**$(a, n-1)$. If the monad has a unit then **leftPower** $(a, 0) := 1$. Otherwise, define **leftPower** $(a, 1) = a$ See Monad and MonadWithUnit for details. See also **leftRecip**.

leftRankPolynomial ()

leftRankPolynomial $()\$F$ calculates the left minimal polynomial of a generic element of an algebra of domain F, a domain of category FramedNonAssociativeAlgebra over a commutative ring R. This generic element is an element of the algebra defined by the same structural constants over the polynomial ring in symbolic coefficients with respect to the fixed basis.

leftRank (*element*)

leftRank (x) returns the number of linearly independent elements in $xb_1, \ldots, xb_n$, where $b = [b_1, \ldots, b_n]$ is a basis. Argument x is an element of a domain of category FramedNonAssociativeAlgebra over a commutative ring R.

leftRecip (*element*)

leftRecip (a) returns an element that is a left inverse of a, or `"failed"`, if there is no unit element, such an element does not exist, or the left reciprocal cannot be determined (see `unitsKnown`).

leftRecip (*element*)

leftRecip (a) returns an element, which is a left inverse of a, or `"failed"` if such an element doesn't exist or cannot be determined (see `unitsKnown`).

leftRegularRepresentation (*element*[, *vectorOfElements*])

This operation is defined on a domain F of category NonAssociativeAlgebra.
$leftRegularRepresentation(a[, [v_1, \ldots, v_n]])$ returns the matrix of the linear map defined by left multiplication by a

with respect to the basis $[v_1, \ldots, v_n]$. If a second argument is missing, the basis is taken to be the fixed basis for F.

leftTraceMatrix (*[vectorOfElements]*)

This operation is defined on a domain F of category NonAssociativeAlgebra.
leftTraceMatrix ($[v]$), where v is an optional vector $[v_1, \ldots, v_n]$, returns the n-by-n matrix M such that $M_{i,j}$ is the left trace of the product $v_i * v_j$ of elements from the basis $[v_1, \ldots, v_n]$. If the argument is missing, the basis is taken to be the fixed basis for F.

leftTrace (*element*)

leftTrace (a) returns the trace of the left regular representation of a, an element of a domain of category FiniteRankNonAssociativeAlgebra.

leftTrim (*string, various*)

leftTrim (s, c) returns string s with all leading characters c deleted. For example, **leftTrim**`("  abc  ", " ")` returns `"abc  "`.
leftTrim (s, cc) returns s with all leading characters in cc deleted. For example, `leftTrim("(abc)", charClass "()")` returns `"abc"`.

leftUnit ()
leftUnits ()

These operations are defined on a domain F of category NonAssociativeAlgebra.
leftUnit ()$F returns a left unit of the algebra (not necessarily unique), or `"failed"` if there is none.
leftUnits ()$F returns the affine space of all left units of an algebra F, or `"failed"` if there is none, where F is a domain of category FiniteRankNonAssociativeAlgebra. The normal result is returned as a record with selector *particular* for an element of F, and *basis* for a list of elements of F.

legendreSymbol (*integer, integer*)

legendreSymbol (a, p) returns the Legendre symbol $L(a/p)$, $L(a/p) = (-1)^{(p-1)/2}$ **mod** p for prime p. This is 0 if $a = 0$, 1 if a is a quadratic residue **mod** p, and -1 otherwise. Note: because the primality test is expensive, use **jacobi** (a, p) if you know that p is prime.

LegendreP (*nonNegativeInteger, element*)

LegendreP (n, x) is the nth Legendre polynomial, $P[n](x)$, defined by $\frac{1}{\sqrt{(1-2xt+t^2)}} = \sum_{n=0}^{\infty} P[n](x)t^n$.

length (*various*)

length (a) returns the length of integer a in digits.

less? (*aggregate, nonNegativeInteger*)

less? (u, n) tests if u has less than n elements.

leviCivitaSymbol ()

leviCivitaSymbol () is the rank *dim* tensor defined by **leviCivitaSymbol** ()$(i_1, \ldots i_{\text{dim}})$, which is $+1$, -1 or 0 according to whether the permutation $i_1, \ldots, i_{\text{dim}}$ is an even permutation, an odd permutation, or not a permutation of $i_0, \ldots, i_0 + \text{dim} - 1$, respectively, where i_0 is the minimum index.

lexGroebner (*listOfPolynomials, listOfSymbols*)

lexGroebner (lp, lv) computes a Gröbner basis for the list of polynomials lp in lexicographic order. The variables lv are ordered by their position in the list lp.

lhs (*equationOrRewriteRule*)

lhs (x) returns the left hand side of an equation or rewrite-rule.

library (*filename*)

library $(name)$ creates a new library file with filename $name$.

lieAdmissible? ()

lieAdmissible? ()$F tests if the algebra defined by the commutators is a Lie algebra. The domain F is a member of the category FiniteRankNonAssociativeAlgebra(R). The property of anticommutativity follows from the definition.

lieAlgebra? ()

lieAlgebra? ()$F tests if the algebra of F is anticommutative and that the Jacobi identity $(a * b) * c + (b * c) * a + (c * a) * b = 0$ is satisfied for all a, b, c in F.

light (*color*)

light (c) sets the shade of a hue c to its highest value.

limit (*expression, equation* [*, direction*])

limit $(f(x), x = a)$ computes the real two-sided limit of f as its argument x approaches a.
limit $(f(x), x = a, \text{"}left\text{"})$ computes the real limit of f as its argument x approaches a from the left.
limit $(f(x), x = a, \text{"}right\text{"})$ computes the corresponding limit as x approaches a from the right.

limitedIntegrate (*rationalFunction, symbol, listOfRationalFunctions*)

limitedIntegrate $(f, x, [g_1, \ldots, g_n])$ returns fractions $[h, [c_i, g_i]]$ such that the g_i's are among $[g_1, \ldots, g_n]$, $dc_i/dx = 0$, and $d(h + \sum_i c_i \mathtt{log} g_i)/dx = f$ if possible, `"failed"` otherwise.

linearDependenceOverZ (*vector*)
linearlyDependentOverZ? (*vector*)

linearlyDependenceOverZ ($[v_1, \ldots, v_n]$) tests if the elements v_i of a ring (typically algebraic numbers or Expressions) are linearly dependent over the integers. If so,

the operation returns $[c_1, \ldots, c_n]$ such that $c_1 v_1 + \cdots + c_n v_n = 0$ (for which not all the c_i's are 0). If linearly independent over the integers, `"failed"` is returned.
linearlyDependentOverZ? $([v1, \ldots, vn])$ returns *true* if the *vi*'s are linearly dependent over the integers, and *false* otherwise.

lineColorDefault ([*palette*])

lineColorDefault () returns the default color of lines connecting points in a two-dimensional viewport.
lineColorDefault (p) sets the default color of lines connecting points in a two-dimensional viewport to the palette p.

linSolve (*listOfPolynomials, listOfVariables*)

linSolve ($lp, lvar$) finds the solutions of the linear system of polynomials $lp = 0$ with respect to the list of symbols $lvar$.

li (*expression*)

$li(x)$ returns the logarithmic integral of x defined by, $\int \frac{dx}{log(x)}$.

list (*element*)

list (x) creates a list consisting of the one element x.

list? (*sExpression*)

list? (s) tests if SExpression value s is a Lisp list, possibly the null list.

listBranches (*listOfListsOfPoints*)

listBranches (c) returns a list of lists of points representing the branches of the curve c.

listRepresentation (*permutation*)

listRepresentation (p) produces a representation *rep* of the permutation p as a list of preimages and images i, that is, permutation p maps (*rep*.`preimage`).k to (*rep*.`image`).k for all indices k.

listYoungTableaus (*listOfIntegers*)

listYoungTableaus (*lambda*), where *lambda* is a proper partition, generates the list of all standard tableaus of shape *lambda* by means of lattice permutations. The numbers of the lattice permutation are interpreted as column labels.

listOfComponents (*threeSpace*)

listOfComponents (*sp*) returns a list of list of list of points for threeSpace object *sp* assumed to be composed of a list of components, each a list of curves, which in turn is each a list of points, or calls **error** if this is not possible.

listOfCurves (*sp*) returns a list of list of subspace component properties for threeSpace object *sp* assumed to be a list of curves, each of which is a list of subspace components, or calls **error** if this is not possible.

lo (*segment*)

lo (s) returns the first endpoint of s. For example, `lo(1..h) = 1`.

log (*expression*)
logIfCan (*expression*)

log (x) returns the natural logarithm of x.
logIfCan (z) returns **log** (z) if possible, and `"failed"` otherwise.

log2 ([*float*])

log2 () returns $ln(2) = 0.6931471805\ldots$.
log2 (x) computes the base 2 logarithm for x.

log10 ([*float*])

log10 () returns $ln(10) = 2.3025809299\ldots$.
log10 (x) computes the base 10 logarithm for x.

logGamma (*float*)

logGamma (x) is the natural log of (x). Note: this can often be computed even if (x) cannot.

lowerCase ([*string*])
lowerCase? (*character*)

lowerCase () returns the class of all characters for which **lowerCase?** is *true*.
lowerCase (c) returns a corresponding lower case alphabetic character c if c is an upper case alphabetic character, and c otherwise.
lowerCase (s) returns the string with all characters in lower case.
lowerCase? (c) tests if character c is an lower case letter, that is, one of $a \ldots z$.

listOfProperties (*threeSpace*)

listOfProperties (*sp*) returns a list of subspace component properties for *sp* of type ThreeSpace, or calls **error** if this is not possible.

listOfPoints (*threeSpace*)

listOfPoints (*sp*), where *sp* is a ThreeSpace object, returns the list of points component contained in *sp*.

mainKernel (*expression*)

mainKernel (f) returns a kernel of f with maximum nesting level, or `"failed"` if f has no kernels (that is, f is a constant).

mainVariable (*polynomial*)

mainVariable (u) returns the variable of highest ordering that actually occurs in the polynomial p, or `"failed"` if no variables are present. Argument u can be either a polynomial or a rational function.

makeFloatFunction (*expression, symbol* [, *symbol*])

Argument *expr* may be of any type that is coercible to type

InputForm (objects of the most common types can be so coerced).
makeFloatFunction $(expr, x)$ returns an anonymous function of type Float $\rightarrow$ Float defined by $x \mapsto$ expr.
makeFloatFunction $(expr, x, y)$ returns an anonymous function of type (Float, Float) $\rightarrow$ Float defined by $(x, y) \mapsto$ expr.

makeVariable (*element*)

makeVariable (s), where s is a symbol, differential indeterminate, or a differential polynomial, returns a function f defined on the non-negative integers such that $f(n)$ returns the nth derivative of s.
makeVariable (s, n) returns the nth derivative of a differential indeterminate s as an algebraic indeterminate.

makeObject (*functions, range*[, *range*])

Arguments f, g, and h appearing below with arguments (for example, $f(x, y)$) denote symbolic expressions involving those arguments.

Arguments f, g, and h appearing below as symbols without arguments denote user-defined functions which map one or more SmallFloat values to SmallFloat values.

Values a, b, c, and d denote numerical values.

makeObject $(curve(f, g, h), a..b)$ returns the space sp of the domain ThreeSpace with the addition of the graph of the parametric curve $x = f(t)$, $y = g(t)$, $z = h(t)$ as t ranges from **min** (a, b) to **max** (a, b).

makeObject $(curve(f(t), g(t), h(t)), t = a..b)$ returns the space sp of the domain ThreeSpace with the addition of the graph of the parametric curve $x = f(t)$, $y = g(t)$, $z = h(t)$ as t ranges from **min** (a, b) to **max** (a, b).

makeObject $(f, a..b, c..d)$ returns the space sp of the domain ThreeSpace with the addition of the graph of $z = f(x, y)$ as x ranges from **min** (a, b) to **max** (a, b) and y ranges from **min** (c, d) to **max** (c, d).

makeObject $(f(x, y), x = a..b, y = c..d)$ returns the space sp of the domain ThreeSpace with the addition of the graph of $z = f(x, y)$ as x ranges from **min** (a, b) to **max** (a, b) and y ranges from **min** (c, d) to **max** (c, d).

makeObject $(surface(f, g, h), a..b, c..d)$ returns the space sp of the domain ThreeSpace with the addition of the graph of the parametric surface $x = f(u, v)$, $y = g(u, v)$, $z = h(u, v)$ as u ranges from **min** (a, b) to **max** (a, b) and v ranges from **min** (c, d) to **max** (c, d).

$makeObject(surface(f(u, v), g(u, v), h(u, v)), u = a..b, v = c..d)$ returns the space sp of the domain ThreeSpace with the addition of the graph of the parametric surface $x = f(u, v)$, $y = g(u, v)$, $z = h(u, v)$ as u ranges from **min** (a, b) to **max** (a, b) and v ranges from **min** (c, d) to **max** (c, d).

makeYoungTableau (*listOfIntegers, listOfIntegers*)

makeYoungTableau $(lambda, gitter)$ computes for a given lattice permutation *gitter* and for an improper partition *lambda* the corresponding standard tableau of shape *lambda*. See **listYoungTableaus**.

mantissa (*float*)

mantissa (x) returns the mantissa part of x.

map (*function, structure*[, *structure*])
map! (*function, structure*)

map (fn, u) maps the one-argument function fn onto the components of a structure, returning a new structure. Most structures allow f to have different source and target domains. Specifically, the function f is mapped onto the following components of the structure as follows. If u is:

- a series: the coefficients of the series.
- a polynomial: the coefficients of the non-zero monomials.
- a direct product of elements: the elements.
- an aggregate, tuple, table, or a matrix: all its elements.
- an operation of the form $op(a_1, \ldots, a_n)$: each a_i, returning $op(f(a_1), \ldots, f(a_n))$.
- a fraction: the numerator and denominator.
- complex: the real and imaginary parts.
- a quaternion or octonion: the real and all imaginary parts.
- a finite or infinite series or stream: all the coefficients.
- a factored object: onto all the factors.
- a segment $a..b$ or a segment binding of the form $x = a..b$: each of the elements from a to b.
- an equation: both sides of the equation.

map (fn, u, v) maps the two argument function fn onto the components of a structure, returning a new structure. Arguments u and v can be matrices, finite aggregates such as lists, tables, and vectors, and infinite aggregates such as streams and series.

map! (f, u), where u is homogeneous aggregate, destructively replaces each element x of u by $f(x)$.

See also **match**.

mapCoef (*function, freeAbelianMonoid*)
mapGen (*function, freeAbelianMonoid*)

mapCoeff (f, m) maps unary function f onto the coefficients of a free abelian monoid of the form $e_1 a_1 + \ldots + e_n a_n$ returning $f(e_1)a_1 + \ldots + f(e_n)a_n$.
mapGen (fn, m) similarly returns $e_1 f(a_1) + \ldots + e_n f(a_n)$. See FreeAbelianMonoidCategory using Browse.

mapDown! (*tree, value, function*)

These operations make a preorder traversal (node then left branch then right branch) of a tree t of type BalancedBinaryTree(S), destructively mapping values of type S from the root to the leaves of the tree, then returning the modified tree as value; p is a value of type S.
mapDown! (t, p, f), where f is a function of type (S, S) $\rightarrow$ S,

replaces the successive interior nodes of t as follows. The root value x is replaced by $q = f(x,p)$. Then **mapDown!** is recursively applied to (l, q, f) and (r, q, f) where l and r are respectively the left and right subtrees of t.
mapDown! (t, p, f), where f is a function of type (S, S, S) → List S, is similar. The root value of t is first replaced by p. Then f is applied to three values: the value at the current, left, and right node (in that order) to produce a list of two values `l` and `r`, which are then passed recursively as the second argument of **mapDown!** to the left and right subtrees.

mapExponents (*function, polynomial*)

mapExponents (fn, u) maps function fn onto the exponents of the non-zero monomials of polynomial u.

mapUp! ([*tree,*]*tree, function*)

These operations make an endorder traversal (left branch then right branch then node) of a tree t of type BalancedBinaryTree(S), destructively mapping values of type S from the leaves to the root of the tree, then returning the modified tree as value; p is a value of type S.
mapUp! (t, f), where f has type (S, S) → S, replaces the value at each interior node by $f(l, r)$, where l and r are the values at the immediate left and right nodes.
mapUp! (t, t_1, f) makes an endorder traversal of both t and t_1 (of identical shape) in parallel. The value at each successive interior node of t is replaced by $f(l, r, l_1, r_1)$, where l and r are the values at the immediate left and right nodes of t, and l_1 and r_1 are corresponding values of t_1.

mask (*integer*)

mask (n) returns $2^n - 1$ (an n-bit mask).

match? (*string, string, character*)

match? $(s, t, char)$ tests if s matches t except perhaps for multiple and consecutive occurrences of character $char$. Typically $char$ is the blank character.

match (*list, list*[*, option*])

match $(la, lb[, u])$, where la and lb are lists of equal length, creates a function that can be used by **map**. The target of a source value x in la is the value y with the corresponding index in lb. Optional argument u defines the target for a source value a which is not in la. If u is a value of the source domain, then a is replaced by u, which must be a member of la. If u is a value of the target domain, the value returned by the map for a is u. If u is a function f, then the value returned is $f(a)$. If no third argument is given, an error occurs when such a a is found.

mathieu11 ([*listOfIntegers*])
mathieu12 ([*listOfIntegers*])
mathieu22 ([*listOfIntegers*])
mathieu23 ([*listOfIntegers*])
mathieu24 ([*listOfIntegers*])

mathieu11 $([li])$ constructs the mathieu group acting on the eleven integers given in the list li. Duplicates in the list will be removed and **error** will be called if li has fewer or more than eleven different entries. The default value of li is $[1, \ldots, 11]$. Operations $mathieu12$, $mathieu22$, and $mathieu23$ and $mathieu24$ are similar. These operations provide examples of permutation groups in AXIOM.

matrix (*listOfLists*)

matrix (l) converts the list of lists l to a matrix, where the list of lists is viewed as a list of the rows of the matrix.
matrix (llo), where llo is a list of list of objects of type OutputForm (normally unexposed), returns an output form displaying llo as a matrix.

max ([*various*])

max () returns the largest small integer.
max (u) returns the largest element of aggregate u.
max (x, y) returns the maximum of x and y relative to a total ordering "<".

maxColIndex (*matrix*)

maxColIndex (m) returns the index of the last column of the matrix or two-dimensional array m.

maxIndex (*aggregate*)

maxIndex (u) returns the maximum index i of indexed aggregate u. For most indexed aggregates (vectors, strings, lists), **maxIndex** (u) is equivalent to $\#u$.

maxRowIndex (*matrix*)

maxRowIndex (m) returns the index of the "last" row of the matrix or two-dimensional array m.

meatAxe (*listOfListsOfMatrices* [*, boolean, integer, integer*])

meatAxe $(aG[, randomElts, numOfTries, maxTests])$ tries to split the representation given by aG and returns a 2-list of representations. All matrices of argument aG are assumed to be square and of equal size. The default values of arguments $randomElts$, $numOfTries$ and $maxTests$ are $false$, 25, and 7, respectively.

member? (*element, aggregate*)

member? (x, u) tests if x is a member of u.
member? (*pp*, *gp*), where *pp* is a permutation and *gp* is a group, tests whether *pp* is in the group *gp*.

merge (*various*)
merge! (*various*)

merge $([s1, s2, \ldots, sn])$ will create a new ThreeSpace object that has the components of all the ones in the list; groupings of components into composites are maintained.
merge $(s1, s2)$ will create a new ThreeSpace object that has the components of $s1$ and $s2$; groupings of components into composites are maintained.

merge ($[p,]a, b$) returns an aggregate c which merges a and b. The result is produced by examining each element x of a and y of b successively. If $p(x, y)$ is $true$, then x is inserted into the result. Otherwise y is inserted. If x is chosen, the next element of a is examined, and so on. When all the elements of one aggregate are examined, the remaining elements of the other are appended. For example, **merge** ($<, [1, 3], [2, 7, 5]$) returns $[1, 2, 3, 7, 5]$. By default, function p is $\leq$.
merge! ($[p], u, v$) destructively merges the elements u and v into u using comparison function p. Function p is $\leq$ by default.

mesh (*u*[, *v, w, x*])

Argument sp below is a ThreeSpace object sp. Argument lc is a list of curves. Each curve is either a list of points (objects of type Point) or else a list of lists of small floats.
mesh (lc) returns a ThreeSpace object defined by lc.
mesh (sp) returns the list of curves contained in space sp.
mesh ($[sp,], lc, close1, close2$) adds the list of curves lc to the ThreeSpace object sp. Boolean arguments $close1$ and $close2$ tell how the curves and surface are to be closed. If $close1$ is $true$, each individual curve will be closed, that is, the last point of the list will be connected to the first point. If $close2$ is $true$, the first and last curves are regarded as boundaries and are connected. By default, the argument sp is empty.

midpoints (*listOfIntervals*)

These operations are defined on "intervals" represented by records with keys $right$ and $left$, and rational number values.
midpoints ($isolist$) returns the list of midpoints for the list of intervals $isolist$.
midpoint (int) returns the midpoint of the interval int.

min ([*u, v*])

min () returns the element of type SmallInteger.
min (u) returns the smallest element of aggregate u.
min (x, y) returns the minimum of x and y relative to total ordering $<$.

minColIndex (*matrix*)

minColIndex (m) returns the index of the "first" column of the matrix or two-dimensional array m.

minimalPolynomial (*element, positiveInteger*)

minimalPolynomial ($x[, n]$) computes the minimal polynomial of x over the field of extension degree n over the ground field F. The default value of n is 1.

minimalPolynomial (*element*)

minimalPolynomial (a) returns the minimal polynomial of element a of a finite rank algebra. See FiniteRankAlgebra using Browse.

minimumDegree (*polynomial, variable*)

minimumDegree (p, v) gives the minimum degree of polynomial p with respect to v, that is, viewed as a univariate polynomial in v.
minimumDegree (p, lv) gives the list of minimum degrees of the polynomial p with respect to each of the variables in the list lv.
See also FiniteAbelianMonoidRing and MonogenicLinearOperator.

minIndex (*aggregate*)

minIndex ($aggregate$) returns the minimum index i of aggregate u. Note: the **minIndex** of most system-defined indexed aggregates is 1. See also PointCategory.

minordet (*matrix*)

minordet (m) computes the determinant of the matrix m using minors, or calls **error** if the matrix is not square.

minPoly (*expression*)

minPoly (k) returns polynomial p such that $p(k) = 0$.

minRowIndex (*matrix*)

minRowIndex (m) returns the index of the "first" row of the matrix or two-dimensional array m.

minusInfinity ()

minusInfinity () returns `%minusInfinity`, the AXIOM name for $-\infty$.

modifyPointData (*space, nonNegativeInteger, point*)

modifyPointData (sp, i, p) changes the point at the indexed location i in the ThreeSpace object sp to p. This operation is useful for making changes to existing data.

moduloP (*integer*)

moduloP (x), such that $p =$ **modulus**(), returns a, where $x = a + bp$ where x is a p-adic integer. See PAdicIntegerCategory using Browse.

modulus ()

modulus ()$\$R$ returns the value of the modulus p of a p-adic integer domain R. See PAdicIntegerCategory using Browse.

moebiusMu (*integer*)

moebiusMu (n) returns the Moebius function $\mu(n)$, defined as -1, 0 or 1 as follows: $\mu(n) = 0$ if n is divisible by a square > 1, and $(-1)^k$ if n is square-free and has k distinct prime divisors.

monicDivide (*polynomial, polynomial*[, *variable*])

monicDivide ($p, q[, v]$) divides the polynomial p by the monic polynomial q, returning the record containing a $quotient$ and $remainder$. For multivariate polynomials, the polynomials are viewed as a univariate polynomials in v. If p and q are univariate polynomials, then the third argument may be omitted. The operation calls **error** if q is not monic with respect to v.

monomial (*coefficient, exponent*[, *option*])

monomial $(coef, exp)$ creates a term of a univariate polynomial or series object from a coefficient $coef$ and exponent exp. The variable name must be given by context (as through a declaration for the result).
monomial $(c, [x_1, \ldots, x_k], [n_1, \ldots, n_k])$ creates a term $cx_1^{n_1} \ldots x_k^{n_k}$ of a multivariate power series or polynomial from coefficient c, variables x_j and exponents n_j.
monomial (c, x, n) creates a term cx^n of a polynomial or series from a coefficient c, variable x, and exponent n.
monomial $(c, [n_1, \ldots, n_k])$ creates a CliffordAlgebra element $ce(n_1), \ldots, ce(n_k)$ from a coefficient c and basis elements $c(i_j)$

monomial? (*polynomialOrSeries*)

monomial? (p) tests if polynomial or series p is a single monomial.

monomials (*polynomial*)

monomials (p) returns the list of non-zero monomials of polynomial p, $[a_1 X^{(1)}, \ldots, a_n X^{(n)}]$.

more? (*aggregate, nonNegativeInteger*)

more? (u, n) tests if u has greater than n elements.

movedPoints (*permutation*)

movedPoints (p) returns the set of points moved by the permutation p.
movedPoints (gp) returns the points moved by the group gp.

mulmod (*integer, integer, integer*)

mulmod (a, b, p), where a, b are non-negative integers both $<$ integer p, returns ab **mod** p.

multiEuclidean (*listOfElements, element*)

multiEuclidean $([f_1, \ldots, f_n], z)$ returns a list of coefficients $[a_1, \ldots, a_n]$ such that $z / \prod_{i=1}^{n} f_i = \sum_{j=1}^{n} a_j / f_j$. If no such list of coefficients exists, `"failed"` is returned.

multinomial (*integer, listOfIntegers*)

multinomial $(n, [m_1, m_2, \ldots, m_k])$ returns the multinomial coefficient $n!/(m_1! m_2! \ldots m_k!)$.

multiple (*expression*)

multiple (x) directs the pattern matcher that x should preferably match a multi-term quantity in a sum or product. For matching on lists, multiple(x) tells the pattern matcher that x should match a list instead of an element of a list. This operation calls **error** if x is not a symbol.

multiplyCoefficients (*function, series*)

multiplyCoefficients (f, s) returns $\sum_{n=0}^{\infty} f(n) a_n x^n$ where s is the series $\sum_{n=0}^{\infty} a_n x^n$.

multiplyExponents (*various, nonNegativeInteger*)

multiplyExponents (p, n), where p is a univariate polynomial or series, returns a new polynomial or series resulting from multiplying all exponents by the non negative integer n.

multiset (*listOfElements*)

multiset (ls) creates a multiset with elements from ls.

multivariate (*polynomial, symbol*)

multivariate (p, v) converts an anonymous univariate polynomial p to a polynomial in the variable v.

name (*various*)

name (f) returns the name part of the file name for file f.
name (op) returns the name of basic operator op.
name (s) returns symbol s without its scripts.

nand (*boolean, boolean*)

nand (a, b) returns the logical negation of a and b, either booleans or bit aggregates. Note: **nand** $(a, b) = true$ if and only if one of a and b is $false$.

nary? (*basicOperator*)

nary? (op) tests if op accepts an arbitrary number of arguments.

ncols (*matrix*)

ncols (m) returns the number of columns in the matrix or two-dimensional array m.

new ([*various*])

new ()R create a new object of type R. When R is an aggregate, new creates an empty object. Other variations are as follows:

> **new** (s), where s is a symbol, returns a new symbol whose name starts with %s.
>
> **new** (n, x) returns **fill!** $(new(n), x)$, an aggregate of n elements, each with value x.
>
> **new** $(m, n, r)$$R$ creates an m-by-n array or matrix of type R all of whose entries are r.
>
> **new** (d, pre, e), where d, smathpre, and smathe are strings, constructs the name of a new writable file with d as its directory, pre as a prefix of its name and e as its extension. When d or e is the empty string, a default is used. The operation calls **error** if a new file cannot be written in the given directory.

newLine ()

newLine () sends a new line command to output. See DisplayPackage.

nextColeman (*listOfIntegers, listOfIntegers, matrix*)

nextColeman $(alpha, beta, C)$ generates the next Coleman-matrix of column sums *alpha* and row sums *beta* according to the lexicographical order from bottom-to-top. The first Coleman matrix is created using $C = \mathbf{new}(1, 1, 0)$. Also, **new** $(1, 1, 0)$ indicates that C is the last Coleman matrix. See SymmetricGroupCombinatoricFunctions for details.

nextLatticePermutation (*integers, integers, boolean*)

nextLatticePermutation $(lambda, lattP, constructNotFirst)$ generates the lattice permutation according to the proper partition $lambda$ succeeding the lattice permutation $lattP$ in lexicographical order as long as $constructNotFirst$ is $true$. If $constructNotFirst$ is $false$, the first lattice permutation is returned. The result nil indicates that $lattP$ has no successor. See SymmetricGroupCombinatoricFunctions for details.

nextPartition (*vectorOfIntegers, vectorOfIntegers, integer*)

nextPartition $(gamma, part, number)$ generates the partition of $number$ which follows $part$ according to the right-to-left lexicographical order. The partition has the property that its components do not exceed the corresponding components of $gamma$. the first partition is achieved by $part = []$. Also, [ ] indicates that $part$ is the last partition. See SymmetricGroupCombinatoricFunctions for details.

nextPrime (*positiveInteger*)

nextPrime (n) returns the smallest prime strictly larger than n.

nil ()

nil ()$\$R$ returns the empty list of type R.

nilFactor (*element, nonNegativeInteger*)

nilFactor $(base, exponent)$ creates a factored object with a single factor with no information about the kind of $base$. See Factored for details.

node? (*aggregate, aggregate*)

node? (u, v) tests if node u is contained in node v (either as a child, a child of a child, etc.).

nodes (*recursiveAggregate*)

nodes (a) returns a list of all the nodes of aggregate a.

noncommutativeJordanAlgebra? ()

noncommutativeJordanAlgebra? ()$\$F$ tests if the algebra F is flexible and Jordan admissible. See FiniteRankNonAssociativeAlgebra.

nor (*boolean, boolean*)

nor (a, b) returns the logical nor of booleans or bit aggregates a and b. Note: **nor** (a, b) = true if and only if both a and b are $false$.

norm (*element*[, *option*])

norm (x) returns:

for complex x: **conjugate** (x) .

for floats: the absolute value.

for quaternions or octonions: the sum of the squares of its coefficients.

for a domain of category FiniteRankAlgebra: the determinant of the regular representation of x with respect to any basis.

norm $(x[, p])$, where p is a positiveInteger and x is an element of a domain of category FiniteAlgebraExtensionField over ground field F, returns the norm of x with respect to the field of extension degree d over the ground field of size. The default value of p is 1. The operation calls **error** if p does not divide the extension degree of x. Note: **norm** $(x, p) = \prod_{i=0}^{n/p} x^{q^{pi}}$

normal? (*element*)

normal? (a), where a is a member of a domain of category $FiniteAlgebraicExtensionField$ over a field F, tests whether the element a is normal over the ground field F, that is, if $a^{q^i}, 0 \le i \le$ **extensionDegree**$() - 1$ is an F-basis, where $q =$ **size**$()$.

normalElement ()

normalElement ()$\$R$, where R is a domain of category $FiniteAlgebraicExtensionField$ over a field F, returns a element, normal over the ground field F, that is, $a^{q^i}, 0 \le i <$ **extensionDegree**$()$ is an F-basis, where $q =$ **size**$()$. At the first call, the element is computed by **createNormalElement** then cached in a global variable. On subsequent calls, the element is retrieved by referencing the global variable.

normalForm (*polynomial, listOfpolynomials*)

normalForm $(poly, gb)$ reduces the polynomial $poly$ modulo the precomputed Gröbner basis gb giving a canonical representative of the residue class.

normalise (*element*)

normalise (v) returns the column vector v divided by its Euclidean norm; when possible, the vector v is expressed in terms of radicals.

normalize (*element* [,] *option*)

normalize (flt) normalizes float flt at current precision. **normalize** $(f[, x])$ rewrites f using the least possible number of real algebraically independent kernels involving symbol x. If no symbol x is given, the operation rewrites f using the least possible number of real algebraically independent kernels.

normalizeAtInfinity (*vectorOfFunctions*)

normalizeAtInfinity (v) makes v normal at infinity, where v is a vector of functions defined on a curve.

not (*boolean*)

not (n) returns the negation of boolean or bit aggregate n.
not (n) returns the bit-by-bit logical *not* of the small integer n.

nrows (*matrix*)

nrows (m) returns the number of rows in the matrix or two-dimensional array m.

nthExponent (*factored, positiveInteger*)

nthExponent (u, n) returns the exponent of the nth factor of u, or 0 if u has no such factor.

nthFactor (*factor, positiveInteger*)

nthFactor (u, n) returns the base of the nth factor of u, or 1 if n is not a valid index for a factor. If u consists only of a unit, the unit is returned.

nthFlag (*factored, positiveInteger*)

nthFlag (u, n) returns the information flag of the nth factor of u, `"nil"` if n is not a valid index for a factor.

nthFractionalTerm (*partialFraction, integer*)

nthFractionalTerm (p, n) extracts the nth fractional term from the partial fraction p, or 0 if the index n is out of range.

nthRoot (*expression, integer*)
nthRootIfCan (*expression, integer*)

Argument x can be of type Expression, Complex, Float and SmallFloat, or a series.

nthRoot (x, n) returns the nth root of x. If x is not an expression, the operation calls **error** if this is not possible.

nthRootIfCan (z, n) returns the nth root of z if possible, and `"failed"` otherwise.

null? (*sExpression*)

null? (s) is *true* if s is the SExpression object ().

nullary ()

nullary (x), where x has type R, returns a function f of type → R such that such that $f()$ returns the value c. See also **constant** for a similar operation.

nullary? (*basicOperator*)

nullary? (*op*) tests if basic operator *op* is nullary.

nullity (*matrix*)

nullity (m) returns the dimension of the null space of the matrix m.

nullSpace (*matrix*)

nullSpace (m) returns a basis for the null space of the matrix m.

numberOfComponents ([*threeSpace*])

numberOfComponents ()$F returns the number of absolutely irreducible components for a domain F of functions defined over a curve.
numberOfComponents (*sp*) returns the number of distinct object components in the ThreeSpace object s such as points, curves, and polygons.

numberOfComputedEntries (*stream*)

numberOfComputedEntries (*st*) returns the number of explicitly computed entries of stream *st*.

numberOfCycles (*permutation*)

numberOfCycles (p) returns the number of non-trivial cycles of the permutation p.

numberOfDivisors (*integer*)

numberOfDivisors (n) returns the number of integers between 1 and n inclusive which divide n. The number of divisors of n is often denoted by $\tau(n)$.

numberOfFactors (*factored*)

numberOfFactors (u) returns the number of factors in factored form u.

numberOfFractionalTerms (*partialFraction*)

numberOfFractionalTerms (p) computes the number of fractional terms in p, or 0 if there is no fractional part.

numberOfHues ()

numberOfHues () returns the number of total hues. See also **totalHues**.

numberOfImproperPartitions (*integer, integer*)

numberOfImproperPartitions (n, m) computes the number of partitions of the nonnegative integer n in m nonnegative parts with regarding the order (improper partitions). Example: `numberOfImproperPartitions (3, 3)` is 10, since `[0, 0, 3], [0, 1, 2], [0, 2, 1], [0, 3, 0], [1, 0, 2], [1, 1, 1], [1, 2, 0], [2, 0, 1], [2, 1, 0], [3, 0, 0]` are the possibilities. Note: this operation has a recursive implementation.

numberOfMonomials (*polynomial*)

numberOfMonomials (p) gives the number of non-zero monomials in polynomial p.

numer (*fraction*)
numerator (*fraction*)

Argument x is from domain Fraction(R) for some domain R, or of type Expression
numer (x) returns the numerator of x as an object of domain R; if x is of type Expression, it returns an object of domain SMP(D, Kernel(Expression R)).
numerator (x) returns the numerator of x as an element of

Fraction(R); if x if of type Expression, it returns an object of domain Expression.

numerators (*continuedFraction*)

numerators (cf) returns the stream of numerators of the approximants of the continued fraction cf. If the continued fraction is finite, then the stream will be finite.

numeric (*expression*[, *n*])

numeric (x, n) returns a float approximation of expression x to n decimal digits accurary.

objectOf (*typeAnyObject*)

objectOf (a) returns a printable form of an object of type Any.

objects (*threeSpace*)

objects (sp) returns the ThreeSpace object sp. The result is returned as record with fields: $points$, the number of points; $curves$, the number of curves; $polygons$, the number of polygons; and $constructs$, the number of constructs.

oblateSpheroidal (*function*)

oblateSpheroidal (a), where a is a small float, returns a function to map the point (ξ, η, ϕ) to cartesian coordinates $x = asinh(\xi) sin(\eta) cos(\phi)$, $y = asinh(\xi) sin(\eta) sin(\phi)$, $z = acosh(\xi) cos(\eta)$.

octon (*element, element* [, *elements*])

octon (q_e, q_E) constructs an octonion whose first 4 components are given by a quaternion q_e and whose last 4 components are given by a quaternion q_E.
octon ($r_e, r_i, r_j, r_k, r_E, r_I, r_J, r_K$) constructs an octonion from scalars.

odd? (x)

odd? (n) tests if integer n is odd.
odd? (p) tests if p is an odd permutation, that is, **sign** (p) is -1.

oneDimensionalArray ([*integer,*]*elements*)

oneDimensionalArray (ls) creates a one-dimensional array consisting of the elements of list ls.
oneDimensionalArray (n, s) creates a one-dimensional array of n elements, each with value s.

one? (*element*)

one? (a) tests whether a is the unit 1.

open (*file*[, *string*])

open (s[, *mode*]) returns the file s open in the indicated mode: `"input"` or `"output"`. Argument *mode* is `"output"` by default.

operator (*symbol*[, *nonNegativeInteger*])

operator (f, n) makes f into an n-ary operator. If the second argument n is omitted, f has arbitary *arity*, that is, f takes an arbitrary number of arguments.

operators (*expression*)

operators (f) returns a list of all basic operators in f, regardless of level.

optional (*symbol*)

optional (x) tells the pattern matcher that x can match an identity (0 in a sum, 1 in a product or exponentiation), or calls **error** if x is not a symbol.

or (*boolean, boolean*)

a **or** b returns the logical *or* of booleans or bit aggregates a and b.
n **or** m returns the bit-by-bit logical *or* of the small integers n and m.

orbit (*group, elements*)

orbit (gp, el) returns the orbit of the element el under the permutation group gp, that is, the set of all points gained by applying each group element to el.
orbit (gp, ls), where ls is a list or unordered set of elements, returns the orbit of ls under the permutation group gp.

orbits (*group*)

orbits (gp) returns the orbits of the permutation group gp.

ord (*character*)

ord (c) returns an integer code corresponding to the character c.

order (*element*)

order (p) returns:

- if p is a float: the magnitude of p (Note: $base^{\mathbf{order}(x)} \leq |x| < base^{(1+\mathbf{order}(x))}$.)
- if p is a differential polynomial: the maximum number of differentiations of a differential indeterminate among all those appearing in p.
- if p is a differential variable: the number of differentiations of the differential indeterminate appearing in p.
- if p is an element of finite field: the order of an element in the multiplicative group of the field (the function calls **error** if p is 0).
- if p is a univariate power series: the degree of the lowest order non-zero term in f. (A call to this operation results in an infinite loop if f has no non-zero terms.)
- if p is a q-adic integer: the exponent of the highest power of q dividing p (see PAdicIntegerCategory).
- if p is a permutation: the order of a permutation p as a group element.
- if p is permutation group: the order of the group.

order (p, q) returns the order of the differential polynomial p in differential indeterminate q.

order (p, q) returns the order of multivariate series p viewed as a series in q (this operation results in an infinite loop if f has no non-zero terms).
order (p, q) returns the largest n such that q^n divides polynomial p, that is, the order of $p(x)$ at $q(x) = 0$.

orthonormalBasis (*matrix*)

orthonormalBasis (M) returns the orthogonal matrix B such that BMB^{-1} is diagonal, or calls **error** if M is not a symmetric matrix.

output (x)

output (x) displays x on the "algebra output" stream defined by)set output algebra.

outputAsFortran (*outputForms*)

outputAsFortran (f) outputs OutputForm object f in FORTRAN format to the destination defined by the system command)set output fortran. If f is a list of OutputForm objects, each expression in f is output in order.
outputAsFortran (s, f), where s is a string, outputs $s = f$, but is otherwise identical.

outputAsTex (*outputForms*)

outputAsTex (f) outputs OutputForm object f in TeX format to the destination defined by the system command)set output tex. If f is a list of OutputForm objects, each expression in f is output in order.

outputFixed ([*nonNegativeInteger*])

outputFixed ([n]) sets the output mode of floats to fixed point notation, that is, as an integer, a decimal point, and a number of digits. If n is given, then n digits are displayed after the decimal point.

outputFloating ([*nonNegativeInteger*])

outputFloating ([n]) sets the output mode to floating (scientific) notation, that is, $m10^e$ is displayed as `mEe`. If n is given, n digits will be displayed after the decimal point.

outputForm (*various*)

outputForm (x) creates an object of type OutputForm from x, an object of type $Integer$, $SmallFloat$, $String$, or $Symbol$.

outputGeneral ([*nonNegativeInteger*])

outputGeneral ([n]) sets the output mode (default mode) to general notation, that is, numbers will be displayed in either fixed or floating (scientific) notation depending on the magnitude. If n is given, n digits are displayed after the decimal point.

outputSpacing (*nonNegativeInteger*)

outputSpacing (n) inserts a space after n digits on output.
outputSpacing (0) means no spaces are inserted. By default, $n = 10$.

over (*outputForm, outputForm*)

over (o_1, o_2), where o_1 and o_2 are objects of type OutputForm (normally unexposed), creates an output form for the vertical fraction of o_1 over o_2.

overbar (*outputForm*)

overbar (o), where o is an object of type OutputForm (normally unexposed), creates the output form o with an overbar.

pack! (*file*)

pack! (f) reorganizes the file f on disk to recover unused space.

packageCall ()

packageCall $(f)\$P$, where P is the package InputFormFunctions1(R) for some type R, returns the input form corresponding to f$R. See also **interpret**.

pade (*integer, integer, series*[*, series*])

pade $(nd, dd, s[, ds])$ computes the quotient of polynomials (if it exists) with numerator degree at most nd and denominator degree at most dd. If a single univariate Taylor series s is given, the quotient approximate must match the series s to order $nd + dd$. If two series s and ds are given, ns is the numerator series of the function and ds is the denominator series.

padicFraction (*partialFraction*)

padicFraction (q) expands the fraction p-adically in the primes p in the denominator of q. For example, **padicFraction** $(3/(2^2)) = 1/2 + 1/(2^2)$. Use **compactFraction** to return to compact form.

pair? (*sExpression*)

pair? (s) tests if $SExpression$ object is a non-null Lisp object.

parabolic (*point*)

parabolic (pt) transforms pt from parabolic coordinates to Cartesian coordinates: the function produced will map the point (u, v) to $x = 1/2(u^2 - v^2)$, $y = uv$.

parabolicCylindrical (*point*)

parabolicCylindrical (pt) transforms pt from parabolic cylindrical coordinates to Cartesian coordinates: the function produced will map the point (u, v, z) to $x = 1/2(u^2 - v^2)$, $y = uv$, z.

paraboloidal (*point*)

paraboloidal (pt) transforms pt from paraboloidal coordinates to Cartesian coordinates: the function produced will map the point (u, v, phi) to $x = uv\mathbf{cos}(\phi)$, $y = uv\mathbf{sin}(\phi)$, $z = 1/2(u^2 - v^2)$.

paren (*expressions*)

paren (f) returns (f) unless f is a list $[f_1, \ldots, f_n]$ in which case it returns $(f_1, \ldots, f_n)$. This prevents f or the constituent f_i from being evaluated when operators are applied to it. For example, `log(1)` returns 0, but `log(paren 1)` returns the formal kernel `log((1))`. Also, **atan**(**paren** [*x, 2*]) returns the formal kernel **atan**((*x, 2*)).

partialDenominators (*continuedFraction*)

partialDenominators (x) extracts the denominators in x. If $x =$ **continuedFraction**$(b_0, [a_1, \ldots], [b_1, \ldots])$, then **partialDenominators** $(x) = [b_1, b_2 \ldots]$.

partialFraction (*element, factored*)

partialFraction $(numer, denom)$ is the main function for constructing partial fractions. The second argument $denom$ is the denominator and should be factored.

partialNumerators (*continuedFraction*)

partialNumerators (x) extracts the numerators in x, if $x =$ **continuedFraction**$(b_0, [a_1, \ldots], [b_1, \ldots], \ldots)$, then **partialNumerators** $(x) = [a_1, \ldots]$.

partialQuotients (*continuedFraction*)

partialQuotients (x) extracts the partial quotients in x, if $x =$ **continuedFraction**$(b_0, [a_1, \ldots], [b_1, \ldots], \ldots)$, then **partialQuotients** $(x) = [b_0, b_1, \ldots]$.

particularSolution (*matrix, vector*)

aSolution (M, v) finds a particular solution x of the linear system $Mx = v$. The result x is returned as a vector, or `"failed"` if no solution exists.

partition (*integer*)

partition (n) returns the number of partitions of the integer n. This is the number of distinct ways that n can be written as a sum of positive integers.

partitions (*integer*[, *integer, integer*])

partitions (i, j) is the stream of all partitions whose number of parts and largest part are no greater than i and j.
partitions (n) is the stream of all partitions of integer n.
partitions (p, l, n) is the stream of partitions of n whose number of parts is no greater than p and whose largest part is no greater than l.

parts (*aggregate*)

parts (u) returns a list of the consecutive elements of u. Note: if u is a list, **parts** $(u) = u$.

pastel (*color*)

pastel (c) sets the shade of a hue c above "bright" but below "light".

pattern (*rewriteRule*)

pattern (r) returns the pattern corresponding to the left hand side of the rewrite rule r.

patternMatch (*expression, expression, patternMatchResult*)

patternMatch $(expr, pat, res)$ matches the pattern pat to the expression $expr$. The argument res contains the variables of pat which are already matched and their matches. Initially, res is the result of **new()**, an empty list of matches.

perfectNthPower? (*integer, nonNegativeInteger*)

perfectNthPower? (n, r) tests if n is an r th power.

perfectNthRoot (*integer*[, *nonNegativeInteger*])

perfectNthRoot (n) returns a record with fields "base" x and "exponent" r such that $n = x^r$ and r is the largest integer such that n is a perfect r th power.
perfectNthRoot (n, r) returns the r th root of n if n is an r th power, and `"failed"` otherwise.

perfectSqrt (*integer*)

perfectSqrt (n) returns the square root of n if n is a perfect square, and `"failed"` otherwise.

perfectSquare? (*integer*)

perfectSquare? (n) tests if n is a perfect square.

permanent (*matrix*)

permanent (x) returns the permanent of a square matrix x, equivalent to the **determinant** except that coefficients have no change of sign.

permutation (*integer, integer*)

permutation (n, m) returns the number of permutations of n objects taken m at a time. Note: **permutation** $(n, m) = n!/(n - m)!$.

permutationGroup (*listPermutations*)

permutationGroup (ls) coerces a list of permutations ls to the group generated by this list.

permutationRepresentation (*permutations*[, *n*])

permutationRepresentation (pi, n) returns the matrix $\delta_{i,pi(i)}$ (Kronecker delta) if the permutation pi is in list notation and permutes $1, 2, \ldots, n$. Argument pi may either be permutation or a list of integers describing a permutation by list notation.
permutationRepresentation $([pi_1, \ldots, pi_k], n)$ returns the list of matrices $[(\delta_{i,pi_1}(i)), \ldots, (\delta_{i,pi_k(i)})]$ (Kronecker delta) for permutations $pi_1, \ldots, pi_k$ of $1, 2, \ldots, n$.

permutations (*integer*)

permutations (n) returns the stream of permutations formed from $1, 2, \ldots, n$.

physicalLength (*flexibleArray*)

physicalLength! (*flexibleArray, positiveInteger*)

These operations apply to a flexible array a and concern the "physical length" of a, the maximum number of elements that a can hold. When a destructive operation (such as **concat!**) is applied that increases the number of elements of a beyond this number, new storage is allocated (generally to be about 50% larger than current storage allocation) and the elements from the old storage are copied over to the new storage area.

physicalLength (a) returns the physical length of a.
physicalLength! (a, n) causes new storage to be allocated for the elements of a with a physical length of n. The **maxIndex** elements from the old storage area are copied. An **error** is called if n is less than **maxIndex**(a).

pi ()

pi () returns π, also denoted by the special symbol `%pi`.

pile (*listOfOutputForms*)

pile (lo), where lo is a list of objects of type OutputForm (normally unexposed), creates the output form consisting of the elements of lo displayed as a pile, that is, each element begins on a new line and is indented right to the same margin.

plenaryPower (*element, positiveInteger*)

Argument a is a member of a domain of category NonAssociativeAlgebra
plenaryPower (a, n) is recursively defined to be **plenaryPower** $(a, n-1)$ * **plenaryPower**$(a, n-1)$ for $n > 1$ and a for $n = 1$.

plusInfinity ()

plusInfinity () returns the constant `%plusInfinity` denoting $+\infty$.

point (*u* [, *option*])

point (p) returns a ThreeSpace object which is composed of one component, the point p. **point** (l) creates a point defined by a list l.

point (sp) checks to see if the ThreeSpace object sp is composed of only a single point and, if so, returns the point, or calls **error** if sp has more than one point.
point (sp, l) adds a point component defined by a list l to the ThreeSpace object sp.
point (sp, i) adds a point component into a component list of the ThreeSpace object sp at the index given by i.
point (sp, p) adds a point component defined by the point p described as a list, to the ThreeSpace object sp.

point? (*space*)

point? (sp) queries whether the ThreeSpace object sp, is composed of a single component which is a point.

pointColor (*palette*)

pointColor (v) specifies a color v for two-dimensional graph points. This option is expressed in the form `pointColor == v` in the **draw** command. Argument p is either a palette or a float.

pointColorDefault ([*palette*])

pointColorDefault () returns the default color of points in a two-dimensional viewport.
pointColorDefault (p) sets the default color of points in a two-dimensional viewport to the palette p.

pointSizeDefault ([*positiveInteger*])

pointSizeDefault () returns the default size of the points in a two-dimensional viewport.
pointSizeDefault (i) sets the default size of the points in a two-dimensional viewport to i.

polarCoordinates (x)

polarCoordinates (x) returns a record with components (r, ϕ) such that $x = re^{i\phi}$.

polar (*point*)

polar (pt) transforms point pt from polar coordinates to Cartesian coordinates. The function produced will map the point (r, θ) to $x = r\mathbf{cos}(\theta)$, $y = r\mathbf{sin}(\theta)$.

pole? (*series*)

pole? (f) tests if the power series f has a pole.

polygamma (*k, x*)

polygamma (k, x) is the kth derivative of **digamma** (x), often written $\psi(k, x)$ in the literature.

polygon ([*sp,*]*listOfPoints*)
polygon? (*space*)

polygon $([sp,]lp)$ adds a polygon defined by lp to the ThreeSpace object sp. Each lp is either a list of points (objects of type Point) or else a list of small floats. If sp is omitted, it is understood to be empty.
polygon (sp) returns ThreeSpace object sp as a list of polygons, or an error if sp is not composed of a single polygon.
polygon? (sp) tests if the ThreeSpace object sp contains a single polygon component.

polynomial (*series, integer* [, *integer*])

polynomial (s, k) returns a polynomial consisting of the sum of all terms of series s of degree $\leq k$ and greater than or equal to 0.
polynomial (s, k_1, k_2) returns a polynomial consisting of the sum of all terms of Taylor series s of degree d with $0 \leq k_1 \leq d \leq k_2$.

pop! (*stack*)

pop! (s) returns the top element x from stack s, destructively removing it from s, or calls **error** if s is empty. Note: Use **top** (s) to obtain x without removing it from s.

position (*aggregate, aggregate* [, *index*])

position ($x, a[, n]$) returns the index i of the first occurrence of x in a where $i \geq n$, and **minIndex** (a) $- 1$ if no such x is found. The default value of n is 1.
position (cc, t, i) returns the position $j >= i$ in t of the first character belonging to character class cc.

positive? (*orderedSetElement*)

positive? (x) tests if x is strictly greater than 0.

positiveRemainder (*integer, integer*)

positiveRemainder (a, b), where $b > 1$, yields r where $0 \leq r < b$ and $r = a$ **rem** b.

possiblyInfinite? (*stream*)

possiblyInfinite? (s) tests if the stream s could possibly have an infinite number of elements. Note: for many datatypes, **possiblyInfinite?** (s) $=$ **not explictlyFinite?**(s).

postfix (*outputForm, outputForm*)

postfix (op, a), where op and a are objects of type OutputForm (normally unexposed), creates an output form which prints as: a op.

powerAssociative? ()

powerAssociative? ()F, where `F` is a domain of category FiniteRankNonAssociativeAlgebra, tests if all subalgebras generated by a single element are associative.

powerSum (*integer*)

powerSum (n) is the n th power sum symmetric function. See CycleIndicators for details.

powmod (*integer, integer, integer*)

powmod (a, b, p), where a and b are non-negative integers, each $< p$, returns a^b **mod** p.

precision ([*positiveInteger*])

precision () returns the precision of Float values in decimal digits.
precision (n) set the precision in the base to n decimal digits.

prefix (*outputForm, listOfOutputForms*)

prefix (o, lo), where o is an object of type OutputForm (normally unexposed) and lo is a list of objects of type OutputForm, creates an output form depicting the $n$ary prefix application of o to a tuple of arguments given by list lo.

prefix? (*string, string*)

prefix? (s, t) tests if the string s is the initial substring of t.

prefixRagits (*listOfIntegers*)

prefixRagits (rx) returns the non-cyclic part of the ragits of the fractional part of a radix expansion. For example, if $x = 3/28 = 0.10714285714285\ldots$, then **prefixRagits** (x) $= [1, 0]$.

presub (*outputForm, outputForm*)

presub (o_1, o_2), where o_1 and o_2 are objects of type OutputForm (normally unexposed), creates an output form for o_1 presubscripted by o_2.

presuper (*outputForm, outputForm*)

presuper (o_1, o_2), where o_1 and o_2 are objects of type OutputForm (normally unexposed), creates an output form for o_1 presuperscripted by o_2.

primaryDecomp (*ideal*)

primaryDecomp (I) returns a list of primary ideals such that their intersection is the ideal I.

prime (*outputForm* [, *positiveInteger*])

prime ($o[, n]$), where o is an object of type OutputForm (normally unexposed), creates an output form for o following by n primes (that is, a prime like " ' "). By default, $n = 1$.

prime? (*element*)

prime? (x) tests if x cannot be written as the product of two non-units, that is, x is an irreducible element. Argument x may be an integer, a polynomial, an ideal, or, in general, any element of a domain of category UniqueFactorizationDomain.

primeFactor (*element, integer*)

primeFactor ($base, exponent$) creates a factored object with a single factor whose $base$ is asserted to be prime (flag = `"prime"`).

primeFrobenius (*finiteFieldElement* [, *nonNegativeInteger*])

Argument a is a member of a domain of category FieldOfPrimeCharacteristic(p).
primeFrobenius ($a[, s]$) returns a^{p^s}. The default value of s is 1.

primes (*integer, integer*)

primes (a, b) returns a list of all primes p with $a \leq p \leq b$.

primitive? (*finiteFieldElement*)

primitive? (b) tests whether the element b of a finite field is a generator of the (cyclic) multiplicative group of the field, that is, is a primitive element.

primitiveElement (*expressions* [, *expression*])

primitiveElement (a_1, a_2) returns a record with four components: a primitive element a with selector $primelt$, and three polynomials q_1, q_2, and q with selectors $pol1$, $pol2$, and $prim$. The prime element a is such that the algebraic extension generated by a_1 and a_2 is the same as that generated by a, $a_i = q_i(a)$ and $q(a) = 0$. The minimal polynomial for a_2 may involve a_1, but the minimal polynomial for a_1 may not involve a_2. This operations uses **resultant**.

primitiveMonomials (*polynomial*)

primitiveMonomials (p) gives the list of monomials of the polynomial p with their coefficients removed. Note: **primitiveMonomials** $(\sum a_i X^{(i)}) = [X^{(1)}, \ldots, X^{(n)}]$.

primitivePart (*polynomial*[, *symbol*])

primitivePart $(p[, v])$ returns the unit normalized form of polynomial p divided by the **content** of p with respect to variable v. If no v is given, the content is removed with respect to all variables.

principalIdeal (*listOfPolynomials*)

principalIdeal $([f_1, \ldots, f_n])$ returns a record whose "generator" component is a generator of the ideal generated by $[f_1, \ldots, f_n]$ whose "coef" component is a list of coefficients $[c_1, \ldots, c_n]$ such that $generator = \sum_i c_i f_i$.

print (*outputForm*)

print (o) writes the output form o on standard output using the two-dimensional formatter.

product (*element*, *element*)

product $(f(n), n = a..b)$ returns $\prod_{n=a}^{b} f(n)$ as a formal product.
product $(f(n), n)$ returns the formal product $P(n)$ verifying $P(n+1)/P(n) = f(n)$.
product (s, t), where s and t are cartesian tensors, returns the outer product of s and t. For example, if $r =$ **product**(s, t) for rank 2 tensors s and t, then r is a rank 4 tensor given by $r_{i,j,k,l} = s_{i,j} t_{k,l}$.
product (a, b), where a and b are elements of a graded algebra returns the degree-preserving linear product. See GradedAlgebra for details.

prolateSpheroidal (*smallFloat*)

prolateSpheroidal (a) returns a function to transform prolate spheroidal coordinates to Cartesian coordinates. This function will map the point (ξ, η, ϕ) to $x = a\sinh(\xi)\sin(\eta)\cos(\phi)$, $y = a\sinh(\xi)\sin(\eta)\sin(\phi)$, $z = a\cosh(\xi)\cos(\eta)$.

prologue (*text*)

prologue (t) extracts the prologue section of a IBM SCRIPT Formula Formatter or TeX formatted object t.

properties (*basicOperator*[, *prop*])

properties (op) returns the list of all the properties currently attached to op.
property (op, s) returns the value of property s if it is attached to op, and `"failed"` otherwise.

pseudoDivide (*polynomial*, *polynomial*)

pseudoDivide (p, q) returns (c, q, r), when $p' := p$ **leadingCoefficient**$(q)^{\deg(p)-\deg(q)+1} = cp$ is pseudo right-divided by q, that is, $p' = sq + r$.

pseudoQuotient (*polynomial*, *polynomial*)

pseudoQuotient (p, q) returns r, the quotient when $p' := p\text{leadingCoefficient}(q)^{\text{deg}p - \text{deg}q + 1}$ is pseudo right-divided by q, that is, $p' = sq + r$.

pseudoRemainder (*polynomial*, *polynomial*)

pseudoRemainder $(p, q) = r$, for polynomials p and q, returns the remainder when $p' := p\text{leadingCoefficient}(q)^{\text{deg}p - \text{deg}q + 1}$ is pseudo right-divided by q, that is, $p' = sq + r$.

puiseux (*expression*[, *options*])

puiseux (f) returns a Puiseux expansion of the expression f. Note: f should have only one variable; the series will be expanded in powers of that variable. Also, if x is a symbol, **puiseux** (x) returns x as a Puiseux series.
puiseux $(f, x = a)$ expands the expression f as a Puiseux series in powers of $(x - a)$.
puiseux (f, n) returns a Puiseux expansion of the expression f. Note: f should have only one variable; the series will be expanded in powers of that variable and terms will be computed up to order at least n.
puiseux $(f, x = a, n)$ expands the expression f as a Puiseux series in powers of $(x - a)$; terms will be computed up to order at least n.
puiseux $(n{+}{-}{>}a(n), x = a, r_0.., r)$ returns $\sum_{n=r_0, r_0+r, r_0+2r, \ldots} a(n)(x-a)^n$.
puiseux $(a(n), n, x = a, r_0.., r)$ returns $\sum_{n=r_0, r_0+r, r_0+2r, \ldots} a(n)(x-a)^n$.
Note: Each of the last two commands have alternate forms whose third argument is the finite segment $r_0..r_1$ producing a similar series with a finite number of terms.

push! (*element*, *stack*)

push! (x, s) pushes x onto stack s, that is, destructively changing s so as to have a new first (top) element x.

pushdown (*polynomial*, *symbol*)
pushdterm (*monomial*, *symbol*)

pushdown (prf, var) pushes all top level occurences of the variable var into the coefficient domain for the polynomial prf.
pushdterm $(monom, var)$ pushes all top level occurences of the variable var into the coefficient domain for the monomial $monom$.

pushucoef (*polynomial*, *variable*)

pushucoef $(upoly, var)$ converts the anonymous univariate polynomial $upoly$ to a polynomial in var over rational functions.

pushuconst (*rationalFunction, variable*)

pushuconst (r, var) takes a rational function and raises all occurences of the variable var to the polynomial level.

pushup (*polynomial, variable*)

pushup (prf, var) raises all occurences of the variable var in the coefficients of the polynomial prf back to the polynomial level.

qelt (*u* [, *options*])

qelt $(u, p[, options])$ is equivalent to a corresponding **elt** form except that it performs no check that indicies are in range. Use Browse to discover if a given domain has this alternative operation.

qsetelt! (*u, x, y* [, *z*])

qsetelt! $(u, x, y[, z])$ is equivalent to a corresponding **setelt** form except that it performs no check that indicies are in range.

quadraticForm (*matrix*)

quadraticForm (m) creates a quadratic form from a symmetric, square matrix m.

quatern (*element, element, element, element*)

quatern (r, i, j, k) constructs a quaternion from scalars.

queue ([*listOfElements*])

queue $()\$R$ returns an empty queue of type R.
queue $([x, y, \ldots, z])$ creates a queue with first (top) element x, second element y, ..., and last (bottom) element z.

quickSort (*predicate, aggregate*)

quickSort (f, agg) sorts the aggregate agg with the ordering predicate f using the quicksort algorithm.

quo (*integer, integer*)

a **quo** b returns the quotient of a and b discarding the remainder.

quoByVar (*series*)

quoByVar $(a_0 + a_1 x + a_2 x^2 + \cdots)$ returns $a_1 + a_2 x + a_3 x^2 + \cdots$ Thus, this function subtracts the constant term and divides by the series variable. This function is used when Laurent series are represented by a Taylor series and an order.

quote (*outputForm*)

quote (o), where o is an object of type OutputForm (normally unexposed), creates an output form o with a prefix quote.

quotedOperators (*rewriteRule*)

quotedOperators (r), where r is a rewrite rule, returns the list of operators on the right-hand side of r that are considered quoted, that is, they are not evaluated during any rewrite, but applied formally to their arguments.

quotient (*ideal, polynomial*)

quotient (I, f) computes the quotient of the ideal I by the principal ideal generated by the polynomial f, $(I : (f))$.
quotient (I, J) computes the quotient of the ideals I and J, $(I : J)$.

radical (*ideal*)

radical (I) returns the radical of the ideal I.

radicalEigenvalues (*matrix*)

radicalEigenvalues (m) computes the eigenvalues of the matrix m; when possible, the eigenvalues are expressed in terms of radicals.

radicalEigenvectors (*matrix*)

radicalEigenvectors (m) computes the eigenvalues and the corresponding eigenvectors of the matrix m; when possible, values are expressed in terms of radicals.

radicalEigenvector (*eigenvalue, matrix*)

radicalEigenvector (c, m) computes the eigenvector(s) of the matrix m corresponding to the eigenvalue c; when possible, values are expressed in terms of radicals.

radicalOfLeftTraceForm ()

radicalOfLeftTraceForm $()\$F$ returns the basis for the null space of **leftTraceMatrix** $()\$F$, where F is a domain of category FramedNonAssociativeAlgebra. If the algebra is associative, alternative or a Jordan algebra, then this space equals the radical (maximal nil ideal) of the algebra.

radicalRoots (*fractions*)

radicalRoots (rf, v) finds the roots expressed in terms of radicals of the rational function rf with respect to the symbol v.
radicalRoots (lrf, lv) finds the roots expressed in terms of radicals of the list of rational functions lrf with respect to the list of symbols lv.

radicalSolve (*eq, x*)

See **solve** (u, v).

radix (*rationalNumber, integer*)

radix (rn, b) converts rational number rn to a radix expansion in base b.

ramified? (*polynomial*)
ramifiedAtInfinity? ()

Domain F is a domain of functions on a fixed curve.
ramified? $(p)\$F$ tests whether $p(x) = 0$ is ramified.
ramifiedAtInfinity? () tests if infinity is ramified.

random ([*u, v*])

random $()\$R$ creates a random element from domain D.
random $(gp[, i])$ returns a random product of maximal i generators of the permutation group *gp*. The value of i is 20

by default.

range (*listOfSegments*)

range (ls), where ls is a list of segments of the form $[a_1..b_1, \ldots, a_n..b_n]$, provides a user-specified range for clipping for the **draw** command. This command may also be expressed locally to the **draw** command as the option $range == ls$. The values a_i and b_i are either given as floats or rational numbers.

ranges (*listOfSegments*)

ranges (l) provides a list of user-specified ranges for the **draw** command. This command may also be expressed as an option to the **draw** command in the form **ranges** $== l$.

rank (*matrix*)

rank (m) returns the rank of the matrix m. Also:
rank (A, B) computes the rank of the complete matrix $(A|B)$ of the linear system $AX = B$.
rank (t), where t is a Cartesion tensor, returns the tensorial rank of t (that is, the number of indices). See also FiniteRankAlgebra and FiniteRankNonAssociativeAlgebra.

rarrow (*outputForm, outputForm*)

rarrow (o_1, o_2), where o_1 and o_2 are objects of type OutputForm (normally unexposed), creates a form for the mapping $o_1 \rightarrow o_2$.

ratDenom (*expression* [, *option*])

ratDenom $(f[, u])$ rationalizes the denominators appearing in f. If no second argument is given, then all algebraic quantities are moved into the numerators. If the second argument is given as an algebraic kernel a, then a is removed from the denominators. Similarly, if u is a list of algebraic kernels $[a_1, \ldots, a_n]$, the operation removes the a_i's from the denominators in f.

rational? (*element*)
rationalIfCan (*element*)
rational (*element*)

rational? (x) tests if x is a rational number, that is, that it can be converted to type Fraction(Integer). Specifically, if x is complex, a quaternion, or an octonion, it tests that all imaginary parts are 0.
rationalIfCan (x) returns x as a rational number if possible, and `"failed"` if it is not.
rational (x) returns x as a rational number if possible, and calls **error** if it is not.

rationalApproximation (*float, nonNegativeInteger* [, *positiveInteger*])

rationalApproximation $(f, n[, b])$ computes a rational approximation r to f with relative error $< b^{-n}$, that is $|(r - f)/f| < b^{-n}$, for some positive integer base b. By default, $b = 10$. The first argument f is either a float or small float.

rationalFunction (*series, integer, integer*)

rationalFunction (f, m, n) returns a rational function consisting of the sum of all terms of f of degree d with $m \leq d \leq n$. By default, n is the maximum degree of f.

rationalPoint? (*value, value*)

rationalPoint? $(a, b)\$F$ tests if $(x = a, y = b)$ is on the curve defining function field F. See FunctionFieldCategory.

rationalPoints ()

rationalPoints $()\$$ returns the list of all the affine rational points on the curve defining function field F. See FunctionFieldCategory.

rationalPower (*puiseuxSeries*)

rationalPower $(f(x))$ returns r where the Puiseux series $f(x) = g(x^r)$.

ratPoly (*expression*)

ratPoly (f) returns a polynomial p such that p has no algebraic coefficients, and $p(f) = 0$.

rdexquo (*lodOperator*)

rdexquo (a, b), where a and b are linear ordinary differential operators, returns q such that $a = bq$, or `"failed"` if no such q exists.

rightDivide (*lodOperator, lodOperator*)
rightQuotient (*lodOperator, lodOperator*)
rightRemainder (*lodOperator, lodOperator*)

rightDivide (a, b) returns the pair q, r such that $a = qb + r$ and the degree of r is less than the degree of b. The pair is returned as a record with fields $quotient$ and $remainder$. This process is called "right division". Also:
rightQuotient (a, b) returns only q. **rightRemainder** (a, b) returns only r.

read! (*file*)
readIfCan! (*file*)

read! (f) extracts a value from file f. The state of f is modified so a subsequent call to **read!** will return the next element.
readIfCan! (f) returns a value from the file f or `"failed"` if this is not possible (that is, either f is not open for reading, or f is at the end of the file).

readable? (*file*)

readable? (f) tests if the named file exists and can be opened for reading.

readLine! (*file*)
readLineIfCan! (*file*)

readLineIfCan! (f) returns a string of the contents of a line from file f, or `"failed"` if this is not possible (if f is not readable or is positioned at the end of file).
readLine! (f) returns a string of the contents of a line from

the file f, and calls **error** if this is not possible.

real (x)
real? (*expression*)

real (x) returns real part of x. Argument x can be an expression or a complex value, quaternion, or octonion.
real? (f) tests if expression $f = real(f)$.

realEigenvectors (*matrix, float*)

realEigenvectors (m, eps) returns a list of records, each containing a real eigenvalue, its algebraic multiplicity, and a list of associated eigenvectors. All these results are computed to precision *eps* as floats or rational numbers depending on the type of *eps*. Argument m is a matrix of rational functions.

realElementary (*expression*[, *symbol*])

realElementary (f, sy) rewrites the kernels of f involving sy in terms of the 4 fundamental real transcendental elementary functions: $log, exp, tan, atan$. If sy is omitted, all kernels of f are rewritten.

realRoots (*rationalfunctions, v*[, *w*])

realRoots (rf, eps) finds the real zeros of a univariate rational function rf with precision given by eps.
realRoots (lp, lv, eps) computes the list of the real solutions of the list lp of rational functions with rational coefficients with respect to the variables in lv, with precision *eps*. Each solution is expressed as a list of numbers in order corresponding to the variables in lv.

realZeros (*polynomial, rationalNumber*[, *option*])

realZeros (*pol*) returns a list of isolating intervals for all the real zeros of the univariate polynomial *pol*.
realZeros (*pol*[, *eps*]) returns a list of intervals of length less than the rational number *eps* for all the real roots of the polynomial *pol*. The default value of *eps* is ???.
realZeros (*pol*, int[, *eps*]) returns a list of intervals of length less than the rational number *eps* for all the real roots of the polynomial *pol* which lie in the interval expressed by the record int. The default value of *eps* is ???.

recip (*element*)

recip (x) returns the multiplicative inverse for x, or `"failed"` if no inverse can be found. See also FiniteRankNonAssociativeAlgebra andMonadWithUnit.

recur (*function*)

recur (f), where f is a function of type (NonNegativeInteger, R) $\rightarrow$ R for some domain R, returns the function g such that $g(n, x) = f(n, f(n-1, \ldots f(1, x) \ldots))$. For related functions, see MappingPackage.

red ()

red () returns the position of the red hue from total hues.

reduce (*op, aggregate*[, *identity, element*])

reduce $(f, u[, ident, a])$ reduces the binary operation f across u. For example, if u is $[x_1, x_2, \ldots, x_n]$ then **reduce** (f, u) returns $f(\ldots f(x_1, x_2), \ldots, x_n)$.

An optional identity element of f provided as a third argument affects the result if u has less than two elements. If u is empty, the third argument is returned if given, and a call to **error** occurs otherwise. If u has one element and the third argument is given, the value returned is $f(ident, x_1)$. Otherwise x_1 is returned. Thus both **reduce** $(+, u)$ and **reduce** $(+, u, 0)$ return $\sum_{i=1}^{n} x_i$. Similarly, **reduce** $(*, u)$ and **reduce** $(*, u, 1)$ return $\prod_{i=1}^{n} x_i$.

An optional fourth argument z acts as an "absorbing element". **reduce** (f, u, x, z) reduces the binary operation f across u, stopping when an "absorbing element" z is encountered. For example **reduce** $(or, u, false, true)$ will stop iterating across u returning $true$ as soon as an $x_i = true$ is found. Note: if u has one element x, **reduce** (f, u) returns x, or calls **error** if u is empty.

reduceBasisAtInfinity (*basis*)

reduceBasisAtInfinity $(b_1, \ldots, b_n)$, where the b_i are functions on a fixed curve, returns $(x^i\, b_j)$ for all i, j such that $x^i\, b_j$ is locally integral at infinity. See FunctionFieldCategory using Browse.

reducedContinuedFraction (*element, stream*)

reducedContinuedFraction (b_0, b) returns a continued fraction constructed as follows. If $b = [b_1, b_2, \ldots]$ then the result is the continued fraction $b_0 + 1/(b_1 + 1/(b_2 + \cdots))$. That is, the result is the same as **continuedFraction** $(b_0, [1, 1, 1, \ldots], [b_1, b_2, b_3, \ldots])$.

reducedForm (*continuedFraction*)

reducedForm (x) puts the continued fraction x in reduced form, that is, the function returns an equivalent continued fraction of the form **continuedFraction** $(b_0, [1, 1, 1, \ldots], [b_1, b_2, b_3, \ldots])$.

reducedSystem (*matrix*[, *vector*])

reducedSystem (A, v) returns a matrix B such that $Ax = v$ and $Bx = v$ have the same solutions. By default, $v = 0$.

reductum (*polynomial*)

reductum (p) returns polynomial p minus its leading monomial, or zero if handed the zero element. See also IndexedDirectProdcutCategory and MonogenicLinearOperator.

refine (*polynomial, interval, precision*)

refine $(pol, int, tolerance)$ refines the interval int containing exactly one root of the univariate polynomial *pol* to size less than the indicated $tolerance$. Argument int is an interval denoted by a record with selectors $left$ and $right$, each with rational number values. The tolerance is either a rational number or another interval. In the latter case, `"failed"` is returned if no such isolating interval exists.

regularRepresentation (*element, basis*).

regularRepresentation ($a, basis$) returns the matrix of the linear map defined by left multiplication by a with respect to basis $basis$. Element a is a complex element or an element of a domain R of category FramedAlgebra. The second argument may be omitted when a fixed basis is defined for R.

reindex (*cartesianTensor, listOfIntegers*)

reindex ($t, [i_1, \ldots, i_{\rm dim}]$) permutes the indices of cartesian tensor t. For example, if $r = $ **reindex**$(t, [4, 1, 2, 3])$ for a rank 4 tensor t, then r is the rank 4 tensor given by $r(i, j, k, l) = t(l, i, j, k)$.

relationsIdeal (*listOfPolynomials*)

relationsIdeal ($polyList$) returns the ideal of relations among the polynomials in $polyList$.

relerror (*float, float*)

relerror (x, y), where x and y are floats, computes the absolute value of $x - y$ divided by y, when $y \neq 0$.

rem (*element, element*)

a **rem** b returns the remainder of a and b.

remove (*predicate, aggregate*)

Argument u is any extensible aggregate such as a list.
remove $(pred, u)$ returns a copy of u removing all elements x such that $p(x)$ is $true$. Argument u may be any homogeneous aggregate including infinite streams. Note: for lists and streams, `remove(p, u) == [x for x in u | not p(x)]`.
remove! $(pred, u)$ destructively removes all elements x of u such that **pred** (x) is $true$. The value of u after all such elements are removed is returned.
remove! (x, u) destructively removes all values x from u.
remove! (k, t), where t is a keyed dictionary, searches the table t for the key k, removing and returning the entry if there. If t has no such key, it returns `"failed"`.

removeCoshSq (*expression*)

removeCoshSq (f) converts every $\cosh(u)^2$ appearing in f into $1 - \sinh(x)^2$, and also reduces higher powers of **cosh** (u) with that formula.

removeDuplicates (*aggregate*)
removeDuplicates! (*aggregate*)

removeDuplicates (u) returns a copy of u with all duplicates removed.
removeDuplicates! (u) destructively removes duplicates from u.

removeSinhSq (*expression*)

removeSinhSq (f) converts every $\sinh(u)^2$ appearing in f into $1 - \cosh(x)^2$, and also reduces higher powers of **sinh** (u) with that formula.

removeSinSq (*expression*)

removeSinSq (f) converts every $\sin(u)^2$ appearing in f into $1 - \cos(x)^2$, and also reduces higher powers of **sin** (u) with that formula.

removeZeroes (*[integer,]laurentSeries*)

removeZeroes $([n,] f(x))$ removes up to n leading zeroes from the Laurent series $f(x)$. If no integer n is given, all leading zeroes are removed.

reopen! (*file, string*)

reopen! $(f, mode)$ returns a file f reopened for operation in the indicated mode: `"input"` or `"output"`. For example, **reopen!** $(f,$ `"input"`$)$ will reopen the file f for input.

repeating (*listOfElements[, stream]*)
repeating? (*stream*)

repeating (l) is a repeating stream whose period is the list l.
repeating? (l, s) tests if a stream s is periodic with period l.

replace (*string, segment, string*)

replace $(s, i..j, t)$ replaces the substring $s(i..j)$ of s by string t.

represents (*listOfElements[, listOfBasisElements]*)

represents $([a^1, .., a^n][, [v^1, .., v^n]])$ returns $a^1v^1 + \cdots + a^nv^n$. Arguments v_i are elements of a domain of category FiniteRankAlgebra or FiniteRankNonAssociativeAlgebra built over a ring R. The a_i are elements of R. In a framed algebra or finite algebra extension field domain with a fixed basis, $[v_1, \ldots, v_n]$ defaults to the elements of the fixed basis. See FramedAlgebra, FramedNonAssociateAlgebra, and FiniteAlgebraicExtensionField.
See also FunctionFieldCategory.

resetNew ()

resetNew () resets the internal counter that **new** () uses.

resetVariableOrder ()

resetVariableOrder () cancels any previous use of **setVariableOrder** and returns to the default system ordering.

rest (*aggregate[, nonNegativeInteger]*)

rest (u) returns an aggregate consisting of all but the first element of u (equivalently, the next node of u).
rest (u, n) returns the $n^{\rm th}$ node of u. Note: **rest** $(u, 0) = u$.

resultant (*polynomial, polynoial[, variable]*)

resultant (p, q, v) returns the resultant of the polynomials p and q with respect to the variable v. If p and q are univariate polynomials, the variable v defaults to the unique variable.

retract (*element*)

retractIfCan (*element*)

retractIfCan $(a)@S$ returns a as an object of type S, or `"failed"` if this is not possible.
retract $(a)@S$ transforms a into an element of S, or calls **error** if this is not possible.

retractable? (*typeAnyObject*)

retractable? $(a)\$S$ tests if object a of type Any can be converted into an object of type S.

reverse (*linearAggregate*)
reverse! (*linearAggregate*)

reverse (a) returns a copy of linear aggregate a with elements in reverse order.
reverse! (a) destructively puts the elements of linear aggregate a in reverse order.

rightGcd (*lodOperator, lodOperator*)

rightGcd (a, b), where a and b are linear ordinary differential operators, computes the value g of highest degree such that $a = g * aa$ and $b = g * bb$ for some values aa and bb. The value g is computed using right-division.

rhs (*rewriteRuleOrEquation*)

rhs (u) returns the right-hand side of the rewrite rule or equation u.

right (*binaryRecursiveAggregate*)

right (a) returns the right child.

rightAlternative? ()

See **leftAlternative?**.

rightCharacteristicPolynomial (*element*)

See **leftCharacteristicPolynomial**.

rightDiscriminant (*basis*)

See **leftDiscriminant**.

rightMinimalPolynomial (*element*)

See **leftMinimalPolynomial**.

rightNorm (*element*)

See **leftNorm**.

rightPower (*monad, nonNegativeInteger*)

See **rightPower**.

rightRankPolynomial ()

See **leftRankPolynomial**.

rightRank (*basis*)

See **leftRank**.

rightRecip (*element*)

See **leftRecip**.

rightRegularRepresentation (*element*[, *basis*])

See **leftRegularRepresentation**.

rightTraceMatrix ([*basis*])

See **leftTraceMatrix**.

rightTrim (*string, various*)

See **leftTrim**.

rightUnits ()

See **leftUnits**.

rischNormalize (*expression, x*)

rischNormalize (f, x) returns $[g, [k_1, \ldots, k_n], [h_1, \ldots, h_n]]$ such that $g =$ **normalize**(f, x) and each k_i was rewritten as h_i during the normalization.

rightLcm (*lodOperator, lodOperator*)

rightLcm (a, b), where a and b are linear ordinary differential operators, computes the value m of lowest degree such that $m = aa * a = bb * b$ for some values aa and bb. The value m is computed using right-division.

roman (*integerOrSymbol*)

roman (x) creates a roman numeral for integer or symbol x.

romberg (*floatFunction, fourFloats, threeIntegers*)
rombergOpen (*floatFunction, fourFloats, twoIntegers*)
rombergClose (*floatFunction, fourFloats, twoIntegers*)

romberg $(fn, a, b, epsrel, epsabs, nmin, nmax, nint)$ uses an adaptive romberg method to numerically integrate function fn over the closed interval from a to b, with relative accuracy $epsrel$ and absolute accuracy $epsabs$; the refinement levels for the checking of convergence vary from $nmin$ to $nmax$. The method is called "adaptive" since it requires an additional parameter $nint$ giving the number of subintervals over which the integrator independently applies the convergence criteria using $nmin$ and $nmax$. This is useful when a large number of points are needed only in a small fraction of the entire interval. Parameter fn is a function of type Float → Float; a, b, $epsrel$, and $epsabs$ are floats; $nmin$, $nmax$, and $nint$ are integers. The operation returns a record containing: `value`, an estimate of the integral; `error`, an estimate of the error in the computation; `totalpts`, the total integral number of function evaluations, and `success`, a boolean value that is *true* if the integral was computed within the user specified error criterion. See NumericalQuadrature for details.

rombergClosed $(fn, a, b, epsrel, epsabs, nmin, nmax)$ similarly uses the Romberg method to numerically integrate function fn over the closed interval a to b, but is not adaptive.

rombergOpen $(fn, a, b, epsrel, epsabs, nmin, nmax)$ is similar to **rombergClosed**, except that it integrates function fn over the open interval from a to b.

root (*outputForm* [, *positiveInteger*])

root $(o[, n])$, where o and n are objects of type OutputForm (normally unexposed), creates an output form for the n th root of the form o. By default, $n = 2$.

rootOfIrreduciblePoly (*polynomial*)

rootOfIrreduciblePoly (f) computes one root of the monic, irreducible polynomial f, whose degree must divide the extension degree of F over GF. That is, f splits into linear factors over F.

rootOf (*polynomial* [, *variable*])

rootOf $(p[, y])$ returns y such that $p(y) = 0$. The object returned displays as $'y$. The second argument may be omitted when p is a polynomial in a unique variable y.

rootSimp (*expression*)

rootSimp (f) transforms every radical of the form $(ab^{qn+r})^{1/n}$ appearing in expression f into $b^q(ab^r)^{1/n}$. This transformation is not in general valid for all complex numbers b.

rootsOf (*polynomialOrExpression* [, *symbol*])

rootsOf $(p[, y])$ returns the value of $[y_1, \ldots, y_n]$ such that $p(y_i) = 0$. The y_i are symbols of the form %y with a suffix number which are bound in the interpreter to respective root values. Argument p is either an expression or a polynomial. Argument y may be omitted in which case p must contain exactly one symbol.

rootSplit (*expression*)

rootSplit (f) transforms every radical of the form $(a/b)^{1/n}$ appearing in f into $a^{1/n}/b^{1/n}$. This transformation is not in general valid for all complex numbers a and b.

rotate! (*queue*)

rotate! (q) rotates queue q so that the element at the front of the queue goes to the back of the queue.

round (*float*)

round (x) computes the integer closest to x.

row (*matrix*, *positiveInteger*)

row (m, i) returns the i th row of the matrix or two-dimensional array m.

rowEchelon (*matrix*)

rowEchelon (m) returns the row echelon form of the matrix m.

rst (*stream*)

rst (s) returns a pointer to the next node of stream s. Caution: this function should only be called after a **empty?** test returns *true* since no error check is performed.

rubiksGroup ()

rubiksGroup () constructs the permutation group representing Rubic's Cube acting on integers $10i + j$ for $1 \leq i \leq 6, 1 \leq j \leq 8$. The faces of Rubik's Cube are labelled: Front, Right, Up, Down, Left, Back and numbered from 1 to 6. The pieces on each face (except the unmoveable center piece) are clockwise numbered from 1 to 8 starting with the piece in the upper left corner. The moves of the cube are represented as permutations on these pieces, represented as a two digit integer ij where i is the number of the face and j is the number of the piece on this face. The remaining ambiguities are resolved by looking at the 6 generators representing 90-degree turns of the faces.

rule (*various*)

See Section 6.21 on page 173.

rules (*ruleset*)

rules (r) returns the list of rewrite rules contained in ruleset r.

ruleset (*listOfRules*)

ruleset $([r1, \ldots, rn])$ creates a ruleset from a list of rewrite rules $r_1, \ldots, r_n$.

rungaKutta (*vector*, *integer*, *fourFloats*, *integer*, *function*)
rungaKuttaFixed (*vector*, *integer*, *float*, *float*, *integer*, *function*)

rungaKutta $(y, n, a, b, eps, h, ncalls, derivs)$ uses a 4–th order Runga-Kutta method to numerically integrate the ordinary differential equation $dy/dx = f(y, x)$ from x_1 to x_2, where y is an n–vector of n variables. Initial and final values are provided by solution vector y. The local truncation error is kept within *eps* by changing the local step size. Argument h is a trial step size and *ncalls* is the maximum number of single steps the integrator is allowed to take. Argument *derivs* is a function of type (Vector Float, Vector Float, Float) → Void, which computes the right-hand side of the ordinary differential equation, then replaces the elements of the first argument by updated elements.

rungaKuttaFixed $(y, n, x_1, x_2, ns, derivs)$ is similar to **rungaKutta** except that it uses *ns* fixed steps to integrate the solution vector y from x_1 to x_2, returning the values in y.

saturate (*ideal*, *polynomial* [, *listOfVariables*])

saturate $(I, f[, lvar])$ is the saturation of the ideal I with respect to the multiplicative set generated by the polynomial f in the variables given by $lvar$, a list of variables. Argument $lvar$ may be omitted in which case $lvar$ is taken to be the list of all variables appearing in f.

say (*strings*)

say (u) sends a string or a list of strings u to output.

sayLength (*listOfStrings*)

sayLength (ls) returns the total number of characters in the list of strings ls.

scalarMatrix (*scalar*[, *dimension*])

scalarMatrix ($r[, n]$) returns an n-by-n matrix with scalar r on the diagonal and zero elsewhere. The dimension may be omitted if the result is to be an object of type **SquareMatrix** (n, R) for some n.

scan (*binaryFunction*, *aggregate*, *element*)

scan (f, a, r) successively applies **reduce** (f, x, r) to more and more leading sub-aggregates x of aggregrate a. More precisely, if a is $[a1, a2, \ldots]$, then **scan** (f, a, r) returns $[reduce(f, [a1], r), reduce(f, [a1, a2], r), \ldots]$. Argument a can be any linear aggregate including streams. For example, if a is a list or an infinite stream of the form $[x_1, x_2, \ldots]$, then `scan(+, a, 0)` returns a list or stream of the form $[x_1, x_1 + x_2, \ldots]$.

scanOneDimSubspaces (*listOfVectors*, *integer*)

scanOneDimSubspaces ($basis, n$) gives a canonical representative of the nth one-dimensional subspace of the vector space generated by the elements of $basis$. Consult RepresentationPackage2 using details.

script (*symbol*, *listOfListsOfOutputForms*)

script ($sy, [a, b, c, d, e]$) returns sy with subscripts a, superscripts b, pre-superscripts c, pre-subscripts d, and argument-scripts e. Omitted components are taken to be empty. For example, **script** ($s, [a, b, c]$) is equivalent to **script** ($s, [a, b, c, [], []]$).

scripted? (*symbol*)

scripted? (sy) tests if sy has been given any scripts.

scripts (*symbolOrOutputForm*[, *listOfOutputForms*])

scripts (o, lo), where o is an object of type OutputForm (normally unexposed) and lo is a list $[sub, super, presuper, presub]$ of four objects of type OutputForm (normally unexposed), creates a form for o with scripts on all four corners.
scripts (s) returns all the scripts of s as a record with selectors sub, sup, $presup$, $presub$, and $args$, each with a list of output forms as a value.

search (*key*, *table*)

search (k, t) searches the table t for the key k, returning the entry stored in t for key k, or `"failed"` if t has no such key.

sec (*expression*)
secIfCan (*expression*)

sec (x) returns the secant of x.
secIfCan (z) returns **sec** (z) if possible, and `"failed"` otherwise.

sec2cos (*expression*)

sec2cos (f) converts every **sec** (u) appearing in f into $1/\cos(u)$.

sech (*expression*)
sechIfCan (*expression*)

sech (x) returns the hyperbolic secant of x.
sechIfCan (z) returns **sech** (z) if possible, and `"failed"` otherwise.

sech2cosh (*expression*)

sech2cosh (f) converts every **sech** (u) appearing in f into $1/\cosh(u)$.

second (*aggregate*)

second (u) returns the second element of recursive aggregate u. Note: **second** (u) = **first**(**rest**(u)).

segment (*integer*[, *integer*])

segment ($i[, j]$) returns the segment $i..j$. If not qualified by a **by** clause, this notation for integers i and j denotes the tuple of integers i, $i + 1$, ..., j. When j is omitted, **segment** (i) denotes the half open segment $i..$, that is, a segment with no upper bound.
segment ($x = bd$), where bd is a binding, returns bd. For example, **segment** ($x = a..b$) returns $a..b$.

select (*pred*, *aggregate*)
select! (*pred*, *aggregate*)

select (p, u) returns a copy of u containing only those elements x such that $p(x)$ is $true$. For a list l, $select(p, l) == [x \texttt{ for } x \texttt{ in } l | p(x)]$. Argument u may be any finite aggregate or infinite stream.
select! (p, u) destructively changes u by keeping only values x such that $p(x)$ is true. Argument u can be any extensible linear aggregate or dictionary.

semicolonSeparate (*listOfOutputForms*)

semicolonSeparate (lo), where lo is a list of objects of type OutputForm (normally unexposed), returns an output form which separates the elements of lo by semicolons.

separant (*differentialPolynomial*)

separant ($polynomial$) returns the partial derivative of the differential polynomial p with respect to its leader.

separate (*polynomial*, *polynomial*)

separate (p, q) returns (a, b) such that polynomial $p = ab$ and a is relatively prime to q. The result produced is a record with selectors $primePart$ and $commonPart$ with value a and b respectively.

separateDegrees (*polynomial*)

separateDegrees (p) splits the polynomial p into factors. Each factor is a record with selector deg, a non-negative integer, and $prod$, a product of irreducible polynomials of degree *deg*.

separateFactors (*listOfRecords, polynomial*)

separateFactors $(lfact, p)$ takes the list produced by **separateDegrees** along with the original polynomial p, and produces the complete list of factors.

separateFactors (*listOfRecords, integer*)

separateFactors (ddl, p) refines the distinct degree factorization produced by **ddFact** to give a complete list of factors.

sequences (*listOfIntegers*)
sequences (*listOfIntegers, listOfIntegers*)

sequences $([l_0, l_1, l_2, .., l_n])$ is the set of all sequences formed from l_0 0's, l_1 1's, l_2 2's, ..., l_n n's.
sequences $(l1, l2)$ is the stream of all sequences that can be composed from the multiset defined from two lists of integers $l1$ and $l2$. For example, the pair $([1, 2, 4], [2, 3, 5])$ represents multiset with 1 2, 2 3's, and 4 5's.

series (*specifications*[, ...])

series $(expression)$ returns a series expansion of the expression f. Note: f must have only one variable. The series will be expanded in powers of that variable.
series (sy), where sy is a symbol, returns sy as a series.
series (st), where t is a stream $[a_0, a_1, a_2, \ldots]$ of coefficients a_i from some ring, creates the Taylor series $a_0 + a_1x + a_2x^2 + \ldots$. Also, if st is a stream of elements of type Record(k:NonNegativeInteger, c:R), where k denotes an exponent and c, a non-zero coefficient from some ring R, it creates a stream of non-zero terms. The terms in st must be ordered by increasing order of exponents.
series $(f, x = a[, n])$ expands the expression f as a series in powers of $(x - a)$ with n terms. If n is missing, the number of terms is governed by the value set by the system command)set streams calculate.
series (f, n) returns a series expansion of the expression f. Note: f should have only one variable; the series will be expanded in powers of that variable and terms will be computed up to order at least n.
series $(i+- >a(i), x = a, m..[n, k])$ creates the series $\sum_{i=m..n \text{ by } k} a(i)(x-a)^i$. Here m, n, and k are rational numbers. Upper-limit n and stepsize k are optional and have default values $n = \infty$ and $k = 1$.
series $(a(i), i, x = a, m..[n, k])$ returns $\sum_{i=m..n\mathbf{by}k} a(n)(x-a)^n$.

seriesSolve (*eq, y, x, c*)

eq denotes an equation to be solved; alternatively, an expression u may be given for *eq* in which case the equation *eq* is defined as $u = 0$.

leq denotes a list $[eq_1 \ldots eq_n]$ of equations; alternatively, a list of expressions $[u_1 \ldots u_n]$ may be given of *leq* in which case the equations eq_i are defined by $u_i = 0$.

seriesSolve $(eq, y, x = a, [y(a) =]b)$ returns a Taylor series solution of *eq* around $x = a$ with initial condition $y(a) = b$. Note: *eq* must be of the form $f(x, y)y'(x) + g(x, y) = h(x, y)$.

seriesSolve $(eq, y, x = a, [b_0, \ldots, b_{(n-1)}])$ returns a Taylor series solution of *eq* around $x = a$ with initial conditions $y(a) = b_0$, $y'(a) = b_1$, ... $y^{(n-1)}(a) = b_{(n-1)}$. Equation *eq* must be of the form $f(x, y, y', \ldots, y^{(n-1)}) * y^{(n)}(x) + g(x, y, x', \ldots, y^{(n-1)}) = h(x, y, y', \ldots, y^{(n-1)})$.

$seriesSolve(leq, [y_1, \ldots, y_n], x = a, [y_1(a) = b_1, \ldots, y_n(a) = b_n])$ returns a Taylor series solution of the equations eq_i around $x = a$ with initial conditions $y_i(a) = b_i$. Note: each eq_i must be of the form $f_i(x, y_1, y_2, \ldots, y_n)y_1'(x) + g_i(x, y_1, y_2, \ldots, y_n) = h(x, y_1, y_2, \ldots, y_n)$.

$seriesSolve(leq, [y_1, \ldots, y_n], x = a, [b_1, \ldots, b_n])$ is equivalent to the same command with fourth argument $[y_1(a) = b_1, \ldots, y_n(a) = b_n]$.

setchildren! (*recursiveAggregate*)

setchildren! (u, v) replaces the current children of node u with the members of v in left-to-right order.

setColumn! (*matrix*)

setColumn! (m, j, v) sets the jth column of matrix or two-dimensional array m to v.

setDifference (*list, list*)

setDifference (l_1, l_2) returns a list of the elements of l_1 that are not also in l_2. The order of elements in the resulting list is unspecified.

setelt (*structure, index, value*[, *option*])

setelt (u, x, y), also written $u.x := y$, sets the image of x to be y under u, regarded as a function mapping values from the domain of x to the domain of y. Specifically, if u is:

a list: $u.first := x$ is equivalent to **setfirst!** (u, x). Also, $u.rest := x$ is equivalent to **setrest!** (u, x), and $u.last := x$ is equivalent to **setlast!** (u, x).

a linear aggregate, $u(i..j) := x$ destructively replaces each element in the segment $u(i..j)$ by x. The value x is returned. Note: This function has the same effect as `for k in i..j repeat u.k := x; x`. The length of u is unchanged.

a recursive aggregate, $u.value := x$ is equivalent to **setvalue!** (u, x) and sets the value part of node u to **x**. Also, if u is a BinaryTreeAggregate, $u.left := x$ is equivalent to **setleft!** (u, x) and sets the left child of u to **x**. Simiarly, $u.right := x$ is equivalent to **setright!** (u, x). See also **setchildren!**.

a table of category TableAggregate(Key, Entry): u(k) := e is equivalent to (**insert**$([k, e], t); e$), where k is a key and

e is an entry.

a library: $u.k := v$ saves the value v in the library u, so that it can later be extracted by $u.k$.

setelt (u, i, j, r), also written, $u(i, j) := r$, sets the element in the i th row and j th column of matrix or two-dimensional array u to r.
setelt $(u, rowList, colList, r)$, also written $u([i_1, i_2, \ldots, i_m], [j_1, j_2, \ldots, j_n]) := r$, where u is a matrix or two-dimensional array and r is another m by n matrix or array, destructively alters the matrix u: the x_{i_k, j_l} is set to $r(k, l)$.

setEpilogue! (*formattedObject, listOfStrings*)

setEpilogue! $(t, strings)$ sets the epilogue section of a formatted object t to $strings$. Argument t is either an IBM SCRIPT Formula Formatted or TEX formatted object.

setfirst! (*aggregate, value*)

setfirst! (a, x) destructively changes the first element of recursive aggregate a to x.

setFormula! (*formattedObject, listOfStrings*)

setFormula! $(t, strings)$ sets the formula section of a formatted object t to $strings$.

setIntersection (*list, list*)

setIntersection (l_1, l_2) returns a list of the elements that lists l_1 and l_2 have in common. The order of elements in the resulting list is unspecified.

setlast! (*aggregate, value*)

setlast! (u, x) destructively changes the last element of u to x. Note: $u.last := x$ is equivalent.

setleaves! (*balancedBinaryTree, listOfElements*)

setleaves! (t, ls) sets the leaves of balanced binary tree t in left-to-right order to the elements of ls.

setleft! (*binaryRecursiveAggregate*)

setleft! (a, b) sets the left child of a to be b.

setPrologue! (*formattedObject, listOfStrings*)

setPrologue! $(t, strings)$ sets the prologue section of a formatted object t to $strings$. Argument t is either an IBM SCRIPT Formula Formatted or TEX formatted object.

setProperties! (*basicOperator, associationList*)

setProperties! (*op, al*) sets the property list of basic operator *op* to association list l. Note: argument *op* is modified "in place", that is, no copy is made.

setProperty! (*basicOperator, string, value*)

setProperty! (*op*, s, v) attaches property s to *op*, and sets its value to v. Argument *op* is modified "in place", that is, no copy is made.

setrest! (*aggregate*[, *integer*], *aggregate*)

Arguments u and v are finite or infinite aggregates of the same type.
setrest! (u, v) destructively changes the rest of u to v.
setrest! (x, n, y) destructively changes x so that **rest** (x, n), that is, x after the n th element, equals y. The function will expand cycles if necessary.

setright! (*binaryRecursiveAggregate*)

setright! (a, x) sets the right child of t to be x.

setRow! (*matrix, integer, row*)

setRow! (m, i, v) sets the i th row of matrix or two-dimensional array m to v.

setsubMatrix! (*matrix, integer, integer, matrix*)

setsubMatrix (x, i_1, j_1, y) destructively alters the matrix x. Here $x(i, j)$ is set to $y(i - i_1 + 1, j - j_1 + 1)$ for $i = i_1, \ldots, i_1 - 1 + \mathbf{nrows}(y)$ and $j = j_1, \ldots, j_1 - 1 + \mathbf{ncols}(y)$.

setTex! (*text, listOfStrings*)

setTex! $(t, strings)$ sets the TeX section of a TeX form t to $strings$.

setUnion (*list, list*)

setUnion (l_1, l_2) appends the two lists l_1 and l_2, then removes all duplicates. The order of elements in the resulting list is unspecified.

setvalue! (*aggregate, value*)

setvalue! (u, x) destructively changes the value of node u to x.

setVariableOrder (*listOfSymbols*[, *listOfSymbols*])

setVariableOrder $([a_1, \ldots, a_m], [b_1, \ldots, b_n])$ defines an ordering on the variables given by $a_1 > a_2 > \ldots > a_m >$ other variables $b_1 > b_2 > \ldots > b_n$.
setVariableOrder $([a1, \ldots, an])$ defines an ordering given by $a_1 > a_2 > \ldots > a_n >$ all other variables.

sFunction (*listOfIntegers*)

sFunction (li) is the S-function of the partition given by list of linteger li, expressed in terms of power sum symmetric functions. See CycleIndicators for details.

shade (*palette*)

shade (p) returns the shade index of the indicated palette p.

shellSort (*sortingFunction, aggregate*)

shellSort (f, a) sorts the aggregate a using the shellSort algorithm with sorting function f. Aggregate a can be any finite linear aggregate which is mutable (for example, lists, vectors, and strings). The sorting function f has type (R, R) → Boolean where R is the domain of the elements of a.

shift (*integerNumber, integer*)

shift (a, i) shifts integer number or float a by i digits.

showAll? ()

showAll? () tests if all computed entries of streams will be displayed according to system command)set streams showall.

showAllElements (*stream*)

showAllElements (s) creates an output form displaying all the already computed elements of stream s. This command will not result in any further computation of elements of s. Also, the command has no effect if the user has previously entered)set streams showall true.

showTypeInOutput (*boolean*)

showTypeInOutput (*bool*) affects the way objects of Any are displayed. If *bool* is $true$, the type of the original object that was converted to Any will be printed. If *bool* is $false$, it will not be printed.

shrinkable (*boolean*)

shrinkable $(b)\$R$ tells AXIOM that flexible arrays of domain R are or are not allowed to shrink (reduce their **physicalLength**) according to whether b is $true$ or $false$.

shufflein (*listOfIntegers, streamOfListsOfIntegers*)

shufflein (li, sli) maps **shuffle** (li, u) onto all members u of sli, concatenating the results. See PartitionsAndPermutations.

shuffle (*listOfIntegers, listOfIntegers*)

shuffle $(l1, l2)$ forms the stream of all shuffles of $l1$ and $l2$, that is, all sequences that can be formed from merging $l1$ and $l2$. See PartitionsAndPermutations.

sign (*various*[, ...])

sign (x), where x is an element of an ordered ring, returns 1 if x is positive, -1 if x is negative, 0 if x equals 0.
sign (p), where p is a permutation, returns 1, if p is an even permutation, or -1, if it is odd.
sign (f, x, a, s) returns the sign of rational function f as symbol x nears a, a real value represented by either a rational function or one of the values `%plusInfinity` or `%minusInfinity`. If s is:

- the string `"left"`: from the left (below).
- the string `"right`: from the right (above).
- not given: from both sides if a is finite.

simplify (*expression*)

simplify (f) performs the following simplifications on f :

- rewrites trigs and hyperbolic trigs in terms of sin, cos, $sinh$, $cosh$.
- rewrites sin^2 and $sinh^2$ in terms of cos and $cosh$.
- rewrites $e^a e^b$ as e^{a+b}.

simplifyExp (*expression*)

simplifyExp (f) converts every product $e^a e^b$ appearing in f into e^{a+b}.

simpson (*floatFunction, fourFloats, threeIntegers*)
simpsonClosed (*floatFunction, fourFloats, twoIntegers*)
simpsonOpen (*floatFunction, fourFloats, twoIntegers*)

simpson $(fn, a, b, epsrel, epsabs, nmin, nmax, nint)$ uses the adaptive simpson method to numerically integrate function fn over the closed interval from a to b, with relative accuracy $epsrel$ and absolute accuracy $epsabs$; the refinement levels for the checking of convergence vary from $nmin$ to $nmax$. The method is called "adaptive" since it requires an additional parameter $nint$ giving the number of subintervals over which the integrator independently applies the convergence criteria using $nmin$ and $nmax$. This is useful when a large number of points are needed only in a small fraction of the entire interval. Parameter fn is a function of type Float → Float; a, b, $epsrel$, and $epsabs$ are floats; $nmin$, $nmax$, and $nint$ are integers. The operation returns a record containing: `value`, an estimate of the integral; `error`, an estimate of the error in the computation; `totalpts`, the total integral number of function evaluations, and `success`, a boolean value which is $true$ if the integral was computed within the user specified error criterion. See NumericalQuadrature for details.

simpsonClosed $(fn, a, b, epsrel, epsabs, nmin, nmax)$ similarly uses the Simpson method to numerically integrate function fn over the closed interval a to b, but is not adaptive.

simpsonOpen $(fn, a, b, epsrel, epsabs, nmin, nmax)$ is similar to **simpsonClosed**, except that it integrates function fn over the open interval from a to b.

sin (*expression*)

Argument x can be a Complex, Float, SmallFloat, or Expression value or a series.
sin (x) returns the sine of x if possible, and calls **error** otherwise.
sinIfCan (x) returns **sin** (x) if possible, and `"failed"` otherwise.

sin2csc (*expression*)

sin2csc (f) converts every **sin** (u) appearing in f into $1/\mathrm{csc}(u)$.

singular? (*polynomialOrFunctionField*)
singularAtInfinity? ()

singular? (p) tests whether $p(x) = 0$ is singular.
singular? $(a)\$F$ tests if $x = a$ is a singularity of the algebraic function field F (a domain of FunctionFieldCategory).
singularAtInfinity? $()\$F$ tests if the algebraic function field F has a singularity at infinity.

sinh (*expression*)

sinhIfCan (*expression*)

Argument x can be a Complex, Float, SmallFloat, or Expression value or a series.
sinh (x) returns the hyperbolic sine of x if possible, and calls **error** otherwise.
sinhIfCan (x) returns **sinh** (x) if possible, and "failed" otherwise.

sinh2csch (*expression*)

sinh2csch (f) converts every **sinh** (u) appearing in f into $1/\text{csch}(u)$.

size ()

size ()$F returns the number of elements in the domain of category Finite. By definition, each such domain must have a finite number of elements. See also FreeAbelianMonoidCategory.

size? (*aggregate, nonNegativeInteger*)

size? (a, n) tests if aggregate a has exactly n elements.

sizeLess? (*element, element*)

sizeLess? (x, y) tests whether x is strictly smaller than y with respect to the **euclideanSize**.

sizeMultiplication ()

sizeMultiplication ()$F returns the number of entries in the multiplication table of the field. Note: The time of multiplication of field elements depends on this size.

skewSFunction (*listOfIntegers, listOfIntegers*)

skewSFunction (li_1, li_2) is the S-function of the partition difference $li_1 - li_2$, expressed in terms of power sum symmetric functions. See CycleIndicators for details.

solve (*u, v*[*, w*])

eq denotes an equation to be solved; alternatively, an expression u may be given for *eq* in which case the equation *eq* is defined as $u = 0$.

leq denotes a list $[eq_1 \ldots eq_n]$ of equations; alternatively, a list of expressions $[u_1 \ldots u_n]$ may be given for *leq* in which case the equations eq_i are defined by $u_i = 0$.

epsilon is either a rational number or a float.

complexSolve $(eq, epsilon)$ finds all the real solutions to precision *epsilon* of the univariate equation *eq* of rational functions with respect to the unique variable appearing in *eq*. The complex solutions are either expressed as rational numbers or floats depending on the type of *epsilon*.

complexSolve $([eq_1 \ldots eq_n], epsilon)$ computes the real solutions to precision *epsilon* of a system of equations eq_i involving rational functions. The complex solutions are either expressed as rational numbers or floats depending on the type of *epsilon*.

radicalSolve $(eq[, x])$ finds solutions expressed in terms of radicals of the equation *eq* involving rational functions. Solutions will be found with respect to a Symbol given as a second argument to the operation. This second argument may be omitted when *eq* contains a unique symbol.

radicalSolve (leq, lv) finds solutions expressed in terms of radicals of the system of equations *leq* involving rational functions. Solutions are found with respect to a list *lv* of Symbols, or with respect to all variables appearing in the equations, if no second argument is given.

solve $(eq[, x])$ finds exact symbolic solutions to equation *eq* involving either rational functions or expressions of type Expression(R). Solutions will be found with respect to a Symbol given as a second argument to the operation. The second argument may be omitted when *eq* contains a unique symbol.

solve (leq, lv) finds exact solutions to a system of equations *leq* involving rational functions or expressions of type **Expression** (R). Solutions are found with respect to a list of *lv* of Symbols, or with respect to all variables appearing in the equations if no second argument is given.

solve $(eq, epsilon)$ finds all the real solutions to precision *epsilon* of the univariate equation *eq* of rational functions with respect to the unique variable appearing in *eq*. The real solutions are either expressed as rational numbers or floats depending on the type of *epsilon*.

solve $([eq_1 \ldots eq_n], epsilon)$ computes the real solutions to precision *epsilon* of a system of equations eq_i involving rational functions. The real solutions are either expressed as rational numbers or floats depending on the type of *epsilon*.

solve (M, v), where M is a matrix and v is a Vector of coefficients, finds a particular solution of the system $Mx = v$ and a basis of the associated homogeneous system $MX = 0$.

solve $(eq, y, x = a, [y_0 \ldots y_m])$ returns either the solution of the initial value problem *eq*, $y(a) = y_0$, $y'(a) = a_1, \ldots$ or "failed" if no solution can be found. Note: an error occurs if the equation *eq* is not a linear ordinary equation or of the form $dy/dx = f(x, y)$.

solve (eq, y, x) returns either a solution of the ordinary diffential equation *eq* or "failed" if no non-trivial solution can be found. If *eq* is a linear ordinary differential equation, a solution is of the form $[h, [b_1, \ldots,]]$ where h is a particular solution and $[b_1, \ldots, b_m]$ are linearly independent solutions of the associated homogeneous equation $f(x, y) = 0$. The value returned is a basis for the solution of the homogeneous equation which are found (note: this is not always a full basis).

See also **dioSolve**, **contractSolve**, **polSolve**, **seriesSolve**, **linSolve**.

solveLinearlyOverQ (*vector*)

solveLinearlyOverQ $([v_1, \ldots, v_n], u)$ returns $[c_1, \ldots, c_n]$ such that $c_1 v_1 + \cdots + c_n v_n = u$, or "failed" if no such rational numbers c_i exist. The elements of the v_i and u can be from any extension ring with an explicit linear dependence test, for example, expressions, complex values, polynomials, rational functions, or exact numbers. See LinearExplicitRingOver.

solveLinearPolynomialEquation (*listOfPolys, poly*)

solveLinearPolynomialEquation $([f_1, \ldots, f_n], g)$, where g is a polynomial and the f_i are polynomials relatively prime to one another, returns a list of polynomials a_i such that $g/\prod_i f_i = \sum_i ai/fi$, or "failed" if no such list of a_i's exists.

sort (*[predicate,]aggregate*)
sort! (*[predicate,]aggregate*)

sort $([p,]a)$ returns a copy of a sorted using total ordering predicate p.
sort! $([p,]u)$ returns u destructively changed with its elements ordered by comparison function p.
By default, p is the operation $\leq$. Thus both **sort** (u) and **sort!** (u) returns u with its elements in ascending order.
Also: **sort** (lp) sorts a list of permutations *lp* according to cycle structure, first according to the length of cycles, second, if S has Finite or S has OrderedSet, according to lexicographical order of entries in cycles of equal length.

spherical (*point*)

spherical (pt) transforms point *pt* from spherical coordinates to Cartesian coordinates, mapping (r, θ, ϕ) to $x = r\sin(\phi)\cos(\theta)$, $y = r\sin(\phi)\sin(\theta)$, $z = r\cos(\phi)$.

split (*element, binarySearchTree*)

split (x, t) splits binary search tree t into two components, returning a record of two components: *less*, a binary search tree whose components are all less than x; and, *greater*, a binary search tree with all the rest of the components of t.

split! (*aggregate, integer*)

split! (u, n) splits u into two aggregates: the first consisting of v, the first n elements of u, and w consisting of all the rest. The value of w is returned. Thus $v = \mathbf{first}(u, n)$ and $w := \mathbf{rest}(u, n)$. Note: afterwards **rest** (u, n) returns **empty** ().

splitDenominator (*listOfFractions*)

splitDenominator (u), where u is a list of fractions $[q_1, \ldots, q_n]$, returns $[[p_1, \ldots, p_n], d]$ such that $q_i = p_i/d$ and d is a common denominator for the q_i's. Similarly, the function is defined for a matrix (respectively, a polynomial) u in which case the q_i are the elements of (respectively, the coefficients of) u.

sqfrFactor (*element, integer*)

sqfrFactor $(base, exponent)$ creates a factored object with a single factor whose *base* is asserted to be square-free (flag = "sqfr").

sqrt (*expression[, option]*)

sqrt (x) returns the square root of x.
sqrt (x, y), where x and y are p-adic integers, returns a square root of x where argument y is a square root of x **mod** p. See also PAdicIntegerCategory.

square? (*matrix*)

square? (m) tests if m is a square matrix.

squareFree (*element*)

squareFree (x) returns the square-free factorization of x, that is, such that the factors are pairwise relatively prime and each has multiple prime factors. Argument x can be a member of any domain of category UniqueFactorizationDomain such as a polynomial or integer.

squareFreePart (*element*)

squareFreePart (p) returns product of all the prime factors of p each taken with multiplicity one. Argument p can be a member of any domain of category UniqueFactorizationDomain such as a polynomial or integer.

squareFreePolynomial (*polynomial*)

squareFreePolynomial (p) returns the square-free factorization of the univariate polynomial p.

squareTop (*matrix*)

squareTop (A) returns an n-by-n matrix consisting of the first n rows of the m-by-n matrix A. The operation calls **error** if $m < n$.

stack (*list*)

stack $([x, y, \ldots, z])$ creates a stack with first (top) element x, second element y, ..., and last element z.

standardBasisOfCyclicSubmodule (*listOfMatrices, vector*)

standardBasisOfCyclicSubmodule (lm, v) returns a matrix representation of cyclic submodule over a ring R, where lm is a list of matrices and v is a vector, such that the non-zero column vectors are an R-basis for Av. See RepresentationPackage2 using Browse.

stirling1 (*integer, integer*)
stirling2 (*integer, integer*)

stirling1 (n, m) returns the Stirling number of the first kind.
stirling2 (n, m) returns the Stirling number of the second kind.

string? (*various*)
string (*sExpression*)

string? (s) tests if SExpression object s is a string.
string (s) converts the symbol s to a string. An **error** is called if the symbol is subscripted.
string (s) returns SExpression object s as an element of String if possible, and otherwise calls **error**.

strongGenerators (*listOfPermutations*)

strongGenerators (*gp*) returns strong generators for the permutation group *gp*.

structuralConstants (*basis*)

structuralConstants ($basis$) calculates the structural constants $[(\gamma_{i,j,k})$ `for` k `in` $1..rank()\$R]$ of a domain R of category FramedNonAssociativeAlgebra over a ring R, defined by: $v_i v_j = \gamma_{i,j,1} v_1 + \cdots + \gamma_{i,j,n} v_n$, where $v_1, \ldots, v_n$ is the fixed R-module basis.

style (*string*)

style (s) specifies the drawing style in which the graph will be plotted by the indicated string s. This option is expressed in the form `style == s`.

sub (*outputForm, outputForm*)

sub (o_1, o_2), where o_1 and o_2 are objects of type OutputForm (normally unexposed), creates an output form for o_1 subscripted by o_2.

subMatrix (*matrix, integer, integer, integer, integer*)

subMatrix (m, i_1, i_2, j_1, j_2) extracts the submatrix $[m(i,j)]$ where the index i ranges from i_1 to i_2 and the index j ranges from j_1 to j_2.

submod (*integerNumber, integerNumber, integerNumber*)

submod (a, b, p), where $0 \leq a < b < p > 1$, returns $a - b$ **mod** p, for integer numbers a, b and p.

subResultantGcd (*polynomial, polynomial*)

subResultantGcd (p, q) computes the gcd of the polynomials p and q using the SubResultant GCD algorithm.

subscript (*symbol, listOfOutputForms*)

subscript $(s, [a1, \ldots, an])$ returns symbol s subscripted by output forms $a_1, \ldots, a_n$ as a symbol.

subset (*integer, integer, integer*)

subSet (n, m, k) calculates the kth m-subset of the set $0, 1, \ldots, (n-1)$ in the lexicographic order considered as a decreasing map from $0, \ldots, (m-1)$ into $0, \ldots, (n-1)$. See SymmetricGroupCombinatoricFunctions.

subset? (*set, set*)

subset? (u, v) tests if set u is a subset of set v.

subspace (*threeSpace*)

subspace (s) returns the space component which holds all the point information in the ThreeSpace object s.

substring? (*string, string, integer*)

substring? (s, t, i) tests if s is a substring of t beginning at index i. Note: `substring?(s, t, 0) = prefix?(s, t)`.

subst (*expression, equations*)

subst $(f, k = g)$ formally replaces the kernel k by g in f.
subst $(f, [k_1 = g_1, \ldots, k_n = g_n])$ formally replaces the kernels $k_1, \ldots, k_n$ by $g_1, \ldots, g_n$ in f.
subst $(f, [k_1, \ldots, k_n], [g_1, \ldots, g_n])$ formally replaces kernels k_i by g_i in f.

suchThat (*symbol, predicates*)

suchThat $(sy, pred)$ attaches the predicate $pred$ to symbol sy. Argument $pred$ may also be a list $[p_1, \ldots, p_n]$ of predicates p_i. In this case, the predicate $pred$ attached to sy is p_1 **and** ... **and** p_n.
suchThat $(r, [a_1, \ldots, a_n], f)$ returns the rewrite rule r with the predicate $f(a1, \ldots, an)$ attached to it.

suffix? (*string, string*)

suffix? (s, t) tests if the string s is the final substring of t.

sum (*rationalFunction, symbolOrSegmentBinding*)

sum $(a(n), n)$, where $a(n)$ is an rational function or expression involving the symbol n, returns the indefinite sum A of a with respect to upward difference on n, that is, $A(n+1) - A(n) = a(n)$.
sum $(f(n), n = a..b)$, where $f(n)$, a, and b are rational functions (or polynomials), computes and returns the sum $f(a) + f(a+1) + \cdots + f(b)$ as a rational function (or polynomial).

summation (*expression, segmentBinding*)

summation $(f, n = a..b)$ returns the formal sum $\sum_{n=a}^{b} f(n)$.

sumOfDivisors (*integer*)

sumOfDivisors (n) returns the sum of the integers between 1 and integer n (inclusive) which divide n. This sum is often denoted in the literature by $\sigma(n)$.

sumOfKthPowerDivisors (*integer, nonNegativeInteger*)

sumOfKthPowerDivisors (n, k) returns the sum of the kth powers of the integers between 1 and n (inclusive) which divide n. This sum is often denoted in the literature by $\sigma_k(n)$.

sumSquares (*integer*)

sumSquares (p) returns the list $[a, b]$ such that $a^2 + b^2$ is equal to the integer prime p, and calls **error** if this is not possible. It will succeed if p is 2 or congruent to 1 **mod** 4.

sup (*element, element*)

sup (x, y) returns the least element from which both x and y can be subtracted. The purpose of **sup** is to act as a supremum with respect to the partial order imposed by the $-$ operation on the domain. See OrderedAbelianMonoidSup for details.

super (*outputForm, outputForm*)

super (o_1, o_2), where o_1 and o_2 are objects of type

OutputForm (normally unexposed), creates an output form for o_1 superscripted by o_2.

superscript (*symbol, listOfOutputForms*)

superscript $(s, [a_1, \ldots, a_n])$ returns symbol s superscripted by output forms $[a_1, \ldots, a_n]$.

supersub (*outputForm, listOfOutputForms*)

supersub (o, lo), where o is an object of type OutputForm (normally unexposed) and lo is a list of output forms of the form $[sub_1, super_1, sub_2, super_2, \ldots, sub_n, super_n]$ creates an output form with each subscript aligned under each superscript.

surface (*function, function, function*)

surface (c_1, c_2, c_3) creates a surface from three parametric component functions c_1, c_2, and c_3.

swap! (*aggregate, index, index*)

swap! (u, i, j) interchanges elements i and j of aggregate u. No meaningful value is returned.

swapColumns! (*matrix, integer, integer*)

swapColumns! (m, i, j) interchanges the ith and j^{th} columns of m returning m which is destructively altered.

swapRows! (*matrix, integer, integer*)

swapRows! (m, i, j) interchanges the ith and j^{th} rows of m, returning m which is destructively altered.

symbol? (*sExpression*)

symbol? (s) tests if SExpression object s is a symbol.

symbol (*sExpression*)

symbol (s) returns s as an element of type Symbol, or calls **error** if this is not possible.

symmetric? (*matrix*)

symmetric? (m) tests if the matrix m is square and symmetric, that is, if each $m(i,j) = m(j,i)$.

symmetricDifference (*set, set*)

symmetricDifference (u, v) returns the set aggregate of elements x which are members of set aggregate u or set aggregate v but not both. If u and v have no elements in common, **symmetricDifference** (u, v) returns a copy of u. Note: $symmetricDifference(u, v) =$ **union(difference**(u, v)**, difference**(v, u)**)**

symmetricGroup (*integers*)

symmetricGroup (n) constructs the symmetric group S_n acting on the integers $1, \ldots, n$. The generators are the n-cycle $(1, \ldots, n)$ and the 2-cycle $(1, 2)$.
symmetricGroup (li), where li is a list of integers, constructs the symmetric group acting on the integers in the list li. The generators are the cycle given by li and the 2-cycle $(li(1), li(2))$. Duplicates in the list will be removed.

symmetricRemainder (*integer, integer*)

symmetricRemainder (a, b), where $b > 1$, yields r where $-b/2 \le r < b/2$.

symmetricTensors (*matrices, positiveInteger*)

symmetricTensors (la, n), where la is a list $[a_1, \ldots, a_k]$ of m-by-m square matrices, applies to each matrix a_i, the irreducible, polynomial representation of the general linear group GL_m corresponding to the partition $(n, 0, \ldots, 0)$ of n.

systemCommand (*string*)

systemCommand (cmd) takes the string cmd and passes it to the runtime environment for execution as a system command. Although various things may be printed, no usable value is returned.

tableau (*listOfListOfElements*)

tableau (ll) converts a list of lists ll to an object of type Tableau.

tableForDiscreteLogarithm (*integer*)

tableForDiscreteLogarithm (n) returns a table of the discrete logarithms of a^0 up to a^{n-1} which, when called with the key **lookup** (a^i), returns i for i in $0..n-1$ for a finite field. This operation calls **error** if not called for prime divisors of order of multiplicative group.

table (*[listOfRecords]*)

table $([p_1, p_2, \ldots, p_n])$ creates a table with keys of type Key and entries of type $Entry$. Each pair p_i is a record with selectors key and $entry$ with values from the corresponding domains Key and $Entry$.
table $()\$T$ creates a empty table of domain T of category TableAggregate.

tail (*aggregate*)

tail (a) returns the last node of recursive aggregate a.

tan (*expression*)
tanIfCan (*expression*)

Argument x can be a Complex, Float, SmallFloat, or Expression value or a series.
tan (x) returns the tangent of x.
tanIfCan (x) returns **tan** (x) if possible, and `"failed"` otherwise.

tan2cot (*expression*)

tan2cot (f) converts every **tan** (u) appearing in f into $1/\cot(u)$.

tan2trig (*expression*)

tan2trig (f) converts every **tan** (u) appearing in f into **sin** $(u)/\cos(u)$.

tanh (*expression*)
tanhIfCan (*expression*)

Argument x can be a Complex, Float, SmallFloat, or Expression value or a series.
tanh (x) returns the hyperbolic tangent of x.
tanhIfCan (x) returns **tanh** (x) if possible, and `"failed"` otherwise.

tanh2coth (*expression*)

tanh2coth (f) converts every **tanh** (u) appearing in f into $1/\mathrm{coth}(u)$.

tanh2trigh (*expression*)

tanh2trigh (f) converts every **tanh** (u) appearing in f into **sinh** $(u)/\mathrm{cosh}(u)$.

taylor (*various, ..*)

taylor (u) converts the Laurent series $u(x)$ to a Taylor series if possible, and if not, calls **error**.
taylor (f) converts the expression f into a Taylor expansion of the expression f. Note: f must have only one variable.
taylor (sy), where sy is a symbol, returns sy as a Taylor series.
taylor $(n+->a(n), x=a)$ returns $\sum_{n=0...} a(n)(x-a)^n)$.
taylor $(f, x=a[, n])$ expands the expression f as a series in powers of $(x-a)$ with n terms. If n is missing, the number of terms is governed by the value set by the system command)set streams calculate.
taylor $(i+->a(i), x=a, m..[n,k])$ creates the Taylor series $\sum_{i=m..n \text{ by } k} a(i)(x-a)^i$. Here m, n and k are integers. Upper-limit n and stepsize k are optional and have default values $n=\infty$ and $k=1$.
taylor $(a(i), i, x=a, m..[n,k])$ returns $\sum_{i=m..n\mathbf{by}k} a(n)(x-a)^n$.

taylorIfCan (*laurentSeries*)

taylorIfCan $(f(x))$ converts the Laurent series $f(x)$ to a Taylor series if possible, and returns `"failed"` if this is not possible.

taylorRep (*laurentSeries*)

taylorRep $(f(x))$ returns $g(x)$, where $f=x^n g(x)$ is represented by $[n, g(x)]$.

tensorProduct (*listOfMatrices*[, *listOfMatrices*])

tensorProduct $([a_1, \ldots, a_k][, [b_1, \ldots, b_k]])$ calculates the list of Kronecker products of the matrices a_i and b_i for $1 \leq i \leq k$. If a second argument is missing, the b_i is defined as the corresponding a_i. Also, **tensorProduct** (m), where m is a matrix, is defined as **tensorProduct** $([m],[m])$. Note: If each list of matrices corresponds to a group representation (representation of generators) of one group, then these matrices correspond to the tensor product of the two representations.

terms (*various*)

terms (s) returns a stream of the non-zero terms of series s. Each term is returned as a record with selectors k and c, which correspond to the exponent and coefficient, respectively. The terms are ordered by increasing order of exponents.
terms (m), where m is a free abelian monoid of the form $e_1 a_1 + \cdots + e_n a_n$, returns $[[a_1, e_1], \ldots, [a_n, e_n]]$. See FreeAbelianMonoidCategory.

tex (*formattedObject*)

tex (t) extracts the TeX section of a TeX formatted object t.

third (*aggregate*)

third (u) returns the third element of a recursive aggregate u. Note: **third** $(u) = first(rest(rest(u)))$.

title (*string*)

title (s) specifies string s as the title for a plot. This option is expressed as a option to the **draw** command in the form `title == s`.

top (*stack*)
top! (*dequeue*)

top (s) returns the top element x from s.
top! (d) returns the element at the top (front) of the dequeue.

toroidal (*value*)

toroidal $(element)$ transforms from toroidal coordinates to Cartesian coordinates: **toroidal** (a) is a function that maps the point (u, v, ϕ) to $x = \mathrm{asinh}(v)\cos(\phi)/(\cosh(v)-\cos(u))$, $y = \mathrm{asinh}(v)\sin(\phi)/(\cosh(v)-\cos(u))$, $z = \mathrm{asin}(u)/(\cosh(v)-\cos(u))$.

toScale (*boolean*)

toScale (b) specifies whether or not a plot is to be drawn to scale. This command may be expressed as an option to the **draw** command in the form $toScale == b$.

totalDegree (*polynomial, listOfVariables*)

totalDegree $(p[, lv])$ returns the maximum sum (over all monomials of polynomial p) of the variables in the list lv. If a second argument is missing, lv is defined to be all the variables appearing in p.

totalfract (*polynomial*)

totalfract (prf) takes a polynomial whose coefficients are themselves fractions of polynomials and returns a record containing the numerator and denominator resulting from putting prf over a common denominator.

totalGroebner (*listOfPolynomials, listOfVariables*)

totalGroebner (lp, lv) computes the Gröbner basis for the list of polynomials lp with the terms ordered first by total degree and then refined by reverse lexicographic ordering. The variables are ordered by their position in the list lv.

tower (*expression*)

tower (f) returns all the kernels appearing in f, regardless of level.

trace (*various, ..*)

trace (m) returns the trace of the matrix m, that is, the sum of its diagonal elements.
trace (a) returns the trace of the regular representation of a, an element of an algebra of finite rank. See FiniteRankAlgebra.
trace $(a[, d])$, where a is an element of a finite algebraic extension field, computes the trace of a with respect to the field of extension degree d over the ground field of size q. This operation calls **error** if d does not divide the extension degree of a. The default value of d is 1. Note:
trace $(a, d) = \sum_{i=0}^{n/d} a^{q^{di}}$.

traceMatrix ([*basis*])

traceMatrix $([v1, .., vn])$ is the n-by-n matrix whose i, j element is $Tr(v_i v_j)$. If no argument is given, the v_i are assumed to be elements of the fixed basis.

tracePowMod (*poly, nonNegativeInteger, poly*)

tracePowMod (u, k, v) returns $\sum_{i=0}^{k} u^{2^i}$, all computed modulo the polynomial v.

transcendenceDegree ()

transcendenceDegree $()\$F$ returns the transcendence degree of the field extension F, or 0 if the extension is algebraic.

transcendent? (*element*)

transcendent? (a) tests whether an element a of a domain that is an extension field over a ground field F is transcendent with respect to F.

transpose (*matrix*[, *options*])

transpose (m) returns the transpose of the matrix m.

transpose $(t[, i, j])$ exchanges the i th and j th indices of t. For example, if $r =$ **transpose**$(t, 2, 3)$ for a rank four tensor t, then r is the rank four tensor given by $r(i, j, k, l) = t(i, k, j, l)$. If i and j are not given, they are assumed the first and last index of t.

tree (*value*[, *listOfChildren*])

tree (x, ls) creates an element of Tree with value x at the root node, and immediate children `ls` in left-to-right order.
tree (x) is equivalent to **tree** $(x, []\$List(S))$ where **x** has type `S`.

trapezoidal (*floatFunction, fourFloats, threeIntegers*)
trapezoidalClosed (*floatFunction, fourFloats, twoIntegers*)
trapezoidalOpen (*floatFunction, fourFloats, twoIntegers*)

trapezoidal $(fn, a, b, epsrel, epsabs, nmin, nmax, nint)$ uses the adaptive trapezoidal method to numerically integrate function fn over the closed interval from a to b, with relative accuracy $epsrel$ and absolute accuracy $epsabs$, where the refinement levels for the checking of convergence vary from $nmin$ to $nmax$. The method is called "adaptive" since it requires an additional parameter $nint$ giving the number of subintervals over which the integrator independently applies the convergence criteria using $nmin$ and $nmax$; this is useful when a large number of points are needed only in a small fraction of the entire interval. Parameter fn is a function of type Float → Float; a, b, $epsrel$, and $epsabs$ are floats; $nmin$, $nmax$, and $nint$ are integers. The operation returns a record containing: `value`, an estimate of the integral; `error`, an estimate of the error in the computation; `totalpts`, the total integral number of function evaluations, and `success`, a boolean value that is $true$ if the integral was computed within the user specified error criterion. See NumericalQuadrature for details.

$trapezoidalClosed(fn, a, b, epsrel, epsabs, nmin, nmax)$ similarly uses the trapezoidal method to numerically integrate function fn over the closed interval a to b, but is not adaptive.

$trapezoidalOpen(fn, a, b, epsrel, epsabs, nmin, nmax)$ is similar to **trapezoidalClosed**, except that it integrates function fn over the open interval from a to b.

triangularSystems (*listOfFractions, listOfSymbols*)

triangularSystems (lf, lv) solves the system of equations defined by lf with respect to the list of symbols lv; the system of equations is obtaining by equating to zero the list of rational functions lf. The result is a list of solutions where each solution is expressed as a "reduced" triangular system of polynomials.

trigs (*expression*)

trigs (f) rewrites all the complex logs and exponentials appearing in f in terms of trigonometric functions.

trim (*string, characterOrCharacterClass*)

trim (s, c) returns s with all characters c deleted from right and left ends. For example, `trim(" abc ", char " ")` returns `"abc"`. Argument c may also be a character class, in which case s is returned with all characters in cc deleted from right and left ends. For example, `trim("(abc)", charClass "()")` returns `"abc"`.

truncate (*various*[, *options*])

truncate (x) returns the integer between x and 0 closest to x.
truncate $(f, m[, n])$ returns a (finite) power series consisting of the sum of all terms of f of degree d with $n \leq d \leq m$. Upper bound m is ∞ by default.

tubePoints (*positiveInteger*)

tubePoints (n) specifies the number of points, n, defining the circle that creates the tube around a three-dimensional curve. The default is 6. This option is expressed in the form `tubePoints == n`.

tubePointsDefault ([*positiveInteger*])

tubePointsDefault (i) sets the number of points to use when creating the circle to be used in creating a three-dimensional tube plot to i.
tubePointsDefault () returns the number of points to be used when creating the circle to be used in creating a three-dimensional tube plot.

tubeRadius (*float*)

tubeRadius (r) specifies a radius r for a tube plot around a three-dimensional curve. This operation may be expressed as an option to the **draw** command in the form `tubeRadius == r`.

tubeRadiusDefault ([*float*])

tubeRadiusDefault (r) sets the default radius for a three-dimensional tube plot to r.
tubeRadiusDefault () returns the radius used for a three-dimensional tube plot.

twist ()

twist (f), where f is a function of type $(A, B)C$, is the function g such that $g(a, b) = f(b, a)$. See MappingPackage for related functions.

unary? (*basicOperator*)

unary? (*op*) tests if basic operator *op* is unary, that is, takes exactly one argument.

union (*set, elementOrSet*)

union (u, x) returns the set aggregate u with the element x added. If u already contains x, **union** (u, x) returns a copy of x.
union (u, v) returns the set aggregate of elements that are members of either set aggregate u or v. See also Multiset.

unit ([*various*])

unit () returns a unit of the algebra (necessarily unique), or `"failed"` if there is none.
unit (u) extracts the unit part of the factored object u.
unit (l) marks off the units on a viewport according to the indicated list l. This option is expressed in the draw command in the form `unit ==`$[f_1, f_2]$.

unit? (*element*)

unit? (x) tests whether x is a unit, that is, if x is invertible.

unitCanonical (*element*)

unitCanonical (x) returns **unitNormal** $(x).canonical$.

unitNormalize (*factored*)

unitNormalize (u) normalizes the unit part of the factorization. For example, when working with factored integers, this operation ensures that the bases are all positive integers.

unitNormal (*element*)

unitNormal (x) tries to choose a canonical element from the associate class of x. If successful, it returns a record with three components "unit", "canonical" and "associate". The attribute `canonicalUnitNormal`, if asserted, means that the "canonical" element is the same across all associates of x. If **unitNormal** $(x) = [u, c, a]$ then $ux = c$, $au = 1$.

unitsColorDefault ([*palette*])

unitsColorDefault (p) sets the default color of the unit ticks in a two-dimensional viewport to the palette p.
unitsColorDefault () returns the default color of the unit ticks in a two-dimensional viewport.

unitVector (*positiveInteger*)

unitVector (n) produces a vector with 1 in position n and zero elsewhere.

univariate (*polynomial*[, *variable*])

univariate $(p[, v])$ converts the multivariate polynomial p into a univariate polynomial in v whose coefficients are multivariate polynomials in all the other variables. If v is omitted, then p must involve exactly one variable.

universe ()

universe ()$\$R$ returns the universal set for finite set aggregate R.

unparse (*inputForm*)

unparse (f) returns a string s such that the parser would transform s to f, or calls **error** if f is not the parsed form of a string.

unrankImproperPartitions0 (*integer, integer, integer*)

unrankImproperPartitions0 (n, m, k) computes the kth improper partition of nonnegative n in m nonnegative parts in reverse lexicographical order. Example: $[0, 0, 3] < [0, 1, 2] < [0, 2, 1] < [0, 3, 0] < [1, 0, 2] < [1, 1, 1] < [1, 2, 0] < [2, 0, 1] < [2, 1, 0] < [3, 0, 0]$. The operation calls **error** if k is negative or too big. Note: counting of subtrees is done by **numberOfImproperPartitions**.

unrankImproperPartitions1 (*integer, integer, integer*)

unrankImproperPartitions1 (n, m, k) computes the kth improper partition of nonnegative n in at most m nonnegative parts ordered as follows: first, in reverse lexicographical order according to their non-zero parts, then according to their positions (i.e. lexicographical order using $subSet$: $[3, 0, 0] < [0, 3, 0] < [0, 0, 3] < [2, 1, 0] < [2, 0, 1] < [0, 2, 1] < [1, 2, 0] < [1, 0, 2] < [0, 1, 2] < [1, 1, 1]$). Note: counting of subtrees is done by **numberOfImproperPartitionsInternal**.

unravel (*listOfElement*)

unravel (t) produces a tensor from a list of components such that **unravel** (**ravel**$(t)) = t$.

upperCase (*string*)
upperCase? (*string*)
upperCase! (*string*)

upperCase! (s) destructively replaces the alphabetic characters in s by upper case characters.
upperCase () returns the class of all characters for which **upperCase?** is $true$.
upperCase (c) converts a lower case letter c to the corresponding upper case letter. If c is not a lower case letter, then it is returned unchanged.
upperCase (s) returns the string with all characters in upper case.
upperCase? (c) tests if c is an upper case letter, that is, one of A..Z.

validExponential (*listOfKernels, expression, symbol*)

validExponential $([k_1, \ldots, k_n], f, x)$ returns g if **exp** $(f) = g$ and g involves only $k_1 \ldots k_n$, and `"failed"` otherwise.

value (*recursiveAggregate*)

value (a) returns the "value" part of a recursive aggregate `a`, typically the root of tree. See, for example, BinaryTree.

var1Steps (*positiveInteger*)

var1Steps (n) indicates the number of subdivisions n of the first range variable. This command may be expressed as an option to the **draw** command in the form **var1Steps** $== n$.

var1StepsDefault ([*positiveInteger*])

var1StepsDefault () returns the current setting for the number of steps to take when creating a three-dimensional mesh in the direction of the first defined free variable (a free variable is considered defined when its range is specified (that is, $x = 0..10$)).
var1StepsDefault (i) sets the number of steps to take when creating a three-dimensional mesh in the direction of the first defined free variable to i (a free variable is considered defined when its range is specified (that is, $x = 0..10$)).

var2Steps (*positiveInteger*)

var2Steps (n) indicates the number of subdivisions, n, of the second range variable. This option is expressed in the form **var2Steps** $== n$.

var2StepsDefault ([*positiveInteger*])

variable (*various*)

variable (f) returns the (unique) power series variable of the power series f.
variable $(segb)$ returns the variable from the left hand side of the SegmentBinding $segb$. For example, if $segb$ is $v = a..b$, then **variable** $(segb)$ returns v.
variable (v) returns s if v is any derivative of the differential indeterminate s.

variables (*expression*)

variables (f) returns the list of all the variables of expression, polynomial, rational function, or power series f.

vconcat (*outputForms*[, *OutputForm*] *(normally unexposed)*)

vconcat (o_1, o_2), where o_1 and o_2 are objects of type OutputForm (normally unexposed), returns an output form for the vertical concatenation of forms o_1 and o_2.
vconcat (lo), where lo is a list of objects of type OutputForm (normally unexposed), returns an output form for the vertical concatenation of the elements of lo.

vector (*listOfElements*)

vector (l) converts the list l to a vector.

vectorise (*polynomial, nonNegativeInteger*)

vectorise (p, n) returns $[a_0, \ldots, a_{n-1}]$ where $p = a_0 + a_1 x + \cdots + a_{n-1} x^{n-1}$ + higher order terms. The degree of polynomial p can be different from $n - 1$.

vertConcat (*matrix, matrix*)

vertConcat (x, y) vertically concatenates two matrices with an equal number of columns. The entries of y appear below the entries of x.

viewDefaults ()

viewDefaults () resets all the default graphics settings.

viewPosDefault ([*listOfNonNegativeIntegers*])

viewPosDefault $([x, y])$ sets the default X and Y position of a viewport window. Unless overridden explicitly, newly created viewports will have the X and Y coordinates x, y.
viewPosDefault () returns the default X and Y position of a viewport window unless overridden explicitly, newly created viewports will have these X and Y coordinate.

viewSizeDefault ([*listOfPositiveIntegers*])

viewSizeDefault $([w, h])$ sets the default viewport width to w and height to h.

viewWriteAvailable ()

viewWriteAvailable () returns a list of available methods for writing, such as BITMAP, POSTSCRIPT, etc.

viewWriteDefault (*listOfStrings*)

viewWriteDefault () returns the list of things to write in a viewport data file; a viewAlone file is always generated.
viewWriteDefault (l) sets the default list of things to write in a viewport data file to the strings in l; a viewAlone file is always generated.

void ()

void () produces a void object.

weakBiRank (*element*)

weakBiRank (x) determines the number of linearly independent elements in the $b_i x b_j$, $i, j = 1, \ldots, n$, where $b = [b_1, \ldots, b_n]$ is the fixed basis of a domain of category FramedNonAssociativeAlgebra.

weight (u)

weight (u) returns

- if u is a differential polynomial: the maximum weight of all differential monomials appearing in the differential polynomial u.
- if u is a derivative: the weight of the derivative u.
- if u is a basic operator: the weight attached to u.

weight (p, s) returns the maximum weight of all differential monomials appearing in the differential polynomial p when p is viewed as a differential polynomial in the differential indeterminate s alone.
weight (op, n) attaches the weight n to op.

weights (*differentialPolynomial, differentialIndeterminated*)

weights (p, s) returns a list of weights of differential monomials appearing in the differential polynomial p when p is viewed as a differential polynomial in the differential indeterminate s alone. If s is missing, a list of weights of differential monomials appearing in differential polynomial p.

whatInfinity (*orderedCompletion*)

whatInfinity (x) returns 0 if x is finite, 1 if x is ∞, and -1 if x is $-\infty$.

wholePart (*various*)

wholePart (x) returns the whole part of the fraction x, that is, the truncated quotient of the numerator by the denominator.
wholePart (x) extracts the whole part of x. That is, if $x = \textbf{continuedFraction}(b_0, [a_1, a_2, \ldots], [b_1, b_2, \ldots])$, then **wholePart** $(x) = b_0$.
wholePart (p) extracts the whole part of the partial fraction p.

wholeRadix (*listOfIntegers*)

wholeRadix (l) creates an integral radix expansion from a list of ragits. For example, **wholeRadix** $([1, 3, 4])$ returns 134.

wholeRagits (*listOfIntegers*)

wholeRagits (rx) returns the ragits of the integer part of a radix expansion.

wordInGenerators (*permutation, permutationGroup*)

wordInGenerators (p, gp) returns the word for the permutation p in the original generators of the permutation group *gp*, represented by the indices of the list, given by **generators**.

wordInStrongGenerators (*permutation, permutationGroup*)

wordInStrongGenerators (p, gp) returns the word for the permutation p in the strong generators of the permutation group *gp*, represented by the indices of the list, given by **strongGenerators**.

wordsForStrongGenerators (*listOfListsOfIntegers*)

wordsForStrongGenerators (gp) returns the words for the strong generators of the permutation group *gp* in the original generators of *gp*, represented by their indices in the list of nonnegative integers, given by **generators**.

wreath (*symmetricPolynomial, symmetricPolynomial*)

wreath (s_1, s_2) is the cycle index of the wreath product of the two groups whose cycle indices are s_1 and s_2, symmetric polynomials with rational number coefficients.

writable? (*file*)

writable? (f) tests if the named file can be opened for writing. The named file need not already exist.

write! (*file, value*)

write! (f, s) puts the value s into the file f. The state of f is modified so that subsequent calls to **write!** will append values to the end of the file.

writeLine! (*textfile*[, *string*])

writeLine! (f) finishes the current line in the file f. An empty string is returned. The call **writeLine!** (f) is equivalent to **writeLine!** $(f, "")$.
writeLine! (f, s) writes the contents of the string s and finishes the current line in the file f. The value of s is returned.

xor (*boolean, boolean*)

xor (a, b) returns the logical *exclusive-or* of booleans or bit aggregates a and b.
xor (n, m) returns the bit-by-bit logical *xor* of the small integers n and m.

xRange (*curve*)

xRange (c) returns the range of the x-coordinates of the points on the curve c.

yCoordinates (*function*)

yCoordinates (f), where f is a function defined over a curve, returns the coordinates of f with respect to the natural basis for the curve. Specifically, the operation returns $[[a_1, \ldots, a_n], d]$ such that $f = (a_1 + \ldots + a_n y^{n-1})/d$.

yellow ()

yellow () returns the position of the yellow hue from total hues.

youngGroup (*various*)

youngGroup $([n_1, \ldots, n_k])$ constructs the direct product of the symmetric groups $Sn_1, \ldots, Sn_k$.
youngGroup $(lambda)$ constructs the direct product of the symmetric groups given by the parts of the partition *lambda*.

yRange (*curve*)

yRange (c) returns the range of the y-coordinates of the points on the curve *c*.

zag (*outputForm, outputForm*)

zag (o_1, o_2), where o_1 and o_2 are objects of type OutputForm (normally unexposed), return an output form displaying the continued fraction form for o_2 over o_1.

zero (*nonNegativeInteger*[, *nonNegativeInteger*])

zero (n) creates a zero vector of length n.
zero (m, n) returns an m-by-n zero matrix.

zero? (*element*)

zero? (x) tests if x is equal to 0.

zeroDim? (*ideal*)

zeroDim? (I) tests if the ideal I is zero dimensional, that is, all its associated primes are maximal.

zeroDimPrimary? (*ideal*)

zeroDimPrimary? (I) tests if the ideal I is 0-dimensional primary.

zeroDimPrime? (*ideal*)

zeroDimPrime? (I) tests if the ideal I is a 0-dimensional prime.

zeroOf (*polynomial*[, *symbol*])

zeroOf $(p[, y])$ returns y such that $p(y) = 0$. If possible, y is expressed in terms of radicals. Otherwise it is an implicit algebraic quantity that displays as $'y$. If no second argument is given, then p must have a unique variable y.

zerosOf (*polynomial*[, *symbol*])

zerosOf (p, y) returns $[y_1, \ldots, y_n]$ such that $p(y_i) = 0$. The y_i's are expressed in radicals if possible. Otherwise they are implicit algebraic quantities that display as y_i. The returned symbols $y_1, \ldots, y_n$ are bound in the interpreter to respective root values. If no second argument is given, then p must have a unique variable y.

zRange (*curve*)

zRange (c) returns the range of the z-coordinates of the points on the curve *c*.

APPENDIX F

Programs for AXIOM Images

This appendix contains the AXIOM programs used to generate the images in the AXIOM Images color insert of this book. All these input files are included with the AXIOM system. To produce the images on page 6 of the AXIOM Images insert, for example, issue the command:

```
)read images6
```

These images were produced on an IBM RS/6000 model 530 with a standard color graphics adapter. The smooth shaded images were made from X Window System screen dumps. The remaining images were produced with AXIOM-generated PostScript output. The images were reproduced from slides made on an Agfa ChromaScript PostScript interpreter with a Matrix Instruments QCR camera.

F.1 images1.input

Read torus knot program.

A (15,17) torus knot.

```
)read tknot                                   1
                                              2
torusKnot(15,17, 0.1, 6, 700)                 3
```

F.2 images2.input

These images illustrate how Newton's method converges when computing the complex cube roots of 2. Each point in the (x, y)-plane represents the complex number $x + iy$, which is given as a starting point for Newton's method. The poles in these images represent bad starting values. The flat areas are the regions of convergence to the three roots.

Read the programs from Chapter 10.
Create a Newton's iteration function for $x^3 = 2$.

```
)read newton                                                    1
)read vectors                                                   2
f := newtonStep(x**3 - 2)                                       3
                                                                4
```

The function f^n computes n steps of Newton's method.

Clip values with magnitude > 4.
The vector field for f^3
The surface for f^3
The surface for f^4

```
clipValue := 4                                                  5
drawComplexVectorField(f**3, -3..3, -3..3)                      6
drawComplex(f**3, -3..3, -3..3)                                 7
drawComplex(f**4, -3..3, -3..3)                                 8
```

F.3 images3.input

```
)r tknot                                                        1
for i in 0..4 repeat torusKnot(2, 2 + i/4, 0.5, 25, 250)        2
```

F.4 images5.input

The parameterization of the Etruscan Venus is due to George Frances.

```
venus(a,r,steps) ==                                             1
  surf := (u:SF, v:SF): Point SF +->                            2
    cv := cos(v)                                                3
    sv := sin(v)                                                4
    cu := cos(u)                                                5
    su := sin(u)                                                6
    x := r * cos(2*u) * cv + sv * cu                            7
    y := r * sin(2*u) * cv - sv * su                            8
    z := a * cv                                                 9
    point [x,y,z]                                               10
  draw(surf, 0..%pi, -%pi..%pi, var1Steps==steps,               11
       var2Steps==steps, title == "Etruscan Venus")             12
                                                                13
```

The Etruscan Venus

```
venus(5/2, 13/10, 50)                                           14
```

The Figure-8 Klein Bottle parameterization is from "Differential Geometry and Computer Graphics" by Thomas Banchoff, in *Perspectives in Mathematics,* Anniversary of Oberwolfasch 1984, Birkhäuser-Verlag, Basel, pp. 43-60.

```
klein(x,y) ==                                                   15
  cx := cos(x)                                                  16
  cy := cos(y)                                                  17
  sx := sin(x)                                                  18
  sy := sin(y)                                                  19
  sx2 := sin(x/2)                                               20
  cx2 := cos(x/2)                                               21
  sq2 := sqrt(2.0@SF)                                           22
  point [cx * (cx2 * (sq2 + cy) + (sx2 * sy * cy)), _           23
```

```
               sx * (cx2 * (sq2 + cy) + (sx2 * sy * cy)), _      24
               -sx2 * (sq2 + cy) + cx2 * sy * cy]                25
                                                                 26
```

Figure-8 Klein bottle

```
draw(klein, 0..4*%pi, 0..2*%pi, var1Steps==50,                   27
     var2Steps==50,title=="Figure Eight Klein Bottle")            28
```

The next two images are examples of generalized tubes.

Rotate a point p by θ around the origin.

```
)read ntube                                                      29
rotateBy(p, theta) ==                                            30
  c := cos(theta)                                                31
  s := sin(theta)                                                32
  point [p.1*c - p.2*s, p.1*s + p.2*c]                           33
                                                                 34
```

A circle in three-space.

```
bcircle t ==                                                     35
  point [3*cos t, 3*sin t, 0]                                    36
                                                                 37
```

An ellipse that twists around four times as `t` revolves once.

```
twist(u, t) ==                                                   38
  theta := 4*t                                                   39
  p := point [sin u, cos(u)/2]                                   40
  rotateBy(p, theta)                                             41
                                                                 42
```

Twisted Torus

```
ntubeDrawOpt(bcircle, twist, 0..2*%pi, 0..2*%pi,                 43
             var1Steps == 70, var2Steps == 250)                  44
                                                                 45
```

Create a twisting circle.

```
twist2(u, t) ==                                                  46
  theta := t                                                     47
  p := point [sin u, cos(u)]                                     48
  rotateBy(p, theta)                                             49
                                                                 50
```

Color function with 21 stripes.

```
cf(u,v) == sin(21*u)                                             51
                                                                 52
```

Striped Torus

```
ntubeDrawOpt(bcircle, twist2, 0..2*%pi, 0..2*%pi,                53
  colorFunction == cf, var1Steps == 168,                         54
  var2Steps == 126)                                              55
```

F.5 images6.input

The height and color are the real and argument parts of the Gamma function, respectively.

```
gam(x,y) ==                                                      1
  g := Gamma complex(x,y)                                        2
  point [x,y,max(min(real g, 4), -4), argument g]                3
                                                                 4
                                                                 5
```

The Gamma Function

```
draw(gam, -%pi..%pi, -%pi..%pi,                                  6
     title == "Gamma(x + %i*y)", _                               7
     var1Steps == 100, var2Steps == 100)                         8
                                                                 9
b(x,y) == Beta(x,y)                                              10
                                                                 11
```

The Beta Function

```
draw(b, -3.1..3, -3.1 .. 3, title == "Beta(x,y)")                12
                                                                 13
atf(x,y) ==                                                      14
  a := atan complex(x,y)                                         15
```

```
    point [x,y,real a, argument a]                                  16
                                                                    17
```

The Arctangent function

```
draw(atf, -3.0..%pi, -3.0..%pi)                                     18
```

F.6 images7.input

First we look at the conformal map $z \mapsto z + 1/z$.

Read program for drawing conformal maps.

```
)read conformal                                                     1
                                                                    2
                                                                    3
```

The coordinate grid for the complex plane.

```
f z == z                                                            4
                                                                    5
```

Mapping 1: Source

```
conformalDraw(f, -2..2, -2..2, 9, 9, "cartesian")                   6
                                                                    7
```

The map $z \mapsto z + 1/z$

```
f z == z + 1/z                                                      8
                                                                    9
```

Mapping 1: Target

```
conformalDraw(f, -2..2, -2..2, 9, 9, "cartesian")                   10
```

The map $z \mapsto -(z+1)/(z-1)$ maps the unit disk to the right half-plane, as shown on the Riemann sphere.

The unit disk.

```
f z == z                                                            11
                                                                    12
```

Mapping 2: Source

```
riemannConformalDraw(f,0.1..0.99,0..2*%pi,7,11,"polar")             13
                                                                    14
```

The map $x \mapsto -(z+1)/(z-1)$.

```
f z == -(z+1)/(z-1)                                                 15
                                                                    16
```

Mapping 2: Target

```
riemannConformalDraw(f,0.1..0.99,0..2*%pi,7,11,"polar")             17
                                                                    18
```

Riemann Sphere Mapping

```
riemannSphereDraw(-4..4, -4..4, 7, 7, "cartesian")                  19
```

F.7 images8.input

```
)read dhtri                                                         1
)read tetra                                                         2
```

Sierpinsky's Tetrahedron

```
drawPyramid 4                                                       3
                                                                    4
)read antoine                                                       5
```

Antoine's Necklace

```
drawRings 2                                                         6
                                                                    7
)read scherk                                                        8
```

Scherk's Minimal Surface

```
drawScherk(3,3)                                                     9
                                                                    10
)read ribbons                                                       11
```

Ribbon Plot

```
drawRibbons([x**i for i in 1..5], x=-1..1, y=0..2)                  12
```

F.8 conformal.input

The functions in this section draw conformal maps both on the plane and on the Riemann sphere.

```
C  := Complex SmallFloat                                  -- Complex Numbers   1
S  := Segment SmallFloat                                  -- Draw ranges       2
R3 := Point SF                                            -- Points in 3-space 3
                                                                               4
```

conformalDraw(*f, rRange, tRange, rSteps, tSteps, coord*) draws the image of the coordinate grid under *f* in the complex plane. The grid may be given in either polar or Cartesian coordinates. Argument *f* is the function to draw; *rRange* is the range of the radius (in polar) or real (in Cartesian); *tRange* is the range of θ (in polar) or imaginary (in Cartesian); *tSteps, rSteps*, are the number of intervals in the *r* and θ directions; and *coord* is the coordinate system to use (either `"polar"` or `"cartesian"`).

```
conformalDraw: (C -> C, S, S, PI, PI, String) -> VIEW3D  5
conformalDraw(f,rRange,tRange,rSteps,tSteps,coord) ==    6
  transformC :=                                          7
    coord = "polar" => polar2Complex                     8
    cartesian2Complex                                    9
  cm := makeConformalMap(f, transformC)                  10
  sp := createThreeSpace()                               11
  adaptGrid(sp, cm, rRange, tRange, rSteps, tSteps)      12
  makeViewport3D(sp, "Conformal Map")                    13
```

Margin notes: Function for changing an (x, y) pair into a complex number. (lines 7–9) Create a fresh space. (line 11) Plot the coordinate lines. (line 12) Draw the image. (line 13)

riemannConformalDraw(*f, rRange, tRange, rSteps, tSteps, coord*) draws the image of the coordinate grid under *f* on the Riemann sphere. The grid may be given in either polar or Cartesian coordinates. Its arguments are the same as those for **conformalDraw**.

```
riemannConformalDraw:(C->C,S,S,PI,PI,String)->VIEW3D        14
riemannConformalDraw(f, rRange, tRange,                     15
                     rSteps, tSteps, coord) ==              16
  transformC :=                                             17
    coord = "polar" => polar2Complex                        18
    cartesian2Complex                                       19
  sp := createThreeSpace()                                  20
  cm := makeRiemannConformalMap(f, transformC)              21
  adaptGrid(sp, cm, rRange, tRange, rSteps, tSteps)         22
  curve(sp,[point [0,0,2.0@SF,0],point [0,0,2.0@SF,0]])     23
  makeViewport3D(sp,"Map on the Riemann Sphere")            24
                                                            25
adaptGrid(sp, f, uRange, vRange,  uSteps, vSteps) ==        26
  delU := (hi(uRange) - lo(uRange))/uSteps                  27
  delV := (hi(vRange) - lo(vRange))/vSteps                  28
  uSteps := uSteps + 1; vSteps := vSteps + 1                29
  u := lo uRange                                            30
  for i in 1..uSteps repeat                                 31
    c := curryLeft(f,u)                                     32
    cf := (t:SF):SF +-> 0                                   33
    makeObject(c,vRange::SEG Float,colorFunction==cf,       34
      space == sp, tubeRadius == .02, tubePoints == 6)      35
    u := u + delU                                           36
```

Margin notes: Function for changing an (x, y) pair into a complex number. (lines 17–19) Create a fresh space. (line 20) Plot the coordinate lines. (line 22) Add an invisible point at the north pole for scaling. (line 23) Plot the coordinate grid using adaptive plotting for coordinate lines, and draw tubes around the lines. (line 26) Draw coordinate lines in the v direction; curve c fixes the current value of u. (line 31) Draw the v coordinate line. (line 34)

```
    v := lo vRange                                                        37
```

Draw coodinate lines in the u direction; curve c fixes the current value of v.

Draw the u coordinate line.

```
    for i in 1..vSteps repeat                                             38
      c := curryRight(f,v)                                                39
      cf := (t:SF):SF +-> 1                                               40
      makeObject(c,uRange::SEG Float,colorFunction==cf,                   41
        space == sp, tubeRadius == .02, tubePoints == 6)                  42
      v := v + delV                                                       43
    void()                                                                44
                                                                          45
```

Map a point in the complex plane to the Riemann sphere.

```
  riemannTransform(z) ==                                                  46
    r := sqrt norm z                                                      47
    cosTheta := (real z)/r                                                48
    sinTheta := (imag z)/r                                                49
    cp := 4*r/(4+r**2)                                                    50
    sp := sqrt(1-cp*cp)                                                   51
    if r>2 then sp := -sp                                                 52
    point [cosTheta*cp, sinTheta*cp, -sp + 1]                             53
                                                                          54
```

Convert Cartesian coordinates to complex Cartesian form.

```
  cartesian2Complex(r:SF, i:SF):C ==                                      55
    complex(r, i)                                                         56
                                                                          57
```

Convert polar coordinates to complex Cartesian form.

```
  polar2Complex(r:SF, th:SF):C ==                                         58
    complex(r*cos(th), r*sin(th))                                         59
                                                                          60
```

Convert complex function f to a mapping: (SF,SF) ↦ R3 in the complex plane.

```
  makeConformalMap(f, transformC) ==                                      61
    (u:SF,v:SF):R3 +->                                                    62
      z := f transformC(u, v)                                             63
      point [real z, imag z, 0.0@SF]                                      64
                                                                          65
```

Convert a complex function f to a mapping: (SF,SF) ↦ R3 on the Riemann sphere.

```
  makeRiemannConformalMap(f, transformC) ==                               66
    (u:SF, v:SF):R3 +->                                                   67
      riemannTransform f transformC(u, v)                                 68
                                                                          69
```

Draw a picture of the mapping of the complex plane to the Riemann sphere.

```
  riemannSphereDraw: (S, S, PI, PI, String) -> VIEW3D                     70
  riemannSphereDraw(rRange,tRange,rSteps,tSteps,coord) ==                 71
    transformC :=                                                         72
      coord = "polar" => polar2Complex                                    73
      cartesian2Complex                                                   74
```

Coordinate grid function.

```
    grid := (u:SF, v:SF): R3 +->                                          75
      z1 := transformC(u, v)                                              76
      point [real z1, imag z1, 0]                                         77
```

Create a fresh space.

Draw the flat grid.

```
    sp := createThreeSpace()                                              78
    adaptGrid(sp, grid, rRange, tRange, rSteps, tSteps)                   79
    connectingLines(sp,grid,rRange,tRange,rSteps,tSteps)                  80
```

Draw the sphere.

```
    makeObject(riemannSphere,0..2*%pi,0..%pi,space==sp)                   81
    f := (z:C):C +-> z                                                    82
    cm := makeRiemannConformalMap(f, transformC)                          83
```

Draw the sphere grid.

```
    adaptGrid(sp, cm, rRange, tRange, rSteps, tSteps)                     84
    makeViewport3D(sp, "Riemann Sphere")                                  85
                                                                          86
```

Draw the lines that connect the points in the complex plane to the north pole of the Riemann sphere.

```
  connectingLines(sp,f,uRange,vRange,uSteps,vSteps) ==                    87
    delU := (hi(uRange) - lo(uRange))/uSteps                              88
    delV := (hi(vRange) - lo(vRange))/vSteps                              89
    uSteps := uSteps + 1; vSteps := vSteps + 1                            90
    u := lo uRange                                                        91
```

For each u.

```
    for i in 1..uSteps repeat                                             92
```

For each v.

Project p1 onto the sphere.
Create a line function.

Draw the connecting line.

```
    v := lo vRange                                                  93
    for j in 1..vSteps repeat                                       94
      p1 := f(u,v)                                                  95
      p2 := riemannTransform complex(p1.1, p1.2)                    96
      fun := lineFromTo(p1,p2)                                      97
      cf := (t:SF):SF +-> 3                                         98
      makeObject(fun, 0..1,space==sp,tubePoints==4,                 99
                 tubeRadius==0.01,colorFunction==cf)                100
      v := v + delV                                                 101
    u := u + delU                                                   102
  void()                                                            103
                                                                    104
```

A sphere sitting on the complex plane, with radius 1.

```
riemannSphere(u,v) ==                                               105
  sv := sin(v)                                                      106
  0.99@SF*(point [cos(u)*sv,sin(u)*sv,cos(v),0.0@SF])+             107
    point [0.0@SF, 0.0@SF, 1.0@SF, 4.0@SF]                          108
                                                                    109
```

Create a line function that goes from p1 to p2

```
lineFromTo(p1, p2) ==                                               110
  d := p2 - p1                                                      111
  (t:SF):Point SF +->                                               112
    p1 + t*d                                                        113
```

F.9 tknot.input

Create a (p, q) torus-knot with radius r around the curve. The formula was derived by Larry Lambe.

Function for the torus knot.

The cross section.

Draw the circle around the knot.

```
)read ntube                                                         1
torusKnot: (SF, SF, SF, PI, PI) -> VIEW3D                           2
torusKnot(p, q ,r, uSteps, tSteps) ==                               3
  knot := (t:SF):Point SF +->                                       4
    fac := 4/(2.2@SF-sin(q*t))                                      5
    fac * point [cos(p*t), sin(p*t), cos(q*t)]                      6
  circle := (u:SF, t:SF): Point SF +->                              7
    r * point [cos u, sin u]                                        8
  ntubeDrawOpt(knot, circle, 0..2*%pi, 0..2*%pi,                    9
               var1Steps == uSteps, var2Steps == tSteps)            10
                                                                    11
```

F.10 ntube.input

The functions in this file create generalized tubes (also known as generalized cylinders). These functions draw a 2-d curve in the normal planes around a 3-d curve.

Points in 3-Space
Points in 2-Space
Draw ranges
Introduce types for functions for:
—the space curve function
—the plane curve function
—the surface function
Frenet frames define a coordinate system around a point on a space curve.
The current Frenet frame for a point on a curve.

```
R3 := Point SF                                                      1
R2 := Point SF                                                      2
S := Segment Float                                                  3
                                                                    4
ThreeCurve := SF -> R3                                              5
TwoCurve := (SF, SF) -> R2                                          6
Surface := (SF, SF) -> R3                                           7
                                                                    8
FrenetFrame :=                                                      9
   Record(value:R3,tangent:R3,normal:R3,binormal:R3)                10
frame: FrenetFrame                                                  11
                                                                    12
```

ntubeDraw(*spaceCurve, planeCurve,* $u_0..u_1$, $t_0..t_1$) draws *planeCurve* in the normal planes of *spaceCurve.* The parameter $u_0..u_1$ specifies the parameter range for *planeCurve* and $t_0..t_1$ specifies the parameter range for *spaceCurve.* Additionally, the plane curve function takes a second parameter: the current parameter of *spaceCurve.* This allows the plane curve to change shape as it goes around the space curve. See Section F.4 on page 692 for an example of this.

```
ntubeDraw: (ThreeCurve,TwoCurve,S,S) -> VIEW3D                      13
ntubeDraw(spaceCurve,planeCurve,uRange,tRange) ==                   14
  ntubeDrawOpt(spaceCurve, planeCurve, uRange, _                    15
               tRange, []$List DROPT)                               16
                                                                    17
ntubeDrawOpt: (ThreeCurve,TwoCurve,S,S,List DROPT)                  18
    -> VIEW3D                                                       19
ntubeDrawOpt(spaceCurve,planeCurve,uRange,tRange,l) ==              20
                                                                    21
  delT:SF := (hi(tRange) - lo(tRange))/10000                        22
  oldT:SF := lo(tRange) - 1                                         23

  fun := ngeneralTube(spaceCurve,planeCurve,delT,oldT)              24
  draw(fun, uRange, tRange, l)                                      25
                                                                    26
```

This function is similar to **ntubeDraw**, but takes optional parameters that it passes to the axiomFundraw command.

nfrenetFrame(*c, t, delT*) numerically computes the Frenet frame about the curve *c* at *t.* Parameter *delT* is a small number used to compute derivatives.

```
nfrenetFrame(c, t, delT) ==                                         27
  f0 := c(t)                                                        28
  f1 := c(t+delT)                                                   29
  t0 := f1 - f0                                                     30
  n0 := f1 + f0                                                     31
  b := cross(t0, n0)                                                32
  n := cross(b,t0)                                                  33
  ln := length n                                                    34
  lb := length b                                                    35
  ln = 0 or lb = 0 =>                                               36
      error "Frenet Frame not well defined"                         37
  n := (1/ln)*n                                                     38
  b := (1/lb)*b                                                     39
  [f0, t0, n, b]$FrenetFrame                                        40
```

The tangent.

The binormal.

The normal.

Make into unit length vectors.

ngeneralTube(*spaceCurve, planeCurve,delT, oltT*) creates a function that can be passed to the system axiomFundraw command. The function is a parameterized surface for the general tube around *spaceCurve. delT* is a small number used to compute derivatives. *oldT* is used to hold the current value of the *t* parameter for *spaceCurve.* This is an efficiency measure to ensure that frames are only computed once for each value of *t.*

Indicate that `frame` is global.

If not already computed, compute new frame.

Project `p` into the normal plane.

```
ngeneralTube: (ThreeCurve, TwoCurve, SF, SF) -> Surface  41
ngeneralTube(spaceCurve, planeCurve, delT, oldT) ==      42
  free frame                                             43
  (v:SF, t: SF): R3 +->                                  44
    if (t ~= oldT) then                                  45
      frame := nfrenetFrame(spaceCurve, t, delT)         46
      oldT := t                                          47
    p := planeCurve(v, t)                                48
    frame.value + p.1*frame.normal + p.2*frame.binormal  49
```

F.11 dhtri.input

Create affine transformations (DH matrices) that transform a given triangle into another.

Compute a DHMATRIX that transforms `t1` to `t2`, where `t1` and `t2` are the vertices of two triangles in 3-space.

Compute a DHMATRIX that transforms `t1` to `t2`, where `t1` and `t2` are the vertices of two tetrahedrons in 3-space.

Put the vertices of a tetrahedron into matrix form.

Compute a vector normal to the given triangle, whose length is the square root of the area of the triangle.

Compute the area of a triangle using Heron's formula.

```
tri2tri: (List Point SF, List Point SF) -> DHMATRIX(SF)   1
tri2tri(t1, t2) ==                                        2
  n1 := triangleNormal(t1)                                3
  n2 := triangleNormal(t2)                                4
  tet2tet(concat(t1, n1), concat(t2, n2))                 5
                                                          6
tet2tet: (List Point SF, List Point SF) -> DHMATRIX(SF)   7
tet2tet(t1, t2) ==                                        8
  m1 := makeColumnMatrix t1                               9
  m2 := makeColumnMatrix t2                              10
  m2 * inverse(m1)                                       11
                                                         12
makeColumnMatrix(t) ==                                   13
  m := new(4,4,0)$DHMATRIX(SF)                           14
  for x in t for i in 1..repeat                          15
    for j in 1..3 repeat                                 16
      m(j,i) := x.j                                      17
    m(4,i) := 1                                          18
  m                                                      19
                                                         20
triangleNormal(t) ==                                     21
  a := triangleArea t                                    22
  p1 := t.2 - t.1                                        23
  p2 := t.3 - t.2                                        24
  c := cross(p1, p2)                                     25
  len := length(c)                                       26
  len = 0 => error "degenerate triangle!"                27
  c := (1/len)*c                                         28
  t.1 + sqrt(a) * c                                      29
                                                         30
triangleArea t ==                                        31
  a := length(t.2 - t.1)                                 32
  b := length(t.3 - t.2)                                 33
  c := length(t.1 - t.3)                                 34
  s := (a+b+c)/2                                         35
  sqrt(s*(s-a)*(s-b)*(s-c))                              36
```

F.12 tetra.input

Bring DH matrices into the environment.

Set up the coordinates of the corners of the tetrahedron.

Some needed points.

The base of the tetrahedron.

The "middle triangle" inscribed in the base of the tetrahedron.

The bases of the triangles of the subdivided tetrahedron.

Create the transformations that bring the base of the tetrahedron to the bases of the subdivided tetrahedron.

Draw a Sierpinsky tetrahedron with n levels of recursive subdivision.

Recursively draw a Sierpinsky tetrahedron.

Draw the 4 recursive pyramids.

Draw a tetrahedron into the given space with the given color, transforming it by the given DH matrix.

```
)set expose add con DenavitHartenbergMatrix                          1
                                                                     2
x1:SF := sqrt(2.0@SF/3.0@SF)                                         3
x2:SF := sqrt(3.0@SF)/6                                              4
                                                                     5
p1 := point [-0.5@SF, -x2, 0.0@SF]                                   6
p2 := point [0.5@SF, -x2, 0.0@SF]                                    7
p3 := point [0.0@SF, 2*x2, 0.0@SF]                                   8
p4 := point [0.0@SF, 0.0@SF, x1]                                     9
                                                                     10
baseTriangle  := [p2, p1, p3]                                        11
                                                                     12
mt  := [0.5@SF*(p2+p1), 0.5@SF*(p1+p3), 0.5@SF*(p3+p2)]              13
                                                                     14
bt1 := [mt.1, p1, mt.2]                                              15
bt2 := [p2, mt.1, mt.3]                                              16
bt3 := [mt.2, p3, mt.3]                                              17
bt4 := [0.5@SF*(p2+p4), 0.5@SF*(p1+p4), 0.5@SF*(p3+p4)]              18
                                                                     19
tt1 := tri2tri(baseTriangle, bt1)                                    20
tt2 := tri2tri(baseTriangle, bt2)                                    21
tt3 := tri2tri(baseTriangle, bt3)                                    22
tt4 := tri2tri(baseTriangle, bt4)                                    23
                                                                     24
drawPyramid(n) ==                                                    25
  s := createThreeSpace()                                            26
  dh := rotatex(0.0@SF)                                              27
  drawPyramidInner(s, n, dh)                                         28
  makeViewport3D(s, "Sierpinsky Tetrahedron")                        29
                                                                     30
drawPyramidInner(s, n, dh) ==                                        31
  n = 0 => makeTetrahedron(s, dh, n)                                 32
  drawPyramidInner(s, n-1, dh * tt1)                                 33
  drawPyramidInner(s, n-1, dh * tt2)                                 34
  drawPyramidInner(s, n-1, dh * tt3)                                 35
  drawPyramidInner(s, n-1, dh * tt4)                                 36
                                                                     37
makeTetrahedron(sp, dh, color) ==                                    38
  w1 := dh*p1                                                        39
  w2 := dh*p2                                                        40
  w3 := dh*p3                                                        41
  w4 := dh*p4                                                        42
  polygon(sp, [w1, w2, w4])                                          43
  polygon(sp, [w1, w3, w4])                                          44
  polygon(sp, [w2, w3, w4])                                          45
  void()                                                             46
```

F.13 antoine.input

Draw Antoine's Necklace. Thank you to Matthew Grayson at IBM's T.J Watson Research Center for the idea.

Bring DH matrices into the environment.
The current transformation for drawing a sub ring.
Draw Antoine's Necklace with n levels of recursive subdivision. The number of subrings is 10^n.
Do the real work.

```
)set expose add con DenavitHartenbergMatrix                          1
                                                                     2
torusRot: DHMATRIX(SF)                                               3
                                                                     4
                                                                     5
drawRings(n) ==                                                      6
  s := createThreeSpace()                                            7
  dh:DHMATRIX(SF) := identity()                                      8
  drawRingsInner(s, n, dh)                                           9
  makeViewport3D(s, "Antoine's Necklace")                           10
                                                                    11
```

In order to draw Antoine rings, we take one ring, scale it down to a smaller size, rotate it around its central axis, translate it to the edge of the larger ring and rotate it around the edge to a point corresponding to its count (there are 10 positions around the edge of the larger ring). For each of these new rings we recursively perform the operations, each ring becoming 10 smaller rings. Notice how the DHMATRIX operations are used to build up the proper matrix composing all these transformations.

Recursively draw Antoine's Necklace.
Angle around ring.
Angle of subring from plane.
Amount to translate subring.
The translation increment.
Subdivide into 10 linked rings.
Transform ring in center to a link.
Draw a single ring into the given subspace, transformed by the given DHMATRIX.
Parameterization of a torus, transformed by the DHMATRIX in torusRot.

```
drawRingsInner(s, n, dh) ==                                         12
  n = 0 =>                                                          13
    drawRing(s, dh)                                                 14
    void()                                                          15
  t := 0.0@SF                                                       16
  p := 0.0@SF                                                       17
  tr := 1.0@SF                                                      18
  inc := 0.1@SF                                                     19
  for i in 1..10 repeat                                             20
    tr := tr + inc                                                  21
    inc := -inc                                                     22
    dh' := dh*rotatez(t)*translate(tr,0.0@SF,0.0@SF)*               23
           rotatey(p)*scale(0.35@SF, 0.48@SF, 0.4@SF)               24
    drawRingsInner(s, n-1, dh')                                     25
    t := t + 36.0@SF                                                26
    p := p + 90.0@SF                                                27
  void()                                                            28
                                                                    29
drawRing(s, dh) ==                                                  30
  free torusRot                                                     31
  torusRot := dh                                                    32
  makeObject(torus, 0..2*%pi, 0..2*%pi, var1Steps == 6,             33
             space == s, var2Steps == 15)                           34
                                                                    35
torus(u ,v) ==                                                      36
  cu := cos(u)/6                                                    37
  torusRot*point [(1+cu)*cos(v),(1+cu)*sin(v),(sin u)/6]            38
```

F.14 scherk.input

Scherk's minimal surface, defined by: $e^z \cos(x) = \cos(y)$. See: *A Comprehensive Introduction to Differential Geometry,* Vol. 3, by Michael Spivak, Publish Or Perish, Berkeley, 1979, pp. 249-252.

Offsets for a single piece of Scherk's minimal surface.

Draw Scherk's minimal surface on an m by n patch.

Draw only odd patches.

Draw a patch.

The first patch that makes up a single piece of Scherk's minimal surface.

The second patch.

The third patch.

The fourth patch.

Draw the surface by breaking it into four patches and then drawing the patches.

```
(xOffset, yOffset):SF                                             1
                                                                  2
                                                                  3
drawScherk(m,n) ==                                                4
  free xOffset, yOffset                                           5
  space := createThreeSpace()                                     6
  for i in 0..m-1 repeat                                          7
    xOffset := i*%pi                                              8
    for j in 0 .. n-1 repeat                                      9
      rem(i+j, 2) = 0 => 'iter                                   10
      yOffset := j*%pi                                           11
      drawOneScherk(space)                                       12
  makeViewport3D(space, "Scherk's Minimal Surface")              13
                                                                 14
scherk1(u,v) ==                                                  15
  x := cos(u)/exp(v)                                             16
  point [xOffset + acos(x), yOffset + u, v, abs(v)]              17
                                                                 18
scherk2(u,v) ==                                                  19
  x := cos(u)/exp(v)                                             20
  point [xOffset - acos(x), yOffset + u, v, abs(v)]              21
                                                                 22
scherk3(u,v) ==                                                  23
  x := exp(v) * cos(u)                                           24
  point [xOffset + u, yOffset + acos(x), v, abs(v)]              25
                                                                 26
scherk4(u,v) ==                                                  27
  x := exp(v) * cos(u)                                           28
  point [xOffset + u, yOffset - acos(x), v, abs(v)]              29
                                                                 30
drawOneScherk(s) ==                                              31
  makeObject(scherk1,-%pi/2..%pi/2,0..%pi/2,space==s,            32
             var1Steps == 28, var2Steps == 28)                   33
  makeObject(scherk2,-%pi/2..%pi/2,0..%pi/2,space==s,            34
             var1Steps == 28, var2Steps == 28)                   35
  makeObject(scherk3,-%pi/2..%pi/2,-%pi/2..0,space==s,           36
             var1Steps == 28, var2Steps == 28)                   37
  makeObject(scherk4,-%pi/2..%pi/2,-%pi/2..0,space==s,           38
             var1Steps == 28, var2Steps == 28)                   39
  void()                                                         40
```

APPENDIX G

Glossary

!
(syntax) Suffix character for *destructive operations*.

,
(syntax) a separator for items in a *tuple*, for example, to separate arguments of a function `f(x,y)`.

=>
(syntax) the expression `a => b` is equivalent to `if a then` *exit* `b`.

?
1. *(syntax)* a suffix character for Boolean-valued **function** names, for example, **odd?**. 2. Prefix character for "optional" pattern variables. For example, the pattern `f(x + y)` does not match the expression `f(7)`, but `f(?x + y)` does, with `x` matching 0 and `y` matching 7. 3. The special type `?` means *don't care*. For example, the declaration: `x : Polynomial ?` means that values assigned to `x` must be polynomials over an arbitrary *underlying domain*.

abstract datatype
a programming language principle used in AXIOM where a datatype definition has defined in two parts: (1) a *public* part describing a set of *exports*, principally operations that apply to objects of that type, and (2) a *private* part describing the implementation of the datatype usually in terms of a *representation* for objects of the type. Programs that create and otherwise manipulate objects of the type may only do so through its exports. The representation and other implementation information is specifically hidden.

abstraction
described functionally or conceptually without regard to implementation.

accuracy
the degree of exactness of an approximation or measurement. In computer algebra systems, computations are typically carried out with complete accuracy using integers or rational numbers of indefinite size. Domain Float provides a function **precision** to change the precision for floating-point computations. Computations using SmallFloat have a fixed precision but uncertain accuracy.

add-chain
a hierarchy formed by *domain extensions*. If domain `A` extends domain `B` and domain `B` extends domain `C`, then `A` has *add-chain* `B-C`.

aggregate
a data structure designed to hold multiple values. Examples of aggregates are List, Set, Matrix and Bits.

AKCL
Austin Kyoto Common LISP, a version of *KCL* produced by William Schelter, Austin, Texas.

algorithm
a step-by-step procedure for a solution of a problem; a program

ancestor
(of a domain or category) a category that is a *parent*, or a *parent* of a *parent*, and so on. See a **Cross Reference** page of a constructor in Browse.

application
(syntax) an expression denoting "application" of a function to a set of *argument* parameters. Appli-

cations are written as a *parameterized form.* For example, the form `f(x,y)` indicates the "application of the function `f` to the tuple of arguments `x` and `y`." See also *evaluation* and *invocation.*

apply
See *application.*

argument
1. (actual argument) a value passed to a function at the time of a *function call*; also called an *actual parameter.* 2. (formal argument) a variable used in the definition of a function to denote the actual argument passed when the function is called.

arity
1. (function) the number of arguments. 2. (operator or operation) corresponds to the arity of a function implementing the operator or operation.

assignment
(syntax) an expression of the form `x := e`, meaning "assign the value of `e` to `x`." After *evaluation,* the *variable* `x` *points* to an object obtained by evaluating the expression `e`. If `x` has a *type* as a result of a previous *declaration,* the object assigned to `x` must have that type. The interpreter must often *coerce* the value of `e` to make that happen. For example, the expression `x : Float := 11` first *declares* `x` to be a float, then forces the interpreter to coerce the integer `11` to `11.0` in order to assign a floating-point value to `x`.

attribute
a name or functional form denoting *any* useful computational or mathematical property. For example, `commutative("*")` asserts that "`*`" is commutative. Also, `finiteAggregate` is used to assert that an aggregate has a finite number of immediate components.

basis
(algebra) `S` is a basis of a module `M` over a *ring* if `S` generates `M`, and `S` is linearly independent.

benefactor
(of a given domain) a domain or package that the given domain explicitly references (for example, calls functions from) in its implementation. See a **Cross Reference** page of a constructor in Browse.

binary
operation or function with *arity* 2.

binding
the association of a variable with properties such as *value* and *type.* The top-level *environment* in the interpreter consists of bindings for all user variables and functions. When a *function* is applied to arguments, a local environment of bindings is created, one for each formal *argument* and *local variable.*

block
(syntax) a control structure where expressions are sequentially *evaluated.*

body
a *function body* or *loop body.*

boolean
objects denoted by the *literals* `true` and `false`; elements of domain Boolean. See also Bits.

built-in function
a *function* in the standard AXIOM library. Contrast *user function.*

cache
1. (noun) a mechanism for immediate retrieval of previously computed data. For example, a function that does a lengthy computation might store its values in a *hash table* using the function argument as the key. The hash table then serves as a cache for the function (see also `)set function cache`). Also, when *recurrence relations* that depend upon `n` previous values are compiled, the previous `n` values are normally cached (use `)set functions recurrence` to change this). 2. (verb) to save values in a cache.

capsule
the part of the *body* of a *domain constructor* that defines the functions implemented by the constructor.

case
(syntax) an operator used to evaluate code conditionally based on the branch of a Union. For example, if value `u` is `Union(Integer, "failed")`, the conditional expression `if u case Integer then A else B` evaluates `A` if `u` is an integer and `B` otherwise.

Category
the distinguished object denoting the type of a category; the class of all categories.

category
(basic concept) types denoting classes of domains. Examples of categories are Ring ("the class of all rings") and Aggregate ("the class of all aggregates"). Categories form a hierarchy (formally, a directed acyclic graph) with the distinquished category Type

at the top. Each category inherits the properties of all its ancestors. Categories optionally provide "default definitions" for operations they export. Categories are defined in AXIOM by functions called *category constructors*. Technically, a category designates a class of domains with common *operations* and *attributes* but usually with different *functions* and *representations* for its constituent *objects*. Categories are always defined using the AXIOM library language (see also *category extension*). See also file **catdef.spad** for definitions of basic algebraic categories in AXIOM, **aggcat.spad** for data structure

category constructor
a function that creates categories, described by an abstract datatype in the AXIOM programming language. For example, the category constructor Module is a function that takes a domain parameter `R` and creates the category "modules over `R`."

category extension
A category `A` *directly extends* a category `B` if its definition has the form `A == B with ...` or `A == Join(...,B,...)`. In this case, we also say that `B` is the *parent* of `A`. We say that a category `A` extends `B` if `B` is an *ancestor* of `A`. A category `A` may also directly extend `B` if `B` appears in a conditional expression within the `Exports` part of the definition to the right of a `with`. See, for example, file **catdef.spad** for definitions of the algebra categories in AXIOM, **aggcat.spad** for data structure categories.

category hierarchy
hierarchy formed by category extensions. The root category is Type. A category can be defined as a *Join* of two or more categories so as to have multiple *parents*. Categories may also be parameterized so as to allow conditional inheritance.

character
1. an element of a character set, as represented by a keyboard key. 2. a component of a string. For example, the `1`st element of the string `"hello there"` is the character *h*.

client
(of a given domain) any domain or package that explicitly calls functions from the given domain. See a **Cross Reference** page of a constructor in Browse.

coercion
an automatic transformation of an object of one *type* to an object of a similar or desired target type. In the interpreter, coercions and *retractions* are done automatically by the interpreter when a type mismatch occurs. Compare *conversion*.

comment
textual remarks imbedded in code. Comments are preceded by a double dash ("`--`"). For AXIOM library code, stylized comments for on-line documentation are preceded by two plus signs ("++").

Common LISP
A version of *LISP* adopted as an informal standard by major users and suppliers of LISP.

compile-time
the time when category or domain constructors are compiled. Contrast *run-time*.

compiler
a program that generates low-level code from a higher-level source language. AXIOM has three compilers. A *graphics compiler* converts graphical formulas to a compiled subroutine so that points can be rapidly produced for graphics commands. An *interpreter compiler* optionally compiles *user functions* when first *invoked* (use `)set functions compile` to turn this feature on). A *library compiler* compiles all constructors (available on an "as-is" basis for Release 1).

computational object
In AXIOM, domains are objects. This term is used to distinguish the objects that are members of domains rather than the domains themselves.

conditional
a *control structure* of the form `if A then B else C`. The *evaluation* of `A` produces `true` or `false`. If `true`, `B` evaluates to produce a value; otherwise `C` evaluates to produce a value. When the value is not required, the `else C` part can be omitted.

constant
(syntax) a reserved word used in *signatures* in AXIOM programming language to signify that an operation always returns the same value. For example, the signature `0: constant -> $` in the source code of AbelianMonoid tells the AXIOM compiler that `0` is a constant so that suitable optimizations might be performed.

constructor
a *function* that creates a *category*, *domain*, or *package*.

continuation
when a line of a program is so long that it must be broken into several lines, then all but the first line are called *continuation lines.* If such a line is given interactively, then each incomplete line must end with an underscore.

control structure
program structures that can specify a departure from normal sequential execution. AXIOM has four kinds of control structures: *blocks, case* statements, *conditionals,* and *loops.*

conversion
the transformation of an object of one *type* to one of another type. Conversions that can be performed automatically by the interpreter are called *coercions.* These happen when the interpreter encounters a type mismatch and a similar or declared target type is needed. In general, the user must use the infix operation "`::`" to cause this transformation.

copying semantics
the programming language semantics used in PASCAL but *not* in AXIOM. See also *pointer semantics* for details.

data structure
a structure for storing data in the computer. Examples are *lists* and *hash tables.*

datatype
equivalent to *domain* in AXIOM.

declaration
(syntax) an expression of the form `x : T` where `T` is some *type.* A declaration forces all values *assigned* to `x` to be of that type. If a value is of a different type, the interpreter will try to *coerce* the value to type `T`. Declarations are necessary in case of ambiguity or when a user wants to introduce an *unexposed* domain.

default definition
a function defined by a *category.* Such definitions appear in category definitions of the form
`C: Category == T add I`
in an optional implementation part `I` to the right of the keyword `add`.

default package
an optional *package* of *functions* associated with a category. Such functions are necessarily defined in terms of other operations exported by the category.

definition
(syntax) 1. An expression of the form `f(a) == b` defining function `f` with *formal arguments* `a` and *body* `b`; equivalent to the statement `f == (a) +-> b`. 2. An expression of the form `a == b` where `a` is a *symbol,* equivalent to `a() == b`. See also *macro* where a similar substitution is done at *parse* time.

delimiter
a *character* that marks the beginning or end of some syntactically correct unit in the language, for example, "`"`" for strings, blanks for identifiers.

dependent
(of a given constructor) another constructor that mentions the given constructor as an argument or among the types of an exported operation. See a **Cross Reference** page of a constructor in Browse.

destructive operation
An operation that changes a component or structure of a value. In AXIOM, destructive operations have names ending with an exclamation mark ("`!`"). For example, domain List has two operations to reverse the elements of a list, one named **reverse** that returns a copy of the original list with the elements reversed, another named **reverse!** that reverses the elements *in place,* thus destructively changing the original list.

documentation
1. on-line or hard-copy descriptions of AXIOM; 2. text in library code preceded by "++" comments as opposed to general comments preceded by "`--`".

domain
(basic concept) a domain corresponds to the usual notion of datatypes. Examples of domains are List Float ("lists of floats"), Fraction Polynomial Integer ("fractions of polynomials of integers"), and Matrix Stream CardinalNumber ("matrices of infinite *streams* of cardinal numbers"). The term *domain* actually abbreviates *domain of computation.* Technically, a domain denotes a class of objects, a class of *operations* for creating and otherwise manipulating these objects, and a class of *attributes* describing computationally useful properties. Domains may also define *functions* for its exported operations, often in terms of some *representation* for the objects. A domain itself is an *object* created by a *function* called a *domain constructor.* The types of the exported operations of a domain are arbitary; this gives rise to a special class of domains called *packages.*

domain constructor
a function that creates domains, described by an abstract datatype in the AXIOM programming language. Simple domains like Integer and Boolean are created by domain constructors with no arguments. Most domain constructors take one or more parameters, one usually denoting an *underlying domain*. For example, the domain Matrix(R) denotes "matrices over R." Domains Mapping, Record, and Union are primitive domains. All other domains are written in the AXIOM programming language and can be modified by users with access to the library source code and the library compiler.

domain extension
a domain constructor A is said to *extend* a domain constructor B if A's definition has the form `A == B add ...`. This intuitively means "functions not defined by A are assumed to come from B." Successive domain extensions form *add-chains* affecting the *search order* for functions not implemented directly by the domain during *dynamic lookup*.

dot notation
using an infix dot (".") for the operation **elt**. If `u` is the list `[7,4,-11]` then both `u(2)` and `u.2` return `4`. Dot notation nests to the left: `f.g.h` is equivalent to `(f.g).h`.

dynamic
that which is done at *run-time* as opposed to *compile-time*. For example, the interpreter may build a domain "matrices over integers" dynamically in response to user input. However, the compilation of all functions for matrices and integers is done during *compile-time*. Constrast *static*.

dynamic lookup
In AXIOM, a *domain* may or may not explicitly provide *function* definitions for all its exported *operations*. These definitions may instead come from domains in the *add-chain* or from *default packages*. When a *function call* is made for an operation in the domain, up to five steps are carried out.

1. If the domain itself implements a function for the operation, that function is returned.
2. Each of the domains in the *add-chain* are searched; if one of these domains implements the function, that function is returned.
3. Each of the *default packages* for the domain are searched in order of the *lineage*. If any of the default packages implements the function, the first one found is returned.
4. Each of the *default packages* for each of the domains in the *add-chain* are searched in the order of their *lineage*. If any of the default packages implements the function, the first one found is returned.
5. If all of the above steps fail, an error message is reported.

empty
the unique value of objects with type Void.

environment
a set of *bindings*.

evaluation
a systematic process that transforms an *expression* into an object called the *value* of the expression. Evaluation may produce *side effects*.

exit
(reserved word) an *operator* that forces an exit from the current *block*. For example, the block `(a := 1; if i > 0 then exit a; a := 2)` will prematurely exit at the second statement with value 1 if the value of `i` is greater than zero. See "`=>`" for an alternate syntax.

explicit export
1. (of a domain `D`) any *attribute*, *operation*, or *category* explicitly mentioned in the *type* exports part `E` for the domain constructor definition `D: E == I` 2. (of a category `C`) any *attribute*, *operation*, or *category* explicitly mentioned in the *type* specification part `E` for the category constructor definition `C:` *Category* `== E`

export
explicit export or *implicit export* of a domain or category

expose
some constructors are *exposed*, others *unexposed*. Exposed domains and packages are recognized by the interpreter. Use `)set expose` to control what is exposed. Unexposed constructors will appear in Browse prefixed by a star ("`*`").

expression
1. any syntactically correct program fragment. 2. an element of domain Expression.

extend
see *category extension* or *domain extension*.

field
(algebra) a *domain* that is a *ring* where every non-zero element is invertible and where `xy=yx`; a

member of category Field. For a complete list of fields, click on **Domains** under **Cross Reference** for Field in Browse.

file
1. a program or collection of data stored on disk, tape or other medium. 2. an object of a File domain.

float
a floating-point number with user-specified precision; an element of domain Float. Floats are *literals* written either without an exponent (for example, `3.1416`), or with an exponent (for example, `3.12E-12`). Use function *precision* to change the precision of the mantissa (`20` digits by default). See also *small float.*

formal parameter
(of a function) an identifier *bound* to the value of an actual *argument* on *invocation.* In the function definition `f(x,y) == u`, for example, `x` and `y` are the formal parameters.

frame
the basic unit of an interactive session; each frame has its own *step number*, *environment*, and *history.* In one interactive session, users can create and drop frames, and have several active frames simultaneously.

free
(syntax) A keyword used in user-defined functions to declare that a variable is a *free variable* of that function. For example, the statement `free x` declares the variable `x` within the body of a function `f` to be a free variable in `f`. Without such a declaration, any variable `x` that appears on the left-hand side of an assignment before it is referenced is regarded as a *local variable* of that function. If the intention of the assignment is to give a value to a *global variable* `x`, the body of that function must contain the statement `free x`. A variable that is a parameter to the function is always local.

free variable
(of a function) a variable that appears in a body of a function but is not *bound* by that function. Contrast with *local variable.*

function
implementation of *operation.* A function takes zero or more *argument* parameters and produces a single return value. Functions are objects that can be passed as parameters to functions and can be returned as values of functions. Functions can also create other functions (see also InputForm). See also *application* and *invocation.* The terms *operation* and *function* are distinct notions in AXIOM. An operation is an abstraction of a function, described by a *name* and a *signature.* A function is created by providing an implementation of that operation by AXIOM code. Consider the example of defining a user-function `fact` to compute the **factorial** of a nonnegative integer. The AXIOM statement `fact: Integer -> Integer` describes the operation, whereas the statement `fact(n) = reduce(*,[1..n])` defines the function. See also *generic function.*

function body
the part of a *function*'s definition that is evaluated when the function is called at *run-time*; the part of the function definition to the right of the "==".

garbage collection
a system function that automatically recycles memory cells from the *heap.* AXIOM is built upon *Common LISP* that provides this facility.

garbage collector
a mechanism for reclaiming storage in the *heap.*

Gaussian
a complex-valued expression, for example, one with both a real and imaginary part; a member of a Complex domain.

generic function
the use of one function to operate on objects of different types. One might regard AXIOM as supporting generic *operations* but not generic functions. One operation `+: (D, D) -> D` exists for adding elements in a ring; each ring however provides its own type-specific function for implementing this operation.

global variable
A variable that can be referenced freely by functions. In AXIOM, all top-level user-defined variables defined during an interactive user session are global variables. AXIOM does not allow *fluid variables*, that is, variables *bound* by a function `f` that can be referenced by functions that `f` calls.

Gröbner basis
(algebra) a special basis for a polynomial ideal that allows a simple test for membership. It is useful in solving systems of polynomial equations.

group
(algebra) a monoid where every element has a

multiplicative inverse.

hash table
a data structure designed for fast lookup of information stored under "keys". A hash table consists of a set of *entries*, each of which associates a *key* with a *value*. Finding the object stored under a key can be fast for a large number of entries since keys are *hashed* into numerical codes for fast lookup.

heap
1. an area of storage used by data in programs. For example, AXIOM will use the heap to hold the partial results of symbolic computations. When cancellations occur, these results remain in the heap until *garbage collected*. 2. an object of a Heap domain.

history
a mechanism that records input and output data for an interactive session. Using the history facility, users can save computations, review previous steps of a computation, and restore a previous interactive session at some later time. For details, issue the system command *)history ?* to the interpreter. See also *frame*.

ideal
(algebra) a subset of a ring that is closed under addition and multiplication by arbitrary ring elements; thus an ideal is a module over the ring.

identifier
(syntax) an AXIOM name; a *literal* of type Symbol. An identifier begins with an alphabetical character, %, ?, or !, and may be followed by any of these or digits. Certain distinguished *reserved words* are not allowed as identifiers but have special meaning in AXIOM.

immutable
an object is immutable if it cannot be changed by an *operation*; it is not a *mutable object*. Algebraic objects are generally immutable: changing an algebraic expression involves copying parts of the original object. One exception is an object of type Matrix. Examples of mutable objects are data structures such as those of type List. See also *pointer semantics*.

implicit export
(of a domain or category) any exported *attribute* or *operation* or *category* that is not an *explicit export*. For example, Monoid and * are implicit exports of Ring.

index
1. a variable that counts the number of times a *loop* is repeated. 2. the "address" of an element in a data structure (see also category LinearAggregate).

infix
(syntax) an *operator* placed between two *operands*; also called a *binary operator*. For example, in the expression `a + b`, "+" is the infix operator. An infix operator may also be used as a *prefix*. Thus `+(a,b)` is also permissible in the AXIOM language. Infix operators have a *precedence* relative to one another.

input area
a rectangular area on a HyperDoc screen into which users can enter text.

instantiate
to build a *category*, *domain*, or *package* at run-time.

integer
a *literal* object of domain Integer, the class of integers with an unbounded number of digits. Integer literals consist of one or more consecutive digits (0-9) with no embedded blanks. Underscores can be used to separate digits in long integers if desirable.

interactive
a system where the user interacts with the computer step-by-step.

interpreter
the part of AXIOM responsible for handling user input during an interactive session. The interpreter parses the user's input expression to create an expression tree, then does a bottom-up traversal of the tree. Each subtree encountered that is not a value consists of a root node denoting an operation name and one or more leaf nodes denoting *operands*. The interpreter resolves type mismatches and uses type-inferencing and a library database to determine appropriate types for the operands and the result, and an operation to be performed. The interpreter next builds a domain to perform the indicated operation, and invokes a function from the domain to compute a value. The subtree is then replaced by that value and the process continues. Once the entire tree has been processed, the value replacing the top node of the tree is displayed back to the user as the value of the expression.

invocation
(of a function) the run-time process involved in

evaluating a *function application.* This process has two steps. First, a local *environment* is created where *formal arguments* are locally *bound* by *assignment* to their respective actual *argument.* Second, the *function body* is evaluated in that local environment. The evaluation of a function is terminated either by completely evaluating the function body or by the evaluation of a `return` expression.

iteration
repeated evaluation of an expression or a sequence of expressions. Iterations use the reserved words `for`, `while`, and `repeat`.

Join
a primitive AXIOM function taking two or more categories as arguments and producing a category containing all of the operations and attributes from the respective categories.

KCL
Kyoto Common LISP, a version of *Common LISP* that features compilation of LISP into the `C` Programming Language.

library
In AXIOM, a collection of compiled modules respresenting *category* or *domain* constructors.

lineage
the sequence of *default packages* for a given domain to be searched during *dynamic lookup.* This sequence is computed first by ordering the category *ancestors* of the domain according to their *level number*, an integer equal to the minimum distance of the domain from the category. Parents have level 1, parents of parents have level 2, and so on. Among categories with equal level numbers, ones that appear in the left-most branches of `Join`s in the source code come first. See a **Cross Reference** page of a constructor in Browse. See also *dynamic lookup.*

LISP
acronym for List Processing Language, a language designed for the manipulation of non-numerical data. The AXIOM library is translated into LISP then compiled into machine code by an underlying LISP system.

list
an object of a List domain.

literal
an object with a special syntax in the language. In AXIOM, there are five types of literals: *booleans, integers, floats, strings,* and *symbols.*

local
(syntax) A keyword used in user-defined functions to declare that a variable is a *local variable* of that function. Because of default assumptions on variables, such a declaration is often not necessary but is available to the user for clarity when appropriate.

local variable
(of a function) a variable *bound* by that function and such that its binding is invisible to any function that function calls. Also called a *lexical* variable. By default in the interpreter:

1. any variable `x` that appears on the left-hand side of an assignment is normally regarded a local variable of that function. If the intention of an assignment is to change the value of a *global variable* `x`, the body of the function must then contain the statement `free x`.
2. any other variable is regarded as a *free variable.*

An optional declaration `local x` is available to declare explicitly a variable to be a local variable. All *formal parameters* are local variables to the function.

loop
1. an expression containing a `repeat`. 2. a collection expression having a `for` or a `while`, for example, `[f(i) for i in S]`.

loop body
the part of a loop following the `repeat` that tells what to do each iteration. For example, the body of the loop `for x in S repeat B` is `B`. For a collection expression, the body of the loop precedes the initial `for` or `while`.

macro
1. *(interactive syntax)* An expression of the form `macro a == b` where `a` is a *symbol* causes `a` to be textually replaced by the expression `b` at *parse* time. 2. An expression of the form `macro f(a) == b` defines a parameterized macro expansion for a parameterized form `f`. This macro causes a form `f(x)` to be textually replaced by the expression `c` at parse time, where `c` is the expression obtained by replacing `a` by `x` everywhere in `b`. See also *definition* where a similar substitution is done during *evaluation.* 3. *(programming language syntax)*

An expression of the form `a ==> b` where `a` is a symbol.

mode
a type expression containing a question-mark ("?"). For example, the mode POLY ? designates *the class of all polynomials over an arbitrary ring.*

mutable
objects that contain *pointers* to other objects and that have operations defined on them that alter these pointers. Contrast *immutable.* AXIOM uses *pointer semantics* as does *LISP* in contrast with many other languages such as PASCAL that use *copying semantics.* See *pointer semantics* for details.

name
1. a *symbol* denoting a *variable,* such as the variable `x`. 2. a *symbol* denoting an *operation,* that is, the operation `divide: (Integer, Integer) -> Integer`.

nullary
a function with no arguments, for example, **characteristic**; operation or function with *arity* zero.

object
a data entity created or manipulated by programs. Elements of domains, functions, and domains themselves are objects. The most basic objects are *literals*; all other objects must be created by *functions.* Objects can refer to other objects using *pointers* and can be *mutable.*

object code
code that can be directly executed by hardware; also known as *machine language.*

operand
an argument of an *operator* (regarding an operator as a *function*).

operation
an abstraction of a *function,* described by a *signature.*
For example, `fact: NonNegativeInteger -> NonNegativeInteger` describes an operation for "the factorial of a (non-negative) integer."

operator
special reserved words in the language such as "+" and "*"; operators can be either *prefix* or *infix* and have a relative *precedence.*

overloading
the use of the same name to denote distinct operations; an operation is identified by a *signature* identifying its name, the number and types of its arguments, and its return types. If two functions can have identical signatures, a *package call* must be made to distinguish the two.

package
a special case of a domain, one for which the exported operations depend solely on the parameters and other explicit domains (contain no $). Intuitively, packages are collections of (*polymorphic*) functions. Facilities for integration, differential equations, solution of linear or polynomial equations, and group theory are provided by packages.

package call
(syntax) an expression of the form `e $ P` where `e` is an *application* and `P` denotes some *package* (or *domain*).

package constructor
same as *domain constructor.*

parameter
see *argument.*

parameterized datatype
a domain that is built on another, for example, polynomials with integer coefficients.

parameterized form
a expression of the form `f(x,y)`, an *application* of a function.

parent
(of a domain or category) a category which is explicitly declared in the source code definition for the domain either to the left of the `with` or as an *export* of the domain. See *category extension.* See also a **Cross Reference** page of a constructor in Browse.

parse
1. (verb) to transform a user input string representing a valid AXIOM expression into an internal representation as a tree-structure; the resulting internal representation is then "interpreted" by AXIOM to perform some indicated action.

partially ordered set
a set with a reflexive, transitive and antisymetric *binary* operation.

pattern matching
1. (on expressions) Given an expression called the "subject" `u`, the attempt to rewrite `u` using a set of "rewrite rules." Each rule has the form `A == B`

where A indicates an expression called a "pattern" and B denotes a "replacement." The meaning of this rule is "replace A by B." If a given pattern A matches a subexpression of u, that subexpression is replaced by B. Once rewritten, pattern matching continues until no further changes occur. 2. (on strings) the attempt to match a string indicating a "pattern" to another string called a "subject", for example, for the purpose of identifying a list of names. In Browse, users may enter *search strings* for the purpose of identifying constructors, operations, and attributes.

pile
alternate syntax for a block, using indentation and column alignment (see also *block*).

pointer
a reference implemented by a link directed from one object to another in the computer memory. An object is said to *refer* to another if it has a pointer to that other object. Objects can also refer to themselves (cyclic references are legal). Also more than one object can refer to the same object. See also *pointer semantics*.

pointer semantics
the programming language semantics used in languages such as LISP that allow objects to be *mutable*. Consider the following sequence of AXIOM statements:

```
x : Vector Integer := [1,4,7]
y := x
swap!(x,2,3)
```

The function **swap!** is used to interchange the second and third value in the list x, producing the value [1,7,4]. What value does y have after evaluation of the third statement? The answer is different in AXIOM than it is in a language with *copying semantics*. In AXIOM, first the vector [1,2,3] is created and the variable x set to *point* to this object. Let's call this object V. Next, the variable y is made to point to V just as x does. Now the third statement interchanges the last 2 elements of V (the "!" at the end of the name **swap!** tells you that this operation is destructive, that is, it changes the elements *in place*). Both x and y perceive this change to V. Thus both x and y then have the value [1,7,4]. In PASCAL, the second statement causes a copy of V to be stored under y. Thus the change to V made by the third statement does not affect y.

polymorphic
a *function* (for example, one implementing an *algorithm*) defined with categorical types so as to be applicable over a variety of domains (the domains which are members of the categorical types). Every AXIOM function defined in a domain or package constructor with a domain-valued parameter is polymorphic. For example, the same matrix "+" function is used to add "matrices over integers" as "matrices over matrices over integers."

postfix
an *operator* that follows its single *operand*. Postfix operators are not available in AXIOM.

precedence
(syntax) refers to the so-called *binding power* of an operator. For example, "*" has higher binding power than "+" so that the expression a + b * c is equivalent to a + (b * c).

precision
the number of digits in the specification of a number. The operation **digits** sets this for objects of Float.

predicate
1. a Boolean-valued function, for example, odd: Integer -> Boolean. 2. a Boolean-valued expression.

prefix
(syntax) an *operator* such as "-" that is written *before* its single *operand*. Every function of one argument can be used as a prefix operator. For example, all of the following have equivalent meaning in AXIOM: f(x), f x, and f.x. See also *dot notation*.

quote
the prefix *operator* "'" meaning *do not evaluate*.

Record
(basic domain constructor) a domain constructor used to create an inhomogeneous aggregate composed of pairs of *selectors* and *values*. A Record domain is written in the form Record(a1: D1, ..., an: Dn) (n > 0) where a1, ..., an are identifiers called the *selectors* of the record, and D1, ..., Dn are domains indicating the type of the component stored under selector an.

recurrence relation
A relation that can be expressed as a function f with some argument n which depends on the value of f at k previous values. In most cases, AXIOM will rewrite a recurrence relation on compilation

so as to *cache* its previous `k` values and therefore make the computation significantly more efficient.

recursion
use of a self-reference within the body of a function. Indirect recursion is when a function uses a function below it in the call chain.

recursive
1. A function that calls itself, either directly or indirectly through another function. 2. self-referential. See also *recursive.*

reference
see *pointer*

relative
(of a domain) A package that exports operations relating to the domain, in addition to those exported by the domain. See a **Cross Reference** page of a constructor in Browse.

representation
a *domain* providing a data structure for elements of a domain, generally denoted by the special identifier *Rep* in the AXIOM programming language. As domains are *abstract datatypes*, this representation is not available to users of the domain, only to functions defined in the *function body* for a domain constructor. Any domain can be used as a representation.

reserved word
a special sequence of non-blank characters with special meaning in the AXIOM language. Examples of reserved words are names such as `for`, `if`, and `free`, operator names such as "+" and **mod**, special character strings such as "==" and "`:=`".

retraction
to move an object in a parameterized domain back to the underlying domain, for example to move the object 7 from a "fraction of integers" (domain Fraction Integer) to "the integers" (domain Integer).

return
when leaving a function, the value of the expression following `return` becomes the value of the function.

ring
a set with a commutative addition, associative multiplication, a unit element, where multiplication is distributive over addition and subtraction.

rule
(syntax) 1. An expression of the form `rule A == B` indicating a "rewrite rule." 2. An expression of the form `rule (R1;...;Rn)` indicating a set of "rewrite rules" `R1`,...,`Rn`. See *pattern matching* for details.

run-time
the time when computation is done. Contrast with *compile-time*, and *dynamic* as opposed to *static*. For example, the decision of the intepreter to build a structure such as "matrices with power series entries" in response to user input is made at run-time.

run-time check
an error-checking that can be done only when the program receives user input; for example, confirming that a value is in the proper range for a computation.

search string
a string entered into an *input area* on a HyperDoc screen.

selector
an identifier used to address a component value of a Record datatype.

semantics
the relationships between symbols and their meanings. The rules for obtaining the *meaning* of any syntactically valid expression.

semigroup
(algebra) a *monoid* which need not have an identity; it is closed and associative.

side effect
action that changes a component or structure of a value. See *destructive operation* for details.

signature
(syntax) an expression describing the type of an *operation*. A signature has the form `name : source -> target`, where `source` is the type of the arguments of the operation, and `target` is the type of the result.

small float
an object of the domain SmallFloat for floating-point arithmetic as provided by the computer hardware.

small integer
an object of the domain SmallInteger for integer arithmetic as provided by the computer hardware.

source
the *type* of the argument of a *function*; the type expression before the `->` in a *signature*. For example, the source of `f : (Integer, Integer) -> Integer` is `(Integer, Integer)`.

sparse
data structure whose elements are mostly identical (a sparse matrix is one filled mostly with zeroes).

static
that computation done before run-time, such as compilation. Contrast *dynamic*.

step number
the number that precedes user input lines in an interactive session; the output of user results is also labeled by this number.

stream
an object of Stream(R), a generalization of a *list* to allow an infinite number of elements. Elements of a stream are computed "on demand." Streams are used to implement various forms of power series.

string
an object of domain String. Strings are *literals* consisting of an arbitrary sequence of *characters* surrounded by double-quotes (""""), for example, `"Look here!"`.

subdomain
(basic concept) a *domain* together with a *predicate* characterizing the members of the domain that belong to the subdomain. The exports of a subdomain are usually distinct from the domain itself. A fundamental assumption however is that values in the subdomain are automatically *coerceable* to values in the domain. For example, if `n` and `m` are declared to be members of a subdomain of the integers, then *any binary* operation from Integer is available on `n` and `m`. On the other hand, if the result of that operation is to be assigned to, say, `k`, also declared to be of that subdomain, a *run-time* check is generally necessary to ensure that the result belongs to the subdomain.

such that clause
(syntax) the use of "`|`" followed by an expression to filter an iteration.

suffix
(syntax) an *operator* that is placed after its operand. Suffix operators are not allowed in the AXIOM language.

symbol
objects denoted by *identifier literals*; an element of domain Symbol. The interpreter, by default, converts the symbol `x` into Variable(x).

syntax
rules of grammar and punctuation for forming correct expressions.

system commands
top-level AXIOM statements that begin with "`)`". System commands allow users to query the database, read files, trace functions, and so on.

tag
an identifier used to discriminate a branch of a Union type.

target
the *type* of the result of a *function*; the type expression following the "`->`" in a *signature*.

top-level
refers to direct user interactions with the AXIOM interpreter.

totally ordered set
(algebra) a partially ordered set where any two elements are comparable.

trace
use of system function `)trace` to track the arguments passed to a function and the values returned.

tuple
an expression of two or more other expressions separated by commas, for example, `4,7,11`. Tuples are also used for multiple arguments both for *applications* (for example, `f(x,y)`) and in *signatures* (for example, `(Integer, Integer) -> Integer`). A tuple is not a data structure, rather a syntax mechanism for grouping expressions.

type
The type of any *category* is the unique symbol *Category*. The type of a *domain* is any *category* to which the domain belongs. The type of any other object is either the (unique) domain to which the object belongs or a *subdomain* of that domain. The type of objects is in general not unique.

Type
a category with no operations or attributes, of which all other categories in AXIOM are *extensions*.

type checking
a system function that determines whether the datatype of an object is appropriate for a given operation.

type constructor
a *domain constructor* or *category constructor*.

type inference
when the interpreter chooses the type for an object based on context. For example, if the user interactively issues the definition `f(x) == (x + %i)**2` then issues `f(2)`, the interpreter will infer the type of `f` to be `Integer -> Complex Integer`.

unary
operation or function with *arity* 1.

underlying domain
for a *domain* that has a single domain-valued parameter, the *underlying domain* refers to that parameter. For example, the domain "matrices of integers" (Matrix Integer) has underlying domain Integer.

Union
(basic domain constructor) a domain constructor used to combine any set of domains into a single domain. A Union domain is written in the form `Union(a1: D1, ..., an: Dn)` (n > 0) where `a1`, ..., `an` are identifiers called the *tags* of the union, and `D1`, ..., `Dn` are domains called the *branches* of the union. The tags `ai` are optional, but required when two of the `Di` are equal, for example, `Union(inches: Integer, centimeters: Integer)`. In the interpreter, values of union domains are automatically coerced to values in the branches and vice-versa as appropriate. See also *case*.

unit
(algebra) an invertible element.

user function
a function defined by a user during an interactive session. Contrast *built-in function*.

user variable
a variable created by the user at top-level during an interactive session.

value
1. the result of *evaluating* an expression. 2. a property associated with a *variable* in a *binding* in an *environment*.

variable
a means of referring to an object, but not an object itself. A variable has a name and an associated *binding* created by *evaluation* of AXIOM expressions such as *declarations*, *assignments*, and *definitions*. In the top-level *environment* of the interpreter, variables are *global variables*. Such variables can be freely referenced in user-defined functions although a `free` declaration is needed to assign values to them. See *local variable* for details.

Void
the type given when the *value* and *type* of an expression are not needed. Also used when there is no guarantee at run-time that a value and predictable mode will result.

wild card
a symbol that matches any substring including the empty string; for example, the search string "`*an*`" matches any word containing the consecutive letters "`a`" and "`n`".

workspace
an interactive record of the user input and output held in an interactive history file. Each user input and corresponding output expression in the workspace has a corresponding *step number*. The current output expression in the workspace is referred to as `%`. The output expression associated with step number `n` is referred to by `%%(n)`. The `k` th previous output expression relative to the current step number `n` is referred to by `%%(- k)`. Each interactive *frame* has its own workspace.

Index

Request for Information

If you are in the USA, Canada, or Mexico and wish to receive further information about AXIOM (including prices, licensing and ordering procedures) then please call NAG Inc. or complete the details below and return to NAG Inc.

Please send me further information about AXIOM - The Scientific Computation System.

Name:______________________________________ Dept:____________

Organization:____

Address:____

Tel:________________ Fax:________________ e-mail:________________

NAG Inc, 1400 Opus Place, Suite 200, Downers Grove, IL 60515-5702, USA, Tel: +1 708 971 2337 Fax: +1 708 971 2706

Request for Information

If you are NOT in the USA, Canada, or Mexico and wish to receive further information about AXIOM (including prices, licensing and ordering procedures) then please call NAG Ltd. or complete the details below and return to NAG Ltd.

Please send me further information about AXIOM - The Scientific Computation System.

Name:______________________________________ Dept:____________

Organization:____

Address:____

Tel:________________ Fax:________________ e-mail:________________

NAG Ltd, Wilkinson House, Jordan Hill Road, OXFORD, OX2 8DR, UK, Tel: +44 865 511245 Fax: +44 865 310139
NAG GmbH, Schleißheimerstr. 5, D-8046 Garching bei München, Deutschland, Tel: +49 89 3207395 Fax: +49 89 3207396

please affix
postage

NAG Inc
1400 Opus Place, Suite 200
Downers Grove, IL 60515-5702
USA

please affix
postage

NAG Ltd
Wilkinson House
Jordan Hill Road
OXFORD
UK OX2 8DR

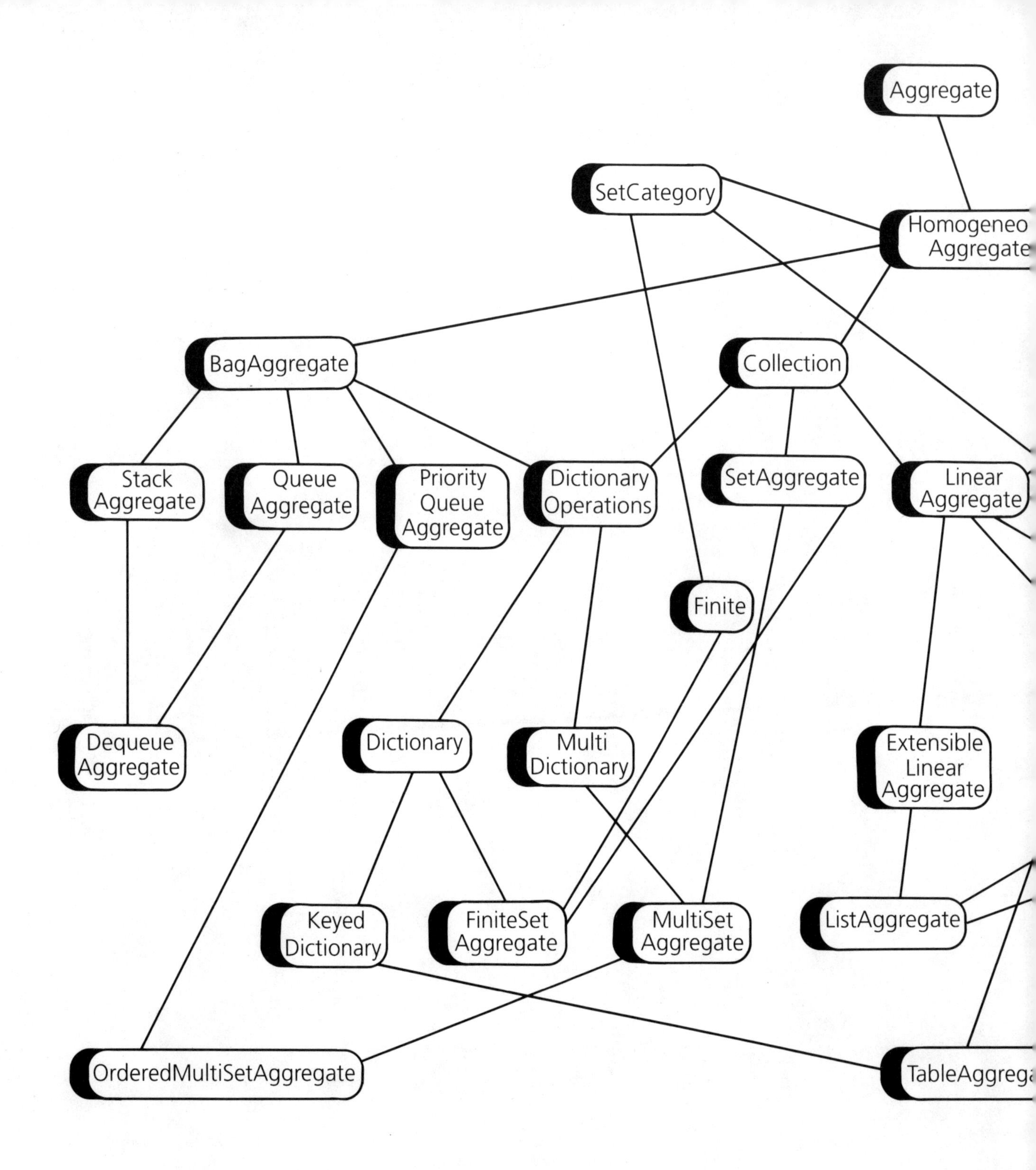

Aggregate
SetCategory
Homogeneo
Aggregate
BagAggregate
Collection
Stack
Aggregate
Queue
Aggregate
Priority
Queue
Aggregate
Dictionary
Operations
SetAggregate
Linear
Aggregate
Finite
Dequeue
Aggregate
Dictionary
Multi
Dictionary
Extensible
Linear
Aggregate
Keyed
Dictionary
FiniteSet
Aggregate
MultiSet
Aggregate
ListAggregate
OrderedMultiSetAggregate
TableAggrega